Hydrogen Energy Technologies

Hydrogen Energy Technologies

Editors

Bahman Shabani
Mahesh Suryawanshi

MDPI

Basel • Beijing • Wuhan • Barcelona • Belgrade • Novi Sad • Cluj • Manchester

Editors
Bahman Shabani
RMIT University
Melbourne, Australia

Mahesh Suryawanshi
UNSW Sydney
Sydney, Australia

Editorial Office
MDPI
St. Alban-Anlage 66
4052 Basel, Switzerland

This is a reprint of articles from the Topic published online in the open access journals *Energies* (ISSN 1996-1073), *Sci* (ISSN 2413-4155), and *Hydrogen* (ISSN 2673-4141) (available at: https://www.mdpi.com/topics/hydrogen).

For citation purposes, cite each article independently as indicated on the article page online and as indicated below:

Lastname, A.A.; Lastname, B.B. Article Title. *Journal Name* **Year**, *Volume Number*, Page Range.

ISBN 978-3-0365-9300-5 (Hbk)
ISBN 978-3-0365-9301-2 (PDF)
doi.org/10.3390/books978-3-0365-9301-2

Contents

Preface

The worldwide shift towards embracing hydrogen technology underscores the need for scientific research and advancement in this field. This publication aims to contribute to this progress by compiling recent cutting-edge studies that explore different facets of hydrogen technology within the various stages of its value chain. The editors extend their appreciation to the Energies editorial and administrative teams for their support and collaboration in ensuring that the quality of this contribution is high.

Bahman Shabani and Mahesh Suryawanshi
Editors

energies

MDPI

Article

Model-Based System Performance Analysis of a Solid Oxide Fuel Cell System with Anode Off-Gas Recirculation

Eun-Jung Choi [1], Sangseok Yu [2], Ji-Min Kim [2] and Sang-Min Lee [1,*]

1 Department of Clean Fuel and Power Generation, Korea Institute of Machinery & Materials (KIMM), 156 Gajeongbuk-ro, Yuseong-gu, Daejeon 34103, Korea; ejchoi@kimm.re.kr
2 School of Mechanical Engineering, Chungnam University, 99, Daehak-ro, Yuseong-gu, Daejeon 34134, Korea; sangseok@cnu.ac.kr (S.Y.); jiminkim@kimm.re.kr (J.-M.K.)
* Correspondence: victlee@kimm.re.kr; Tel.: +82-42-868-7833

Abstract: Designing proper solid oxide fuel cell (SOFC) system configurations is essential for their high efficiency. The present study analyzes the performance improvement of the SOFC system with anode off-gas recirculation (AOGR). Two AOGR configurations are suggested. Depending on the heat flows of off gases, the configurations are called AOGR #1 and #2, respectively. Additionally, a reference system is examined for comparison. This study aims to numerically evaluate the characteristics and performance of each system under various operating conditions such as fuel and air utilization factors. The operating current density and steam to carbon ratio are fixed at 0.3 A/cm^2 and 2.5, respectively. The results indicate that the system performance shows a large difference depending on the system configurations. The SOFC system with AOGR has better performance than the reference system under the operating conditions considered in this paper. However, it is also revealed that depending on the system configuration and operating conditions, AOGR can be effective or ineffective for system performance. Therefore, a deliberate operating strategy for AOGR systems needs to be developed based on the load conditions.

Keywords: SOFC; AOGR; hydrogen recirculation; system performance; power generation system

Citation: Choi, E.-J.; Yu, S.; Kim, J.-M.; Lee, S.-M. Model-Based System Performance Analysis of a Solid Oxide Fuel Cell System with Anode Off-Gas Recirculation. *Energies* **2021**, *14*, 3607. https://doi.org/10.3390/en14123607

Academic Editor: Bahman Shabani

Received: 23 May 2021
Accepted: 14 June 2021
Published: 17 June 2021

1. Introduction

Solid oxide fuel cells (SOFCs) have recently received attention as an alternative power source since they have especially high electrical efficiency, low emission and fuel flexibility. Challenging issues for SOFC commercialization are having long-term durability and enhancing economic efficiency. Due to the long operation time of a SOFC, it is mainly utilized as a stationary power plant. Therefore, achieving high system efficiency and fuel utilization becomes important because it is directly linked to economic efficiency [1,2]. To improve system efficiency, designing an appropriate system configuration is necessary in addition to applying highly efficient components. The overall system efficiency varies according to the layout of the system components.

Generally, thermal energy from stack off-gas and system exhaust gas are utilized in order to improve system efficiency. The heat is recovered at heat recovery heat exchangers (HR-HEs) and often supplied to the fuel/air preheater and reformer or used to generate steam necessary for the reforming reaction [3,4]. The SOFC combined heat and power (CHP) system has also been widely suggested for efficient SOFC systems [5–9]. In utilizing exhaust heat from the stack and system, the system composition and its configuration highly affect the system's overall efficiency. Therefore, many studies on designing system configurations have been conducted [6,7].

In order to improve system efficiency, anode off-gas recirculation (AOGR) can be adopted for the SOFC system. Anode off-gas (AOG) contains unreacted hydrocarbons and a high content of steam. The recirculated AOG reacts as fuel inside a fuel cell, leading to an increase in the fuel utilization efficiency. The efficiency of the steam methane reforming (SMR) reaction is also promoted by additionally supplied steam from AOG [10,11].

In the study of Powell et al. [12], a 2 kW class SOFC system with AOGR was experimentally verified. Heat and steam from AOG were delivered to an adiabatic steam reformer. As a result, the overall fuel utilization efficiency was increased up to 93%, and the system achieved a maximum net LHV efficiency of 57% at 1.7 kW. Additionally, a parametric study of the SOFC system with AOGR was conducted by Lee et al. [10]. In this study, a turbocharger and an ejector were suggested to supply air and recirculated AOG. The effects of the external reforming (ER) ratio, fuel utilization and steam to carbon (S/C) ratio were examined. The suggested system showed electrical efficiency of 64.6% when the ER ratio, fuel utilization and S/C ratio were 0.4, 0.75 and 2.5, respectively. Through the sensitivity analysis, it was revealed that fuel utilization was the most influential factor in the system efficiency. Table 1 presents a literature summary of the SOFC system with AOGR analyzed by many researchers.

Table 1. A literature summary of the SOFC system with AOGR.

Authors	SOFC Power	η_{ele} (Max.)	AOGR Device	Comments
Lee et al. [10]	5 kW	64.6%	Ejector	The turbocharger and ejector are used to blow the cathode air and AOG. Sensitivity analysis has been conducted to determine the optimal operating schemes.
Powell et al. [12]	1.7–2.2 kW	56.6% (LHV)	Blower	The system uses adiabatic ESR and AOGR system. Required heat and steam for SMR are provided by recirculated AOG.
Koo et al. [13]	113.8 kW	66% (LHV)	Blower	A cascade system having a double SOFC system and a single SOFC system with AOGR was analyzed using the exergy-based analysis method.
Wagner et al. [14]	6 kW	66% (LHV)	Fan	A novel micro AOGR fan has been introduced and experimentally coupled to SOFC system.
Baba et al. [15]	1 kW	-	Ejector	SOFC system with a variable flow ejector was examined under partial load and full load conditions.
Tanaka et al. [16]	10 kW	58.7% (LHV)	Blower	Tow AOGR blowers were developed and coupled with SOFC system simulator.
Dietrich et al. [17]	0.3 kW	41% (LHV)	Injector	SOFC running on propane with AOGR was experimentally examined and compared to a partial oxidation system.

Additionally, the method that can generate additional electricity by utilizing SOFC exhaust gas has been widely studied. SOFC hybridization with other power generation systems allows the system efficiency to be effectively improved [7,18,19]. Kuchonthara et al. [19] evaluated a combined power generation system with a SOFC and various gas turbine (GT) cycles. The results indicated that the humid air turbine promoted the thermal efficiency of the overall system. The effectiveness of a SOFC-engine hybrid system was experimentally demonstrated by Kim et al. [18]. A 5 kW class SOFC stack and internal combustion engine were combined, and the electrical efficiency of the hybrid system increased by up to 26%.

In this paper, two SOFC system configurations with AOGR were developed. The differences between the first AOGR system (AOGR #1) and the second AOGR system (AOGR #2) were the heat flows of AOG, cathode off-gas (COG) and catalytic combustor off-gas (CCOG). The details were described in Section 2. The performances of each system were evaluated in comparison with those of a reference system. For comparison, component-level mathematical modeling was conducted. Analysis of the effect of the fuel/air utilization factor and recirculation ratio was conducted based on the simulation result. In particular, the temperature of each component, net power, and electrical and thermal efficiency were examined.

2. System Configurations

To evaluate the system efficiency of various SOFC system configurations, the reference SOFC system and the system with AOGR were analyzed. The reference system shown in Figure 1a consists of an external steam reformer (ESR), a SOFC stack, a catalytic combustor (CC), air blowers, a steam generator, an air preheater and an HR-HE. For an efficient system, a direct internal reforming (DIR) SOFC was suggested. The DIR stack enables hydrocarbons to be reformed directly inside a stack [20,21]. As the steam reforming reaction is highly endothermic, the ESR needs a large amount of heat. In the reference system, ESR thermally integrated with CC was used, in other words the heat of combustion from CC was directly provided to the ESR. This ESR is called an allothermal reformer [4]. Due to the thermal stability of CC, additional air to the CC blows when needed to keep the temperature of the CC lower than 1123.15 K [22]. Thermal energy from COG was used to generate the steam required for the steam reforming process. Cooled COG has the effect on preventing the excessive temperature increase of the CC. CCOG supplies heat to the ESR and stack air flow. The rest of the CCOG thermal energy was recovered from the HR-HE.

The first concept of the SOFC system with AOGR is suggested in Figure 1b, and it is named the AOGR #1 system. AOG flowed into the fuel preheater through a recirculation blower. The recirculation blower was generally able to withstand hot gas up to 1073.15 K, so recirculated AOG needed to be cooled at the fuel preheater. The recirculated AOG is able to warm the ESR and simultaneously provide the additional steam required for the SMR reactions. Distinctive flows of AOG and fuel are marked with blue lines in Figure 1b.

Figure 1c shows the second concept of the SOFC system with AOGR, which is called the AOGR #2 system. While the ESRs shown in Figure 1a,b were thermally integrated to the CC, the ESR shown in Figure 1c obtained the required heat only from the reactant flows, namely, the adiabatic reformer. It has been demonstrated that the SMR reaction can occur using only the sensible heat of the inlet gas [4,12]. The inlet fuel to the ESR was heated in two stages by the heat from recirculated AOG and COG. The heat of CCOG was used to warm up the stack supplied air and was recovered at the HR-HE. The differences of the AOGR #2 system compared to the AOGR #1 system are highlighted with green lines in Figure 1b.

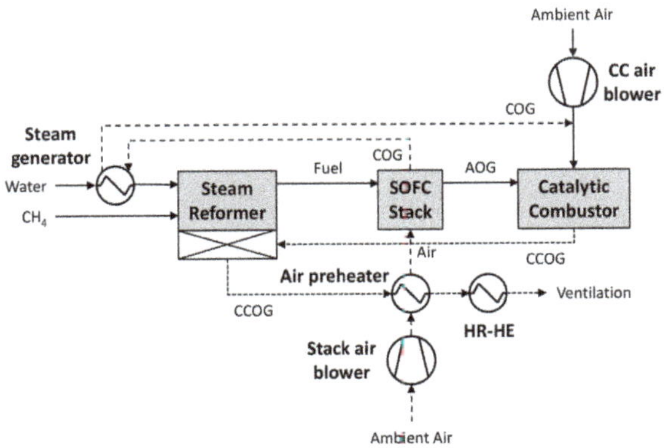

(a)

Figure 1. *Cont.*

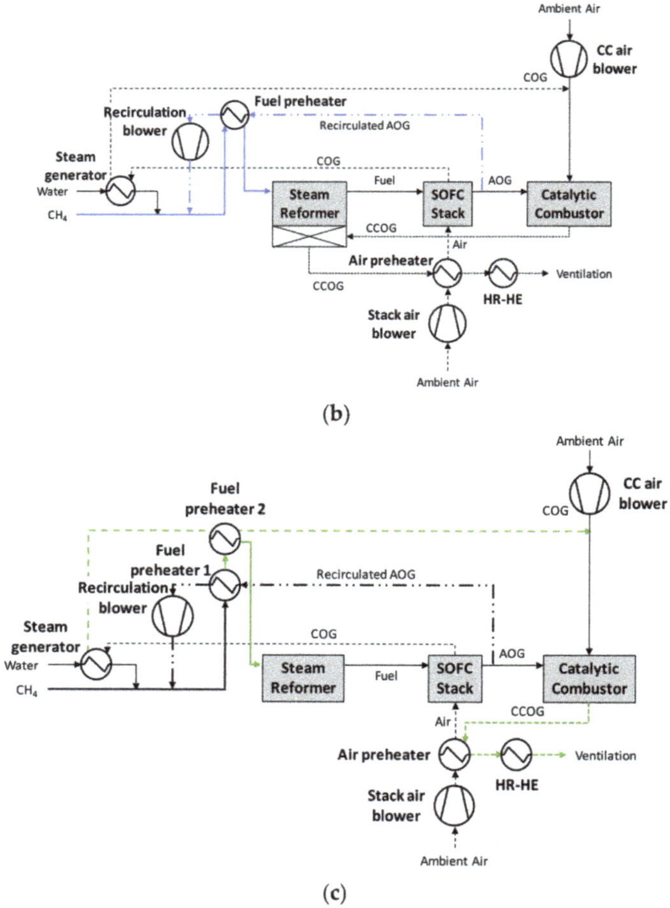

Figure 1. Schematic diagram of the system configuration; (**a**) reference system, (**b**) AOGR #1 system, and (**c**) AOGR #2 system.

3. Model Description

The purpose of the model developed in this paper was firstly to compare the performance and efficiency of various configurations and then to develop a control strategy in subsequent research. To check the system response according to the control design, the model has to be dynamic and not too complex [1,23]. In addition, the needs of dynamic models have been addressed by several authors because of the high operating temperature of SOFCs [24,25]. Therefore, a dynamic and lumped component model was developed in the SIMULINK environment. First, the steady-state characteristics of each system were examined to check the effectiveness of the systems. Dynamic characteristics and control strategies will be analyzed in future papers. In the following sections, the descriptions of each component model are discussed.

3.1. A SOFC Stack

In this model, a 1 kW class DIR stack has been adopted for an efficient SOFC system, and it is a planar SOFC stack with an anode-supported cell. The conditions inside a SOFC stack are appropriate for the SMR reaction because of its proper operating temperature and electrochemically generated steam. Additionally, internal reforming has several advantages, such as reducing the size of the external reformer and air blower power consumption by means of decreasing the stack air flow rate for stack cooling [23,26,27].

The SOFC stack is composed of reactant channels and the positive-electrolyte-negative (PEN) structure. The model accounts for mass balances, thermal balances and electrochemical reactions. Specifications of the stack are presented in Table 2. Fuel and air mixtures are assumed to follow the ideal gas law.

Table 2. A SOFC stack specifications.

Parameter	Values
Cell length (m)	0.1
Cell width (m)	0.1
Number of cells	40
Anode electrode thickness (m)	1×10^{-3}
Cathode electrode thickness (m)	3×10^{-4}
Electrolyte thickness (m)	2×10^{-5}
Anode channel height (m)	3×10^{-4}
Cathode channel height (m)	5×10^{-4}
Channel width (m)	1×10^{-3}
Number of channels	60
Anode porosity	0.4
Density of PEN (kg m^{-3})	5300
Specific heat of PEN (J kg^{-1} K^{-1})	500

3.1.1. Mass Balance Model

In the mass transfer model, the species in the anode channel are considered as CH_4, H_2O, CO, H_2 and CO_2, and those in the cathode channel are only N_2 and O_2. Four reactions inside a stack are considered. All the reactions are presented in Table 3.

Table 3. Reactions considered inside a stack.

Reaction	Equation	
Steam methane reforming reaction (SMR)	$CH_4 + H_2O \rightarrow CO + 3H_2$	(R1)
Water gas shift reaction (WGS)	$CO + H_2O \rightarrow H_2 + CO_2$	(R2)
H_2 oxidation reaction (Ox)	$H_2 + O^{2-} \rightarrow H_2O + 2e^-$	(R3)
O_2 reduction reaction (Red)	$0.5O_2 + 2e^- \rightarrow O^{2-}$	(R4)
Overall reaction	$H_2 + 0.5O_2 \rightarrow H_2O$	(R5)

Mass balance dynamics for i species in fuel and air channels are given in Equations (1) and (2). In the fuel channel, (R1)—(R3) reactions described in Table 3 take place. (R4) reaction occurs in the air channel. r_i is the molar rate of formation for i species, and is obtained from the equations below.

$$\frac{\partial C_{i,f}}{\partial t} = -u_f \frac{\partial C_{i,f}}{\partial x} + r_{i,f} \quad (i = CH_4,\ H_2O,\ CO,\ H_2,\ CO_2) \tag{1}$$

$$\frac{\partial C_{i,a}}{\partial t} = -u_a \frac{\partial C_{i,a}}{\partial x} + r_{i,a} \quad (i = N_2,\ O_2) \tag{2}$$

For calculating the reaction rates of (R1) and (R2), there are various relations accounting for the reaction rates of SMR and WGS [1,20,28]. Among the equations, Chinda et al. [29] model presented in Equations (3)–(8) has been used. Chinda et al. derived the reaction rate from Arhenius' curve fits using the data by Lehnert et al. [30]. R_k and k_k represent the reaction rate and forward reaction rate constant of reaction k, respectively.

Chinda et al. considered that the SMR reaction occurs at the surface of the anode and the WGS reaction takes place inside the void volume of the anode.

$$R_{R1} = k_{R1}\left(P_{CH_4}P_{H_2O} - \frac{P_{H_2}^3 P_{CO}}{K_{R1}}\right)\left[\text{mol m}^{-2}\,\text{s}^{-1}\right] \tag{3}$$

$$R_{R2} = k_{R2}\left(P_{H_2O}P_{CO} - \frac{P_{H_2}P_{CO_2}}{K_{R2}}\right)\left[\text{mol m}^{-3}\,\text{s}^{-1}\right] \tag{4}$$

$$k_{R1} = 2395\exp\left(-\frac{231266}{RT}\right)\left[\text{mol m}^{-2}\,\text{Pa}^{-2}\,\text{s}^{-1}\right] \tag{5}$$

$$k_{R2} = 0.0171\exp\left(-\frac{103191}{RT}\right)\left[\text{mol m}^{-3}\,\text{Pa}^{-2}\,\text{s}^{-1}\right] \tag{6}$$

K_k indicates the equilibrium constant of reaction k and can be calculated by Equations (7) and (8) [2], Z is defined as (1000/T(K)-1), The unit of K_{R1} is Pa^2, and K_{R2} is a dimensionless constant.

$$K_{R1} = 1.0267\times 10^{10} exp\left(-0.2513Z^4 + 0.3665Z^3 + 0.5810Z^2 - 27.13Z + 3.2770\right) \tag{7}$$

$$K_{R2} = exp\left(-0.2935Z^3 + 0.6351Z^2 + 4.1788Z + 0.3169\right) \tag{8}$$

The reaction rates of (R3) and (R4) are equal to the electrochemical reaction rate of the fuel cell and can be obtained by Faraday's law as shown in Equation (9).

$$R_{R3} = R_{R4} = R_{R5} = \frac{J}{2F}\left[\text{mol s}^{-1}\right] \tag{9}$$

From Equations (3)–(9), the molar rate of formation per volume (r_i) can be inferred. Nitrogen does not react, so r_{N_2} is zero.

$$r_{CH_4} = -\frac{1}{t_f}R_{R1} \tag{10}$$

$$r_{H_2O} = -\frac{1}{t_f}R_{R1} - \varepsilon_a R_{R2} + \frac{1}{V_a}R_{R3} \tag{11}$$

$$r_{CO} = \frac{1}{t_f}R_{R1} - \varepsilon_a R_{R2} \tag{12}$$

$$r_{H_2} = 3\frac{1}{t_f}R_{R1} + \varepsilon_a R_{R2} - \frac{1}{V_a}R_{R3} \tag{13}$$

$$r_{CO_2} = \varepsilon_a R_{R2} \tag{14}$$

$$r_{O_2} = -0.5\frac{1}{V_c}R_{R3} \tag{15}$$

The pressure decrease (Δp) through the anode and cathode channels is calculated using Equation (16). Equations (17) and (18) represent the friction factor (f) for laminar and turbulent flow, respectively. φ is the aspect ratio of the channel, and v_g is the gas velocity.

$$f = \frac{24}{Re\left[1 - 1.3553\varphi + 1.9467\varphi^2 - 1.7012\varphi^3 + 0.9564\varphi^4 - 0.2537\varphi^5\right]}\ (Re < 2000) \tag{16}$$

$$f = \frac{0.0791}{Re^{0.25}}\ (Re \geq 2000) \tag{17}$$

$$\Delta p = f\frac{4L_{ch}}{D_h}\frac{1}{2}\rho_g u_g^2 \tag{18}$$

3.1.2. Electrochemical Reaction Model

Cell potential can be computed by Equation (19). The thermodynamic reversible potential (V_{rev}) is determined by the Nernst equation in Equation (20) [31]. A voltage drop from the thermodynamic reversible potential exists because of overpotentials summarized in Table 4.

$$V_{cell} = V_{rev} - \eta_{act,a} - \eta_{act,c} - \eta_{ohm} - \eta_{conc} \tag{19}$$

$$V_{rev} = E_0 - \frac{RT_{PEN}}{2F} ln \frac{P_{H_2O}}{P_{H_2} P_{O_2}^{0.5}} \tag{20}$$

$$E_0 = 1.2723 - 2.7645 \times 10^{-4} T_{PEN} \tag{21}$$

Table 4. Equations for the overpotentials of the stack based on Ref. [29].

Equation
Activation overpotential
$\eta_{act,a} = \frac{RT}{\alpha_a F} \sinh^{-1}\left(\frac{j}{j_{0,a}}\right)$
$j_{0,a} = 1.5 \times 10^{10} \exp\left(-\frac{137000}{RT}\right)$
$\eta_{act,c} = \frac{RT}{\alpha_c F} shih^{-1}\left(\frac{j}{2j_{0,c}}\right)$
$j_{0,c} = 8 \times 10^{10} \exp\left(-\frac{140000}{RT}\right)$
Ohmic overpotential
$\eta_{ohmic} = j(ASR_{ohmic}) = j\frac{t_e}{\sigma}$
$\sigma = 3.34 \times 10^4 \exp\left(-\frac{10300}{T_{PEN}}\right)$
Concentration overpotential
$\eta_{conc,c} = \frac{RT}{2F} \ln\left(1 - \frac{j}{j_{\lim,c}}\right)$

3.1.3. Thermal Balance Model

For thermal energy calculations, the lumped capacitance method was used. Homogeneous temperature among the fuel cell components was assumed, so the temperature of the fuel cell was considered the same as T_{PEN}. Heat absorption and release related to the reactions inside the fuel cell and convective heat transfer between the fuel cell and fuel/air bulk flows were considered in this model. Consequently, three temperatures were achieved from the thermal balance model: air, fuel bulk flow temperature and PEN temperature. The relevant equation for thermal balances in the PEN structure appears as follows.

$$\begin{aligned} \frac{\partial T_{PEN}}{\partial t} &= -\frac{q_{g,a}}{\rho_{PEN} c_{p,PEN} L_c w_c t_{PEN}} - \frac{q_{g,c}}{\rho_{PEN} c_{p,PEN} L_c w_c t_{PEN}} \\ &+ \frac{1}{\rho_{PEN} c_{p,PEN}}\left(-\frac{1}{t_{PEN}} \Delta H_{R1} R_{R1} - \Delta H_{R2} \varepsilon_a R_{R2} - \frac{1}{t_{PEN} \omega_c L_c} \Delta H_{R5} R_{R5}\right) \end{aligned} \tag{22}$$

For the calculation of heat transfer at gas channels, the Nusselt number (Nu) was considered to be a constant value of 3.39. Anode and cathode channels can act as fins of uniform rectangular cross-sectional areas, so relations for heat transfer at extended surfaces are used, as shown below, where η_f is the fin efficiency and the convective heat transfer coefficient h_{ch} is obtained from Nu.

$$q_g = \eta_f h_{ch} \left(N_{ch} A_f + A_b\right) \left(T_{PEN} - T_g\right) \tag{23}$$

3.2. *Steam Methane Reformer*

The methane steam reforming process requires a large amount of heat; thus, exhaust gas from the system is generally used as a heat source for the ESR, namely, an allothermal reformer. The reference and AOGR #1 systems suggested in this paper adopted this type of ESR. However, DIR stacks can mitigate the demand for high-performance ESR [4]. ESR can operate at a lower temperature range than the normal operating temperature

(973.15–1073.15 K) [32]. Hence, an adiabatic reformer was also examined in the AOGR #2 system. A shell and tube type catalytic steam reformer was selected in this paper, and the suggested specifications of the reformer are detailed in Table 5. The model consists of mass and thermal balance models. Gas mixtures are considered ideal gases, and the porosity of the bed is constant.

Table 5. Reformer specifications.

Parameter	Values
Length (m)	0.3
Diameter (m)	0.2
Particle diameter (m)	1.2×10^{-3}
Catalyst pore radius (m)	1×10^{-8}
Bed porosity	0.4
Density of reformer (kg m^{-3})	3970
Specific heat of reformer (J kg^{-1} K^{-1})	765

For catalytic steam reformers, it is important to find the most appropriate catalyst because it directly affects the performance of the reformer. Among various active metals, nickel (Ni) is widely used because of its high reactivity and long durability [33,34]. Xu and Froment examined the kinetics of the SMR process with a $Ni/MgAl_2O_4$ catalyst under an operating range of approximately 675–1000 K. The results are the most widely used for SMR kinetcis [35]. The SMR model in this paper has been developed based on the Xu and Froment model. While various reactions take place in catalytic SMR, only the SMR reaction, the WGS reaction, and the direct steam reforming (DSR) reaction are considered. The SMR (R1) and WGS (R2) reactions are defined in Table 3, and the DSR (R6) reaction is presented below. It is assumed that the species in the reactant flow are CH_4, H_2O, CO, H_2 and CO_2. The SMR kinetics based on Xu and Froment model are organized in Table 6 [35].

$$\text{Direct steam reforming reaction (DSR)} \quad CH_4 + 2H_2O \rightarrow CO_2 + 3H_2 \text{ (R6)} \tag{24}$$

Table 6. The equations for SMR kinetics based on Xu and Froment model based on Ref. [35].

Equation
Reaction rate[mol kg$_{\text{cat}}^{-1}$s^{-1}]
$R_{SMR,\text{R1}} = \frac{k_{\text{R1}}}{p_{H_2}^{2.5}} \frac{\left(p_{CH_4} p_{H_2O} - p_{H_2}^3 p_{CO}/K_{e,\text{R1}}\right)}{Den^2}$
$R_{SMR,\text{ R2}} = \frac{k_{\text{R2}}}{p_{H_2}} \frac{\left(p_{CO} p_{H_2O} - p_{H_2} p_{CO_2}/K_{e,\text{R2}}\right)}{Den^2}$
$R_{SMR,\text{R6}} = \frac{k_{\text{R3}}}{p_{H_2}^{3.5}} \frac{\left(p_{CH_4} p_{H_2O}^2 - p_{H_2}^4 p_{CO_2}/K_{e,\text{R3}}\right)}{Den^2}$
$Den = 1 + K_{CO} p_{CO} + K_{H_2} p_{H_2} + K_{CH_4} p_{CH_4} + K_{H_2O} p_{H_2O}/p_{H_2}$
Kinetic rate constant
$k_{\text{R1}} = 4.225 \times 10^{15} \exp\left(-\frac{240100}{RT_{SMR}}\right)$
$k_{\text{R2}} = 1.955 \times 10^{6} \exp\left(-\frac{67130}{RT_{SMR}}\right)$
$k_{\text{R6}} = 1.020 \times 10^{15} \exp\left(-\frac{243900}{RT_{SMR}}\right)$
Equilibrium constant
$K_{e,\text{R1}} = \exp(-26830/T + 30.114) bar^2$
$K_{e,\text{R2}} = \exp(4400/T - 4.036)$
$K_{e,\text{R6}} = \exp(-22430/T + 26.078) bar^3$
Rate of formation or consumption of each species[mol kg$_{\text{cat}}^{-1}$s^{-1}]
$r_{SMR,CH_4} = -\eta_{\text{R1}} R_{SMR,\text{ R1}} - \eta_{\text{R6}} R_{SMR,\text{ R6}}$
$r_{SMR,CO_2} = \eta_{\text{R2}} R_{SMR,\text{ R2}} + \eta_{\text{R6}} R_{SMR,\text{ R6}}$
$r_{SMR,H_2O} = -\eta_{\text{R1}} R_{SMR,\text{ R1}} - \eta_{\text{R2}} R_{SMR,\text{ R2}} - 2\eta_{\text{R6}} R_{SMR,\text{ R6}}$
$r_{SMR,H_2} = 3\eta_{\text{R1}} R_{SMR,\text{ R1}} + \eta_{\text{R2}} R_{SMR,\text{ R2}} + 4\eta_{\text{R6}} R_{SMR,\text{ R6}}$
$r_{SMR,CO} = \eta_{\text{R1}} R_{SMR,\text{ R1}} - \eta_{\text{R2}} R_{SMR,\text{ R2}}$

Mass and thermal balances in both the gas and solid phases are presented below. Equations (25) and (26) represent the mass balance in the gas and solid phases for each species i, respectively [28]. The mass transfer coefficient ($k_{g,i}$) is presented in Equations (27) and (28) [36].

$$\varepsilon_b \frac{\partial C_{g,i}}{\partial t} = -v_g \frac{\partial C_{g,i}}{\partial x} - k_{g,i} a_v (C_{g,i} - C_{s,i}) \tag{25}$$

$$\frac{dC_{s,i}}{dt} = -k_{g,i} a_v (C_{g,i} - C_{s,i}) + (1 - \varepsilon_b) \rho_{cat} r_{SMR,i} \tag{26}$$

$$k_{g,i} = j_{D,i} Re Sc_i^{1/3} \frac{D_i}{d_p} \tag{27}$$

$$\varepsilon_b j_{D,i} = 0.765 \mathrm{Re}^{-0.82} + 0.365 \cdot Sc_i^{-0.398} \tag{28}$$

Equations (29) and (30) are the thermal balances in the gas and solid phases. The convective heat transfer coefficient ($h_{g,SMR}$) in Equation (30) is known from the Chilton–Colburn j-factor (j_H). For forced convection through a packed bed, Yoshida et al. [37] suggested empirical correlations of j_H. Ψ is an empirical coefficient depending on the particle shape, and its value is 1 for a sphere.

$$\varepsilon_b \frac{\partial T_g}{\partial t} = -v_g \frac{\partial T_g}{\partial x} + \frac{h_{g,SMR} a_v}{\rho_g c_{p,g}} (T_s - T_g) \tag{29}$$

$$\frac{\partial T_s}{\partial t} = -\frac{h_{g,SMR} a_v}{\rho_{bed} c_{p,b}} (T_s - T_g) + \frac{(1 - \varepsilon_b) \rho_{cat}}{\rho_b c_{p,b}} \sum_k (-\Delta H_k \eta_k R_k)\ (k = \mathrm{R1,\ R2,\ R6}) \tag{30}$$

$$h_{g,SMR} = j_H \frac{c_{p,g} \rho_g v_g}{Pr^{2/3}} \tag{31}$$

$$j_H = 0.91 \mathrm{Re}^{-0.51} \Psi\ 0.01 < Re < 50 \tag{32}$$

$$j_H = 0.61 \mathrm{Re}^{-0.41} \Psi\ 50 < Re < 7000 \tag{33}$$

3.3. Catalytic Combustor

To compute the temperature and species in the CC, a mathematical model including mass and thermal balances was developed. A Pt-catalyzed monolithic combustor is analyzed. Specifications of CC are given in Table 7. The oxidation reactions over Pt considered in this model are CO, CH_4 and H_2 oxidation. The rate expressions and reaction rate per Pt surface area are organized in Table 8 based on Ref. [38]. Chemical reactions are assumed to occur only on the external surface of the catalytic wall.

Table 7. Catalytic combustor specifications.

Parameter	Values
Length (m)	0.25
Width (m)	0.25
Height (m)	0.1
Number of cells	10,000
Hydraulic diameter (m)	1.2×10^{-3}
Catalyst pore radius (m)	1×10^{-8}
Tortuosity	14
Monolith porosity	0.63
Density of solid (kg m^{-3})	430
Catalyst (m^2 m^{-3})	26,895
Mass (m^2 m^{-3})	25

Table 8. Equations for the reaction rate of CC based on Ref. [38].

Equation	
Oxidation reactions of CC	
$CO + 0.5O_2 \rightarrow CO_2$	(R1)
$CH_4 + 2O_2 \rightarrow CO_2 + H_2O$	(R2)
$H_2 + 0.5O_2 \rightarrow H_2O$	(R3)
Reaction rate[mol m^{-2}s^{-1}]	
$R_{CC,CO} = k_1 y_{CO} y_{O_2} / G$	
$R_{CC,CH_4} = k_2 y_{CH_4} y_{O_2} / G$	
$R_{CC,H_2} = k_1 y_{H_2} y_{O_2} / G$	
$G = (1 + K_{CO} c_{CO})^2$	
Rate constant[molK m^{-2}s^{-1}]	
$k_1 = 6.699 \times 10^9 \exp(-12556/T)$	
$k_2 = 7.326 \times 10^6 \exp(-19000/T)$	
Adsorption equilibrium constant	
$K_{CO} = 65.5 \exp(961/T)$	

Homogeneous temperature, concentration and velocity within the channel are assumed for mass and thermal balance computations. Equations (34) and (35) describe the mass balances for the gas and solid phases for each species i, where i refers to CH_4, H_2O, H_2, CO, CO_2, O_2 and N_2. The corresponding equations for thermal balances for the gas and solid phases are presented in Equations (36) and (37). For the heat transfer coefficient ($h_{g,CC}$) calculation, Nu is considered to have a constant value of 3.39.

$$\varepsilon_m \frac{\partial y_{g,i}}{\partial t} = -v_g \frac{\partial y_{g,i}}{\partial x} - k_{m,i} S (y_{g,i} - y_{s,i}) \tag{34}$$

$$(1 - \varepsilon_m) \frac{dy_{s,\,i}}{dt} = k_{m,i} S (y_{g,i} - y_{s,i}) + \frac{RT_{CC}}{P_{tot}} \alpha R_{CC,i} \tag{35}$$

$$\varepsilon_m \rho_g c_{p,g} \frac{\partial T_g}{\partial t} = h_{g,CC} S (T_s - T_g) - v_g \rho_g c_{p,g} \frac{\partial T_g}{\partial x} \tag{36}$$

$$(1 - \varepsilon_m) \rho_s c_{p,s} \frac{\partial T_s}{\partial t} = h_{g,CC} S (T_g - T_s) + \alpha \left(\sum_i (-\Delta H_i) R_i \right) \tag{37}$$

3.4. Balance of Plant

3.4.1. Air and Recirculation Blower

A model of air and recirculation blowers was employed to examine the outlet temperature and power consumption. The stack air flow rate was determined by the air utilization factor, and the CC air flow rate was controlled to maintain a CC temperature below 1123.15 K. In terms of the recirculation blower, the target recirculation ratio determines the flow rate. The blower outlet temperature and power consumption are obtained from Equations (38) and (39). In the computation, the values of η_{isen}, η_{motor} and η_{mech} were 0.8, 0.9 and 0.9, respectively.

$$T_{blower,o} = T_{blower,i} - T_{blower,i} \frac{1 - (p_o/p_i)^{\kappa - 1/\kappa}}{\eta_{isen}} \tag{38}$$

$$P_{blower} = \dot{m}_{blower} c_{p,air} (T_{blower,o} - T_{blower,i})\, \eta_{motor} \eta_{mech} \tag{39}$$

3.4.2. Heat Exchanger

Heat exchangers are employed to recover heat from AOG, COG and exhaust gas. They act as fuel/air preheaters, a steam generator and a HR-HE. The outlet temperature of each gas and heat transfer rate are defined by the effectiveness-NTU method, as shown in

Equation (40) since the information is insufficient to use the LMTD method. ε_{HE} indicates the effectiveness of a heat exchanger and was set to 0.75.

$$q_{HE} = \varepsilon_{HE} C_{min} (T_{h,i} - T_{c,i}) \quad C_{min} = \min\{C_{hot}, C_{cool}\} \tag{40}$$

3.5. Performance Factor

U_{fuel} and U_{air} indicate the fuel and air utilization factors, respectively. Methane is only supplied as fuel in this model. If there are other hydrocarbons in fuel, they should be added to the denominator in Equation (41). The fuel and air flow rates of the stack are determined from the target utilization factor. Recirculated hydrogen, however, is not considered when the fuel flow rate is calculated.

$$U_{fuel} = \frac{J}{8F\dot{n}_{CH_4}} \tag{41}$$

$$U_{air} = \frac{J}{4Fy_{O2}\dot{n}_{air}} \tag{42}$$

The electrical, thermal and total efficiencies are shown in Equations (43)–(45). The electrical efficiency of the system is the ratio of the net generated energy of the system to the chemical energy of the supplied fuel. To estimate the thermal efficiency, the temperature of the final exhaust gas from the CC (T_{vent}) is assumed to become 393.15 K after the HR-HE.

$$\eta_{ele} = \frac{P_{stack} - P_{FCBlower} - P_{CCBlower}}{\dot{m}_{CH_4} LHV_{CH_4}} \tag{43}$$

$$\eta_{th} = \frac{C_{p,CCOG}\dot{m}_{CCOG}(T_{CCOG,O} - T_{vent})}{\dot{m}_{CH_4} LHV_{CH_4}} \tag{44}$$

$$\eta_{tot} = \frac{\eta_{ele} + \eta_{th}}{\dot{m}_{CH_4} LHV_{CH_4}} \tag{45}$$

4. Results and Discussion

4.1. Model Validation

To demonstrate the reliability of the developed model, a model validation process is essential. Comparing the simulation results to the experimental data of the whole system is the best way to validate a model, however, the purpose of this paper was to estimate the performance depending on the system configuration. Thus, only the model of the most complex component, which is a SOFC stack, is demonstrated in this paper. A 1 kW class stack as described in Table 1 was used for the experiment. During the experiments, the fuel and air flow rates were constant regardless of the current. H_2 and CO_2 were supplied at 27.96 and 10.61 lpm, respectively, as fuel. The air flow rate was fixed at 66.10 lpm. The load current was varied from 0 to 29.92 A. The stack temperature was maintained at 1023.15 K during the experiment. The model validation results are displayed in Figure 2. As a result, the model was able to accurately predict the performance of the stack.

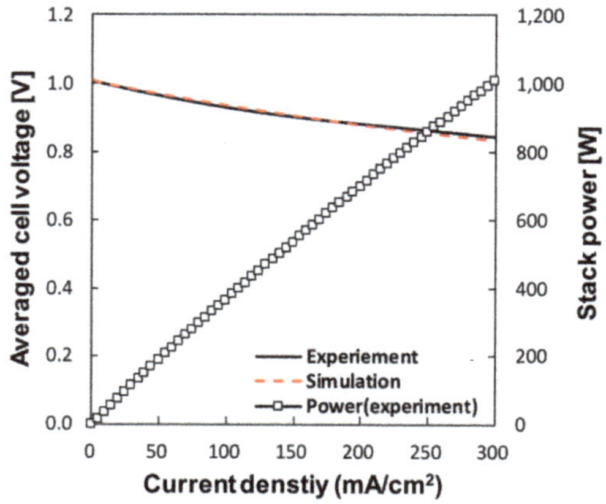

Figure 2. Comparison between the predictions and experimental data.

4.2. Operating Conditions for Simulation

The operating conditions for the simulation are presented in Table 9. Methane was used as fuel. The fuel and air flow rates were determined depending on the fuel/air utilization factor and operating current density. Steam flow rate was calculated based on the fuel flow rate and S/C ratio. The steam was generated at steam generator using thermal energy from off-gas. The PI controller adjusted the CC air flow rate so that the temperature of the CC did not exceed 1123.75 K. In this study, the performance of each system was investigated with various fuel/air utilization factors and recirculation ratios. We define the reference condition of the simulation as the condition with both a utilization factor of 0.6 and a recirculation ratio of 0.2.

Table 9. Operating conditions for the simulation.

Parameter	Values
Inlet fuel	CH_4
Operating pressure of the system (bar)	1.2
Operating current density (A cm^{-2})	0.3
Maximum temperature of CC (K)	1123.15
Exhaust gas temperature (K)	393.15
S/C ratio at ESR	2.5
Ambient temperature (K)	298.15
Fuel/Air utilization factor (Reference value)	0.3–0.8 (0.6)
Recirculation ratio (Reference value)	0–0.8 (0.2)

4.3. Result 1: The Effect of the Fuel/Air Utilization Factors

4.3.1. Reference System

The fuel flow rate changed from 6.42 to 2.41 lpm in accordance with the fuel utilization factor of 0.3 to 0.8. The air flow rate was fixed at 30.57 lpm. Figure 3a,b indicate the temperature of each component, the CC air flow rate and the heat transfer rate of each heat exchanger. As the fuel utilization factor increased, less fuel was supplied to the system. This decreased the combustion energy at the CC and the additional CC air flow rate for cooling. When the fuel utilization factor was 0.8, the temperature of CC became lower than 1123.15 K without additional air flow to the CC. The ESR was thermally integrated with CCOG in the reference system; therefore, the temperature of the ESR decreased. In Figure 3b, it was observed that the heat transfer rate from CCOG to ESR and the amount of recovered heat were rapidly reduced. On the other hand, the stack temperature increased due to the lowered internal reforming rate. The SMR process is a strong endothermic process, so the more abundance the internal reforming reactions are, the lower the temperature of the stack.

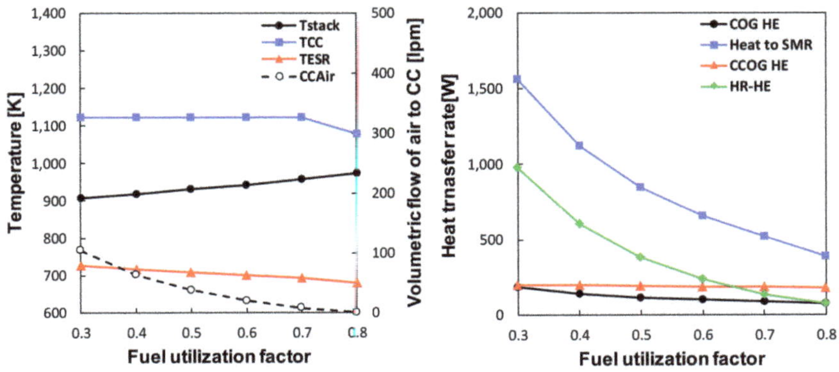

(**a**) (**b**)

Figure 3. The effect of fuel utilization factor on the reference system; (**a**) temperature of each component, (**b**) heat transfer rate at each heat exchanger (COG HE: steam generator, CCOG HE: air pre-heater, HR HE: heat recovery heat exchanger).

To examine the effect of the air utilization factor, the fuel utilization factor was fixed at 0.6 and the air utilization factor was changed from 0.3 to 0.8. The results are depicted in Figure 4a,b. A decreased stack air flow rate led to a temperature rise in the stack, and the raised temperature accelerated the internal reforming reaction. Consequently, the hydrogen molar flow rate to the CC increased, causing larger heat generation in the CC. Although extra air was supplied to the CC, the total amount of air diminished because the stack air flow rate decreased further. Less heat was transferred from the CCOG to the ESR, thus, the temperature of the ESR slightly decreased.

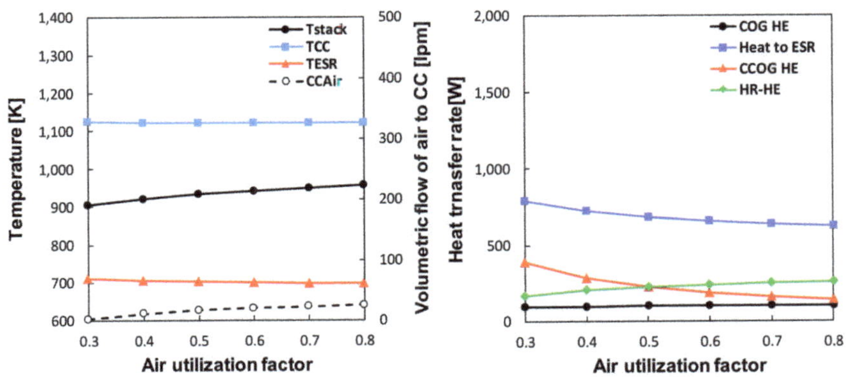

(**a**) (**b**)

Figure 4. The effect of air utilization factor on the reference system; (**a**) temperature of each component, (**b**) heat transfer rate at each heat exchanger (COG HE: steam generator, CCOG HE: air pre-heater, HR HE: heat recovery heat exchanger).

4.3.2. SOFC System with AOGR #1

In this section, the AOGR #1 system is examined. As the condition described in Section 4.3.1, the fuel or air utilization factor was changed from 0.3 to 0.8. For the calculation, the recirculation ratio was held constant at 0.2. The results of the fuel and air utilization factors are presented in Figure 5a,b, respectively. Overall, the results are similar to those presented in Figures 3 and 4. By comparing the CC air flow rate, the generated heat of the CC of this system was slightly lower than that of the reference system. The AOGR system enhanced the efficiency of fuel utilization, so the available energy in CC becomes reduced. The temperature of the recirculation blower is also shown in Figure 5a,b, and it falls within the appropriate operating temperature. The recirculation blower temperature was between 691.79 and 745.86 K.

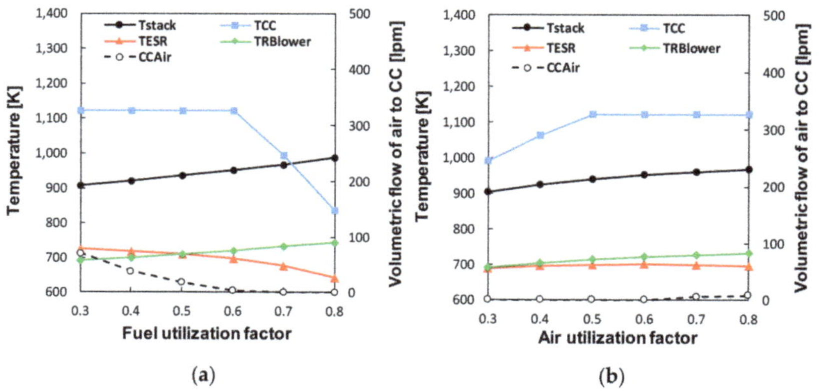

Figure 5. The effect of (**a**) fuel and (**b**) air utilization factors on the AOGR #1 system.4.3.3. SOFC System with AOGR #2.

The difference between the AOGR #1 and AOGR #2 systems is the heat supply method for the ESR. As mentioned in Section 2, the ESR applied in the system with AOGR #1 was thermally integrated with the CC to directly absorb heat from the CCOG. Meanwhile, the ESR used in the AOGR #2 system received the required heat only from the reactants. Fuel preheater 2 was also added to recover additional heat from the COG, as shown in Figure 1c. The effect of fuel and air utilization factors on component temperatures is presented in Figure 6a,b. The changes in temperature of the stack, ESR and CC were similar to those presented in Section 4.3.2. However, CCOG did not directly flow to the ESR, and the temperatures of the stack and CC in the AOGR #2 system became higher than those in the AOGR #1 system. When the fuel utilization factor was 0.8, the stack temperature slightly decreased. Under this condition, the effect of the stack temperature reduction because of the lowered inlet air temperature became greater than the effect of the stack temperature increase caused by the weakened internal reforming reaction.

With regard to the recirculation blower, the temperature of the recirculation blower increased to 728.02 K at the fuel utilization factor of 0.8. The heat supply amount at the fuel preheater diminished in accordance with the increase in the fuel utilization factor. On the other hand, the temperature of the recirculation blower decreased as the air utilization factor increased because of the reduced heat transfer rate in the steam generator.

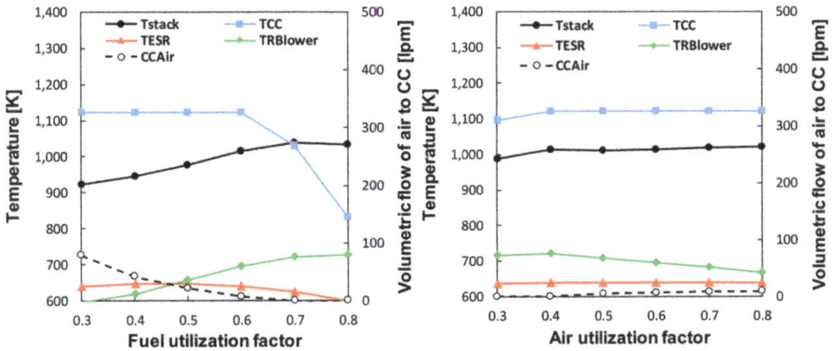

(a) (b)

Figure 6. The effect of (**a**) fuel and (**b**) air utilization factors on the AOGR #2 system.

4.4. Result 2: Performance Analysis

4.4.1. Comparison of Net Power

Figure 7a represents the net power of each system with various fuel utilization factors. For the computation of net power, the power consumptions of the stack air blower, CC air blower and recirculation blower were considered. According to the result of the reference system, the net power continuously increases with increasing fuel utilization factor. An increased stack temperature affected the enhancement of net power. For the AOGR #1 and AOGR #2 systems, however, the net power decreased when the fuel utilization factor reached a certain value. The maximum net power of AOGR #1 and AOGR #2 was 774.92 and 848.88 W, respectively, at a fuel utilization factor of 0.7. An increase in the stack temperature had a positive effect on power enhancement, on the other hand, the power was simultaneously negatively affected since a relatively high fuel utilization factor caused a fuel dilution problem at the anode. The effect of the air utilization factor on the net power is depicted in Figure 7b. The weakened cooling effect by lower stack air flow resulted in performance improvement for all systems.

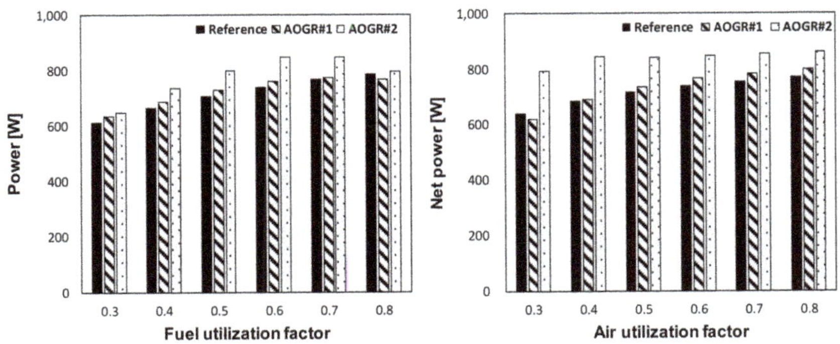

(a) (b)

Figure 7. Net power of each configuration with various (**a**) fuel and (**b**) air utilization factors.

When comparing the results of each system, the net power of the AOGR #2 system was higher than that of the other systems. This was because the AOGR #2 system applied the heat of CCOG only to the stack inlet air, increasing the stack temperature. The performance improvement became noticeable with a high fuel utilization factor (except for 0.8) and low air utilization factor. The net power of the AOGR #2 system was on average 9.48% and

7.24% higher than that of the reference system and the AOGR #1 system in Figure 7a and on average 17.70% and 15.55% higher in Figure 7b, respectively. This result explains that when a DIR stack is used, a high temperature at the ESR is not necessarily required. DIR reactions sufficiently compensate for lowered ESR performance, and a temperature rise in the stack develops the performance.

4.4.2. Comparison of Efficiency

The electrical and thermal efficiency of each system depending on the fuel and air utilization factor is described in Figure 8a,b. In Figure 8a, decreased amount of input fuel energy resulted in an increase in electrical efficiency for all systems when the fuel utilization factor increased. Meanwhile, the thermal efficiency was reduced as the amount of generated heat at the CC diminished because of the lower fuel supply. The total efficiency of the reference system and AOGR #1 system was enhanced from 38.25 to 55.35% and from 31.94 to 52.58%, respectively. Despite the higher net power of the AOGR #1 system than the reference system, the reference system had higher total efficiency due to the larger heat recovery amount. The AOGR #2 system showed relatively higher total efficiency with little change. The maximum total efficiency was 60.81% when the electrical and thermal efficiencies were 40.83 and 19.98%, respectively. When the stack air flow rate decreased, the electrical efficiency of the stack improved owing to the rise in the stack temperature for all systems, as shown in Figure 8b. The maximum total efficiency was 49.35% for the reference system, 44.90% for the AOGR #1 system and 61.36% for the AOGR #2 system when the air utilization factor was 0.8.

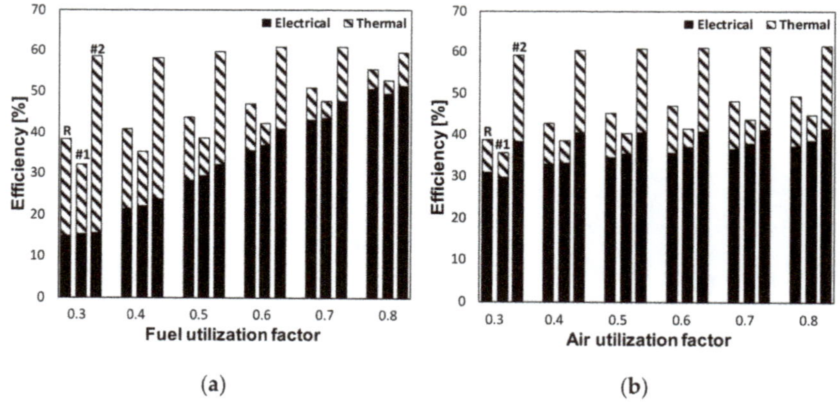

Figure 8. Efficiency of each configuration with various (**a**) fuel and (**b**) air utilization factors. (R: reference system, #1: AOGR #1 system, #2: AOGR #2 system).

In Figure 8a,b, the AOGR #2 system shows the highest total efficiency among the system configurations. Unlike the other two systems, a large amount of heat can be recovered at the HR-HE of AOGR #2. Therefore, the thermal efficiency of this system was superior to that of the other systems. By comparing the results shown in Figures 7 and 8, the fuel flow rate was a relatively influential factor in the system performance.

4.5. Result 3: The Effect of Recirculation Ratio

4.5.1. The SOFC System with AOGR #1

Figure 9a shows the effect of the recirculation ratio in the AOGR #1 system. Both the air and fuel utilization factors were 0.6, and the recirculation ratio changed from 0 to 0.8. As the recirculation ratio increased, the inlet fuel flow rate to the CC decreased, causing a reduction in the temperature of the ESR and CC. The recirculated fuel diluted the anode fuel at a high recirculation ratio and then weakened the internal reforming reaction in the stack. This phenomenon increased the temperature of the stack.

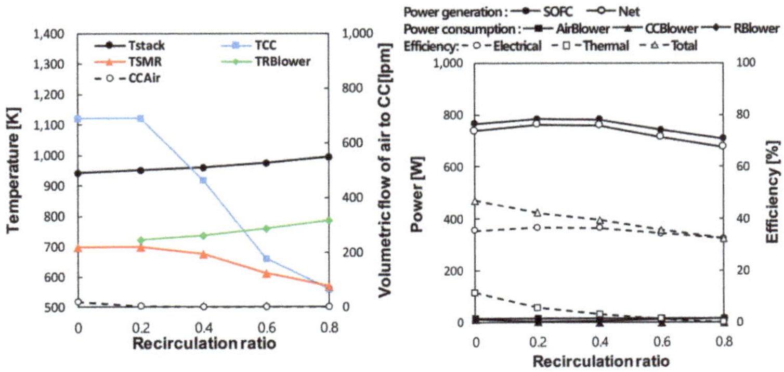

(a) (b)

Figure 9. The effect of recirculation ratio on the AOGR #1 system; (**a**) temperature of each component and (**b**) power and efficiency.

Power and efficiency changes are presented in Figure 9b. The maximum net power and electrical efficiency were 762.6 W and 36.7% at a recirculation ratio of 0.2. The maximum thermal efficiency was observed when AOG was not recirculated. Because heat generation of CC kept decreasing as the recirculation ratio increased, the AOGR system was unable to achieve higher thermal efficiency. Therefore, the total efficiency fell as more hydrogen was recirculated.

4.5.2. The SOFC System with AOGR #2

The effect of the recirculation ratio varied according to the operation condition. In this paper, the fuel and air flow rates to the system were considered as the main parameters. It was revealed that the effect of the air utilization factor on the performance of the AOGR #2 system was relatively small in the previous section, thus, power and efficiency changes with various fuel utilization factors and recirculation ratios were investigated in this section with a fixed air utilization factor of 0.6. Recirculation did not exceed 0.6 to prevent CC temperatures that were too low at high fuel utilization factors. Figure 10 shows the generated power of the stack, and the black asterisk marker indicates the maximum power points at each fuel utilization factor. The result shows that when inlet fuel flow was high, a high recirculation ratio has the advantage of power generation. However, a low recirculation ratio was better at an excessively high fuel utilization factor. An optimized point generating the maximum power of 860.82 W was achieved when the fuel utilization factor was 0.61 and the recirculation ratio was 0.26.

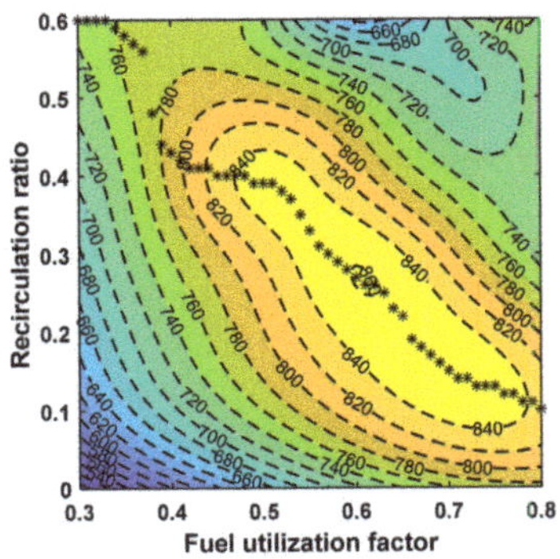

Figure 10. Generated power of AOGR #2 system with various fuel utilization factors and recirculation ratios.

Electrical and thermal efficiency changes are presented in Figure 11. The change in electrical efficiency is similar to the result of Figure 9b. The black asterisk marker in Figure 11a represents the point of the maximum electrical efficiency at each fuel utilization factor. At a fixed fuel utilization factor, the electrical efficiency increased along the raised recirculation ratio until fuel dilution occurred. The maximum value was 53.44% when the fuel utilization factor and recirculation ratio were 0.80 and 0.10, respectively. In Figure 11b, the thermal efficiency always showed a maximum value as there was no recirculated hydrogen. The thermal efficiency continued decreasing when the fuel flow rate decreased.

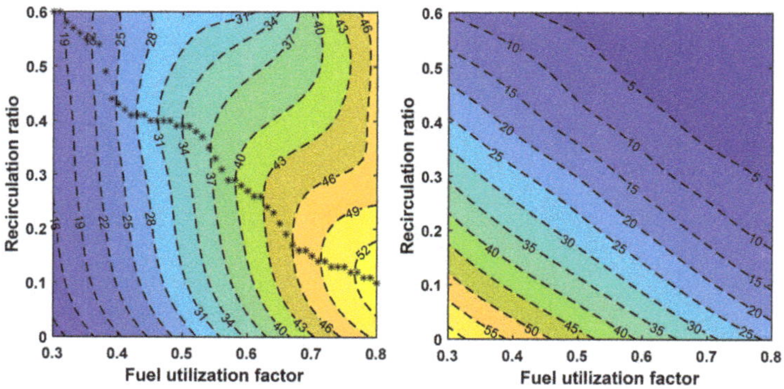

(a) (b)

Figure 11. Efficiency of the AOGR #2 system with various fuel utilization factors and recirculation ratios; (**a**) electrical and (**b**) thermal efficiency.

5. Conclusions

To validate the effectiveness of the SOFC system with AOGR, three configurations of the SOFC system were numerically evaluated. Two different system configurations with AOGR systems using recirculation blowers were examined, and the results were compared to those of a reference system. With a developed dynamic model, the temperature of each component, power, and system efficiency were analyzed under various fuel/air utilization factors and recirculation ratios.

As the fuel and air utilization factors increased, the net power and total efficiency of the systems were enhanced because of the rise in the stack temperature. For the SOFC system with AOGR, the performance began to decrease after a certain fuel utilization factor because of fuel dilution by recirculated AOG. The AOGR #1 system showed greater electricity generation than the reference system, but the total efficiency of the reference system was higher than that of the AOGR #1 system since the reference system can recover more heat from the exhaust gas. Both the electrical and total efficiency of AOGR #2 had the highest values among the three systems. The ESR of the AOGR #2 system did not absorb additional heat from CCOG, so the thermal energy utilization of the AOGR #2 system was much higher than that of the other systems.

Anode off-gas recirculation can improve fuel utilization efficiency but also diminish the performance because of fuel dilution. The performance change with various recirculation ratios and fuel utilization factors was examined for AOGR #1 and AOGR #2 systems, and the results indicate that there is an optimum recirculation ratio depending on the operating conditions. The maximum power and electrical efficiency of AOGR #2 were 860.82 W with a fuel utilization factor of 0.61 and recirculation ratio of 0.26 and 53.44% with a fuel utilization factor of 0.80 and recirculation ratio of 0.10, respectively. In terms of thermal efficiency, however, the maximum value was achieved when the recirculation ratio was 0. While operating the SOFC system, the requirements of electricity and heat varied according to the situation. Therefore, considering the electrical and thermal load, the system operating conditions need to be controlled. Based on this study, research analyzing the dynamic characteristics of each system and developing optimized control strategies are in progress.

Author Contributions: Conceptualization, E.-J.C. and S.-M.L.; simulation, E.-J.C. and S.Y.; investigation, E.-J.C. and J.-M.K.; writing, E.-J.C.; supervision, S.-M.L. and E.-J.C. All authors have read and agreed to the published version of the manuscript.

Funding: This work was supported by the Technology Innovation Program funded by the Korea Evaluation Institute of Industrial Technology (KEIT) and the Ministry of Trade, Industry & Energy (MOTIE) of the Republic of Korea (No. 20004963).

Institutional Review Board Statement: Not applicable.

Informed Consent Statement: Not applicable.

Conflicts of Interest: The authors declare no conflict of interest.

Nomenclature

A	area (m^2)
a_v	external surface area per unit volume of catalyst bed (m^2 m^{-3})
C	molar concentration (mole m^{-3})
c_p	specific heat capacity (J kg^{-1} K^{-1})
D	effective diffusion coefficient (m^2 s^{-1})
D_h	hydraulic diameter (m)
d_p	particle diameter (m)
E_0	Nernst voltage (V)
F	Faraday constant (C $mole^{-1}$)
f	friction factor
$\cdot H$	heat of reaction (kJ $mole^{-1}$)
H	convective heat transfer coefficient (W m^{-2} K^{-1})
J	current (A)
j	current density (A cm^{-2})
j_0	exchange current density (A cm^{-2})
j_D, j_H	Chilton-Colburn factor for mass and heat transfer
j_{lim}	limit current density (A cm^{-2})
K	equilibrium constant (described in the paper)
k	forward reaction rate constant (described in the paper)
k_g	gas to solid mass transfer coefficient (m^3 m^{-2} s^{-1})
k_m	mass transfer coefficient (m s^{-1})
L_c	cell length (m)
m	mass flow rate (kg s^{-1})
N_{ch}	number of channels
N_u	Nusselt number
n	molar flow rate (mole s^{-1})
q	convective heat transfer rate (W m^{-2})
P	partial pressure (pa), power (W)
Pr	Prandtl number
R	ideal gas constant 8.314 (J $mole^{-1}$ K^{-1})
R_k	reaction rate ($k = R1,\ R2,\ R3,\ R4,\ R5$, described in the paper)
Re	Reynolds number
r	molar rate of formation or consumption (mole m^{-3} s^{-1})
r_{SMR}	molar rate of formation or consumption at SMR (mole kg_{cat}^{-1} s^{-1})
S	geometric surface area per unit reactor volume (m^2 m^{-3})
Sc	Schmidt number
T	temperature (K)
t	time (s), thickness (m)
t_f	anode electrode thickness (m)
U	utilization factor

u	velocity (m s^{-1})
V	voltage (V)
V_a	anode volume (m^3)
V_c	cathode volume (m^3)
v	superficial velocity of the gas (m s^{-1})
w	cell width (m)
y	molar fraction
α	catalytic surface area per unit reactor volume (m^2 m^{-3}), ion transfer coefficients
ε	porosity
η	overpotential (V), efficiency
η_k	effectiveness factor of reaction k
κ	specific heat ratio
ρ	density (kg m^{-3})
σ	electrode conductivity (S m^{-1})
φ	aspect ratio
Subscripts	
a	air, anode
b	bed, base
c	cathode
cat	catalyst
ch	channel
e	electrode
ele	electrical
f	fuel, fin
g	gas
i	species, inlet
k	reaction
m	monolith
o	outlet
s	solid
Abbreviations	
AOG	anode off-gas
AOGR	anode off gas recirculation
CC	catalytic combustor
COG	cathode off-gas
CCOG	catalytic combustor off-gas
DIR	direct inter reforming
ESR	external steam reformer
ER	external reforming
HE	heat exchanger
HR-HE	heat recovery heat exchanger
PEN	positive-electrolyte-negative
RBlower	recirculation blower
S/C	steam to carbon
SMR	steam methane reforming
SOFC	solid oxide fuel cell
WGS	water gas shift

References

1. Xi, H.; Sun, J.; Tsourapas, V. A control oriented low order dynamic model for planar SOFC using minimum Gibbs free energy method. *J. Power Sources* **2007**, *165*, 253–266. [CrossRef]
2. Haberman, B.A.; Young, J.B. Three-dimensional simulation of chemically reacting gas flows in the porous support structure of an integrated-planar solid oxide fuel cell. *Int. J. Heat Mass Transf.* **2004**, *47*, 3617–3629. [CrossRef]
3. Yin, F.; Ji, S.; Chen, B.; Zhao, L.; Liu, H.; Li, C. Preparation and characterization of $LaFe_{1-x}MgxO_3/Al_2O_3/FeCrAl$: Catalytic properties in methane combustion. *Appl. Catal. B Environ.* **2006**, *66*, 265–273. [CrossRef]
4. Van Biert, L.; Visser, K.; Aravind, P.V. A comparison of steam reforming concepts in solid oxide fuel cell systems. *Appl. Energy* **2020**, *264*, 114748. [CrossRef]

5. Farhad, S.; Hamdullahpur, F.; Yoo, Y. Performance evaluation of different configurations of biogas-fuelled SOFC micro-CHP systems for residential applications. *Int. J. Hydrogen Energy* **2010**, *35*, 3758–3768. [CrossRef]
6. Braun, R.J.; Klein, S.A.; Reindl, D.T. Evaluation of system configurations for solid oxide fuel cell-based micro-combined heat and power generators in residential applications. *J. Power Sources* **2006**, *158*, 1290–1305. [CrossRef]
7. Zhang, X.; Chan, S.H.; Li, G.; Ho, H.K.; Li, J.; Feng, Z. A review of integration strategies for solid oxide fuel cells. *J. Power Sources* **2010**, *195*, 685–702. [CrossRef]
8. Suzuki, M.; Sogi, T.; Higaki, K.; Ono, T.; Takahashi, N.; Shimazu, K.; Shigehisa, T. Development of SOFC residential cogeneration system at Osaka gas and Kyocera. *ECS Trans.* **2007**, *7*, 27–30. [CrossRef]
9. Riensche, E.; Stimming, U.; Unverzagt, G. Optimization of a 200 kW SOFC cogeneration power plant: Part I: Variation of process parameters. *J. Power Sources* **1998**, *73*, 251–256. [CrossRef]
10. Lee, K.; Kang, S.; Ahn, K.Y. Development of a highly efficient solid oxide fuel cell system. *Appl. Energy* **2017**, *205*, 822–833. [CrossRef]
11. Lisbona, P.; Corradetti, A.; Bove, R.; Lunghi, P. Analysis of a solid oxide fuel cell system for combined heat and power applications under non-nominal conditions. *Electrochim. Acta* **2007**, *53*, 1920–1930. [CrossRef]
12. Powell, M.; Meinhardt, K.; Sprenkle, V.; Chick, L.; McVay, G. Demonstration of a highly efficient solid oxide fuel cell power system using adiabatic steam reforming and anode gas recirculation. *J. Power Sources* **2012**, *205*, 377–384. [CrossRef]
13. Koo, T.; Kim, Y.S.; Lee, D.; Yu, S.; Lee, Y.D. System Simulation and Exergetic Analysis of Solid Oxide Fuel Cell Power Generation System with Cascade Configuration. *Energy* **2021**, *214*, 119087. [CrossRef]
14. Wagner, P.H.; Wuillemin, Z.; Constantin, D.; Diethelm, S.; van Herle, J.; Schiffmann, J. Experimental Characterization of a Solid Oxide Fuel Cell Coupled to a Steam-Driven Micro Anode off-Gas Recirculation Fan. *Appl. Energy* **2020**, *262*. [CrossRef]
15. Baba, S.; Takahashi, S.; Kobayashi, N.; Hirano, S. Performance of Anodic Recirculation by a Variable Flow Ejector for a Solid Oxide Fuel Cell System under Partial Loads. *Int. J. Hydrogen Energy* **2020**, *45*, 10039–10049. [CrossRef]
16. Tanaka, Y.; Sato, K.; Yamamoto, A.; Kato, T. Development of Anode Off-Gas Recycle Blowers for High Efficiency SOFC Systems. *ECS Trans.* **2013**, *57*, 443–450. [CrossRef]
17. Dietrich, R.-U.; Lindermeir, A.; Immisch, C.; Spieker, C.; Spitta, C.; Stenger, S.; Leithner, R.; Kuster, T.; Oberland, A. SOFC System Using a Hot Gas Ejector for Offgas Recycling for High Efficient Power Generation from Propane. *ECS Trans.* **2013**, *57*, 171–184. [CrossRef]
18. Kim, Y.S.; Lee, Y.D.; Ahn, K.Y. System integration and proof-of-concept test results of SOFC–engine hybrid power generation system. *Appl. Energy* **2020**, *277*, 115542. [CrossRef]
19. Kuchonthara, P.; Bhattacharya, S.; Tsutsumi, A. Combinations of solid oxide fuel cell and several enhanced gas turbine cycles. *J. Power Sources* **2003**, *124*, 65–75. [CrossRef]
20. Aguiar, P.; Adjiman, C.S.; Brandon, N.P. Anode-supported intermediate temperature direct internal reforming solid oxide fuel cell I: Model-based steady-state performance. *J. Power Sources* **2004**, *138*, 120–136. [CrossRef]
21. Kendall, K.; Singhal, S.C. *High-Temperature Solid Oxide Fuel Cells: Fundamentals, Design and Applications*, 1st ed.; Elsevier Science: Amsterdam, The Netherlands, 2013.
22. Kong, Q.; Yin, Y.; Xue, B.; Jin, Y.; Feng, W.; Chen, Z.G.; Su, S.; Sun, C. Improved catalytic combustion of methane using CuO nanobelts with predominantly (001) surfaces. *Beilstein J. Nanotechnol.* **2018**, *9*, 2526–2532. [CrossRef]
23. Van Biert, L.; Godjevac, M.; Visser, K.; Aravind, P.V. Dynamic modelling of a direct internal reforming solid oxide fuel cell stack based on single cell experiments. *Appl. Energy* **2019**, *250*, 976–990. [CrossRef]
24. Ota, T.; Koyama, M.; Wen, C.J.; Yamada, K.; Takahashi, H. Object-based modeling of SOFC system: Dynamic behavior of micro-tube SOFC. *J. Power Sources* **2003**, *118*, 430–439. [CrossRef]
25. Achenbach, E. Three-dimensional and time-dependent simulation of a planar solid oxide fuel cell stack. *J. Power Sources* **1994**, *49*, 333–348. [CrossRef]
26. Nakagawa, N.; Sagara, H.; Kato, K. Catalytic activity of Ni–YSZ–CeO_2 anode for the steam reforming of methane in a direct internal-reforming solid oxide fuel cell. *J. Power Sources* **2001**, *92*, 88–94. [CrossRef]
27. Lee, A.L.; Zabransky, R.; Huber, W. Internal reforming development for solid oxide fuel cells. *Ind. Eng. Chem. Res.* **1990**, *29*, 766–773. [CrossRef]
28. Abbas, S.Z.; Dupont, V.; Mahmud, T. Kinetics study and modelling of steam methane reforming process over a NiO/Al_2O_3 catalyst in an adiabatic packed bed reactor. *Int. J. Hydrogen Energy* **2017**, *42*, 2889–2903. [CrossRef]
29. Chinda, P.; Chanchaona, S.; Brault, P.; Wechsatol, W. A planar anode-supported solid oxide fuel cell model with internal reforming of natural gas. *Eur. Phys. J. Appl. Phys.* **2011**, *54*, 23405. [CrossRef]
30. Lehnert, W.; Meusinger, J.; Thom, F. Modelling of gas transport phenomena in SOFC anodes. *J. Power Sources* **2000**, *87*, 57–63. [CrossRef]
31. Campanari, S.; Iora, P. Definition and sensitivity analysis of a finite volume SOFC model for a tubular cell geometry. *J. Power Sources* **2004**, *132*, 113–126. [CrossRef]
32. Mbodji, M.; Commenge, J.M.; Falk, L.; di Marco, D.; Rossignol, F.; Prost, L.; Valentin, S.; Joly, R.; Del-Gallo, P. Steam methane reforming reaction process intensification by using a millistructured reactor: Experimental setup and model validation for global kinetic reaction rate estimation. *Chem. Eng. J.* **2012**, *207–208*, 871–884. [CrossRef]
33. Rostrup-Nielsen, J.R. *Catalytic Steam Reforming*; Springer: Berlin, Germany, 1984.

34. Twigg, M.V. *Catalyst Handbook*; Wolfe Publishing: London, UK, 1989.
35. Xu, J.; Froment, G.F. Methane steam reforming, methanation and water-gas shift: I. Intrinsic kinetics. *AIChE J.* **1989**, *35*, 88–96. [CrossRef]
36. Geankoplis, C.J. *Transport Processes and Unit Operations*; Pearson: London, UK, 1993.
37. Yoshida, F.; Ramaswami, D.; Hougen, O.A. Temperatures and partial pressures at the surfaces of catalyst particles. *AIChE J.* **1962**, *8*, 5–11. [CrossRef]
38. Oh, S.H.; Cavendish, J.C. Transients of monolithic catalytic converters. Response to step changes in feed stream temperature as related to controlling automobile emissions. *Ind. Eng. Chem. Prod. Res. Dev.* **1982**, *21*, 29–37. [CrossRef]

energies

MDPI

Article

Heat Transfer Analysis in Supercritical Hydrogen of Decoupled Poisoned Hydrogen Moderator with Non-Uniform Heat Source of Chinese Spallation Neutron Source

Jianfei Tong [1,2,3], Lingbo Zhu [2,3,4], Yiping Lu [4], Tianjiao Liang [2,3], Youlian Lu [2,3], Songlin Wang [2,3], Chaoju Yu [2,3], Shikui Dong [1,*] and Heping Tan [1]

1 School of Energy Science and Engineering, Harbin Institute of Technology, Harbin 150001, China; tongjf@ihep.ac.cn (J.T.); tanhp@hit.edu.cn (H.T.)
2 Institute of High Energy Physics, Chinese Academy of Sciences, Beijing 100049, China; zhulingbo@ihep.ac.cn (L.Z.); liangtj@ihep.ac.cn (T.L.); luyl@ihep.ac.cn (Y.L.); wangsl@ihep.ac.cn (S.W.); yucj@ihep.ac.cn (C.Y.)
3 Spallation Neutron Source Science Center, Dongguan 523803, China
4 Department of Mechanical and Power Engineering, Harbin University of Science and Technology, Harbin 150001, China; luyiping@hrbust.edu.cn
* Correspondence: dongsk@hit.edu.cn

Abstract: The flow field distribution and thermal properties of supercritical hydrogen are crucial factors affecting the quality of neutrons output from spallation neutron source, which may contribute to the optimization design of the moderator. Several sensitivity studies on affecting heat transfer characteristics of liquid hydrogen inside a moderator were executed, and a choice was made to use a computational fluid dynamics method for numerical simulation. The sensitivity degree of factors affecting the heat transfer characteristics of liquid hydrogen are in sequence of inlet mass flow, beam power and operating pressure. Especially when the beam power is 500 kW (the temperature range of liquid hydrogen is about 20~30 K); where the effect of mass flow rate is remarkable, the cooling effect is best in the range of 60~90 g/s $\times$ 394 mm^2. Meanwhile, the maximum temperature of liquid hydrogen is close to the bottom recirculation zone due to the influence of the flow field and the heat deposition distribution of the poisoned plate. The effect of variable pressure on the temperature of liquid hydrogen is not significant, whereas the sudden rise of wall temperature is observed near the large specific heat region of 15 bar.

Keywords: Chinese spallation neutron source; decoupled poisoned hydrogen moderator; non-uniform heat source; numerical simulation; supercritical pressure

Citation: Tong, J.; Zhu, L.; Lu, Y.; Liang, T.; Lu, Y.; Wang, S.; Yu, C.; Dong, S.; Tan, H. Heat Transfer Analysis in Supercritical Hydrogen of Decoupled Poisoned Hydrogen Moderator with Non-Uniform Heat Source of Chinese Spallation Neutron Source. *Energies* **2021**, *14*, 4547. https://doi.org/10.3390/en14154547

Academic Editor: Vladislav A. Sadykov

Received: 15 June 2021
Accepted: 24 July 2021
Published: 27 July 2021

1. Introduction

The Chinese Spallation Neutron Source (CSNS) generates intensive neutrons due to the impact of 1.6 GeV high-energy pronton on heavy metal targets. Thermal and cold neutrons after deceleration can be used to study the atomic structure and motion of certain objects [1,2]. The current beam power of a CSNS target station is 100 kW; the phase II target is plans to upgrade this to 500 kW. As the core component of the target station, the moderator slows down the leaking neutrons generated in the spall target for neutron scattering experiments [3]. The Decouple Poison Hydrogen Moderator (DPHM) is taken as the research object owing to its complex internal structure. At the same time, the existence of a poisoned plate inside the container leads to uneven distribution of flow and heat transfer.

During the operation of the target station, special working conditions should be considered, such as power off mode, distortion mode and refrigerator failure, etc., which will lead to a sharp increase in the thermal load of the hydrogen circulation system and a rise in temperature and pressure. In order to protect the cryogenic equipment, liquid hydrogen must be discharged in an emergency. In the above process, the ambient pressure decreases from 1.5 MPa to 0.1 MPa, and the liquid hydrogen in the DPHM changes from a supercritical state to supercooled state [4,5]. Due to the transcritical process, the change of pressure and temperature has a strong influence on the thermophysical properties of liquid hydrogen, for instance density, viscosity and thermal conductivity, which all make the flow heat transfer in the container more complicated [6,7]. In addition, thermal deposition increases significantly with the enhancement of proton beam power at the target station, thus leading to the further rise of the overall temperature of DPHM. To ensure the stable operation of the system, the average temperature of liquid hydrogen and the temperature difference between the inlet and outlet should be maintained in certain extent.

The heat transfer characteristics of supercritical fluid have been investigated extensively, especially the investigation of in tubes have become a focal point of research content [8–11]. Considering the influence of different factors on heat transfer, Wang et al. [12] found that the buoyancy effect under the non-uniform heat source can effectively alleviate the degree of Heat Transfer Deterioration (HTD). Meanwhile, a modified turbulent model, which can accurately predict the influence of semicircular heating condition on the flow and heat transfer, was also proposed. Zhu et al. [13] conducted a numerical study about the effects of gravity on the heat transfer performance of supercritical CO_2 flowing within a vertical tube. They found that the effect of gravity on heat transfer is pronounced and closely related to the variations of thermophysical properties, particularly with low mass flux condition. The experiments on turbulent heat transfer via supercritical CO_2 in a vertical tube were carried out by Kim et al. [14]. It is indicated that the distribution of wall temperature, which had a noticeable peak value when the wall heat flux was moderated and the mass flux was low, varied with buoyancy effect and flow acceleration. Nevertheless, it seems that issues such as the heat transfer characteristics of supercritical liquid hydrogen have not received sufficient attention. Youn et al. [15] obtained the variable rules of thermophysical properties of supercritical hydrogen by changing inlet conditions, for example the temperature, pressure, mass flow rate, etc. The evaluations of numerous correlations for heat transfer to supercritical hydrogen flowing turbulently in circular tubes were presented by Locke and Landrum [16]. Furthermore, In comparision with other correlations of supercritical hydrogen, by modifying the relevant parameters of correlations about non-hydrogen supercritical fluid with variable properties, this method can be applied to liquid hydrogen and obtain more accurate prediction results. James [17] described the entire validation process for a model of the heat transfer coefficient and the expected content required to complete the validation. By utilizing the above model, the convection heat transfer coefficient between the supercritical, cryogenic hydrogen and the moderator vessel walls was verified. Xie and Zhang [18–20] have carried out a number of studies on enhanced heat transfer of supercritical liquid hydrogen. Specifically, the rib structure was not only added to the wall, but also included some surface grooves and bulges, which were conducive to the enhancement of cooling performance and reducing the influence of the non-uniform distribution on heat transfer. The numerical simulation method of convective heat transfer is also important, which determines the reliability and accuracy of the simulation results. Mosavati et al. [21,22] proposed a novel calculation method for convective heat transfer in a closed cavity, specifically for the distribution factors using backward Monte Carlo method. Moreover, the effects of Rayleigh number, temperature ratio, radiation conductivity and other parameters on heat flux were studied, and the results were compared and analyzed. They conclude that lowering the temperature ratio will make the heat source surface temperature distribution smoother and lead to greater radiation heat transfer flux. At the same time, by increasing the Rayleigh number, the convective heat transfer flux can be significantly improved, and the heat flux on the heat source surface becomes uneven.

Therefore, it is necessary to explore the heat flow characteristics of liquid hydrogen in DPHM under the conditions of changing pressure, heat source and mass flow rate. The study reported here was undertaken to validate the accuracy of the supercritical model by taking the liquid hydrogen in the moderator as the research object. Meanwhile, the corresponding heat sources under different beam power of DPHM were investigated based on the neutron physical-thermal coupling method. Finally, the Computational Fluid Dynamics (CFD) method was used to simulate the flow and heat transfer characteristics of liquid hydrogen by changing the boundary conditions of DPHM.

2. Mathematical and Physical Model

2.1. Physical Model

Due to the internal complexity of the DPHM model, it was simplified for the convenience of numerical simulation analysis, as shown in Figure 1. The main structure of DPHM is made of the inlet-pipe of hydrogen, the outlet-pipe of hydrogen, container and poisoned plate, in which the structure of the pipeline is coaxial multi-layer casing. The distance between the exit of the inlet-pipe of hydrogen and the bottom surface of the container is 5 mm. After the liquid hydrogen flows along inlet-pipe of hydrogen for a distance, the poisoned plate separates the flow of it into two uneven sides, which are injected vertically to the bottom surface of the inner cavity under the action of pressure difference to form a circular jet for cooling [23]. The pressure generated by the impact forces the liquid hydrogen to flow radially along the wall, and then, due to the limitation of the internal structure, it is concentrated upward at the neck of the container and finally discharged by the hydrogen return tube. Due to the thermal deposition caused by neutron collisions, liquid hydrogen is mainly heated by pipelines, poisoned plate and container during flow.

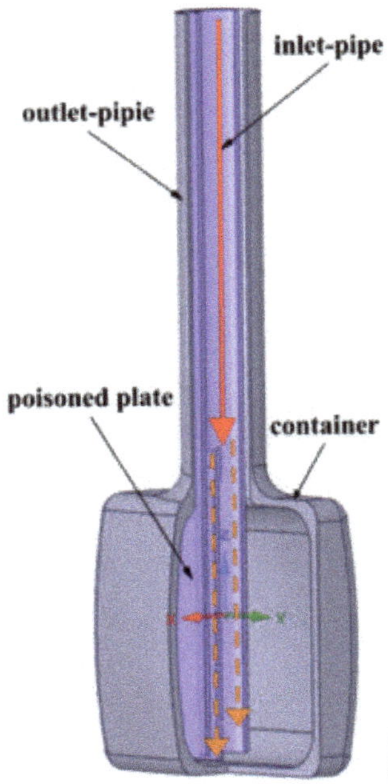

(a)

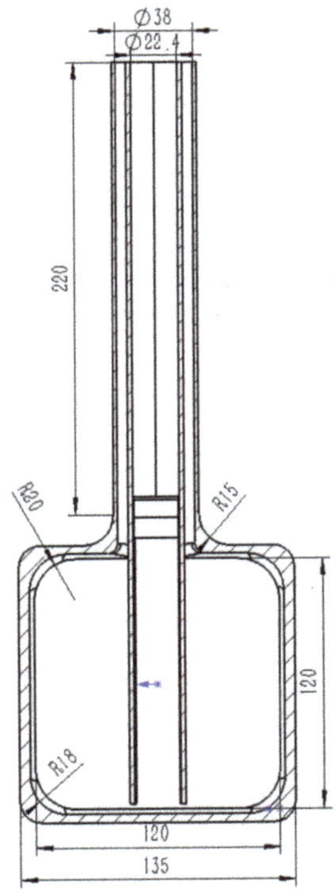

(b)

Figure 1. The Computational Model of DPHM: (**a**) Model Diagram; (**b**) Detailed Size (mm).

The MCNPX is a universal Monte Carlo transport program developed by Los Alamos National Laboratory in the United States, which has been proved to be able to accurately simulate the scattering reaction of high-energy particles inside the moderator [24]. In the current work, the external coupling method was used to modify the common parameters of MCPNX 2.5 and CFX 11.0 software [25,26], the heat source distribution of the moderator calculated by the former was taken as the thermal boundary condition of the input of the latter.

2.2. Meshing

The mesh of DPHM generated by commercial software ICEM 14.0 is shown in Figure 2, where the unstructured mesh with strong adaptability was selected to discretize the fluid domain. The overall quality of the grid is detailed; the maximum and average skewness are 0.652 and 0.223 respectively, and the average aspect ratio is 3.124, which proves the reliability of grid division. Considering the influence of boundary layer on the main flow area, the local mesh of areas with large curvature and complex flow, such as inlet, outlet and wall boundary area, was encrypted so as to accurately capture the flow characteristics. In order to accurately predict the turbulent flow field, the height of the first layer of the boundary layer grid is set to be small enough to meet the criterion of y^+ value close to 1. Five inflation layers are divided in the sensitive region, so the estimation of the thermal gradients can be improved. The selection height of the first layer grid in the calculation model is calculated by the following Equation [27],

$$Re = \frac{\rho u L}{\mu} \tag{1}$$

$$C_f = 0.058 Re^{-0.2} \tag{2}$$

$$\tau_w = \frac{1}{2} C_f \rho u^2 \tag{3}$$

$$U_\tau = \sqrt{\frac{\tau_w}{\rho}} \tag{4}$$

$$y = \frac{y^+ \mu}{U_\tau \rho} \tag{5}$$

where Re, ρ, u, L, μ, C_f, τ_w and U_τ are the dimensionless number, density, bulk fluid velocity, characteristic length, dynamic viscosity of fluid, boundary layer friction coefficient, boundary layer shear stress and shear velocity. Finally, after calculation, the value of the first layer grid height y^+ is 2 mm.

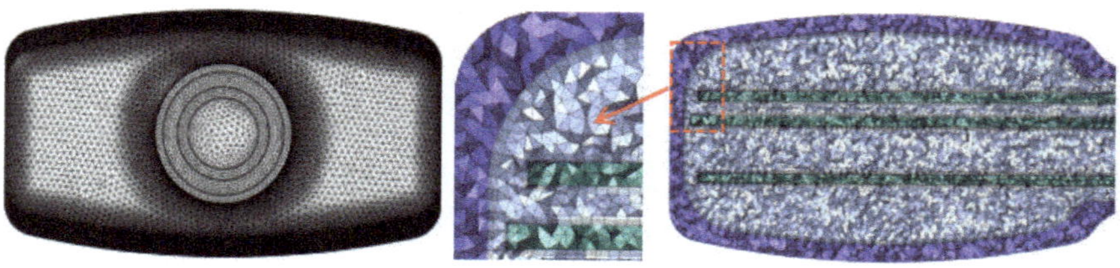

Figure 2. Generation mesh in the computational domain.

2.3. Governing Equations

The liquid hydrogen jet cooling in the moderator belongs to high-speed flow, so the influence of gravity on heat transfer flow can be ignored. In the transcritical process, the pressure has a great influence on the density of liquid hydrogen [28]. Therefore, this article assumes that the liquid hydrogen in the moderator is a compressible fluid, and only the steady flow is studied. The specific governing equations are as follows:

The continuity equation [29]:

$$\nabla \cdot (\rho \mathbf{u}) = 0 \tag{6}$$

The momentum equation:

$$\begin{cases} \nabla \cdot (\rho u \mathbf{u}) = -\frac{\partial p}{\partial x} + \frac{\partial \tau_{xx}}{\partial x} + \frac{\partial \tau_{yx}}{\partial y} + \frac{\partial \tau_{zx}}{\partial z} \\ \nabla \cdot (\rho v \mathbf{u}) = -\frac{\partial p}{\partial x} + \frac{\partial \tau_{xy}}{\partial x} + \frac{\partial \tau_{yy}}{\partial y} + \frac{\partial \tau_{zy}}{\partial z} \\ \nabla \cdot (\rho w \mathbf{u}) = -\frac{\partial p}{\partial x} + \frac{\partial \tau_{xz}}{\partial x} + \frac{\partial \tau_{yz}}{\partial y} + \frac{\partial \tau_{zz}}{\partial z} \end{cases} \tag{7}$$

The energy equation [30]:

$$\nabla \cdot [(\rho E + P)\mathbf{u}] = \nabla \cdot \left(\lambda_{eff} \nabla T - \left(\tau_{eff} \cdot \mathbf{u}\right)\right) + S_h \tag{8}$$

where $\mathbf{u}$, $\boldsymbol{\tau}$, p, P, E, T, λ_{eff}, τ_{eff} and S_h denote the velocity vector, stress tensor, bulk fluid pressure, static pressure, total energy, bulk fluid temperature, effective thermal conductivity, effective stress tensor and volumetric heat source, respectively.

In this paper, the steady-state heat specifies and the material is isotropic, so the Fourier law was employed to describe the heat conduction within the solid wall [31].

$$\nabla \cdot (\lambda \nabla T) = 0 \tag{9}$$

where λ is the thermal conductivity of the fluid. The shear stress transport (SST) k-ω turbulence model, which takes into account the transport of turbulent shear stress and has a high accuracy in predicting the complex flow, is used to solve the three-dimensional flow of liquid hydrogen.

2.4. Boundary Conditions

The boundary conditions for each structure of the computational model, including fluid domain and solid domain, should be set before the simulation begins. Moreover, the non-uniform heat source of DPHM is imported into CFX software by compiling User Defined Function (UDF).

1. The inlet temperature ranges from 18 to 30 K, the inlet mass flow rate 30 to 150 g/s, the pressure from 11 to 15 bar. It is assumed that the flow at the inlet has been fully developed, and the boundary condition of the entrance is set as the average mass. The outlet boundary condition was set as the pressure outlet according to the standard atmospheric pressure, and the outlet calculation domain was appropriately extended to avoid the backflow phenomenon.
2. The standard wall function is used for wall treatment, and the no-slip boundary condition is adopted. The container is set as an adiabatic wall surface, the second-order upwind format is used for the discrete equation, and Semi-Implicit Method for Pressure Linked Equations (SIMPLE) algorithm is selected for the pressure-velocity coupling method. The maximum number of convergent iterations is 8000, and the convergence residual Root Mean Square of the residual values (RMS) value is set to 10^{-6} to obtain a stable convergent solution.
3. The non-uniform heat source of the moderator was obtained by external coupling of MCNPX and CFX software and applied to the moderator. Table 1 lists the comparison of the corrected heat sources between MCNPX and CFX when the proton beam power is 500 kW. Since the models used in the two softwares are slightly different, there are some errors in the calculation values of thermal deposition of materials, but they are within the allowable range, which proves the accuracy and reliability of the coupling results.

Table 1. Comparison of heat deposition calculation results between MCNPX and CFX (W).

Structure	CFX	MCNPX
Liquid hydrogen	332	324
Poisoned plate	43	40
Aluminum coating	44	43
The container	389	385

2.5. *Verification of Grid Independence*

In this paper, the grid independence of DPHM is verified under the conditions of beam power of 100 kW, pressure of 15 bar and inlet flow of 30 g/s. Five groups of grids were selected for comparison, and the three parameters of maximum temperature, maximum pressure of liquid hydrogen and maximum container temperature were detected. The results are shown in Figure 3. It can be seen that the temperature variation tends to be stable when the number of grid elements exceeds 1.68 million, indicating that the independent solution can be obtained for the grid number at this condition. The Grid Convergence Index (GCI) was used to quantify the grid independence [32]. The GCI_{34} for fine, and medium grids was 1.76%. The GCI_{45} for medium and coarse grids was 4.61%. The value of $GCI_{45}/(r^{P}GCI_{34})$ was 1.03, which was approximately 1 and indicates that the solutions were well within the asymptotic range of convergence. Therefore, considering the computational efficiency and accuracy, the fourth group of grids was finally employed for calculation.

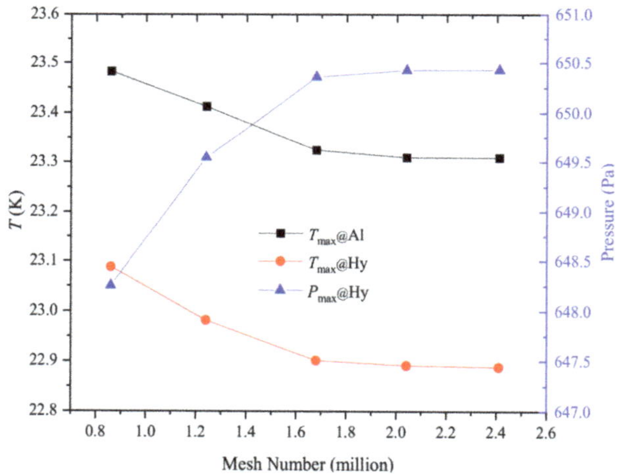

Figure 3. Grid independence validation of DPHM.

2.6. *Verification of Physical Parameters*

Pseudo-critical temperature refers to the temperature corresponding to the maximum specific heat capacity of fluid at constant pressure in a supercritical state (Critical pressure and temperature of liquid hydrogen are 12.9 bar and 33.15 K, respectively). In the vicinity of this temperature range, the thermophysical properties of the fluid will change dramatically, thus affecting the flow and heat transfer. As shown in Figure 4a, the density ρ and dynamic viscosity μ decrease with the gradual increase of temperature under 15 bar but converge to a constant value as the temperature is far beyond the critical value. Meanwhile, the change in specific heat c_p is most significant, which is further reflected in Figure 4b. There is a prominent phenomenon called the "thermal spike phenomenon", which is not conducive to the stability of heat transfer. This primary peak in λ and c_p of hydrogen disappears with the augment of pressure due to the corresponding Pseudo-critical temperature increasing roughly from 32 to 34 K [33].

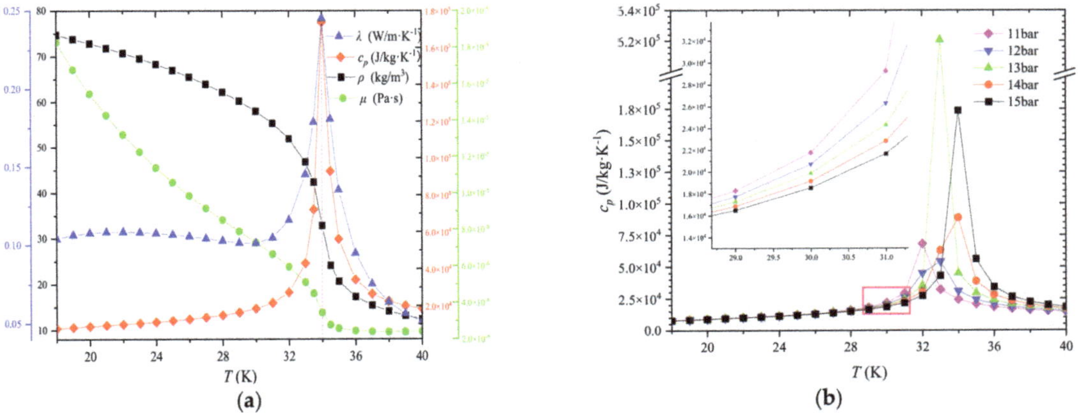

Figure 4. Variations of thermophysical properties of liquid hydrogen under different pressure: (**a**) Pressure = 15 bar; (**b**) Specific Heat (different pressure).

Accurate prediction of the thermal properties of hydrogen is crucial to the reliability of the simulation results. Based on equations of state (EoSs) theory [34,35], the physical parameters of hydrogen, in which the α parameter and pressure–volume–temperature (PVT) relationships have an important influence, can be accurately predicted. In order to more accurately simulate the flow heat transfer of non-ideal fluids under supercritical conditions, under the premise of considering the calculation accuracy and efficiency, this paper adopts the Peng–Robinson (PR-EOS) equation of state [31], which is universal in engineering, to calculate the physical parameters of working fluids:

$$p = \frac{RT}{v_{PR} - b} - \frac{a(T)}{v_{PR}(v_{PR} + b) + b(v_{PR} - b)} \tag{10}$$

where

$$a(T) = 0.45724\frac{R^2T_c^2}{p_c}\left[1 + \kappa\left(1 - \sqrt{(T/T_c)}\right)\right]^2 \tag{11}$$

$$b = 0.0778\frac{RT_c}{p_c} \tag{12}$$

$$\kappa = 0.37464 + 1.54226\omega - 0.26992\omega^2 \tag{13}$$

where R, v_{PR}, T_c, p_c, and ω are the molecular gas constant, the molar volume, the critical temperature, the critical pressure and the Pitzer acentric factor.

Whereas the prediction results of liquid density obtained by the PR-EOS formula have poor accuracy, especially when the pressure is close to the critical region and in a small temperature range. Khashayar [36] evaluated 11 correlations for predicting hydrogen density and found that the modified Redlich–Kwong EoS by Mathias and Copeman (RKMC) [37] was widely applicable to predict liquid hydrogen properties in various ranges. The deviation between the experimental results and the RKMC EoS equation was small in the temperature range of 14~32 K, which proved the rationality and accuracy of the RKMC EoS equation. Since the liquid hydrogen in this paper belongs to the supercritical state, and the temperature change is located in the above range, the RKMC EOS equation is used for the subsequent calculation in this paper. The specific equation are as follows:

$$p = \frac{RT}{v - b} - \frac{a_c\alpha(T_r)}{v(v + b)} \tag{14}$$

$$b = 0.08664\frac{RT_c}{p_c} \tag{15}$$

$$a_c = 0.42748\frac{R^2T_c^2}{p_c} \tag{16}$$

where

$$\alpha = \begin{cases} \left[1 + c_1(1-\sqrt{T_r}) + c_2(1-\sqrt{T_r})^2 + c_3(1-\sqrt{T_r})^3\right]^2, T_r \le 1 \\ \left[1 + c_1(1-\sqrt{T_r})\right]^2, T_r > 1 \end{cases} \tag{17}$$

where a_c is the critical attractive parameter, b is the molar co-volume, α is the temperature dependence function (alpha function), c_1,c_2,c_3 are the coefficients of the Mathias and Copeman alpha function, and T_r is the reference temperature.

Figure 5 shows the density and viscosity distributions at temperatures of 15~40 K at pressures of 5, 10 and 15 bar. In order to verify the accuracy of the above model, a comparison was made with the data of GASPAK [38], the results showing that the two sets of data were well in agreement and that the error was within the allowable range.

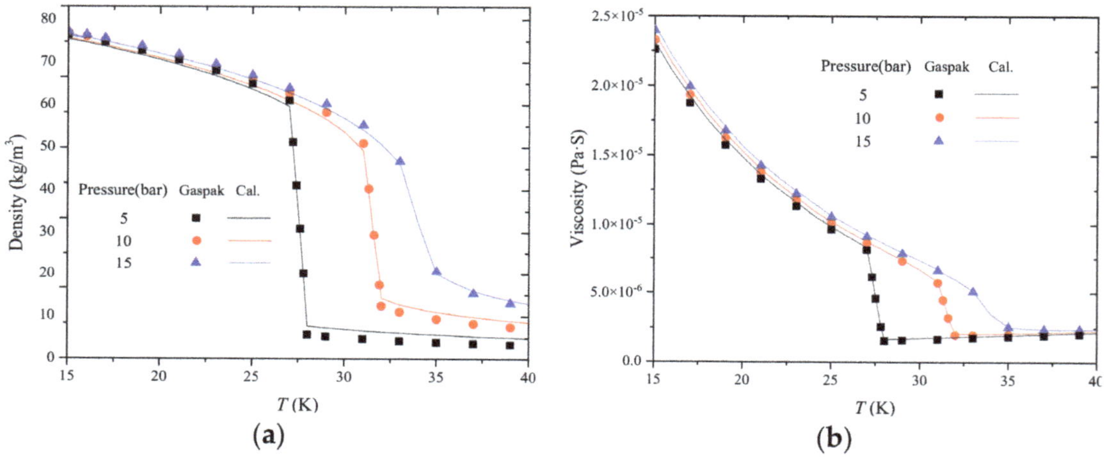

Figure 5. Comparison of simulation results with GSPAK data of thermophysical properties: (**a**) Density; (**b**) Viscosity.

3. Results and Discussion

3.1. The Heat Source Distribution

The cross section of nuclear data, which were used in the simulation calculation with the MCNPX of this paper, are mainly from the databases and form the basis to ensure the correctness of simulation results. A large number of variance reduction techniques are used in the calculation to improve the calculation accuracy and reduce the calculation time. In the process of liquid hydrogen flow heat transfer, its heat source mainly comes from the container (which is all generated by the nuclear reaction on the wall of the moderator), poisoned plate and the background radiation.

Figure 6 shows the heat source distribution of DPHM at 100 kW power obtained by coupling calculation of CFX and MCNPX simulation software, which are, respectively, container (Figure 6a) and poisoned plate (Figure 6b). It can be seen from the contours that the values calculated by the two softwares are slightly different, but the general trend distribution is consistent. A large number of long-wave neutrons elastic scattering reactions, which owing to the narrowing effect of the poisoned plate, result in an energy concentration in the poisoned plate, thus leading to the highest heat deposition at this point (with a maximum value of 3.1×10^6 W/m^3).

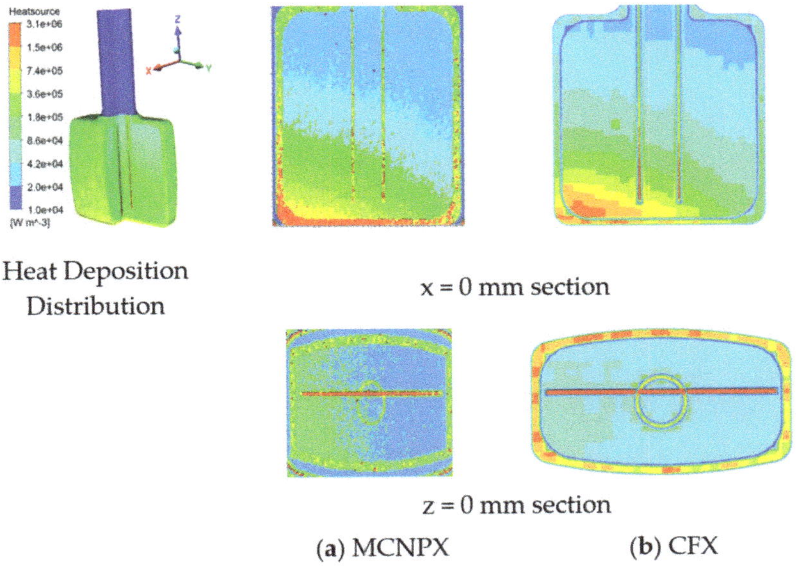

Figure 6. Comparison of heat deposition calculation results between MCNPX and CFX: (**a**) MCNPX; (**b**) CFX.

3.2. *Effect of Beam Power on Heat Transfer*

The influence of beam power increase on the heat transfer characteristics of liquid hydrogen is characterized by the change of temperature. The temperature distribution with large beam power from 100 kW to 500 kW is shown in Figure 7. As is evident from Figure 7, the temperature of liquid hydrogen in the cavity tends to be symmetrically distributed at different beam power. On the contrary, the temperature at different wall positions of the external aluminum container is quite different, which is manifested in the relatively high temperature values in four corners. The maximum value appears at the lower left corner of the hydrogen container at 500 kW, which is specifically 33 K. And as power increases, the difference becomes more pronounced, due to the irregular heat source of the moderator itself. As shown in Figure 6, the temperature at the wall position corresponding to the high heat deposition is also higher, and the increase of power further highlights this phenomenon. In addition, it can be seen that the liquid hydrogen temperature in the cavity has a partial fluctuation under the power of 300~400 kW, which is thanks to the increase of the overall heat source affecting the physical properties of liquid hydrogen. The overall temperature increases from about 18 K to 20 K corresponding to 500 kW in this range, which is in a transition state, so the liquid hydrogen temperature is unevenly distributed.

Considering the neutron performance and hydrogen system safety, the wall temperature of the container is forbidden to exceed the vaporization temperature of the working fluid; thus, the maximum temperature value is the focus of attention and ensures that it is within the safe range. The specific relationship between maximum temperature of container, poisoned plate and hydrogen with beam power is studied in Figure 8; it is found that there is an approximate linear growth relation between them. For the local thermal deposition of DPHM, due to the increase of the Footprint size of the beam, the growth multiple will be smaller than that of the increase of beam power. The latent heat in steady flow is negligible, so the heat change in DPHM is mainly caused by the scattering reaction, which means that the energy generated by per unit volume is proportional to the temperature difference. In this case, the heat deposition increases linearly, resulting in a linear increase in maximum temperature.

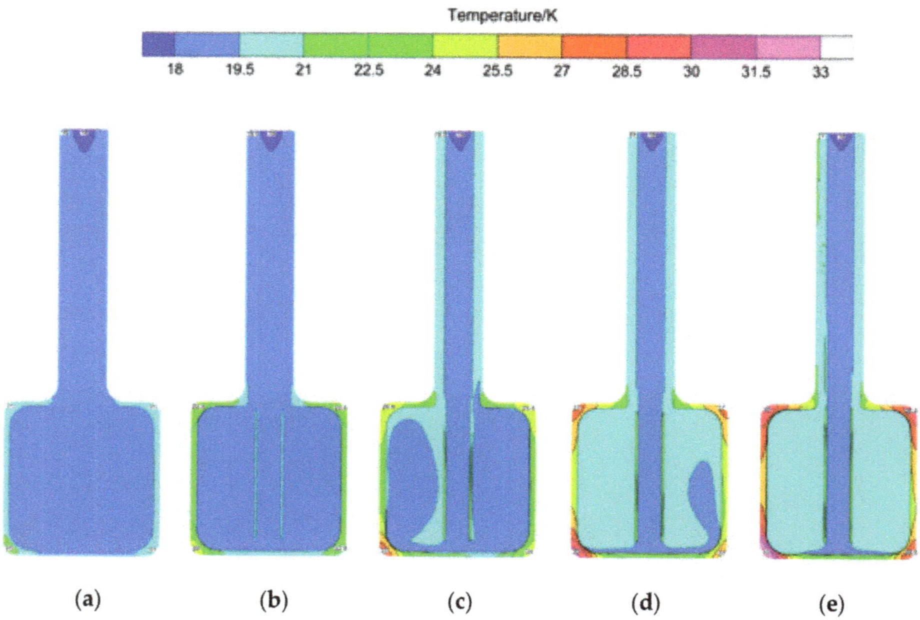

Figure 7. The temperature distribution of DPHM with various beam power (q = 60 g/s, x = 0 mm): (**a**) 100 kW; (**b**) 200 kW; (**c**) 300 kW; (**d**) 400 kW; (**e**) 500 kW.

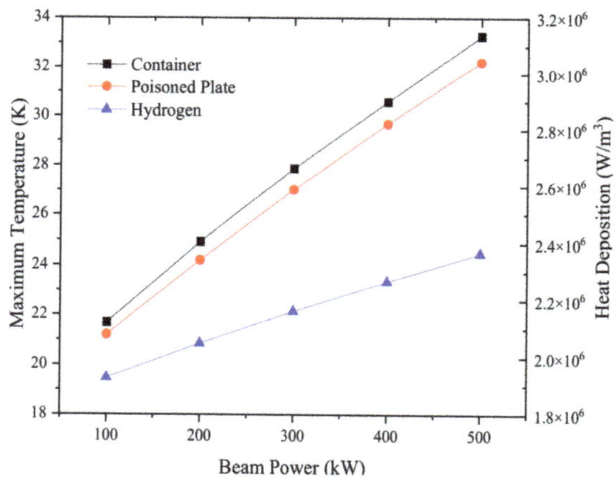

Figure 8. The variation of maximum temperature with different beam power.

3.3. Effect of Mass Flow on Heat Transfer

The analysis of flowing process mainly includes the selection of flow range and the analysis of pressure loss. The higher the mass flow, the smaller overall temperature rise of liquid hydrogen, and the temperature distribution inside the container is relatively uniform, which is also beneficial to the neutron moderating effect. At the cross section of x = 0 mm, it can be observed, from Figure 9, that the average temperature of the side with lower liquid hydrogen flow is higher due to the uneven segmentation of the cross-sectional area of the import pipeline by the poisoned plate. The velocity contours under different mass flow are also listed, and it can be seen that the temperature distribution is basically unchanged but that the overall temperature decreases gradually with the increase of flow rate.

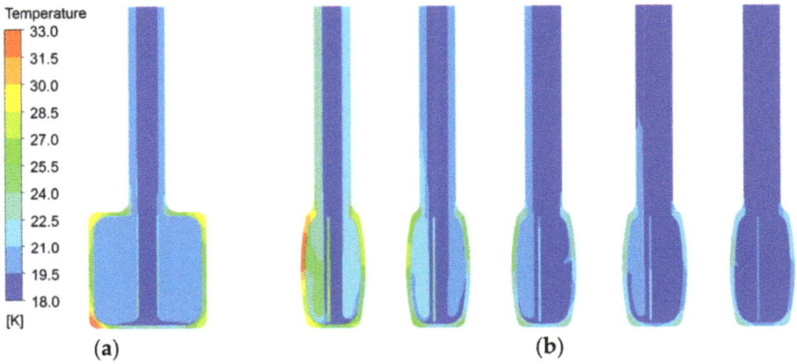

Figure 9. Temperature distribution of DPHM (beam power = 500 kW): (**a**) $x = 0$ mm (60 g/s); (**b**) $y = 0$ mm (30, 60, 90, 120, 150 g/s).

Correspondingly, Figure 10 presents the detailed flow of liquid hydrogen inside the cavity under the 60 g/s mass flow when the turbulence viscosity ratio is about 5, which is also one of the important factors affecting heat transfer. There are two distinct symmetrical vortices on both sides of the bottom of the container according to Figure 10a; part of the liquid hydrogen from the exit is involved in the vortex, and the other part is gathered up at the neck by the inertial force. Moreover, it can be seen that due to the separation of the poisoned plate, the size and position of the vortex generated by different flow rates are different but all located near the inner wall of the container, where the intensity of vortex dominates the heat transfer.

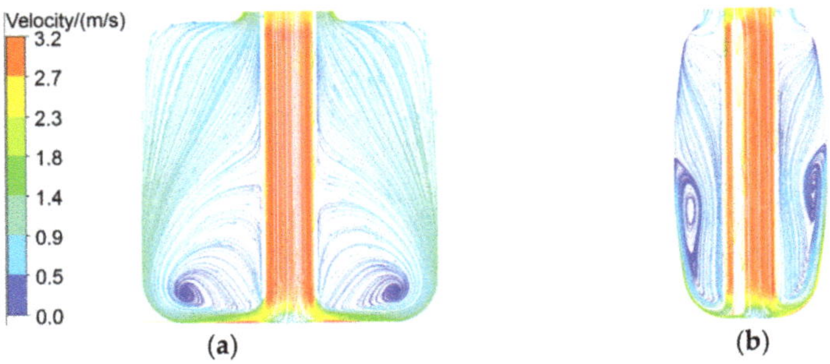

Figure 10. The streamline in the DPHM: (**a**) $x = 0$ mm; (**b**) $y = 0$ mm.

The effect of mass flow on maximum temperature under 500 kW beam power is shown in Figure 11. Specifically, the heat transfer effect is significantly improved by distinct increment of turbulence effect in the flow range of 30~60 g/s, while the slope decreases with the increase of mass flow, which proves that the cooling effect reduces gently. It also should be noted that the increase of mass flow is accompanied by the increase of pressure loss, which will be detrimental to the stability of the flow. The intersection point between the temperature curve of liquid hydrogen and the corresponding pressure drop curve is noted, which is in the range of 60~90 g/s, indicating that the cooling effect is the best and the flow is relatively stable.

The liquid hydrogen carries out jet impact on the bottom surface, which was taken as the research object result of its good heat transfer effect, after flowing out of the hydrogen intake pipe with a vertical distance of 5 mm, as shown in Figure 12. For a smooth surface, with the flow of liquid hydrogen reaching the wall, the pressure forces the jet to flow axially along the wall. Combined with the pressure distribution, it can be seen that the target surface located in the impact zone is under the maximum pressure, and then diffuses and decreases along the periphery, leading to a gradual decrease in velocity. The temperature in the center area of the target surface increases slowly with a similar regularity. This is because the viscous boundary layer on the wall gradually thickens and the surface heat transfer coefficient decreases when the liquid hydrogen flows in the radial direction.

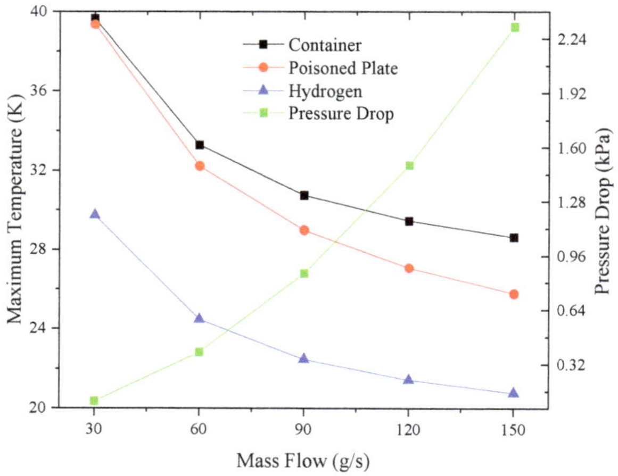

Figure 11. The variation of maximum temperature with different mass flow.

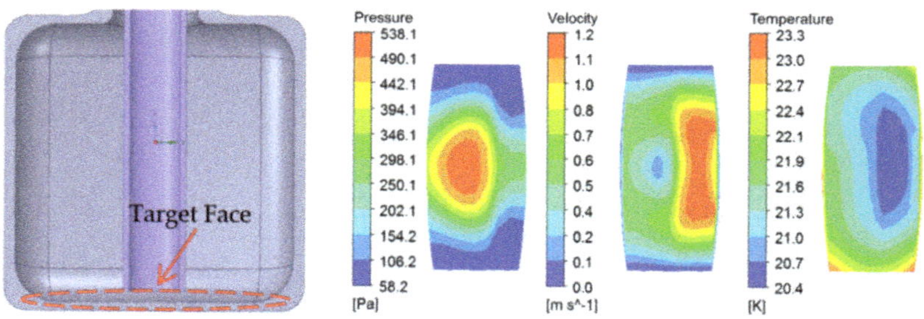

(**a**) Pressure (**b**) Velocity (**c**) Temperature

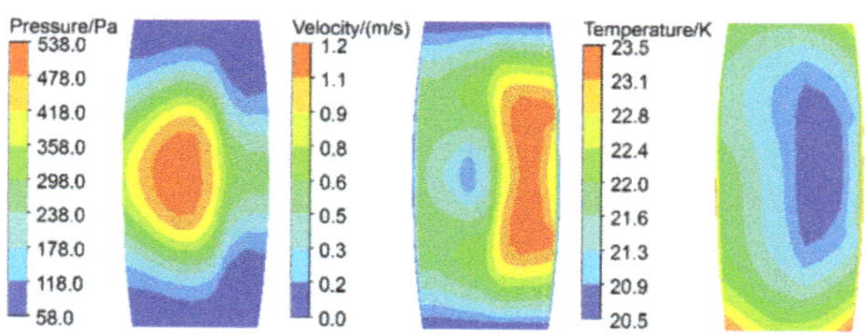

Figure 12. Physical parameters distribution of target surface: (**a**) Pressure; (**b**) Velocity; (**c**) Temperature.

The local heat transfer coefficient is obtained by the following equations,

$$h = \frac{q_w}{T_w - T_b} \tag{18}$$

$$T_b = f(H_b, P) \tag{19}$$

$$H_b = \frac{\int_A \rho w H dA}{\int_A \rho w dA} \tag{20}$$

where q_w, T_w, T_b, w and H are the local heat flux of wall, averaging temperature of wall, bulk fluid temperature, bulk fluid axial velocity and bulk fluid enthalpy.

Figure 13 shows the distribution of the flow heat transfer coefficient h at the bottom target face along the x-axis, corresponding to the temperature distribution. It reaches the maximum value at the position of poisoned plate and then gradually decreases, which is due to the poor heat transfer at this place due to the flow dead zone generated by the shunting phenomenon. After that, the heat transfer coefficient increases gradually with the flow recovery.

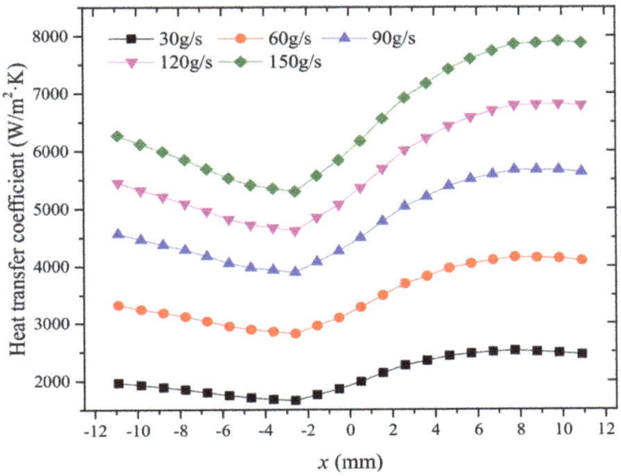

Figure 13. The distribution of the heat transfer coefficient of the target surface along the x axis.

3.4. Effect of Pressure on Heat Transfer

To explore the influence of law of pressure change in a pseudo-critical region, the inlet temperature of hydrogen is set to 30 K. Figure 14 shows the influence of changing mass flow on the axial distribution of the wall temperature, T_w, of the hydrogen intake tube and the bulk temperature, T_b, of liquid hydrogen inside the tube. The results show that the temperature of both the wall and the liquid hydrogen increases gradually with the deepening of the flow distance. This is because there is no heat source around when the liquid hydrogen flows into the pipeline at the beginning, so the temperature change is not obvious. After that, due to the influence of thermal deposition of the poisoned plate and container, the two kinds of temperature increase significantly, and the temperature rise range of the wall surface is about 3 K. When the pressure is 15 bar, the temperature value will change at the mutation position corresponding to the physical property of liquid hydrogen in Figure 4, while the temperature will not change at a distance from the critical value. Moreover, different pressures have little effect on the average temperature of the fluid in the tube. It can be concluded that under supercritical pressure, the drastic change of physical properties in the critical region will have a great impact on turbulent heat transfer, resulting in the deterioration of convective heat transfer.

The basic formula of jet impingement heat transfer is the Newton cooling Equation,

$$q = h(T_w - T_j) \tag{21}$$

where T_j is jet temperature and h is local impact convective heat transfer coefficient. In order to obtain the average h of the whole target surface, the local h curve of the target surface must be obtained, and then the h of each point can be calculated through a surface integral along the radius [39]. The local Nusselt number can be expressed as

$$Nu = \frac{h_j d}{\lambda_j} \tag{22}$$

where h_j, d and λ_j are local impact convective heat transfer coefficient, diameter of nozzle and thermal conductivity of jet.

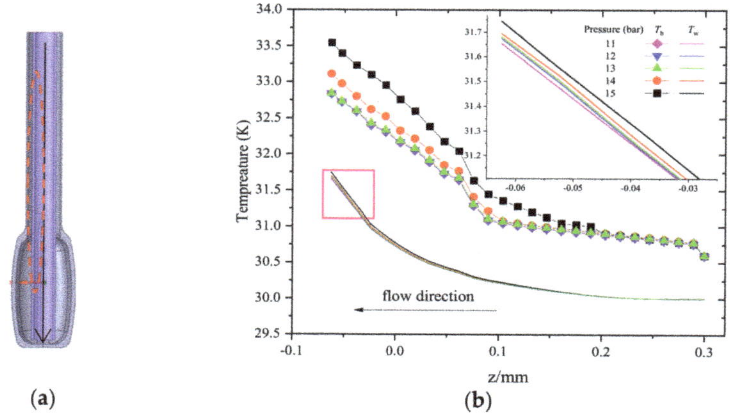

Figure 14. Variations of temperature along axial direction under different pressures: (**a**) $x = 0$ mm; (**b**) $y = 0$ mm.

In order to more specifically reflect the heat flow characteristics of liquid hydrogen inside the vessel, the relationship between the average Nusselt number on the target surface and the resistance f along the hydrogen inlet pipe and the jet Reynolds number Re was explored under different pressure conditions. As can be seen from Figure 15, Nu increases significantly with the increase of Re and tends to change linearly. Simultaneously, the flow boundary layer in the inlet pipe segment gradually becomes thinner, the flow resistance decreases, the resistance coefficient decreases, and the pressure on the target surface also increases. The intersection point of Nu and f represents the inlet flow of liquid hydrogen corresponding to different pressures when the optimal cooling effect is achieved under the premise of considering the comprehensive characteristics of flow resistance and heat transfer characteristics. On the other hand, with the continuous increase of Re, the variation trend of Nu corresponding to different pressures is the same. At the pressure of 15 bar, the corresponding overall curve of Nu moves down and the heat transfer capacity decreases, which further proves that near the critical region, while the physical property variation caused by the pressure change worsens the turbulent heat transfer.

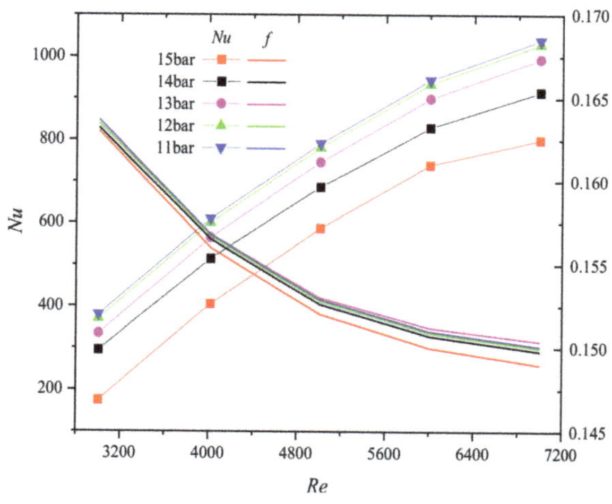

Figure 15. The variation of Nu with Re at the target surface.

4. Conclusions

In this paper, the heat transfer performance of hydrogen in a decoupled poison hydrogen moderator under supercritical pressure is investigated by numerical method, and the effects of different boundary conditions on flow and heat transfer are compared. The major conclusions are summarized as follows:

1. The sensitivity degree of factors affecting the heat transfer characteristics of liquid hydrogen are in sequence of inlet mass flow, beam power and operating pressure.
2. The temperature tends to be stable and pressure loss increases gradually when the mass flow rate exceeds a certain range, especially when the beam power is 500 kW (the temperature range of liquid hydrogen is about 20~30 K); the cooling effect is best in the range of 60~90 g/s × 394 mm^2.
3. With the beam power increasing, the maximum temperature of the container, poisoned plate and hydrogen maintain the linear growth trend. The maximum temperature of liquid hydrogen is close to the bottom recirculation zone due to the influence of the flow field and the heat deposition distribution of the poisoned plate.
4. As the pressure increased, there was no significant difference in trends of the bulk temperature, T_b, of liquid hydrogen, whereas, near the large specific heat region of 15 bar, the wall temperature T_w exhibited a sudden enhancement.

Therefore, the concrete influence law of the specific heat of liquid hydrogen on heat transfer is expected for future studies. Meanwhile, it is necessary to establish a 2D model of a moderator for avoiding random errors.

Author Contributions: J.T. and L.Z. are the main authors of this manuscript. All the authors contributed to this manuscript. J.T. and L.Z. conceived the novel idea, and Y.L. (Yiping Lu) and S.D. performed the analysis; J.T. and Y.L. (Youlian Lu) analyzed the data and contributed analysis tools; J.T. and L.Z. wrote the entire paper. J.T., T.L. and S.D. checked, reviewed and revised the paper; J.T., C.Y. and S.W. performed final proofreading and supervised this research. All authors have read and agreed to the published version of the manuscript.

Funding: This article Supported by the Program for Guangdong Introducing Innovative and Entrepreneurial Teams. (Project number: 2017ZT07S225).

Institutional Review Board Statement: Not applicable.

Informed Consent Statement: Not applicable.

Conflicts of Interest: The authors declare no conflict of interest.

Nomenclature

T	bulk fluid temperature, K
E	total energy, kJ/kg
P	static pressure, Pa
R	molecular gas constant, J/(mol·K)
L	characteristic length, mm
H	bulk fluid enthalpy, kJ/kg
Nu	Nussel number,
C_f	boundary layer friction coefficient
Re	Reynolds number
S_h	volumetric heat source, W/m^3
T_w	averaging temperature of wall, K
T_c	critical temperature, K
T_b	bulk temperature of fluid, K
T_r	reference temperature, K
U_τ	shear velocity, m/s
α	temperature dependence function (alpha function)
a_c	critical attractive parameter, MPa·m^6·k·mol^{-2}

b	molar co-volume, $m^3 \cdot k \cdot mol^{-1}$
p	ideal gas pressure, Pa
u	bulk fluid velocity, m/s
h	heat transfer coefficient, $W/m^2 \cdot K$
f	resistance coefficient, -
w	bulk fluid axial velocity, m/s
ρ	density, kg/m^3
λ	thermal conductivity of the fluid, $W/m \cdot K^{-1}$
ω	acentric factor, -
μ	dynamic viscosity, Pa·S
τ	shear stress, N/m^2
p_c	critical pressure, K
C_p	Specific heat capacity, $J/kg \cdot K^{-1}$
v_{PR}	the molar volume, L
λ_{eff}	effective thermal conductivity, $W/m \cdot K^{-1}$
τ_{eff}	effective stress tensor, N/m^2
τ_w	boundary layer shear stress, N/m^2
$c1,c2,c3$	coefficients of the Mathias and Copeman alpha function

References

1. Wei, J.; Chen, H.S.; Chen, Y.W.; Chen, Y.B.; Chi, Y.L.; Deng, C.D.; Dong, H.Y.; Dong, L.; Fang, S.X.; Feng, J.; et al. China Spallation Neutron Source: Design, R&D, and outlook. *Nucl. Instrum. Methods Phys. Res. A* **2009**, *600*, 10–13.
2. Wang, F.W.; Jia, X.J.; Liang, T.J.; He, C.H.; Yin, W.; Zhang, S.Y.; Zhu, T.; Yu, Q.Z.; Wang, P. Physical design of the target station and spectrometers for a spallati-on neutron source. *Physics* **2008**, *34*, 449–453.
3. Wang, F.W.; Yan, Q.W.; Liang, T.J.; Yin, W.; Zhang, P.L.; Fu, S.N.; Tang, J.Y.; Fang, S.S.; Zhang, Z.; Zhang, J. Netron Scattering and Spallation Neutron Source. *Physics* **2005**, *34*, 731–738.
4. Wang, G.P.; Xiao, J.; He, K.; Li, S.P. Conceptual design of cryogenic system for Chinese Spallation Neutron Source. *Cryogenics* **2009**, *5*, 27–30.
5. He, C.C.; He, K.; Wang, G.P. Hydrogen safety design of cryogenic system for Chinese Spallation Neutron Source. *Cryogenics* **2010**, *6*, 1–5.
6. Zhao, K.; Yang, G.P.; Shen, J.; Zhang, Z.; Chen, J. Experimental study on adjustment system of liquid hydrogen density. *Cryogenics* **2018**, *2*, 45–50.
7. Wang, G.P.; Xiao, J.; Zhang, Y.; Qiu, Y.N.; Hu, Z.J.; Ding, M.Y.; Sang, M.J.; Li, S.P. Pressure control design of Chinese Spallation Neutron Source supercritical hydrogen circulation system. *Vac. Cryog.* **2011**, *1*, 251–256.
8. Huang, D.; Wu, Z.; Sunden, B.; Li, W. A brief review on convection heat transfer of fluids at supercritical pressures in tubes and the recent progress. *Appl. Energ.* **2016**, *162*, 494–505. [CrossRef]
9. Mohseni, M.; Bazargan, M. A New Correlation for the Turbulent Prandtl Number in Upward Rounded Tubes in Supercritical Fluid Flows. *J. Heat Transf.* **2016**, *138*, 081701. [CrossRef]
10. Du, X.; Lv, Z.; Yu, X.; Cao, M.; Zhou, J.; Ren, Y.; Qiu, Q.; Zhu, X. Heat transfer of supercritical CO2 in vertical round tube: A considerate turbulent Prandtl number modification. *Energy* **2020**, *192*, 116612. [CrossRef]
11. Qiu, Q.; Du, X.; Zhao, S.; Zhu, X.; Shen, S. Numerical study on heat transfer of SCW near the pseudo-critic-al temperature in a hexagon sub-channel. *Nucl. Eng. Des.* **2018**, *331*, 263–273. [CrossRef]
12. Wang, Z.W.; Qi, G.L.; Li, M.J. Numerical Investigation of Heat Transfer to Supercritical Water in Vertical Tube under Semicircular Heating Condition. *Energies* **2019**, *12*, 3958. [CrossRef]
13. Zhu, X.J.; Zhang, R.Z.; Yu, X.; Cao, M.G.; Ren, Y.X. Numerical Study on the Gravity Effect on Heat Transfer of Supercritical CO_2 in a Vertical Tube. *Energies* **2020**, *13*, 3502. [CrossRef]
14. Kim, D.E.; Kim, M.H. Experimental study of the effects of flow acceleration and buoyancy on heat transfer in a supercritical fluid flow in a circular tube. *Nucl. Eng. Des.* **2010**, *240*, 3336–3349. [CrossRef]
15. Youn, B.; Mills, A.F. Flow of Supercritical Hydrogen in a Uniformly Heated Circular Tube. *Numer. Heat Transfer Part A Appl.* **1993**, *24*, 1–24. [CrossRef]
16. Locke, J.M.; Landrum, D.B. Study of Heat Transfer Correlations for Supercritical Hydrogen in Regenerative Cooling Channels. *J. Propul. Power* **2008**, *24*, 94–102. [CrossRef]
17. Freels, J.D. COMSOL Validation Progress on Supercritical Hydrogen Heat Transfer. In Proceedings of the COMSOL Multiphysics Conference 2007 Proceedings, COMSOL, Boston, MA, USA, October 2007; Dravid, V., Ed.; pp. 341–347.
18. Xie, P.Y.; Zhang, X.B. A method of rib-bed plate enhancing heat transfer in hydrogen rocket engine chamber wall. *Int. J. Hydrogen Energy* **2019**, *44*, 20504–20515. [CrossRef]
19. Xie, P.Y.; Zhang, X.B. Enhancement heat transfer analysis of supercritical hydrogen fuel in small-scale channels with spherical concave. *Int. J. Therm. Sci.* **2020**, *152*, 106287. [CrossRef]

20. Xie, P.Y.; Zhang, X.B. Turbulent heat transfer analysis in supercritical hydrogen fuel flow considering thermal stratification. *Numer. Heat Transfer Part A Appl.* **2020**, *77*, 913–929. [CrossRef]
21. Mosavati, B.; Mosavati, M.; Kowsary, F. Solution of radiative inverse boundary design problem in a combined radiating-free convecting furnace. *Int. Commun. Heat Mass* **2013**, *45*, 130–136. [CrossRef]
22. Mosavati, B.; Mosavati, M.; Kowsary, F. Inverse boundary design solution in a combined radiating-free convecting furnace filled with participating medium containing specularly reflecting walls. *Int. Commun. Heat Mass* **2016**, *76*, 69–76. [CrossRef]
23. Bolek, A.; Bayraktar, S. Flow and heat transfer investigation of a circular jet issuing on different types of surfaces. *Sadhana Acad. Proc. Eng.* **2019**, *44*, 242–252. [CrossRef]
24. Tong, J.F.; Wang, J.L.; Zhang, H.C.; Ji, Y.; Tan, H.P. Three dimensional numerical simulation of thermal-hydraulic behaviors of CSNS decoupled poisoned hydrogen moderator with non-uniform heat source. In Proceedings of the ASME 2016 24th International Conference on Nuclear Engineering, Charlotte, NC, USA, 26–30 June 2016; pp. 1–6.
25. Pelowitz, D.B. *MCNPX User' S Manual Version 2.5.0*; Los Alamos National Laboratory: Los Alamos, NM, USA, 2005.
26. ANSYS Inc. *ANSYS CFX Reference Guide, Release 11.0*; ANSYS Inc.: Canonsburg, PA, USA, 2006.
27. Versteeg, H.; Malalasekra, W. *An Introduction to Computational Fluid Dynamics: The Finite Volume Method*, 2nd ed.; Prentice Hall: Hoboken, NJ, USA, 2007.
28. Bae, J.H.; Yoo, J.Y. Direct numerical simulation of turbulent supercritical flows with heat transfer. *Phys. Fluids* **2005**, *17*, 105104. [CrossRef]
29. Hirsch, C. Numerical Computation of Internal and External Flows. In *Fundamentals of Computational Fluid Dynamics*, 2nd ed.; Hirsch John Wiley & Sons Ltd.: New York, NY, USA, 2007; Volume 1.
30. Tao, W.Q. *Numerical Heat Transfer*, 2nd ed.; Xi'an Jiaotong University Press: Xi'an, China, 2001.
31. Yan, X.; Baugn, J.W.; Mesbath, M. The effect of reynolds number on the heat transfer distribution from a flat Plate to an impinging jet. In Proceedings of the ASME Winter Annual Meeting, Anaheim, CA, USA, 1992; Volume 226, pp. 1–7.
32. Roache, P.J. Perspective: A method for uniform reporting of grid refinement studies. *J. Fluids Eng.* **1994**, *116*, 405–413. [CrossRef]
33. Tong, J.F.; Wang, S.L.; Yin, W.; Lu, Y.L.; Yu, Q.Z.; Hu, C.M.; Yu, C.J.; Du, W.T.; Yao, C.J.; Liang, T.J. Neutronics/Thermal-Hydraulic Coupling for CSNS Decoupled Posioned Hydrogen Moderator. *J. Eng. Thermophysrus.* **2013**, *34*, 905–909.
34. Boublik, T. The BACK equation of state for hydrogen and related compounds. *Fluid Phase Equilib* **2005**, *240*, 96–100. [CrossRef]
35. Peng, D.Y.; Robinson, D.P. A new two-constant equation of state. *Ind. Eng. Chem. Fundam.* **1976**, *15*, 59–64. [CrossRef]
36. Khashayar, N. Comparative study of eleven equations of state in predicting the thermodynamic properties of hydrogen. *Int. J. Hydrogen Eng.* **2010**, *35*, 3802–3811.
37. Mathias, P.M.; Copeman, T.W. Extension of the Peng-Robinson equation-of-state to complex mixtures: Evaluation of the various forms of the local composition concept. *Fluid Phase Equilib* **1983**, *13*, 91–108. [CrossRef]
38. McCarty, R.D.; Arp, V.; Fox, J.R. *GASPAK Version 3.35/3.45*; CRYODATA Inc.: Littleton, CO, USA, 2007.
39. Flourusse, L.J.; Peters, C.J.; Pamies, J.C.; Vega, L.F.; Meijer, H. Solubility of hydrogen in heavy n-alkanes: Experiments and SAFT modeling. *AIChE. J.* **2003**, *49*, 3260–3269. [CrossRef]

energies

Article

Thermal Efficiency and Economics of a Boil-Off Hydrogen Re-Liquefaction System Considering the Energy Efficiency Design Index for Liquid Hydrogen Carriers

Minsoo Choi [1], Wongwan Jung [1], Sanghyuk Lee [2], Taehwan Joung [3] and Daejun Chang [1,*]

[1] Department of Mechanical Engineering, Korea Advanced Institute of Science and Technology (KAIST), Daejeon 34141, Korea; minsoo.choi@kaist.ac.kr (M.C.); wgjung@kaist.ac.kr (W.J.)
[2] Energy Plant Research Center, Samsung Heavy Industries, Seongnam-si 13486, Korea; sh.91.lee@samsung.com
[3] International Maritime Research Center (IMRC), Korea Research Institute of Ships & Ocean Engineering (KRISO), Daejeon 34103, Korea; thjoung@kriso.re.kr
* Correspondence: djchang@kaist.ac.kr; Tel.: +82-42-350-1514

Abstract: This study analyzes the thermodynamic, economic, and regulatory aspects of boil-off hydrogen (BOH) in liquid hydrogen (LH_2) carriers that can be re-liquefied using a proposed re-liquefaction system or used as fuel in a fuel cell stack. Five LH_2 carriers sailing between two designated ports are considered in a case study. The specific energy consumption of the proposed re-liquefaction system varies from 8.22 to 10.80 kWh/kg as the re-liquefaction-to-generation fraction (R/G fraction) is varied. The economic evaluation results show that the cost of re-liquefaction decreases as the re-liquefied flow rate increases and converges to 1.5 $/kg at an adequately large flow rate. Three energy efficient design index (EEDI) candidates are proposed to determine feasible R/G fractions: an EEDI equivalent to that of LNG carriers, an EEDI that considers the energy density of LH_2, and no EEDI restrictions. The first EEDI candidate is so strict that the majority of the BOH should be used as fuel. In the case of the second EEDI candidate, the permittable R/G fraction is between 25% and 33%. If the EEDI is not applied for LH_2 carriers, as in the third candidate, the specific life-cycle cost decreases to 67% compared with the first EEDI regulation.

Keywords: liquid hydrogen carrier; boil-off hydrogen; specific energy consumption; exergy efficiency; economics; energy efficiency design index

Citation: Choi, M.; Jung, W.; Lee, S.; Joung, T.; Chang, D. Thermal Efficiency and Economics of a Boil-Off Hydrogen Re-Liquefaction System Considering the Energy Efficiency Design Index for Liquid Hydrogen Carriers. *Energies* **2021**, *14*, 4566. https://doi.org/10.3390/en14154566

Academic Editor: Bahman Shabani

Received: 21 June 2021
Accepted: 20 July 2021
Published: 28 July 2021

1. Introduction

Due to the current global attention concerning the reduction of carbon dioxide (CO_2) emissions, which is the largest contributor to global warming [1], the demand for renewable energy is increasing. A large portion of CO_2 emissions is attributed to the combustion of the fossil fuels, which provide approximately 80% of the total world energy supply [2]. From 2009 to 2018, global CO_2 emissions increased by 16% [3]. To reduce CO_2 emissions, power generation using renewable energy, which is not accompanied by CO2 emissions, has been increasing. According to the International Energy Agency, power generation achieved using renewable energy has increased by 57% from 2010 to 2018 [4]. The portion of renewable energy used for electricity generation has been projected to consistently increase in the future [5].

Although renewable energy can be used to reduce CO_2 emissions, its production varies from region to region. This uneven distribution of renewable energy requires the use of energy carriers that can transport surplus renewable energy. Hydrogen is regarded as a potential energy carrier candidate for storing and transporting surplus renewable energy over long distances [6]. The delivered hydrogen generates energy from an oxidation process in fuel cells, which produces no CO_2 and only pure water [7]. Hydrogen also has other principal advantages as an energy carrier, as noted by Rosen et al. [8]. It is free from exhaustion because it can be obtained by electrolyzing water into oxygen and hydrogen. Additionally, unlike electricity itself, it can be stored in a variety of forms such as compressed gas, liquid, and in chemical compounds including ammonia and metal hydrides [9].

Considering the large-scale transportation of hydrogen, it is essential to develop liquid hydrogen (LH_2) carriers. LH_2 has a density of 70 kg/m^3, which is higher than the density of compressed hydrogen (39 kg/m^3) at 70 MPa [10]. Compared to ammonia, LH_2 is safer, as it is nontoxic. Moreover, because it is a pure substance that is not combined with other elements, it does not require any additional energy or processes to break its chemical bonds.

The main challenge concerning the transportation of hydrogen in a liquid state is its cryogenic boiling temperature, which is 20 K at atmospheric pressure. Although LH_2 storage tanks on land are heavily equipped with an insulation layer, heat ingress is inevitable [11], resulting in the generation of boil-off hydrogen (BOH) and increase in the pressure of the storage tank if no preventative measures are used. This issue also exists for LH_2 cargo tanks used in LH_2 carriers.

In the case of relatively large-scale liquefied natural gas (LNG) carriers, whose cargo temperature is around 110 K, a boil-off gas (BOG) re-liquefaction system is usually installed. This system extracts the BOG from the cargo tanks, re-liquefies it using a refrigeration cycle, and returns the re-liquefied BOG to the cargo tank. Many studies have been conducted concerning this BOG re-liquefaction system [12–15]. Romero et al. analyzed and evaluated the conditions, parameters, and energy consumption of a reverse Brayton cycle-based BOG re-liquefaction system [16]. Kwak et al. investigated a small-scale BOG re-liquefaction system for use on LNG-fueled ships. They optimized the nitrogen reverse Brayton cycle using a genetic algorithm and compared two cases with and without BOG compressors [17]. Yin et al. compared a reverse Brayton cycle-based BOG re-liquefaction system with two expanders to such a system with a single expander. They simulated two cycles using parallel and serial nitrogen expanders and compared them with the base case that possessed a single expander [18]. Sayyaadi et al. used a genetic algorithm to optimize the price of the BOG re-liquefaction system products [19]. In these previous studies of BOG re-liquefaction systems, the specific energy consumption (SEC), which indicates the energy required to re-liquify 1 kg of BOG, was found to be between 0.73 to 1.41 kWh/kg. This value varied depending on the inlet conditions of the BOG system. Moreover, the precooling and precompression processes affected the SEC. The reverse helium Brayton cycle for BOG re-liquefaction was preferred because it was easy to operate. Additionally, this cycle did not require high pressures and was more appropriate for onboard systems in terms of safety [17]. These studies can be utilized to design appropriate BOH re-liquefaction systems.

Many studies have been conducted to liquefy hydrogen from ambient temperatures [20–22]. The main results of these investigations may also be applied towards designing BOH re-liquefaction systems. Chang et al. analyzed and compared various figures of hydrogen liquefaction systems in terms of exergy efficiency [23]. In their study, the reverse Brayton cycle with two expanders showed an exergy efficiency of 24.2%, while the Claude cycle had an exergy efficiency of 27.2%. Asadnia et al. proposed a hydrogen liquefaction system using combined mixed refrigerants [24]. They separated the cycles into two sections, in which the first section was used for precooling and the second was used for cryogenic liquefaction. In the precooling section, 11 materials including hydrocarbons were used as a mixed refrigerant. In the cryogenic section, hydrogen, helium and neon were used as mixed refrigerants. The simulation results showed an SEC of 7.69 kWh/kg. Sadaghiani et al. also proposed a hydrogen liquefaction system that used mixed refrigerants [25]. The proposed system was composed of two cycles, one of which used a mixture of hydrocarbons while the other used a mixture of hydrogen, neon and helium as a mixed refrigerant. This system demonstrated an SEC of 4.41 kWh/kg. Chang et al. proposed a hydrogen liquefaction system that utilized the cold energy of LNG [26]. In that study, the LNG precooled the hydrogen and was vaporized after precooling. Following the design of a hydrogen liquefaction system using LNG precooling, three configurations of the liquefaction cycles were compared in terms of their energy consumption. Considering the required energy for the liquefaction process described in their paper, the SEC for the most efficient configuration was calculated as 17.37 kWh/kg, which includes the energy consumed for ortho-para conversion. Yuksel et al. also proposed a hydrogen liquefaction system with four serial helium turbo expanders and analyzed this system [27]. These previous studies showed that the SECs of hydrogen liquefaction systems vary with the configuration of the system and the inlet conditions of the hydrogen.

BOH re-liquefaction systems, however, differ from liquefaction systems for hydrogen at ambient temperatures. The temperature of BOH is much lower than the ambient temperature. Additionally, such a system does not require an ortho-para conversion process because the LH_2 in the cargo tank has already been converted into para-hydrogen. Lee et al. proposed a partial BOH re-liquefaction system for use on an LH_2 carrier [28]. In their study, some of the BOH from the LH_2 cargo tank was used as a fuel in a proton-exchange membrane fuel cell (PEMFC). The remainder was precompressed to 40 bar using a cold compressor with an inlet temperature of 120 K. This compressed BOH was cooled down using helium in a reverse Brayton cycle and re-liquefied by expansion. This BOH re-liquefaction system exhibited an SEC of 3.30 kWh/kg and exergy efficiency of 74.9%.

Previous studies of BOH re-liquefaction systems lacked an investigation of the economic feasibility of these systems that considered the installation and operation costs. A comparison with the production costs of LH_2 can be used to investigate the economic feasibility of BOH re-liquefaction systems. Moreover, considering that existing LNG carriers vary in terms of their capacity, economic case studies of BOH re-liquefaction systems with varying capacities are required.

Another critical factor to consider in the design and operation of the BOH re-liquefaction system is the energy efficiency design index (EEDI) regulation of LH_2 carriers. The EEDI is the design indicator regulated by the International Maritime Organization (IMO), which restricts the CO_2 emissions of ships during the design process. Ship of certain types as indicated by the IMO must obey this index to operate. The EEDI also affects the electricity generation of BOH re-liquefaction systems. However, such a regulation for LH_2 carriers has not yet been established. LH_2 is similar to conventional liquefied gases such as LNG and liquefied petroleum gas (LPG) in that it is liquefied. However, liquid hydrogen differs from other conventional liquefied gases because it is an extremely low-density liquefied gas that is free from CO_2 emissions. Consequently, the establishment of EEDI regulations for LH_2 carriers is difficult in comparison with that of other liquefied gases.

The objective of this study is to investigate the thermodynamic, economic, and regulatory aspects of BOH in LH_2 carriers in which the BOH can be either re-liquefied using the proposed re-liquefaction system or used as fuel for the fuel cells. A case study is conducted that considers five LH_2 carriers sailing between two designated ports. A thermodynamic analysis is carried out, followed by an economic assessment considering the re-liquefaction-to-generation (R/G) fraction. As these evaluations are meaningful in terms of the allowable EEDI, three EEDI candidates are proposed for estimating a feasible R/G fraction.

2. Proposed Boil-Off Hydrogen Re-Liquefaction System Combined with Fuel Cells

Figure 1 shows a process flow diagram of the proposed system. The system consists of a re-liquefaction cycle and PEMFC stack. The proposed re-liquefaction cycle is modified from a reverse Brayton cycle to utilize the cold energy of BOH heading to the PEMFC stack. The BOH to be utilized in the PEMFC stacks is first heated to the operating temperature of the PEMFC system. HX 1 and HX 3 are introduced to transfer the cold energy of the BOH to the helium refrigerant and increase the BOH temperature to the ambient temperature. Stream 101 indicates the BOH generation from the cargo tank. This stream is divided into two streams, which are labelled as 102 and 106. Stream 102 is cooled down and re-liquefied to a subcooled liquid state. Stream 106 cools the helium refrigerant and is utilized for propulsion. Stream 201 indicates the low-pressure helium stream. This stream is pressurized to a high pressure through Comp 1 and Comp 2. After AC 1, the helium is cooled down along with the BOH heading to the PEMFC system. After the helium is compressed, it is cooled down using the cold helium in HX 4. It is then primarily expanded to an intermediate pressure via EXP 1 and cooled down via HX 5. After passing through HX 5, the helium is expanded to a low pressure through EXP 2 and the BOH is liquefied in HX 6. After the BOH has been re-liquefied, this cold helium then cools down the hot helium and hydrogen.

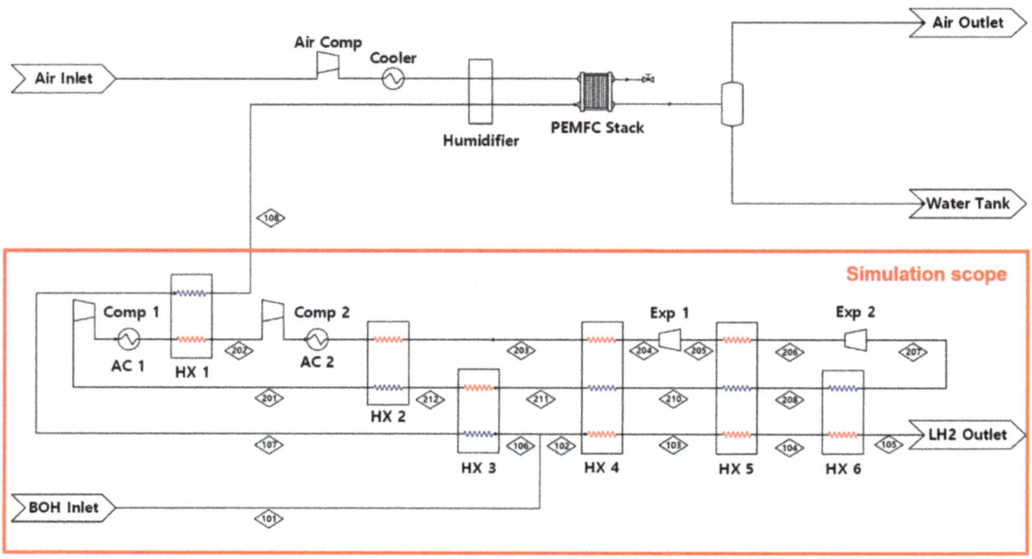

Figure 1. Process flow diagram of the proposed system.

Aspen Hysys V11 was used to simulate the proposed system. The modified Benedict-Webb-Rubin equation was applied for the states of ortho- and para-hydrogen. The Aspen properties (RefProp) were used for helium. Table 1 shows the boundary conditions for the process simulation.

Table 1. Boundary conditions of the BOH re-liquefaction cycle.

Item	Unit	Value
Cargo LH_2 density	kg/m^3	70.83
Boil off rate	%/day	0.2 [29]
BOH inlet temperature	K	80
BOH outlet temperature	K	20
BOH pressure	bar	1.013

The following assumptions were made for this simulation:

- The dead-state temperature and pressure are 298 K and 1.01 bar, respectively.
- The temperature at the exit of the aftercooler is 313 K.
- The minimum temperature at the inlet of the compressor is 240 K.
- The high and low pressures of the helium cycle are 10 and 1.2 bar, respectively.
- The adiabatic efficiencies of the compressors are 75%.
- The adiabatic efficiencies of the expanders are 75%.
- The pressure drop in the heat exchangers is negligible.
- In the heat exchangers, the minimum temperature difference is larger than 1% of the hottest stream temperature. If the temperature of the hottest stream in the heat exchanger is greater than 300 K, the minimum temperature difference is 3 K [23,26].

3. Method of Evaluation

3.1. Energy and Exergy Efficiency

The total energy required to re-liquefy BOH is calculated using Equation (1).

$$\dot{W}_{net} = \dot{W}_{Comp\ 1} + \dot{W}_{Comp\ 2} - \dot{W}_{Exp\ 1} - \dot{W}_{Exp\ 2} \quad (1)$$

$\dot{W}_{net}$: Total work required to re-liquefy BOH
$\dot{W}_{Comp\ 1}$: Work input for Comp 1
$\dot{W}_{Comp\ 2}$: Work input for Comp 2
$\dot{W}_{Exp\ 1}$: Work output for Exp 1
$\dot{W}_{Exp\ 2}$: Work output for Exp 2

The R/G fraction, which is the ratio of the re-liquefied flow rate to the BOH generation flow rate, is estimated using Equation (2). The SEC is defined by Equation (3) to evaluate the energy required to re-liquefy the 1 kg of BOH. Because the re-liquefaction system utilizes the cold energy of the hydrogen, the SEC varies with the R/G fraction.

$$\text{R/G fraction} = \frac{\dot{m}_{re-liqeufaction}}{\dot{m}_{BOH-generation}} \times 100\ \% \quad (2)$$

$$\text{SEC} = \frac{\dot{W}_{net}}{\dot{m}_{re-liqeufaction}} \quad (3)$$

$\dot{m}_{BOH-generation}$: Mass flow rate of the generated BOH
$\dot{m}_{re-liqeufaction}$: Mass flow rate of the re-liquefied BOH

In thermodynamics, physical flow exergy refers to the maximum useful work delivered to an external user as the stream reaches the dead state [30]. Considering refrigeration systems, it refers to the reversible and minimum work required for refrigeration to occur at a certain state. The physical flow exergy of stream can be estimated using Equation (4) [31]. In Equation (4), subscripts S and 0 refer to present state of the stream and dead state, respectively. Subscript ** means state that has same temperature with the dead state and same pressure with the present state. The first two terms in Equation (4) corresponds to thermal exergy, which is the physical exergy from the temperature difference of the stream with the dead state. The last two terms represent mechanical exergy, which is the physical exergy from pressure difference of the stream with the dead state.

$$E_S = E_T + E_M = (H_S - H_{**}) - T_0(S_S - S_{**}) + T_0(S_0 - S_{**}) - (H_0 - H_{**}) \quad (4)$$

E_S: Physical flow exergy of stream
E_T: Thermal exergy of stream
E_M: Mechanical exergy of stream
H_S: Enthalpy of stream at present state
H_{**}: Enthalpy of stream at state **
H_0: Enthalpy of stream at dead state
S_S: Entropy of stream at present state
S_{**}: Entropy of stream at state **
S_0: Entropy of stream at dead state
T_0: Ambient temperature

During re-liquefaction, the irreversibility between processes causes exergy loss. To calculate this exergy loss, the physical exergy difference between inlets and outlets of a component can be used [32]. This exergy loss makes the system less efficient and require more work than an ideal system. From this point of view, the system exergy efficiency can be estimated using the numerical indicator η_{ex} via Equation (5).

$$\eta_{ex} = \frac{\dot{E}_{re-liquefaction-in} - \dot{E}_{re-liquefaction-outlet}}{\dot{W}_{net} + \dot{E}_{BOH\ to\ PEMFC-in} - \dot{E}_{BOH\ to\ PEMFC-out}} \quad (5)$$

$\dot{E}_{re-liquefaction-in}$: Physical flow exergy of Stream 102
$\dot{E}_{re-liquefaction-outlet}$: Physical flow exergy of Stream 105
$\dot{E}_{BOH\ to\ PEMFC-in}$: Physical flow exergy of Stream 106
$\dot{E}_{BOH\ to\ PEMFC-out}$: Physical flow exergy of Stream 108

3.2. Economics

CAPEX is defined as the initial investment required to construct a plant [33], and it consists of the direct project expenses, indirect project expenses, contingency and fee as depicted in Figure 2. The direct expenses encompass the equipment costs, material costs, and labor costs required to install the equipment. The indirect project expenses include the freight costs, insurance, and taxes. They also include the overhead costs required to construct the plant. The contingency is the cost that covers unforeseen circumstances, while the fee is related to the contractors. Among these costs, the sum of the direct and indirect costs is called the bare module cost. The contingency and fee are assumed as 15% and 3% of the bare module cost, respectively. The bare module cost for each component is estimated using the Aspen Capital Cost Estimator V11.

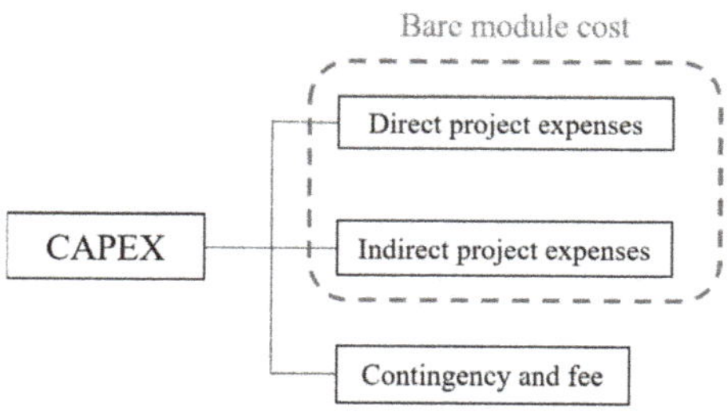

Figure 2. Composition of CAPEX.

OPEX is defined as the costs associated with the day-to-day operations of a plant [33]. OPEX consists of direct manufacturing costs, fixed manufacturing costs, and general manufacturing expenses as depicted in Figure 3. The direct manufacturing costs are the operating expenses, which vary with the production rate. They include raw materials costs, utilities costs, and operational labor. The fixed costs are independent of changes in the production rate. They include taxes and insurance. The general expenses are overhead costs that are necessary to carry out business functions. They include administration, distribution and selling costs, as well as costs for research and development. Equation (6) [33] is used to estimate OPEX. Table 2 shows the specific values used to estimate CAPEX and OPEX.

$$OPEX = 0.18\ CAPEX + 2.73\ C_{OL} + 1.23\ (C_{UT} + C_{WT}) \quad (6)$$

C_{OL}: Cost of the operator salary
C_{UT}: Cost of utilities
C_{WT}: Cost of the cooling water

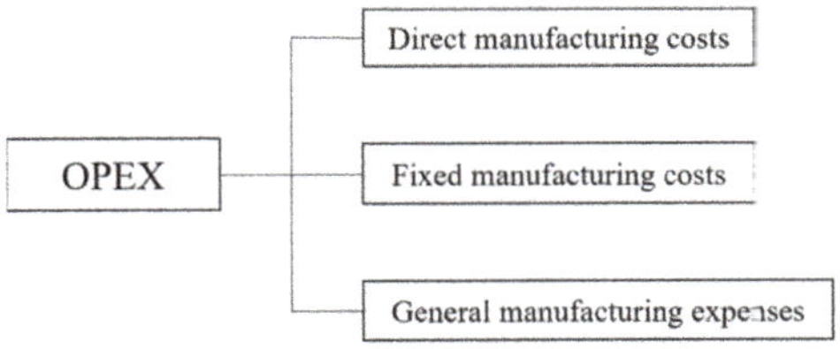

Figure 3. Composition of OPEX.

Table 2. Cost values for CAPEX and OPEX estimation.

Item	Unit	Value
Operator salary	\$/yr	59,580 [33]
Cooling water cost	\$/MWh	243.68 [33]
Electricity	\$/MWh	26.59 [34]
PEMFC system	\$/kW	100 [35]

The life cycle cost (LCC) is defined as the total costs required to install and operate the system during the life cycle [33]. It is estimated using Equation (7). The specific life cycle cost (SLCC) is defined as the LCC required for 1 kg of BOH, which is estimated using Equation (8). Additionally, the cost difference is defined as the difference between the LH_2 production cost and SLCC, as expressed by Equation (9).

$$\text{LCC} = \text{CAPEX} + (\text{Life Cycle}) \times \text{OPEX} \tag{7}$$

$$\text{SLCC} = \frac{\text{LCC}}{\dot{m}_{re-liqeufaction}} \tag{8}$$

$$(\text{Cost difference}) = (LH_2\ \text{production cost}) - \text{SLCC} \tag{9}$$

3.3. Restrictions on CO_2 Emissions from LH_2 Carriers

The attained EEDI indicates the CO_2 emissions per unit of deadweight divided by the ship speed, which is calculated using Equation (10) for each ship [36].

$$\text{EEDI}_{att} = \frac{P_{ME} \cdot C_{ME} \cdot SFC_{ME} + P_{AE} \cdot C_{AE} \cdot SFC_{AE}}{DWT \cdot V_{ref} \cdot} \tag{10}$$

P_{ME}: Power of the main engine
P_{AE}: Power of the auxiliary engine
C_{ME}: Conversion factor of the main engine between the fuel consumption and CO_2 emissions
C_{AE}: Conversion factor of the auxiliary engine between the fuel consumption and CO_2 emissions
SFC_{ME}: Specific fuel consumption of the main engine
SFC_{AE}: Specific fuel consumption of the auxiliary engine
DWT: Deadweight of the ship
V_{ref}: Speed of the ship

A diesel electric propulsion obtained using LNG is assumed for LH_2 carriers. Specific fuel consumption is assumed as 175 g/kWh. The conversion factor between the fuel consumption and CO_2 emissions is 2.75 [37]. According to the Marine Environment Protection Committee (MEPC) issued by the IMO, the power of the main engine for diesel electric propulsion is calculated using Equation (11). The parameter η is taken as 91.3 %, which indicates the product of the electrical efficiencies of the generators, transformers, converters and motors. Considering ships whose rated output of the motor is larger than 10,000 kW, the power of the auxiliary engine is calculated using Equation (12) [36].

$$P_{ME} = 0.83 \times \frac{MPP_{motor}}{\eta} \tag{11}$$

$$P_{AE} = 0.025 \times MPP_{motor} + 250\ \text{kW} \tag{12}$$

MPP_{motor}: Rated output of the motor
η: Product of the electrical efficiencies of the generator, transformer, converter, and motor

The PEMFC system uses the BOH to generate electricity, which is then utilized for propulsion in conjunction with the electricity from the main engine. Therefore, the required power of the main engine is calculated using Equation (13). In this study, the efficiency of the PEMFC system is assumed to be 42% compared with lower heating value of hydrogen. In the case of LNG carriers with a BOG re-liquefaction system, the power required for the BOG re-liquefaction is added to the auxiliary engine power, as shown in Equation (14).

$$P_{ME} = 0.83 \times \frac{MPP_{motor}}{\eta} - P_{PEMFC} \tag{13}$$

$$P_{AE} = 0.025 \times MPP_{motor} + 250\ \text{kW} + P_{re-liq} \tag{14}$$

P_{PEMFC}: Electricity generated from the PEMFC
P_{re-liq}: Power required for re-liquefaction

The required EEDI indicates the criteria that the ship under EEDI regulations must satisfy. Equations (15)–(17) show the methodology for calculating the required EEDI [36].

$$EEDI_{ref} = a \cdot b^{-c} \tag{15}$$

$$EEDI_{req} = (1 - X) \times EEDI_{ref} \tag{16}$$

$$EEDI_{att} \le EEDI_{req} \tag{17}$$

$EEDI_{ref}$: Reference EEDI
$EEDI_{req}$: Required EEDI

The parameters a and c in the required EEDI equation are determined based on the type of ships. The variable b is the deadweight of the ship. X, which is between 0 and 1, is a reduction factor that indicates the reinforcement of the regulations over time. The time factor (referred to as the 'phase') represents the reinforcement of the regulations over time, which is determined using the value of X. For example, phase 3 indicates the year after 2025 and the factor X in this time is 0.3.

Because the EEDI regulations for LH_2 carriers have not yet been designated, various perspectives should be considered before determining the final designation. This study considers the following EEDI candidates:

- EEDI Candidate 1: EEDI equivalent to that of LNG carriers
- EEDI Candidate 2: EEDI considering the energy density of LH_2
- EEDI Candidate 3: No restrictions using LNG as a fuel

3.4. EEDI Candidate 1: EEDI Equivalent to That of LNG Carriers

The concept behind EEDI candidate 1 is to utilize the required EEDI of LNG carriers for LH_2 carriers. Table 3 shows the parameters used for the evaluation of the required EEDI of LNG carriers. In this EEDI candidate, the parameters in Table 3 and the deadweight of the LH_2 carrier are used to calculate the required EEDI for LH_2 carriers. Therefore, the required EEDI of an LNG carrier with the same deadweight as the LH_2 carrier is calculated and compared with the attained EEDI for the LH_2 carrier.

Table 3. Parameters and variables for the required EEDI of LNG carriers.

Ship Type	a	b	c	X
LNG carrier	2253.70	Deadweight (tons)	0.474	0.3 (Phase 3)

3.5. EEDI Candidate 2: EEDI Considering the Energy Density of LH_2

EEDI candidate 2 considers the energy density difference between LH_2 and LNG. As shown in Table 4, the density of LH2 is 16% of that of LNG. This low density of LH_2 makes the attained EEDI of an LH_2 carrier calculated using Equation (10) smaller than that of an LNG carrier with the same volumetric capacity. Conversely, LH_2 has a heating value that is 2.58 times that of LNG. This indicates that LH_2 can carry more energy within the same mass as LNG. EEDI candidate 2, therefore, considers this energy density factor. The energy density is used to introduce the "re-scaled deadweight" concept shown in Equation (19) in place of the mass density. Using this rescaled deadweight, Equation (20) defines the "energy-based EEDI", which applies the energy density concept to the attained EEDI. In EEDI candidate 2, it is compared with the required EEDI of LNG carriers with the same volumetric capacity.

$$DWT = DWT_{cargo} + DWT_{other} \tag{18}$$

$$DWT_{re-scaled} = \frac{LHV_{LH2}}{LHV_{LNG}} DWT_{cargo} + DWT_{other} \tag{19}$$

$$EEDI_{energy-based} = \frac{P_{ME} \cdot C_{ME} \cdot SFC_{ME} + P_{AE} \cdot C_{AE} \cdot SFC_{AE}}{DWT_{re-scaled} \cdot V_{ref}} \tag{20}$$

DWT_{cargo}: Deadweight of cargo
DWT_{other}: Deadweight without cargo
$DWT_{re-scaled}$: Rescaled deadweight
LHV_{LH2}: Lower heating value of LH_2
LHV_{LNG}: Lower heating value of LNG
$EEDI_{energy-based}$: Energy-based EEDI

Table 4. Densities and lower heating values of LH_2 and LNG.

Item	Units	LNG	LH_2	Ratio of LH_2/LNG
Density	kg/m^3	437.89	70.83	0.16
Gravimetric lower heating value	MWh/kg	12.92	33.33	2.58
Volumetric lower heating value	MWh/m^3	5660.94	2360.41	0.42

3.6. EEDI Candidate 3: No Restrictions Using LNG as a Fuel

EEDI candidate 3 refers to the case in which LH_2 carriers have no restrictions regarding their CO_2 emissions provided that conventional clean fuels such as LNG are used. Unlike other gas carriers that carry CO_2-rich fuels, such as LPG and LNG, LH_2 carriers are used to transport CO_2-free hydrogen. The strong regulations on CO_2 emissions from LH_2 carriers, such as those considered in EEDI candidates 1 and 2, may thereby be impartial to liquid hydrogen, ultimately preventing the shipping of this clean fuel. It would therefore be fair to impose no restrictions on CO_2 emissions if these ships utilize relatively clean fuels such as LPG or LNG. In this case, the practicality of BOH re-liquefaction can be determined purely on an economic basis.

4. Case Study

4.1. Target Ship Descriptions

LH_2 carriers with five different sizes are considered for the subsequent case studies. The data of LNG carriers from the Clarksons database is used to assume the cargo capacity, rated output of motors for propulsion, and deadweight of LH_2 carriers [38]. The deadweight of an LH_2 carrier is assumed to be sum of the LH_2 cargo weight and the deadweight of an LNG carrier without cargo and with the same capacity. The rated output of the motor (MPP) is calculated to have the same propulsion power of an LNG carrier with the same capacity. Table 5 shows the specifications of the LH_2 carriers based on these assumptions.

Table 5. Specifications of the ship assumptions.

Ship	Cargo Capacity (m^3)	Speed (knots)	MPP (MW)	Dead Weight (tons)
Ship #1	50,000	19	11.49	6300
Ship #2	74,000	17.5	14.02	9400
Ship #3	154,000	19	23.40	19,800
Ship #4	174,000	19	26.00	23,000
Ship #5	210,000	19.5	27.68	24,400

4.2. Voyage Conditions

The LH_2 export terminal is assumed to located be at Darwin, Australia, while the import terminal is assumed to located be at Pyeongtaek, South Korea. The boil-off rate for a laden voyage is assumed to be 0.2%/day [29]. The BOH generation for a ballast voyage is assumed to be 40% of that of the laden voyage. Table 6 shows the voyage conditions between Darwin and Pyeongtaek. Table 7 shows the BOH generation rates for the laden and ballast voyages. Table 8 shows the total amounts of BOH generated during one-way trips.

Table 6. Summary of the voyage conditions from Darwin to Pyeongtaek.

Item	Unit	Value
Distance	km	5600
Voyage time	days	7
Loading time	days	2
Unloading time	days	2
BOR	%/day	0.2

Table 7. BOH generation during voyage.

Voyage	Ship	Unit	Value
Laden voyage	Ship #1	kg/day	6400
	Ship #2	kg/day	9500
	Ship #3	kg/day	19,900
	Ship #4	kg/day	22,400
	Ship #5	kg/day	27,000
Ballast voyage	Ship #1	kg/day	2600
	Ship #2	kg/day	3800
	Ship #3	kg/day	7900
	Ship #4	kg/day	9000
	Ship #5	kg/day	10,800

Table 8. Total amount of BOH during voyage.

Voyage	Case	Unit	Value
Laden voyage	Ship #1	Tons	58
	Ship #2	Tons	86
	Ship #3	Tons	179
	Ship #4	Tons	201
	Ship #5	Tons	243
Ballast voyage	Ship #1	Tons	23
	Ship #2	Tons	34
	Ship #3	Tons	72
	Ship #4	Tons	81
	Ship #5	Tons	97

4.3. Liquid Hydrogen Production Cost

An LH_2 production cost is assumed for comparison with the LCC of the BOH re-liquefaction system. The Fuel Cells Program Records from the Department of Energy provides the costs for hydrogen production and liquefaction. These documents also provide the terminal cost of LH_2 [39,40]. As a result, the total cost for LH_2 production and shipping is assumed to be 6.5 $/kg as described in Table 9.

Table 9. LH_2 production cost assumption.

Item	Unit	Value
LH_2 production cost	$/kg	6.50

5. Results and Discussion

5.1. Energy and Exergy Efficiency Results

Figure 4 shows the SEC of the proposed BOH re-liquefaction system, which varies from 8.22 to 10.80 kWh/kg as the R/G fraction varies from 10% to 100%. The BOH that is diverted to the PEMFC cools down the helium refrigerant through HX 1 and HX 3. As the temperature of the helium at the inlet of the compressors decreases, the specific volume of the helium also decreases. The compressor work required to achieve a specific pressure ratio decreases as this specific volume decreases. In the 100% re-liquefaction case, the temperature of the helium increases from 311 to 486 K during compression from 1.20 to 2.89 bar in Comp 1. In this case, a specific compressor work of 907.94 kJ/kg is required. Conversely, in the 10% re-liquefaction case, the cold BOH heading to the PEMFC stacks cools down the helium refrigerant in HX 3. The inlet temperature of Comp 1 is 240 K and increases to 375 K during compression from 1.20 to 2.89 bar in Comp 1. In this case, the specific compressor work is 700.91 kJ/kg. By comparing the 100% to the 10% R/G fraction, the cold energy from BOH reduces 23% of the required compressor work. As noted for the compression at Comp 1, the cold energy from the BOH reduces the compressor work of Comp 2. In the 100% re-liquefaction case, the inlet and outlet temperatures are 313 and 582 K at Comp 2, respectively, where the helium refrigerant is compressed from 2.89 to 10 bar via 1396.85 kJ/kg of specific compressor work. Similarly, the BOH cools down the helium in the 10% re-liquefaction case. The inlet and outlet temperatures are 240 and 446 K, respectively, with same pressure ratio in the 10% re-liquefaction case, and the specific compressor work is 1071.37 kJ/kg. By comparing the effects of 100% and 10% R/G fractions at Comp 2, the cold energy from the BOH reduces 23% of the required compressor work.

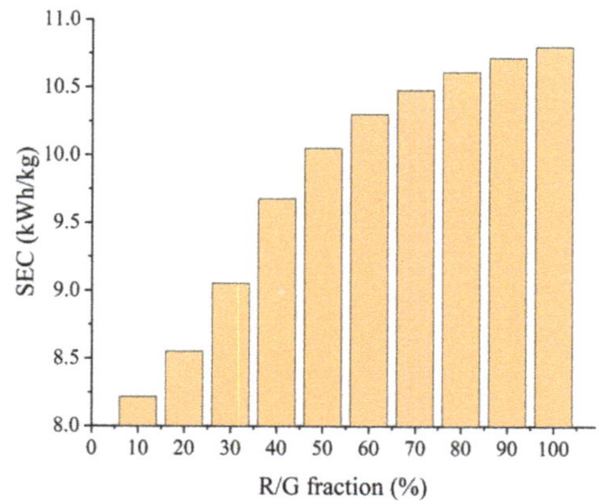

Figure 4. SEC with varying R/G fraction.

Figure 5 shows the exergy efficiency of the BOH re-liquefaction system with varied R/G fractions. The exergy efficiency increases from 0.209 to 0.258 as the R/G fraction increases from 10% to 30%, and it then converges after an R/G fraction of 30%. Figure 6 shows the exergy loss at each component in the re-liquefaction system. The exergy loss in the expanders and compressors is caused by mechanical irreversibility. The exergy loss in the after-coolers and heat exchangers is caused by the heat transfer between a finite temperature difference. It should be mentioned that the exergy loss due to heat transfer decreases as the R/G fraction increases from 10% to 30%, while it converges after 30%. When the R/G fraction is lower than 30%, the excess cold energy is provided by the BOH heading to the PEMFC. The excess cold energy enlarges the temperature difference between the helium and BOH heading to the PEMFC, and this large temperature difference causes a large amount of exergy loss.

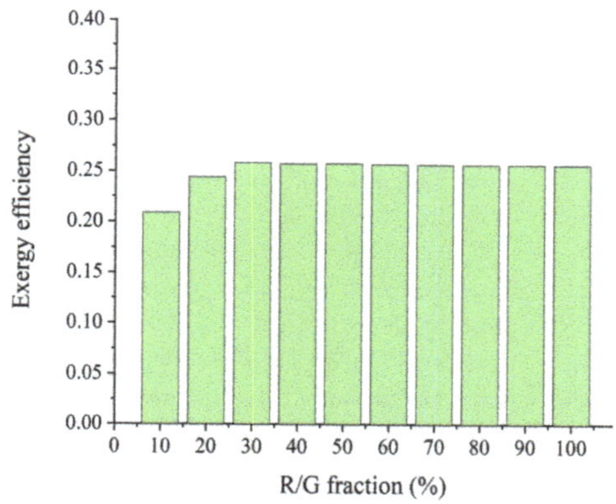

Figure 5. Exergy efficiency with varied R/G fractions.

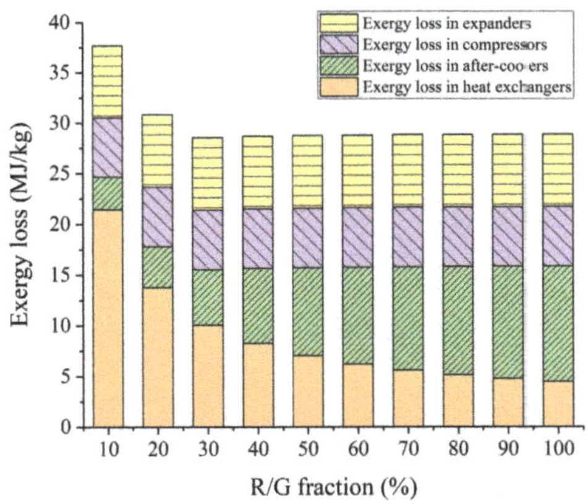

Figure 6. Specific exergy loss with varied R/G fractions.

In the process flow diagram depicted in Figure 1, Stream 108 indicates the BOH diverted to the PEMFC stacks. This stream provided cold energy through HX 3 and HX 1 and is designed to be 310 K, which is the ambient temperature. However, in the cases of 10% and 20% re-liquefaction, the excess cold energy is not fully utilized, and the temperature of Stream 108 is lower than 310 K. Because of this low temperature of Stream 108, the temperature differences in HX 1 and HX 2 are larger than those in the higher R/G fraction cases. As a result, increased exergy losses of 58% and 15% are generated by the heat transfer at the 10% and 20% R/G fractions, respectively, compared to the other R/G fraction cases.

5.2. Economic Evaluation Results

Figure 7 shows the structures of the LCCs for the BOH re-liquefaction systems. It is indicated that OPEX, which includes the operation and maintenance expenses, more influences the LCC than CAPEX, which contains the initial investment of the system. It is obvious that the total LCC increases with the increasing LH_2 capacity of the ship. However, the SLCC, which is the LCC per 1 kg of BOH to be re-liquefied, decreases because the increase of the LCC is lower than the increase of the mass of the re-liquefied BOH. Figure 8a shows the SLCC of the BOH re-liquefaction system as it varies with the capacity of the ship and R/G fraction. It is indicated that at the same R/G fraction, the SLCC decreases as the capacity of the ship increases. Moreover, the SLCC decreases as the R/G fraction increases for the same ship. It can be deduced that as the mass of BOH re-liquefaction increases, the SLCC of the BOH re-liquefaction system decreases. Figure 8b shows the SLCC results with the varied mass of the re-liquefied BOH. It is indicated that as the re-liquefied mass increases, the specific LCC decreases. The slope of the graph in Figure 8b decreases as the re-liquefied mass increases. After the re-liquefied mass is greater than 7.2 ton/day, the SLCC converges at 1.5 $/kg.

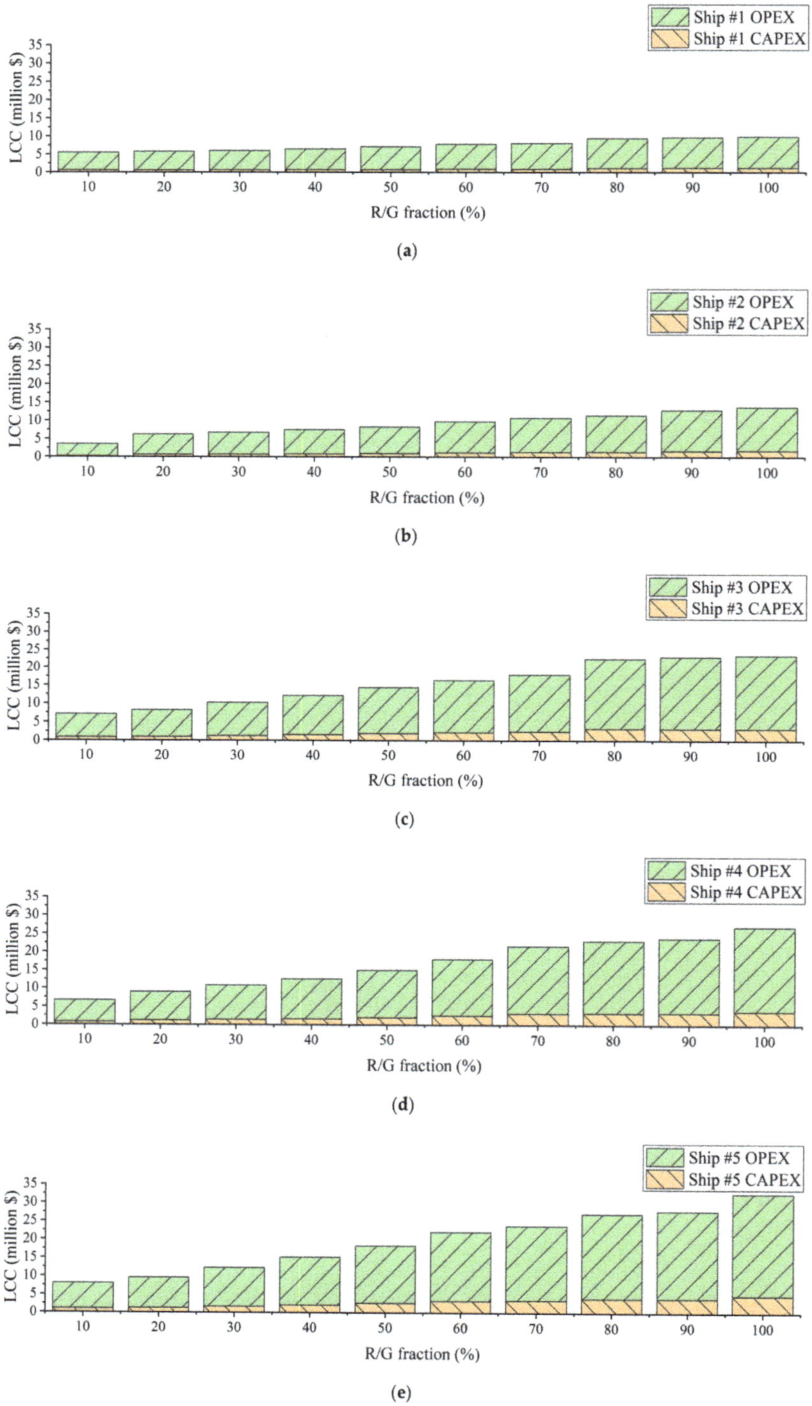

Figure 7. Life cycle cost structures of (**a**) Ship #1, (**b**) Ship #2, (**c**) Ship #3, (**d**) Ship #4, and (**e**) Ship #5.

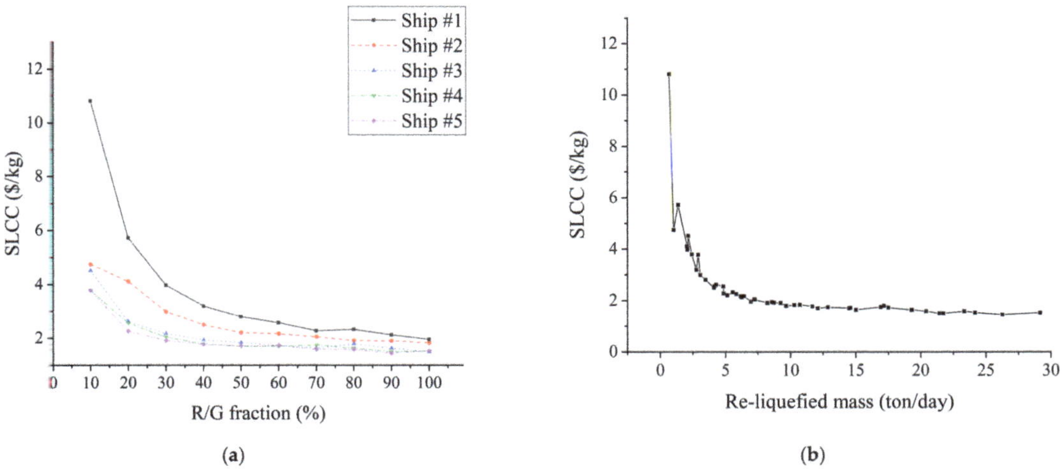

Figure 8. SLCC for each ship with varied (**a**) R/G fractions and (**b**) re-liquefied masses.

Compared to the LH_2 production cost of 6.50 \$/kg (as mentioned in Section 4.3), the BOH re-liquefaction system is considered to be beneficial for 20% to 100% R/G fractions. The production cost and SLCC can be used to estimate the economic benefit obtained by using such a system. During the voyage from Darwin to Pyeongtaek described in Section 4.2, the cost difference between the LH_2 production cost and SLCC for a round-trip is estimated in Figure 9.

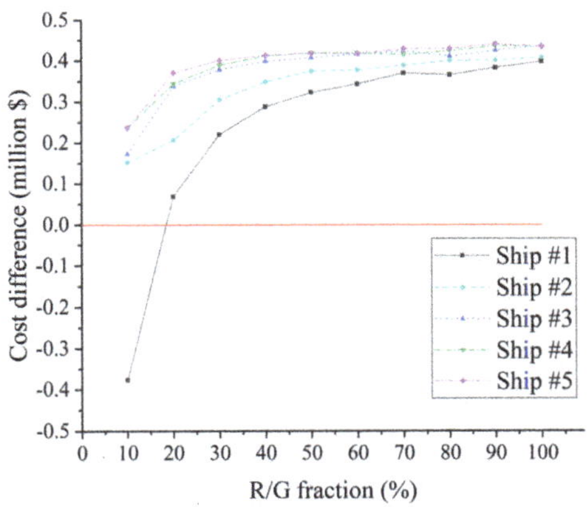

Figure 9. Cost difference with varied R/G fractions during the voyage.

5.3. Consequences of the EEDI

5.3.1. EEDI Candidate 1

Figure 10 shows the attained EEDI results of the ships with varied R/G fractions. Each graph includes the required EEDI phase 3 line of the LNG carrier with the same volumetric capacity. As shown in Equation (10), the calculation results obtained using the same R/G fraction for each ship tend to decrease as the volume capacity of the ships increase. This indicates that as the volumetric capacity of the ship increases, the attained EEDI of the LH_2 carriers tends to decrease and becomes more similar to EEDI candidate 1. As the R/G fraction increases, the BOH utilized in the PEMFC decreases and the required power from the main engine increases. The more power the main engine generates using the LNG fuel, the more CO_2 the ship emits.

Table 10 shows the permittable R/G fractions according to EEDI candidate 1, indicating that only a small amount of BOH is permittable for re-liquefaction. In the cases of Ships #1 and #2, whose capacity is relatively smaller than the other LH_2 carriers, additional hydrogen is required to satisfy the EEDI candidate 1. Additionally, in the cases of Ships #3 to #5 with larger capacities, less than 15% of the generated BOH is permittable for re-liquefaction.

Table 10. Permittable R/G fractions of the LH_2 carriers according to EEDI candidate 1.

Case	Permittable R/G Fraction (%)
Ship #1	No re-liquefaction
Ship #2	No re-liquefaction
Ship #3	9.37
Ship #4	8.20
Ship #5	14.81

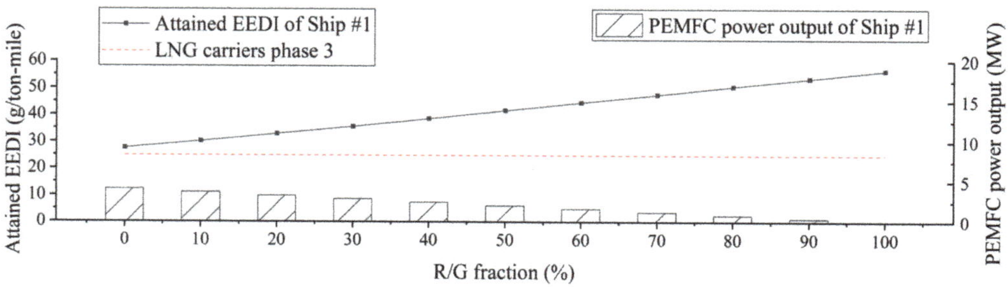

(**a**)

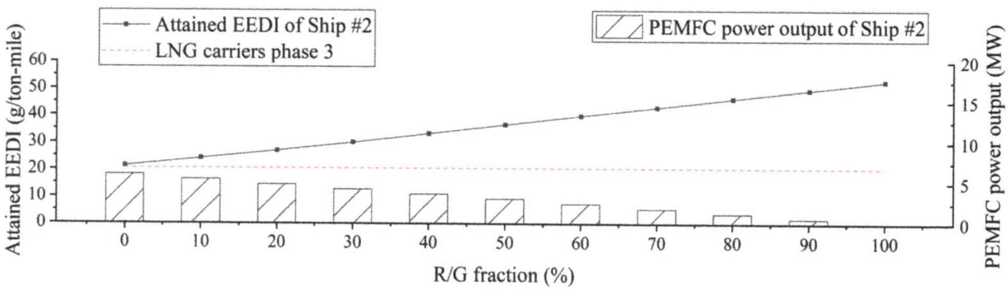

(**b**)

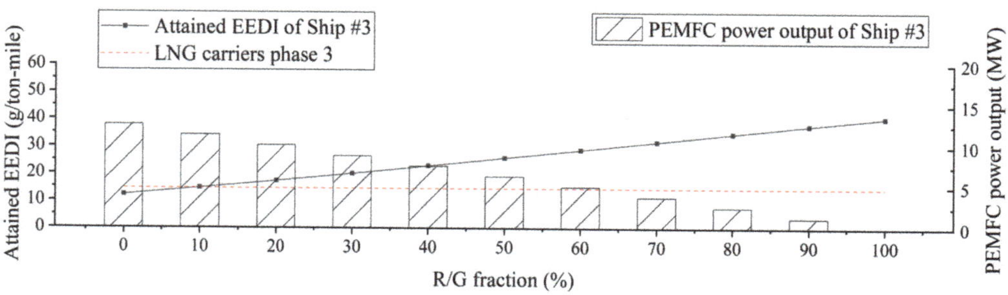

(**c**)

Figure 10. *Cont.*

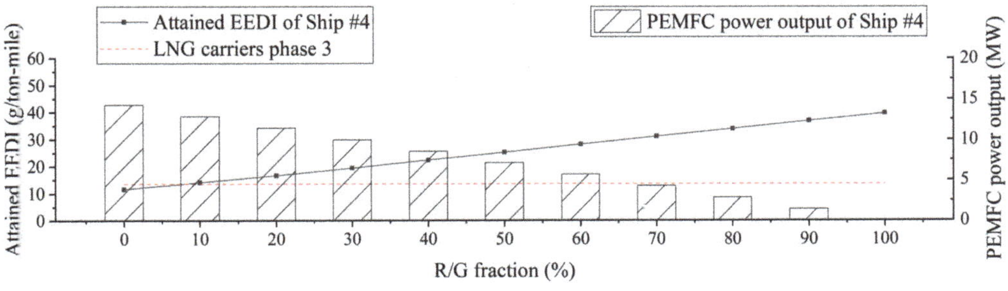

(**d**)

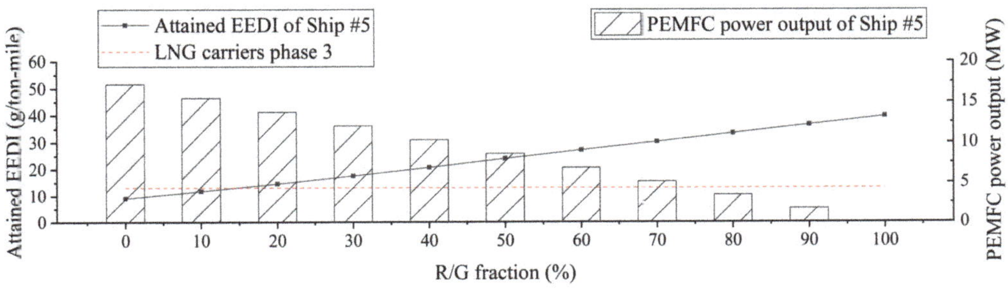

(**e**)

Figure 10. Attained EEDIs for (**a**) Ship #1, (**b**) Ship #2, (**c**) Ship #3, (**d**) Ship #4, and (**e**) Ship #5 using EEDI candidate 1.

5.3.2. EEDI Candidate 2

Figure 11 shows the energy-based EEDI calculation results defined for EEDI candidate 2. Each graph for Ships #1 to #5 presents the results of this energy-based EEDI with varied R/G fractions. The graphs also exhibit the required EEDI phase 3 for LNG carriers with the same rescaled deadweight as each LH_2 carrier. Similar to the attained EEDI shown in Figure 10, as the R/G fraction increases, the energy-based EEDI increases. Moreover, the energy-based EEDI tends to decrease as the volumetric capacity of the ships increases. However, unlike the attained EEDI results shown in Figure 10, every ship is able to reliquefy a ratio of BOH between 25% and 33% such that the energy-based EEDI is less than the required EEDI phase 3 of LNG carriers. These results were obtained due to the rescaled deadweight that was increased from the original deadweight considering the differing heating values of LH_2 and LNG. Table 11 shows the permittable R/G fractions of Ships #1 to #5. The permittable R/G fraction tends to increase as the volumetric capacity of the ships increases.

Table 11. Permittable R/G fractions of the LH_2 carriers according to EEDI candidate 2.

Case	Permittable R/G Fraction (%)
Ship #1	26.20
Ship #2	25.45
Ship #3	30.00
Ship #4	27.79
Ship #5	33.35

The differences between Tables 10 and 11 indicate how the mass and energy densities of LH_2 differ from those of LNG. Because LH_2 has a lower density but larger heating value than LNG, the permittable R/G fraction is larger in EEDI candidate 2 than in EEDI candidate 1. The cargo of the currently used energy carriers under EEDI regulations is mainly hydrocarbon materials such as oil and LNG. These materials have different densities and heating values compared to hydrogen. The existing EEDI regulation for energy carriers, which is calculated using the mass-based deadweight, is used due to the properties of these hydrocarbons. Therefore, the application of this regulation directly to LH_2 carriers without considering the properties of LH_2 is inappropriate. The large heating value of hydrogen should be reflected in these regulations such that the energy carrier may carry energy efficiently.

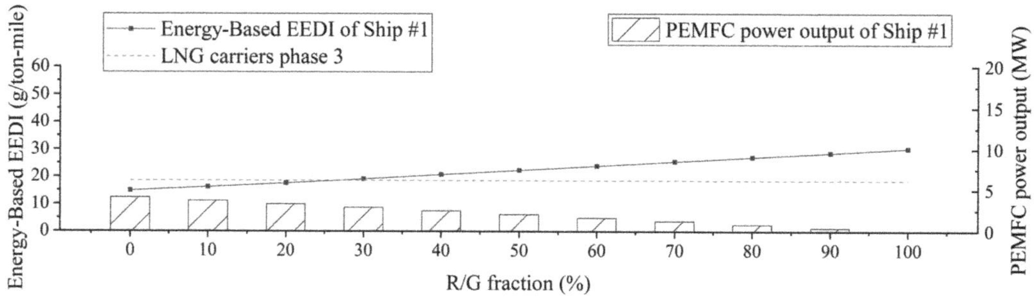

(a)

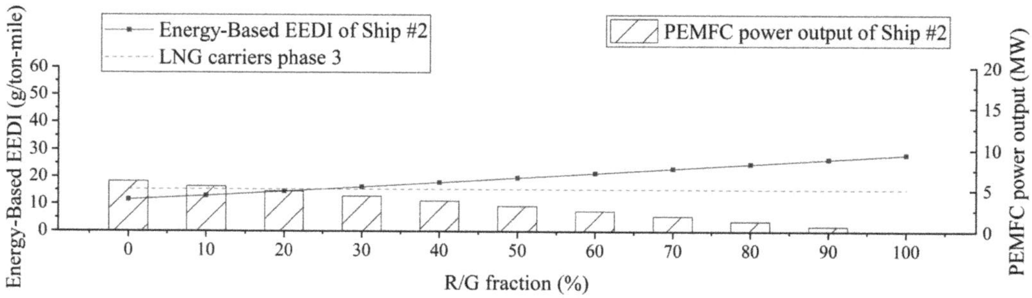

(b)

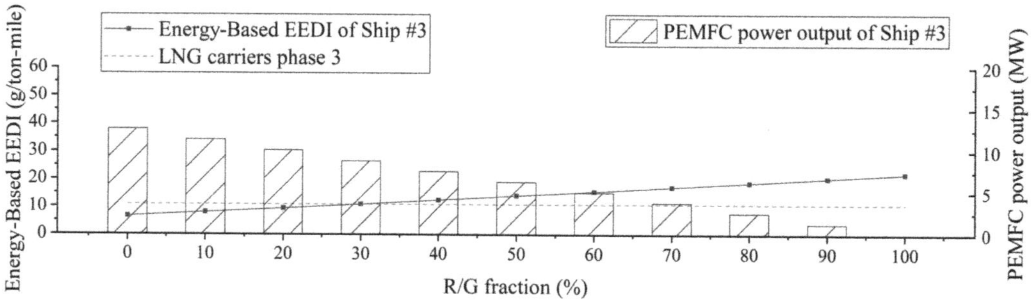

(c)

Figure 11. *Cont.*

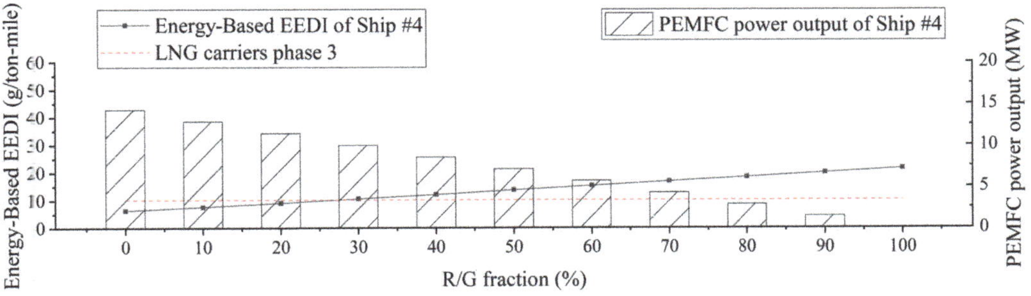

(**d**)

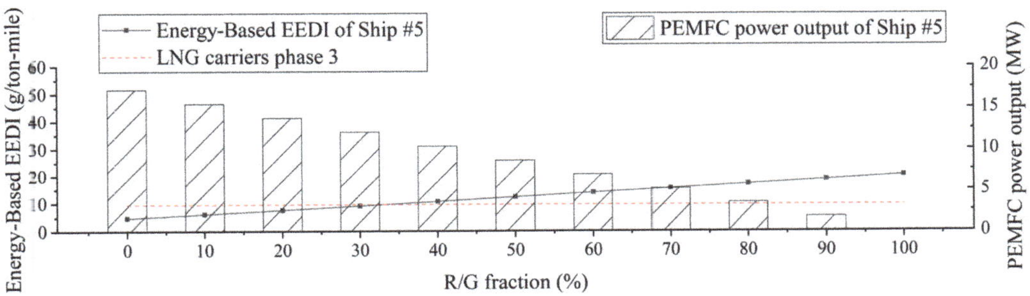

(**e**)

Figure 11. Energy-based EEDIs for (**a**) Ship #1, (**b**) Ship #2, (**c**) Ship #3, (**d**) Ship #4, and (**e**) Ship #5 using EEDI candidate 2.

5.3.3. EEDI Candidate 3

EEDI candidate 3 exempts the LH_2 carriers from the EEDI regulations. The LH_2 carriers deliver LH_2 cargo, which emits no CO_2, unlike other fuels. In addition, it is highly likely that only CO_2-free LH_2 will be allowed for international trading. Therefore, although regulations on CO_2 emissions may not be imposed on LH_2 carriers, LH_2 is far less CO_2-intensive than other liquefied cargos such as LNG and LPG considering the entire supply chain.

In this case, the BOH R/G fraction is determined mainly via economic motivations. As discussed in Section 5.2, the SLCC of the BOH re-liquefaction system decreases as the R/G fraction increases. Consequently, all BOH may be re-liquefied considering the economic results obtained using EEDI candidate 3.

Table 12 shows the SLCCs for the permittable R/G fractions obtained using each EEDI candidate. As described in Section 5.3.1, the permittable R/G fraction indicates the amount satisfying the EEDI restrictions for each candidate. In the case of EEDI candidate 3, this ratio is 100% because there is no EEDI restriction. Compared with EEDI candidate 1, the SLCC for the permittable R/G fraction decreases from 50% to 68% depending on the capacity of the LH_2 carriers in EEDI candidate 3. Likewise, the SLCC decreases from 18% to 48% compared to the EEDI candidate 2. These results indicate the economic advantages that may be obtained when LH_2 carriers are not subjected to EEDI restrictions. Considering this advantage and the CO_2-free characteristic of LH_2, the EEDI-free regulation of LH_2 carriers can be considered, which exempts LH_2 carriers with LNG fuels from the CO_2 emissions restrictions.

Table 12. SLCCs for the permittable R/G fractions obtained using each EEDI candidate.

SLCC with EEDI Candidate ($/kg)	Ship #1	Ship #2	Ship #3	Ship #4	Ship #5
EEDI Candidate 1	-	-	4.64	4.01	3.05
EEDI Candidate 2	4.64	3.51	2.18	2.16	1.87
EEDI Candidate 3	1.96	1.83	1.50	1.53	1.53

6. Conclusions

This study proposed a partial BOH re-liquefaction system based on the reverse Brayton helium cycles. This system divides the generated BOH into two streams, one of which is to be re-liquefied and the other is utilized to generate electricity in PEMFC stacks. Various evaluations for the system were performed based on an assumed voyage route, five different LH_2 carrier specifications, and an assumed LH_2 production cost.

The SEC increased from 8.22 to 10.80 kWh/kg as the R/G fraction increased from 10% to 100%. The exergy efficiency was increased from 0.209 to 0.258 as the R/G fraction increased from 10% to 30%, and it converged to 0.258 when the R/G fraction was larger than 30%. The exergy loss in heat transfer occupied the largest portion of all. Due to the excessive cold energy of the BOH heading to the PEMFC stacks, compared to other R/G fraction cases, 58% and 15% more exergy loss occurred in 10% and 20% cases, respectively.

The system economics indicated that the re-liquefied mass of BOH is inversely proportional to the SLCC. The gradient of this decrease became smoother as the re-liquefied mass of BOH increased. When the re-liquefied mass of BOH was larger than 7200 kg/day, the SLCC was almost unchanged from 1.5 $/kg; this value is much lower than 6.50 $/kg, which is the assumed LH_2 production cost.

Considering EEDI candidate 1, the attained EEDI demonstrated that most of the BOH should not be re-liquefied when the required EEDI was evaluated based on the parameters of the LNG carrier for the required EEDI phase 3 with the same volumetric capacity. However, for EEDI candidate 2, it was shown that the permittable R/G fraction was between 25% and 33% considering energy-based EEDI and required EEDI phase 3. Finally, for EEDI candidate 3, the EEDI-free regulation of LH_2 carriers was discussed considering the CO_2-free characteristic of LH_2. If the EEDI regulation is not used for LH_2 carriers, the SLCC of the BOH re-liquefaction system decreases up to 68% compared to LNG carriers with equivalent EEDI regulations.

Author Contributions: Conceptualization, M.C., W.J. and S.L.; methodology, W.J. and T.J.; software, M.C.; validation, M.C. and W.J.; formal analysis, M.C.; investigation, M.C.; resources, T.J.; data curation, M.C.; writing—original draft preparation, M.C.; writing—review and editing, W.J.; visualization, M.C.; supervision, W.J. and D.C.; project administration, D.C.; funding acquisition, D.C. All authors have read and agreed to the published version of the manuscript.

Funding: This research was a part of the project titled 'Development of Safety and Control Standards for Hydrogen Ships: Cargo Handling and Fuel Gas Supply Systems' (Grant number: 20200456), funded by the Ministry of Oceans and Fisheries, Korea.

Conflicts of Interest: The authors declare no conflict of interest.

Nomenclature

Scripts	Description
$\dot{W}_{net}$	Total work required to re-liquefy BOH (kW)
$\dot{W}_{Comp\ 1}$	Work input for Comp 1 (kW)
$\dot{W}_{Comp\ 2}$	Work input for Comp 2 (kW)
$\dot{W}_{Exp\ 1}$	Work output for Exp 1 (kW)
$\dot{W}_{Exp\ 2}$	Work output for Exp 2 (kW)
$\dot{m}_{BOH-generation}$	Mass flow of the generated BOH (kg/s)

$\dot{m}_{re-liqeufaction}$	Mass flow of the re-liquefied BOH (kg/s)
E_S	Physical flow exergy of stream (kJ)
E_T	Thermal exergy of stream (kJ)
E_M	Mechanical exergy of stream (kJ)
H_S	Enthalpy of stream at present state (kJ)
H_{**}	Enthalpy of stream at state ** (kJ)
H_0	Enthalpy of stream at dead state (kJ)
S_S	Entropy of stream at present state (kJ/K)
S_{**}	Entropy of stream at state ** (kJ/K)
S_0	Entropy of stream at dead state (kJ/K)
T_0	Ambient temperature (K)
η_{ex}	Exergy efficiency
$\dot{E}_{BOH\ to\ PEMFC-in}$	Physical flow exergy of Stream 106 (kW)
$\dot{E}_{BOH\ to\ PEMFC-out}$	Physical flow exergy of Stream 108 (kW)
$\dot{E}_{re-liquefaction-in}$	Physical flow exergy of Stream 102 (kW)
$\dot{E}_{re-liquefaction-out}$	Physical flow exergy of Stream 105 (kW)
C_{OL}	Cost of the operator salary ($/yr)
C_{UT}	Cost of utilities ($/GJ)
C_{WT}	Cost of the cooling water ($/GJ)
$EEDI_{att}$	Attained EEDI (g/ton-mile)
P_{ME}	Power of the main engine (kW)
P_{AE}	Power of the auxiliary engine (kW)
C_{ME}	Conversion factor of the main engine between fuel consumption and CO_2 emissions
C_{AE}	Conversion factor of the auxiliary engine between fuel consumption and CO_2 emissions
SFC_{ME}	Specific fuel consumption of the main engine (g/kWh)
SFC_{AE}	Specific fuel consumption of the auxiliary engine (g/kWh)
DWT	Deadweight of the ship (ton)
V_{ref}	Speed of the ship (knot)
MPP_{motor}	Rated output of the motor (kW)
η	Product of the electrical efficiencies of the generator, transformer, converter, and motor
P_{PEMFC}	Electricity generated from the PEMFC (kW)
P_{re-liq}	Power required for re-liquefaction (kW)
$EEDI_{ref}$	Reference EEDI (g/ton-mile)
$EEDI_{req}$	Required EEDI (g/ton-mile)
DWT_{cargo}	Deadweight of cargo (ton)
DWT_{other}	Deadweight without cargo (ton)
$DWT_{re-scaled}$	Re-scaled deadweight (ton)
LHV_{LH2}	Lower heating value of LH2 (MJ/kg)
LHV_{LNG}	Lower heating value of LNG (MJ/kg)
$EEDI_{energy-based}$	Energy based EEDI (ton)
Abbreviation	Description
BOG	Boil-off gas
BOH	Boil-off hydrogen
CO_2	Carbon dioxide
IMO	International Maritime Organization
LCC	Life cycle cost ($)
LH_2	Liquid hydrogen
LNG	Liquified natural gas
MEPC	Marine Environment Protection Committee
PEMFC	Proton-exchange membrane fuel cell
R/G fraction	Re-liquefaction-generation fraction (%)
SEC	Specific energy consumption (kWh/kg)
SLCC	Specific life cycle cost ($/kg)

References

1. NOAA's Annual Greenhouse Gas Index. Available online: https://www.esrl.noaa.gov/gmd/aggi/ (accessed on 3 May 2021).
2. Total Energy Supply (TES) by Source. Available online: https://www.iea.org/data-and-statistics/?country=WORLD&fuel=Energysupply&indicator=TPESbySource (accessed on 3 May 2021).
3. Total CO2 Emissions. Available online: https://www.iea.org/data-and-statistics/?country=WORLD&fuel=CO2emissions&indicator=TotCO2 (accessed on 3 May 2021).
4. Renewable Electricity Generation by Source (Non-Combustible). Available online: https://www.iea.org/data-and-statistics/?country=WORLD&fuel=Energysupply&indicator=RenewGenBySource (accessed on 3 May 2021).
5. Share of Renewables in Power Generation in the Sustainable Development Scenario, 2000–2030. Available online: https://www.iea.org/data-and-statistics/charts/share-of-renewables-in-power-generation-in-the-sustainable-development-scenario-2000-2030 (accessed on 6 May 2021).
6. Pethaiah, S.S.; Sadasivuni, K.K.; Jayakumar, A.; Ponnamma, D.; Tiwary, C.S.; Sasikumar, G. Methanol Electrolysis for Hydrogen Production Using Polymer Electrolyte Membrane: A Mini-Review. *Energies* **2020**, *13*, 5879. [CrossRef]
7. Jayakumar, A. *An Assessment on Polymer Electrolyte Membrane (PEM) Fuel Cell Stack Components*; Apple Academic Press: Boca Raton, FL, USA, 2017; ISBN 9781771886062.
8. Rosen, M.A.; Koohi-Fayegh, S. The prospects for hydrogen as an energy carrier: An overview of hydrogen energy and hydrogen energy systems. *Energy Ecol. Environ.* **2016**, *1*, 10–29. [CrossRef]
9. Niaz, S.; Manzoor, T.; Pandith, A.H. Hydrogen storage: Materials, methods and perspectives. *Renew. Sustain. Energy Rev.* **2015**, *50*, 457–469. [CrossRef]
10. Yanxing, Z.; Maoqiong, G.; Yuan, Z.; Xueqiang, D.; Jun, S. Thermodynamics analysis of hydrogen storage based on compressed gaseous hydrogen, liquid hydrogen and cryo-compressed hydrogen. *Int. J. Hydrog. Energy* **2019**, *44*, 16833–16840. [CrossRef]
11. Notardonato, W.U.; Swanger, A.M.; Fesmire, J.E.; Jumper, K.M.; Johnson, W.L.; Tomsik, T.M. Zero boil-off methods for large-scale liquid hydrogen tanks using integrated refrigeration and storage. *IOP Conf. Ser. Mater. Sci. Eng.* **2017**, *278*, 012012. [CrossRef]
12. Tan, H.; Zhao, Q.; Sun, N.; Li, Y. Enhancement of energy performance in a boil-off gas re-liquefaction system of LNG carriers using ejectors. *Energy Convers. Manag.* **2016**, *126*, 875–888. [CrossRef]
13. Choi, J. Development of partial liquefaction system for liquefied natural gas carrier application using exergy analysis. *Int. J. Nav. Archit. Ocean Eng.* **2018**, *10*, 609–616. [CrossRef]
14. Sayyaadi, H.; Babaelahi, M. Multi-objective optimization of a joule cycle for re-liquefaction of the Liquefied Natural Gas. *Appl. Energy* **2011**, *88*, 3012–3021. [CrossRef]
15. Kim, D.; Hwang, C.; Gundersen, T.; Lim, Y. Process design and economic optimization of boil-off-gas re-liquefaction systems for LNG carriers. *Energy* **2019**, *173*, 1119–1129. [CrossRef]
16. Romero, J.; Orosa, J.A.; Oliveira, A.C. Research on the Brayton cycle design conditions for reliquefaction cooling of LNG boil off. *J. Mar. Sci. Technol.* **2012**, *17*, 532–541. [CrossRef]
17. Kwak, D.H.; Heo, J.H.; Park, S.H.; Seo, S.J.; Kim, J.K. Energy-efficient design and optimization of boil-off gas (BOG) re-liquefaction process for liquefied natural gas (LNG)-fuelled ship. *Energy* **2018**, *148*, 915–929. [CrossRef]
18. Yin, L.; Ju, Y.L. Comparison and analysis of two nitrogen expansion cycles for BOG Re-liquefaction systems for small LNG ships. *Energy* **2019**, *172*, 769–776. [CrossRef]
19. Sayyaadi, H.; Babaelahi, M. Thermoeconomic optimization of a cryogenic refrigeration cycle for re-liquefaction of the LNG boil-off gas. *Int. J. Refrig.* **2010**, *33*, 1197–1207. [CrossRef]
20. Krasae-in, S.; Stang, J.H.; Neksa, P. Development of large-scale hydrogen liquefaction processes from 1898 to 2009. *Int. J. Hydrogen Energy* **2010**, *35*, 4524–4533. [CrossRef]
21. Krasae-In, S.; Stang, J.H.; Neksa, P. Exergy analysis on the simulation of a small-scale hydrogen liquefaction test rig with a multi-component refrigerant refrigeration system. *Int. J. Hydrog. Energy* **2010**, *35*, 8030–8042. [CrossRef]
22. Ratlamwala, T.A.H.; Dincer, I.; Gadalla, M.A. Thermodynamic analysis of a novel integrated geothermal based power generation-quadruple effect absorption cooling-hydrogen liquefaction system. *Int. J. Hydrog. Energy* **2012**, *37*, 5840–5849. [CrossRef]
23. Chang, H.M.; Ryu, K.N.; Baik, J.H. Thermodynamic design of hydrogen liquefaction systems with helium or neon Brayton refrigerator. *Cryogenics* **2018**, *91*, 68–76. [CrossRef]
24. Asadnia, M.; Mehrpooya, M. A novel hydrogen liquefaction process configuration with combined mixed refrigerant systems. *Int. J. Hydrog. Energy* **2017**, *42*, 15564–15585. [CrossRef]
25. Sadaghiani, M.S.; Mehrpooya, M. Introducing and energy analysis of a novel cryogenic hydrogen liquefaction process configuration. *Int. J. Hydrog. Energy* **2017**, *42*, 6033–6050. [CrossRef]
26. Chang, H.M.; Kim, B.H.; Choi, B. Hydrogen liquefaction process with Brayton refrigeration cycle to utilize the cold energy of LNG. *Cryogenics* **2020**, *108*, 103093. [CrossRef]
27. Yuksel, Y.E.; Ozturk, M.; Dincer, I. Analysis and assessment of a novel hydrogen liquefaction process. *Int. J. Hydrog. Energy* **2017**, *42*, 11429–11438. [CrossRef]
28. Lee, H.; Shao, Y.; Lee, S.; Roh, G.; Chun, K.; Kang, H. Analysis and assessment of partial re-liquefaction system for liquefied hydrogen tankers using liquefied natural gas (LNG)and H2 hybrid propulsion. *Int. J. Hydrog. Energy* **2019**, *44*, 15056–15071. [CrossRef]

29. Ishimoto, Y.; Voldsund, M.; Nekså, P.; Roussanaly, S.; Berstad, D.; Gardarsdottir, S.O. Large-scale production and transport of hydrogen from Norway to Europe and Japan: Value chain analysis and comparison of liquid hydrogen and ammonia as energy carriers. *Int. J. Hydrog. Energy* **2020**, *45*, 32865–32883. [CrossRef]
30. Bejan, A. *Advanced Engineering Thermodynamics*, 3rd ed.; John Wiley & Sons: Hoboken, NJ, USA, 2006; ISBN 9781119245964.
31. Morosuk, T.; Tsatsaronis, G. Splitting physical exergy: Theory and application. *Energy* **2019**, *167*, 698–707. [CrossRef]
32. Tsatsaronis, G.; Morosuk, T. Advanced exergetic analysis of a novel system for generating electricity and vaporizing liquefied natural gas. *Energy* **2010**, *35*, 820–829. [CrossRef]
33. Turton, R.; Bailie, R.C.; Whiting, W.B.; Shaeiwitz, J.A. *Analysis, Synthesis and Design of Chemical Processes*, 4th ed.; Prentice Hall: Upper Saddle River, NJ, USA, 2012.
34. Federal Energy Regulatory Commission Market Assessments. Available online: https://cms.ferc.gov/sites/default/files/2020-11/ngas-ovr-archive.pdf (accessed on 10 May 2021).
35. Moreno, N.G.; Molina, M.C.; Gervasio, D.; Robles, J.F.P. Approaches to polymer electrolyte membrane fuel cells (PEMFCs) and their cost. *Renew. Sustain. Energy Rev.* **2015**, *52*, 897–906. [CrossRef]
36. 2018 Guidelines on The Method of Calculation of The Attained Energy Efficiency Design Index (Eedi) for New Ships. Available online: https://wwwcdn.imo.org/localresources/en/KnowledgeCentre/IndexofIMOResolutions/MEPCDocuments/MEPC.308(73).pdf (accessed on 6 May 2021).
37. Jeong, J.; Seo, S.; You, H.; Chang, D. Comparative analysis of a hybrid propulsion using LNG-LH2 complying with regulations on emissions. *Int. J. Hydrog. Energy* **2018**, *43*, 3809–3821. [CrossRef]
38. Clarksons Research. Available online: https://www.clarksons.net/portal/ (accessed on 20 July 2021).
39. Connelly, E.; Penev, M.; Elgowainy, A.; Hunter, C. Current Status of Hydrogen Liquefaction Costs. Available online: https://www.hydrogen.energy.gov/pdfs/19001_hydrogen_liquefaction_costs.pdf (accessed on 6 May 2021).
40. Vickers, J.; Peterson, D.; Randolph, K. Cost of Electrolytic Hydrogen Production with Existing Technology. Available online: https://www.hydrogen.energy.gov/pdfs/20004-cost-electrolytic-hydrogen-production.pdf (accessed on 6 May 2021).

energies

MDPI

Article

The Impact of Economic, Energy, and Environmental Factors on the Development of the Hydrogen Economy

Justyna Cader *, Renata Koneczna and Piotr Olczak

Mineral and Energy Economy Research Institute, Polish Academy of Sciences, 31-261 Cracow, Poland; rkoneczna@min-pan.krakow.pl (R.K.); olczak@min-pan.krakow.pl (P.O.)
* Correspondence: jcader@min-pan.krakow.pl

Abstract: This article attempts to model interdependencies between socio-economic, energy, and environmental factors with selected data characterizing the development of the hydrogen economy. The study applies Spearman's correlation and a linear regression model to estimate the influence of gross domestic product, population, final energy consumption, renewable energy, and CO_2 emission on chosen hydrogen indicators—production, patents, energy technology research, development, and demonstration budgets. The study was conducted in nine countries selected for their actions towards a hydrogen economy based on analyses of national strategies, policies, research and development programs, and roadmaps. The results confirm the statistically significant impact of the chosen indicators, which are the drivers for the development of the hydrogen economy from 2008 to 2018. Moreover, the empirical results show that different characteristics in each country contribute to the development of the hydrogen economy vision.

Keywords: linear regression model; Spearman's correlation; national hydrogen strategies; hydrogen economy implementation

Citation: Cader, J.; Koneczna, R.; Olczak, P. The Impact of Economic, Energy, and Environmental Factors on the Development of the Hydrogen Economy. *Energies* **2021**, *14*, 4811. https://doi.org/10.3390/en14164811

Academic Editor: Bahman Shabani

Received: 8 July 2021
Accepted: 4 August 2021
Published: 7 August 2021

1. Introduction

In recent years, hydrogen production and fuel cell technologies have attracted the attention of the authorities in many countries. Moreover, the number of research studies related to these technologies increases significantly every year. Such technologies are perceived as breakthrough solutions with the potential to become clean and sustainable energy carriers [1]. Hence, many countries include the possibility of supporting the development of hydrogen and fuel cell vehicle production (along with the necessary infrastructure) in their energy policy scenarios for the coming years. These policies are connected to the national net-zero emissions target declarations in European [2–7] and Asian [8–10] countries, as well as the USA [11] and Australia [12], among others. Hydrogen is predicted to play a significant role in the energy transformation of global economies [13].

The demand for hydrogen is mainly linked to oil refining and chemical industries [14,15]. Despite the limited consumption of hydrogen in other sectors of the economy, it has significant potential in the power industry through its use in fuel cells. Currently, most hydrogen applications are focused on generator cooling and hydrogen burning in boilers or CHP units onsite. In transport, hydrogen can be used as a fuel, both directly (fuel cells, internal combustion engines) or indirectly (complex synthetic fuels) [16].

Various technologies are involved in hydrogen production [17,18]. Hydrogen can be produced by electrolysis using nuclear or renewable electricity (green hydrogen). The proportion of hydrogen produced from renewable energy is still insignificant, mainly due to cost. Hydrogen can also be produced using fossil fuels, e.g., hard or lignite coal and natural gas, with CO_2 emissions reduced via carbon capture storage (blue hydrogen). Fossil fuels are still the main source of fuel for hydrogen production (natural gas, approximately 75%; coal, approximately 23%) [18]. Hydrogen production is also possible without using the CCS method of Steam Methane Reforming (SMR) or the gasification method (grey or brown hydrogen). The pyrolysis process of methane is an alternative production method (turquoise hydrogen).

The development of hydrogen energy is currently related to the search for ways to implement economic production of green hydrogen by reducing the cost of large-scale production [1,19]. Moreover, the implementation of hydrogen technology is particularly important in transport since a continuous increase in road traffic is expected [20]. Transport is predicted to be the main target for the application of hydrogen energy in the near future [21].

The implementation of new hydrogen energy technologies depends, inter alia, on the state of the economy of individual countries and the long-term goals and scenarios included in their national strategies, policies, research and development (R&D) programs, and roadmaps [14,22]. The processes of decarbonization of the economy and energy transformation are also associated with many socio-economic, environmental, and legal factors [23,24]. Most of the studies of the hydrogen economy involve prognostic analyses. Such analyses consider the production and demand for hydrogen together with the development of fuel cell vehicles and the accompanying infrastructure.

Current policies for a hydrogen economy based on a Romanian example were described by Iordache et al. [25]. They referred to energy (energy dependency, RES, net import of electricity, combined heat and power electricity production), transport (road density, roads fuel consumption, motor vehicles), and environmental indicators (CO_2 emissions). Xu et al. [26] studied the influence of factors such as CO_2 emissions, per capita income, the scale of the labor input, the added value in the industry, and European governmental mechanisms for the production of renewable energy based on hydrogen. National economic variables have been found to have a positive impact on hydrogen-based renewable energy. The number and type of patents in selected countries were used as tools to map the development of the hydrogen economy by Sinigaglia et al. [27]. In terms of technological progress, Japan and the United States were found to be the most advanced. The level of application of hydrogen technologies in selected economies for 2008 was analyzed by Leben and Hočevar (2008) [28]. They used correlations of national development indicators (gross domestic product, public expenditure on education, R&D expenditure, number of researchers, science and technology doctorates, general patents, greenhouse gas emissions, total final energy consumption) together with a group of hydrogen indicators (number of hydrogen refueling stations, fuel cell vehicles, and hydrogen production). The study supported the hypothesis that the implementation of hydrogen technologies was dependent on selected national development indicators.

In addition, the analysis of environmental (CO_2 emissions) and economic (GDP, oil prices) factors in relation to renewable energy sources was applied by Sadorsky [29], showing a significant impact of emissions and GDP on the consumption of renewable energy. Similar conclusions were reported by Wang et al. [30], describing the significant impact of GDP on RES consumption. Marques and Fuinhas [31] showed that market dependencies such as fossil fuel prices and incomes were not significant for the development of RES at the turn of the 20th and 21st centuries. Mendonça et al. [32] demonstrated the positive impact of GDP and population on CO_2 emissions and renewable energy production as a way to reduce emissions.

This research attempts to correlate the following factors:

- socio-economic (gross domestic product, population)
- energy (gross final energy consumption including transport and industry, the share of renewable energy of primary energy supply)
- environmental (total CO_2 emissions including transport and industry emissions)
- hydrogen (volume of hydrogen production, the number of patents for hydrogen production and fuel cells, research, development, and demonstration subsidies for energy technology).

Using Spearman's correlation and a linear regression approach, the relationship between indicators was examined in this research. These methods were previously used in various studies related to the analysis of environmental, economic, and energy factors in the areas of renewable energy. Directly in the field of the hydrogen economy, Spearman's method for national development and hydrogen was used by Leben and Hočevar [28]. Spearman's rank correlation was also used by Durmuşoğlu et al. [33] to visualize the factors that influence environmental performance. The correlations were sought for the variables of GDP, CO_2 emissions, and renewable energy consumption as well as for environmental, energy, and economic indexes. Spearman's correlation was also used to search for indicators covering the economy, society, and the environment to present the level of sustainable development [34]. The linear regression approach was applied by Asumadu-Sarkodie and Owusu [35], showing a positive correlation between CO_2 emissions, energy use, GDP, and population. Menz and Vachon [36] discussed the contribution of various policies and systems in US states to the development of wind energy. For this purpose, they used linear regression equations.

In this article the comparison was made for the following nine countries: China, the United States, Japan, the Republic of Korea, the Netherlands, France, the United Kingdom, Germany, Australia. These countries were selected because of their commitment to implementation of hydrogen strategies and overall progress towards a hydrogen economy (see Section 1.1). Time-series data from 2008 to 2018 were analyzed.

Multiple factors, such as socio-economic, energy, and environmental ones, strongly influence the development of the countries with zero-emission policies. Because of that, there is a need to distinguish which specific indicators affect the development of the hydrogen economy in order to create better and more effective strategies in the coming years. The presented methodology, as well as multidimensional factors, will confirm the hypothesis:

> *"There is a correlation between selected economic, energy, and environmental indicators and the development of the hydrogen economy in countries involved in the implementation of hydrogen technologies".*

This study adds value to the extant literature in two novel ways. The first is the analysis of the possibilities for the development of hydrogen technologies and modeling interdependencies between chosen indicators. This is considered by monitoring the number of subsidies and the number of patents dedicated to these solutions, as well as observations of the hydrogen market in the country (through the volume of hydrogen production and consumption) concerning a selected group of indicators essential to strategic policy creation. Secondly, the study improves the global debate on the steps taken so far to evolve the low-carbon economy and the energy transition associated with the vision of the future of hydrogen.

1.1. Hydrogen Strategy in Chosen Countries

Hydrogen is a crucial factor in most strategies employed by 75 countries to achieve net-zero greenhouse gas emissions. By the beginning of 2021, over 30 countries had created hydrogen roadmaps or strategies at a national level. An additional six are currently drafting their strategies. Among the countries analyzed, four have hydrogen strategies and the rest have hydrogen roadmaps (Figure 1). The main focus of these policies is transport and industry.

The United States is the world leader in stocks of FCEVs, with approximately one in three FCEVs on USA roads. The USA is closely followed by China, Japan, Korea [37], Germany, and France [38]. The situation in China and Korea is particularly dynamic, with new sales climbing from a few units in 2017 to almost 4400 in China and 4100 in Korea, in 2019. The global forecast is for 4.5 million FCEVs by 2030 [37].

Figure 1. The latest hydrogen-related government documents cited in this paper.

In 2019, there were 470 hydrogen refueling stations in the world. Compared to 2018, there was a significant increase in their number by over 20%. Most of these stations are located in Japan (113). Germany has 81, the United States has 64, and China has 61 [37]. The number of FCEVs is projected to continue to increase, e.g., in China to one million by 2030 [39], in Korea, to 800,000 by 2030 [8], in the Netherlands to 15,000 by 2030 [5], in France to 22,500 by 2028 [40], and in Japan to 800,000 by 2030 [9].

Much attention has also been given to buses, trains, inland and coastal navigation, car fleets, and airplanes. For example, the EU indicated that around 45,000 fuel cell trucks and buses could be put on the road by 2030. Fuel cell trains could also replace approximately 570 diesel trains by 2030 [41]. The French National Railway Company has already taken the first step in this direction by ordering 15 hydrogen trains [40]. Japan and China are also developing hydrogen railway technology [42].

Germany will allocate EUR 0.6 billion for the purchase of buses with an alternative drive system as part of the H2 Mobility program [43]. Japan plans to purchase 1200 hydrogen buses by 2030 [9]. Since air traffic will continue to run on liquid fuel, it makes sense to require suppliers to use electric jet fuel, the production of which uses green hydrogen. In the interest of an ambitious market ramp-up, a 2% minimum for 2030 is being discussed [2].

The priorities of hydrogen strategies also include the need to decarbonize construction, industry (e.g., steel, chemical), and the power system. Korea is already a leader in hydrogen-based micro-CHP plants (mCHP) in new buildings. It is predicted that by 2030, there will be 20,000 mCHP in buildings, consuming 150,000 tons of hydrogen per year. In addition, as much as 3.5 GW of power generated from hydrogen fuel cells will be installed [8]. Japan forecasts that there will be 5.3 million stationary fuel cells in households by 2030 [9].

In the EU, clusters can play a major role in helping countries in the decarbonizing industry. Their goal is to facilitate cooperation and help create an energy-saving industry, e.g., the Rotterdam cluster [3]. In the Netherlands, the chemical industry can play an important role in capturing and utilizing carbon dioxide with green hydrogen. In Germany, attention is focused on the steel industry. Planned investments will focus on alternative processes, i.e., hydrogen injection into blast furnaces and direct hydrogen reduction in dedicated installations. Such solutions are also promoted by Posco, the dominant steel producer in South Korea [44].

Another priority for hydrogen strategies is the power system. For example, Japan is promoting the installation of gas supply systems to store surplus electricity from renewable energy as part of the Fukushima demonstration project [9].

Certain international initiatives should also be mentioned. In Germany, the coalition committee's package for the future provides EUR 2.0 billion to intensify international cooperation in the field of hydrogen at all levels [2]. Globally, there are 228 hydrogen projects across the value chain which result from the assumed strategic goals [38]. Europe is the global leader in the number of proposed hydrogen projects, with Australia, Japan, and Korea. China and the USA are following as additional hubs [45]. Over half of the announced projects (55%) are located in Europe [38]. One of the most active EU member states in expanding hydrogen technologies is Germany [2].

If all projects are successful, the total investment will exceed USD 300 billion in hydrogen spending by 2030 [38]. Particular attention is given to R&D projects. Several countries are developing ambitious research programs as part of their national hydrogen strategy (e.g., Australia, South Korea, and several EU member states) [46]. In 2019, significant funding was allocated to these projects by Japan (USD 281.7 million), the USA (USD 120 million), and Germany (USD 50.7 million) according to an IEA estimation [47]. Germany has launched a research campaign called 'Hydrogen Technologies 2030'. Its key elements are technologies dedicated to the transport sector, steel and chemical industries, the green hydrogen production industry, technologies for export, and the creation of a new research network.

Patents are the result of many R&D projects in this field. In 2019, significant progress was made in the filing of patents for hydrogen technologies and fuel cells: China (1493), Japan (682), and Korea (444) [48]. These three countries account for over 55% of all global fuel cell patents and over 65% of all hydrogen-related patents. EU countries have only issued around 16% of all patents. Among the European countries analyzed, Germany filed the most patents (136 in 2019) [48].

Growing demand for hydrogen in various sectors will depend on innovative solutions to increase hydrogen production. In Europe, Germany is the largest producer of hydrogen with an estimated annual volume of over 2.4 billion m^3 in 2019. The second-largest producer is the Netherlands (2.1 billion m^3) [37]. However, China ranks first in the world in the production of hydrogen (4.3 billion m^3), which accounts for 18% of the total world production [49]. Second in the world is the USA (2.6 billion m^3). These countries produce hydrogen from fossil fuels (natural gas, oil, coal), and to a much lesser extent, via electrolysis. Replacing them with renewable energy sources is a priority.

Australia, California, and the United Kingdom have great potential in the development of renewable energy. However, Germany aims to become the lead supplier of green hydrogen technology to the global market [40]. Geoscience Australia estimates that, based on the quality of its wind, solar, and water resources alone, about 11% (872,000 km^2) may be highly suitable for hydrogen production [12]. However, the United Kingdom has one of the largest offshore wind farm markets in the world. The state of California in the USA has large resources of renewable energy (31.7% of the energy mix in 2019 [50], 50% in 2030 [51]). It should also be mentioned that Japan has built a hydrogen plant in Namie, Fukushima, to implement full-scale power-to-gas technology. The facility, called Fukushima Hydrogen Energy Research Field (FH2R), uses a 20 MW solar power plant on a 180,000 m^2 site along with grid energy to electrolyze water in a renewable 10 MW hydrogen production unit, the world's largest [52]. Australia is also planning a project on a similar scale as part of the ARENA program [42].

The development of hydrogen technologies contributes to the establishment of international partnerships by various countries, increasing the prospects of creating a hydrogen economy. The example of such cooperation is The International Partnership for Hydrogen and Fuel Cells in the Economy (IPHE). The intergovernmental organization was created to facilitate the transition to clean energy and mobility systems based on hydrogen technologies. Apart from the analyzed and described countries (China, the United States, Japan, Republic of Korea, the Netherlands, France, the United Kingdom, Germany, Australia), the members of this organization also include Chile, Italy, Austria, Brazil, Costa Rica, Iceland, Canada, India, and European Commission [53]. All member states are obliged to accelerate the development of hydrogen technologies, which directly increases the prospects for the development of the hydrogen economy in these countries. It is visible among others by dominating the global market in terms of the distribution of fuel cell vehicles and the number of hydrogen refueling stations [54].

The contribution of individual countries to the development of the hydrogen economy is also visible through the global collaboration The Hydrogen Valley Platform founded by the Fuel Cells and Hydrogen Joint [55]. The platform collects flagship hydrogen projects. Currently, 36 Hydrogen Valleys in 19 countries are described. In addition to the aforementioned countries, Denmark, French Guiana, Portugal, Romania, Slovakia, Spain, and Thailand stand out in terms of planned investments in hydrogen technology.

The foundations for the creation of hydrogen economies and societies in African countries are established thanks to the pancontinental association The Africa Hydrogen Partnership. Particular efforts in this direction are visible in Morocco (partnership with Germany to develop the first green hydrogen plant in Africa) and in the Republic of South Africa (expanding knowledge and innovation in hydrogen technology via the Hydrogen South Africa—HySA initiative) [56].

2. Data and Methodology

2.1. Indicators

The study correlates selected economic, energy, and environmental indicators along with a group of specific factors connected with the development of the hydrogen economy.

2.1.1. Hydrogen Indicators (HyInd)

The group of hydrogen indicators consisted of hydrogen production volume (HPV, in billion m^3), the number of patents (PAT, numbers of patents), and energy RD&D budget (RDD, in USD million) in the hydrogen production and fuel cells category. Additionally, data for hydrogen consumption were also compiled (HCV, in billion m^3).

The hydrogen production and consumption data could be used to monitor trends in the hydrogen market at national levels. They could also be used to illustrate hydrogen use. The hydrogen volume indicator is the estimated amount of hydrogen produced in a particular market and refers to production in physical terms. The data come from the AI-powered statistical database for market analysis, IndexBox. They are shown after raw and mirror information is combined, and after performing IB AI algorithms to eliminate any anomalies and to complete missing data [57]. To a large extent, the data for hydrogen consumption overlap with production volumes due to the current use of hydrogen and the complexity of hydrogen storage and distribution [17]. Most hydrogen is produced and consumed on-site [58,59]. For this reason, only production data were used in the correlation analysis and modeling.

The process of developing new methods of production and application of hydrogen is associated with its potential funding. Government subsidies support research and development, and further technological changes relevant to the industry, energy, and transport sectors. The allocated subsidies reflect an assessment of the state's efforts to increase competitiveness in a given technology and are important in accelerating the implementation of hydrogen technology [60]. Collective data on RD&D budgets (for hydrogen and fuel cell technology in particular) were used. Collective data were used firstly because different countries use different methodologies to allocate subsidies and secondly because the data are incomplete for several years. These data are compiled by the International Energy Agency (IEA) and include central or federal government budgets and expenditures by state-owned companies. The database reflects the expenditure allocated to basic and applied research, experimental development programs, and energy-related and fundamental research programs in selected countries [45]. The statistics are available only for IEA member countries.

A good indicator of the level of innovation is the number of patents filed [61]. This enables the preferred directions of technological progress for a given country to be determined [62,63]. Patents are considered to be key in accelerating the development of the hydrogen economy [27]. The number of patents per country per year (from 2014 to 2018) in the hydrogen production and fuel cells category according to an established classification system (the Cooperative Patent Classification—CPC) were obtained from the Fuel Cells and Hydrogen Observatory (FCHO) as extracted from the PatBase database [48].

2.1.2. Economic, Energy, and Environmental Indicators

The following significant national indicators were identified:

- economic (gross domestic product—GPD, in USD trillion)
- energy (total final energy consumption—TFC, including transport TFC_T and industry TFC_I, in Mtoe; proportion of primary energy supply that is renewable—RES, in %)
- environmental (total carbon dioxide emissions—CO_2, including transport CO_2_T and industry emissions CO_2_I, in Mt).

The population (POP, in mln) of the country was also taken into account.

Population and GDP are the variables used in the vast majority of the studies related to the environment and energy [26,28,32,64]. GDP is a synthetic and objective measure of economic performance, but it is essential to policy creation [65]. For comparison, data normalization in terms of gross domestic product and population was also used, eliminating the influence of the size of the countries concerned. GDP data were obtained from the Organization for Economic Cooperation and Development (OECD) database. Population statistics were compiled by the World Bank.

Carbon dioxide emissions contribute to climate change and environmental degradation. Currently, hydrogen is produced from fossil fuels, with significant CO_2 emissions. Since the 1930s, the vision of a hydrogen economy has been associated with the reduction of emissions [66]. Reducing carbon emissions across the economy is essential to achieve carbon neutrality. The impact of greenhouse gases in the context of a hydrogen economy has been considered by various authors [25,26,28]. The reduction in CO_2 emissions is an indicator of the level of decarbonization of a given country. This may contribute to supporting hydrogen technologies.

TFC data help to estimate the environmental impact of energy use. The indicator can be used to monitor and evaluate the success of key policies that have been designed to influence energy consumption and energy efficiency [67]. A significant relationship exists between energy consumption and economic growth in the long term [68,69]. In addition, economic growth drives energy consumption in the end-use sectors of transport and industry. The share of the industry and transport sectors in the total final energy consumption in the selected countries is significant. The statistics reports and database compiled by the IEA were used to provide data for carbon dioxide emissions and total final energy consumption [70,71].

The last factor is renewable energy defined as the proportion of the total primary energy supply that is renewable. The source of the RES indicator was the OECD. The RES indicator illustrates the commitment of a given country to search for clean and ecological energy sources as an effective solution to increasing energy production, taking into account environmental constraints (e.g., greenhouse emissions). Access to clean, modern, and more efficient energy in all countries is also important in the context of sustainable development [72], while the use of renewable energy sources in the hydrogen production process is defined as the long-term goal of a developed hydrogen economy [73]. Interest in green hydrogen production solutions facilitates the development of renewable energy technologies on a large scale [13]. Countries with a large share of RES with the possibility of further renewable-cost reduction, have the potential to develop a clean hydrogen economy [1].

2.2. Methods

The analysis involved two stages. First, Spearman's correlation was used, the second, one-parameter and multi-parameter linear regression models (ordinary least squares method) were used. The basis for choosing Spearman's correlation is that it is more general using than the Pearson correlation (which is only for a linear relationship). Additionally, Spearman's correlation is more resistant to outliers in trials than the Pearson correlation. Spearman's method was used in many articles concerning the area of renewable energy [28,33,34,74–79].

2.2.1. Spearman's Correlation

Spearman's correlation was used to investigate the relationship between the two selected parameters, one from each of the hydrogen indicator groups and one from the economic, energy, and environmental indicator groups—Equation (1).

$$r_{xy} = \frac{\frac{1}{n}\sum_i^n (R(x_i) - \overline{R(x)})(R(y_i) - \overline{R(y)})}{\sqrt{\left(\frac{1}{n}\sum_i^n (R(x_i) - \overline{R(x)})^2\right)\left(\frac{1}{n}\sum_i^n (R(y_i) - \overline{R(y)})^2\right.}} \tag{1}$$

where:

x—parameter
y—parameter
R(x) and R(y)—ranks of the x and y variables
$\overline{R(x)}$ and $\overline{R(y)}$—mean ranks
n—total number of observations
i—number of observations

2.2.2. The Linear Regression Models

In the second part of the calculation, the regression tool from the RStudio and Analysis ToolPak (MS Excel) was used to perform a linear regression analysis using the least squares method. This enables analysis of the influence of independent variables on the dependent variable. To select parameters potentially best suited to the parameters closely related to the hydrogen economy, a single-parameter regression model was built. A linear regression model has also been used many times in the energy area, for example, in [35,36,80–84].

Linear regression Equation (2) is as follows:

$$y_{io}(cn) = c1(cn, y_o, x_{d0}) + c2(cn, y_o, x_{d0}) \times x_{id0}(cn) \tag{2}$$

where:

$x_{d0}(cn) \in DI(cn),\ y_o \in DO(cn)$
c1, c2—regression coefficients, values depend on the country and parameters used in the calculations, including their values
cn—country
years of analysis: from 2008 to 2018
DI—data input (raw values in Supplementary Data, Table S1)
DO—data output (raw values in Supplementary Data, Table S1)
DO = (HPV, RDD); DI = (POP, GDP, TFC, TFC_T, TFC_I, RES, CO2, CO2_T, CO2_I)

The purpose of the analysis of various parameters is to maximize the value of the linear regression coefficient R^2 depending on the country and y_0, x_{d0}. For each parameter, the c1 and c2 coefficients were calculated.

An example of the application of Formula (3) is presented below:

$$HPV_i(JPN) = c1(JPN, HPV, GDP) + c2(JPN, HPV, GDP) \times GDP_i(cn) \tag{3}$$

where:

JPN—Japan
GDP—Gross Domestic Product as parameter
GDP_i—GDP value in year i

Next, a parametric regression model was built to select a pair of parameters potentially best suited to the parameters closely related to the hydrogen economy (hydrogen production and share of technology energy research, development, and demonstration (RDD) budget).

Overall multiple (two-parameter) linear regression Equation (4):

$$y_{io}(cn) = c1(cn, y_o, x_{d1}, x_{d2}) + c2(cn, y_o, x_{d1}, x_{d2}) \times x_{id1}(cn) + c3(cn, y_o, x_{d1}, x_{d2}) \times x_{id2}(cn) \quad (4)$$

where:

$x_{d1}(cn), x_{d2}(cn) \in DI(cn),\ x_{d1} \neq x_{d2}$
$y_o \in DO(cn)$
c1, c2, c3—regression coefficients; values depend on the country and parameters used in the calculations, including their values
cn—country
i—years of analysis: from 2008 to 2018
DI—data input (raw values in Supplementary Data, Table S1)
DO—data output (raw values in Supplementary Data, Table S1)

Analyzed combinations in terms of DO and DI:

$$DO = (HPV, RDD);\ DI = (POP, GDP, TFC, TFC\_T, TFC\_I, RES, CO2, CO2\_T, CO2\_I)$$

$$DO = (HPV, RDD);\ DI = \left(\frac{POP}{GDP}, \frac{TFC}{GDP}, \frac{TFC\_T}{GDP}, \frac{TFC\_I}{GDP}, RES, \frac{CO2}{GDP}, \frac{CO2\_T}{GDP}, \frac{CO2\_I}{GDP}\right)$$

$$DO = (HPV, RDD);\ DI = \left(\frac{GDP}{POP}, \frac{TFC}{POP}, \frac{TFC\_T}{POP}, \frac{TFC\_I}{POP}, RES, \frac{CO2}{POP}, \frac{CO2\_T}{POP}, \frac{CO2\_I}{POP}\right)$$

$$DO = (HPV, RDD);\ DI = \log(POP, GDP, TFC, TFC\_T, TFC\_I, RES, CO2, CO2\_T, CO2\_I)$$

$$DO = \log(HPV, RDD);\ DI = \log\left(\frac{POP}{GDP}, \frac{TFC}{GDP}, \frac{TFC\_T}{GDP}, \frac{TFC\_I}{GDP}, RES, \frac{CO2}{GDP}, \frac{CO2\_T}{GDP}, \frac{CO2\_I}{GDP}\right)$$

$$DO = \log(HPV, RDD);\ DI = \log\left(\frac{GDP}{POP}, \frac{TFC}{POP}, \frac{TFC\_T}{POP}, \frac{TFC\_I}{POP}, RES, \frac{CO2}{POP}, \frac{CO2\_T}{POP}, \frac{CO2\_I}{POP}\right)$$

$$DO = (HPV, RDD);\ DI = \mathrm{sqrt}(POP, GDP, TFC, TFC\_T, TFC\_I, RES, CO2, CO2\_T, CO2\_I)$$

$$DO = (HPV, RDD);\ DI = \mathrm{sqrt}\left(\frac{POP}{GDP}, \frac{TFC}{GDP}, \frac{TFC\_T}{GDP}, \frac{TFC\_I}{GDF}, RES, \frac{CO2}{GDP}, \frac{CO2\_T}{GDP}, \frac{CO2\_I}{GDP}\right)$$

$$DO = (HPV, RDD);\ DI = \mathrm{sqrt}\left(\frac{GDP}{POP}, \frac{TFC}{POP}, \frac{TFC\_T}{POP}, \frac{TFC\_I}{POP}, RES, \frac{CO2}{POP}, \frac{CO2\_T}{POP}, \frac{CO2\_I}{POP}\right)$$

$$DO = (HPV, RDD);\ DI = (POP, GDP, TFC, TFC\_T, TFC\_I, RES, CO2, CO2\_T, CO2\_I)^2$$

$$DO = (HPV, RDD);\ DI = \left(\frac{POP}{GDP}, \frac{TFC}{GDP}, \frac{TFC\_T}{GDP}, \frac{TFC\_I}{GDP}, RES, \frac{CO2}{GDP}, \frac{CO2\_T}{GDP}, \frac{CO2\_I}{GDP}\right)^2$$

$$DO = (HPV, RDD);\ DI = \left(\frac{GDP}{POP}, \frac{TFC}{POP}, \frac{TFC\_T}{POP}, \frac{TFC\_I}{POP}, RES, \frac{CO2}{POP}, \frac{CO2\_T}{POP}, \frac{CO2\_I}{POP}\right)^2$$

The purpose of the analysis of the combination of different pairs of parameters is to maximize the value of the linear regression coefficient of determination R^2 depending on the country and selected cn, y_0, x_{d1}, x_{d2}. For each selected combination, the following coefficients were calculated: c1, c2, and c3 (using the least squares method).

An example of the application of Formula (5) is presented below:

$$HPV_i(JPN) = c1(JPN, HPV, GDP, TFC) + c2(JPN, HPV, GDP, TFC) \times GDP_i(cn) + c3(JPN, HPV, GDP, TFC) \times TFC_i(cn) \quad (5)$$

where:

JPN—Japan
GDP—Gross Domestic Product as parameter
GDP_i—GDP value in year i

3. Results and Discussion

3.1. Spearman's Correlation

The study aimed to determine whether there is a correlation between the variables (HPV, RDD, PAT) and the indices (POP, GDP, TFC, TFC_T, TFC_I, RES, CO2, CO2_T, CO2_I) and to see which variables influence hydrogen indicators. The study used Spearman's correlation test for nine countries.

In Spearman's correlations of the matrices, it is clear that correlations occur in all the analyzed dependencies (Figures 2 and 3; Supplementary Data, Figures S1–S18). The reason for their association strength is different. The following thresholds were adopted in the analyses [85]: very strong (the correlation coefficient ≥0.80), strong (the correlation coefficient ≥0.60; <0.80), moderate (the correlation coefficient ≥0.40; <0.60), weak (the correlation coefficient ≥0.20; <0.40), and very weak (the correlation coefficient >0.0; <0.20). The relationships between the two variables can be positive or negative.

In 2008 (Figure 2), HPV was most strongly correlated with POP (0.60). The remaining correlation values are at a similar level in the range (0.45–0.57). These are moderate dependencies. In the case of RDD, the strongest correlation is seen for TFC_I, and CO2_I (0.90). The other correlations are also stronger compared to the relationship between the chosen indicators and HPV.

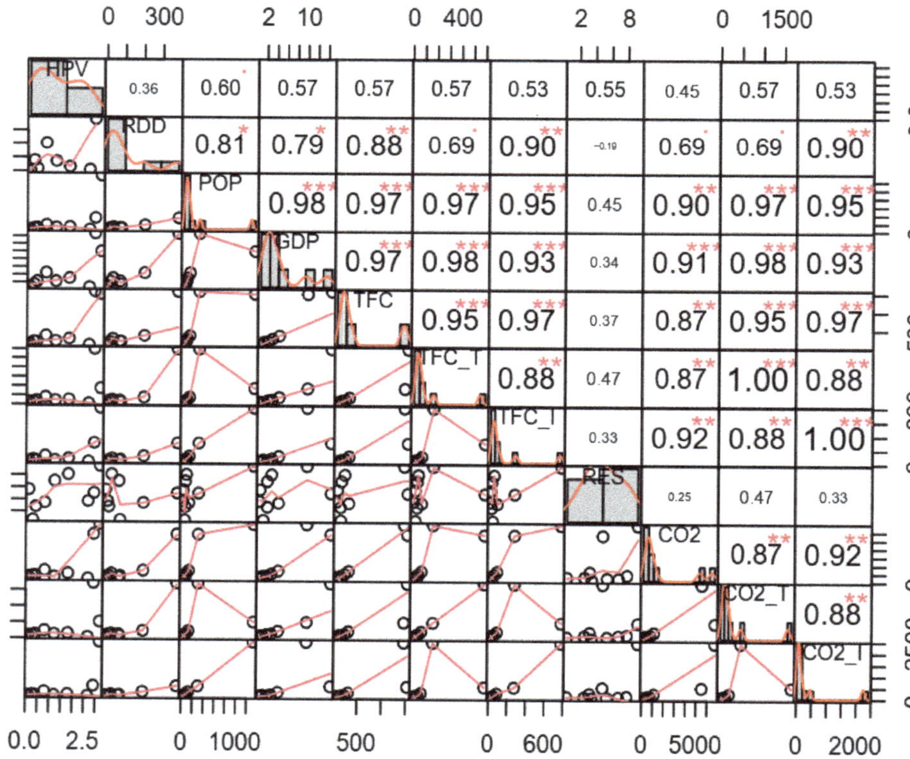

Figure 2. Spearman's correlation and compilation of data for all analyzed countries in 2008. *, **, ***—the more symbols, the stronger the correlation. Abbreviations: HPV—hydrogen production volume; RDD—energy RD&D in the hydrogen production and fuel cells category; POP—population; GPD—gross domestic product; TFC—total final energy consumption; TFC_T—total final energy consumption in transport; TFC_I—total final energy consumption in industry; RES—share of renewable energy of primary energy supply; CO2—total carbon dioxide emissions; CO2_T—total carbon dioxide emissions in transport; CO2_I—total carbon dioxide emissions in industry.

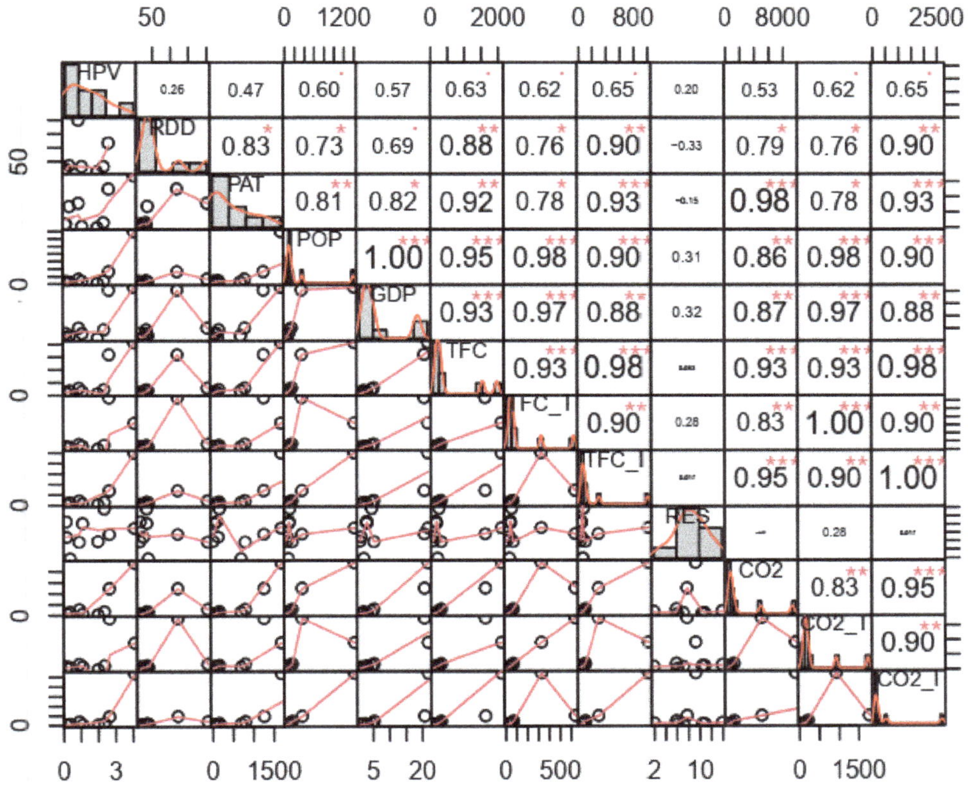

Figure 3. Spearman's correlation and compilation of data for all analyzed countries in 2018. *, **, ***—the more symbols, the stronger the correlation. Abbreviations: HPV—hydrogen production volume; RDD—energy RD&D in the hydrogen production and fuel cells category; PAT—number of patents; POP—population; GPD—gross domestic product; TFC—total final energy consumption; TFC_T—total final energy consumption in transport; TFC_I—total final energy consumption in industry; RES—share of renewable energy of primary energy supply; CO2—total carbon dioxide emissions; CO2_T—total carbon dioxide emissions in transport; CO2_I—total carbon dioxide emissions in industry.

In 2018 (Figure 3), HPV was most strongly correlated with TFC_I and CO2_I (0.65). The relationship between HPV and POP remained at the same level.

The very strong correlations in 2018 for RDD were maintained with TFC_I and CO2_I (0.90), as well as with TFC (0.88). Patents are most significantly correlated with CO2 (0.98). Significant relationships are also seen for the PAT relationship with TFC_I, CO2_I, TFC, GDP, and POP.

The research has shown that throughout the period analyzed (2008–2018) in addition to a correlation with economic indicators, HPV also strongly correlates with environmental and energy indicators (Supplementary Data, Figures S1–S18). The strongest correlations $\geq$0.90 are seen in three non-European countries. HPV-CO2 correlations, including in transport (CO2_T) and industry (CO2_I) sectors, lead to the conclusion that to achieve environmental sustainability, hydrogen production is a key factor and cannot be ignored. At the same time, hydrogen should be produced using renewable energy sources and should be completely free of CO_2 emissions. It was also observed that there is a correlation regarding HPV-TFC including TFC_I and TFC_T. There are very strong correlations in all Asian countries, the United Kingdom, and Australia. In addition, it is strong in the Netherlands, Germany, and the United States. This is evidence that the production of hydrogen not only has an environmental impact but is also oriented to final energy consumption

(Supplementary Data, Figures S1–S18). Secondly, most of the correlations between RDD and the selected indicators were very strong or moderate negative correlations. The same is true for the correlation between patents (which are often the result of research and development works) and the analyzed indicators. Among the very strong and strong correlations (≥0.60), negative correlations for RDD occur with final energy consumption (KOR, UK,), CO_2 emission (KOR, UK, JPN), and renewable energy sources (KOR, FRA, USA). Patents strongly correlate with CO2_T (KOR, GER, AUS), RES (USA, AUS), and TFC (USA, AUS). It might be suggested that public funding is still very important, however, the level of funding in some countries is declining and the share of private funding is increasing to achieve a sustainable environment.

Attention should be given to the increase in hydrogen production, where many strong correlations with selected indicators have been demonstrated. This may be a major factor in achieving environmental goals that will accelerate the process of decarbonization of the economies of sustainable countries. However, in order to achieve this, hydrogen must be produced from renewable energy sources.

3.2. Regression Models

A total of 92 × 3 parameter combinations for HPV were analyzed, and separately, a total of 92 × 3 parameter combinations for RDD were also analyzed. Table 1 shows the results for HPV in the form of pairs for which the mean R^2 values were the highest. Similarly, the RDD results are shown in Table 2.

Table 1. The values of the coefficient of determination for the linear regression (R^2) for selected combinations of HPV and parameters x_{d1} and x_{d2}—four pairs with the highest mean R^2 for individual countries were presented. Color agenda: red—the highest value, green—the lower value; the best combination is bold.

x_{d1}	x_{d2}	R^2/Country										
		CHN	USA	JPN	KOR	NLD	FRA	UK	GER	AUS	mean	min
TFC_T/GDP	**CO2_T/GDP**	**0.995**	**0.991**	**0.889**	**0.906**	**0.689**	**0.664**	**0.912**	**0.884**	**0.892**	**0.869**	**0.664**
TFC_T/GDP	RES	0.988	0.994	0.873	0.929	0.569	0.381	0.900	0.953	0.845	0.826	0.381
GDP/POP	CO2_T/POP	0.996	0.999	0.902	0.933	0.641	0.339	0.868	0.901	0.839	0.824	0.339
TFC_T/GDP	CO2/GDP	0.995	0.993	0.879	0.927	0.704	0.279	0.866	0.931	0.820	0.821	0.279

Abbreviations: POP—population; GPD—gross domestic product; TFC_T—total final energy consumption in transport; RES—share of renewable energy of primary energy supply; CO2—total carbon dioxide emissions; CO2_T—total carbon dioxide emissions in transport; CHN—China; USA—United States; JPN—Japan; KOR—Republic of Korea; NLD—Netherlands; FRA—France; UK—United Kingdom; GER—Germany; AUS—Australia.

Table 2. The values of the coefficient of determination for the linear regression R^2 for selected combinations of RDD and parameters x_{d1} and x_{d2}—four pairs with the highest mean R^2 for individual countries were presented. Color agenda: red—the highest value, green—the lower value; the best combination is bold.

x_{d1}	x_{d2}	R^2/Country									
		USA	JPN	KOR	NLD	FRA	UK	GER	AUS	mean	min
POP/GDP	RES	0.960	0.801	0.836	0.766	0.947	0.501	0.715	0.781	0.788	0.501
RES	**CO2_T/GDP**	**0.944**	**0.791**	**0.842**	**0.794**	**0.959**	**0.568**	**0.544**	**0.775**	**0.777**	**0.544**
TFC_T/GDP	RES	0.940	0.788	0.845	0.807	0.965	0.544	0.499	0.777	0.771	0.499
TFC_I/GDP	RES	0.933	0.813	0.853	0.847	0.952	0.431	0.401	0.808	0.755	0.401

Abbreviations: POP—population; GPD—gross domestic product; TFC_T—total final energy consumption in transport; RES—share of renewable energy of primary energy supply; CO2_T—total carbon dioxide emissions in transport; USA—United States; JPN—Japan; KOR—Republic of Korea; NLD—Netherlands; FRA—France; UK—United Kingdom; GER—Germany; AUS—Australia.

In two-parameter linear regression, the highest value of the mean R^2 was achieved for HPV and the following parameter pair: TFC_T/GDP and CO2_T/GDP. In the category of countries, the highest value of R^2 for this pair of parameters was seen for China and the USA, and the lowest value was seen for France (0.664).

In two-parameter linear regression, the highest value of the mean R^2 was achieved for RDD and the following parameter pair: PCP/GDP and RES. In the category of countries, the highest value of R^2 for this pair of parameters applies to the USA and France, and the lowest value was seen for the United Kingdom (0.501). However, for the second pair (bold pair) in Table 2, the minimum value is four percentage points higher than for the first pair, therefore the following pair of parameters were used to present the results: RES and CO2_T/GDP. In this case, the maximum value of the R^2 coefficient is also observed for the USA and France, and the lowest for Germany (0.544).

The results for the values of the c1, c2, c3 coefficients for the chosen pairs from Tables 1 and 2 are presented in Table 3.

Table 3. The values of the regression coefficients c1, c2, c3 (according to Formula (4)) for pairs x_{d1} and x_{d2} were selected based on Tables 1 and 2.

Country	HPV = c1 + c2*TFC_T/GDP + c3*CO2_T/GDP				RDD = c1 + c2*RES + c3*CO2_T/GDP			
	c1, bn m^3	c2	c3	*p*-Value	c1, USD mln	c2	c3	*p*-Value
CHN	0.983	2.952	−0.974	0.338	Lack of data			
USA	1.626	0.064	−0.011	0.443	−2716.3	123.2	21.4	0.265
JPN	1.026	0.196	−0.071	0.373	−554.1	60.9	9.2	0.281
KOR	0.708	−0.015	−0.002	0.908	−89.8	−4.8	2.6	0.77
NLD	0.765	0.017	0.032	0.97	−31.6	8.8	1.2	0.226
FRA	−2.976	3.618	−1.204	0.254	98.1	−9.2	0.6	0.679
UK	−0.253	0.022	0.005	0.955	131.3	−4.5	−1.8	0.065
GER	3.973	2.984	−1.114	0.009	173.0	−5.7	−1.9	0.273
AUS	0.401	0.044	−0.018	0.074	14.0	2.5	−0.3	0.545

Abbreviations: HPV—hydrogen production volume; RDD—energy RD&D in the hydrogen production and fuel cells category; GPD—gross domestic product; TFC_T—total final energy consumption in transport; RES—share of renewable energy of primary energy supply; CO2_T—total carbon dioxide emissions in transport; CHN—China; USA—United States; JPN—Japan; KOR—Republic of Korea; NLD—Netherlands; FRA—France; UK—United Kingdom; GER—Germany; AUS—Australia.

For all countries except South Korea, an increase in energy consumption in transport has a positive effect on hydrogen production. However, for all countries except the Netherlands and the UK, CO_2 emissions from transport had a negative influence on hydrogen production. In the USA, Japan, the Netherlands, and Australia, the positive effect of renewable energy sources is reflected in the proportion of hydrogen and fuel cells included in the energy technology budget. With the exception of the UK, Germany, and Australia, subsidies decrease as a result of the negative impact of CO_2 emissions.

Despite the good fit of the model to empirical data (according to the mean coefficient of determination in the above approach), the *p*-value is too high. For this reason, it was decided to calculate the best-fit regression model for each country. The results are presented in Table 4 and show the highest R^2 coefficients for one-parametric regression, with a *p*-value < 0.05. The multiple regression models presented show that different characteristics in each country influence the level of hydrogen production or RD&D budget (Table 4; Supplementary Data, Table S2).

Table 4. Selected parameters x_{d0}, x_{d1}, and x_{d2} for which R^2 is the highest under the condition p-value < 0.05.

Country	HPV					RDD				
	x_{d1}	x_{d2}	R^2	x_{d0}	R^2	x_{d1}	x_{d2}	R^2	x_{d0}	R^2
CHN	CO2/POP	CO2_I/POP	0.9934	POP	0.91	Lack of data				
USA	TFC_I/POP	CO2_I/POP	0.9995	POP	0.99	GDP/POP	TFC/POP	0.9785	RES	0.84
JPN	TFC/POP	CO2/POP	0.8868	GDP	0.86	TFC	CO2_T	0.6178	CO2	0.36
KOR	POP	TFC_I	0.9189	GDP	0.90	TFC/GDP	TFC_I/GDP	0.9891	CO2	0.73
NLD	Min. p-value 0.09			Min. p-value 0.07		TFC_I/POP	RES	0.8375	Min. p-value 0.23	
FRA	Min. p-value 0.16			Min. p-value 0.57		RES	CO2_T	0.9494	RES	0.86
UK	POP	RES	0.9384	RES	0.82	TFC	CO2_T	0.6759	CO2_T	0.46
GER	CO2/GDP	CO2_T/GDP	0.9355	Min. p-value 0.25		Min. p-value 0.11			CO2_T	0.37
AUS	CO2/POP	CO2_I/POP	0.9418	POP	0.81	Min. p-value 0.18			GDP	0.46

Abbreviations: HPV—hydrogen production volume; RDD—energy RD&D in the hydrogen production and fuel cells category; POP—population; GPD—gross domestic product; TFC—total final energy consumption; TFC_I—total final energy consumption in industry; RES—share of renewable energy of primary energy supply; CO2—total carbon dioxide emissions; CO2_T—total carbon dioxide emissions in transport; CO2_I—total carbon dioxide emissions in industry; CHN—China; USA—United States; JPN—Japan; KOR—Republic of Korea; NLD—Netherlands; FRA—France; UK—United Kingdom; GER—Germany; AUS—Australia.

For three countries, the HPV regression analysis gives the highest coefficient of determination (R^2) for POP, whilst for two of the countries, it is a GDP indicator. For the data observed in NLD and FRA, no regression equation with an appropriate degree of significance was determined. In the case of two-parameter regression, the minimum R^2 value equals 0.89. However, similar to univariate regression for NLD and FRA, the results were not statistically significant.

In the univariate regression for the explanatory variable RDD, the dependencies on CO2 and CO2_T showed the greatest R^2 factor with a p-value < 0.05. For NLD the p-value achieved is too high. For other countries, the coefficient of determination was reached at the level of 0.36 (for JPN) and 0.86 (for FRA).

In the case of two-parameter regression, the minimum fit factor was 0.62 for JPN, and the highest was achieved for the GDP/POP and TFC/POP model for the USA. Causal relationships were established between the dependent and explanatory variables, which indicates that most of the variables can be used to predict each other.

In the case of China, industrial CO_2 emissions positively affected the level of hydrogen production. The main hydrogen production process in the industry, i.e., steam methane reforming, has a significant carbon footprint. The high level of CO_2 emissions (almost 7 kg CO_2/kg of H_2), comes from fuel consumption and the process reactions [86,87]. According to Soltani et al. [86], it is estimated that about 3% of global industrial CO_2 emissions come from this process. In China, on average 33% of CO_2 emissions came from this process over the period analyzed. The indicator of total CO_2 emissions from all sectors is less significant but has a negative impact on hydrogen production.

The final energy consumption in the USA industry sector has a positive influence on hydrogen production. In 2018, record-breaking energy consumption was observed in the end-user sector (industry) in the USA. In recent years, fluctuations have been noticeable, inter alia, with the Great Recession in 2008 and a gradual return to average energy consumption levels seen before 2008. The increasing energy demand in the growing USA economy is not compatible with the ambitions for climate [88]. Today, 95% of hydrogen in the USA is produced via endothermic processes involving natural gas reforming, more specifically steam reforming. As heat must be supplied for the production of hydrogen, energy consumption in these processes and related industrial sectors is increasing. Simultaneously, the production of hydrogen is negatively correlated with CO_2 emissions in industry. The main drawback of hydrogen production processes is the fact that carbon dioxide is released into the atmosphere. Carbon dioxide capture and sequestration (CCS), as well as the modernization of systems for capturing CO_2 from large industrial SMR installations [89–91], will significantly reduce emissions in the industrial sector. The reduction of emissions from the SMR in the future together with the development of other green methods of hydrogen production, is compatible with the climate neutrality goal.

Further analysis found that the GDP had a significant negative effect on the USA RD&D budget for hydrogen technologies. Currently, energy efficiency and renewables are the top priorities in the energy technology RD&D budget (according to the IEA). This is in contrast with the decreasing total RD&D spending on hydrogen and fuel cells. Solutions involving cheap and effective energy sources are required. It should be noted that expenditure on fossil fuel technologies is still nearly five times greater than that spent on hydrogen technologies (USD 581.4 million in 2018). In 2019, it was estimated that hydrogen and fuel cell research was allocated only 1.5% of energy technology RD&D budgets. Despite the constantly growing budget for energy technologies and the increase in gross domestic spending on R&D [92] current subsidies for hydrogen technologies negatively correlate with GDP growth. However, there are plans for significant investment in these solutions in the coming years.

As energy consumption in Japan increases, the level of hydrogen production also increases. However, CO_2 emissions have a negative effect on hydrogen production. Japan has set a goal to reduce current hydrogen production emissions by 60% by 2030. Net-zero CO_2 emissions across the production-to-final-use hydrogen cycle should be achieved in the long term [93]. Emissions from the hydrogen production process will be reduced further thanks to CO_2 capture and storage technologies. The impact of total energy consumption is related to the fact that hydrogen is an energy carrier. It must be produced from another substance before it can be used as a source of energy or fuel. In Japan, the development of the supply chain and the diversification of uses (mobility, electric energy production, industry, and the housing sector), mean that hydrogen will become a key energy source.

This study also found that the total energy consumption in Japan negatively correlates with the number of subsidies allocated to hydrogen technologies. In contrast, CO_2 emissions from transport positively correlate with the RD&D budget for hydrogen technologies. Japan has identified hydrogen as the solution to the problem of clean energy for transport. Fuel cell vehicles are recognized as one of the key drivers of the hydrogen economy, especially in the future of transport where green hydrogen could completely replace traditional fossil fuels. As with other new developments, making hydrogen fuel cell technology as efficient and profitable as possible requires investment. Moreover, Japanese automakers Toyota, Nissan, and Honda are global leaders in the development of fuel cell vehicles. In addition, they work with Air Liquide to strengthen the national refueling network [94]. The potential of hydrogen to create emission-free transport, industry, and energy generation is recognized, which will initiate a new phase of growing economy in Japan [95].

The population growth in Korea has had a positive impact on hydrogen production. The population increase is correlated with the need to meet the current energy demand. The total energy consumption in industry also has a positive effect on hydrogen production, but to a lesser extent. In Korea, several interrelations can be identified between the production of hydrogen and energy consumption in industry. Firstly, most of the hydrogen produced is still intended as feedstock for petrochemical plants [96]. In the future, hydrogen could play a significant role as a cleaner feedstock for chemicals [44]. It is predicted that hydrogen will play a significant role in transforming the energy power system. Korea, a hydrogen frontrunner, has set leadership priorities not only with respect to fuel cell cars, but also in the field of large-scale stationary fuel cells for power generation. Driven by population growth together with economic growth and industrial competition, hydrogen will play a particular role in contributing to the total final energy consumption.

Moreover, with energy consumption in the Korean industry allowing for the impact of GDP, an increase in RD&D subsidies for hydrogen and fuel cells can be seen. In contrast, total energy consumption by all sectors has a smaller but negative impact on subsidies for hydrogen technologies. South Korea is known for its innovative prowess. The hydrogen industry was worth USD 12.54 billion in 2020 according to the Korea Energy Economics Institute [97]. The hydrogen economy is seen as a key contributor to economic growth (USD 38.54 billion) as well as providing many thousands of new jobs (420,000 jobs) by 2040. It is estimated that hydrogen could account for 5% of the projected energy consumption in 2040, provided that roadmap targets are met [44]. South Korea plans to strengthen its energy base in all sectors. The effect that energy consumption in industry has on the number of subsidies for hydrogen technology is taken into consideration and is the most significant factor in the regression model presented here.

In the Netherlands, total energy consumption has a significantly positive impact on the energy technology budget in the hydrogen technologies category. The share of RES in the total primary energy supply is less important since the value of subsidies decreases as it grows. Research work in the field of energy efficiency is carried out in a multidirectional and multifaceted manner. In the last two years of the period under consideration, a significant increase in expenditure on hydrogen and fuel cells can be seen. The share of the budget increased almost 38 times from 2016 to 2018 (USD 0.344 million in 2016, USD 13.225 million in 2018). However, subsidies for the development of RES were even higher and in 2018 the share of the renewables budget was over eight times higher than the expenditure on hydrogen. Research expenditure on hydrogen technologies is justified by the need to increase energy efficiency. The amount of energy obtained from the combustion of hydrogen (about 118 MJ/kg at 298 K) is much higher than that obtained from gasoline (about 44 MJ/kg) [98]. Hydrogen energy is converted directly into electricity with high efficiency and low power losses. Hence, industry sees potential for adopting hydrogen as an energy source for heating. Currently, the production of hydrogen in the industry is estimated to be 180 PJ per year. The industrial hydrogen system has a significant impact on the energy system in the Netherlands [99].

The proportion of the French energy technology budget allocated to hydrogen and fuel cells is negatively affected by increases in renewable energy sources, and to a lesser extent, by CO_2 emissions in transport. The main goal of hydrogen production in the French strategy is to decarbonize industrial processes, for which the demand for hydrogen is currently the highest [6]. To produce carbon-neutral hydrogen, access to clean energy is essential. Due to the additional power demand and the departure from nuclear energy, renewable energy sources are being used. The decrease in the overall share of RES in the primary energy supply may increase RD&D expenditure, not only on renewables, but also on the development of hydrogen technologies.

Further analysis found that the increase in the proportion of renewable energy sources in the UK has a negative effect on hydrogen production, whilst the population growth is positively correlated with the amount of hydrogen generated. Currently, renewable energy sources account for less than 5% of hydrogen production. However, this situation is forecast to change in the future [100]. A decrease in CO_2 emissions in transport positively influences the subsidies for hydrogen technologies. Simultaneously, with increasing total energy consumption, the RD&D budget for hydrogen and fuel cells increases. Currently, the level of carbon-free hydrogen production is insignificant. However, since over a third of industrial energy consumption is for high-temperature processes [4], the UK government needs to provide subsidies over the next decade to make hydrogen technology the low-emission energy solution ideal for generating this type of energy.

In Germany, the production of hydrogen is positively related to total CO_2 emissions. However, the growth in CO_2 emissions from transport is negatively related to the amount of hydrogen generated. Whilst transport emissions account for a significant proportion of total emissions, fossil fuels remain the main source. Germany has a very large industrial sector, consuming vast amounts of energy. Demand for hydrogen is expected to remain particularly noticeable in industry, while its growth in transport will be driven by a smaller-scale market growth impulse until 2030. The long-term goal of the German economy is for a gradual increase in the use of hydrogen in transport, especially green hydrogen; this will result in a decrease in CO_2 emissions in this sector. However, it should be noted that the hydrogen strategy has to start with blue hydrogen (produced from natural gas with CO_2 emission), due to insufficient volumes of green hydrogen in the near future [2]. The environmental policy focused on hydrogen technologies and cooperation of the government with individual sectors of the economy is necessary for the aspect of eliminating barriers to the hydrogen economy. In Germany, such challenges are visible, among others, in the transport sector, e.g., high costs of production and purchase of FCEVs, their limited availability and the lack of modern German models, low development of refuelling network infrastructure, and their profitability [101].

The last finding of this research was that the industrial CO_2 emissions in Australia are positively correlated with hydrogen production, whilst total CO_2 emissions have a negative and less significant influence on the level of hydrogen generated. Industry accounts for a significant proportion of hydrogen production compared with the volumes dedicated to other sectors. This is reflected in the lower impact of total CO_2 emissions compared to industrial emissions. Mineral refining, chemical production, and steel manufacturing are currently emission-intensive industries in the Australian economy [12]. The use of hydrogen may allow low-carbon products to be obtained in these sectors. Australia has a great potential to produce low-emission hydrogen thanks to the opportunity of using large coal and natural gas resources in combination with the use of carbon capture and storage technologies. The possibility of producing clean hydrogen is also noticeable due to the intensely increasing share of renewable sources such as solar and wind in energy generation.

4. Conclusions

The growth rate of all socio-economic and environmental variables changes over time. There are numerous reasons for these fluctuations. The coming years will be decisive with respect to decarbonization, energy transformation, and the development of the hydrogen ecosystem.

The future of hydrogen will not only have environmental, energy, and economic dimensions, but will also be a cross-cutting topic with far-reaching consequences for foreign policy, security of supply, and geo-economic cooperation. Hydrogen is expected to play a key role in a future climate-neutral economy enabling emission-free transport and energy storage as well as energy-saving industry. All analyzed countries recognize the important role of hydrogen in their national energy and climate plans up to 2030. However, they still need instructions on what economic, social, and environmental factors are conducive to the development of hydrogen.

The USA, China, Japan, South Korea, the Netherlands, France, the United Kingdom, Germany, and Australia are all strongly committed to decarbonization of the economy. National strategies, roadmaps, financial support, and targets for hydrogen have generated unprecedented momentum for the hydrogen industry. All activities will require continuation in the form of new partnerships and the creation of hydrogen communities.

The posed hypothesis *There is a correlation between selected economic, energy, and environmental indicators and the development of the hydrogen economy in countries involved in the implementation of hydrogen technologies* was proved by the presented dependencies. As this article shows, several factors influence the hydrogen economy. Five key indicators have been identified: population, GDP, CO_2 emissions (including CO_2 from industry and transport), RES, and TFC (including industrial and transport sectors). These should be considered when modeling and analyzing the future role of hydrogen. Hydrogen production volumes, the share of RD&D budget, and the number of patents filed were selected as indicators of selected aspects of the hydrogen economy.

The hydrogen and fuel cells category is a small proportion of the total RD&D budget (1.5–16%). However, hydrogen RD&D is growing in most countries. In 2018, growth was positive in all countries except Korea and France. In contrast, the dynamics of hydrogen production are negative. Only in selected years are the dynamics positive for a few countries. The exception is China, where the dynamics of hydrogen production are positive throughout the 2008–2019 period.

The multiple regression models and correlations presented here show that, to a great extent, the different characteristics in each country contribute to the development of the vision of the hydrogen economy. The increase in the share of renewable energy sources in a given country can significantly contribute to strengthening the country's future hydrogen production market, whilst at the same time limiting the harmful impact of CO_2 emissions. Hydrogen production, driven by an increasing population and economic growth, will play a crucial role in contributing to total final energy consumption. The amount of energy obtained from hydrogen is much higher than that from fossil fuels and can be efficiently converted directly into electricity.

Important synergies exist between hydrogen production and CO_2 emissions. Hydrogen, as an energy carrier, must be produced from another substance. Unfortunately, this still tends to involve fossil fuels. Reducing emissions along the entire value chain of hydrogen technologies may be achieved in the distant future. However, today, in certain industrial sectors such as steelmaking, blue hydrogen could be used to reduce carbon emissions. It is necessary to properly target environmental policies in order to reward low-emission and zero-emission technologies and hence the fuels produced in these processes.

The proportion of the energy technology RD&D budget allocated to hydrogen and fuel cells is strongly linked to CO_2 emissions. Fuel cells are a promising technology and compare favorably with internal combustion engine technology. Subsidies for hydrogen technologies are necessary to develop energy-efficient solutions in many sectors of the economy. Subsidies are influenced by total energy consumption.

It can be seen that the research variables affect a selected group of hydrogen indicators. This study contributes to the further development of the hydrogen economy. The analysis could be extended using additional indicators, e.g., the size and value of hydrogen imports and exports.

Supplementary Materials: The following are available online at https://www.mdpi.com/article/10.3390/en14164811/s1, Figures S1–S18: Matrices for analyzed countries (Figure S1 Matrix for China part1, Figure S2 Matrix for China part2, Figure S3 Matrix for South Korea part1, Figure S4 Matrix for South Korea part2, Figure S5 Matrix for the United Kingdom part1, Figure S6 Matrix for the United Kingdom part2, Figure S7 Matrix for Germany part1, Figure S8 Matrix for Germany part2, Figure S9 Matrix for France part1, Figure S10 Matrix for France part2, Figure S11 Matrix for Japan part1, Figure S12 Matrix for Japan part2, Figure S13 Matrix for the USA part1, Figure S14 Matrix for the USA part2, Figure S15 Matrix for the Netherlands part1, Figure S16 Matrix for the Netherlands part2, Figure S17 Matrix for the Australia part1, Figure S18 Matrix for Australia part2, Table S1: Raw data for statistical analysis, Table S2: Multiple regression models in chosen countries.

Author Contributions: Conceptualization, J.C. and R.K.; methodology, P.O.; formal analysis, J.C. and P.O; validation, J.C, R.K. and P.O.; investigation, J.C., R.K. and P.O.; resources, J.C., R.K. and P.O.; writing—original draft preparation, J.C., R.K. and P.O.; writing—review and editing, J.C., R.K. and P.O.; visualization, J.C. and P.O.; project administration, J.C. and R.K.; funding acquisition, J.C., R.K. and P.O. All authors have read and agreed to the published version of the manuscript.

Funding: The research was funded within the frame of statutory works of the Mineral and Energy Economy Research Institute, Polish Academy of Sciences.

Institutional Review Board Statement: Not applicable.

Informed Consent Statement: Not applicable.

Data Availability Statement: Not applicable.

Conflicts of Interest: The authors declare no conflict of interest.

Abbreviations

CHP	cogeneration or combined heat and power
CCS	carbon capture storage
FCEV	fuel cell electric vehicles
HPV	hydrogen production volume
PAT	number of patents
RDD	energy RD&D in the hydrogen production and fuel cells category
HCV	hydrogen consumption
GPD	gross domestic product
TFC	total final energy consumption
TFC_T	total final energy consumption in transport
TFC_I	total final energy consumption in industry
RES	share of renewable energy of primary energy supply
CO2	total carbon dioxide emissions
CO2_T	total carbon dioxide emissions in transport
CO2_I	total carbon dioxide emissions in industry
POP	population
CHN	China
USA	United States
JPN	Japan
KOR	Republic of Korea
NLD	Netherlands
FRA	France
UK	United Kingdom
GER	Germany
AUS	Australia

References

1. IRENA. *Green Hydrogen Cost Reduction: Scaling Up Electrolysers to Meet the 1.5 °C Climate Goal*; International Renewable Energy Agency: Abu Dhabi, United Arab Emirates, 2020; ISBN 9789292602956.
2. Federal Ministry for Economic Affairs and Energy the National Hydrogen Strategy 2020. Available online: https://www.bmwi.de/Redaktion/EN/Publikationen/Energie/the-national-hydrogen-strategy.pdf?__blob=publicationFile&v=6 (accessed on 10 January 2021).
3. Visman, B. Roadmaps to a Hydrogen Future in the Netherlands by 2050. Available online: https://www.semanticscholar.org/paper/Roadmaps-to-a-hydrogen-future-in-the-Netherlands-by-Visman/cdba1a8591613944d5b406c307d98b218e418fe2 (accessed on 10 January 2021).
4. HM Government Industrial Decarbonisation Strategy. Available online: https://assets.publishing.service.gov.uk/government/uploads/system/uploads/attachment_data/file/970229/Industrial_Decarbonisation_Strategy_March_2021.pdf (accessed on 2 February 2021).
5. Ministry of Economic Affairs and Climate Policy Netherlands Government Strategy on Hydrogen. Available online: https://www.government.nl/documents/publications/2020/04/06/government-strategy-on-hydrogen (accessed on 10 January 2021).
6. Ministère de la Transition Écologique Stratégie Nationale Pour Le Développement de l'hydrogène Décarboné En France. Available online: https://www.economie.gouv.fr/plan-de-relance/profils/entreprises/strategie-nationale-developpement-hydrogene-decarbone-appels-projets-territoriaux (accessed on 1 February 2021).
7. HM Government Energy Department Powering Our Net Zero Future. Energy White Paper. Available online: https://assets.publishing.service.gov.uk/government/uploads/system/uploads/attachment_data/file/945899/201216_BEIS_EWP_Command_Paper_Accessible.pdf (accessed on 20 January 2021).
8. Study Task Force Hydrogen Roadmap Korea. A Vision, Roadmap and Recommendation to Develop Korea's Hydrogen Economy. Available online: http://www.h2eva.org (accessed on 3 February 2021).
9. Hydrogen and Fuel Cell Strategy Council the Strategic Road Map for Hydrogen and Fuel Cells. Industry-Academia-Government Action Planto Realize a "Hydrogen Society". Available online: https://www.meti.go.jp/english/press/2019/pdf/0312_002b.pdf (accessed on 2 February 2021).
10. China Hydrogen Alliance. *White Paper on Hydrogen Energy and Fuel Cell Industry in China*; China Hydrogen Alliance: Beijing, China, 2019.
11. FCHEA Road Map to a US Hydrogen Economy. Reducing Emissions and Driving Growth across the Nation. Available online: https://www.fchea.org/us-hydrogen-study (accessed on 2 February 2021).
12. COAG Energy Council Hydrogen Working Group Australia's National Hydrogen Strategy. Available online: https://www.industry.gov.au/sites/default/files/2019-11/australias-national-hydrogen-strategy.pdf (accessed on 2 February 2021).

13. Li, L.; Lin, J.; Wu, N.; Xie, S.; Meng, C.; Zheng, Y.; Wang, X.; Zhao, Y. Review and Outlook on the International Renewable Energy Development. *Energy Built Environ.* **2020**. [CrossRef]
14. IEA The Future of Hydrogen. Available online: https://www.iea.org/reports/the-future-of-hydrogen (accessed on 2 February 2021).
15. Jovan, D.J.; Dolanc, G. Can Green Hydrogen Production Be Economically Viable under Current Market Conditions. *Energies* **2020**, *13*, 6599. [CrossRef]
16. FCHO Hydrogen Molecule Market. Available online: https://www.fchobservatory.eu/reports (accessed on 2 August 2021).
17. NRC. NAE Transportation, Distribution, and Storage of Hydrogen. In *The Hydrogen Economy: Opportunities, Costs, Barriers, and R&D Needs*; National Research Council and National Academy of Engineering, Ed.; The National Academies Press: Washington, DC, USA, 2004.
18. Umbach, F.; Pfeiffer, J. Germany and the EU's Hydrogen Strategies in Perspective—The Need for Sober Analyses. *Perisc. Occas. Anal. Br. Ser.* **2020**, *2020*, 1–12.
19. O'Brien, J.E.; Hartvigsen, J.L.; Boardman, R.D.; Hartvigsen, J.J.; Larsen, D.; Elangovan, S. A 25 KW High Temperature Electrolysis Facility for Flexible Hydrogen Production and System Integration Studies. *Int. J. Hydrogen Energy* **2020**, *45*, 15796–15804. [CrossRef]
20. Hydrogen Europe Papers: Hydrogen Europe's Position Paper on the Sustainable and Smart Mobility Strategy. Available online: https://www.hydrogeneurope.eu/wp-content/uploads/2021/04/Hydrogen-Europe_SSMS_Paper-compressed.pdf (accessed on 2 February 2021).
21. Meng, X.; Gu, A.; Wu, X.; Zhou, L.; Zhou, J.; Liu, B.; Mao, Z. Status Quo of China Hydrogen Strategy in the Field of Transportation and International Comparisons. *Int. J. Hydrogen Energy* **2020**, in press. [CrossRef]
22. Hydrogen Europe Factsheets & Infographics: Market and Industry Ambitions. Available online: https://hydrogeneurope.eu/publications-0 (accessed on 2 February 2021).
23. Singh, H.V.; Bocca, R.; Gomez, P.; Dahlke, S.; Bazilian, M. The Energy Transitions Index: An Analytic Framework for Understanding the Evolving Global Energy System. *Energy Strateg. Rev.* **2019**, *26*, 100382. [CrossRef]
24. World Economic Forum Fostering Effective Energy Transition 2019 Edition Insight Report. Available online: http://www3.weforum.org/docs/WEF_Fostering_Effective_Energy_Transition_2019.pdf (accessed on 12 January 2021).
25. Iordache, I.; Gheorghe, A.V.; Iordache, M. Towards a Hydrogen Economy in Romania: Statistics, Technical and Scientific General Aspects. *Int. J. Hydrogen Energy* **2013**, *38*, 12231–12240. [CrossRef]
26. Xu, R.; Chou, L.C.; Zhang, W.H. The Effect of CO^2 Emissions and Economic Performance on Hydrogen-Based Renewable Production in 35 European Countries. *Int. J. Hydrogen Energy* **2019**, *44*, 29418–29425. [CrossRef]
27. Sinigaglia, T.; Freitag, T.E.; Kreimeier, F.; Martins, M.E.S. Use of Patents as a Tool to Map the Technological Development Involving the Hydrogen Economy. *World Pat. Inf.* **2019**, *56*, 1–8. [CrossRef]
28. Leben, J.; Hočevar, S. Correlation between National Development Indicators and the Implementation of a Hydrogen Economy in Slovenia. *Int. J. Hydrogen Energy* **2012**, *37*, 5468–5480. [CrossRef]
29. Sadorsky, P. Renewable Energy Consumption, CO_2 Emissions and Oil Prices in the G7 Countries. *Energy Econ.* **2009**, *31*, 456–462. [CrossRef]
30. Wang, B.; Mi, Z.; Nistor, I.; Yuan, X.C. How Does Hydrogen-Based Renewable Energy Change with Economic Development? Empirical Evidence from 32 Countries. *Int. J. Hydrogen Energy* **2018**, *43*, 11629–11638. [CrossRef]
31. Marques, A.C.; Fuinhas, J.A. Drivers Promoting Renewable Energy: A Dynamic Panel Approach. *Renew. Sustain. Energy Rev.* **2011**, *15*, 1601–1608. [CrossRef]
32. de Souza Mendonça, A.K.; de Andrade Conradi Barni, G.; Moro, M.F.; Bornia, A.C.; Kupek, E.; Fernandes, L. Hierarchical Modeling of the 50 Largest Economies to Verify the Impact of GDP, Population and Renewable Energy Generation in CO_2 Emissions. *Sustain. Prod. Consum.* **2020**, *22*, 58–67. [CrossRef]
33. Durmuşoğlu, B.; Selam, A.A.; Fırat, S.Ü.O. A Comparative Study on the Relations among Sustainability Focused Indexes with REC, CO2, and GDP. *Am. Sci. Res. J. Eng. Technol. Sci.* **2017**, *34*, 81–94.
34. Frugoli, P.A.; Almeida, C.M.V.B.; Agostinho, F.; Giannetti, B.F.; Huisingh, D. Can Measures of Well-Being and Progress Help Societies to Achieve Sustainable Development? *J. Clean. Prod.* **2015**, *90*, 370–380. [CrossRef]
35. Asumadu-Sarkodie, S.; Owusu, P.A. Recent Evidence of the Relationship between Carbon Dioxide Emissions, Energy Use, GDP, and Population in Ghana: A Linear Regression Approach. *Energy Sources Part B Econ. Plan. Policy* **2017**, *12*, 495–503. [CrossRef]
36. Menz, F.C.; Vachon, S. The Effectiveness of Different Policy Regimes for Promoting Wind Power: Experiences from the States. *Energy Policy* **2006**, *34*, 1786–1796. [CrossRef]
37. IEA Hydrogen. IEA, Paris. Available online: https://www.iea.org/reports/hydrogen (accessed on 10 January 2021).
38. The Hydrogen Council and McKinsey & Company Hydrogen Insights a Perspective on Hydrogen Investment, Market Development and Cost Competitiveness. Available online: www.hydrogencouncil.com (accessed on 14 January 2021).
39. SAE China Hydrogen Fuel Cell Vehicle Technology Roadmap. Available online: http://www.sae-china.org (accessed on 2 February 2021).
40. Fuel Cells and Hydrogen 2 Joint Undertaking Opportunities for Hydrogen Energy Technologies Considering the National Energy & Climate Plans. Available online: https://www.fch.europa.eu/publications/opportunities-hydrogen-energy-technologies-considering-national-energy-climate-plans (accessed on 2 February 2021).
41. FCH. *Hydrogen Roadmap Europe*; Fuel Cells and Hydrogen: Brussels, Belgium, 2019; ISBN 9789292463328.

42. PwC Embracing Clean Hydrogen for Australia How the Journey towards Decarbonisation Can Be Fuelled by Hydrogen. Available online: https://www.pwc.com.au/infrastructure/embracing-clean-hydrogen-for-australia-270320.pdf (accessed on 12 January 2021).
43. H2 MOBILITY. We Are Building the Filling Station Network of the Future. H2 MOBILITY Deutschland GmbH & Co KG. Available online: https://h2.live/en/h2mobility (accessed on 2 March 2021).
44. Stangarone, T. South Korean Efforts to Transition to a Hydrogen Economy. *Clean Technol. Environ. Policy* **2020**. [CrossRef]
45. IEA Hydrogen Projects Database. Available online: https://www.iea.org/reports/hydrogen-projects-database (accessed on 2 February 2021).
46. EC. *Communication from the Commission to the European Parliament, the Council, the European Economic and Social Committee and the Committee of the Regions. A Hydrogen Strategy for a Climate-Neutral Europe*; European Commission: Brussels, Belgium, 2020.
47. IEA Energy Technology RD&D Budgets 2020. Database. Available online: https://www.iea.org/reports/energy-technology-rdd-budgets-2020 (accessed on 2 February 2021).
48. FCHO Total Patent Registrations. Available online: https://www.fchobservatory.eu/observatory/patents (accessed on 8 February 2021).
49. Tu, K.J. Prospects of a Hydrogen Economy with Chinese Characteristics. *Études de l'Ifri*, 21 October 2020; p. 62.
50. The California Energy Commission Total System Electric Generation. Available online: https://www.energy.ca.gov/data-reports/energy-almanac/california-electricity-data/2019-total-system-electric-generation (accessed on 1 May 2021).
51. The California Air Resources Board California's 2030 Climate Commitment Renewable Resources for Half of the State's Electricity by 2030. Available online: https://ww3.arb.ca.gov/html/fact_sheets/2030_renewables.pdf (accessed on 3 April 2021).
52. Hydrogen Power & Clean Energy Community a Consortium in the Country Has Launched the Fukushima Hydrogen Energy Research Field. Available online: https://www.hydrogenfuelnews.com/japan-opens-the-largest-hydrogen-production-unit-on-the-globe/8539553/ (accessed on 30 March 2021).
53. IPHE International Partnership for Hydrogen and Fuel Cells in the Economy—Partners. Available online: https://www.iphe.net/partners (accessed on 30 July 2021).
54. Samsun, R.C.; Antoni, L.; Rex, M. Mobile Fuel Cell Application: Tracking Market Trends. In *Advanced Fuel Cells Technology*; Collaboration Programme by International Energy Agency: Paris, France, 2020.
55. FCH Hydrogen Valleys. Available online: https://www.h2v.eu/about-us (accessed on 30 July 2021).
56. Clifford Chance Focus on Hydrogen: A New Energy Frontier for Africa. Available online: https://www.cliffordchance.com/briefings/2021/01/focus-on-hydrogen--a-new-energy-frontier-for-africa.html (accessed on 30 July 2021).
57. IndexBox Hydrogen Data. Available online: https://app.indexbox.io/meta#p_cov (accessed on 2 January 2021).
58. Abe, I. Statistics on Hydrogen Production and Consumption. In *Energy Carriers and Conversion Systems. Encyclopedia of Life Support Systems*; Ohta, T., Nejat Veziroglu, T., Eds.; Developed under the Auspices of the UNESCO; Eolss Publishers: Paris, France, 2008; pp. 131–135.
59. IRENA Hydrogen: A Renewable Energy Perspective. Available online: https://irena.org/publications/2019/Sep/Hydrogen-A-renewable-energy-perspective (accessed on 12 January 2021).
60. Van Benthem, A.A.; Kramer, G.J.; Ramer, R. An Options Approach to Investment in a Hydrogen Infrastructure. *Energy Policy* **2006**, *34*, 2949–2963. [CrossRef]
61. Granstrand, O. Innovation and intellectual property rights. In *The Oxford Handbook of Innovation*; Fagerberg, J., Mowery, D.C., Nelson, R.R., Eds.; Oxford University Press: Oxford, UK, 2005.
62. Moreira, R.; De Carvalho, F.M.S.; Bergamaschi, V.S.; Politano, R. Patentes Depositadas em Âmbito Nacional Como Indicador de Desenvolvimento Das Tecnologias de Produção de Hidrogênio. *Quim. Nova* **2013**, *36*, 748–751. [CrossRef]
63. Raghupathi, V.; Raghupathi, W. Innovation at Country-Level: Association between Economic Development and Patents. *J. Innov. Entrep.* **2017**, *6*, 4. [CrossRef]
64. Lin, B.; Raza, M.Y. Analysis of Energy Related CO2 Emissions in Pakistan. *J. Clean. Prod.* **2019**, *219*, 981–993. [CrossRef]
65. Fraumeni, B. Gross Domestic Product: Are Other Measures Needed? *IZA World Labor* **2017**, *368*, 1–11. [CrossRef]
66. Turner, J.A. Sustainable Hydrogen Production. *Science* **2004**, *305*, 972–974. [CrossRef]
67. UNECE G-1: Final Energy Consumption. Available online: https://unece.org/DAM/env/europe/monitoring/Indicators/G-1-en-final.pdf (accessed on 8 February 2021).
68. Esen, Ö.; Bayrak, M. Does More Energy Consumption Support Economic Growth in Net Energy-Importing Countries? *J. Econ. Financ. Adm. Sci.* **2017**, *22*, 75–98. [CrossRef]
69. Mukhtarov, S.; Mikayilov, J.I.; Ismayilov, V. The Relationship between Electricity Consumption and Economic Growth: Evidence from Azerbaijan. *Int. J. Energy Econ. Policy* **2020**, *7*, 32–38.
70. IEA CO2 Emissions from Fuel Combustion. Database. Available online: https://www.iea.org/data-and-statistics (accessed on 2 February 2021).
71. IEA Total Final Consumption (TFC) by Sector. World Energy Balances 2020. Database. Available online: https://www.iea.org/data-and-statistics/data-product/world-energy-balances (accessed on 2 February 2021).
72. United Nations the Sustainable Development Goals. Available online: https://sdgs.un.org/goals/goal7 (accessed on 15 January 2021).
73. Turner, J.A. A Realizable Renewable Energy Future. *Science* **1999**, *285*, 687–689. [CrossRef]

74. Yadav, A.; Kumar, N. Solar Resource Estimation Based on Correlation Matrix Response for Indian Geographical Cities. *Int. J. Renew. Energy Res.* **2016**, *6*, 695–701.
75. Jurasz, J.; Canales, F.A.; Kies, A.; Guezgouz, M.; Beluco, A. A Review on the Complementarity of Renewable Energy Sources: Concept, Metrics, Application and Future Research Directions. *Sol. Energy* **2020**, *195*, 703–724. [CrossRef]
76. Liczmańska-Kopcewicz, K.; Pypłacz, P.; Wiśniewska, A. Resonance of Investments in Renewable Energy Sources in Industrial Enterprises in the Food Industry. *Energies* **2020**, *13*, 4285. [CrossRef]
77. Mehedintu, A.; Sterpu, M.; Soava, G. Estimation and Forecasts for the Share of Renewable Energy Consumption in Final Energy Consumption by 2020 in the European Union. *Sustainability* **2018**, *10*, 1515. [CrossRef]
78. Kryzia, D.; Olczak, P.; Wrona, J.; Kopacz, M.; Kryzia, K.; Galica, D. Dampening Variations in Wind Power Generation Through Geographical Diversification. In Proceedings of the IOP Conference Series: Earth and Environmental Science, Krakow, Poland, 14–17 November 2019; Volume 214.
79. Cader, J.; Olczak, P.; Koneczna, R. Regional Dependencies of Interest in the "My Electricity" Photovoltaic Subsidy Program in Poland. *Energy Policy J.* **2021**, *24*, 97–116. [CrossRef]
80. El Kafazi, I.; Bannari, R.; Abouabdellah, A.; Aboutafail, M.O.; Guerrero, J.M. Energy Production: A Comparison of Forecasting Methods Using the Polynomial Curve Fitting and Linear Regression. In Proceedings of the 2017 International Renewable and Sustainable Energy Conference (IRSEC), Tangier, Morocco, 4–7 December 2017; IEEE: Piscataway, NJ, USA, 2017; pp. 1–5.
81. Mohammed, A.; Alshibani, A.; Alshamrani, O.; Hassanain, M. A Regression-Based Model for Estimating the Energy Consumption of School Facilities in Saudi Arabia. *Energy Build.* **2021**, *237*, 110809. [CrossRef]
82. Ferreira, Â.P.; Ramos, J.G.; Fernandes, P.O. A Linear Regression Pattern for Electricity Price Forecasting in the Iberian Electricity Market. *Rev. Fac. Ing. Univ. Antioq.* **2019**, 117–127. [CrossRef]
83. Olczak, P.; Olek, M.; Matuszewska, D.; Dyczko, A.; Mania, T. Monofacial and Bifacial Micro PV Installation as Element of Energy Transition—The Case of Poland. *Energies* **2021**, *14*, 499. [CrossRef]
84. Olczak, P.; Kryzia, D.; Matuszewska, D.; Kuta, M. "My Electricity" Program Effectiveness Supporting the Development of PV Installation in Poland. *Energies* **2021**, *14*, 231. [CrossRef]
85. Statstutor Spearman's Correlation. Available online: https://www.statstutor.ac.uk/resources/uploaded/spearmans.pdf (accessed on 30 July 2021).
86. Soltani, R.; Rosen, M.A.; Dincer, I. Assessment of CO2 Capture Options from Various Points in Steam Methane Reforming for Hydrogen Production. *Int. J. Hydrogen Energy* **2014**, *39*, 20266–20275. [CrossRef]
87. da Costa Labanca, A.R. Carbon Black and Hydrogen Production Process Analysis. *Int. J. Hydrogen Energy* **2020**, *45*, 25698–25707. [CrossRef]
88. Energy Information Administration Today in Energy. Available online: https://www.eia.gov/todayinenergy/detail.php?id=39092 (accessed on 2 October 2020).
89. Foundation for Research and Technology Hellas Final Report Summary—HY2SEPS-2 Project (Hybrid Membrane-Pressure Swing Adsorption (PSA) Hydrogen Purification Systems). Available online: https://cordis.europa.eu/project/id/278538/pl (accessed on 12 January 2021).
90. CACHET Carbon Dioxide Capture and Hydrogen Production from Gaseous Fuels. Available online: https://cordis.europa.eu/project/id/19972 (accessed on 3 February 2021).
91. Busse, A.; Power, G.; MacMurray, J. *Demonstration of Carbon Capture and Sequestration of Steam Methane Reforming Process Gas Used for Large-Scale Hydrogen Production*; Air Products and Chemicals, Inc.: Allentown, PA, USA, 2018.
92. OECD Gross Domestic Spending on R&D (Indicators). Available online: https://data.oecd.org/rd/gross-domestic-spending-on-rd.htm (accessed on 2 February 2021).
93. METI Hydrogen and Fuel Cell Strategy Roadmap. An Action Plan for Achieving a Hydrogen Society. Available online: www.meti.go.jp (accessed on 12 January 2021).
94. Sen, S. Japan: Taking a Lead in Hydrogen. Available online: https://www.thechemicalengineer.com/features/japan-taking-a-lead-in-hydrogen/ (accessed on 2 March 2021).
95. Chaube, A.; Chapman, A.; Shigetomi, Y.; Huff, K.; Stubbins, J. The Role of Hydrogen in Achieving Long Term Japanese Energy System Goals. *Energies* **2020**, *13*, 4539. [CrossRef]
96. Kan, S. South Korea's Hydrogen Strategy and Industrial Perspectives. *Édito Énergie Ifri*, 25 March 2018; pp. 1–7.
97. The UK's Department for International Trade. *The Hydrogen Economy South Korea. Market Intelligence Report*; The UK's Department for International Trade: London, UK, 2011.
98. Revankar, S.T.; Brown, N.R.; Kane, C.; Oh, S. *Development of Efficient Flowsheet and Transient Modeling for Nuclear Heat Coupled Sulfur Iodine Cyclefor Hydrogen Production*; Technical Report; Purdue University: West Lafayette, IN, USA, 2010.
99. Weeda, M.; Segers, R. *The Dutch Hydrogen Balance, and the Current and Future Representation of Hydrogen in the Energy Statistics. TNO 2020 P10915*; TNO: Amsterdam, The Netherlands, 2020.
100. Ballard; Deloitte China. *Fueling the Future of Mobility—Hydrogen and Fuel Cell Solutions for Transportation—Report*; Deloitte: New York, NY, USA, 2020; Volume 1, 103p.
101. Trencher, G.; Edianto, A. Drivers and Barriers to the Adoption of Fuel Cell Passenger Vehicles and Buses in Germany. *Energies* **2021**, *14*, 833. [CrossRef]

energies

MDPI

Article

An Experimental Study of the Possibility of In Situ Hydrogen Generation within Gas Reservoirs

Pavel Afanasev [1,*], Evgeny Popov [1], Alexey Cheremisin [1], Roman Berenblyum [2], Evgeny Mikitin [3], Eduard Sorokin [3], Alexey Borisenko [3], Viktor Darishchev [4], Konstantin Shchekoldin [4] and Olga Slavkina [4]

1 Skolkovo Institute of Science and Technology, 121205 Moscow, Russia; e.popov@skoltech.ru (E.P.); a.cheremisin@skoltech.ru (A.C.)
2 Hydrogen Source AS, 0114 Oslo, Norway; roman.berenblyum@hydrogen-source.com
3 Lukoil Engineering LLC, 109028 Moscow, Russia; Evgenij.Mikitin@lukoil.com (E.M.); Eduard.Sorokin@lukoil.com (E.S.); Alexey.Borisenko@lukoil.com (A.B.)
4 Ritek LLC, 400048 Volgograd, Russia; Viktor.Darischev@lukoil.com (V.D.); Konstantin.Schekoldin@lukoil.com (K.S.); olga.slavkina@lukoil.com (O.S.)
* Correspondence: pavel.afanasev@skoltech.ru

Abstract: Hydrogen can be generated in situ within reservoirs containing hydrocarbons through chemical reactions. This technology could be a possible solution for low-emission hydrogen production due to of simultaneous CO_2 storage. In gas fields, it is possible to carry out the catalytic methane conversion (CMC) if sufficient amounts of steam, catalyst, and heat are ensured in the reservoir. There is no confirmation of the CMC's feasibility at relatively low temperatures in the presence of core (reservoir rock) material. This study introduces the experimental results of the first part of the research on in situ hydrogen generation in the Promyslovskoye gas field. A set of static experiments in the autoclave reactor were performed to study the possibility of hydrogen generation under reservoir conditions. It was shown that CMC can be realized in the presence of core and ex situ prepared Ni-based catalyst, under high pressure up to 207 atm, but at temperatures not lower than 450 °C. It can be concluded that the crushed core model improves the catalytic effect but releases carbon dioxide and light hydrocarbons, which interfere with the hydrogen generation. The maximum methane conversion rate to hydrogen achieved at 450 °C is 5.8%.

Keywords: hydrogen production; steam methane reforming; in situ hydrogen generation

Citation: Afanasev, P.; Popov, E.; Cheremisin, A.; Berenblyum, R.; Mikitin, E.; Sorokin, E.; Borisenko, A.; Darishchev, V.; Shchekoldin, K.; Slavkina, O. An Experimental Study of the Possibility of In Situ Hydrogen Generation within Gas Reservoirs. *Energies* **2021**, *14*, 5121. https://doi.org/10.3390/en14165121

Academic Editor: Bahman Shabani

Received: 21 July 2021
Accepted: 16 August 2021
Published: 19 August 2021

1. Introduction

The growing demand for clean energy resources stimulates the development of unconventional and alternative energy. Renewable energy is a promising and developing field, but hydrogen has a number of benefits as an energy source. According to the world's long-term programs for developing hydrogen technologies, hydrogen can ensure 12% of the world's total primary energy demand in 2050 [1]. Besides that, hydrogen is a valuable chemical product required for the refining industry and the production cycle of ammonia, methanol, and others. However, there is no cheap way for sustainable hydrogen production without greenhouse gas emissions.

Hydrogen can be obtained from natural gas through catalytic steam methane reforming (SMR), partial oxidation, autothermal reforming, and methane cracking. It also can be produced from water through electrolysis of water or from coal through coal gasification [2]. However, all these hydrogen production methods are very energy consuming. In addition, energy for hydrogen production is usually produced by burning hydrocarbons with carbon dioxide emissions. Most hydrogen is produced mainly via the SMR process, which also produces up to 10 kg of CO_2 per kg of generated hydrogen [3]. Greenhouse gases are produced during energy generation, as direct products of the chemical reactions, as well as during stages of compression and transportation of reagents and products. It is essential to realize carbon capture and storage (CCS) technology to make synthesized hydrogen "blue". These actions increase the cost of hydrogen significantly [4,5].

A promising, energy-efficient, and cost-effective technology for producing low-emission (almost "green," or without any greenhouse gas emissions) hydrogen is an in situ hydrogen generation in hydrocarbon reservoirs. Hydrogen can be generated in situ in oil/bitumen fields (for example through bitumen gasification) [6–10], coal deposits (underground coal gasification) [11–13] or gas fields [14–16]. These feedstock types imply different hydrogen generation mechanisms: oil aquathermolysis and thermolysis, coke gasification, methane cracking, steam methane reforming, and water-gas shift reaction.

The chemical transformations occurring in the gas reservoir include mainly steam methane reforming, water-gas shift reaction, and methane cracking (at temperatures higher than 500 °C) in the presence of a metal-based catalyst [17,18], according to the following forward reactions:

$$CH_4 + H_2O \leftrightarrow CO + 3H_2 - 206\ \text{kJ/mol}, \tag{1}$$

$$CO + H_2O \leftrightarrow CO_2 + H_2 + 41\ \text{kJ/mol}, \tag{2}$$

$$CH_4 \leftrightarrow C + 2H_2 - 75\ \text{kJ/mol}. \tag{3}$$

At the same time, side reactions take place that consume generated hydrogen. These reactions mostly include methanation reactions: reverse reactions (1) and (3) and forward reaction [7]:

$$2CO + 2H_2 \leftrightarrow CH_4 + CO_2 + 59.0\ \text{kcal/mol}. \tag{4}$$

The generated hydrogen can be stored underground and produced at any time. Moreover, it is expected that hydrogen will rise to the geological uplifts of the reservoir under the influence of gravitational forces. Simultaneously, the environmentally undesirable greenhouse gases, such as carbon and nitrogen oxides, having a higher density than hydrogen, will sink to the bottom of the field under the influence of gravity. These gases are also more soluble in water, compared with hydrogen. In addition, carbon oxides also can react with rocks, forming insoluble compounds such as carbonates. So, greenhouse gases may not be produced at all during hydrogen production from the gas reservoir [16].

Technology considered in this study, implies pure hydrogen production from gas fields with simultaneous CO_2 storage [14,15]. It can be implemented even in depleted or abandoned fields or fields in a late stage of exploration because the main process proceeds with an increase in the amount of gaseous components (up to four volumes of hydrogen can be generated from one volume of methane). The existing infrastructure (wells, pipeline network) can be used in hydrogen production, leading to a significant decrease in the produced hydrogen cost. For example, the produced hydrogen can be transported using a modern gas pipeline through mixing with natural gas in concentrations up to 20 and even 70% (for the Nord Stream) [19].

In this research, the idea of in situ hydrogen generation within gas fields supposes the implementation of the CMC (catalytic methane conversion) in the porous medium of the reservoir. The technology implies the injection of a catalyst precursor (aqueous solution of Ni-containing salt) or an active catalyst (particles of Ni-based catalyst) into a hydrocarbon-containing zone on the first stage. Since the reducing conditions are in the reservoir, active phase of catalyst can be formed from the precursor in situ. Then the temperature in the reaction zone should be raised to a temperature, at which catalyzed SMR and methane cracking occur.

The study introduces the results of laboratory experiments performed in an autoclave reactor at initial conditions the same as in the Promyslovskoye gas field, using core material taken from this target field. The Promyslovskoye gas field is located 96 km southwest of Astrakhan city, Russia. It contains about 1700 mln m^3 of natural gas, the reservoir temperature is 48 °C, the initial pressure is 8.9 MPa, and the current reservoir pressure is 2.3 MPa. The porosity of the target layer is about 29%, residual gas saturation is 77%, and residual water saturation is 23%. The depth of gas-bearing layers is about 730 m.

The temperature range from 300 to 450 °C was discovered during the experiments to estimate the possibility of hydrogen generation from methane in situ within the target field. These temperatures can be achieved in a porous medium of rock due to steam/overheated water injection into the reservoir (up to 350 °C) or due to in situ combustion of saturating hydrocarbons (up to 700 °C for oil combustion) [20,21]. The effects of different forms of catalyst and steam/methane ratios on CMC were also investigated during the experiments. The obtained data can help conclude the expediency of the new stage of field exploration and manage the process of CMC to intensify and speed up of in situ hydrogen generation processes.

This is the first publication from the planned series of publications devoted to the in situ hydrogen generation from methane under gas reservoir conditions. The concept, feasibility, and regularities of the considered process are investigated in the current study. The results of experiments performed at more favorable conditions (higher temperature and dynamic mode) will be presented in further publications.

2. Materials and Methods

Experiments were designed to study the possibility of methane conversion into hydrogen at relatively low temperatures in the presence of different types of catalysts: in situ synthesized (precursor is nickel nitrate hexahydrate) and ex situ synthesized Ni-based. The influences of the type of porous medium and steam/methane ratio on the process were also investigated.

2.1. Porous Medium

Several different porous media were investigated, varying from crushed alumina to crushed ceramics, to river sand and crushed core. These types of porous medium simulated different types of reservoir rock, including the target gas field. Industrial alumina (Al_2O_3) pellets, Alumac 5D® (Salindres, France), were used as an inert porous medium. Alumac 5D® has a high specific surface area of about 335 m^2/g, is very hydrophilic, inert to most liquids and gases and, is stable at temperatures up to roughly 2000 °C. Granules of Al_2O_3 were crushed to 0.8–1.2 mm before use. River sand with granules size 0.8–1.2 mm was used as filler in some experiments to model sandstone rock samples. Its composition can be roughly approximated as SiO_2. One more option for the porous medium was crushed ceramics. The mineralogical composition of this filler is presented in Table 1.

Table 1. Mineralogical composition of crushed ceramic filler.

Mineral	Value, wt.%
Mullite	68.1
Quartz	31.9

In other experiments, non-extracted core (rock) samples from the Promyslovskoye gas field were used to recreate reservoir conditions and investigate the influence of the real core on the process of hydrogen generation. The average content of total organic carbon determined with the rock-eval method [22,23] is 1.35 wt.% Data for the averaged composition of the mineral matrix is demonstrated in Table 2.

Table 2. Averaged mineralogical composition of the core for laboratory experiments.

Mineral	Value, wt.%
Albite	12.4
Anhydrite	0.8
Calcite	10.3
Halite	4.1
Illite	1.2
Pyrite	0.1
Quartz	71.1

Experimental studies observe slightly overestimated results of the methane conversion since the packed model may not exactly repeat the properties of the consolidated core. For example, the porosity and permeability of the consolidated core cannot be reproduced with high accuracy.

2.2. Catalyst Preparation Procedure

There were two types of monometallic Ni-based catalysts used in the experiments. The first one was an in situ prepared catalyst, which can be delivered into the reservoir in the form of a water solution of the catalyst precursor, then obtained through chemical transformations directly at the reservoir [24]. So, this catalyst was prepared in the reactor during experiments from catalyst precursor solution. The second one was the ex situ prepared catalyst which was nickel oxide particles supported on alumina. This catalyst can be delivered into the reservoir in the form of suspension together with steam or overheated water. In this case, the catalyst was prepared in advance and loaded into the reactor before the experiments.

The catalyst precursor, used for in situ prepared catalyst, was water-soluble nickel nitrate hexahydrate ($Ni(NO_3)_2{\cdot}6H_2O$, chemically pure), which had to be decomposed under high temperature according to the summary equation [25,26]:

$$2Ni(NO_3)_2{\cdot}6H_2O = 2NiO + O_2 + 4NO_2 + 12H_2O. \quad (5)$$

This salt solution in deionized water was put into the reactor before the experiment with other reactants (water and methane). The catalyst here is the particles of nickel oxide, which have a catalytic effect themselves or can be reduced to a more active metallic phase by interaction with hydrogen or a mixture of steam and methane at a high temperature [27–30] according to the equation:

$$NiO + H_2 = Ni + H_2O. \quad (6)$$

The second type of catalyst was the ex situ prepared catalyst by wet impregnation of α-Al_2O_3 (granular size of 0.5–1.0 mm, the specific surface area of 174 m^2/g) with a water solution of nickel salt. This catalyst was obtained through heat treatment of the carrier, and soaked in 31.42% nickel nitrate solution in a muffle furnace. Catalyst's preparing procedure included treating 100 g of α-Al_2O_3 particles with 150 g of the catalyst precursor solution (soak period-2 h), drying the carrier with the precursor solution in the air at 110 °C, while water was not evaporated. Next, the heat treatment of the carrier particles coated with precursor salt particles was necessary. Heat treatment was carried out in a muffle furnace, in the air atmosphere, for 3 h at 150 °C and then 3 h at 450 °C. The decomposition of $Ni(NO_3)_2{\cdot}6H_2O$ occurred and nickel oxide particles formed on the substrate's surface because of the last operation. The catalyst can be used in the experiments after this procedure. Such supported catalyst contains 16.16% of the active component, calculated in terms of nickel oxide.

2.3. Experimental Setup

An autoclave reactor used for static experiments is a reactor by Parr (USA), fabricated of Inconel 600 alloy, designed for experiments at max temperature 600 °C and max pressure

~408 atm. The reactor volume is 1 L. It has a control block, external heaters, magnetic stirrer, thermocouple, check valve, manometer, and bursting disc (as a safety measure). High-pressure, high-temperature tubes with the dimensions 1/8″ and 1/16″, and vessels by Swagelok were used for connection lines. The appearance of the reactor and hydrodynamic scheme of installation used in the experiments are shown in Figures 1 and 2, respectively.

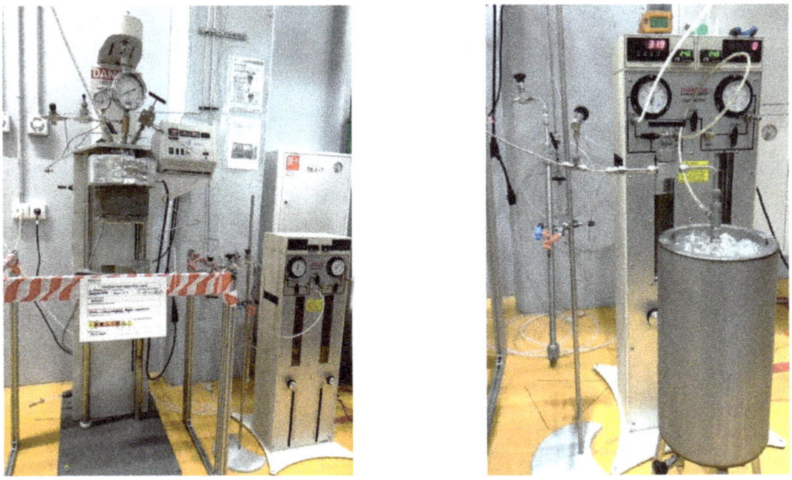

(a) (b)

Figure 1. (**a**) The appearance of autoclave installation and (**b**) sampling system used in experiments.

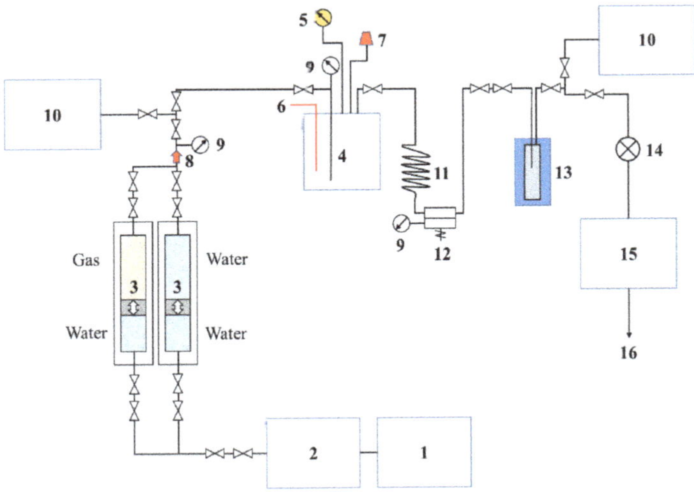

Figure 2. The scheme of autoclave installation used in experiments: 1—computer; 2—pump (Quizix); 3—piston column (V = 1 L, with gas); 4—autoclave (reactor); 5—digital pressure gauge; 6—thermocouple; 7—bursting disc; 8—check valve; 9—manometer; 10—vacuum pump; 11—condenser (cooler); 12—back pressure regulator; 13—separator (V = 0.25 L); 14—gas meter (0.5 L); 15—gas chromatograph; 16—ventilation system with gas afterburning.

2.4. Experimental Procedure

The experiments focused on studying the activity of CMC at considered conditions, the concept of in situ nanoscale active catalyst synthesis, the feasibility of CMC in the presence of real core, and the influence of steam amount and type of porous medium on methane conversion. The set of experiments was performed by subsequently changing the type of catalyst (in situ and ex situ prepared in the form of nickel oxide), values of steam/methane ratio (from 2 to 21), temperatures (from 300 to 450 °C), and types of porous media (without filler, sandpack model, crushed ceramics, crushed alumina, and crushed core). So, the temperature of the CMC initialization, product gas composition, methane conversion rates, as well as optimal parameters for the CMC can be revealed.

In the first stage, the experimental procedure included the preparation of the catalyst (or catalyst precursor), filler for simulation of porous medium, assembly, and pressurization of autoclave installation. Then, the exact quantities of water, catalyst, and filler were placed into the reactor. After that, the whole hydrodynamic system was vacuumed, and the exact quantity of methane was injected (the quantities of loaded reagents in different tests are presented in Table 3). In the next stage, heat treatment of the reagents was performed (operational temperatures for the different tests are also presented in Table 3) with periodic gas sampling (for gas chromatography analysis by Agilent 7890 B). The experimental procedure is similar to the literature described [16].

Table 3. Reagent loads and experimental parameters.

Exp. No.	Water, mL	Methane, L	Catalyst, g	Porous Medium	T, °C
1	30.0	18.5	34.5 [1]	-	350
2	29.3	7.3	1.0 [1]	-	350–450
3	20.1	2.5	5.9	Ceramics	450
4	20.0	2.5	4.5 [1]	River sand	450
5	33.3	2.5	37.5	Alumina	450
6	42.4	2.5	5.4	Core	300–450
7	88.7	7.4	7.2	Core	450

[1] Precursor (catalyst was prepared in situ in the experiment).

If the hydrogen content in product gases is low, additional water can be injected into the reactor. This action is aimed to shift the thermodynamic equilibrium of the system to the products, as water is one of the reagents and can possibly create additional mixing of gas components. It should be noticed that water and methane are injected from the bottom of the reactor by a high-pressure pump (Quizix), and gas samples for gas chromatography analysis are taken from the top. It is also assumed that the injected water is vaporized right in front of the reactor because the inlet tube has a high temperature. At the end of the heat treatment period, the heaters of the experimental setup are turned off, the pressure decreases, and nitrogen injection begins until the reactor cools down. In the final stage, the methane conversion rate was calculated, and conclusions were made. The methane conversion rate can be calculated using the following equation, considering only the SMR process:

$$\text{Methane conversion rate } (\%) = \frac{n_{CH4,inj}(\text{mol}) - n_{CH4,rem}(\text{mol})}{n_{CH4,inj}(\text{mol})} \times 100\ \%, \quad (7)$$

where $n_{CH4,inj}$ is the amount of methane injected during the whole experimental time, and $n_{CH4,rem}$ is the amount of methane remaining after the heat treatment period and collected simultaneously with the pressure decreasing and reactor cooling down. Then the calculated value should be compared with the methane conversion rate, calculated with respect to the amount of synthesized hydrogen, directly detected on the chromatograph.

Experiment No. 1 implied heat treatment of methane and water in an autoclave at the relatively low temperature of 350 °C (Table 3). This limitation in the temperature range is determined by the maximum value that can be achieved in the reservoir by injecting steam or superheated water into the reservoir. In this experiment, in situ synthesized catalyst was obtained in the reactor because of the decomposition of the catalyst precursor nickel nitrate hexahydrate. The experiment was carried out in the absence of a porous medium with a steam/methane ratio of 2.

Within experiment No. 2, heat treatment was performed in two stages. The first stage of heat treatment was carried out at a temperature of 350 °C, but then the temperature was increased to 450 °C. The amount of initially loaded methane was reduced. It led to an increase in the steam/methane ratio to an optimal value of five [31,32]. The amount of initially loaded catalyst precursor was also reduced. The experiment was carried out without a porous medium.

In experiment No. 3, ex situ prepared catalyst with a large specific surface area was used. At the same time, the main volume of the reactor was filled using a crushed ceramics model. In this experiment, water was not loaded into the reactor before heat treatment but was injected in several stages after the temperature had risen to 450 °C. It is believed that the injection of new portions of steam and rapid diffusion of components at this temperature ensured an even distribution of components in the pore volume of the crushed ceramics model.

In experiment No. 4, a sandpack model was used as a filler. A new attempt to synthesize an active catalyst in situ from a catalyst precursor-nickel nitrate hexahydrate, in the presence of a sandpack model at a temperature of 450 °C, was made.

Experiment No. 5 was carried out in the presence of inert alumina granules in the reactor (as a porous medium) and under conditions of an increased amount of ex situ prepared catalyst. The increased steam/methane ratio in the experiment makes it possible to create more favorable conditions for the SMR process due to the displacement of the equilibrium of the main reactions (Equations (1) and (2)) to the right, towards the products. The heat treatment of the reactor was carried out at a temperature of 450 °C. The gas sample was taken only once, at the end of the heat treatment period.

In experiments No. 6 and 7 crushed core samples from the target gas field were used to study the process of CMC in conditions close to reservoir conditions and investigate the effect of the core. Taking into account the results of previous experiments, experiment No. 6 was designed with heat treatment at two temperatures: at 300 °C and then at 450 °C, with an ex situ prepared catalyst and corepack model. In turn, experiment No. 7 repeated conditions of experiment No. 6 (with the ex situ prepared catalyst and corepack model), but was performed at lower steam/methane ratio and a heat treatment of only 450 °C.

3. Results

3.1. Determining of Thermodynamic Constraints

First of all, the thermodynamic calculations for the primary catalytic SMR process were performed. It is assumed that the SMR is the main mechanism of generating hydrogen at considered experimental conditions. Thus, this thermodynamic model is a simplification of the CMC process, which does not describe the system correctly at high temperatures (at which, for example, catalytic methane cracking occurs). At the same time, the approach overestimates methane conversion under conditions, at which other processes of hydrogen generation besides the SMR, do not yet play a significant role. It is because kinetic limitations are not taken into consideration.

The model considers only reversible reactions (Equations (1) and (2)) in a plug flow reactor for simplicity. Then, numerical methods can be used to calculate the equilibrium composition of product gases and the methane conversion rate at the reactor outlet for any values of temperature (T), pressure (p), and steam/methane ratio (ß). It is assumed that temperature and pressure are constant along the reactor's entire length. Let χ be the methane fraction converted to carbon dioxide after the whole reaction time, ξ is the methane conversion rate and K_{p1}, K_{p2} is the reaction equilibrium constants for reactions (Equations (1) and (2)), respectively. It is possible to write an expression for K_{p2} through mole fractions of hydrogen, carbon dioxide, carbon monoxide, and water (taking into account quasi-equilibrium for the reaction—Equation (2)). At the same time, these mole fractions can be written through χ and ξ, according to the material balance equations for each chemical element and mathematical expression for χ and ξ (by definitions). Then the expression for K_{p2} could be written as:

$$K_{p2}(T) \times (\xi - \chi) \times (ß - \xi - \chi) = (3\xi + \chi) \times \chi. \quad (8)$$

The expression for K_{p1} also could be written through component's mole fractions (hydrogen, carbon monoxide, methane, and water) and overall pressure:

$$K_{p1}(T) \times (1 - \xi) \times (ß - \xi - \chi) \times (1 + ß + 2\xi)^2 = (3\xi + \chi)^3 \times (\xi - \chi) \times p^2. \quad (9)$$

If we analytically solve Equation (8) for χ, then put the answer into the Equation (9), we have the cubic equation for ξ, where $K_{p1}(T)$, $K_{p2}(T)$, ß and p are parameters. This cubic equation could be solved numerically for each value of T, p, and ß, if put here expressions for $K_{p1}(T)$ and $K_{p2}(T)$. These expressions can be written through thermodynamic functions ($_{\Delta r}G^o_T$, $_{\Delta r}H^o_T$, $_{\Delta r}S^o_T$, $_{\Delta r}C_p$) and approximation formulas for the reduced Gibbs energies for both of the reactions separately [33]. The result of calculating expressions for $K_{p1}(T)$ and $K_{p2}(T)$ is provided in the study [34].

In this study, the methane conversion rates and the product gas mixture's equilibrium composition were calculated using the processing of code written in the Python programming language. The calculation results for the steam/methane ratio equal to 5 are shown in Figure 3. The interval of conditions considered in experiments within this study is highlighted with a frame.

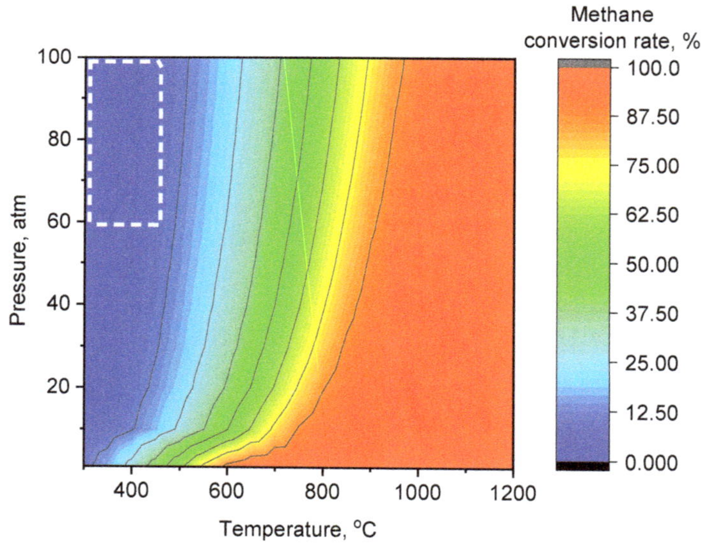

Figure 3. The dependence of methane conversion rate on temperature and pressure at constant steam/methane ratio equals 5 as a result of thermodynamic calculations for the primary catalytic SMR (steam methane reforming) process.

Also, the dependencies of methane conversion rates on temperature were plotted at various pressures (in the range from 1 to 100 atm), but at a constant steam/methane ratio of 5 (Figure 4a). They demonstrate the effect of external pressure on the catalytic SMR process. The dependencies of methane conversion rates on the temperature at various steam/methane ratios (in the range from 1 to 20) but at constant pressure 100 atm were also plotted (Figure 4b) and show the effect of additional portions of steam on the catalytic SMR process. The intervals of conditions considered in experiments within this study are highlighted with frames.

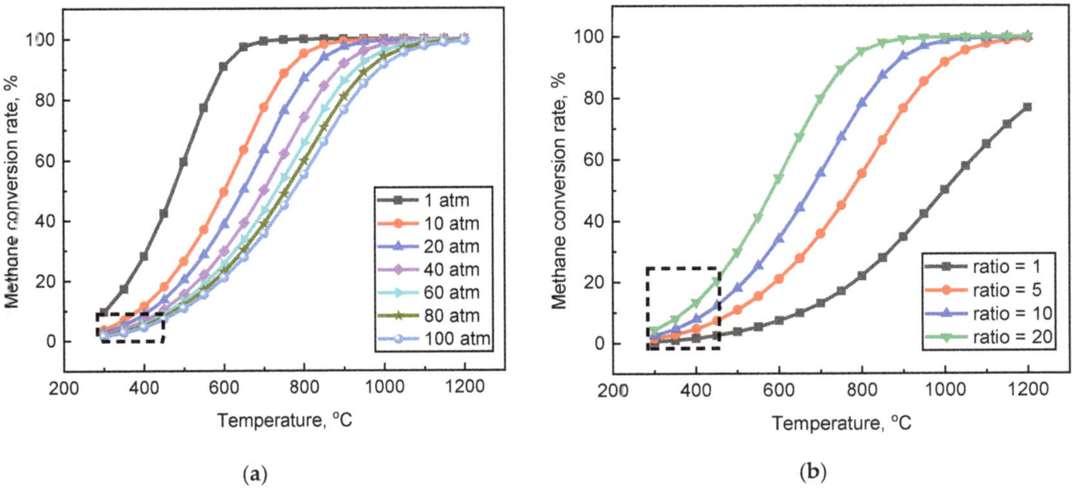

Figure 4. (**a**) The dependencies of methane conversion rates on the temperature at different pressures and constant steam/methane ratio 5 and (**b**) at different steam/methane ratios and constant pressure 100 atm, as a result of thermodynamic calculations for the primary catalytic SMR process.

Thermodynamic calculations for the SMR process allow for optimizing the experimental design and adjusting the operating parameters (initial loading of reagents, heating conditions, and others). The thermodynamic approach allows researchers an understanding of the maximum achievable methane conversion rates under specific conditions. It also provides information about the equilibrium product gas composition and about the completeness of the processes in a specific experiment. It also allows a conclusion to be made about catalyst activity, the influence of external factors, and the possible mechanism of the hydrogen generation process.

It can be seen from the obtained dependencies that the highest methane conversions are attainable at high temperatures, low pressures, and at high values of the steam/methane ratio. However, high values of steam/methane ratio and low pressures are practically unachievable in gas field conditions. One possible way of increasing the methane conversion rate is to increase the temperature inside the reservoir up to 800 °C and higher. This study describes the results of experiments, performed at relatively low temperatures, achievable in the gas reservoir due to steam or overheated water injection. In subsequent publications, the results of high-temperature experiments will be presented.

3.2. Experimental

The experimental parameters: maximum pressures in the reactor achieved during the experiments and the product gas mixture's composition for each of the experiments are shown in Table 4.

As a result of experiment No. 1, only trace amounts of hydrogen were detected in the gas samples. This may mean that the CMC is not active at the considered temperature of 350 °C, or some factors interfere with the process (for example, the active phase of the catalyst may not yet form at this temperature). However, in addition to the methane and water, significant amounts of carbon dioxide, up to 13 mol.%., and nitrogen dioxide, up to 5 mol.%., were detected in the product gases. At the same time, the methane fraction in the reactor decreased to 78 mol.%. Nitrogen monoxide and nitrogen were also detected in the product gas mixture as minor gas components.

Table 4. Summary of experimental parameters and product gas composition.

Exp. No.	Max P, atm	Max Concentration of Main Product Gas Components, mol.%.			Other Gas Components
		Hydrogen	Methane	Carbon Dioxide	
1	115	0.002	78.00	13.00	NO, NO_2, N_2
2	64	0.011	98.24	0.97	CO, NO_2, N_2
3	138	35.300	56.90	8.00	CO, C_2H_6
4	103	0.100	98.50	0.26	CO, N_2
5	120	3.100	93.00	0.40	N_2
6	207	53.500	21.42	15.67	CO, H_2S, C_2H_4, C_2H_6, C_3H_8
7	140	6.970	39.74	47.70	CO, C_2H_4, C_2H_6, C_3H_6, C_3H_8, C_4H_{10}, C_5H_{12}

The next experiment, No. 2, was optimized compared with the previous one. However, this did not lead to a significant increase in the hydrogen concentration in the reactor. The maximum achieved hydrogen fraction in the product gas mixture was only 0.011 mol.%. Simultaneously, the methane fraction in the reactor was about 98.2 mol.%, and the carbon dioxide fraction was about 1% vol. Carbon monoxide, nitrogen dioxide, and nitrogen were also detected in product gases in trace amounts. The higher temperature of heat treatment, up to 450 °C did not lead to the activation of hydrogen generation processes, since the methane concentration in the product gas mixture was almost 100%. It is more likely that the active phase of the catalyst cannot be obtained in situ with the considered conditions.

During experiment No. 3, water (steam) was injected into the reactor in several stages, with simultaneous gas composition monitoring. The maximum value of the steam/methane ratio ~10 was achieved due to additional water injections (4 injection cycles in total). As a result of the experiment, the hydrogen fraction in the product gases was about 18.5 mol.%. after one hour of heat treatment (Figure 5a).

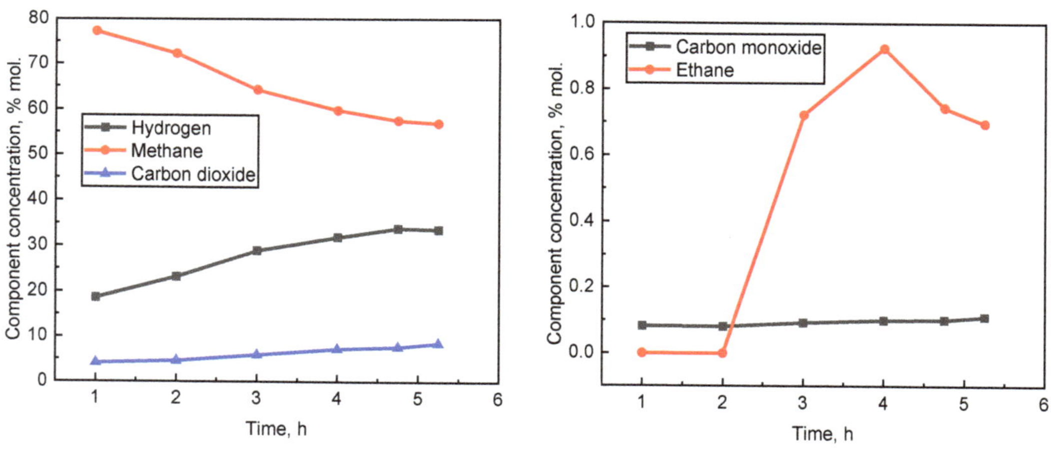

(a) (b)

Figure 5. (**a**) The dependencies of major gas component concentrations and (**b**) minor gas component concentrations in product gas mixture in experiment No. 3 on time.

Additional portions of injected steam led to an increase of hydrogen fraction reaching the maximum value of 33.9 mol.%. The fraction of carbon dioxide in the reactor also increased consistently, up to 8.0 mol.%., and the concentration of methane decreased to ~57 mol.%. Probably, the main effects that led to a significant increase in the hydrogen concentration are the presence of a porous model in the experiment and increased values of the steam/methane ratio. Interestingly, the ethane component appeared in the reactor, with a concentration of 0.93 mol.%. The dependencies of major and minor gas component fractions in the product gas mixture on experimental time are presented in Figure 5.

The replacement of a crushed ceramics model with a sandpack model in experiment No. 4 and the loading of a catalyst precursor into the reactor instead of the ex situ prepared catalyst led to a significant decrease in hydrogen amounts. In this experiment, the Ni-based catalyst particles (nickel oxide) should have been formed from the particles of the precursor during the thermal decomposition of nickel nitrate hexahydrate, according to the reaction Equation (4). As a result of the experiment, hydrogen was detected in the product gas mixture only in trace amounts (~0.1 mol.%). At the same time, initial methane and carbon dioxide fractions almost did not change, reaching 98.5 and 0.26 mol.%., respectively.

As a result of experiment No. 5, performed in the presence of the ex situ prepared catalyst and crushed alumina, a hydrogen concentration in the product gases of 3.1 mol.%. was achieved. In this case, the steam/methane ratio in the experiment was high and equal to 17. The methane fraction in the reactor decreased to 93.0 mol.%. At the same time, the carbon dioxide fraction in the product gas mixture reached 0.4 mol.%.

Experiment No. 6 on implementing the CMC in the presence of core material from a real gas field and an ex situ prepared catalyst was carried out for almost 9 h with periodic gas sampling. In this case, the first 2 h reactor was heated to a temperature of 300 °C, and then to a temperature of 450 °C. Additional portions of water were injected into the reactor during the experiment (4 injection cycles, 42.4 mL in total), and excess pressure was released from the system if the value was higher than ~100 atm. However, these actions did not lead to activation of the CMC process. As a result of heat treatment at 300 °C, only trace amounts of hydrogen were detected in the gas samples (Figure 6a).

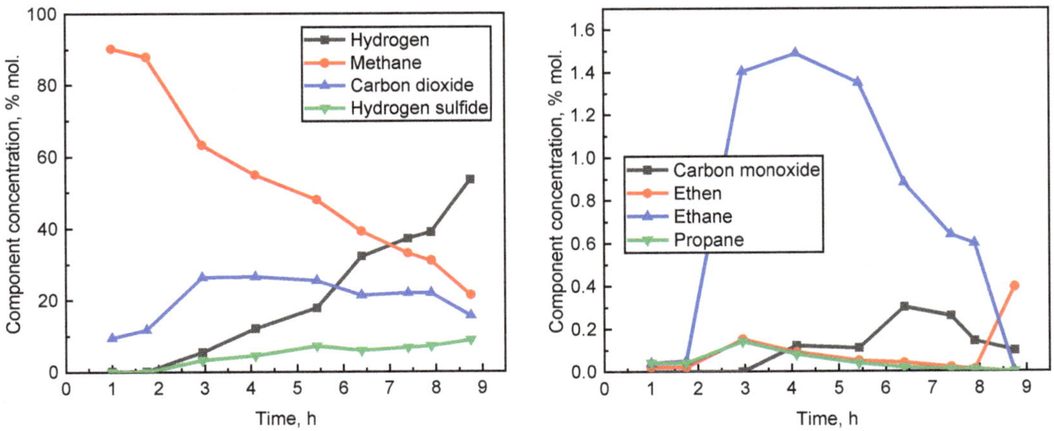

(a) (b)

Figure 6. (**a**) The dependencies of major gas component concentrations and (**b**) minor gas component concentrations in product gas mixture in experiment No. 6 on time.

The maximum hydrogen fraction in the product gas mixture, achieved during the heat treatment at 450 °C, was 53.5 mol.%. In this case, the methane fraction in the product gas mixture was only 21.42 mol.%, and carbon dioxide was 15.67 mol.%. Hydrogen sulfide, carbon monoxide, ethane, and ethene constituted a large total fraction in the product gas mixture. Thus, it can be concluded that some transformations occur with the core material during the experiment, which leads to the appearance of fat gas components and an increase in the concentration of carbon dioxide in the product gas mixture. The dependencies of the major and minor gas component concentrations in the product gas mixture on time are presented in Figure 6. It can be expressed from the graphs that the dependencies for the major gas components have a monotonic nature. It indicates that the system has not reached an equilibrium state at the time of the experiment completion. The methane conversion rate for the whole experiment, calculated through material balance equations, is equal to 5.80. The methane conversion rate calculated for the shorter period corresponding to the active methane conversion process is possibly much higher.

Experiment No. 7 had a similar design to experiment No. 6. However, initially, 28.6 mL of water was loaded into the reactor, then another 60.1 mL of water was injected during the heat treatment (3 injection cycles). As a result of the additional injected water, the pressure in the system rose to 153 atm. Therefore, excess pressure was released from the reactor to about 100 atm. Three pressure relief cycles were made during the whole experimental time. Even with the achievement of high values of the steam/methane ratio (up to 15), the product gas mixture contained only up to 6.97 mol.%. of hydrogen. At the same time, the methane fraction decreased to 39.74 mol.%, and the carbon dioxide fraction increased to 47.7 mol.%. The major and minor gas components' dependencies in the product gas mixture on time are presented in Figure 7. It can be expressed from the graphs that the dependencies for the major gas components reach a plateau at the end of the heat treatment period, which indicates the approach to the equilibrium state of the system. The methane conversion rate for the whole experiment, calculated through material balance equations, is equal to 3.71.

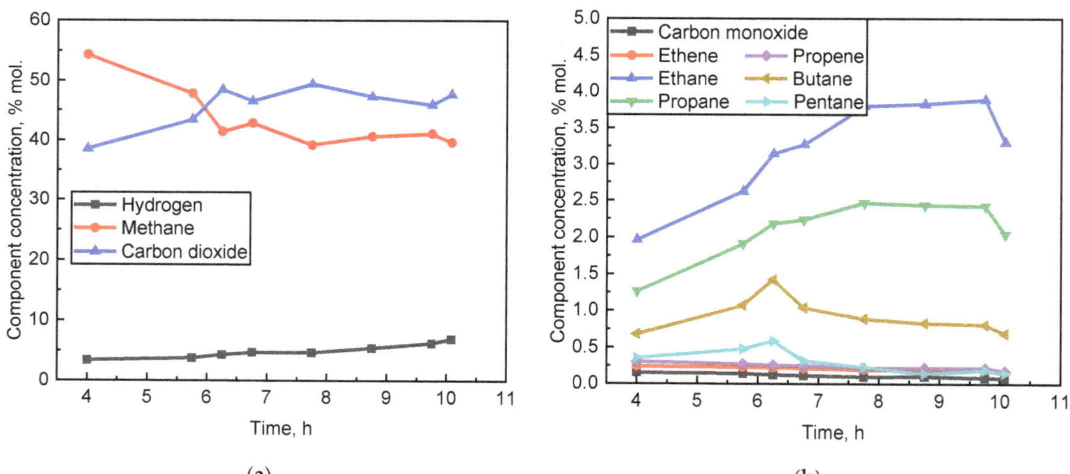

Figure 7. (**a**) The dependencies of major gas component concentrations and (**b**) minor gas component concentrations in product gas mixture in experiment No. 7 on time.

Besides the main gas components, the product gas mixture contains hydrocarbons with a carbon chain length up to C_5. Ethane was detected as a minor gas component in a concentration of up to 3.89 mol.%; ethene, up to 0.24 mol.%; propane, up to 2.46 mol.%; propene, up to 0.3 mol.%; butane and isobutane with a total concentration of up to 1.42 mol.%; and pentane, up to 0.59 mol.% (Figure 7). These components are present in the product gas mixture, most likely as a result of the organic matter of core decomposition (during the processes of thermolysis, aquathermolysis, and others).

4. Discussion

As a result of the experiments, the highest hydrogen concentration of 53.5 mol.%. was achieved in some gas samples. Although the methane conversion rate at individual stages can be quite high, the value for the whole experiment was only 5.8%. The achieved value of methane conversion is somewhat lower than those described in the literature for the SMR process carried out under similar conditions. Table 5 shows a comparative analysis of experimental data obtained by other researchers in similar methane conversion processes.

Table 5. Comparison of achieved methane conversion rate with literature data.

Catalyst	Methane Conv., %	T, °C	P, atm	Steam to Methane Ratio	Reference
10 wt.% Ni/Al_2O_3 [1]	0.0	500	1	2	[35]
10 wt.% Ni/Al_2O_3	15.0	500	1	2	
10 wt.% Ni/Al_2O_3	9.0	400	1	1	[36]
10 wt.% Ni/Al_2O_3	31.0	500	1	1	
7 wt.% Ni/Al_2O_3 + 1 wt.% Ag	75.0	500	1	4	[37]
Ni/Al_2O_3 [2]	25.0	450	1	2	[38]
10 wt.% Ni/Al_2O_3	32.0	500	1	3	[39]
16.2 wt.% Ni/Al_2O_3	5.8	450	207	21	Current study

[1] Catalyst reduction at 500 °C; [2] nanoparticle clusters.

The SMR process in the works mentioned above was carried out in a dynamic mode in continuous flow reactors, at a much lower pressure, in the presence of a catalyst previously reduced at high temperature. This study describes experiments carried out in a static mode, which imposes restrictions, for example, on the mixing of reagents. At the same time, this study describes experiments carried out simultaneously in the low-temperature and high-pressure ranges, in the presence of a core from a real gas field. Besides that, the nickel-based catalyst used in this study was activated directly in the reactor during the experiment. There was no preliminary reduction treatment stage, and the catalyst reduction took place in a steam-methane atmosphere at a temperature not higher than the experimental temperature. As a result, the hydrogen concentrations and the methane conversion rates obtained are lower than those described in the literature. The effects of temperature, the type of porous medium, and the steam to methane ratio on the catalyst activity and methane conversion are discussed below.

4.1. Applicability of Different Forms of Catalyst

In the above series of experiments, the use of two types of catalysts was considered: in situ synthesized from a water-soluble precursor-nickel nitrate hexahydrate (during the heat treatment process) and ex situ synthesized. The last type is nickel oxide particles deposited on a porous substrate from α-Al_2O_3, obtained separately from the main process.

The reason for the low hydrogen yield in experiment No. 1 may lie in the absence of an active phase of catalyst in the system. The experimental temperature of 350 °C, might be insufficient for the complete decomposition of the precursor, according to the reaction Equation (5), or for the conversion of the oxide form of the catalyst into a more active–metallic form. At the same time, the gas components NO, NO_2, and N_2 are present in the product gas mixture, indicating changes in the catalyst precursor. An excess amount of catalyst precursor was taken in the experiment in order to avoid the influence of the amount of catalyst on the activity of the process (Table 3).

Besides, a significant decrease in methane fraction from the initial 100% and an increase in the carbon dioxide fraction in the reactor should be noted. This indicates the hydrogen generation process's occurrence according to Equations (1)–(3). However, synthesized hydrogen can enter into secondary processes of catalyst or nitrogen oxides reduction.

Which is probably the main reason hydrogen was detected in experiment No. 1 only in trace amounts.

In the next experiment, No. 2, the CMC conditions were changed to determine the efficiency of the in situ prepared catalyst. An increase in temperature at the second stage of heat treatment up to 450 °C, an increase in the steam/methane ratio to 5, and a decrease in the loaded precursor amount did not allow significant methane conversion. According to the thermodynamics of the primary SMR process, an increase in temperature and steam/methane ratio leads to an increase in hydrogen concentration. Besides, a decreased amount of catalyst precursor could possibly reduce secondary reactions consuming hydrogen. Nevertheless, as in the previous experiment, only trace amounts of hydrogen were detected in the product gas mixture. In this case, the product gas consisted almost entirely of unreacted methane. The experimental results indicate the impossibility of obtaining an active catalyst in situ in the reactor during the CMC at temperatures up to 450 °C.

Experiment No. 4 was performed to look at in situ catalyst generation again by repeating successful experiment No. 3, but replacing the ex situ catalyst with the in situ one. As a result of the experiment, only trace amounts of hydrogen were obtained. However, the sandpack model could also contribute to the hydrogen yield (this influence has yet to be studied), besides the type of the catalyst.

Replacing the in situ synthesized catalyst with an ex situ prepared catalyst in experiments No. 3, 6, and 7 made it possible to obtain significant hydrogen concentrations in the product gas mixture after heat treatment of a steam-methane mixture at the temperature of 450 °C. The ex situ prepared, Ni-based catalyst supported on a porous substrate has a large specific surface area coated with nano and microparticles of nickel oxide (Figure 8). The use of an ex situ prepared catalyst ensures the oxide form of the catalyst in the reactor in the absence of nitrogen oxides.

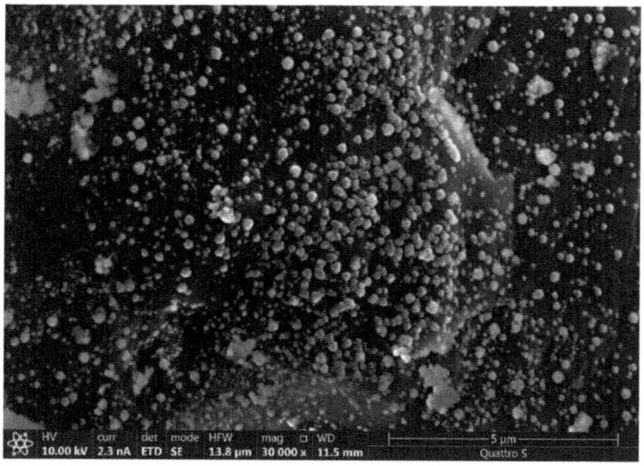

Figure 8. SEM image of ex situ prepared catalyst's surface with nickel oxide particles (light color).

Thus, experiments indicate that the active phase of the nickel-based catalyst does not form from a water-soluble nickel nitrate hexahydrate, at temperatures up to 450 °C. Other conditions or catalyst precursors should be used for active catalyst in situ generation. Nevertheless, for the implementation of the SMR process at considerable temperatures, ex situ prepared nickel-based catalyst can be used. The delivery of such a catalyst into the formation is possible in the form of a suspension, together with the injected steam or water.

4.2. *Effect of Temperature on the CMC*

The temperature at which the process of CMC is carried out primarily determines the process's thermodynamic constraints. A higher methane conversion rate can be achieved with higher process temperatures (Figure 3), keeping other parameters the same. The temperature also determines in what form the catalyst could be used. For example, the insufficiently high heat treatment temperature was probably why the active phase of the catalyst was not formed during in situ in experiments No. 1 and 2. Additionally, the insufficiently high heat treatment temperature is also the reason for the hydrogen formation process's low activity.

Experiment No. 6, carried out in the presence of a real core from a gas field and an ex situ prepared catalyst, demonstrates the effect of temperature on the hydrogen generation process's activity. As a result of reagents heat treatment at 300 °C, only trace amounts of hydrogen were detected in the gas samples (Figure 6). The temperature of 300 °C is not enough for the active CMC. However, heat treatment at 450 °C led to significant hydrogen concentrations in the product gas mixture (5.47 mol.% in the first gas sample taken at a given temperature). It should be noted that the obtained high concentrations of hydrogen as a result of heat treatment at 450 °C could also be achieved with an increase of the steam/methane ratio to 21 (since temperature and steam/methane ratio were increased simultaneously) and a pressure drop. Such actions led to the shift of the thermodynamic equilibrium of the main reactions (Equations (1) and (2)), and could affect the experiment's result.

4.3. Effect of a Packed Model on the CMC

The above series of experiments also examined the effect of a porous medium composition on the CMC. Experiments were carried out in the reactor's bulk in the absence of filler, as well as in the presence of a sandpack model, crushed ceramic, crushed alumina, and a corepack model, simulating porous medium. As a result, it can be noted that significant hydrogen fractions in product gas mixtures can be obtained in the case of crushed ceramics, alumina, and core as a porous media.

For example, in experiment No. 3, the hydrogen concentration in a specific gas sample was 35.3 mol.%, and some of the ceramic granules were covered with coke (Figure 9b). This fact indicates that hydrogen generation is proceeded not only by the mechanism of the catalytic SMR but also by the mechanism of the catalytic cracking of methane (according to the reaction Equation (3)).

(a)

(b)

Figure 9. (**a**) The appearance of the ceramic granules of the packed model before the experiment No. 3 and (**b**) after the experiment No. 3.

The crushed ceramics model probably contributed to the activity of hydrogen generation processes since it could contain catalyst promoters that increase catalyst activity. Probably, the crushed ceramics model also contains additional acidic catalytic sites, on which the methane cracking reaction took place, leading to the deposition of coke on some of the granules.

Experiment No. 5 was carried out in the presence of a crushed alumina model. In this case, alumina is an inert material and pure substance of constant composition, Al_2O_3. Probably, in the absence of catalyst promoters in the packed model, the catalytic effect was significantly reduced compared with that observed in experiments No. 3, 6, 7. Therefore, the achieved hydrogen concentration in the product gases of experiment No. 5 was significantly lower than that obtained in experiments with crushed ceramics and core models. These models, which are mixtures of substances, may contain catalyst promoters (transition metal atoms), which increased catalytic effect and led to significant hydrogen concentrations.

It can also be seen from the results of gas chromatography for experiment No. 6 that the product gas mixture contains small amounts of ethane, methene, and propane. These components are cracking products (thermal cracking, hydrocracking, and aquathermolysis) of the organic matter of the core and can be detected because the core used in the experiment was not previously cleaned of the original organic matter.

Besides that, hydrogen sulfide has a significant fraction (up to 8.91 mol.%.) in the product gases in experiment No. 6. Hydrogen sulfide is most likely formed due to the decomposition of sulfur-containing components of the core (organic and inorganic) and the interaction of decomposition products with hydrogen synthesized during the experiment.

Hydrogen sulfide is also known as a catalyst poison [40] and, even in small amounts, can significantly reduce the activity of the catalyst. In turn, wet natural gas, as well as the decomposition products of the organic matter of the core, can lead to rapid coking of the catalyst, and therefore, significantly reduce the catalyst's activity. However, the experimental results indicate that the activity of the catalyst did not decrease during the experiment. This behavior can be explained by the significant concentrations of hydrogen and steam in the reactor and general reducing conditions.

Based on the results of gas chromatography for experiment No. 6, it can also be assumed that a significant contribution to the total amount of synthesized carbon dioxide was made by carbon dioxide from the core. Carbon dioxide was also formed due to the decomposition of the carbonate minerals from the core. So, the mass of the crushed core model significantly decreased. The calculation of the material balance confirms this assumption.

Similar to experiment No. 6, the carbon dioxide fraction in the product gas mixture of experiment No. 7 reaches a high value, up to 47.7 mol.%. This value is significantly higher than the equilibrium value calculated based on the SMR process's thermodynamic modeling at the considered parameters and calculated methane conversion rate. Such high carbon dioxide fraction can be explained by the decomposition of carbonate minerals of the core [41], which leads to the release of significant amounts of carbon dioxide. So, the decomposition temperature of calcium carbonate can be significantly reduced in the presence of water and carbon dioxide [42,43].

Pyrolysis results of two core samples, taken before and after experiment No. 7, confirm the previous conclusion. There is a peak on the pyrolytic spectrum corresponding to the active release of carbon dioxide from the unprocessed core sample (Figure 10, red curve). It starts at a temperature of about 450 °C at the 15th minute of heating and has a maximum at temperature of 558 °C. This peak is almost absent on the pyrolytic spectrum of the core sample taken after heat treatment during the experiment (Figure 10, green curve). The decomposition of carbonate minerals here occurs in the absence of steam and the excess pressure of carbon dioxide, which means the defined decomposition temperatures can be higher than in the real experiments.

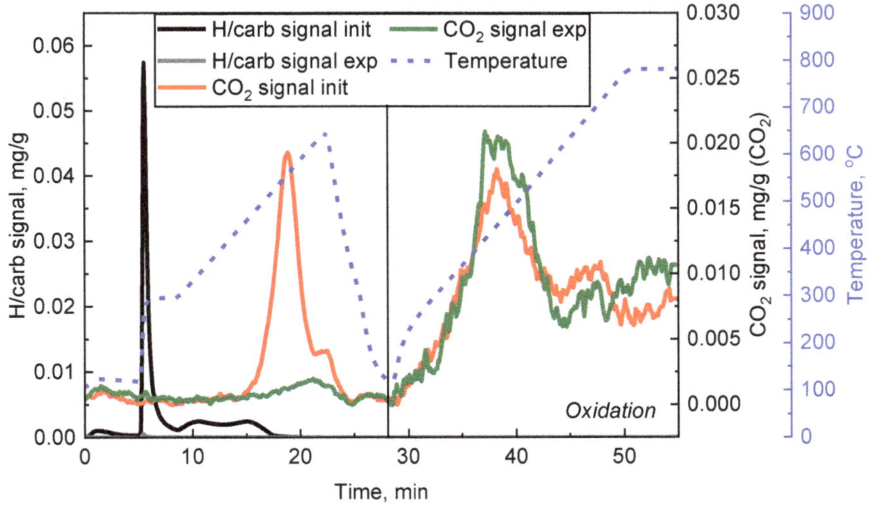

Figure 10. Pyrolytic spectra of core samples before (black and red curves) and after (grey and green curves) experiment. No. 7.

It is also seen from the pyrolysis data that the main part of the organic matter of the core decomposes in the temperature range of 135–295 °C. The curve obtained for hydrocarbons released from the core sample taken before the experiment has a peak with a maximum at a temperature of 264 °C (Figure 10, black curve). At the same time, the curve obtained for hydrocarbons from the core sample after heat treatment doesn't have this peak (Figure 10, grey curve). This fact confirms that the organic matter of the core is completely decomposed during the experiments at 450 °C.

The presence in the reactor of carbon dioxide released from the core leads to the underestimated hydrogen and other gas component fractions, compared with the equilibrium values calculated based on thermodynamic modeling and the methane conversion rate. Besides that, additional amounts of carbon dioxide in the reactor lead to a shift in the equilibrium of the main reactions of the SMR process (Equations (1) and (2)) towards the reactants, interfering with the hydrogen generation.

4.4. Effect of Steam to Methane Ratio on the CMC

Within the experiments, the amount of initially loaded water and the amount of water (steam) injected during the experiment were also changed to study the effect of the steam/methane ratio on the CMC. Based on the results of thermodynamic modeling (Figure 4b), it can be concluded that a higher methane conversion rate can be achieved with a higher amount of steam in the initial gas mixture (with an increase in the steam/methane ratio), keeping other parameters the same.

For example, the steam/methane ratio of ~10 achieved in experiment No. 3 as a result of additional water (steam) injections had a positive effect on the thermodynamic equilibrium. Additional water (steam) injections shifted the equilibrium of the main reactions (Equations (1) and (2)) towards the products and contributed to the achievement of a higher methane conversion rate, compared with experiments No. 1 and 2, in which the steam/methane ratios were 2 and 5, respectively. Additional water (steam) injections could also possibly lead to more active mixing of reagents.

A decrease in the steam/methane ratio from 21 (in experiment No. 6) to 15 (in experiment No. 7) led only to a slight decrease in the activity of the hydrogen generation processes. This conclusion can be made by comparing both experiments in terms of the calculated methane conversion rate obtained from the material balance equations for each of the components.

Considering the experimental results and thermodynamic calculations, we can conclude that the high value of the steam/methane ratio has a positive effect on the amount of synthesized hydrogen. However, providing a steam/methane ratio higher than 10 is not realistic (especially under the reservoir conditions). Nevertheless, these results confirm that the technology has great opportunities for applying in gas reservoirs with high water saturation.

4.5. Difference between Hydrogen Concentration and Methane Conversion Rate

It should be noted that the values of the gas component's fractions obtained in experiments are fair only for specific gas samples. It is better also to compare different static experiments in terms of the methane conversion rate calculated from the material balance equations for each of the components. In this case, the methane conversion rate calculations according to the detectable amounts of hydrogen should be performed. For example, the calculated value of the methane conversion rate for experiment No. 6 is 5.80%. This value corresponds to a constant equilibrium concentration of hydrogen in the reactor, about 18.8 mol.%. (based on the thermodynamic modeling of the SMR process). This value is lower than the actual observed value in some gas samples.

In experiment No. 7, product gases contained up to 6.97 mol.%. of hydrogen, and this value is much less than the value achieved in experiment No. 6 with a similar design. In turn, the methane conversion rate in experiment No. 7 is 3.71%, and this value is close to the methane conversion rate achieved in experiment No. 6. This value corresponds to the equilibrium hydrogen concentration in the product gas mixture of about 12.9 mol.%. It indicates an underestimated hydrogen concentration in gas samples (probably due to the large amounts of carbon dioxide released from the core) compared with the equilibrium value.

Thus, the results of further experiments devoted to studying the CMC process under the reservoir conditions should be compared in terms of the gas component concentrations and the calculated parameters of the methane conversion rate.

5. Conclusions

Within this study, the CMC in application to the gas reservoir for in situ hydrogen generation was investigated. A series of static experiments were then carried out in an autoclave reactor at temperatures of 300–450 °C and pressures of ~65–200 atm, representing gas field reservoir conditions under steam or overheated water injection. The effects of the heat treatment temperature, the form of catalyst used, the type of porous medium, and the steam/methane ratio on the CMC were studied. In particular, the CMC was implemented in the presence of core material, taken from the target gas field, with the initial fluid saturations representing reservoir conditions.

It was found that the CMC can be implemented under reservoir conditions in the presence of a crushed core of the target gas field at temperatures above 450 °C. At this temperature the ex situ prepared Ni-based catalyst should be used. The highest hydrogen concentration achieved during the current research was 53.5 mol.%. (in a separate gas sample), that corresponds to the methane conversion rate to hydrogen for the whole experiment of 5.8%.

Based on the obtained data and results, the following conclusions can be made:

1. The experimental results prove the activity of used Ni-based catalyst supported on Al_2O_3 substrate in the CMC. At the same time, the possibility of reducing the oxide phase of the catalyst with the formation of an active metal phase directly in the reactor during the experiment is confirmed;
2. The temperature of 350 °C is insufficient for realizing the CMC in the presence of the considered catalyst and porous media. The process becomes possible at a temperature of 450 °C, with the achievement of methane conversion rates of the order of 4–6%;
3. The packed model, which is a simulated reservoir rock, plays a key role in the process. It increases the catalytic surface area. It also includes transition metal atoms, which

can promote the main catalytic effect. Thus, the highest hydrogen concentrations were detected in experiments with crushed ceramics and crushed core models. In turn, the absence of a porous media negatively affected the hydrogen yield;

4. Based on the results of experiment No. 3, at a temperature of 450 °C, hydrogen generation from methane can occur both by the mechanism of the catalytic SMR and by the mechanism of catalytic methane cracking;
5. An ex situ prepared catalyst in an amount of 0.3 wt.% successfully catalyzed the CMC. The catalyst remained active during the whole experiment, even in the presence of relatively high amounts of hydrogen sulfide in the reactor (8.91 mol.%. in experiment No. 6);
6. The heat treatment of core material of the target gas field at the temperature of 450 °C leads to the decomposition of the mineral (carbonate) and organic matter with the release of additional amounts of carbon dioxide and light hydrocarbons, respectively;
7. An increase in the steam/methane ratio leads to a shift in the thermodynamic equilibrium of the component system towards the products and, consequently, to an increase in the amount and concentration of synthesized hydrogen. In this case, an increase in the steam/methane ratio above 10 is impractical.

The obtained results indicate the potential prospects of in situ hydrogen generation from the methane of depleted gas fields. This technology requires heating of the reservoir to 450 °C and above. It is possible due to the implementation of in situ combustion of hydrocarbons that saturate the reservoir (bitumen/oil or even natural gas). The outcomes of the dynamic high-temperature experiments showing the achievement of a methane conversion rate to hydrogen of about 40% (in the presence of crushed core material of target gas field) will be presented in following publications.

Author Contributions: Conceptualization, P.A., E.P., A.C. and R.B.; methodology, P.A., E.P., A.C. and R.B.; software, P.A.; validation, P.A., E.P., A.C. and R.B.; formal analysis, P.A.; investigation, P.A., E.P.; resources, E.M. and V.D.; data curation, P.A.; writing—original draft preparation, P.A.; writing—review and editing, E.P., A.C. and R.B.; visualization, P.A.; supervision, A.C., E.M. and V.D.; project administration, E.S., A.B., K.S. and O.S.; funding acquisition, E.M., E.S. and A.B. All authors have read and agreed to the published version of the manuscript.

Funding: This research received no external funding.

Acknowledgments: The authors would like to thank the Skolkovo Institute of Science and Technology and Integrated Center for Hydrocarbon Recovery for supporting and assisting this research. Authors thank Lukoil-Engineering LLC and Ritek LLC for providing the core material. The research was carried out in conjunction with Ritek LLC, which develops of advanced low-carbon energy technologies. The authors would like to show their gratefulness to Nikolay Taraskin, Sergey Buzov and Kirill Maerle for help in performing experiments and analyzing experimental data.

Conflicts of Interest: The authors declare no conflict of interest.

References

1. Hydrogen Council. Hydrogen Scaling Up. United Nations. 2017. Available online: https://hydrogencouncil.com/wp-content/uploads/2017/11/Hydrogen-scaling-up-Hydrogen-Council.pdf (accessed on 13 November 2017).
2. T-raissi, A. Hydrogen: Automotive fuel of the future. *IEEE Power Energy Mag.* **2004**, *2*, 40–45. [CrossRef]
3. Balcombe, P.; Speirs, J.; Johnson, E.; Martin, J.; Brandon, N.; Hawkes, A. The carbon credentials of hydrogen gas networks and supply chains. *Renew. Sustain. Energy Rev.* **2018**, *91*, 1077–1088. [CrossRef]
4. IEA. The future of Fuel: The Future of Hydrogen; France. 2019. Available online: https://iea.blob.core.windows.net/assets/9e3a3493-b9a6-4b7d-b499-7ca48e357561/The_Future_of_Hydrogen.pdf (accessed on 15 June 2019).
5. IEAGHG. Techno-Economic Evaluation of SMR Based Standalone (Merchant) Hydrogen Plant with CCS. UK. 2017. Available online: https://ieaghg.org/exco_docs/2017-02.pdf (accessed on 15 February 2017).
6. Hallam, R.J.; Hajdo, L.E.; Donnelly, J.K. Thermal Recovery of Bitumen at Wolf Lake. *SPE Reserv. Eng.* **1989**, *4*, 178–186. [CrossRef]
7. Hajdo, L.E.; Hallam, R.J.; Vorndran, L.D.L. Hydrogen Generation During In-Situ Combustion. In Proceedings of the SPE 1985 California Regional Meeting, Bakersfield, CA, USA, 27–29 March 1985; pp. 675–689. [CrossRef]

8. Kapadia, P.R.; Kallos, M.S.; Gates, I.D. A Comprehensive Kinetic Theory to Model Thermolysis, Aquathermolysis, Gasification, Combustion, and Oxidation of Athabasca Bitumen. In Proceedings of the SPE Improved Oil Recovery Symposium, Tulsa, OK, USA, 24–28 April 2010; pp. 1–31. [CrossRef]
9. Kapadia, P.R.; Kallos, M.S.; Leskiw, C.; Gates, I.D. Potential for Hydrogen Generation during In situ Combustion of Bitumen. In Proceedings of the SPE EUROPEC/EAGE Annual Conference and Exhibition, Amsterdam, The Netherlands, 8–11 June 2009; pp. 1–14. [CrossRef]
10. Kapadia, P.R.; Wang, J.J.; Kallos, M.S.; Gates, I.D. Practical process design for in situ gasification of bitumen. *Appl. Energy* **2013**, *107*, 281–296. [CrossRef]
11. Self, S.J.; Reddy, B.V.; Rosen, M.A. Review of underground coal gasification technologies and carbon capture. *Int. J. Energy Environ. Eng.* **2012**, *3*, 1–8. [CrossRef]
12. Cui, Y.; Liang, J.; Wang, Z.; Zhang, X.; Fan, C.; Wang, X.; Syngas, H.Á. Experimental forward and reverse in situ combustion gasification of lignite with production of hydrogen-rich syngas. *Int. J. Coal Sci. Technol.* **2014**, *1*, 70–80. [CrossRef]
13. Scott, E. Production of Hydrogen from Underground Coal Gasification. Patent WO 2008/033268 Al, 7 October 2008.
14. Surguchev, L.; Berenblym, R.; Dmitrievsky, A. Process for Generating Hydrogen. U.S. Patent 8763697 B2, 2014.
15. Gates, I.; Davidson, S. In-Situ Process to Produce Hydrogen from Underground Hydrocarbon Reservoirs. Patent WO 2017/136924 Al, 2017.
16. Surguchev, L.; Berenblyum, B. In-situ H_2 generation from hydrocarbons and CO_2 storage in the reservoir. In Proceedings of the Fourth EAGE CO_2 Geological Storage Workshop, Stavanger, Norway, 22–24 April 2014.
17. Zhang, Y.; Smith, K.J. Carbon formation thresholds and catalyst deactivation during CH4 decomposition on supported Co and Ni catalysts. *Catal. Lett.* **2004**, *95*, 7–12. [CrossRef]
18. Amin, A.; Epling, W.; Croiset, E. Reaction and Deactivation Rates of Methane Catalytic Cracking over Nickel. *Ind. Eng. Chem. Res.* **2011**, *50*, 12460–12470. [CrossRef]
19. Gazprom Export. Blue Fuel-Gazprom Export Global Newsletter. 48. 2018. Available online: http://www.gazpromexport.ru/files/BLUE_FUEL_48326.pdf (accessed on 18 December 2018).
20. Zhang, J.; Chen, Z. Chapter 9: Formation Damage by Thermal Methods Applied to Heavy Oil Reservoirs. In *Formation Damage during Improved Oil Recovery: Fundamentals and Applications*; Yuan, B., Wood, A.D., Eds.; Elsevier Inc.: Amsterdam, The Netherlands, 2018; ISBN 9780128137833. [CrossRef]
21. Ahmed, T.; Meehan, D.N. Chapter 6: Introduction to Enhanced Oil Recovery. In *Advanced Reservoir Management and Engineering*, 2nd ed.; Elsevier Inc.: Amsterdam, The Netherlands, 2012; pp. 541–585. ISBN 9780123855480. [CrossRef]
22. Espitalie, J.; Bordenave, M.L. Rock-Eval pyrolysis. In *Applied Petroleum Geochemistry*; Technip: Paris, France, 1993; pp. 237–361.
23. Espitalie, J.; Drouet, S.; Marquis, F. Petroleum evaluation by using the petroleum evaluation workstation (a Rock-Eval connected with computer). *Geol. Oil Gas* **1994**, *1*, 23–32.
24. Amrollahi Biyouki, A.; Hosseinpour, N.; Bahramian, A.; Vatani, A. In-situ upgrading of reservoir oils by in-situ preparation of NiO nanoparticles in thermal enhanced oil recovery processes. *Colloids Surf. A Physicochem. Eng. Asp.* **2017**, *520*, 289–300. [CrossRef]
25. Brockner, W.; Ehrhardt, C.; Gjikaj, M. Thermal decomposition of nickel nitrate hexahydrate, $Ni(NO_3)_2{\cdot}6H_2O$, in comparison to $Co(NO_3)_2{\cdot}6H_2O$ and $Ca(NO_3)_2{\cdot}4H_2O$. *Thermochim. Acta* **2007**, *456*, 64–68. [CrossRef]
26. Małecka, B.; Łącz, A.; Drozdz, E.; Małecki, A. Thermal decomposition of d-metal nitrates supported on alumina. *J. Therm. Anal. Calorim.* **2015**, *119*, 1053–1061. [CrossRef]
27. Rashidi, H.; Ebrahim, H.A.; Dabir, B. Reduction kinetics of nickel oxide by methane as reducing agent based on thermogravimetry. *Thermochim. Acta* **2013**, *561*, 41–48. [CrossRef]
28. Kharatyan, S.L.; Chatilyan, H.A.; Manukyan, K.V. Kinetics and Mechanism of Nickel Oxide Reduction by Methane. *J. Phys. Chem. C* **2019**, *123*, 21513–21521. [CrossRef]
29. Rodriguez, J.A.; Hanson, J.C.; Frenkel, A.I.; Kim, J.Y.; Pérez, M. Experimental and theoretical studies on the reaction of H2 with NiO: Role of O vacancies and mechanism for oxide reduction. *J. Am. Chem. Soc.* **2002**, *124*, 346–354. [CrossRef] [PubMed]
30. Jeangros, Q.; Hansen, T.W.; Wagner, J.B.; Damsgaard, C.D.; Dunin-Borkowski, R.E.; Hébert, C.; Van Herle, J.; Hessler-Wyser, A. Reduction of nickel oxide particles by hydrogen studied in an environmental TEM. *J. Mater. Sci.* **2013**, *48*, 2893–2907. [CrossRef]
31. Hou, K.; Hughes, R. The kinetics of methane steam reforming over a Ni/α-Al2O catalyst. *Chem. Eng. J.* **2001**, *82*, 311–328. [CrossRef]
32. Xu, J.; Froment, G.F. Methane Steam Reforming, Methanation and Water-Gas Shift: 1. Intrinsic Kinetics. *AIChE J.* **1989**, *35*, 88–96. [CrossRef]
33. Glushko, V.P.; Gurvich, L.V.; Weitz, I.V.; Medvedev, V.A.; Hachkuruzov, G.A.; Jungmann, V.S.; Bergman, G.F.; Baibuz, V.F.; Iorish, V.S. *Thermodynamic Properties of Substances in 6 Volumes*; Nauka: Moscow, Russia, 1979.
34. Zhavoronkov, N.M.; Kisil, I.M.; Olevskiy, V.M.; Kharlamov, V.V. *Nitrogenman's Handbook*, 2nd ed.; Khimia: Moscow, Russia, 1986.
35. Matsumura, Y.; Nakamori, T. Steam reforming of methane over nickel catalysts at low reaction temperature. *Appl. Catal. A Gen.* **2004**, *258*, 107–114. [CrossRef]
36. Kho, E.T.; Scott, J.; Amal, R. Ni/TiO_2 for low temperature steam reforming of methane. *Chem. Eng. Sci.* **2016**, *140*, 161–170. [CrossRef]

37. Dan, M.; Mihet, M.; Biris, A.R.; Marginean, P.; Almasan, V.; Borodi, G.; Watanabe, F.; Biris, A.S.; Lazar, M.D. Supported nickel catalysts for low temperature methane steam reforming: Comparison between metal additives and support modification. *React. Kinet. Mech. Catal.* **2012**, *105*, 173–193. [CrossRef]
38. Lai, G.H.; Lak, J.H.; Tsai, D.H. Hydrogen Production via Low-Temperature Steam-Methane Reforming Using Ni-CeO_2-Al_2O_3 Hybrid Nanoparticle Clusters as Catalysts. *ACS Appl. Energy Mater.* **2019**. [CrossRef]
39. Khzouz, M.; Gkanas, E.I. Experimental and numerical study of low temperature methane steam reforming for hydrogen production. *Catalysts* **2018**, *8*, 5. [CrossRef]
40. Rostrup-Nielsen, J.R. Sulfur poisoning. In *Progress in Catalyst Deactivation. NATO Advanced Study Institutes Series*; Figueiredo, J.L., Ed.; Martinus Nijhoff Publisher: Leiden, The Netherlands, 1982; pp. 209–227. [CrossRef]
41. Luo, Y.H.; Zhu, D.Q.; Pan, J.; Zhou, X.L. Thermal decomposition behaviour and kinetics of Xinjiang siderite ore. *Trans. Inst. Min. Metall. Sect. C Miner. Process. Extr. Metall.* **2016**, *125*, 17–25. [CrossRef]
42. Warne, S.S.J.; French, D.H. The decomposition of anhydrous carbonate minerals in coal and oil shale ashes produced at temperatures of 400 and 575 C. *Thermochim. Acta* **1984**, *75*, 139–149. [CrossRef]
43. Giammaria, G.; Lefferts, L. Catalytic effect of water on calcium carbonate decomposition. *J. CO_2 Util.* **2019**, *33*, 341–356. [CrossRef]

energies

Article

A Highly Selective Novel Green Cation Exchange Membrane Doped with Ceramic Nanotubes Material for Direct Methanol Fuel Cells

Marwa H. Gouda , Tamer M. Tamer * and Mohamed S. Mohy Eldin *

Polymer Materials Research Department, Advanced Technology and New Materials Research Institute (ATNMRI), City of Scientific Research and Technological Applications (SRTA-City), Alexandria 21934, Egypt; mgouda@srtacity.sci.eg
* Correspondence: tmahmoud@srtacity.sci.eg (T.M.T.); mmohyeldin@srtacity.sci.eg (M.S.M.E.)

Abstract: Herein, a pair of inexpensive and eco-friendly polymers were blended and formulated based on poly (ethylene oxide) (PEO) and poly (vinyl alcohol) (PVA). FTIR, XRD, EDX and TEM techniques were used to describe a Phosphated titanium oxide (PO_4TiO_2) nanotube synthesised using a straightforward impregnation-calcination procedure. For the first time, the produced nanoparticles were inserted as a doping agent into this polymeric matrix at a concentration of (1–3) wt.%. FTIR, TGA, DSC and XRD were used to identify the formed composite membranes. Furthermore, because there are more hydrogen bonds generated between the polymer's functional groups and oxygen functional groups PO_4TiO_2, oxidative stability and tensile strength are improved with increasing doping addition and obtain better results than Nafion117. The permeability of methanol reduced as the weight % of PO_4TiO_2 increased. In addition, the ionic conductivity of the membrane with 3 wt.% PO_4-TiO_2 is raised to (28 mS cm^{-1}). The optimised membrane (PVA/PEO/PO_4TiO_2-3) had a higher selectivity (6.66×10^5 S cm^{-3} s) than Nafion117 (0.24×10^5 S cm^{-3} s) and can be used as a proton exchange membrane in the development of green and low-cost DMFCs.

Keywords: proton exchange membrane; poly (vinyl alcohol); poly (ethylene oxide); titanium oxide; direct methanol fuel cell; fuel cell

Citation: Gouda, M.H.; Tamer, T.M.; Mohy Eldin, M.S. A Highly Selective Novel Green Cation Exchange Membrane Doped with Ceramic Nanotubes Material for Direct Methanol Fuel Cells. *Energies* **2021**, *14*, 5664. https://doi.org/10.3390/en14185664

Academic Editor: Bahman Shabani

Received: 20 May 2021
Accepted: 3 September 2021
Published: 9 September 2021

1. Introduction

Chemical energy is instantly converted into electrical energy by the fuel cell. It is a sort of power-producing equipment that can efficiently convert and store energy. Hydrocarbons such as methanol or ethanol can be used as fuel in those cells. It produces zero emissions or minimum pollution [1]. As a type of proton exchange membrane fuel cell (PEMFC), the direct methanol fuel cell (DMFC) is widely utilised in home appliances, vehicles, aerospace and other fields. [2].

A membrane separates the fuel and oxidant compartments in a fuel cell, allowing for efficient ion transport and charge balance. Due to its chemical stability, mechanical properties and ionic conductivity, the Nafion membranes are the most perfluorinated PEMs utilised in DMFCs [1,3]. However, nafion membrane manufacture is expensive and time-consuming, which limits its commercialisation [4,5]. As a result, replacing them with ecologically benign and cost-efficient polymeric films is crucial and essential [6–8].

To replace Nafion membranes, sulfonation or blending of polymers [9] and/or doping agents, such as porous and functionalised inorganic materials and functionalised carbon materials, are inserted into the polymeric matrix [10–15]. The most prevalent non-perfluorinated polymers utilised to build novel alternative polymeric membranes are poly(styrene) (PS), poly (ether ether ketone) (PEEK), poly(benzimidazole) (PBI) and poly (arylene ether sulfone) (PSU). However, the use of toxic chemical time, solvents and temperature in the preparing of these non-degradable polymers makes membrane synthesis expensive, complicated and environmentally unfriendly. Therefore, using biodegradable, cheap and green polymers such as polyethene oxide (PEO) and polyvinyl alcohol (PVA) is a more appealing strategy from an economic and technological standpoint than inventing innovative complicated polymers or adapting existing commercial membranes [10,16–18]. In addition, the catalysts and film are critical components of a DMFC. As a result, building a cost-effective membrane brings DMFC systems closer to widespread use.

In addition to its chemical stability, hydrophilicity, adhesive properties and film-forming abilities, PVA is also environmentally friendly and low cost [19–21]. Polyvinyl alcohol is therefore commonly employed in medicinal, commercial and industrial settings. Polyvinyl alcohol's proton conductivity and, as a result, its stiff and semi-crystalline structure hampers its use as a proton exchange membrane in fuel cells. As a result, adding doping agents or mixing with another polymer electrolyte to correct this flaw is a viable option [19,22,23]. Since hydrogen connections develop between the -OH groups of PVA and the ether linkage of polyethene oxide, blending PVA with PEO is preferred [20,24]. On the other hand, PEO is an environmentally acceptable polymer that is utilised to synthesise polymer electrolyte systems in various energy devices due to its improved ionic conductivity, low toxicity and flexibility [25,26].

To increase membrane properties, many researchers adopted the conventional practise of incorporating doping compounds into polymer matrix to create nanocomposite barrier membranes [27–31]. Due to its huge surface area, mechanical toughness, chemical resistance, barrier to fuel crossing, cheap price and low level of toxicity, phosphated titania (PO_4TiO_2) in a polymer matrix has been studied for use in fuel cell applications [19,20]. PO_4TiO_2 also includes oxygen-containing hydrophilic functional groups, which enhance water sorption and produce proton conduction channels [20]. When PO_4TiO_2 nanotubes are embedded into polymers, the hydrogen bonds will be generated between hydroxyl groups along the polymer backbone and oxygen groups of PO_4TiO_2. These hydrogen bonds will reflect on the membranes' mechanical properties, strengthening them and limiting extreme swelling and water sorption [20,26], enhancing the ionic conductivity of formulated membranes containing PO_4TiO_2 nanotubes.

This project aims to develop innovative nanocomposite membranes constructed from mild processing of environmentally safe and economic polymers compatible with water as the principal solvent to further DMFC commercialisation. Due to its exceptional capacity to build films with PEO polymer, polyvinyl alcohol was selected as the key polymer for the membrane. The polymers were crosslinked completely and concurrently converted to sulfonated PVA using crosslinkers such as 4-sulfophithalic acid (SPA) and glutaraldehyde (GA). SPVA/PEO/PO_4TiO_2 nanotubes were synthesised and injected as a doping agent in the PVA matrix at various ratio to create new nanocomposite membranes. The parameters such as oxidative chemical stability, proton conductivity, mechanical resistance and restriction of the methanol permeability will be controlled due to the formation of hydrogen bond of formulated matrix and oxygen groups of PO_4-TiO_2, which could be improved DMFC performance employing such membranes.

2. Materials and Methods

PEO (MW: 900,000 g mol^{-1}, Acros Organics) and PVA (99% hydrolysis and medium MW, USA). Glutaraldehyde (GA) (Alfa Aesar, 25 wt.% in H_2O) and 4-sulphophthalic acid (SPA) (Sigma-Aldrich, 99.9 wt.% in H_2O) were used as covalent and ionic cross-linkers, respectively [30]. Titanium (IV) oxide rutile (TiO_2, <5 μm, ≥99.9%, Sigma-Aldrich, Darmstadt, Germany) and H_3PO_4 (Fisher Chemical, 85 wt.%)

2.1. Synthesis

2.1.1. Synthesis of Phosphated Titanium Oxide Nanotube (PO_4-TiO_2)

TiO_2 nanotubes were synthesised as mentioned in the previous work [19]. TiO_2 nanotubes were mixed to 0.1 mol/L^{-1} phosphoric acid in a molar ratio 1:1 and the suspension was shaken in hot water (80 °C). The mixture was rinsed with H_2O and dried overnight at 110 °C. The powder was then burned in a muffle furnace at 450 degrees Celsius.

2.1.2. Preparation of SPVA/PEO/ PO_4TiO_2 Membranes

Here, 100 mL of PVA solution (10%) was prepared, and PEO (2 g) was dissolved in 100 mL deionizedH_2O: ethanol (80:20) vol percent at 50 °C for 1 h, before blending PVA: PEO (85:15) wt. percent and covalent crosslinking the polymers blend with glutaraldehyde (0.5 g, 50 wt. percent). The inorganic-organic nanocomposite (structure illustrated in Figure 1) was then made by incorporating varying concentrations of PO_4-TiO_2 nanotubes (1, 2, 3 wt. percent relative to PVA) in the polymeric mix, and they were given the names PVA/PEO, PVA/PEO/PO_4TiO_2-1, PVA/PEO/PO_4TiO_2-2 and PVA/PEO/PO_4TiO_2-3 accordingly.

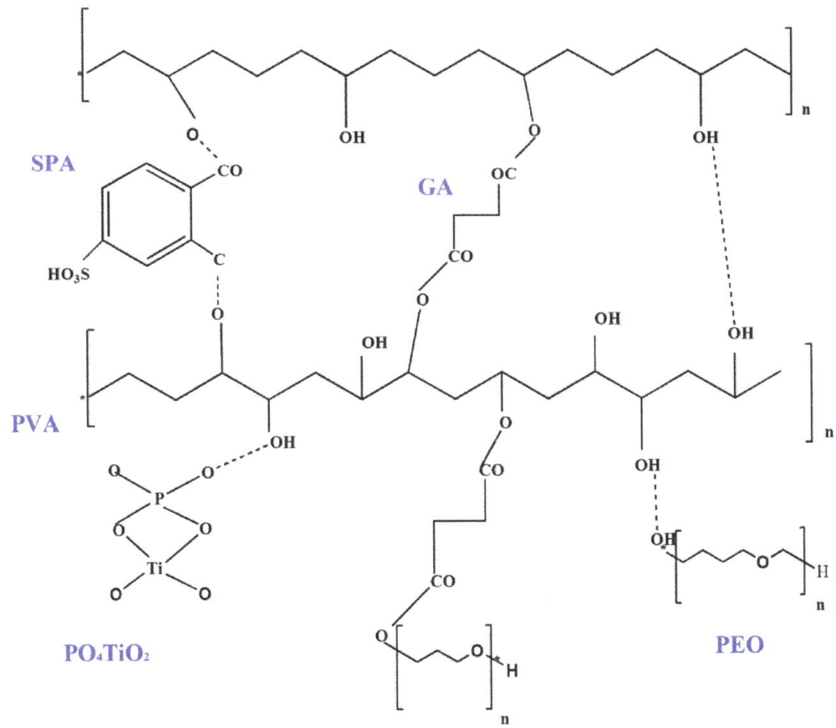

Figure 1. Probable structure of the SPVA/ PEO/ PO_4TiO_2membrane.

2.2. Characterisation

A Fourier transform infrared spectrophotometer (Shimadzu FTIR-8400 S- Japan) was used to monitor the functional groups of PO_4TiO_2 nanotubes and composite membranes, while an X-ray diffractometer was used to analyze the structures (Schi-madzu7000-Japan). A thermo-gravimetric analyser (Shimadzu TGA-50, Tokyo, Japan) was used to track SPVA/PEO/PO_4TiO_2 membranes; the temperature range was 25–800 °C, with the heating rate was 10 °C/min under nitrogen environment. The membranes were also evaluated using differential scanning calorimetry (DSC) (Shimadzu DSC-60, Japan) at temperatures ranging from 25 to 300 °C. The SPVA/PEO/PO_4-TiO_2-1 membrane's morphological structure was revealed using a scanning electron microscope (SEM). Transmission electron microscopy (TEM, JEM 2100 electron microscope) and energy-dispersive X-ray (EDX) were used to visualise the PO_4-TiO_2 nanotube (Joel Jsm 6360LA-Japan).

The hydrophilicity of membranes was determined by measuring the contact angles between membrane surfaces and water drops. a Rame-Hart Instrument Co. model 500-FI contact-angle analyser was used to analyse the measurements. To determine swelling ratio (SR) and water uptake, a certain weight of membrane with actual dimensions was soaked in deionised water for 24 h then gently dried on tissue paper to remove surface water before analysis again. Finally, the composite membranes' SR and WU were calculated using Equations (1) and (2).

$$SR(\%) = \frac{L_{wet} - L_{dry}}{L_{dry}} \times 100 \quad (1)$$

$$WU(\%) = \frac{W_{wet} - W_{dry}}{W_{dry}} \times 100 \quad (2)$$

where L_{dry} and L_{wet} denote the length of dry and wet of tested membranes, respectively, and W_{dry} and W_{wet} denote the weight of dry and wet tested sample.

The nanocomposite membranes' ion exchange capacity (IEC) was estimated by acid-base titration [32]. The weighted membranes were submerged in a 50 cm^3 2M NaCl solution for two days before titrating with a 0.01 N NaOH solution. The IEC was calculated using Equation (3) below:

$$IEC\left(\frac{meq}{g}\right) = \frac{V_{NaOH} \times C_{NaOH}}{W_d} \quad (3)$$

The volume of sodium hydroxide consumed in titration, the concentration of sodium hydroxide solution, and the dry sample weight, respectively, are represented by V_{NaOH}, C_{NaOH} and W_d.

To investigate the proton conductivity of formulated films, the electrochemical impedance spectroscopy (EIS) will be utilised using PAR 273A potentiostat (Princeton Applied Research, Inc., Oak Ridge, TN, USA) and a SI 1255 HF frequency response analyser (FRA, Schlumberger Solartron). according to the published method in the literature with modification [1]. the ionic conductivity of the membranes was estimated using Equation (4),

$$\sigma = \frac{d}{RA} \quad (4)$$

where σ (S cm^{-1}) is the membrane's ionic conductivity, R (Ω) is its resistance, A (cm^2) is its area and d (cm) is its thickness.

To estimate the methanol permeability, The tested membrane was seated within two vessels in a glass diffusion chamber to assess its methanol permeability. The receptor vessel (B) was charged with water, while the other vessel (A) was filled with 2 M methanol [29]. the crossing of methanol through membrane as a function of time was calculated according to Equation (5),

$$C_B(t) = \frac{A}{V_B}\frac{P}{L}C_A(t - t_0) \quad (5)$$

where A (cm^2) is the active membrane area, VB (cm^3) is the capacity of the receptor vessel, L (cm) is the crosssection film thickness, CB and CA (mol L^{-1}) are the concentrations of methanol in vessels B and A, respectively, and the period (t–t_0) is the time of the methanol crossover (cm^2 s^{-1}). The selectivity of the membranes (the ratio of ionic conductivity to methanol permeability) was determined since it can provide vital information about the fuel cell's performance.

The oxidative stability of tested membranes was measured gravimetrically as a function of membrane weight soaked in oxidative solution [Fenton's reagent (3 wt.% H_2O_2 containing 2 ppm $FeSO_4$)] at 68 °C for 24 h [19].

The dry nano-composite membranes were put through a tensile strength test at room temperature until they broke, using Lloyd Instruments LR10k [32].

3. Results

3.1. Characterisation of PO_4-TiO_2 Nanotube and Nanocomposite Membranes

The FT-IR spectra of prepared nanoparticle TiO_2 and PO_4-TiO_2 were presented separately in Figure 2. For TiO_2 nanoparticles, Ti-O bonds are responsible for the bands at 715 cm^{-1} and 1025 cm^{-1}. the bands at 1622 cm^{-1} refer to the bending vibration of the Ti-OH band. The band at 3387 cm^{-1} are assigned to O-H stretching vibration bonds due to moisture adsorption on the material's surface [20,33]. For the chart of PO_4-TiO_2 particles, the band at 690 cm^{-1} corresponds to the stretching of the Ti-O bond. The bands at 890, 1085 and 1270 cm^{-1} are referred to as P-O bonds vibration. The band located at 1425 cm^{-1} is attributed to the stretching vibration band of the P=O bond. The O-H bonds from H_2O molecules adsorption are proofed by the bands at 1630 and 3117 cm^{-1}. The band located at 2374 cm^{-1} is related to the presence of CO_2 [34,35].

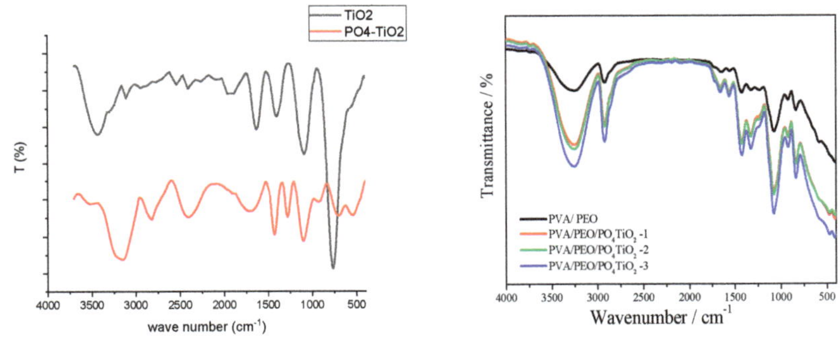

Figure 2. FTIR spectra of PO4TiO$_2$ (**left chart**), PVA/PEO/PO4TiO$_2$ membranes (**right chart**).

For the membranes, the Figure 2 shows that the bands around 3250 cm^{-1} refer to the characteristic stretching vibration band of hydroxyl groups on PVA and PEO. the bands at 1650 cm^{-1} are attributed to the bending vibration O-H bonds. The band at 1112 cm^{-1} is the characteristic band of PEO [36]. Bands at 2840 cm^{-1} can be assigned to the vibration of methylene C-H bonds in the polymer's structure. The characteristic peak for sulfate groups of sulfophithalic acid (SPA) was cited at 900 cm^{-1}, while the small bands at 1700 cm^{-1} indicate C=O bonds of the sulfophithalic acid (SPA), which confirms the crosslinking process. The band at 1100 cm^{-1} is assigned to P-O bonds of phosphate titanium oxide.

In Figure 3, show the XRD pattern of TiO_2 and PO_4TiO_2 (on the left side) and composite membranes on the right side. The constructed membranes' amorphous shape shows good ion conduction [37], while the titanium dioxide rutile characteristic peaks intensity at two angles of 28, 36, 41 and 54 [38]. This is because the phosphate entering the titanium oxide lattice changed its original crystalline phase due to the different synthesis processes for PO_4TiO_2. Therefore, the intensity of the sharp peak of the original titanium oxide at 28° is disappeared in the diffractogram of phosphate titanium oxide. In comparison, the ridge at 54° of the TiO_2 is absent in the diffractograms of PO_4TiO_2.

Morphological analysis of membranes was studied using SEM and presented in Figure 4. Figure 4a, b demonstrates SEM images of membranes that show a smooth surface with no defects for the undoped crosslinked membrane. At the same time, particles of phosphate titanium oxide tubes appeared in the doped membrane, which was further confirmed by EDX spectra as shown in Figure 4e. However, the SEM image in Figure 4c illustrate the porous structure of the cross-sectional of the doped membrane. Consequently, these voids lead to an improvement in the ionic conductivity of the films [39]. While TEM image of phosphate titanium oxides shown in Figure 4e proofed the forming of nanotubes shape with nanoscale size as illustrated in Figure 4f.

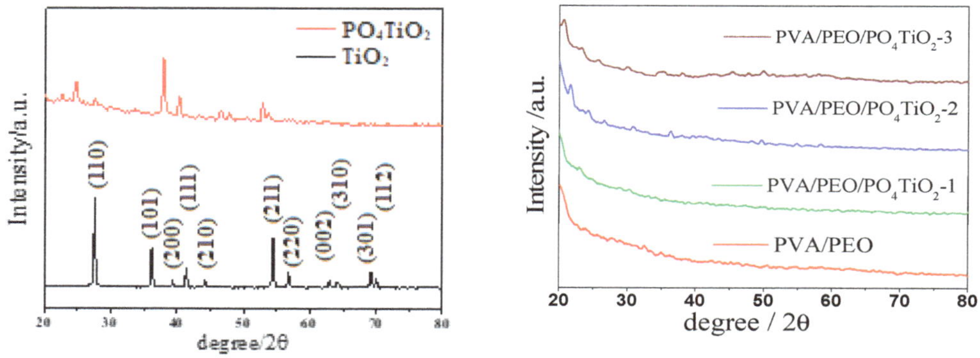

Figure 3. XRD patterns of PO_4TiO_2 (**left chart**) and PVA/PEO/PO4TiO_2 membranes (**right chart**).

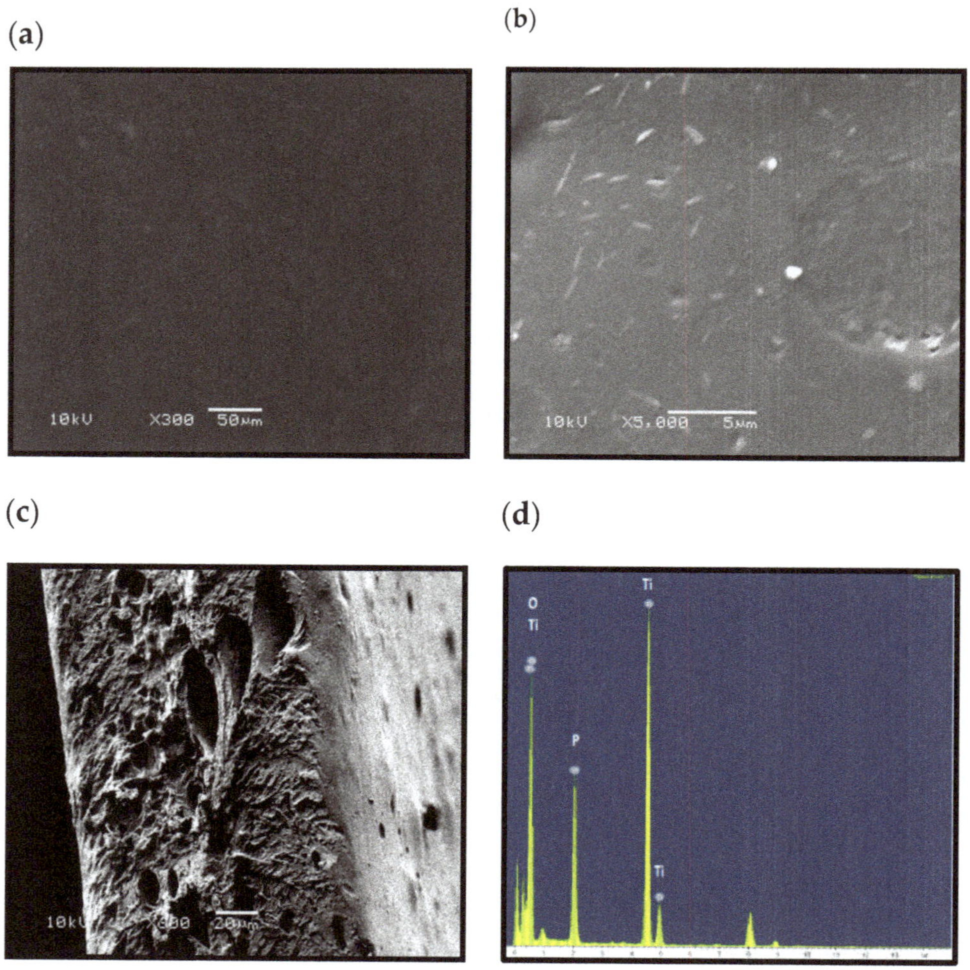

Figure 4. *Cont.*

(e) **(f)**

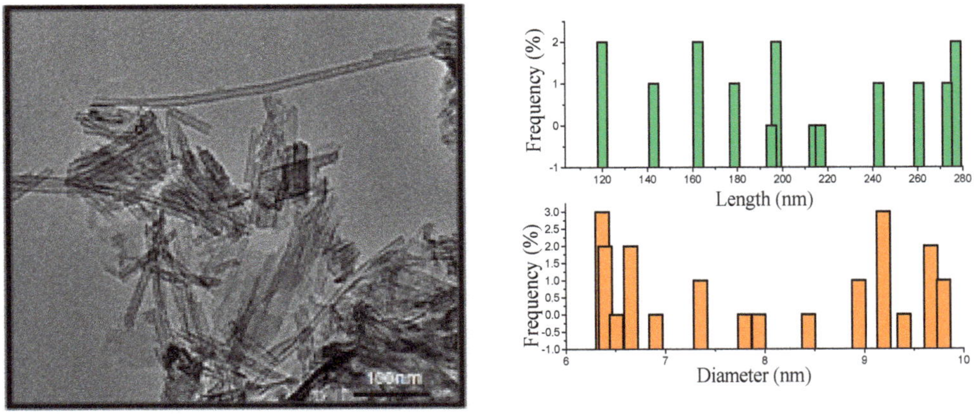

Figure 4. SEM images for (**a**) undoped membrane, (**b**,**c**) doped membrane (PVA/PEO/PO_4TiO_2-1) surface and cross-section (**d**) EDX analysis for PO_4TiO_2, (**e**) TEM image for PO_4TiO_2nanotubes and (**f**) the frequency distribution plot of PO_4TiO_2nanotubes size from TEM image.

3.2. Mechanical and Thermal Properties

The addition of TiO_2 or functionalisation of TiO_2 develops the mechanical tensile characteristics of the polymeric form significantly [19–21]. For example, as displayed in Table 1, increasing the amount of PO_4TiO_2 in the polymeric film enhanced the tensile strength of the composite films by improving their compatibility. This behaviour can be explained by improving the interaction between functional groups along two polymer backbones, such as ether linkages, hydroxyl groups and the various phosphate groups of PO_4TiO_2 nanoparticles, via ionic, covalent and hydrogen interactions interfacial adhesion, as compared to the neat membrane.

Table 1. Physicochemical parameters of the formulated composite membranes compare to Nafion 117 [1,24].

Membrane	Thickness (μm)	WU (%)	SR (%)	Contact Angle (°)	Tensile Strength (MPa)	Oxidative Stability (RW, %) *
SPVA/PEO	130	95	90	65.36	15.5	90
SPVA/PEO/PO_4TiO_2-1	150	40	42	67.23	24.9	94
SPVA/PEO/PO_4TiO_2-2	175	22	13	70.36	32.5	98
SPVA/PEO/PO_4TiO_2-3	184	16	10	72.30	40.3	99
Nafion 117	170	9.5	13	102	25	92

* The retained weight of membranes (RW) after immersion for a day in Fenton's reagent.

The TGA of formulated composite films in the presence or absence of PO_4TiO_2 nanoparticles is shown in Figure 5. Moisture evaporation in all membranes can be defined as the initial weight loss of all manufactured membranes at 150 °C (10%) [40]. The following weight loss of composite membranes was demonstrated between (150–300) °C range, possibly due to the breakdown of functional groups [41,42]. Finally, from 300 to 580 °C, all samples show a significant decomposition, which could be connected to polymeric chain decomposition [43], which began at 250 °C for the undoped membrane and began at 310 °C with a lower weight % for the doped membranes, with 3 wt percent doping. According to these findings, the addition of PO_4TiO_2 to composite membranes increases their thermal stability by increasing hydrogen bonding in the composite. Furthermore, the presence of only one endothermic peak in DSC, as shown in Figure 5, demonstrates flawless membrane miscibility, and the removal of this peak at PO_4TiO_2 (3 wt.%) may be attributed to constructing new physical bonds (i.e., hydrogen bonds) between the nanoparticles and the polymeric matrix [29]. When a result, as the concentration of the doping agent increased, the melting temperature of the membranes fell. This behaviour may be described by hydrogen bond interactions that partially degrade membrane crystallinity, lowering the melting point and increasing ionic conductivity [29].

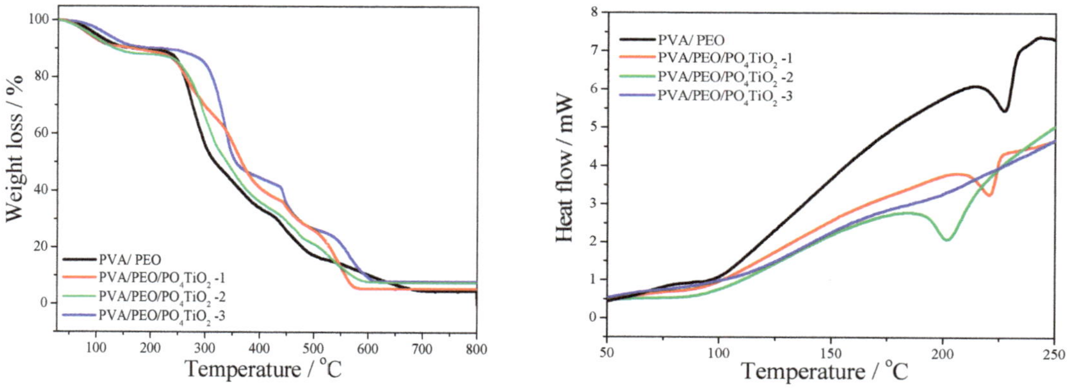

Figure 5. TGA of PVA/PEO/PO_4TiO_2 membranes (**left chart**) and DSC of PVA/PEO/PO_4TiO_2 membranes (**right chart**) curves of nanocomposite membranes.

Table 1 depicts the behaviour of nanocomposite membranes in contact with deionised water. When contact angles are less than 90 degrees, membrane surfaces are deemed hydrophobic, and when they are greater than 90 degrees, they are considered hydrophilic. However, as the doping agent content increases, the composite membranes become less hydrophilic and have a lower hydrophilic quality [26,44]. When the amount of PO_4TiO_2 in the polymeric blend was increased from 1% to 3%, the swelling ratio and water sorption of the composite membranes were lowered, which is vital because water overload may be avoided [45]. To put it another way, increasing the doping agent in the membrane matrix makes the structure more compact, reducing water overload in the polymeric matrix channels [46,47].

3.3. Oxidative Stability

The chemical stability parameters of the formulated nanocomposite membranes were described in Table 1. The SPVA/PEO membrane has the lowest stability against the oxidation condition; however, adding PO_4TiO_2 as a dopant improves polymeric composite protection against OOH and OH radical. The PVA/PEO/PO_4TiO_2-3 membrane was the most chemically stable, with weight retention of nearly 100%, suggesting that adding a doping agent such as TiO_2 or functionalising TiO_2 improves the oxidation chemical stability of formulated membranes [20,48].

3.4. Ionic Conductivity, IEC and Methanol Crossover

As the composite membrane contains further acidic exchangeable groups from PO_4TiO_2, the IEC values increase as PO_4TiO_2 in the composite membranes increases. This is due to the acidic (phosphate) sites of PO_4TiO_2 increasing the charges in the membranes, which promotes ionic conduction [19,20]. This is due to the SPVA/PEO/ PO_4TiO_2- 3 membrane's superior ionic conductivity (28 mS cm^{-1}) when compared to an undoped membrane (12 mS cm^{-1}). Adding PO_4TiO_2 to the polymeric matrix avoids methanol crossing when it comes to the fuel permeability of composite membranes. The undoped polymeric membrane exhibited a methanol permeability of 4.5×10^{-7} cm^2 s^{-1}, but the SPVA/PEO/PO_4TiO_2-3 membrane permeability of 0.42×10^{-7} cm^2 s^{-1} when PO_4TiO_2 was added to the membrane matrix, as indicated in Table 2. The capacity of the doping agent to restrict the polymeric matrix channels, reducing water uptake and hence fuel permeability, may be the cause of the membrane containing the doping agent's decreased methanol permeability [19,20,49,50]. When compared to undoped SPVA/PEO membrane (0.26×10^5 S cm^{-3} s) and Nafion 117 (0.24×10^5 S cm^{-3} s), SPVA/PEO/ PO_4TiO_2-3 (6.66×10^5 S cm^{-3} s) showed higher selectivity, indicating that the nanocomposite membranes produced are suitable for use in DMFCs [49].

Table 2. Ionic conductivity, methanol permeability, IEC and selectivity of the fabricated membranes and Nafion 117 [1].

Membrane	IEC (meq g^{-1})	Ionic Conductivity (mS cm^{-1})	Methaanol Permeability (10^{-7} cm^2 s^{-1})	Selectivity (10^5 S cm^{-3} s)
SPVA/PEO	0.20	12	4.5	0.26
SPVA/PEO/PO_4TiO_2-1	0.35	17.7	2.10	0.84
SPVA/PEO/PO_4TiO_2-2	0.45	20.5	1.51	1.35
SPVA/PEO/PO_4TiO_2-3	0.60	28	0.42	6.66
Nafion 117	0.89	34.0	14.1	0.24

4. Conclusions

Using eco-friendly and readily available polymers, a simple blending and solution casting approach created a more economical nanocomposite membrane. Furthermore, incorporating PO_4TiO_2 nanotubes into the polymeric blend improves the membrane's physicochemical parameters, such as ionic conductivity, mechanical properties, oxidative stability, reducing water sorption and limiting methanol permeability, especially in the composite membrane with 3 percent PO_4TiO_2. that also demonstrate the most suitable oxidative chemical stability and methanol crossover limiting. Finally, the manufactured membrane with the best characteristics (PVA/PEO/PO_4TiO_2-3) could be used as a cation exchange composite membrane to construct environmentally friendly and low-cost DMFCs.

Author Contributions: M.H.G. performed the experiments; M.H.G. and M.S.M.E. analysed the data; M.H.G., T.M.T. and M.S.M.E. wrote the paper. All authors have read and agreed to the published version of the manuscript.

Funding: This research received no external funding.

Institutional Review Board Statement: Not applicable.

Informed Consent Statement: Not applicable.

Data Availability Statement: Not applicable.

Acknowledgments: The authors would like to express their grateful thanks to the City of Scientific Research and Technological Applications (SRTA-City) for its support in preparation and characterization and evaluation.

Conflicts of Interest: The authors declare no conflict of interest.

References

1. Yuan, C.; Wang, Y. Synthesis and characterization of a crosslinked membrane based on sulfonated poly(aryl ether sulfone) and sulfonated polyvinyl alcohol applied in direct methanol fuel cells. *J. Polym. Res.* **2020**, *27*, 329. [CrossRef]
2. Beydaghi, H.; Javanbakht, M.; Salarizadeh, P.; Bagheri, A.; Amoozadeh, A. Novel proton exchange membrane nanocompo-sites based on sulfonated tungsten trioxide for application in direct methanol fuel cells. *Polymers* **2017**, *119*, 253–262. [CrossRef]
3. Zhiwei, W.; Hao, Z.; Qiang, C.; Sumei, Z.; Feng, Y.; Jian, K.; Jinyao, C.; Ya, C.; Ming, X. Preparation and characterization of PVA proton exchange membranes containing phosphonic acid groups for direct methanol fuel cell applications. *J. Polym. Res.* **2019**, *26*, 200. [CrossRef]
4. Pandey, R.P.; Shukla, G.; Manohar, M.; Shahi, V.K. Graphene oxide based nanohybrid proton exchange membranes for fuel cell applications: An overview. *Adv. Colloid Interface Sci.* **2017**, *240*, 15–30. [CrossRef] [PubMed]
5. Ye, Y.-S.; Rick, J.; Hwang, B.-J. Water Soluble Polymers as Proton Exchange Membranes for Fuel Cells. *Polymers* **2012**, *4*, 913–963. [CrossRef]
6. Ma, J.; Choudhury, N.A.; Sahai, Y. A comprehensive review of direct borohydride fuel cells. *Renew. Sustain. Energy Rev.* **2010**, *14*, 183–199. [CrossRef]

7. Merino-Jiménez, I.; León, C.P.; Shah, A.A.; Walsh, F.C. Developments in direct borohydride fuel cells and remaining chal-lenges. *J. Power Sources* **2012**, *219*, 339–357. [CrossRef]
8. Gouda, M.H.; Elnouby, M.; Aziz, A.N.; Youssef, M.E.; Santos, D.M.F.; Elessawy, N.A. Green and Low-Cost Membrane Electrode Assembly for Proton Exchange Membrane Fuel Cells: Effect of Double-Layer Electrodes and Gas Diffusion Layer. *Front. Mater.* **2020**, *6*, 337. [CrossRef]
9. Mohy Eldin, M.S.; Hashem, A.E.; Tamer, T.M.; Omer, A.M.; Yossuf, M.E.; Sabet, M.M. Development of cross-linked chi-tosan/alginate polyelectrolyte proton exchanger membranes for fuel cell applications. *Int. J. Electrochem. Sci.* **2017**, *12*, 53840–53858.
10. Tamer, T.M.; Omer, A.M.; Sabet, M.M.; Youssef, M.E.; Hashem, A.I.; Mohy Eldin, M.S. Development of polyelectrolyte sulfonated chitosan-alginate as an alternative methanol fuel cell membrane. *Desalin. Water Treat.* **2021**, *227*, 132–148. [CrossRef]
11. Pourzare, K.; Mansourpanah, Y.; Farhadi, S. Advanced nanocomposite membranes for fuel cell applications: A comprehen-sive review. *Biofuel Res. J.* **2016**, *12*, 496–513. [CrossRef]
12. Mohy Eldin, M.S.; Abd Elmageed, M.H.; Omer, A.M.; Tamer, T.M.; Yossuf, M.E.; Khalifa, R.E. Novel proton exchange mem-branes based on sulfonated cellulose acetate for fuel cell applications: Preparation and characterization. *Int. J. Electrochem. Sci.* **2016**, *11*, 10150–10171. [CrossRef]
13. Mohy Eldin, M.S.; Omer, A.M.; Tamer, T.M.; Abd Elmageed, M.H.; Youssef, M.E.; Khalifa, R.E. Novel aminated cellulose acetate membranes for direct methanol fuel cells (DMFCs). *Int. J. Electrochem. Sci.* **2017**, *12*, 4301–4318. [CrossRef]
14. Mohy Eldin, M.S.; Abd Elmageed, M.H.; Omer, A.M.; Tamer, T.M.; Yossuf, M.E.; Khalifa, R.E. Development of novel phospho-rylated cellulose acetate polyelectrolyte membranes for direct methanol fuel cell application. *Int. J. Electrochem. Sci.* **2016**, *11*, 3467–3491. [CrossRef]
15. Bakangura, E.; Wu, L.; Ge, L.; Yang, Z.; Xu, T. Mixed matrix proton exchange membranes for fuel cells: State of the art and perspectives. *Prog. Polym. Sci.* **2016**, *57*, 103–152. [CrossRef]
16. Wei, Q.; Zhang, Y.; Wang, Y.; Chai, W.; Yang, M. Measurement and modeling of the effect of composition ratios on the properties of poly(vinyl alcohol)/poly(vinyl pyrrolidone) membranes. *Mater. Des.* **2016**, *103*, 249–258. [CrossRef]
17. Maarouf, S.; Tazi, B.; Guenoun, F. Preparation and characterization of new composite membranes containing polyvinylpyrrolidone, polyvinyl alcohol, sulfosuccinic acid, silicotungstic acid and silica for direct methanol fuel cell applications. *J. Mater. Environ. Sci.* **2017**, *8*, 2870–2876.
18. Pintauro, P. Perspectives on Membranes and Separators for Electrochemical Energy Conversion and Storage Devices. *Polym. Rev.* **2015**, *55*, 201–217. [CrossRef]
19. Gouda, M.H.; Elessawy, N.A.; Santos, D.M.F. Synthesis and Characterization of Novel Green Hybrid Nanocomposites for Application as Proton Exchange Membranes in Direct Borohydride Fuel Cells. *Energies* **2020**, *13*, 1180. [CrossRef]
20. Gouda, M.H.; Gouveia, W.; El Essawy, N.A.; Šljukić, B.; Nassr, A.A.; Santos, D.M.F. Simple design of PVA-based blend doped with $SO_4(PO_4)$-functionalised TiO_2 as an effective membrane for direct borohydride fuel cells. *Int. J. Hydrog. Energy* **2020**, *45*, 15226–15238. [CrossRef]
21. Gouda, M.H.; Gouveia, W.; Afonso, M.L.; Šljukić, B.; Elessawy, N.A.; Santos, D.M.F. Novel ternary polymer blend membranes doped with SO4/PO4-TiO2 for low temperature fuel cells; Paper No. ICCPE 106. In Proceedings of the 5th World Congress on Mechanical, Chemical, and Material Engineering (MCM'19), Lisbon, Portugal, 15–17 August 2019. [CrossRef]
22. Abu-Saied, M.A.; Soliman, E.A.; Abualnaj, K.M.; El Desouky, E. Highly Conductive Polyelectrolyte Membranes Polyvinyl Alcohol)/Poly (2-Acrylamido-2-Methyl Propane Sulfonic Acid) (PVA/PAMPS) for Fuel Cell Application. *Polymers* **2021**, *13*, 2638. [CrossRef]
23. Eldin, M.S.M.; Abu-Saied, M.A.; Elzatahry, A.A.; El-Khatib, K.M.; Hassan, E.A.; El-Sabbah, M.M. Novel acid-base poly vinyl chloride-doped ortho-phosphoric acid membranes for fuel cell applications. *Int. J. Electrochem. Sci.* **2011**, *6*, 5417–5429.
24. Eldin, M.S.M.; Elzatahry, A.A.; El-Khatib, K.M.; Hassan, E.A.; El-Sabbah, M.M.; Abu-Saied, M.A. Novel grafted nafion membranes for proton-exchange membrane fuel cell applications. *J. Appl. Polym. Sci.* **2011**, *119*, 120–133. [CrossRef]
25. Rochliadi, A.; Bundjali, B.; Arcana, I.M. Polymer electrolyte membranes prepared by blending of poly(vinyl alcohol)-poly(ethylene oxide) for lithium battery application. In Proceedings of the Joint International Conference on Electric Vehicular Technology and Industrial, Mechanical, Electrical and Chemical Engineering (ICEVT & IMECE), Surakarta, Indonesia, 4–5 November 2015; pp. 370–373.
26. Gouda, M.H.; Gouveia, W.; Afonso, M.L.; Šljukić, B.; El Essawy, N.A.; Nassr, A.A.; Santos, D.M.F. Poly(vinyl alcohol)-based crosslinked ternary polymer blend doped with sulfonated graphene oxide as a sustainable composite membrane for direct borohydride fuel cells. *J. Power Sources* **2019**, *432*, 92–101. [CrossRef]
27. Khalifa, R.E.; Omer, A.M.; Abd Elmageed, M.H.; Mohy Eldin, M.S. Titanium Dioxide/Phosphorous-Functionalized Cellulose Acetate Nanocomposite Membranes for DMFC Applications: Enhancing Properties and Performance. *ACS Omega* **2021**, *6*, 17194–17202. [CrossRef] [PubMed]
28. Abu-Saied, M.A.; Elzatahry, A.A.; El-Khatib, K.M.; Hassan, E.A.; El-Sabbah, M.M.; Drioli, E.; Mohy Eldin, M.S. Preparation and characterization of novel grafted cellophane-phosphoric acid-doped membranes for proton exchange membrane fuel-cell applications. *J. Appl. Polym. Sci.* **2012**, *123*, 3710–3724. [CrossRef]

29. Gouda, M.H.; Konsowa, A.H.; Farag, H.A.; Elessawy, N.A.; Tamer, T.M.; Mohy Eldin, M.S. Novel nanocomposite membranes based on cross-linked eco-friendly polymers doped with sulfated titania nanotubes for direct methanol fuel cell application. *Nanomater. Nanotechnol.* **2020**, *10*, 1847980420964368. [CrossRef]
30. Sedesheva, Y.S.; Ivanov, V.S.; Wozniak, A.I.; Yegorov, A.S. Proton-exchange membranes based on sulfonated polymers. *Orient. J. Chem.* **2016**, *32*, 2283–2296. [CrossRef]
31. Awang, N.; Ismail, A.F.; Jaafar, J.; Matsuura, T.; Junoh, H.; Othman, M.H.D.; Rahman, M.A. Functionalization of polymeric materials as a high-performance membrane for direct methanol fuel cell: A review. *React. Funct. Polym.* **2015**, *86*, 248–258. [CrossRef]
32. Mohy Eldin, M.S.; Farag, H.A.; Tamer, T.M.; Konsowa, A.H.; Gouda, M.H. Development of novel iota carrageenan-g-polyvinyl alcohol polyelectrolyte membranes for direct methanol fuel cell application. *Polym. Bull.* **2020**, *779*, 4895–4916. [CrossRef]
33. Benito, H.E.; Sánchez, T.; Alamilla, R.G.; Enríquez, J.M.; Robles, G.S.; Delgado, F.P. Synthesis and physicochemical characterization of titanium oxide and sulfated titanium oxide obtained by thermal hydrolysis of titanium tetrachloride. *Braz. J. Chem. Eng.* **2014**, *31*, 737–745. [CrossRef]
34. Lu, M.; Wang, F.; Liao, Q.; Chen, K.; Qin, J.; Pan, S. FTIR spectra and thermal properties of TiO_2-doped iron phosphate glasses. *J. Mol. Struct.* **2015**, *1081*, 187–192. [CrossRef]
35. Goswami, P.; Ganguli, J. Synthesis characterization and photocatalytic reactions of phosphate mesoporous titania. *Bull. Mater. Sci.* **2012**, *35*, 889–896. [CrossRef]
36. Pucic, I.; Jurkin, T. FTIR assessment of poly(ethylene oxide) irradiated in solid state, melt and aqeuous solution. *Radiat. Phys. Chem.* **2012**, *81*, 1426–1429. [CrossRef]
37. Yu, X.; Qiang, L. Preparation for graphite materials and study on electrochemical degradation of phenol by graphite cathodes. *Adv. Mater. Phys. Chem.* **2012**, *2*, 63–68. [CrossRef]
38. Venkatesan, P.; Dharmalingam, S. Effect of cation transport of SPEEK-Rutile TiO_2 electrolyte on microbial fuel cell performance. *J. Membr. Sci.* **2015**, *492*, 518–527. [CrossRef]
39. Ngai, K.S.; Ramesh, S.; Ramesh, K.; Juan, J.C. A review of polymer electrolytes: Fundamental, approaches and applications. *Ionics* **2016**, *22*, 1259–1279. [CrossRef]
40. Kowsari, E.; Zare, A.; Ansari, V. Phosphoric acid-doped ionic liquid-functionalized graphene oxide/sulfonated polyimide composites as proton exchange membrane. *Int. J. Hydrog. Energy* **2015**, *40*, 13964–13978. [CrossRef]
41. Bayer, T.; Cunning, B.V.; Selyanchyn, R.; Daio, T.; Nishihara, M.; Fujikawa, S.; Sasaki, K.; Lyth, S.M. Alkaline anion exchange membranes based on KOH-treated multilayer graphene oxide. *J. Membr. Sci.* **2016**, *508*, 51–61. [CrossRef]
42. Pandey, R.; Shahi, V. Sulphonated imidized graphene oxide (SIGO) based polymer electrolyte membrane for improved water retention, stability and proton conductivity. *J. Power Sources* **2015**, *299*, 104–113. [CrossRef]
43. Deshmukh, K.; Ahamed, M.B.; Sadasivuni, K.K.; Ponnamma, D.; Deshmukh, R.R.; Pasha, S.K.K.; AlMaadeed, M.A.; Chidambaram, K. Graphene oxide reinforced polyvinyl alcohol/polyethylene glycol blend composites as high-performance dielectric material. *J. Polym. Res.* **2016**, *23*, 159. [CrossRef]
44. Shirdast, A.; Sharif, A.; Abdollahi, M. Effect of the incorporation of sulfonated chitosan/sulfonated graphene oxide on the proton conductivity of chitosan membranes. *J. Power Sources* **2016**, *306*, 541–551. [CrossRef]
45. Beydaghi, H.; Javanbakht, M.; Kowsari, E. Synthesis and Characterization of Poly(vinyl alcohol)/Sulfonated Graphene Oxide Nanocomposite Membranes for Use in Proton Exchange Membrane Fuel Cells (PEMFCs). *Ind. Eng. Chem. Res.* **2014**, *53*, 16621–16632. [CrossRef]
46. Cheng, T.; Feng, M.; Huang, Y.; Liu, X. SGO/SPEN-based highly selective polymer electrolyte membranes for direct methanol fuel cells. *Ionics* **2017**, *23*, 2143–2152. [CrossRef]
47. Luo, T.; Xu, H.; Li, Z.; Gao, S.; Fang, Z.; Zhang, Z.; Wang, F.; Ma, B.; Zhu, C. Novel proton conducting membranes based on copolymers containing hydroxylated poly(ether ether ketone) and sulfonated polystyrenes. *J. Appl. Polym. Sci.* **2017**, *134*, 45205. [CrossRef]
48. Yuan, C.; Wang, Y. The preparation of novel sulfonated poly(aryl ether ketone sulfone)/TiO_2 composite membranes with low methanol permeability for direct methanol fuel cells. *High Perform. Polym.* **2021**, *33*, 326–337. [CrossRef]
49. Yang, C.C.; Chien, W.C.; Li, Y.J. Direct methanol fuel cell based on poly(vinyl alcohol)/titanium oxide nanotubes/poly(styrene sulfonic acid) (PVA/nt-TiO_2/PSSA) composite polymer membrane. *J. Power Sources* **2010**, *195*, 3407–3415. [CrossRef]
50. Ahmad, H.; Kamarudin, S.K.; Hasran, U.A.; Daud, W.R.W. A novel hybrid Nafion-PBI-ZP membrane for direct methanol fuel cells. *Int. J. Hydrog. Energy* **2011**, *36*, 14668–14677. [CrossRef]

energies

MDPI

Review

Intermetallic Compounds Synthesized by Mechanical Alloying for Solid-State Hydrogen Storage: A Review

Yuchen Liu *, Djafar Chabane and Omar Elkedim

FEMTO-ST Institute, University Bourgogne Franche-Comté, UTBM, CNRS, 90000 Belfort, France; djafar.chabane@utbm.fr (D.C.); omar.elkedim@utbm.fr (O.E.)
* Correspondence: yuchen.liu@utbm.fr

Abstract: Hydrogen energy is a very attractive option in dealing with the existing energy crisis. For the development of a hydrogen energy economy, hydrogen storage technology must be improved to over the storage limitations. Compared with traditional hydrogen storage technology, the prospect of hydrogen storage materials is broader. Among all types of hydrogen storage materials, solid hydrogen storage materials are most promising and have the most safety security. Solid hydrogen storage materials include high surface area physical adsorption materials and interstitial and non-interstitial hydrides. Among them, interstitial hydrides, also called intermetallic hydrides, are hydrides formed by transition metals or their alloys. The main alloy types are A_2B, AB, AB_2, AB_3, A_2B_7, AB_5, and BCC. A is a hydride that easily forms metal (such as Ti, V, Zr, and Y), while B is a non-hydride forming metal (such as Cr, Mn, and Fe). The development of intermetallic compounds as hydrogen storage materials is very attractive because their volumetric capacity is much higher (80–160 kgH_2m^{-3}) than the gaseous storage method and the liquid storage method in a cryogenic tank (40 and 71 kgH_2m^{-3}). Additionally, for hydrogen absorption and desorption reactions, the environmental requirements are lower than that of physical adsorption materials (ultra-low temperature) and the simplicity of the procedure is higher than that of non-interstitial hydrogen storage materials (multiple steps and a complex catalyst). In addition, there are abundant raw materials and diverse ingredients. For the synthesis and optimization of intermetallic compounds, in addition to traditional melting methods, mechanical alloying is a very important synthesis method, which has a unique synthesis mechanism and advantages. This review focuses on the application of mechanical alloying methods in the field of solid hydrogen storage materials.

Keywords: mechanical alloying; intermetallic compounds; solid hydrogen storage

Citation: Liu, Y.; Chabane, D.; Elkedim, O. Intermetallic Compounds Synthesized by Mechanical Alloying for Solid-State Hydrogen Storage: A Review. *Energies* **2021**, *14*, 5758. https://doi.org/10.3390/en14185758

Academic Editor: Attilio Converti

Received: 3 August 2021
Accepted: 10 September 2021
Published: 13 September 2021

1. Introduction

Currently, as fossil energy is on the verge of disappearing [1] and pollution caused by fossil fuels is becoming more serious [2], it is urgent to develop clean energy. As an important type of clean energy, hydrogen (H_2) energy has the advantages of being non-polluting, being easy to be produced, and having extremely high energy density. Compared with other clean energy sources, such as geothermal energy, wind energy, and tidal energy, it is the best choice. The utilization of H_2 energy involves many aspects, such as the production, transportation, storage, and utilization of H_2 energy. Among them, what restricts the use of H_2 energy is the storage technology of H_2 energy. The storage of H_2 energy can be roughly divided into two types: physical storage methods and chemical storage methods.

Physical storage includes high-pressure compressed gas storage and ultra-low temperature storage. Compressed H_2 requires high-pressure vessels to be loaded, and these vessels should be strong, light, and anti-explosive under special circumstances. Actually, the volume density of H_2 increases with pressure, but the weight density decreases at the same time. Therefore, as the weight density in the compressed gas system decreases, improvements in the volume storage density are sacrificed. At the same time, the future pressure vessel is expected to consist of three layers: a polymer lining, a carbon fiber composite material (which is a stress-bearing component), and an outer aramid layer that can withstand mechanical and chemical damage [3]. This also drive the cost higher. In summary, although compressed gas H_2 storage technology is simple and mature, lighter, safer, and low-cost containers still need to be researched. For ultra-low temperature liquid H_2 storage, liquefaction and continuous evaporation of H_2 require a relatively large amount of energy, which limits the application of liquid H_2 storage in mobile equipment, such as vehicles. Liquid H_2 storage is suitable for H_2 in a relatively short period of time. Applications include being consumed, such as in the aerospace industry [4].

Chemical storage refers to the combination of H_2 and materials through van der Waals forces or chemical bonds to store H_2 in the H_2 storage material in the form of atoms. This H_2 storage method is the most attractive method because, for chemical H_2 storage, reversible H_2 absorption and desorption reactions can be carried out in a mild environment, and its H_2 storage density is high, and there is no safety risk. There are many types of H_2 storage materials, which can be divided into solid H_2 storage materials and Liquid Organic H_2 Carriers (LOHCs) according to their different material states.

The theoretical H_2 storage capacity of LOHCs is extremely high, but the production of these liquid materials often requires extremely complicated steps, and a wide variety of catalysts are required to assist in the reaction when the reaction is at room temperature, so it is not suitable for the engines of vehicles; liquid materials are more suitable for H_2 storage at fixed sites. Solid H_2 storage materials are more reliable and safer than liquid and gaseous materials in ultra-low temperature storage [5].

Solid H_2 storage materials can be divided into many types, such as intermetallic compounds, complex hydrides, chemical hydrides, etc. [6]. In this review, the focus is on intermetallic compounds.

In recent years, intermetallic compounds have attracted great attention because of their wide application in the development of H_2 storage alloys [7]. There are many applications for intermetallic compounds, including H_2 storage systems, Nickel Metal Hydride (NiMH) battery electrodes, H_2 sensors and catalysts, and cooling systems [8]. Intermetallic compounds are attractive in the development of H_2 storage alloys because they can absorb large amounts of H_2. In addition, they are abundant and have diverse ingredients. Generally, H_2 reacts with intermetallic compounds to produce a solid solution of H_2 in the respective compound or the formed hydride. The hydride produced by the intermetallic compound is called the intermetallic hydride. The general formula is $A_mB_nH_x$. The interactions between metal atoms and interstitial H_2 atoms determine its properties, so it largely depends on the crystal structure of the compound [9].

The materials used as solid-state H_2 storage intermetallic compounds can be divided into A_2B, AB, AB_2, AB_3, AB_5, and BCC according to their composition. According to the different main component, it can be divided into Mg-based H_2 storage alloy, Ti-based H_2 storage alloy, RE-based H_2 storage alloy, etc.

Mg and Mg-based alloys have long been considered potential H_2 storage materials. Fifty years ago, scholars discovered that Mg metal and its alloys can be used as H_2 storage materials. As a potential commercial H_2 storage material, this material has some special advantages: (i) high abundance (2% of the composition of the surface of the Earth and almost unlimited in seawater), (ii) being non-toxic compared with other light elements and their hydrides that rapidly exothermic oxidation in the air, and (iii) being highly safe. In addition, the production technology of Magnesium (Mg) is very mature, and the cost of raw materials is very low.

Research on the synthesis of hydride nanoparticles of Mg and Mg-based alloys by the ball milling has been studied in depth. This method has a low cost, has a simple and convenient procedure, and is a technology that has been studied extensively by scholars. It is suitable for the synthesis of various low-particle-size powder materials and can even synthesize nanomaterials [10]. For H_2 storage material synthesized by mechanical alloying, its particles have a nanometre diameter and, due to its higher specific surface area and larger number of defects, provide hydrogenation nucleation sites and shorter H_2 diffusion distances. Therefore, a fine powder has better H_2 reaction kinetics than big particles [11]. The effect of particle size on the absorption kinetics of Mg-based H_2 storage alloys was shown in [12].

As an active metal, there is always an oxide layer on the surface of Mg. The activation process is to decompose the surface oxide layer that inhibits H_2 absorption. This process is usually carried out at high temperatures. However, mechanical alloying methods can overcome these difficulties because the powder particles undergo severe plastic deformation during mechanical alloying. The oxide layer on the particle surface is then destroyed. Additionally, by synthesizing these alloys by mechanical alloying, the most significant improvement in hydrogenation/dehydration kinetics can be achieved [13]. A catalyst is usually needed to further accelerate the H_2 and dehydrogenation kinetics of the alloy.

Some factors may restrict the hydrogenation kinetics of Mg and Mg alloys. These factors include (i) the formation of oxide layers on the surface that inhibits the penetration of H_2 into the alloys [14], (ii) the slow dissociation rate of H_2 molecules on the Mg surface, (iii) the low speed of the MgH_2/Mg interface [15], and (iv) the slow diffusion of H_2 through magnesium hydride [15]. Most of these factors can be solved by nanostructured Mg using mechanical alloying with the existing catalytic additives.

Titanium H_2 storage alloys mainly include $TiMn_2$ [16], Ti_2Ni [17], TiFe, etc. The TiFe intermetallic compound and its doped derivatives are a potential H_2 storage alloy because it is inexpensive and has a relatively high H_2 storage capacity, about 1.8 wt.%; rapidly absorbs and desorbs H_2; moderates H_2 absorption conditions; and has abundant reserves, so it has a very high potential as a H_2 storage material [18]. The most common method of preparing TiFe is to melt the element mixture of Ti and Fe at high temperatures in an inert atmosphere and then to place it at 800 to 900 °C for a long time. The activation of TiFe obtained in this way requires a high-temperature heat treatment (about 400 °C) under high vacuum conditions and high H_2 pressure (about 5 MPa) [19]; at the same time, the materials are prone to poisoning. TiFe alloy also has many other shortcomings, that is, the poor absorption and desorption kinetics, the obvious lag of hydride formation and decomposition, and the sensitivity to gas impurities [20].

In the past 30 years, great efforts have been made to modify TiFe alloys, such as alloying with one or several elements to partially replace Fe or Ti; chemical surface modification; and mechanical alloying to reduce particle size or alloying with catalysts, additives and other methods. Surface modifications can catalyze the adsorption and dissociation of H_2 molecules on the surface of the alloy by a certain transition metal or alloy deposited on the surface to promote the hydrogenation of the material or can use acids or alkalis to eliminate stable oxide layers. These methods also help to improve the anti-poisoning ability of TiFe [21]. The study by Edalati et al. showed that, through the process of High Pressure Torsion (HPT) [22,23] and Roll Groove (GR) [24], TiFe is activated after severe plastic deformation. TiFe samples activated by HPT and GR processes can absorb and release H_2 even after being stored in the air for several months. The study on the activation mechanism of HPT-treated TiFe shows that various defects such as nanoparticle grain boundaries and micro-cracks are the pathways for H_2 to pass through the oxide layer that hinders material transport on the surface. The main shortcoming of GR is that the sample is not easy to be fully activated immediately. The biggest disadvantage of the HPT process is that the sample amount is usually less than 1 g [25]. In order to expand the production of TiFe for H_2 storage, it is necessary to develop another severe plastic deformation technology that improves activation. This technology must have simple procedures, low cost, and high efficiency in producing severe plastic deformation. Mechanical alloying may be the ideal technology.

RE-based H_2 storage alloys can be divided into AB_5 type alloys and superstructure alloys (AB_3, A_2B_7, A_5B_{19}, AB_4, etc.) according to their compositions. AB_5-type H_2 storage alloy is one of the first H_2 storage alloys discovered, and it has also been used as a commercial H_2 storage alloy for a long time in nickel–H_2 batteries. The AB_5 type H_2 storage alloy has good H_2 storage performance. Its typical alloy is $LaNi_5$, which was first developed in 1969 by the Philips laboratory in the Netherlands.

After discovering the H_2 storage capacity of AB_5 alloy [26], an important research area focused on the engineering application of these compounds [27]. However, the AB_5 alloys used are synthesized by the high-temperature equilibrium method [28].

At the beginning of the $LaNi_5$ study, it was found that the capacity of this alloy decreases significantly during the cycles. After 100 charge and discharge cycles, the capacity drops by 40%, which also determines that pure $LaNi_5$ is not suitable for battery applications. Microscopic studies have shown that, during the cycle, the alloy particles change from an initial size of 7 to 2 mm to fine powder after 25 cycles. This is caused by the repeated strain of the crystal lattice in the process of H_2 absorption and desorption. The decrease in particle size leads to a longer exposure of the alloy surface to the electrolyte and increases in alloy corrosion due to the high affinity of $LaNi_5$ for water. The needle-like crystals of $La(OH)_3$ were detected by scanning transmission electron microscopy. During the continuous pulverization of $LaNi_5$ particles, La atoms are thrown to the grain boundary, where La is oxidized to form $La(OH)_3$. Therefore, the electrode is consumed by the corrosion reaction, and the H_2 storage capacity is reduced [29].

Extensive research on H_2 storage alloy $LaNi_5$ is to partially replace La and Ni with other elements to reduce the volume expansion ratio of the alloy to its hydride [30]. The equilibrium pressure of the alloy can be reduced by introducing many elements: Cr, Co, Cu, Al, and Mn, etc. [31].

At the beginning of this century, RE–magnesium–nickel (RE–Mg–Ni)-based H_2 storage alloys gradually emerged. This type of alloy has a unique structure with a long-period one-dimensional superstructure, in which AB_5 units ($CaCu_5$ type structure) and A_2B_4 units (Laves type structure) are accumulated in a rhombus at a ratio of 1:1 along the c-axis. Kadir et al. [32] showed this structure. Due to its unique structure, this type of alloy has the advantages of a high capacity and a good activation performance. Therefore, this type of alloy is expected to make up for the shortcomings of the low discharge capacity of AB_5 type alloys, which has become one of the research hotspots.

Mechanical alloying is a kind of "non-equilibrium processing" technology. Since Libowitz et al. [33] first reported the synthesis of a metal alloy hydride $ZrNiH_3$ by mechanical alloying in 1958, this method has always been a research hotspots in the field of H_2 storage material synthesis. Unlike traditional melting methods, mechanical alloying has some unique advantages. First, the mechanical alloying procedure is simple, has a low cost, and uses simple equipment. Second, under the heat of positive mixing, mechanical alloying can also alloy two metals and can synthesize metastable alloys, amorphous alloys, or quasicrystals, etc. This is to expand the scope of additives and to develop new H_2 storage alloys to lay a good foundation, which is difficult to achieve using melting methods. Mechanical alloying can realize the nanometerization of the particle size, which often greatly improves the H_2 storage performance of the alloy. However, this method also has some drawbacks, such as powder contamination, difficulty in precise control of the composition, etc.

After the metal or alloy undergoes the mechanical alloying process, the particle size and crystal grain size are greatly reduced while the micro-strain and lattice distortion in the crystal increase. This has a great impact on the H_2 storage performance of the material, mainly in terms of absorption/desorption kinetics and thermodynamics [34,35] as well as H_2 storage capacity. First, the micro-strain and lattice distortion act on thermodynamics and kinetics, which can reduce the temperature of H_2 desorption, etc. Second, for kinetics, grain refinement reduces the diffusion distance, and micro-strain and distortion increase diffusion and reduce the apparent activation energy. Additionally, the free energy of the material after ball milling is increased, which is released in the subsequent heating recovery process. For the H_2 storage capacity, in the AB, AB_5, and AB_2 materials [36,37], a significant decrease in capacity was observed after ball milling, while the capacity of Mg_2Ni and Mg materials did not change significantly [38]. The reason may be the formation of amorphous or disordered structure and the intervention of impurities.

This review focuses on the application of mechanical alloying in the synthesis of intermetallic compounds as solid-state H_2 storage materials.

2. Application of Mechanical Alloying in Mg-Based Hydrogen Storage Alloys

Magnesium in the pure form can absorb up to 7.6 wt.% of H_2 (above 400 °C), but it has low stability and low H_2 absorption/desorption kinetics [39]. Some known compounds, such as Mg–Ni compounds, Mg–Fe compounds, etc., seem to be possible alternatives to magnesium hydride. These materials sacrifice volume or mass H_2 capacity to obtain a better balance of pressure, stability, or cost [40].

Mg–Ni series alloys can form $MgNi_2$ alloy and Mg_2Ni alloy as well as the newly emerged MgNi alloy. Among them, Mg_2Ni alloy form Mg_2NiH_4 after H_2 absorption. The H_2 content is 3.6 wt.%, which is much higher than the commercial H_2 storage alloy $LaNi_5$, which has a H_2 absorption of 1.4 wt.%. This alloy also has good H_2 absorption and desorption kinetics at low temperatures. The theoretical electrochemical capacity of Mg_2Ni can reach 1000 mAh/g. On the basis of binary alloys, researchers often add third elements such as Fe, Ti, La, etc. into Mg–Ni materials [41]. The addition of new elements changes the composition and content of the alloy phase and can also play a catalytic role, thus improving the H_2 storage performance of the alloys. Due to the significant difference between the high vapor pressure of Mg and the different melting temperature of the constituent elements (melting temperature for Mg is 650 °C, and that for Ni is 1452 °C), it is difficult to produce Mg_2Ni alloys through the traditional smelting method. The mechanical alloying technology can easily synthesize these alloys because it is a completely solid state reaction process. In 1995, Singh and Zaluski [42,43] used mechanical alloying to produce Mg_2Ni. Mg_2Ni prepared by mechanical alloying was found to be able to react with H_2 at a relatively low temperature, even room temperature, while the alloy synthesized by the melting method requires a hydrogenation temperature of 250–350 °C and a pressure of 15–50 bar [44]. The H_2 absorption and desorption properties of some Mg–Ni-based alloys are listed in [45].

N. Cui et al. [46] believe that the Ni in the Mg–Ni alloy has high electrocatalytic activity for the H_2 desorption reaction of the alloy in the alkaline solution. Cold welding occurs during the mechanical alloying process. During this period, Ni as the second phase is evenly distributed. The Ni particles on the surface of the Mg alloy are the electrocatalytic active centres of the H_2 evolution reaction. This is because they can reduce the charge transfer resistance, resulting in a significant increase in electrochemical capacity. At the same time, the mechanical alloying reduces the particle size and increases defect density caused by crystallization, which can improve the diffusion of H_2 in the alloy. The Mg and Ni powder in the mechanical alloying process is not only a mixture of two components but also a new composite system with evenly distributed electrocatalytic active centers, large interface, small particle size, and high reactivity due to high defect density. However, he also proposed that excessive grinding damages the crystal lattice and causes a sharp drop in electrochemical capacity.

L. Zaluski et al. [43] prepared a nano-sized Mg_2Ni alloy using the mechanical alloying method under the conditions of a ball-to-alloy ratio of 5 and a grinding time of 60 h. He found that the nanocrystalline Mg_2Ni alloy synthesized by mechanical alloying showed better H_2 adsorption performance than the alloy prepared by conventional methods and that the produced powder can easily absorb H_2 without activation because the mechanically alloyed nanocrystalline produces many very active fresh surfaces in the ball milling process. Conventional polycrystalline Mg_2Ni can react with H_2 at a temperature higher than 250 °C, while nanocrystalline Mg_2Ni can also absorb H_2 at a lower temperature (for example, at 200 °C, which is below the structural transition temperature of Mg_2NiH_4 hydride). No activation is required. Pd can catalyze the H_2 absorption kinetics of nanocrystalline Mg_2Ni

at 200 °C. Nanocrystalline Mg_2Ni with a small amount of Pd can absorb H_2 even at room temperature, does not require activation, and has good kinetics.

The nanocrystalline Mg_2Ni intermetallic compound formed by mechanically alloying Mg and Mg_2Ni can quickly absorb H_2 without activation. Mg + Mg_2Ni composites need to be activated, but their H_2 absorption rate is faster than Mg_2Ni at 150 °C and 12 bar, and the capacity is 4.2 wt.% [47]. Some other additive were also investigated [48–51].

The nanocrystalline and amorphous MgNi alloy prepared by mechanical alloying is also a potential alloy as a negative electrode for NiMH batteries. For example, after 10 hours of ball milling, the MgNi alloy had an initial discharge capacity of 522 mAh/g [52]. In addition, unlike traditional AB_5 and AB_2 H_2 storage materials, this MgNi alloy does not require any activation and absorbs H_2 directly. They also have the advantages of being almost nontoxic and low cost. However, their H_2 absorption and desorption kinetics still needs to be further improved, and the actual discharge capacity is difficult to reach its theoretical value. In addition, from the point of view of commercial applications, this alloy has poor cycle stability as a hydride electrode. For example, after only 20 charge and discharge cycles, the discharge capacity of the MgNi electrode decays by more than 70%. Such cycle stability prevents MgNi from being used as a battery electrode, so it must be improved by other methods. The decrease in capacity is related to the irreversible corrosion of the alloy electrode by the KOH in the battery. This reaction forms a $Mg(OH)_2$ layer [53–55] on the surface of particles. This not only consumes the alloy itself but also greatly increases the charge transfer resistance at the alloy/electrolyte interface and may hinder the diffusion of H_2 into and out of the alloy body [56]. The pulverization of the alloy during the H_2 absorption and desorption cycle exacerbates this harmful phenomenon because the pulverization produces a new active surface and thus forms new additional $Mg(OH)_2$ layers after making contact with the electrolyte.

Mustafa Anik et al. [57] studied the electrochemical properties of Mg_2Ni and MgNi synthesized by Mechanical Alloying (MA). The results showed that the charge and discharge capacity of Mg_2Ni alloy increased sharply with the increase in grinding time within 40 h. The capacity of the alloy for which the grinding time exceeds 40 h no longer increases. They also found that the electrochemical performance of MgNi is much better than that of Mg_2Ni, and the charge–discharge reversibility of the Mg_2Ni alloy is very poor. The lower initial discharge capacity and cycle stability of the Mg_2Ni alloy are not only due to the blocking effect of the $Mg(OH)_2$ layer but also maybe owing to the highly irreversible reaction of the alloy. The author believes that the existence of free electrocatalytically active Ni particles on the surface of the MgNi particles is the main factor that promotes the H_2 transfer reaction on the surface of the alloy.

Chiaki Iwakura et al. [58] dissolved Ti and V into MgNi alloy by mechanical alloying. They found that the two-element solid-solution amorphous $Mg_{0.9}Ti_{0.06}V_{0.04}Ni$ alloy prepared from MA is better than the single-element solid-solution $Mg_{0.9}Ti_{0.1}Ni$ or $Mg_{0.9}V_{0.1}Ni$ alloys show better cycle performance. The AES depth distribution shows that, after the charge–discharge cycle in an alkaline solution, the oxide layer on the surface of the $Mg_{0.9}Ti_{0.06}V_{0.04}Ni$ alloy is thinner than the surface of the $Mg_{0.9}Ti_{0.1}Ni$ or $Mg_{0.9}V_{0.1}Ni$ alloy. The XRD data show that a composite oxide layer composed of Ti and V species precipitates on the surface of the alloy particles, which may be the reason for the synergistic effect of the solid solution of the two elements to promote the charge–discharge cycle performance.

Stephane Ruggeri et al. [59] synthesized a MgNiTi ternary alloy by adding Ti through ball milling on the basis of MgNi alloy. It was found that the initial discharge capacity of $MgNi_{0.95}Ti_{0.05}$ (C1 = 575 mAh/g) was significantly increased compared with the MgNi electrode (C1 = 522 mAh/g). The author believes that this promotion of initial discharge capacity may be related to the formation of a multi-phase structure. The initial discharge capacity of $Mg_{0.5}Ti_{0.5}Ni$ is 338 mAh/g, and after 10 cycles, the capacity stability is 75%. This cycle life is better than that of the initial MgNi alloy. The improvement in cycle life seems to be attributed to the formation of TiO_2 that limits the formation of $Mg(OH)_2$ on the alloy surface. However, the corrosion resistance of the electrode still needs further enhancement.

Jian-Jun Jiang et al. [60] believed that the modification of MgNi materials is very complicated. Substituting other elements for Mg does not significantly change its performance. Therefore, they used MA to synthesize amorphous MgNi alloy electrodes. He found that amorphous MgNi alloys do not require an additional activation process, and the mass transfer reaction largely depends on the oxide on the surface of the alloy particles, not on the electrode itself. The reaction rate is controlled by the film thickness. In the initial cycle, the film thickness is very low, so the electron exchange process is relatively smooth. In further charge and discharge cycles, the thickness of the oxide film increases and begins to hinder the transport of electrons. Additionally, this layer of film cannot provide an effective anti-corrosion effect due to its open structure. For the alloy electrode covered by the oxide film, the exchange current density greatly decreases as the thickness of the oxide film increases. Therefore, if this alloy electrode is used in a strong alkaline solution, surface modification must be considered to improve performance.

O Elkedim and L Huang et al. [61] researched the substitution of Mn and Ti in the Mg_2Ni phase by first principles Density Functional Theory (DFT) calculations. The results show that the doping of Mn reduces the thermodynamic stability of Mg_2Ni. When Ti is doped into Mg_2Ni, with the increase in Ti content, the thermodynamic stability of Mg_2Ni gradually decreases. That is, the doping of Ti and Mn can promote the thermodynamics of H_2 release of Mg_2Ni.

Amirkhiz et al. [62] pointed out that the addition of Single Wall Carbon Nanotube (SWCNT) can be used as a grinding aid in mechanical alloying to prevent the powder from consolidating on the surface of the ball and the container. Yao et al. [63] also believed that carbon materials have high dispersibility and catalytic activity, so they can promote the mechanical alloying process of ductile metals to produce finer particles. They believed that SWCNT can penetrate the thin hydroxide layer on the surface of MgH_2 and can act as a 'H_2 pump' to move H_2 atoms to the surface. Multiwall Carbon Nanotube (MWCNT) also has a similar function. L.W. Huang et al. [64] studied the effects of Al substitution and the addition of MWCNTs about their structure and electrochemical properties of Mg_2Ni alloys. They found that the ground alloy particles showed smaller particle size, agglomeration and better dispersibility, indicating that MWCNT can act as a grinding aid. After ball milling, most MWCNTs retain their tubular structure. All ball milled alloys with additives show excellent activation properties. However, adding MWCNTs is difficult to enhance the cycle life of the electrode. On the one hand, the discharge capacity is improved by the refinement of alloy particles. On the other hand, due to the enhanced dispersibility, more alloy particle surfaces are exposed to the KOH solution, which weakens the corrosion resistance of the ball milled alloy.

Based on the high H_2 capacity of Mg and certain Ti-based alloys, scholars have developed new compounds of Mg–Ti–Ni–Fe alloys. Guo et al. [65] used mechanical alloying to synthesize $Mg_{76}Ti_{12}Fe_{12-x}Ni_x(x = 0, 4, 8, 12)$ alloy and studied the effect of ball milling time on the H_2 storage performance of the alloy. The results show that increasing the grinding time up to 80 hours increases the amount of Mg amorphous phase, which reduces the H_2 storage performance. They also compared the H_2 storage performance of alloys of various compositions after grinding for 40 hours and found that the storage capacity of alloys doped with Ni and Fe at the same time is much higher than that of alloys doped with single elements. In addition, the H_2 absorption plateau pressures of $Mg_{76}Ti_{12}Fe_8Ni_4$ and $Mg_{76}Ti_{12}Fe_4Ni_8$ are reduced. The author explains that this phenomenon is due to the simultaneous formation of Mg_2Ni and NiTi phases during the mechanical alloying process. They also observed that, as the Ni content increases, the Fe content decreases and that the hysteresis between the H_2 absorption and desorption curves gradually decreases.

Mechanical alloying has also been applied to the synthesis and optimization of some binary magnesium-rich intermetallic compounds, such as $LaMg_{12}$, La_2Mg_{17}, etc. X. P. Gao et al. [66] used mechanical alloying to add Ni to the $LaMg_{12}$ alloy and studied its electrochemical H_2 storage performance. When the weight ratio of $LaMg_{12}$ to Ni is 1:3, the maximum discharge capacity of the alloy can reach 1010 mAh/g. However, its cycle performance needs to be further improved. Zhang Yanghuan et al. [67] synthesized $LaMg_{11}Ni$ + x wt.% Ni(x = 100, 200) alloys by mechanical alloying and studied the gaseous and electrochemical H_2 storage kinetics. They found that increasing the Ni content can improve the kinetics, which is attributed to the decrease in activation energy and enthalpy. The milling time also affects two kinetics; the gaseous H_2 absorption kinetic and electrochemical kinetic have maximum values with the change in milling time. LI Xia et al. [68] put Ni and La_2Mg_{17} alloy together for ball milling and obtained La_2Mg_{17}-x wt.%Ni(x = 0, 50, 100, 150, and 200) alloy. They found that the increase in Ni increases the proportion of the amorphous phase. This leads to a decrease in H_2 absorption capacity and cycle stability but an increase in discharge capacity. The H_2 absorption capacity of La_2Mg_{17}-50 wt.%Ni is 5.796 wt.% (3 MPa), and the maximum discharge capacity is 353.1 mAh/g.

MY Song et al. [69] believed that, under certain conditions, the nucleation process controls the hydrogenation and dehydrogenation reactions of Mg, so nucleation can be promoted by creating a large number of defects, and shorten the H_2 diffusion distance by reducing the particle size of Mg particles. Therefore, the hydrogenation and dehydrogenation kinetics of magnesium can be improved by mechanical alloying. This method generates many defect nuclei on the surface and/or inside of magnesium through severe plastic deformation or can be added by adding additives act as active sites for nucleation and, at the same time, greatly reduces the particle size of magnesium to shorten the diffusion distance. They also believe that mechanical alloying changes the rate control step of the hydrogenation reaction because it promotes the diffusion of H_2, so the control step becomes the gas-phase mass transfer and the chemical adsorption of H_2.

Some complex additives have also been applied to improve the H_2 storage performance of pure Mg. Mykhaylo Lotoskyy et al. [70] found an outstanding effect of graphite and TiH_2 adding to Mg by high-energy reactive ball milling. This composite can reversely absorb 6 wt.% H_2. The addition of graphite greatly increases the cycle stability of the material. The author believes that graphite not only reduces the recrystallization of particles during the cycle but also helps to further refine the particles. C Zhou et al. [71] studied the effects of TiH_2, $TiMn_2$, and VTiCr on the kinetics of Mg hydrogenation. They used ball milling to produce alloy powder. The results show that the cycle kinetics of the material with VTiCr is better. The author found that the cycle kinetics of the material is relatively good at 300 °C but that the kinetics deteriorates severely at low temperatures (25–150 °C). The author attributed it to the growth process of microscopic grains during the reaction. Pavel Rizo-Acosta et al. [72] reported the effect of Early Transition Metal (ETM) hydride ($ETMH_x$) as a catalyst on the performance of cycling H_2 storage of Mg. They used reactive ball milling to mix the hydrides of Sc, Y, Ti, Zr, V, and Nb with Mg and found that the presence of $ETMH_x$ is beneficial to the decomposition of H_2 molecules. Later, due to the high diffusion coefficient, H can diffuse rapidly in $ETMH_x$, resulting in The rapid nucleation of MgH_2. For cycle life, $ETMH_x$ can limit grain growth and improve structural stability. Among all of the additives studied, TiH_2 has the best performance. The reversible H_2 capacity for 20 cycles is 4.8 wt.%.

3. Ti-Based Hydrogen Storage Alloys

In 2000, Chiang, C.H., et al. [73] studied the hydrogenation performances of TiFe, $TiFe_2$, and Ti during mechanical alloying in the H_2 gas atmosphere. They found that, through reaction ball milling, TiFe can directly absorb H_2 without activation. Single-phase TiH is produced during the mechanical alloying. In addition, $TiFe_{1.924}$ can also be hydrogenated by mechanical alloying in H_2 because the alloy decomposes to form TiFeH, TiH, and Fe during this process. Based on the hydrides of Ti, TiFe and $TiFe_2$ and the thermal stability of the product powder during the ball milling process, they proposed that the TiFe milling reaction in H_2 includes four steps: (1) The particles are broken to produce a fresh surface. (2) The powder absorbs H_2. (3) H_2 supersaturates in the powder. (4) TiFeH decomposes into TiH and Fe.

The research results of Hoda Emami et al. [74] showed there exists a close relationship between the activation of TiFe alloy and its particle size. Annealed TiFe with a grain size of micrometers does not absorb H_2. Rolled samples with submicron to micron grain sizes are partially activated. The samples processed by HPT have nano- to sub-micron grain sizes and are fully activated under 3 MPa H_2 pressure. The ball-milled sample with a nanometer particle size is fully activated and absorbs H_2 at a pressure as low as 1 MPa.

The results of the literature [75] also show that the grinding of TiFe alloy and the transformation of alloy particles to nanometer sizes greatly simplify the activation process. In 2012, V. Zadorozhnyi et al. [76] demonstrated the possibility of directly synthesizing nano-size TiFe by solid-state reaction method compounds from a single component Fe and Ti using a mechanical alloying method. They discovered the exothermic effect of TiFe during the mechanical alloying process and believed that it was caused by the accumulation of excessive internal energy in the form of defects during the synthesis process. This characteristic plays a crucial role in the powder compaction process after mechanical alloying. It enhances the adhesion of the powder particles, thus ensuring that they are more easily compressed into blocks. Their other experimental results [77] show that the compacted sample of TiFe powder synthesized by mechanical alloying maintains the H_2 adsorption performance of powdered TiFe and still maintains a considerable capacity after 20 absorption–adsorption cycles. After the first absorption–absorption cycle, the TiFe sample that was not compacted by mechanical alloying was immediately destroyed. They proposed that the improved cycle stability of TiFe samples produced by ball milling can be attributed to the formation of a special microstructure that resembles a sponge and acts as a specific bridge between powder particles. The formation of these contact bridges may be due to the material accumulating a great amount of energy in the process of mechanical alloying.

Later, they proposed that, although the nanocrystalline state of the intermetallic compound formed by mechanical alloying plays a crucial role in the activation stage, it is not so important in the subsequent absorption–desorption cycle because of the lattice repetition caused by the reaction itself. Expansion and contraction also form nano-sized grains and introduce a large number of defects. Inui et al. [78] added the importance of lattice defects generated during the MA treatment of TiFe. This partially disordered structure leads to the expansion of the solid solution area in the pressure–composition–isotherm curve, the reduction of the plateau pressure, and it also makes it easier for H_2 to enter the crystal grains.

They have conducted a lot of research on synthesizing TiFe by mechanical alloying and adding various elements for doping. They are doped with Mg and S, Co and Ni, and Al and Cr. The concentrations of Mg and S are as high as 2 and 1 at.%, respectively. These alloys TiFe + 1% S and TiFe + 2% Mg show 0.6–0.7 wt.% reversible H_2 capacity. The S-containing alloy has a very simple activation procedure, namely heating to 100 °C in H_2 and placing it for about 20 minutes [79]. The results of doping Co and Ni also showed that the content of the two elements is as high as 2 at. %, for the mechanical alloying ones, and the extension of the H solid solution region and the reduction in the $(\alpha + \beta)$ plateau have been found [80]. Their work also showed the results of doping Al and Cr. The concentrations of the two are 20 and 6 at.%. Mechanical alloying doped TiFe + 5 % Al and TiFe + 4 % Cr showed an extremely simple activation procedure and a reversible H_2 capacity of 0.7 wt.%, and the alloys containing Al had a higher plateau pressure. Compared with unalloyed TiFe, the hysteresis was significantly reduced. This characteristic is related to the smaller lattice expansion when the b hydride phase is formed. They also reported the data of TiFe–Mn, TiFe–Zr, and TiFe–Cu alloys produced by ball milling [81].

Liang et al. [82] listed some of the ball milling parameters and the corresponding H_2 storage performance.

G. K. Sujan et al. [83] mentioned some problems in the preparation of TiFe powder by mechanical alloying in his review. The main disadvantage is that, for mechanical alloying, the raw material must be powder and the production cost of titanium powder is very high.

Reducing the adhesion of powder on the container wall caused by cold welding is very important to ensure the output and quality of the product. The literatures use process control agents, mainly cyclohexane and benzene, to effectively reduce powder adhesion. Falcao et al. [84] developed an alternative way to produce nanocrystalline TiFe, using TiH_2 in the raw material instead of Ti powder, resulting in a higher yield of TiFe. Contamination during mechanical alloying (for example, carbides formed by organic grinding aids and chromium in stainless steel containers and balls) is also a serious problem.

Tohru Nobuki et al. [85] used a mechanical alloying method to quickly synthesize a TiNi alloy within 20 min. They demonstrated through the cross section of the powder sticking to the grinding ball that the formation of the alloy occurs through the inter-diffusion between thin layers of co-laminated pure elements. The hydrogenation thermodynamics and kinetics of short-term mechanical alloying TiNi are similar to TiNi obtained by melting.

Z. Zhang et al. [86] used mechanical alloying to dope Mg into TiNi alloy to produce $TiMgNi_x$ (x = 0.1, 0.5, 1, 2) alloy and studied its structure and H_2 storage performance. The results show that the average discharge capacity decay of the samples is very low, less than 1.1 % per cycle. Among all of the samples, TiMgNi showed the highest discharge capacity. The author believes that this is related to the MgNi amorphous phase. They have conducted a lot of research on this ternary alloy synthesized by mechanical alloying, including adding additive [87,88] and its application in NiMH batteries [89]. They also used mechanical alloying to prepare alloy $TiMgNi_x$ (x = 0.2, 0.4, 0.6, 0.8, 1) samples with different Ni content and explored the effect of Ni content [88]. They found that, when the Ni content increases, the discharge capacity and activation performance of $TiMgNi_x$ alloy increase linearly. Other researchers also explored the properties of Ti–Mg–Ni alloy produced by ball milling [90,91].

B. Hosni et al. [92] synthesized Ti_2Ni by mechanical alloying at room temperature, and studied its structure and H_2 storage performance. The results show that the activation of the alloy is very simple, requiring only one charge and discharge cycle. As for the cycle life, with the temperature gradually increases, the electrochemical discharge capacity loss after several cycles increases. The maximum capacity increases with the increase in temperature, and at the same time, the corrosion current density decreases. The author believes that the decrease in the oxide surface layer caused this phenomenon. Additive can also be added into Ti_2Ni by mechanical alloying [93].

Hailiang Chu et al. [94] used mechanical alloying to mix $Ti_{0.9}Zr_{0.2}Mn_{1.5}Cr_{0.3}V_{0.3}$ alloy based on $TiMn_2$ with $LaNi_{3.8}Mn_{0.3}Al_{0.4}Co_{0.5}$ (AB_5) and $La_{0.7}Mg_{0.25}Zr_{0.05}Ni_{2.975}Co_{0.525}$ ($AB_{3.5}$) additives to study the structure and electrochemical performance. The addition of additives did not change the AB_2 structure of the main body but significantly increased the electrochemical discharge capacity, reaching 310.4 mAh/g and 314.0 mAh/g, respectively. The author believes that the $AB_{3.5}$ alloy can reduce the charge transfer resistance and that the AB_5 alloy can improve the H_2 diffusion of the AB_2 alloy. Myong JinChoi et al. [95] mixed $TiMn_2$ and TiFe by the mechanical alloying method, then added different content of Ni, and studied its electrochemical performance. The results show that the H_2 absorption capacity of the $TiMn_{0.9}Fe_{0.55}$ composite alloy is about 0.9 wt.% and that the alloy containing 30 wt.% Ni has the highest discharge capacity of 110 mAh/g, and this value was maintained for 20 cycles. The author attributed the improvement in performance to Ni catalyzing the electrochemical reaction and at the same time improving the corrosion resistance of the alloy.

4. RE-Based Hydrogen Storage Alloys

4.1. AB_5 Type Hydrogen Storage Alloy

In 2001, G. Liang et al. [96] began to synthesize $LaNi_5$ using two MA pathways. The first one is the mechanical alloying of La powder and Ni powder, and the second one is the mechanical milling of LaH with Ni. They all result in the formation of the α-$LaNi_5$ hydride phase. After testing, the H_2 storage performance of the two types of $LaNi_5$ is similar to that of the melted bulk $LaNi_5$. Another disadvantage of the AB_5 alloy is that it is susceptible to all kinds of pollution. Improving the toxicity resistance of AB_5 alloy and its hydride is generally achieved through surface modification methods, which (i) provide surface catalytic activity for H_2 dissociation/recombination and/or (ii) introduce catalytic active centers to protect the surface of the alloy, with the active center replacing the alloy and reacting with impurities [97]. The most effective way to form new catalytic active centers on the surface of MH is to introduce platinum group elements. At this time, mechanical alloying is a good introduction method [98].

KD Modibane et al. [99] introduced Pd into AB_5 alloys by mechanical alloying combined with conventional electroless plating technology, with the composition being $[La, Ce, Pr, Nd]\ [Ni, Co, Al, Mn]_5$, and studied the H_2 absorption performance of Pd on the alloy and the impact of poisoning tolerance. The results show that the alloy with 1 wt.% Pd has the best H_2 absorption kinetic performance.

4.2. AB_3 and A_2B_7 Hydrogen Storage Alloy

Kadir et al. [100] reported the study of a new type of ternary alloy, the general formula of which is RMg_2Ni_9 (R = RE, Ca, Y), $PuNi_3$-type structure. It has been found that some ternary alloys based on R–Mg–Ni can reversibly absorb/desorb H_2 at 1.8–1.87 wt.% and are therefore considered potential candidates for H_2 storage alloys [100]. However, its cycle stability and overall performance must be further improved.

Many scholars study high-capacity alloys with AB_3 components (A: RE metals, Mg, Ca, etc.; B: transition metals) [101–103]. Most scholars synthesize AB_3 alloy by melting method, but MA also has unique advantages in synthesizing AB_3 alloy. For example, Hassen Jaafar et al. [104] successfully synthesized the AB_3 alloy $LaMg_2Ni_5Al_4$ using MA.

Mouna Elghali et al. [105] produced AB_3 alloy according to the equation $AB_5 + 2AB_2 = 3AB_3$. They produced $LaZr_2Mn_4Ni_5$. Due to the immiscibility of La and Zr elements, this compound cannot be obtained by melting methods. Therefore, mechanical alloying methods are used. It is found that the cell volume is increased compared with the $LaMg_2Ni_9$ alloy [105]. They also used the above methods to produce alloys such as $LaTi_2Cr_4Ni_5$ [106], $LaZr_2Cr_4Ni_5$ [107], and $CeTi_2Cr_4Ni_5$ [108], with the results showing that these alloys have good reversibility, high discharge capacity, and a good cycle stability.

In the AB_3 alloy, Ca and Mg can be used instead of A. Such an alloy has a H_2 storage capacity of about 1.9 wt.% and a discharge capacity of about 370 mAh/g [109]. $LaCaMgNi_9$ compounds are always produced by smelting or sintering method. However, it is difficult to produce stoichiometric $LaCaMgNi_9$ using melting technology due to the vapor pressures of Ca and Mg being very high, and it is very easy to evaporate during the melting process. Another difficulty is that Ca and Mg are easily contaminated by oxygen at high temperatures [110]. The use of mechanical alloying (MA) can avoid these difficulties. S. Chebab et al. [111] used MA to treat pure La, Ni, Ca, and Mg powders for 30 hours, and the weight percent of this phase reached 67 %. This alloy can absorb 6 H/f.u. of H_2.

In the past few years, people have studied the possibility of La_2Ni_7 alloy as a H_2 storage alloy because of its good H absorption capacity. However, the La_2Ni_7 phase is hindered by its poor cycle stability because of poor corrosion resistance [112]. Different RE elements are usually used to replace La, while Co, Mn, Al, and some other elements are used to replace Ni to improve the electrochemical properties of such alloy electrodes [113].

In the work of M. Balcerzak et al. [114], MA technology was used to manufacture La_2Ni_7 alloy, and then, Mg element was also incorporated to produce the ternary alloy $La_{2-x}Mg_xNi_7$ (x = 0–1). It was found that the electrochemical and thermodynamic properties of this alloy increased with the rise content of Mg, and the alloy with the best performance was $La_{1.5}Mg_{0.5}Ni_7$. Ni element forms a film on the surface of the alloy particles, and the film is very dense, which can effectively protect the material from corrosion by strong alkaline solutions.

In their another work [115], they also synthesized $La_{1.5-x}Pr_xMg_{0.5}Ni_7$ and $La_{1.5-x}Nd_x Mg_{0.5}Ni_7$ alloys (x = 0, 0.25, 0.5, 1) with MA. It was found that replacing La with Pr or Nd elements resulted in increased cycle stability of the alloy and optimized H_2 absorption kinetics.

Martyna Dymek et al. [116] also doped the above materials. La is partially replaced by Mg (5.6 at.%) and Ni is partially replaced by Co. The doped material shows easy activation characteristics (maximum capacity is reached after the second cycle) and excellent H_2 capacity (14% larger than Co-free materials). The exchange current density of the Co-modified $La_{1.5}Mg_{0.5}Ni_7$ alloy is increased by 15%, and the H_2 diffusivity is also improved.

Marek Nowak et al. [117] used MA to synthesize $La_{1.5}Mg_{0.5}Ni_7$ alloy and doped Al or Mn into it. It is found that both elements can improve its H_2 adsorption. The time required for the third cycle to reach 95% of the maximum H_2 capacity was respectively reduced to 5 and 6 minutes and enhanced the stability of the discharge capacity.

5. Body-Centered Cubic (BCC) Alloys

In addition to the abovementioned H_2 storage alloys that have been extensively studied, MA can also be used in the synthesis and performance optimization of BCC structure H_2 storage alloys.

Y.Q. Hu et al. [118] used mechanical alloying to synthesize $TiCr_2$ with BCC structure and compared the performance with the alloy of the same composition produced by mechanical grinding. They found that the overall performance of the alloy produced by mechanical alloying is better than that of the mechanically crushed alloy. The H_2 absorption capacity of the MA sample is 1.0 wt.% (52 °C, 2.5 MPa), and the desorption capacity is 0.6 wt.%. Nobuhiko Takeichi et al. [119] studied the effect of different Cr content on the performance of $TiCr_{2-x}$(x = 0, 0.2 and 0.5). The results show that the sample can react with H_2 under the conditions of 5 MPa and 250 °C. $TiCr_{1.5}$ has the highest H_2 content, reaching 0.47 H/M (40 °C, 8 MPa)

Compared with other intermetallic compounds, the V-rich solid solution with BCC structure has attracted great attention because of its relatively high H_2 storage capacity (up to 4 wt.%) [120]. The volumetric H_2 storage capacity of the BCC phase (VH_2 is 0.16 g/cm^3) exceeds that of liquid H_2 (0.07 g/cm^3) [121]. In addition, due to their high reactivity, V-BCC alloys can absorb H_2 at relatively low temperatures without catalysts.

M. Balcerzak et al. [120] synthesized the V-based BCC solid solution $Ti_{0.5}V_{1.4-x}Ni_{0.1}Cr_x$ (x = 0, 0.1, 0.2, 0.3) using mechanical alloying, studied its structure and electronic properties, and studied the addition of Cr atom pairs The effect of vanadium-rich body core-alloy on H_2 storage performance. X-ray photoelectron spectroscopy measurements show that the addition of Cr has a significant impact on the oxidation resistance of V-BCC alloys. The cyclic charge and discharge method proves that the Cr-doped V-BCC alloy significantly improves the cycle life of the material stability.

They also synthesized $Ti_{0.5}V_{1.5-x}Mn_x$ (x = 0, 0.1, 0.2, 0.3) [120], $Ti_{0.5}V_{1.5-x}Ni_x$ (x = 0, 0.1, 0.2, 0.3) [122] by mechanical alloying.

$Ti_{0.5}V_{1.5-x}Co_x$ and $Ti_{0.5}V_{1.4-x}Ni_{0.1}Co_x$($x$ = 0, 0.1, 0.2, 0.3) [123] solid solutions synthesized by mechanical alloying can absorb H_2 with no activation. Their H_2 storage capacities decrease as Co atoms number increases. However, Co raises the hydrogenation kinetics, lowers the hysteresis, and improves the reversibility of the H_2 adsorption.

Toshihiko Kondo et al. [124] mechanically alloyed $CaMg_2$ with V and synthesized Mg_2CaV_3 ternary BCC alloy. This alloy is activated by graphite grinding and then can reach a H_2 storage capacity of 3.3 wt.% (10 h) at 25 °C. Unlike the classic V-based BCC alloy, the alloy still maintains the BCC structure after hydrogenation. The desorption process starts at 270 °C and 0.1 MPa argon atmosphere.

Huaiyu Shao et al. [125] synthesized the Mg-based BCC alloy $Mg_{60}Ni_5Co_mX_{35-m}$(X = Co, B, Al, Cr, V, Pd, and Cu) using the mechanical alloying method and studied the relationship between its lattice parameters and H_2 absorption performance. The results show that the alloys with lattice parameters in the range of 0.300–0.308 nm absorb more H_2 while the alloys with lattice parameters greater than 0.313 nm have difficulty in absorbing H_2. The geometric effect is one of the main influencing factors.

The details of the H_2 storage performance of the intermetallic compounds synthesized by mechanical alloying for H_2 storage in the past five years are listed in Table 1. The literature [126] shows the thermodynamic data of some materials.

Table 1. H_2 absorption/desorption properties of intermetallic compounds produced by mechanical alloying in the last 5 years.

Type	Material	Method	Temperature (°C)	Pressure (bar)	Kinetics (min)	Cycling (Cycle/Stability%)	Max of H_2 (wt.%)	Discharge Capacity (mAh*g^{-1})	Ref.
A_2B	$Mg_2Ni_{0.92}M_{0.08}$	BM	300		12 abs		4.2		[127]
	$Mg_{67}Ni_{33}$	BM 20 h				20/17		380	[128]
	$Mg_2Ni - 5wt\%V$ and $Mg_2Ni - 10wt\%V$	BM	325	0.8			2.75		[129]
	$Mg2Ni$+ graphite	BM 40 h				20/50.8		351.1	[50]
	$Mg2Ni$	BM 48 h					2.8		[130]
	$Mg_2Ni(Fe)$	BM 4 h	400	29			3.9		[131]
	$Mg2.1Ni0.7V0.3 -5wt\%ACAP$	BM 3 h	300		20 abs 10 des		2.8	400	[132]
	$Mg_2(Ni_{1-x}Cux)$ (x = 0–0.15)	wet-milling + BM	320	60			3.55		[133]
	Mg_3GeNi_2	BM 80 h				20/63.8		69.9	[134]
AB	$TiFe$	BM 40 h				50/80		240	[135]
	$(TiFe)_{100-x}Ni_x$	BM 2 h	227 (x = 5)	50 (x = 5)			1 (x = 5)		[136]
	$(TiFe)$	BM 2 h	22	33			1.07		[18]
	$(TiFe)$	BM 36 h	30	100			1.5		[74]
	$(TiFe)$	BM 3 h	40	25	3 abs		0.75		[137]
	$(TiFe) + 4wt\%Zr$	BM 30 min	20	20	60 abs		0.75		[138]
	$(TiFe)_{100-x}Mx$ (M = Co,Nb)	BM 2 h	35	80			0.875 H/M (x = Co) 0.9H/M (x = Nb)		[80]
	$TiFe + 5wt\%Al$	BM	22	50			0.98		[81]
	$Ti_{1.09}Mg_{0.01}Fe_{0.9}Ni_{0.1}$	BM 1 h	80	40	400s des	10/70	0.8	80.4	[139]
	$TiFe + 4wt\%(Zr + 2Mn)$	BM 15 min	22	4	120 abs		1.05		[140]
AB_2	$Mg(Ni_xMn_{1-x})_2$ (x = 0.1,0.25,0.5,0.75,0.9)	BM 10–20 h	300	2	8 abs(x = 0.1) 3 des(x = 0.1)		1.3		[141]
	$LaMgNi_4$	BM 40 h				9/77.4		385	[142]
	$ZrTiV_{0.8}Ni_2Cr_{0.52}Mn_{0.56}Co_{0.08}Al_{0.04}$	BM 30 min				90/44.4		225	[143]
AB_3	$LaZr_2Mn_4Ni_5$	BM 5 h	20	10		30/99.3	1.1	300	[144]
	$LaTi_2Ni_5Al_4$	BM 20 h					4 H/f.u		[145]
	$LaCaMgNi_9$	BM 20 h	25	9.5		30 /73.3	5 H/f.u	150	[146]
	$CeTi_2Ni_{4.5}Al_{0.2}Mn_{0.3}Cr_4$	BM 5 h				30 /93		140	[147]

Table 1. *Cont.*

Type	Material	Method	Temperature (°C)	Pressure (bar)	Kinetics (min)	Cycling (Cycle/Stability%)	Max of H_2 (wt.%)	Discharge Capacity ($mAh*g^{-1}$)	Ref.
A_2B_7	$La_{1.5-x}Gd_xMg_0.5Ni_7$			16	4 abs	50/65 (x = 0.25)	1.48 (x = 0.25)	304 (x = 0.25)	[148]
	$(0 \leqslant x \leqslant 1.5)$					50/94.9 (x = 0.5)			
	$La_{1.5}Mg_{0.5}Ni_7 + Ni - P$	BM 48 h				60/89.3		280	[149]
	$La2 - xMgxNi7$ (x = 0:0.25:1)	BM 48 h	30	1000	2 des	50/60	1.53 (x = 0.5)	248	[114]
AB_5	$LaNi_{3-x}Mn_xCr_2$ (x = 0,0.1,0.3)	BM 10 h			4	30/90		300 (x = 0.3)	[150]
	$[La, Ce, Pr, Nd][Ni, Co, Al, Mn]_5$ +1 wt% Pd	BM 1 h			78 abs		0.86		[99]
	$MmNi_5$	BM 1 h	27	70	2 abs		1.5		[151]
	$CaNi_{4.8}Mn_{0.2}$	BM 40 h				50/ 61		96	[152]
	$LaNi_5 - MWCNTs$	BM	25	40			1.53		[153]
BCC	$Mg_{55}Co_{45}$	BM 125 h	−15	100			3.24		[154]
	$MgVCr$	BM 100 min	30	60			0.9		[155]
	$Ti_{0.5}V_{1.4}Cr_{0.1}$	BM 14 h	30			50/71.43	2.9	70	[156]
	$Mg_{64}Pd_3Co_{33}$	BM 120 h				20/8		624	[157]
	$Ti_{0.5}V_{1.3}Ni_{0.1}Co_{0.1}$	BM 14 h	30			50/50		75	[123]
	$Ti_{0.7}Nb_{0.3}$	BM 6 h	300	20			3.2		[158]
	$Ti_{0.9}Zr_{0.1}Mn_{1.2}V_{0.4}Cr_{0.4}$	BM 10 h				10/ 24.1		409.5	[159]

6. Conclusions

As a pure solid-state synthesis method, mechanical alloying has many applications in the synthesis and optimization of H_2 storage materials. For different alloy systems, the specific effects of mechanical alloying are different. However, in general, mechanical alloying can reduce the particle size (even down to the nanometer scale), can increase the specific surface area, and can reduce the diffusion distance of H_2. For some elements that cannot form a solid solution or alloy for various reasons, mechanical alloying makes the alloying of these elements possible because it is a pure solid-state method. This is not only a unique advantage compared to traditional melting but also an important basis for the synthesis of new H_2 storage alloys and the addition of various additives to optimize performance. However, many scholars have also proposed mechanical alloying defects, the most important of which is the consolidation of the powder on the inner wall of the container and the surface of the ball. This causes an uneven composition, waste of raw materials, etc., and more importantly, it is difficult to finely control the composition of the material. Several methods have been proposed, such as adding grinding aids (carbon material, etc.) and grinding aid liquids (such as alcohol, etc.), but they still cannot be completely resolved. Mechanical alloying is also a problem without uniform parameters for each alloy system. Before performing mechanical alloying, various scholars or organizations usually conduct a series of experiments to determine the optimal parameters, but this not only consumes a lot of time and materials but may only obtained the best within the set parameter range.

In summary, for solid H_2 storage materials, mechanical alloying is an extremely attractive synthesis method. Many scholars have used this method to synthesize H_2 storage materials with great potential. However, for possible large-scale commercial applications in the future, further optimization of the technology still needs to be considered.

Author Contributions: Supervision, O.E., D.C. and Y.L.; Validation, O.E. and D.C.; Writing—Original draft, Y.L., D.C. and O.E.; Writing—Review and editing, Y.L., D.C. and O.E. All authors have read and agreed to the published version of the manuscript.

Funding: This research received no external funding.

Institutional Review Board Statement: Not applicable.

Informed Consent Statement: Not applicable.

Data Availability Statement: Not applicable.

Acknowledgments: The authors of the paper would like to thank the China Scholarship Council(CSC).

Conflicts of Interest: The authors declare no conflict of interest.

Abbreviations

The following abbreviations are used in this manuscript:

H_2	Hydrogen
LOHCs	Liquid Organic Hydrogen Carriers
NiMH	Nickel Metal Hydride
RE	Rare Earth
Mg	Magnesium
HPT	High Pressure Torsion
GR	Roll Groove
HER	Hydrogen Evolution Reaction
DFT	Density Function Theory
SWCNT	Single Wall Carbon Nanotube
MWCNT	Multiwall Carbon Nanotube
ETM	Early Transition Metal
MH	Metal Hydride
MA	Mechanical Alloying
BCC	Body Centered Cubic

References

1. Shafiee, S.; Topal, E. When will fossil fuel reserves be diminished? *Energy Policy* **2009**, *37*, 181–189. [CrossRef]
2. Vohra, K.; Vodonos, A.; Schwartz, J.; Marais, E.A.; Sulprizio, M.P.; Mickley, L.J. Global mortality from outdoor fine particle pollution generated by fossil fuel combustion: Results from GEOS-Chem. *Environ. Res.* **2021**, *195*, 110754. [CrossRef]
3. Park, G.; Jang, H.; Kim, C. Design of composite layer and liner for structure safety of hydrogen pressure vessel (type 4). *J. Mech. Sci. Technol.* **2021**, 1–11. [CrossRef]
4. Züttel, A. Hydrogen storage methods. *Naturwissenschaften* **2004**, *91*, 157–172. [CrossRef] [PubMed]
5. Mahendra Y.; Xu, Q. Liquid-phase chemical hydrogen storage materials. *Energy Environ. Sci.* **2012**, *5*, 9698–9725. [CrossRef]
6. Rusman, N.A.; Dahari, M. A review on the current progress of metal hydrides material for solid-state hydrogen storage applications. *Int. J. Hydrog. Energy* **2016**, *41*, 12108–12126. [CrossRef]
7. Gasiorowski, A.; Iwasieczko, W.; Skoryna, D.; Drulis, H.; Jurczyk, M. Hydriding properties of nanocrystalline Mg_2-xMxNi alloys synthesized by mechanical alloying (M = Mn, Al). *J. Alloys Compd.* **2004**, *364*, 283–288. [CrossRef]
8. Falahati, H.; Barz, D.P. Evaluation of hydrogen sorption models for AB_5-type metal alloys by employing a gravimetric technique. *Int. J. Hydrog. Energy* **2013**, *38*, 8838–8851. [CrossRef]
9. Palewski, T.; Tristan, N.V.; Drulis, H.; Ćwik, J. Hydrogenation process of the Gd_3M (M = Ni or Co) intermetallics compound. *J. Alloys Compd.* **2005**, *404-406*, 584–587. [CrossRef]
10. Bhuyan, R.K.; Mohapatra, R.K.; Nath, G.; Sahoo, B.K.; Das, D.; Pamu, D. Influence of high-energy ball milling on structural, microstructural, and optical properties of Mg_2TiO_4 nanoparticles. *J. Mater. Sci. Mater. Electron.* **2019**, *31*, 628–636. [CrossRef]
11. Xie, X.B.; Hou, C.; Chen, C.; Sun, X.; Pang, Y.; Zhang, Y.; Yu, R.; Wang, B.; Du, W. First-principles studies in Mg-based hydrogen storage Materials: A review. *Energy* **2020**, *211*, 118959. [CrossRef]
12. Luo, Q.; Li, J.; Li, B.; Liu, B.; Shao, H.; Li, Q. Kinetics in Mg-based hydrogen storage materials: Enhancement and mechanism. *J. Magnes. Alloy.* **2019**, *7*, 58–71. [CrossRef]
13. Jurczyk, M.; Smardz, L.; Okonska, I.; Jankowska, E.; Nowak, M.; Smardz, K. Nanoscale Mg-based materials for hydrogen storage. *Int. J. Hydrog. Energy* **2008**, *33*, 374–380. [CrossRef]
14. Khan, D.; Panda, S.; Ma, Z.; Ding, W.; Zou, J. Formation and hydrogen storage behavior of nanostructured Mg_2FeH_6 in a compressed 2MgH2–Fe composite. *Int. J. Hydrog. Energy* **2020**, *45*, 21676–21686. [CrossRef]
15. Antiqueira, F.J.; Leiva, D.R.; Zepon, G.; de Cunha, B.F.; Figueroa, S.J.; Botta, W.J. Fast hydrogen absorption/desorption kinetics in reactive milled Mg-8 mol% Fe nanocomposites. *Int. J. Hydrog. Energy* **2020**, *45*, 12408–12418. [CrossRef]
16. Liu, B.H.; Kim, D.M.; Lee, K.Y.; Lee, J.Y. Hydrogen storage properties of $TiMn_2$-based alloys. *J. Alloys Compd.* **1996**, *240*, 214–218. [CrossRef]
17. Hosni, B.; Khaldi, C.; ElKedim, O.; Fenineche, N.; Lamloumi, J. Electrochemical properties of Ti_2Ni hydrogen storage alloy. *Int. J. Hydrog. Energy* **2017**, *42*, 1420–1428. [CrossRef]
18. Zadorozhnyy, V.Y.; Milovzorov, G.S.; Klyamkin, S.N.; Zadorozhnyy, M.Y.; Strugova, D.V.; Gorshenkov, M.V.; Kaloshkin, S.D. Preparation and hydrogen storage properties of nanocrystalline TiFe synthesized by mechanical alloying. *Prog. Nat. Sci. Mater. Int.* **2017**, *27*, 149–155. [CrossRef]
19. Reilly, J.J.; Wiswall, R.H. Formation and properties of iron titanium hydride. *Inorg. Chem.* **1974**, *13*, 218–222. [CrossRef]
20. Sandrock, G.D.; Goodell, P.D. Surface poisoning of $LaNi_5$, FeTi and (Fe,Mn)Ti by O_2, Co and H_2O. *J. Less Common Met.* **1980**, *73*, 161–168. [CrossRef]
21. Davids, M.W.; Lototskyy, M.; Nechaev, A.; Naidoo, Q.; Williams, M.; Klochko, Y. Surface modification of TiFe hydrogen storage alloy by metal-organic chemical vapour deposition of palladium. *Int. J. Hydrog. Energy* **2011**, *36*, 9743–9750. [CrossRef]
22. Edalati, K.; Matsuda, J.; Iwaoka, H.; Toh, S.; Akiba, E.; Horita, Z. High-pressure torsion of TiFe intermetallics for activation of hydrogen storage at room temperature with heterogeneous nanostructure. *Int. J. Hydrog. Energy* **2013**, *38*, 4622–4627. [CrossRef]
23. Edalati, K.; Matsuda, J.; Arita, M.; Daio, T.; Akiba, E.; Horita, Z. Mechanism of activation of TiFe intermetallics for hydrogen storage by severe plastic deformation using high-pressure torsion. *Appl. Phys. Lett.* **2013**, *103*, 143902. [CrossRef]
24. Edalati, K.; Matsuda, J.; Yanagida, A.; Akiba, E.; Horita, Z. Activation of TiFe for hydrogen storage by plastic deformation using groove rolling and high-pressure torsion: Similarities and differences. *Int. J. Hydrog. Energy* **2014**, *39*, 15589–15594. [CrossRef]
25. Valiev, R.Z.; Estrin, Y.; Horita, Z.; Langdon, T.G.; Zechetbauer, M.J.; Zhu, Y.T. Producing bulk ultrafine-grained materials by severe plastic deformation. *JOM* **2006**, *58*, 33–39. [CrossRef]
26. Vucht, J.V.; Kuijpers, F.; Bruning, H.C.A.M. Reversible room-temperature absorption of large quantities of hydrogen by intermetallic compounds. *Philips Res. Rep.* **1970**, *25*, 133–140.
27. Huston, E.L.; Sandrock, G.D. Engineering properties of metal hydrides. *J. Less Common Met.* **1980**, *74*, 435–443. [CrossRef]
28. Kim, J.K.; Park, I.S.; Kim, K.J.; Gawlik, K. A hydrogen-compression system using porous metal hydride pellets of $LaNi_5$-xAlx. *Int. J. Hydrog. Energy* **2008**, *33*, 870–877. [CrossRef]
29. Simičić, M.V.; Zdujić, M.; Jelovac, D.M.; Rakin, P.M. Hydrogen storage material based on $LaNi_5$ alloy produced by mechanical alloying. *J. Power Sources* **2001**, *92*, 250–254. [CrossRef]
30. Van Mal, H.H.; Buschow, K.H.; Kuijpers, F.A. Hydrogen absorption and magnetic properties of $LaCo_{5x}Ni_{5-5x}$ compounds. *J. Less Common Met.* **1973**, *32*, 289–296. [CrossRef]
31. Sakai, T.; Oguro, K.; Miyamura, H.; Kuriyama, N.; Kato, A.; Ishikawa, H.; Iwakura, C. Some factors affecting the cycle lives of $LaNi_5$-based alloy electrodes of hydrogen batteries. *J. Less Common Met.* **1990**, *161*, 193–202. [CrossRef]

32. Kadir, K.; Sakai, T.; Uehara, I. Synthesis and structure determination of a new series of hydrogen storage alloys; RMg_2Ni_9 (R = La, Ce, Pr, Nd, Sm and Gd) built from $MgNi_2$ Laves-type layers alternating with AB5 layers. *J. Alloys Compd.* **1997**, *257*, 115–121. [CrossRef]
33. Libowitz, G.G.; Hayes, H.F.; Gibb, T.R. The system zirconium-nickel and hydrogen. *J. Phys. Chem.* **1958**, *62*, 76–79. [CrossRef]
34. Hoang, K.; Van De Walle, C.G. Mechanism for the decomposition of lithium borohydride. *Int. J. Hydrog. Energy* **2012**, *37*, 5825–5832. [CrossRef]
35. Lal, C.; Jain, I.P. Effect of ball milling on structural and hydrogen storage properties of Mg-x wt% FeTi(x = 2 & 5) solid solutions. *Int. J. Hydrog. Energy* **2012**, *37*, 3761–3766. [CrossRef]
36. Tessier, P.; Schulz, R.; Ström-Olsen, J.O. Elastic stress in composite FeTi hydrogen storage materials. *J. Mater. Res.* **1998**, *13*, 1538–1547. [CrossRef]
37. Aymard, L.; Lenain, C.; Courvoisier, L.; Salver-Disma, F.; Tarascon, J. Effect of Carbon Additives on the Electrochemical Properties of AB 5 Graphite Composite Electrodes Prepared by Mechanical Milling. *J. Electrochem. Soc.* **1999**, *146*, 2015. [CrossRef]
38. Jianfeng Zhang.; Zhinian Li.; Yuanfang Wu.; Xiumei Guo.; Jianhua Ye.; Baolong Yuan.; Shumao Wang.; Lijun Jiang. Recent advances on the thermal destabilization of Mg-based hydrogen storage materials. *RSC Adv.* **2019**, *9*, 408–428. [CrossRef]
39. Yartys, V.A.; Lototskyy, M.V.; Akiba, E.; Albert, R.; Antonov, V.E.; Ares, J.R.; Baricco, M.; Bourgeois, N.; Buckley, C.E.; Bellosta von Colbe, J.M.; et al. Magnesium based materials for hydrogen based energy storage: Past, present and future. *Int. J. Hydrog. Energy* **2019**, *44*, 7809–7859. [CrossRef]
40. Baran, A.; Materials, M.P. Magnesium-based materials for hydrogen storage—A scope review. *Materials* **2020**, *13*, 3993 . [CrossRef]
41. Ivanov, E.; Konstanchuk, I.; Stepanov, A.; Boldyrev, V. Magnesium mechanical alloys for hydrogen storage. *J. Less Common Met.* **1987**, *131*, 25–29. [CrossRef]
42. Singh, A.K.; Singh, A.K.; Srivastava, O.N. On the synthesis of the Mg_2Ni alloy by mechanical alloying. *J. Alloys Compd.* **1995**, *227*, 63–68. [CrossRef]
43. Zaluski, L.; Zaluska, A.; Ström-Olsen, J.O. Hydrogen absorption in nanocrystalline Mg_2Ni formed by mechanical alloying. *J. Alloys Compd.* **1995**, *217*, 245–249. [CrossRef]
44. Orimo, S.; Fujii, H. Materials science of Mg-Ni-based new hydrides. *Appl. Phys. A* **2001**, *72*, 167–186. [CrossRef]
45. Sakintuna, B.; Lamari-Darkrim, F.; Hirscher, M. Metal hydride materials for solid hydrogen storage: A review. *Int. J. Hydrog. Energy* **2007**, *32*, 1121–1140. [CrossRef]
46. Cui, N.; He, P.; Luo, J.L. Magnesium-based hydrogen storage materials modified by mechanical alloying. *Acta Mater.* **1999**, *47*, 3737–3743. [CrossRef]
47. Liang, G.; Boily, S.; Huot, J.; Van Neste, A.; Schulz, R. Mechanical alloying and hydrogen absorption properties of the Mg–Ni system. *J. Alloys Compd.* **1998**, *267*, 302–306. [CrossRef]
48. Huang, L.W.; Elkedim, O.; Moutarlier, V. Synthesis and characterization of nanocrystalline Mg_2Ni prepared by mechanical alloying: Effects of substitution of Mn for Ni. *J. Alloys Compd.* **2010**, *504*, S311–S314. [CrossRef]
49. Huang, L.W.; Elkedim, O.; Hamzaoui, R. First principles investigation of the substitutional doping of Mn in Mg_2Ni phase and the electronic structure of Mg_3MnNi_2 phase. *J. Alloys Compd.* **2011**, *509*, S328–S333. [CrossRef]
50. Du, Q.; Li, S.; Huang, G.; Feng, Q. Enhanced electrochemical kinetics of magnesium-based hydrogen storage alloy by mechanical milling with graphite. *Int. J. Hydrog. Energy* **2017**, *42*, 21871–21879. [CrossRef]
51. Gkanas, E.I.; Damian, A.; Ioannidou, A.; Stoian, G.; Lupu, N.; Gjoka, M.; Makridis, S.S. Synthesis, characterisation and hydrogen sorption properties of mechanically alloyed $Mg(Ni_{1-x}Mn_x)_2$. *Mater. Today Energy* **2019**, *13*, 186–194. [CrossRef]
52. Ruggeri, S.; Lenain, C.; Roué, L.; Alamdari, H.; Liang, G.X.; Huot, J.; Schulz, R. Optimization of the Ball-Milling Parameters for the Synthesis of Amorphous MgNi Alloy Used as Negative Electrode in Ni-MH Batteries. *Mater. Sci. Forum* **2001**, *377*, 63–70. [CrossRef]
53. Liu, W.; Lei, Y.; Sun, D.; Wu, J.; Wang, Q. A study of the degradation of the electrochemical capacity of amorphous $Mg_{50}Ni_{50}$ alloy. *J. Power Sources* **1996**, *58*, 243–247. [CrossRef]
54. Goo, N.H.; Woo, J.H.; Lee, K.S. Mechanism of rapid degradation of nanostructured Mg_2Ni hydrogen storage alloy electrode synthesized by mechanical alloying and the effect of mechanically coating with nickel. *J. Alloys Compd.* **1999**, *288*, 286–293. [CrossRef]
55. Lenain, C.; Aymard, L.; Electrochemistry, J.T.J.S.S. Electrochemical properties of Mg_2Ni and Mg_2Ni_2 prepared by mechanical alloying. *Electrochemistry* **1998**, *2*, 285–290.
56. Abe, T.; Tachikawa, T.; Hatano, Y.; Watanabe, K. Electrochemical behavior of amorphous MgNi as negative electrodes in rechargeable Ni–MH batteries. *J. Alloys Compd.* **2002**, *330-332*, 792–795. [CrossRef]
57. Anik, M. Electrochemical hydrogen storage capacities of Mg_2Ni and MgNi alloys synthesized by mechanical alloying. *J. Alloys Compd.* **2010**, *491*, 565–570. [CrossRef]
58. Iwakura, C.; Shin-ya, R.; Miyanohara, K.; Nohara, S.; Inoue, H. Effects of Ti–V substitution on electrochemical and structural characteristics of MgNi alloy prepared by mechanical alloying. *Electrochim. Acta* **2001**, *46*, 2781–2786. [CrossRef]
59. Ruggeri, S.; Roué, L.; Huot, J.; Schulz, R.; Aymard, L.; Tarascon, J.M. Properties of mechanically alloyed Mg–Ni–Ti ternary hydrogen storage alloys for Ni-MH batteries. *J. Power Sources* **2002**, *112*, 547–556. [CrossRef]
60. Jiang, J.J.; Gasik, M. An electrochemical investigation of mechanical alloying of MgNi-based hydrogen storage alloys. *J. Power Sources* **2000**, *89*, 117–124. [CrossRef]

61. Elkedim, O.; Huang, L.; Bassir, D. Advanced study of hydrogen storage by substitutional doping of Mn and Ti in Mg_2Ni phase. *Int. J. Simul. Multidiscip. Des. Optim.* **2014**, *5*, A24. [CrossRef]
62. Amirkhiz, B.S.; Danaie, M.; Mitlin, D. The influence of SWCNT–metallic nanoparticle mixtures on the desorption properties of milled MgH_2 powders. *Nanotechnology* **2009**, *20*, 204016. [CrossRef]
63. Yao, X.; Wu, C.; Du, A.; Lu, G.Q.; Cheng, H.; Smith, S.C.; Zou, J. Mg-Based Nanocomposites with High Capacity and Fast Kinetics for Hydrogen Storage. *J. Phys. Chem. B* **2006**, *110*, 11697–11703. [CrossRef]
64. Huang, L.W.; Elkedim, O.; Nowak, M.; Jurczyk, M.; Chassagnon, R.; Meng, D.W. Synergistic effects of multiwalled carbon nanotubes and Al on the electrochemical hydrogen storage properties of Mg_2Ni-type alloy prepared by mechanical alloying. *Int. J. Hydrog. Energy* **2012**, *37*, 1538–1545. [CrossRef]
65. Guo, J.; Yang, K.; Xu, L.; Liu, Y.; Zhou, K. Hydrogen storage properties of $Mg_{76}Ti1_2Fe_{12-x}Nix$(x = 0,4,8,12) alloys by mechanical alloying. *Int. J. Hydrog. Energy* **2007**, *32*, 2412–2416. [CrossRef]
66. Gao, X. P.; Wang, Y.; Lu, Z.W.; Hu, W. K.; Wu, F.; Song, D.Y.; Shen, P.W. Preparation and Electrochemical Hydrogen Storage of the Nanocrystalline $LaMg_{12}$ Alloy with Ni Powders. *Chem. Mater.* **2004**, *16*, 2515–2517. [CrossRef]
67. Zhang, Y.; Wang, J.; Zhang, P.; Zhu, Y.; Hou, Z.; Shang, H. An Investigation on Hydrogen Storage Kinetics of the Nanocrystalline and Amorphous $LaMg_{12}$-type Alloys Synthesized by Mechanical Milling. *J. Wuhan Univ. Technol.-Mater. Sci. Ed.* **2018**, *33*, 278–287. [CrossRef]
68. Li, X.; Zhao, D.; Zhang, Y.; Xu, J.; Zhang, G.; Zhang, Y. Hydrogen storage properties of mechanically milled La_2Mg_{17-x} wt.(x = 0, 50, 100, 150 and 200) composites. *J. Rare Earths* **2013**, *31*, 694–700. [CrossRef]
69. Song, M.Y. Improvement in hydrogen storage characteristics of magnesium by mechanical alloying with nickel. *J. Mater. Sci.* **1995**, *30*, 1343–1351. [CrossRef]
70. Lotoskyy, M.; Denys, R.; Yartys, V.A.; Eriksen, J.; Goh, J.; Nyamsi, S.N.; Sita, C.; Cummings, F. An outstanding effect of graphite in nano-MgH_2–TiH_2 on hydrogen storage performance. *J. Mater. Chem. A* **2018**, *6*, 10740–10754. [CrossRef]
71. Zhou, C.; Fang, Z.Z.; Robert C. Bowman, J. Stability of Catalyzed Magnesium Hydride Nanocrystalline During Hydrogen Cycling. Part I: Kinetic Analysis. *J. Phys. Chem. C* **2015**, *119*, 22261–22271. [CrossRef]
72. Rizo-Acosta, P.; Cuevas, F.; Latroche, M. Hydrides of early transition metals as catalysts and grain growth inhibitors for enhanced reversible hydrogen storage in nanostructured magnesium. *J. Mater. Chem. A* **2019**, *7*, 23064–23075. [CrossRef]
73. Chiang, C.H.; Chin, Z.H.; Perng, T.P. Hydrogenation of TiFe by high-energy ball milling. *J. Alloys Compd.* **2000**, *307*, 259–265. [CrossRef]
74. Emami, H.; Edalati, K.; Matsuda, J.; Akiba, E.; Horita, Z. Hydrogen storage performance of TiFe after processing by ball milling. *Acta Mater.* **2015**, *88*, 190–195. [CrossRef]
75. Zaluski, L.; Zaluska, A.; Ström-Olsen, J.O. Nanocrystalline metal hydrides. *J. Alloys Compd.* **1997**, *253–254*, 70–79. [CrossRef]
76. Zadorozhnyi, V.Y.; Skakov, Y.A.; Milovzorov, G.S. Appearance of metastable states in Fe-Ti and Ni-Ti systems in the process of mechanochemical synthesis. *Met. Sci. Heat Treat.* **2008**, *50*, 404–410. [CrossRef]
77. Zadorozhnyy, V.; Klyamkin, S.; Zadorozhnyy, M.; Bermesheva, O.; Kaloshkin, S. Hydrogen storage nanocrystalline TiFe intermetallic compound: Synthesis by mechanical alloying and compacting. *Int. J. Hydrog. Energy* **2012**, *37*, 17131–17136. [CrossRef]
78. Inui, H.; Yamamoto, T.; Hirota, M.; Yamaguchi, M. Lattice defects introduced during hydrogen absorption–desorption cycles and their effects on P–C characteristics in some intermetallic compounds. *J. Alloys Compd.* **2002**, *330–332*, 117–124. [CrossRef]
79. Zadorozhnyy, V.Y.; Klyamkin, S.N.; Zadorozhnyy, M.Y.; Gorshenkov, M.V.; Kaloshkin, S.D. Mechanical alloying of nanocrystalline intermetallic compound TiFe doped with sulfur and magnesium. *J. Alloys Compd.* **2014**, *615*, S569–S572. [CrossRef]
80. Berdonosova, E.A.; Zadorozhnyy, V.Y.; Zadorozhnyy, M.Y.; Geodakian, K.V.; Zheleznyi, M.V.; Tsarkov, A.A.; Kaloshkin, S.D.; Klyamkin, S.N. Hydrogen storage properties of TiFe-based ternary mechanical alloys with cobalt and niobium. A thermochemical approach. *Int. J. Hydrog. Energy* **2019**, *44*, 29159–29165. [CrossRef]
81. Zadorozhnyy, V.Y.; Klyamkin, S.N.; Zadorozhnyy, M.Y.; Bermesheva, O.V.; Kaloshkin, S.D. Mechanical alloying of nanocrystalline intermetallic compound TiFe doped by aluminum and chromium. *J. Alloys Compd.* **2014**, *586*, S56–S60. [CrossRef]
82. Liang, L.; Wang, F.; Rong, M.; Wang, Z.; Yang, S.; Wang, J.; Zhou, H.; Liang, L.; Wang, F.; Rong, M.; et al. Recent Advances on Preparation Method of Ti-Based Hydrogen Storage Alloy. *J. Mater. Sci. Chem. Eng.* **2020**, *8*, 18–38. [CrossRef]
83. Sujan, G.K.; Pan, Z.; Li, H.; Liang, D.; Alam, N. An overview on TiFe intermetallic for solid-state hydrogen storage: microstructure, hydrogenation and fabrication processes. *Crit. Rev. Solid State Mater. Sci.* **2019**, *45*, 410–427. [CrossRef]
84. Falcão, R.B.; Dammann, E.D.; Rocha, C.J.; Durazzo, M.; Ichikawa, R.U.; Martinez, L.G.; Botta, W.J.; Leal Neto, R.M. An alternative route to produce easily activated nanocrystalline TiFe powder. *Int. J. Hydrog. Energy* **2018**, *43*, 16107–16116. [CrossRef]
85. Nobuki, T.; Crivello, J.C.; Cuevas, F.; Joubert, J.M. Fast synthesis of TiNi by mechanical alloying and its hydrogenation properties. *Int. J. Hydrog. Energy* **2019**, *44*, 10770–10776. [CrossRef]
86. Zhang, Z.; Elkedim, O.; Balcerzak, M.; Jurczyk, M. Structural and electrochemical hydrogen storage properties of MgTiNix(x = 0.1, 0.5, 1, 2) alloys prepared by ball milling. *Int. J. Hydrog. Energy* **2016**, *41*, 11761–11766. [CrossRef]
87. Zhang, Z.; Elkedim, O.; Ma, Y.Z.; Balcerzak, M.; Jurczyk, M. The phase transformation and electrochemical properties of TiNi alloys with Cu substitution: Experiments and first-principle calculations. *Int. J. Hydrog. Energy* **2017**, *42*, 1444–1450. [CrossRef]
88. Zhang, Z.; Elkedim, O.; Balcerzak, M.; Jurczyk, M.; Chassagnon, R. Effect of Ni content on the structure and hydrogenation property of mechanically alloyed TiMgNix ternary alloys. *Int. J. Hydrog. Energy* **2017**, *42*, 23751–23758. [CrossRef]

89. Zhang, Z.; Elkedim, O.; Zhang, M.; Bassir, D. Systematic investigation of mechanically alloyed Ti-Mg-Ni used as negative electrode in Ni-MH battery. *J. Solid State Electrochem.* **2017**, *22*, 1669–1676. [CrossRef]
90. Li, X.D.; Elkedim, O.; Cuevas, F.; Chassagnon, R. Structural and hydrogenation study on the ball milled TiH_2–Mg–Ni. *Int. J. Hydrog. Energy* **2015**, *40*, 4212–4218. [CrossRef]
91. Li, X.D.; Elkedim, O.; Nowak, M.; Jurczyk, M. Characterization and first principle study of ball milled Ti–Ni with Mg doping as hydrogen storage alloy. *Int. J. Hydrog. Energy* **2014**, *39*, 9735–9743. [CrossRef]
92. Hosni, B.; Li, X.; Khaldi, C.; Elkedim, O.; Lamloumi, J. Structure and electrochemical hydrogen storage properties of Ti2Ni alloy synthesized by ball milling. *J. Alloys Compd.* **2014**, *615*, 119–125. [CrossRef]
93. Li, X.D.; Elkedim, O.; Nowak, M.; Jurczyk, M.; Chassagnon, R. Structural characterization and electrochemical hydrogen storage properties of Ti2-xZrxNi(x = 0, 0.1, 0.2) alloys prepared by mechanical alloying. *Int. J. Hydrog. Energy* **2013**, *38*, 12126–12132. [CrossRef]
94. Chu, H.; Zhang, Y.; Sun, L.; Qiu, S.; Qi, Y.; Xu, F.; Yuan, H. Structure and electrochemical properties of composite electrodes synthesized by mechanical milling Ni-free $TiMn_2$-based alloy with La-based alloys. *J. Alloys Compd.* **2007**, *446-447*, 614–619. [CrossRef]
95. Choi, M.J.; Hong, H.S.; Lee, K.S. Electrochemical characteristics of the composite metal hydride of TiFe and $TiMn_2$ synthesized by mechanical alloying. *J. Alloys Compd.* **2003**, *358*, 306–311. [CrossRef]
96. Liang, G.; Huot, J.; Schulz, R. Hydrogen storage properties of the mechanically alloyed LaNi5-based materials. *J. Alloys Compd.* **2001**, *320*, 133–139. [CrossRef]
97. Lototsky, M.V.; Williams, M.; Yartys, V.A.; Klochko, Y.V.; Linkov, V.M. Surface-modified advanced hydrogen storage alloys for hydrogen separation and purification. *J. Alloys Compd.* **2011**, *509*, S555–S561. [CrossRef]
98. Bratanich, T.I.; Skorokhod, V.V. Reversible Hydriding of $LaNi_{5-x}Al_x$—Pd Composites in the Presence of Carbon Monoxide. *Powder Metall. Met. Ceram.* **2000**, *39*, 575–583. [CrossRef]
99. Modibane, K.D.; Lototskyy, M.; Davids, M.W.; Williams, M.; Hato, M.J.; Molapo, K.M. Influence of co-milling with palladium black on hydrogen sorption performance and poisoning tolerance of surface modified AB5-type hydrogen storage alloy. *J. Alloys Compd.* **2018**, *750*, 523–529. [CrossRef]
100. Kadir, K.; Sakai, T.; Uehara, I. Structural investigation and hydrogen storage capacity of $LaMg_2Ni_9$ and $(La_{0.65}Ca_{0.35})(Mg_{1.32}Ca_{0.68})Ni_9$ of the AB2C9 type structure. *J. Alloys Compd.* **2000**, *302*, 112–117. [CrossRef]
101. Liu, W.; Webb, C.J.; Gray, E.M.A. Review of hydrogen storage in AB3 alloys targeting stationary fuel cell applications. *Int. J. Hydrog. Energy* **2016**, *41*, 3485–3507. [CrossRef]
102. Belgacem, Y.B.; Khaldi, C.; Lamloumi, J. The effect of the discharge rate on the electrochemical properties of AB3-type hydrogen storage alloy as anode in nickel–metal hydride batteries. *Int. J. Hydrog. Energy* **2017**, *42*, 12797–12807. [CrossRef]
103. Fang, F.; Chen, Z.; Wu, D.; Liu, H.; Dong, C.; Song, Y.; Sun, D. Subunit volume control mechanism for dehydrogenation performance of AB3-type superlattice intermetallics. *J. Power Sources* **2019**, *427*, 145–153. [CrossRef]
104. Jaafar, H.; Aymard, L.; Dachraoui, W.; Demortière, A.; Abdellaoui, M. Preparation and characterization of mechanically alloyed AB_3-type based material $LaMg_2Ni_5A_{14}$ and its solid-gaz hydrogen storage reaction. *J. Solid State Chem.* **2018**, *260*, 73–79. [CrossRef]
105. Elghali, M.; Abdellaoui, M.; Paul-Boncour, V.; Latroche, M. Synthesis and structural characterization of mechanically alloyed AB_3-type based material: $LaZr_2Mn_4Ni_5$. *Intermetallics* **2013**, *41*, 76–81. [CrossRef]
106. Ayari, M.; Ghodbane, O.; Abdellaoui, M. Elaboration and electrochemical characterization of $LaTi_2Cr_4Ni_5$ -based metal hydride alloys. *Int. J. Hydrog. Energy* **2015**, *40*, 10934–10942. [CrossRef]
107. Sahli, I.; Ghodbane, O.; Abdellaoui, M. Elaboration and electrochemical characterization of $LaZr_2Cr_4Ni_5$ -based metal hydride alloys. *Ionics* **2016**, *22*, 1973–1983. [CrossRef]
108. Ayari, M.; Ghodbane, O.; Abdellaoui, M. Electrochemical study of the reversible hydrogen storage in CeTi2Cr4Ni5-based metal hydride alloys. *Int. J. Hydrog. Energy* **2016**, *41*, 18582–18591. [CrossRef]
109. Kadir, K.; Kuriyama, N.; Sakai, T.; Uehara, I.; Eriksson, L. Structural investigation and hydrogen capacity of $CaMg_2Ni_9$: A new phase in the AB_2C_9 system isostructural with $LaMg_2Ni_9$. *J. Alloys Compd.* **1999**, *284*, 145–154. [CrossRef]
110. Miraglia, S.; Girard, G.; Fruchart, D.; Liang, G.; Huot, J.; Schulz, R. Structural characterization and some hydrogen absorption properties of (MgxCa1-x)Ni2.6: A new phase in the Mg–Ca–Ni system. *J. Alloys Compd.* **2009**, *478*, L33–L36. [CrossRef]
111. Chebab, S.; Abdellaoui, M.; Latroche, M.; Paul-Boncour, V. Structural and hydrogen storage properties of $LaCaMgNi_9$-type alloy obtained by mechanical alloying. *Mater. Renew. Sustain. Energy* **2015**, *4*, 1–10. [CrossRef]
112. Akiba, E.; Hayakawa, H.; Kohno, T. Crystal structures of novel La–Mg–Ni hydrogen absorbing alloys. *J. Alloys Compd.* **2006**, *408-412*, 280–283. [CrossRef]
113. Szajek, A.; Jurczyk, M.; Rajewski, W. The electronic and electrochemical properties of the $LaNi_5$, LaNi4Al and LaNi3AlCo systems. *J. Alloys Compd.* **2000**, *307*, 290–296. [CrossRef]
114. Balcerzak, M.; Nowak, M.; Jurczyk, M. Hydrogenation and electrochemical studies of La–Mg–Ni alloys. *Int. J. Hydrog. Energy* **2017**, *42*, 1436–1443. [CrossRef]
115. Balcerzak, M.; Nowak, M.; Jurczyk, M. The Influence of Pr and Nd Substitution on Hydrogen Storage Properties of Mechanically Alloyed $(La,Mg)_2Ni_7$-Type Alloys. *J. Mater. Eng. Perform.* **2018**, *27*, 6166–6174. [CrossRef]

116. Dymek, M.; Nowak, M.; Jurczyk, M.; Bala, H. Electrochemical characterization of nanocrystalline hydrogen storage $La_{1.5}Mg_{0.5}Ni_{6.5}Co_{0.5}$ alloy covered with amorphous nickel. *J. Alloys Compd.* **2019**, *780*, 697–704. [CrossRef]
117. Nowak, M.; Balcerzak, M.; Jurczyk, M. Effect of Substitutional Elements on the Thermodynamic and Electrochemical Properties of Mechanically Alloyed La1.5Mg0.5Ni7-xMx alloys (M = Al, Mn). *Metals* **2020**, *10*, 578. [CrossRef]
118. Hu, Y.Q.; Zhang, H.F.; Yan, C.; Ye, L.; Ding, B.Z.; Hu, Z.Q. Preparation and hydrogenation of body-centered-cubic TiCr2 alloy. *Mater. Lett.* **2004**, *58*, 783–786. [CrossRef]
119. Takeichi, N.; Takeshita, H.T.; Oishi, T.; Kaneko, T.; Tanaka, H.; Kiyobayashi, T.; Kuriyama, N. Hydrogenation of Body-Centered-Cubic Titanium-Chromium Alloys Prepared by Mechanical Grinding. *Mater. Trans.* **2002**, *43*, 2161–2164. [CrossRef]
120. Balcerzak, M. Hydrogenation properties of nanocrystalline TiVMn body-centered-cubic alloys. *Int. J. Hydrog. Energy* **2020**, *45*, 15521–15529. [CrossRef]
121. Lototsky, M.V.; Yartys, V.A.; Zavaliy, I.Y. Vanadium-based BCC alloys: Phase-structural characteristics and hydrogen sorption properties. *J. Alloys Compd.* **2005**, *404–406*, 421–426. [CrossRef]
122. Balcerzak, M. Effect of Ni on electrochemical and hydrogen storage properties of V-rich body-centered-cubic solid solution alloys. *Int. J. Hydrog. Energy* **2018**, *43*, 8395–8403. [CrossRef]
123. Balcerzak, M. Structural, Electrochemical and Hydrogen Sorption Studies of Nanocrystalline Ti-V-Co and Ti-V-Ni-Co Alloys Synthesized by Mechanical Alloying Method. *J. Mater. Eng. Perform.* **2019**, *28*, 4838–4844. [CrossRef]
124. Kondo, T.; Sakurai, Y. Hydrogen absorption–desorption properties of Mg–Ca–V BCC alloy prepared by mechanical alloying. *J. Alloys Compd.* **2006**, *417*, 164–168. [CrossRef]
125. Shao, H.; Asano, K.; Enoki, H.; Akiba, E. Correlation study between hydrogen absorption property and lattice structure of Mg-based BCC alloys. *Int. J. Hydrog. Energy* **2009**, *34*, 2312–2318. [CrossRef]
126. Djellouli, A.; Benyelloul, K.; Aourag, H.; Bekhechi, S.; Adjadj, A.; Bouhadda, Y.; ElKedim, O. A datamining approach to classify, select and predict the formation enthalpy for intermetallic compound hydrides. *Int. J. Hydrog. Energy* **2018**, *43*, 19111–19120. [CrossRef]
127. Zhang, Y.; Zhang, H.; Ding, X.; Liu, D.; Zhang, Q.; Si, T. Microstructure characterization and hydrogen storage properties study of $Mg_2Ni_{0.92}M_{0.08}$ (M = Ti, V, Fe or Si) alloys. *Prog. Nat. Sci. Mater. Int.* **2018**, *28*, 464–469. [CrossRef]
128. Hapçı Ağaoğlu, G.; Orhan, G. Elaboration and electrochemical characterization of Mg–Ni hydrogen storage alloy electrodes for Ni/MH batteries. *Int. J. Hydrog. Energy* **2017**, *42*, 8098–8108. [CrossRef]
129. Kumar, S.; Miyaoka, H.; Ichikawa, T.; Dey, G.K.; Kojima, Y. Micro-alloyed Mg_2Ni for better performance as negative electrode of Ni-MH battery and hydrogen storage. *Int. J. Hydrog. Energy* **2017**, *42*, 5220–5226. [CrossRef]
130. Nobuki, T.; Okuzumi, Y.; Hatate, M.; Crivello, J.C.; Cuevas, F.; Joubert, J.M. Mechanosynthesis and Reversible Hydrogen Storage of Mg_2Ni and Mg_2Cu Alloys. *Mater. Trans.* **2019**, *60*, 441–449. [CrossRef]
131. Chen, X.; Zou, J.; Huang, S.; He, G.; Zhao, N.; Zeng, X.; Ding, W. Hydrogen storage in Mg_2 Ni(Fe)H_4 nano particles synthesized from coarse-grained Mg and nano sized Ni(Fe) precursor. *RSC Adv.* **2018**, *8*, 18959–18965. [CrossRef]
132. Grigorova, E.; Tzvetkov, P.; Todorova, S.; Markov, P.; Spassov, T. Facilitated Synthesis of Mg_2Ni Based Composites with Attractive Hydrogen Sorption Properties. *Materials* **2021**, *14*, 1936. [CrossRef] [PubMed]
133. Si, T.; Ma, Y.; Li, Y.; Liu, D. Solid solution of Cu in Mg_2NiH_4 and its destabilized effect on hydrogen desorption. *Mater. Chem. Phys.* **2017**, *193*, 1–6. [CrossRef]
134. Hou, J.; Liu, Z.; Zhu, Y.; Fei, J.; Song, Y.; Zhang, Y.; Zhang, J.; Liu, Y.; Chen, Q.; Li, L. Electrochemical hydrogen storage performance of Mg_3GeNi_2 alloy. *Intermetallics* **2020**, *127*, 106961. [CrossRef]
135. Abrashev, B.; Spassov, T.; Pandev, M.; Vassilev, S.; Popov, A. Hydrogen sorption and electrochemical properties of Ti-Fe based alloys synthesized by mechanical alloying. *Bulg. Chem. Commun.* **2017**, *49*, 247–253.
136. Zadorozhnyy, V.; Berdonosova, E.; Gammer, C.; Eckert, J.; Zadorozhnyy, M.; Bazlov, A.; Zheleznyi, M.; Kaloshkin, S.; Klyamkin, S. Mechanochemical synthesis and hydrogenation behavior of $(TiFe)_{100-x}Ni_x$ alloys. *J. Alloys Compd.* **2019**, *796*, 42–46. [CrossRef]
137. Nobuki, T.; Moriya, T.; Hatate, M.; Crivello, J.C.; Cuevas, F.; Joubert, J.M. Synthesis of tife hydrogen absorbing alloys prepared by mechanical alloying and SPS treatment. *Metals* **2018**, *8*, 264. [CrossRef]
138. Manna, J.; Tougas, B.; Huot, J. Mechanical activation of air exposed TiFe + 4 wthydrogenation by cold rolling and ball milling. *Int. J. Hydrog. Energy* **2018**, *43*, 20795–20800. [CrossRef]
139. Shang, H.; Zhang, Y.; Li, Y.; Qi, Y.; Guo, S.; Zhao, D. Investigation on gaseous and electrochemical hydrogen storage performances of as-cast and milled $Ti_{1.1}Fe_{0.9}Ni_{0.1}$ and $Ti_{1.09}Mg_{0.01}Fe_{0.9}Ni_{0.1}$ alloys. *Int. J. Hydrog. Energy* **2018**, *43*, 1691–1701. [CrossRef]
140. Romero, G.; Lv, P.; Huot, J. Effect of ball milling on the first hydrogenation of TiFe alloy doped with 4 wt% (Zr + 2Mn) additive. *J. Mater. Sci.* **2018**, *53*, 13751–13757. [CrossRef]
141. Gkanas, E.I.; Khzouz, M.; Donac, A.; Ioannidou, A.; Stoian, G.; Lupu, N.; Gjoka, M.; Makridis, S.S. Synthesis and Hydrogen Sorption Characteristics of Mechanically Alloyed $Mg(Ni_xMn_{1-x})_2$ Intermetallics. *arXiv* **2017**, arXiv:1702.04807.
142. Wang, Z.M.; Zhou, H.Y.; Gu, Z.F.; Cheng, G.; Yu, A.B. Preparation of LaMgNi4 alloy and its electrode properties. *J. Alloys Compd.* **2004**, *377*, L7–L9. [CrossRef]
143. Humana, R.M.; Ruiz, F.C.; Thomas, J.E.; Peretti, H.A.; Castro, E.B.; Visintin, A. Properties of composites of metal hydride alloys synthesized by mechanical milling. *J. Solid State Electrochem.* **2016**, *21*, 153–160. [CrossRef]
144. M, E.; M, A. Structural and Hydriding Properties of the $LaZr_2Mn_4Ni_5$-AB3 Type Based Alloy Prepared by Mechanical Alloying from the $LaNi_5$ and $ZrMn_2$ Binary Compounds. *J. Mater. Sci. Eng.* **2016**, *05*. [CrossRef]

145. Jaafar, H.; Dhiab, A.; Slama, C.; Sahli, I.; Abdellaoui, M. Synthesis of quaternary nano-intermetallic $LaTi_2Ni_5Al_4$ with its hydrogen encapsulation and inspection using lithium-hydride cell. *Micro Nano Lett.* **2020**, *15*, 201–205. [CrossRef]
146. Chebab, S.; Abdellaoui, M.; Latroche, M.; Paul-Boncour, V. Electrochemical characterization of mechanically alloyed $LaCaMgNi_9$ compound. *Mater. Renew. Sustain. Energy* **2016**, *5*, 1–7. [CrossRef]
147. Ayari, M.; Sahli, I.; Elghali, M.; Ghodbane, O.; Jaafar, H.; Abdellaoui, M. Synthesis and characterizations of structural and electrochemical properties of $CeTi_2Ni_{4.5}Al_{0.2}Mn_{0.3}Cr_4$ AB_3 type compound. *J. Alloys Compd.* **2021**, *884*, 161017. [CrossRef]
148. Nowak, M.; Balcerzak, M.; Jurczyk, M. Hydrogen storage and electrochemical properties of mechanically alloyed $La_{1.5\text{-}x}Gd_xMg_{0.5}Ni_7$ ($0 \leq x \leq 1.5$). *Int. J. Hydrog. Energy* **2018**, *43*, 8897–8906. [CrossRef]
149. Dymek, M.; Nowak, M.; Jurczyk, M.; Bala, H. Encapsulation of $La_{1.5}Mg_{0.5}Ni_7$ nanocrystalline hydrogen storage alloy with Ni coatings and its electrochemical characterization. *J. Alloys Compd.* **2018**, *749*, 534–542. [CrossRef]
150. Dhaouadi, H.; Ajlani, H.; Zormati, H.; Abdellaoui, M. Elaboration and electrochemical characterization of two hydrogen storage alloy types: $LaNi_{3-x}Mn_xCr_2$(x = 0, 0.1, and 0.3) and La2Ni7. *Ionics* **2018**, *24*, 2017–2027. [CrossRef]
151. Srivastava, S.; Panwar, K. Investigations on Microstructures of Ball-milled $MmNi_5$ Hydrogen Storage Alloy. *Mater. Res. Bull.* **2016**, *73*, 284–289. [CrossRef]
152. Dabaki, Y.; Khaldi, C.; Fenineche, N.; ElKedim, O.; Tliha, M.; Lamloumi, J. Electrochemical Studies on the Ca-Based Hydrogen Storage Alloy for Different Milling Times. *Met. Mater. Int.* **2019**, *27*, 1005–1024. [CrossRef]
153. -ullah Rather, S.; Ullah Rather, S. Hydrogen uptake of mechanically milled lani 5-mwcnts nanocomposite. *Artic. Int. J. Adv. Sci. Eng. Inf. Technol.* **2018**, *6*, 2321–9009.
154. Li, J.; Li, B.; Yu, X.; Zhao, H.; Shao, H. Geometrical effect in Mg-based metastable nano alloys with BCC structure for hydrogen storage. *Int. J. Hydrog. Energy* **2019**, *44*, 29291–29296. [CrossRef]
155. Fujiwara, K.; Uehiro, R.; Edalati, K.; Li, H.W.; Floriano, R.; Akiba, E.; Horita, Z. New MgVCr BCC alloys synthesized by high-pressure torsion and ball milling. *Mater. Trans.* **2018**, *59*, 741–746. [CrossRef]
156. Balcerzak, M.; Wagstaffe, M.; Robles, R.; Pruneda, M.; Noei, H. Effect of Cr on the hydrogen storage and electronic properties of BCC alloys: Experimental and first-principles study. *Int. J. Hydrog. Energy* **2020**, *45*, 28996–29008. [CrossRef]
157. Zhan, Y.; Zhang, Y.; Zhu, Y. F.; Zhuang, X.Y.; Neng, W.A.N.; Yi, Q.U.; Guo, L.; Wang, M.; Li, Q. Electrochemical performances of $Mg_{45}M_5Co_{50}$ (M = Pd, Zr) ternary hydrogen storage electrodes. *Trans. Nonferrous Met. Soc. China* **2016**, *26*, 1388–1395. [CrossRef]
158. de Araujo-Silva, R.; Neves, A.M.; Vega, L.E.; Triques, M.R.; Leiva, D.R.; Kiminami, C.S.; Ishikawa, T.T.; Jorge Junior, A.M.; Botta, W.J. Synthesis of β-Ti-Nb alloys from elemental powders by high-energy ball milling and their hydrogenation features. *Int. J. Hydrog. Energy* **2018**, *43*, 18382–18391. [CrossRef]
159. Kazemipour, M.; Salimijazi, H.; Arefarjmand, A.; Saidi, A.; Saatchi, A. Electrochemical Hydrogen Storage Capacity of $Ti_{0.9}Zr_{0.1}Mn_{1.2}V_{0.4}Cr_{0.4}$ Alloy Synthesized by Ball Milling and Annealing. *Trans. Indian Inst. Met.* **2015**, *69*, 1327–1333. [CrossRef]

energies

MDPI

Review

A Review of Enhancement of Biohydrogen Productions by Chemical Addition Using a Supervised Machine Learning Method

Yiyang Liu [1,†], Jinze Liu [1,†], Hongzhen He [1], Shanru Yang [2], Yixiao Wang [1], Jin Hu [1], Huan Jin [3,*], Tianxiang Cui [3], Gang Yang [4] and Yong Sun [1,5,*]

1 Key Laboratory of Carbonaceous Wastes Processing and Process Intensification of Zhejiang Province, University of Nottingham Ningbo, Ningbo 315100, China; yiyangliu0904@163.com (Y.L.); jinze_Lau@163.com (J.L.); Hongzhen.he20@imperial.ac.uk (H.H.); shyyw12@nottingham.edu.cn (Y.W.); ssyjh3@nottingham.edu.cn (J.H.)
2 Centre for English Language Education (CELE), University of Nottingham Ningbo, Ningbo 315100, China; shanru.yang@nottingham.edu.cn
3 School of Computer Science, University of Nottingham Ningbo, Ningbo 315100, China; tianxiang.cui@nottingham.edu.cn
4 Institute of Process Engineering, Chinese Academy of Sciences, Beijing 100864, China; yanggang@ipe.ac.cn
5 School of Engineering, Edith Cowan University, 70 Joondalup Drive, Perth, WA 6027, Australia
* Correspondence: huan.jin@nottingham.edu.cn (H.J.); y.sun@ecu.edu.au or yong.sun@nottingham.edu.cn (Y.S.)
† Authors have equal contributions.

Abstract: In this work, the impact of chemical additions, especially nano-particles (NPs), was quantitatively analyzed using our constructed artificial neural networks (ANNs)-response surface methodology (RSM) algorithm. Fe-based and Ni-based NPs and ions, including Mg^{2+}, Cu^{2+}, Na^+, NH_4^+, and K^+, behave differently towards the response of hydrogen yield (HY) and hydrogen evolution rate (HER). Manipulating the size and concentration of NPs was found to be effective in enhancing the HY for Fe-based NPs and ions, but not for Ni-based NPs and ions. An optimal range of particle size (86–120 nm) and Ni-ion/NP concentration (81–120 mg L^{-1}) existed for HER. Meanwhile, the manipulation of the size and concentration of NPs was found to be ineffective for both iron and nickel for the improvement of HER. In fact, the variation in size of NPs for the enhancement of HY and HER demonstrated an appreciable difference. The smaller (less than 42 nm) NPs were found to definitely improve the HY, whereas for the HER, the relatively bigger size of NPs (40–50 nm) seemed to significantly increase the H_2 evolution rate. It was also found that the variations in the concentration of the investigated ions only statistically influenced the HER, not the HY. The level of response (the enhanced HER) towards inputs was underpinned and the order of significance towards HER was identified as the following: $Na^+ > Mg^{2+} > Cu^{2+} > NH_4^+ > K^+$.

Keywords: biohydrogen ($BioH_2$); nanoparticles; quantitative assessment; artificial neuron networks; process intensifications

Citation: Liu, Y.; Liu, J.; He, H.; Yang, S.; Wang, Y.; Hu, J.; Jin, H.; Cui, T.; Yang, G.; Sun, Y. A Review of Enhancement of Biohydrogen Productions by Chemical Addition Using a Supervised Machine Learning Method. *Energies* **2021**, *14*, 5916. https://doi.org/10.3390/en14185916

Academic Editor: Bahman Shabani

Received: 25 August 2021
Accepted: 14 September 2021
Published: 17 September 2021

1. Introduction

The further rollback of globalization will ultimately reshape the current supply chain block, especially as more and more countries have realized how pivotal it is to have self-sufficient industries to produce strategic products such as medicine, energy, and even toilet paper rolls [1]. Aside from the public health emergency, energy security is another draconian challenge that countries across the world are reluctantly facing, although the price of crude oil did once plunge to USD 25 per barrel (158.98 L) in the middle of 2020 during the COVID-19 pandemic [2]. Whether to take bolder steps in the energy reliance transition from fossil fuel to renewable energy will make a great difference in the world that our children will be able to inherit in the future [3]. Consequently, by 2021, several developed countries already started to restrict the use of fossil fuels in order to eventually achieve a shift in fuel type [4,5].

Among all sources of energy, hydrogen (H_2) is one of the most favorable candidates due to its inherent appealing features: (1) high energy yield (122 kJ kg^{-1}), (2) generation of water as a result of combustion, and (3) electricity generation through the fuel cell [6,7]. However, the current predominant H_2 generation still comes from fossil-based materials via existing mature industrial chemical processes such as natural gas steam reforming (NGSR), nature gas thermal cracking (NGTC), auto-thermal reforming (ATR), coal gasification, and partial oxidation of heavier-than-naphtha hydrocarbons [8]. Consequently, the paradox of sustainability of H_2 utilization and the non-renewability of H_2 generation will be encountered, although the development of carbon capture storage and utilization (CCSU) such as via a mature catalytic process like Fischer–Tropsch synthesis might alleviate environmental impacts from H_2 generation [9–12].

Apart from the thermal process, the biological hydrogen ($BioH_2$) generation process also plays a supplementary role in H_2 generation due to features such as versatile feedstock (lignocellulose, wet kitchen organic waste, and wastewater) and no green-house gas emissions (GHE). Despite the appealing advantages that are mentioned above, $BioH_2$ production is hampered by its relatively lower process performance [13]. To implement $BioH_2$ in different applications either on a decentralized or centralized basis or both, different process intensification approaches have been proposed, such as hydrolysate detoxification, mixed continuous and batch operations, co-fermentation, process optimization, and chemical addition. Among these approaches, chemical addition is considered to be one of the most attractive and practical ones because of its operational simplicity (without any additional modifications) and relatively low energy consumption [14]. However, current reports are limited to focusing on the facilitation of $BioH_2$ production by all types of chemical additives. In contrast, the nanoparticles (NPs) as a potential type of chemical additive still lack research on their addition and the corresponding quantitative relationships, such as hydrogen yield (HY) and hydrogen evolution rate (HER) with detailed incubation conditions, especially the concentration of different metal elements.

In this paper, instead of making a simple $BioH_2$ production enhancement comparison using the addition of NPs across literature reports, the collected data (such as HY, HER, and the substrate concentrations from literature works) were used to construct the data matrix for supervised machine learning algorithm using the developed artificial neural networks (ANNs) coupled with statistical analysis using response surface methodology (RSM) for more insightful and quantitative correlations and analysis. The review of assessing the impact of NPs additions on $BioH_2$ production in form of HY and HER using a developed ANNs-RSM algorithm, to the best of our knowledge, has not been reported before.

2. Materials and Methods

The literature used in this review was mainly collected from the scientific databases from Web of Science, Google Scholar and Science Direct via keyword search. Various keyword groups were comprised of several words, including "dark fermentation," "biohydrogen," and "nanoparticles." With regard to the possible missing relevant literature, by using the abovementioned searching strategy, an extensive additional search process was conducted with more detailed keywords, including "trace metal," "transitional metal," "iron," "nickel," "gold," "copper," and "metal oxide." During the additional search, these mentioned keywords were also combined with the keyword "biohydrogen."

The ANNs (based on Python 2.7 platform) was deployed for data analysis. The detailed schematic diagram of the construction of the ANNs and data collection is shown in Figure S1. In this work, the widely used feed-forward three-layer networks were used. The simplified cross-out method was used for cross-validation during the data training step. The detailed descriptions of the standard procedures for this methodology can be found in our previous works [15]. During the data training, the mean square error (*MSE*) and mean average relative residual (*MARR*) were computed as follows:

$$MSE\% = \frac{1}{N_{sam}} \sum_{j=1}^{N_{sam}} \left(r_i^{sam} - r_i^{cal} \right)^2 \times 100\% \quad (1)$$

$$MARR\% = \frac{1}{N_{sam}} \sum_{j=1}^{N_{exp}} \left(\frac{\left| r_i^{sam} - r_i^{cal} \right|}{r_i^{sam}} \right) \times 100\% \quad (2)$$

where N_{sam} is the number of data, and r_i^{sam} and r_i^{cal} are actual and calculated values, respectively. The setting for allowable accuracy was 95%. For the ANNs prediction data matrix, the widely used Box–Behnken design (BBD) and the central composite design (CCD) were used to predict the data matrix generation [16]. Once the supervised data learning was complete, the analysis of variation (ANOVA) based on commercial Design Expert® Version 11 software package (Stat-Ease, Inc., Minneapolis, MN, USA) was used for statistical analysis.

3. Literature Survey Comparisons

In this paper, for the convenience of discussion, four different types of NPs (Fe-based, Au-based, Cu-based, and Ni-based) were surveyed across different studies and the results are shown in Figure 1.

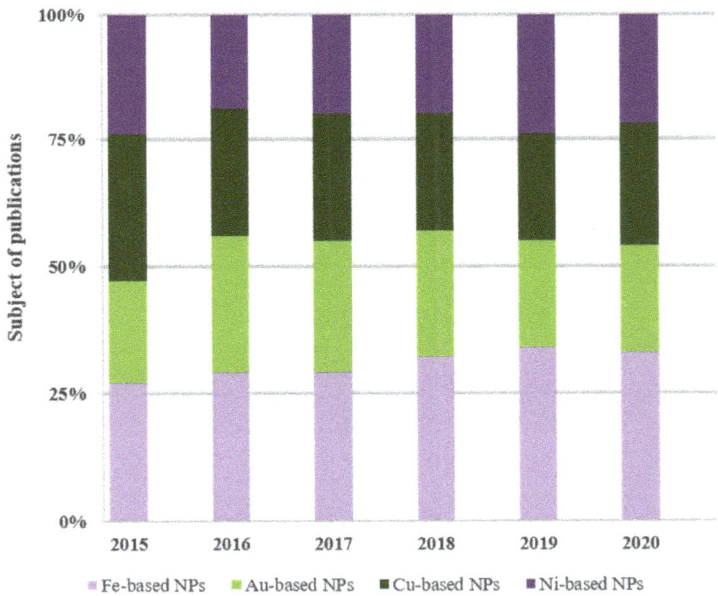

Figure 1. Statistics of publications from Scopus and Google Scholar in regard to $BioH_2$ production by chemical nanoparticle additions.

For each type of NPs, taking Ni-based NPs, for instance, all nickel-related species were included, such as nanoparticles such as zero-valent particles, metal oxide NiO_2, etc. The number of reports on the topic of $BioH_2$ enhancement by NPs additions has been increasing steadily since 2015. Among different NPs, the number of reports using iron-based NPs has presented a discernible trend in recent years. The impetus underlining this trend is possibly associated with its inherent appealing cost-effective feature compared to other NPs such as gold or nickel. Apart from Fe-based NPs, Ni-based NPs have experienced an appreciable increase in recent years, with an exception in 2015 [17,18]. The research interests that focus on Ni-based NPs might be pertinent to the metal cluster of hydrogenase [19]. According to recent classifications, there are three different hydrogenases, namely, [Fe], [NiFe], and [FeFe] [20,21]. During biological chemical reactions, these enzyme active centers play a pivotal role in the metabolism of proton ion-associated redox reactions. Studies have shown that [FeFe] hydrogenase catalyzes H_2 generation, whereas [NiFe] hydrogenase catalyzes the consumption of H_2. [NiFe] hydrogenase presents a relatively higher tolerance to the existence of oxygen and it widely exists in various types of microbial strains, whereas [FeFe] hydrogenase is relatively strict to the presence of oxygen and only exists in some algae and bacteria [22,23]. Regarding [Fe] hydrogenase, it only strictly exists in some methanogen strains [24–26].

4. Underlying Mechanisms of Metal Ions and Metal-Based Nanoparticles

Many extensively studied metal ions and metal-based nanoparticles are regarded as effective additives in culture medium to facilitate $BioH_2$ production in the dark fermentation process, including Na^+, K^+, NH_4^+, Mg^{2+}, Ca^{2+}, Co^{2+}, Zn^{2+}, Cu^{2+}, Fe^{2+}/Fe^{3+}, and Ni^{2+}/Ni^{3+}, among others [27–29]. Extensive studies have found that even small changes in the latter may have a significant impact on $BioH_2$ production; hence, many strategies have been proposed based on them, such as concentration regulation, including concentration manipulation [30], size regulation [17,31], composites fabrication [23,32], and heteroatom doping [33]. In general, the enhancement of NPs addition lies in a few important facts: (i) the controllable release of mental ions that facilitates the passive transport across the membrane [34]; (ii) nanodots that facilitate the electron transport chain during metabolism, such as glycolysis [35]; and (iii) the appropriate level of NPs favorable to the hydrogenase activities (co-enzymes often contain the metal ions in the catalysis center, which ultimately enhances the rate of hydrogen generation [36]. The potential mechanisms of $BioH_2$ enhancement are summarized in Figure 2. Therefore, in this part, this review will focus on the impact of the latter on $BioH_2$ production and its mechanisms.

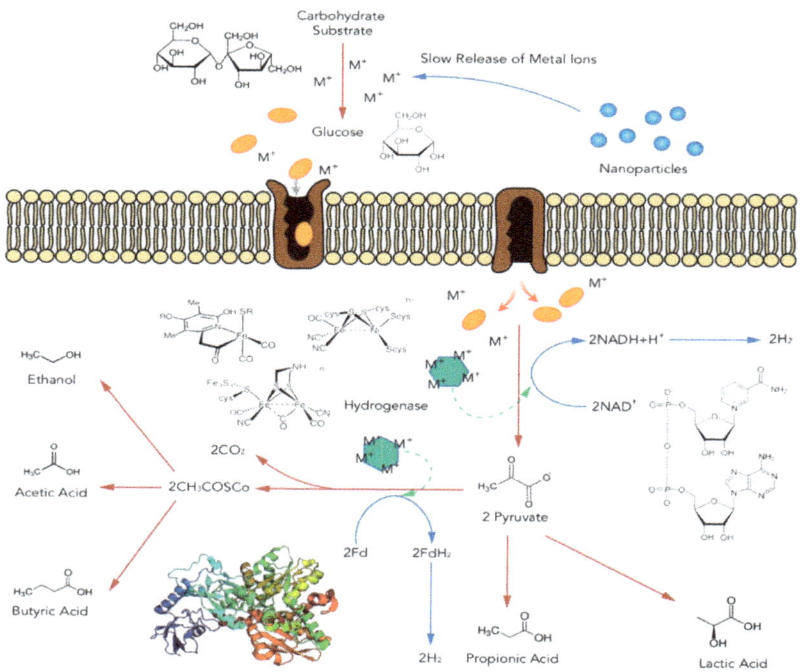

Figure 2. Potential mechanism of $BioH_2$ enhancement by NPs addition.

4.1. Fe-Based Ions and Nanoparticles

Iron is an important trace element in the formation of hydrogenases and other enzymes. The pre-addition of Fe in the culture medium is a widely used strategy to enhance $BioH_2$ production in dark fermentation [37]. As illustrated in Figure 2, first, Fe is the essential element to form the metal content at the active sites of hydrogenase ([FeFe], [FeNi], and [Fe]), thus catalyzing the reduction reaction of H^+ to H_2 [38]. Second, the presence of Fe-based NPs improves the activity of ferredoxin oxidoreductase by reducing the dissolved oxygen (DO) level and enhancing electron transfer due to the surface and quantum size effects [39,40]. In addition, Fe-based components could participate in enriching the microbial community and enhancing the growth of H_2-producing bacteria [41]. The oxidative stress increases when there is a higher Fe concentration, which results in the formation of abundant oxidative radicals, thus leading to the deactivation or decomposition of enzymes [17,30].

4.2. Ni-Based Ions and Nanoparticles

Similarly, nickel ions or Ni-based nanoparticles are another widely studied substance that can significantly enhance $BioH_2$ production in dark fermentation. The mechanisms between Ni-ion/Ni-based nanoparticles and Fe-ion/Fe-based nanoparticles are largely identical but with minor differences. The key mechanisms for Ni include (a) facilitating the synthesis of [FeNi] hydrogenase [42], (b) improving the activity of ferredoxin oxidoreductase [43], and (c) Ni NPs controlling the concentration of Ni^{2+} at the optimum level. In addition, it is worth noting that [NiFe] hydrogenase exists in more bacteria than [FeFe] hydrogenase. Therefore, Ni can promote H_2-producing bacteria in the dark fermentation process to a certain extent [44].

5. Results

5.1. Impact of Fe-Based Ions and NP Addition

To quantitatively unveil the impact of the concentration of Fe-ion/Fe NPs and size effects upon the HY and HER in $BioH_2$ generation, the collected values from the literature (Table 1) were statistically analyzed through our previously established ANN-RSM method and the results are shown in Figure 3.

Table 1. Comparison of $BioH_2$ production with the addition of Fe-based nanoparticles.

NPs	Opt/mg L^{-1}	Substrate	SC/g L^{-1}	Size/nm	HY/mmol g^{-1}	HER/mmol $L^{-1}h^{-1}$	Reference
Fe (NPs)	400	Grass	10.7	50	2.9	5.4	[45]
Fe (NPs)	25	Starch	5	35	3	-	[18]
Fe (NPs)	300	Malate	3	16	20	0.4	[46]
Fe (NPs)	50	Xylose	30	75	13.3	2	[47]
Fe (NPs)	200	MSJ	10	50	0.9	2.4	[48]
Fe (NPs)	200	Sucrose	7.5	50	15.9	10.1	[27]
Fe (NPs)	175	Glucose	7.5	59	12.9	5.69	[28]
Fe (NPs)	50	Starch	6	35	5	-	[43]
Fe (NPs)	250	Malate	4	12	24.2	0.8	[44]
Fe_2O_3 (NPs)	50	Glucose	5	50	1.92	2.5	[49]
Fe_2O_3 (NPs)	50	CDW	15.3	33	16.75	102.5	[17]
Fe_2O_3 (NPs)	200	DW	56	23	7.85	62.4	[30]
Fe_2O_3 (NPs)	50	Wastewater	110	6.5	1.9	49.4	[50]
Fe_2O_3 (NPs)	200	MEG	4	100	8.4	0.6	[51]
Fe_2O_3 (NPs)	300	CAS	10	20	3.875	1.92	[52]
Fe_2O_3 (NPs)	200	Glucose	10	20	9.2	3.1	[52]
Fe_2O_3 (NPs)	60	Glucose	6	60	1.92	2.5	[49]
Fe_3O_4(NPs)	10	Glucose	2.5	100	10.1	0.23	[53]
Fe_3O_4(A-C-NPs)	250	Glucose	5	30	11.656	3.2	[38]
GT-INP (Fe_2O_4 and FeO(OH)(NPs)	1000	CO	1.008	70	1.58	0.0662	[54]
Magnetite (NPs)	200	SJ	3	50	6.7	0.23	[55]
Hematite (NPs)	200	Sucrose	12.5	55	10.4	6	[56]

In this table, MEG refers to mono ethylene glycol, SC refers to substrate concentration, MSJ denotes Marcroalgea Saccharina Japonica, NMBL refers to *R. sphaeroides* NMBL-02 and *E. coli* NMBL-04, MC refers to mixed consortia, BA refers to *Bacillus anthracis* PUNAJAN 1, CP refers to *C. pasteurianum*, EA refers to *E. aerogenes* ATCC13408, EC refers to *E. cloacae*, Cl refers to Clostridium, Ca refers to *C. acetobutylicum* NCIM 2337, SJ refers to sugarcane juice, CAS refers to cassava starch.

Hydrogen Yield (HY)

Compounds	*p*-value	Significance
Diameter	0.0319	Yes
Diameter-NPs	0.0681	Yes
NPs	0.8607	No
$Fe^{2+/3+}$	0.9200	No

Hydrogen Evolution Rate (HER)

Compounds	*p*-value	Significance
Diameter	0.0088	Yes
Diameter-NPs	0.3707	No
NPs	0.0011	Yes
$Fe^{2+/3+}$	0.0784	Yes

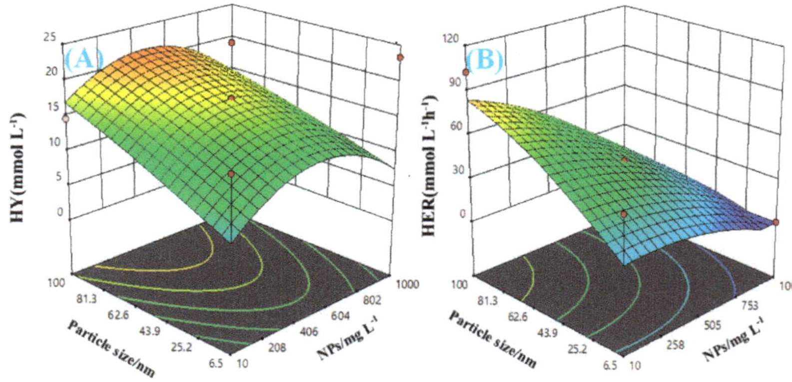

Figure 3. Statistical analysis of HY and HER. HY refers to H_2 yield; HER refers to the H_2 evolution rate. (**A**) Particle size and NP concentrations versus HY, (**B**) particle size and NP concentrations versus HER.

The effects of NPs size and NPs concentration together with the binary combined impact upon the HY and HER were extensively explored. Regarding HY, it was found that the size of the NPs together with the concentration of NPs were both statistically significant to the H_2 yield amongst the surveyed literature's reports of experimental conditions. From Figure 3A, it is indicated that the HY tended to approach the highest value in the range of NP size (81–100 nm) and NP concentration (406–604 mg L^{-1}). For HER, it was found that the size of NPs, the concentration of NPs, and Fe^{2+}/Fe^{3+} were all significant to HER. For the combined effects (NP size and concentration), on the other hand, these effects were found to be statistically insignificant to HER. The 3D plot of HER versus NPs size and NPs concentration (Figure 3B) also tended to show the highest region of HER located at the size range of 81–100 nm. Among the collected literature reports, the HER seemed to be more appreciably and directly related to the relatively larger size of the particle, which might be quite contradictory to some findings. This indicates that the manipulation of NPs ideally in size range of 81–100 nm is favorable for both high HY and HER. Reducing the size of NPs could improve the quantum dot effect, thus improving the electron transport. In contrast, the electron transport phenomena in extracellular media during cultivation is quite complicated and some factors such as osmosis condition and the activity of the fermentation broth might be counter-effective to the nanoparticle size effect for enhancing BioH_2 generation. Currently, very few works have been done to elucidate the mechanisms of this size impact upon selective enhancement of HY and HER. From our statistical analysis, a reasonable explanation for the ideal size effect is that the nanoparticle size of 81–100 nm is more thermodynamically stable than NPs with a smaller size during fermentation, since Fe-based NPs with smaller size are easier to agglomerate and form large Fe-based particles and deteriorate the electron transport performance in extracellular conditions. The fabrication of composites (e.g., Fe@graphene) is a promising strategy to enable the stable existence of small-sized nanoparticles; however, it has not been widely investigated.

5.2. Impact of Ni-Based Ions and NP Addition

The impact of Ni-based ions and NPs upon HY and HER is summarized in Table 2 and the statistical analysis results are shown in Figure 4.

Table 2. Comparison of $BioH_2$ production with the addition of Ni-based nanoparticles.

Nanoparticles	Opt/mg L^{-1}	Substrate	SC/g L^{-1}	Size/nm	HY/mmol g^{-1}	HER/mmol $L^{-1}h^{-1}$	Reference
Ni (NPs)	5.7	Glucose	14.01	13.6	14.1	11.5	[57]
Ni (NPs)	32	Starch	8	80	2.4	10.3	[18]
Ni (NPs)	60	MEG	4.7	60	1.11	1.5	[23]
Ni (NPs)	10	Glucose	1	25	9.5	30	[32]
Ni (NPs)	1	Glucose	2.5	100	11.7	0.28	[53]
Ni (NPs)	4.3	Glucose	13.92	28	12.7	10.4	[57]
Ni (NPs)	2.5	Glucose	5	42.5	10.8	1.3	[58]
Ni (NPs)	25	Starch	10	40	2.7	11.5	[18]
Ni (NPs)	11	Glucose	2.7	120	1.21	0.22	[59]
NiO (NPs)	20	MEG	4	100	7.25	0.5	[51]
NiO (NPs)	10	CDW	15.3	23	15.7	44.9	[17]
NiO (NPs)	1.5	Wastewater	9.6	23.6	0.5	12	[31]
Ni (NPs)	100	CS	20	50	20	0.27	[60]

In this table, MEG refers to mono ethylene glycol, CS: cornstalk.

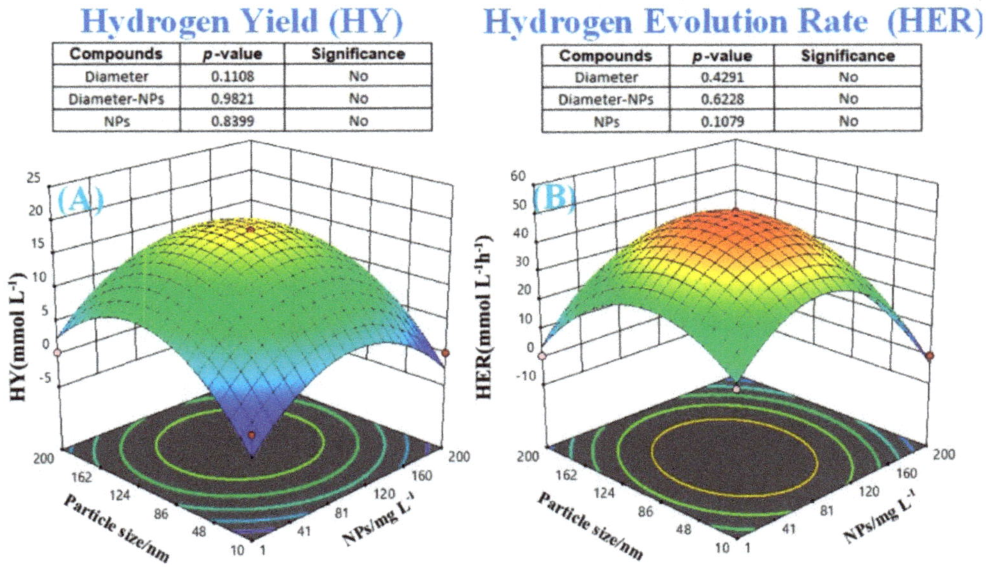

Figure 4. Statistical analysis of HY and HER. (**A**) Particle size and nanoparticle concentration versus HY, (**B**) particle size and nanoparticles concentration versus HER.

Among the collected literature reports, the size and concentration of NPs together with their combined effect were not statistically significant to either HY or HER according to the calculated *p*-value. Regarding HY (Figure 4A), it was found that both too low and too high levels of NPs size and concentration were not favorable. Indeed, an optimal range existed if the NP size and concentration were manipulated within 86–120 nm and (81–120 mg L^{-1}, respectively. Similarly, the HER also presented the same variation patterns as those of HY. An optimal range of particle size (86–120 nm) and Ni-ion/NPs concentration (81–120 mg L^{-1}) existed for HER. Unlike Fe, Ni presented more consistent responding patterns between HY and HER in regards to the variation in the size and concentration of NPs. In addition, studies have indicated that Ni-based ions and NPs tend to selectively enhance some $BioH_2$ generation pathways, such as enhancing the acetate pathway while suppressing or inhibiting butyrate and propionate pathways. However, discrepancies still exist due to different strains of microbes inoculated, cultivation medium, experimental uncertainties, etc. Although the size of NPs was significant to the HER, the combined effects (NP size and concentration) were found to be insignificant. Among the collected literature reports, the HER seemed to be directly related to the relatively larger size of the particles. This indicates that the manipulation of NPs ideally in size range of 81–100 nm is favorable for both HY and HER. This might contradict the first impression that the reduction of NPs size significantly enhances the quantum dot effect that subsequently boosts electron transport. However, the preparation and large-scale deployment of small-sized NPs that can stably exist in the cultivation medium has always been a substantial challenge, which will inevitably increase fixing and operating costs. Fortunately, the enhancement of $BioH_2$ generation seems to be linked to an ideal range of NPs at the size of 81–100 nm; therefore, blindly pursuing small nanoparticles may be meaningless.

5.3. Impact of Other Metal and Non-Metal Nanoparticle Addition

The impact of other metal and non-metal NPs addition upon $BioH_2$ generation is summarized in Table 3.

Table 3. Comparison of $BioH_2$ production with the addition of other nanoparticles, where POME: palm oil mill effluent.

NPs	Opt/mg L^{-1}	Substrate	SC/g L^{-1}	Size/nm	HY/mmol g^{-1}	HER/mmol $L^{-1}h^{-1}$	Reference
Ag	0.002	Glucose	12.5	15	13.8	10.5	[61]
Cu	2.5	Glucose	2.5	97	2.8	5.4	[62]
Pd	5	Glucose	10	100	8.1	6.7	[63]
Au	0.002	Sucrose	15	5	7.5	7.3	[64]
Co	1	Glucose	2.5	100	4.85	0.16	[53]
CoO	1	POME	76.5	17	22.5	0.7	[31]
TiO_2	100	Xylose	30	30	12	1.8	[47]
ZnO	10	MEG	4	100	7.3	0.58	[51]
MgO	1	Glucose	100	100	4.3	0.1	[53]
Cu/SiO_2	0.064	Glucose	5	2.5	5.8	0.54	[65]
Ag/SiO_2	0.107	Glucose	5	2.5	5.4	0.5	[65]
Pd/SiO_2	0.207	Glucose	5	2.5	5.4	0.52	[65]

The addition of NPs was found to be effective at improving $BioH_2$ generation due to the fact that NPs can facilitate electron transport in extracellular cultivation medium during fermentation [66,67]. With regard to the HY and HER, it was quite hard to find one individual NPs that positively enhanced both HY and HER simultaneously. This reflects the complex features of the $BioH_2$ generation process, which generally involves many different steps of sub-metabolic pathways [43,68]. Among the investigated collected literature, CoO-NPs addition was among the most appreciable enhancement for HY and Ag-NPs addition was the most influential factor for HER enhancement. In addition, the impact of adding NPs prepared from hybrid approaches such as combining two different kinds of NPs, i.e., Cu and SiO_2, was marginal. The correlation between $BioH_2$ generation values (HER and HY) and the corresponding size of the NPs added to the fermentation broth was constructed and is plotted in Figure 5. The corresponding HY and HER varied from 0–30 (mmol g^{-1}) and 0–80 (mmol L^{-1} h^{-1}), respectively. Regarding to the enhancement of HY, some reported that smaller size (less than 42 nm) surely increased HY from 10 to 20–25 mmol g^{-1}. On the other hand, for the enhancement of HER, some reported that a relatively bigger size of 40–50 nm seemed to significantly increase the H_2 evolution rate. However, by considering the numbers of reports, the majority of works showed (i) the size of NPs seems to be more effective in enhancing HY than HER, and (ii) the rate of H_2 evolution seems to be less responsive to the size of NPs, though some literature reported exceptionally higher values of HER after NPs (40–50 nm) addition.

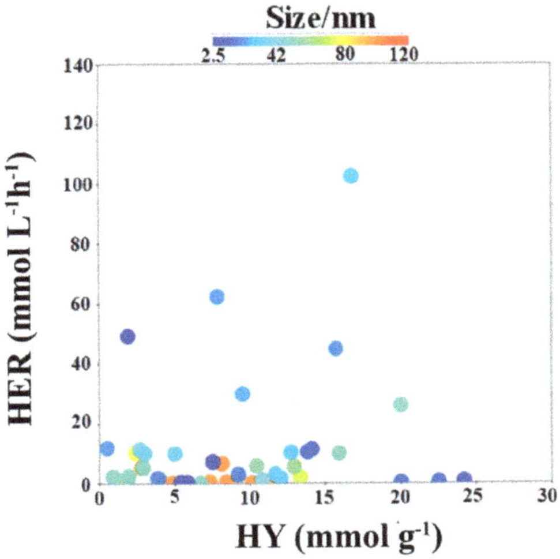

Figure 5. Impact of size of NPs upon HER and HY.

5.4. Impact of Ion Addition

In this work, in order to assess the concentration impacts of different ions upon HY and HER $BioH_2$ generation, ions including Mg^{2+}, Cu^{2+}, Na^+, $NH_4{}^+$, and K^+ were selected and all data are summarized in Table S1.

It is worth noting that some metal ions inevitably introduced into the culture medium due to the use of NPs addition are not in the scope of discussion. It was quite challenging to find out the detailed concentration ranges in each study due to the factor that many reports did not specify the detailed cultivation steps. Although this could be difficult for estimating the level of those ions during the cultivation, the type of defined and undefined cultivation media used in the studies could be utilized to indirectly estimate the range of those different ions accordingly. The level of different ions upon HY and HER $BioH_2$ generation are summarized in Tables S2 and 4, respectively, and the collected values from the literature were statistically analyzed through our previously established ANNs-RSM method.

By comparing the p-values, the impact of the variations in ion concentrations upon HY and HER of $BioH_2$ generation could be identified accordingly [16,69]. It was found that the variations in the investigated ions only statistically influenced HER, but not HY. This suggests pivotal guidance for process intensification for $BioH_2$ generation. The manipulations of ion concentrations in cultivation media can effectively improve or inhibit the rate but not the potential limit of $BioH_2$ generation. In other words, the kinetics of $BioH_2$ generation can be altered by varying some level of ionic concentration. The statistically significant impact of metal ion addition on HER is shown in Figure 6. Among the investigated ions, the single factor included Mg^{2+}, Cu^{2+}, and Na^+ (Figure 6A,B) and the combined factor included Mg^{2+}/Cu^{2+}, Cu^{2+}/Na^+, $Na^+/NH_4{}^+$, Na^+/K^+, and $NH_4{}^+/K^+$ (Figure 6C–E) as the most influential factors for HER. The responding patterns of HER towards different kinds of ions appeared to be appreciably different. These effects can be broadly classified as counter-effective and synergistic. For instance, for the counter-effective impact, the binary Mg^{2+}/Cu^{2+} belongs to this category, as does the binary $NH_4{}^+/K^+$ (Figure 6A,E). For the synergistic effect, the binary Cu^{2+}/Na^+, $Na^+/NH_4{}^+$, and Na^+/K^+ fall into this category (Figure 6B–D). These different ions will act as essential nutritious elements during metabolism at different stages of the growth of microbes [70–72]. For the growth pattern of microbes, there will normally be lagging, exponential, stationary, and death phases [73–75]. After inoculation, the microbes will experience a lagging phase with different duration [76,77]. The length of the lagging phase depends on many factors, such as the harshness of cultivation media, which contains lignocellulosic precursors and high levels of salt concentration [78–80].

Table 4. ANOVA analysis for the effect of ion concentration upon HER.

Source	Sum of Squares	DF	Mean Square	F-Value	*p*-Value
Model	38,286.08	20	1914.30	4.16	0.0005
A-Mg^{2+}	2467.73	1	2467.73	5.36	0.0291
B-Cu^{2+}	1729.50	1	1729.50	3.75	0.0640
C-Na^{+}	7543.84	1	7543.84	16.38	0.0004
D-NH^{4+}	496.57	1	496.57	1.08	0.3091
E-K^{+}	261.49	1	261.49	0.5677	0.4582
AB	7903.35	1	7903.35	17.16	0.0003
AC	1957.27	1	1957.27	4.25	0.0498
AD	513.91	1	513.91	1.12	0.3009
AE	1109.51	1	1109.51	2.41	0.1332
BC	41.84	1	41.84	0.0908	0.7656
BD	330.26	1	330.26	0.7170	0.4052
BE	16.50	1	16.50	0.0358	0.8514
CD	4919.66	1	4919.66	10.68	0.0031
CE	2100.83	1	2100.83	4.56	0.0427
DE	1719.79	1	1719.79	3.73	0.0647
A^2	801.66	1	801.66	1.74	0.1990
B^2	3897.09	1	3897.09	8.46	0.0075
C^2	2148.80	1	2148.80	4.67	0.0406
D^2	387.39	1	387.39	0.8410	0.3679
E^2	1539.54	1	1539.54	3.34	0.0795
Residue	11,515.22	25	460.61		
Lack of fit	11,515.22	20	575.76		
Pure Error	0.0000	5	0.0000		
Cor total	49,801.31	45			

In this table, $r^2 = 0.94$, adjusted $r^2 = 0.93$, predicted $r^2 = 0.93$, and adequate precision (AP) = 15.

The strategies for how to improve and shorten the length of the lagging phase will contribute to the improvement of the duration of the lagging phase [81]. For microbes to initiate their metabolism, elements such as Mg^{2+}, Na^{+}, NH_4^{+}, and K^{+} are essential [82–84]. These elements usually act as the major components of active centers in many enzymes [85–87]. Ensuring a sufficient amount of these necessary elements will facilitate the smooth and fast transition from the lagging phase to the growth phase [88–90]. It is commonly accepted that $BioH_2$ generation will occur mainly in the exponential and stationary phases [91,92]. Clearly, these investigated literature reports provide useful guidance for the levels of these necessary ion elements in the cultivation media. More importantly, through statistical analysis from our developed ANN-RSM algorithm, the level of the response (the enhanced HER) for those inputs was underpinned. In addition, the order of significance for HER was also identified as the following: $Na^{+} > Mg^{2+} > Cu^{2+} > NH_4^{+} > K^{+}$. From a holistic point of view, all the steps involved in $BioH_2$ generation metabolism could be targeted as steps to enhance $BioH_2$ generation (HY and HER). Two major metabolic pathways, namely, butyrate and acetate, are mainly associated with the activities of hydrogenase and the generation of H_2 during dark fermentation [93–95]. From a stoichiometric perspective, the metabolic route towards acetate generates two times that of butyrate pathways [96,97]. From a process intensification point of view, the facilitation of the metabolic pathway towards an acetate pathway is favorable. From our statistical analysis, of all the investigated ions among the literature reports, the yield of $BioH_2$ generation for these chemical additions of ions is not significant, suggesting that the enhancement of $BioH_2$ generation by simple chemical additions of ions might be ineffective at further improving the ceiling value of $BioH_2$ generation yield. For the sake of skewing the delicate balance between butyrate and acetate pathways, the combination of other chemical additions such as activated carbon, biochars, or porous adsorbents will be more effective in enhancing $BioH_2$ generation [98–100].

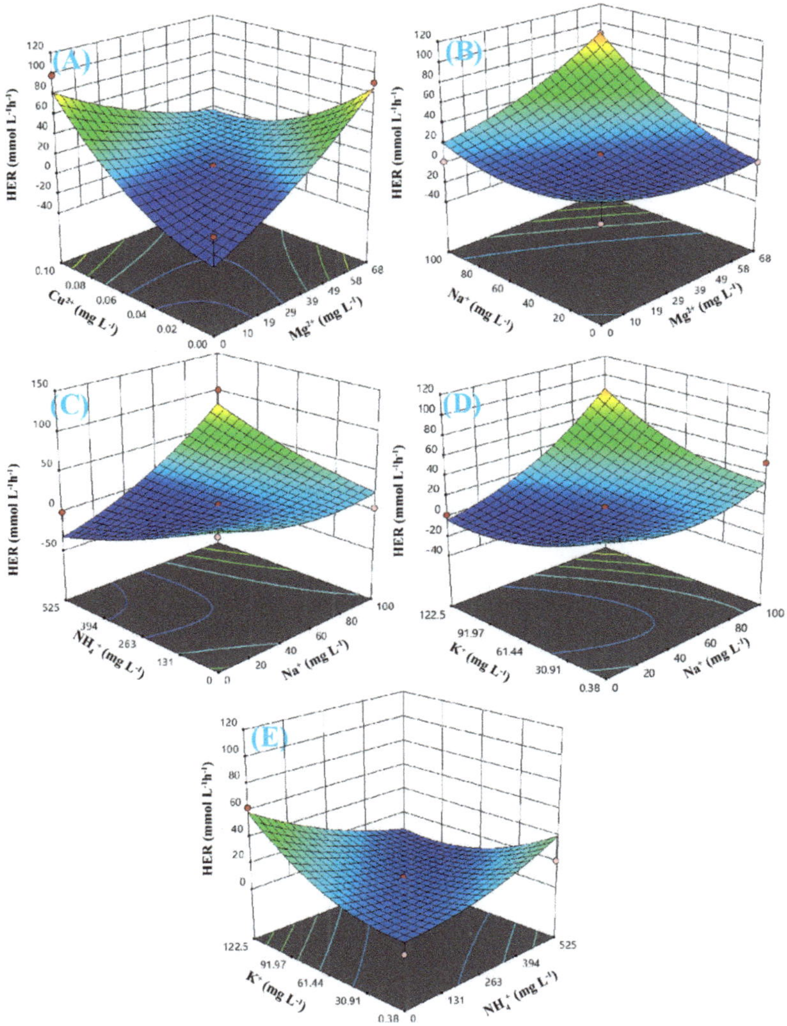

Figure 6. ANNs-RSM analysis of statically significant ion concentrations for HER. (**A**) Mg^{2+}/Cu^{2+} nanoparticle concentration versus HER, (**B**) Na^+/Mg^{2+} nanoparticle concentration versus HER, (**C**) Na^+/NH_4^+ nanoparticle concentration versus HER, (**D**) Na^+/K^+ nanoparticle concentration versus HER, (**E**) NH_4^+/K^+ nanoparticle concentration versus HER.

6. Conclusions

The statistical significance of these different NPs and ion additions were rigorously and quantitatively analyzed through a well-developed ANNs-RSM algorithm. As a result, this work provided effective guidance for the size optimization of NP additions and concentration regulation of ion additives in practice. For Fe-based NPs and ions, both the size of NPs and their corresponding concentration are statistically significant to HY. For HER, it was found that the combined effect of NP size and concentration is insignificant to HER. For Ni-based NPs and ions, neither size nor concentration is statistically significant to HY and HER, respectively. The variation in the size of NPs for the enhancement of

HY and HER behaved differently. The smaller (less than 42 nm) were found to definitely improve HY. Simultaneously, for HER, most reported literature indicated that manipulating the size of NPs is ineffective. It was found that variations in the investigated ions only statistically influenced HER, but not HY. This discovery suggests very pivotal guidance for process intensification for $BioH_2$ generation. Using the constructed algorithm, the level of responses (enhanced HER) towards inputs (other ion additions) was underpinned, and the order of significance towards HER was also identified as the following: $Na^+ > Mg^{2+} > Cu^{2+} > NH_4{}^+ > K^+$. However, the number of relevant literature reports is currently limited; with the support of more experimental data, the results predicted by the ANNs-RSM algorithm will be more credible.

Supplementary Materials: The following are available online at https://www.mdpi.com/article/10.3390/en14185916/s1, Figure S1: Schematic diagram of methodology: (A) The procedures flowchart, (B) ANNs construction: feed forward three layers networks; Table S1: Ion comparison upon $BioH_2$ generation—refers to all data missing as, for convenience of calculation, the missing value was replaced by the averaged value during the artificial neuron network learning process; Table S2: ANOVA analysis for the effect of ion concentration on HY, where $r^2 = 0.94$, adjusted $r^2 = 0.93$, predicted $r^2 = 0.93$, and adequate precision (AP) = 15.

Author Contributions: Drafting and data collection, Y.L., J.L. and Y.W.; paper writing and data collection, H.H.; proofreading, S.Y.; programming and modelling, J.H.; supervision, H.J. and T.C.; drafting, G.Y.; funding acquisition and project management, Y.S. All authors have read and agreed to the published version of the manuscript.

Funding: This work was supported by the Key Laboratory of Carbonaceous Wastes Processing and Process Intensification of Zhejiang Province (2020E10018), the Qianjiang Talent Scheme (QJD1803014), the Ningbo Science and Technology Innovation 2025 Key Project (2020Z100), and the Ningbo Municipal Commonweal Key Program (2019C10033 & 2019C10104), UNNC FoSE Researchers Grant 2020 (I01210100011).

Institutional Review Board Statement: Not applicable.

Informed Consent Statement: Not applicable.

Acknowledgments: The authors would like to express their sincere appreciation for the critical and insightful comments raised by anonymous reviewers, which significantly improved the quality of this work.

Conflicts of Interest: The authors declare no conflict of interest.

Abbreviations

ANOVA	Analysis of variation
ANNs	Artificial neural networks
ATR	Auto-thermal reforming
$BioH_2$	Biological hydrogen
BBD	Box–Behnken design
CCSU	Carbon capture storage and utilization
CCD	Central composite design
DO	Dissolved oxygen
GHE	Greenhouse gas emission
H_2	Hydrogen
HER	Hydrogen evolution rate
HY	Hydrogen yield
MSE	Mean square error
NGSR	Natural gas steam reforming
NGTC	Nature gas thermal cracking
NPs	Nanoparticles
RSM	Response surface methodology

References

1. Economist, T. Building up the pillars of state. *The Economist*, 28 March 2020.
2. Bloomberg, Crude Oil (Nymex). In *Bloomberg Energy Index*, 2020/04/06 ed.; Bloomberg: New York, NY, USA, 2020.
3. Carson, R. *Silent Spring*; Houghton Mifflin Harcourt: New York, NY, USA, 1962.
4. Canadell, P.; Quéré, C.L.; Peters, G.; Korsbakken, J.I.; Andrew, R. Eighteen countries showing the way to carbon zero. *The Conversation*, 26 February 2019.
5. Wang, Y.; Tang, M.; Yusuf, A.; Wang, Y.; Zhang, X.; Yang, G.; He, J.; Jin, H.; Sun, Y. Preparation of Catalyst from Phosphorous Rock Using an Improved Wet Process for Transesterification Reaction. *Ind. Eng. Chem. Res.* **2021**, *60*, 22. [CrossRef]
6. Liu, Z.; Wang, K.; Chen, Y.; Tan, T.; Nielsen, J. Third-generation biorefineries as the means to produce fuels and chemicals from CO_2. *Nat. Catal.* **2020**, *3*, 274–288. [CrossRef]
7. Glenk, G.; Reichelstein, S. Economics of converting renewable power to hydrogen. *Nat. Energy* **2019**, *4*, 216–222. [CrossRef]
8. Sun, Y.; He, J.; Yang, G.; Sun, G.; Sage, V. A review of the enhancement of bio-hydrogen generation by chemicals addition. *Catalysts* **2019**, *9*, 353. [CrossRef]
9. Sun, Y.; Yang, G.; Zhang, L.; Sun, Z. Fischer-Tropsch synthesis in a microchannel reactor using mesoporous silica supported bimetallic Co-Ni catalyst: Process optimization and kinetic modeling. *Chem. Eng. Process. Process. Intensif.* **2017**, *119*, 44–61. [CrossRef]
10. Sun, Y.; Jia, Z.; Yang, G.; Zhang, L.; Sun, Z. Fischer-Tropsch synthesis using iron based catalyst in a microchannel reactor: Performance evaluation and kinetic modeling. *Int. J. Hydrogen Energy* **2017**, *42*, 29222–29235. [CrossRef]
11. Sun, Y.; Yang, G.; Wen, C.; Zhang, L.; Sun, Z. Artificial neural networks with response surface methodology for optimization of selective CO_2 hydrogenation using K-promoted iron catalyst in a microchannel reactor. *J. CO_2 Util.* **2018**, *24*, 10–21. [CrossRef]
12. Sun, Y.; Wang, Y.; He, J.; Yusuf, A.; Wang, Y.; Yang, G.; Xiao, X. Comprehensive kinetic model for acetylene pretreated mesoporous silica supported bimetallic Co-Ni catalyst during Fischer-Trospch synthesis. *Chem. Eng. Sci.* **2021**, *246*, 116828–116844. [CrossRef]
13. Wang, Y.X.; Tang, M.; Ling, J.; Wang, Y.; Liu, Y.; Jin, H.; He, J.; Sun, Y. Modeling biohydrogen production using different data driven approaches. *Int. J. Hydrogen Energy* **2021**, *46*, 29822–29833. [CrossRef]
14. Kumar, G.; Mathimani, T.; Rene, E.R.; Pugazhendhi, A. Application of nanotechnology in dark fermentation for enhanced biohydrogen production using inorganic nanoparticles. *Int. J. Hydrogen Energy* **2019**, *44*, 13106–13113. [CrossRef]
15. Wang, Y.; Yang, G.; Sage, V.; Xu, J.; Sun, G.; He, J.; Sun, Y. Optimization of dark fermentation for biohydrogen production using a hybrid artificial neural network (ANN) and response surface methodology (RSM) approach. *Environ. Prog. Sustain. Energy* **2021**, *40*, e13485. [CrossRef]
16. Sun, Y.; Yang, G.; Xu, M.; Xu, J.; Sun, Z. A simple coupled ANNs-RSM approach in modeling product distribution of Fischer—Tropsch synthesis using a microchannel reactor with Ru-promoted Co/Al_2O_3 catalyst. *Int. J. Energy Res.* **2020**, *44*, 1046–1061. [CrossRef]
17. Gadhe, A.; Sonawane, S.S.; Varma, M.N. Enhancement effect of hematite and nickel nanoparticles on biohydrogen production from dairy wastewater. *Int. J. Hydrogen Energy* **2015**, *40*, 4502–4511. [CrossRef]
18. Taherdanak, M.; Zilouei, H.; Karimi, K. Investigating the effects of iron and nickel nanoparticles on dark hydrogen fermentation from starch using central composite design. *Int. J. Hydrogen Energy* **2015**, *40*, 12956–12963. [CrossRef]
19. Trofanchuk, O.; Stein, M.; Geßner, C.; Lendzian, F.; Higuchi, Y.; Lubitz, W. Single crystal EPR studies of the oxidized active site of [NiFe] hydrogenase from Desulfovibrio vulgaris Miyazaki F. *JBIC J. Biol. Inorg. Chem.* **2000**, *5*, 36–44. [CrossRef]
20. Morra, S.; Arizzi, M.; Allegra, P.; La Licata, B.; Sagnelli, F.; Zitella, P.; Gilardi, G.; Valetti, F. Expression of different types of [FeFe]-hydrogenase genes in bacteria isolated from a population of a bio-hydrogen pilot-scale plant. *Int. J. Hydrogen Energy* **2014**, *39*, 9018–9027. [CrossRef]
21. Peters, J.W.; Schut, G.J.; Boyd, E.S.; Mulder, D.W.; Shepard, E.M.; Broderick, J.B.; King, P.W.; Adams, M.W. [FeFe]-and [NiFe]-hydrogenase diversity, mechanism, and maturation. *Biochim. Biophys. Acta (BBA)-Mol. Cell Res.* **2015**, *1853*, 1350–1369. [CrossRef]
22. Kothari, R.; Singh, D.; Tyagi, V.; Tyagi, S. Fermentative hydrogen production–An alternative clean energy source. *Renew. Sustain. Energy Rev.* **2012**, *16*, 2337–2346. [CrossRef]
23. Elreedy, A.; Ibrahim, E.; Hassan, N.; El-Dissouky, A.; Fujii, M.; Yoshimura, C.; Tawfik, A. Nickel-graphene nanocomposite as a novel supplement for enhancement of biohydrogen production from industrial wastewater containing mono-ethylene glycol. *Energy Convers. Manag.* **2017**, *140*, 133–144. [CrossRef]
24. Pohorelic, B.K.; Voordouw, J.K.; Lojou, E.; Dolla, A.; Harder, J.; Voordouw, G. Effects of deletion of genes encoding Fe-only hydrogenase of Desulfovibrio vulgaris Hildenborough on hydrogen and lactate metabolism. *J. Bacteriol.* **2002**, *184*, 679–686. [CrossRef] [PubMed]
25. Vignais, P.M.; Billoud, B. Occurrence, classification, and biological function of hydrogenases: An overview. *Chem. Rev.* **2007**, *107*, 4206–4272. [CrossRef]
26. Kucharska, K.; Hołowacz, I.; Konopacka-Łyskawa, D.; Rybarczyk, P.; Kamiński, M. Key issues in modeling and optimization of lignocellulosic biomass fermentative conversion to gaseous biofuels. *Renew. Energy* **2018**, *129*, 384–408. [CrossRef]
27. Mohanraj, S.; Kodhaiyolii, S.; Rengasamy, M.; Pugalenthi, V. Phytosynthesized iron oxide nanoparticles and ferrous iron on fermentative hydrogen production using Enterobacter cloacae: Evaluation and comparison of the effects. *Int. J. Hydrogen Energy* **2014**, *39*, 11920–11929. [CrossRef]
28. Mohanraj, S.; Kodhaiyolii, S.; Rengasamy, M.; Pugalenthi, V. Green synthesized iron oxide nanoparticles effect on fermentative hydrogen production by Clostridium acetobutylicum. *Appl. Biochem. Biotechnol.* **2014**, *173*, 318–331. [CrossRef] [PubMed]

29. Patel, S.K.; Lee, J.-K.; Kalia, V.C. Nanoparticles in biological hydrogen production: An overview. *Indian J. Microbiol.* **2018**, *58*, 8–18. [CrossRef] [PubMed]
30. Gadhe, A.; Sonawane, S.S.; Varma, M.N. Influence of nickel and hematite nanoparticle powder on the production of biohydrogen from complex distillery wastewater in batch fermentation. *Int. J. Hydrogen Energy* **2015**, *40*, 10734–10743. [CrossRef]
31. Mishra, P.; Thakur, S.; Mahapatra, D.M.; Ab Wahid, Z.; Liu, H.; Singh, L. Impacts of nano-metal oxides on hydrogen production in anaerobic digestion of palm oil mill effluent–A novel approach. *Int. J. Hydrogen Energy* **2018**, *43*, 2666–2676. [CrossRef]
32. Sun, Y.; Yang, G.; Zhang, J.; Wen, C.; Sun, Z. Optimization and kinetic modeling of an enhanced bio-hydrogen fermentation with the addition of synergistic biochar and nickel nanoparticle. *Int. J. Energy Res.* **2019**, *43*, 983–999. [CrossRef]
33. Kodhaiyolii, S.; Mohanraj, S.; Rengasamy, M.; Pugalenthi, V. Phytofabrication of bimetallic Co–Ni nanoparticles using Boerhavia diffusa leaf extract: Analysis of phytocompounds and application for simultaneous production of biohydrogen and bioethanol. *Mater. Res. Express* **2019**, *6*, 095051. [CrossRef]
34. Jiang, X.C.; Hu, J.S.; Lieber, A.M.; Jackan, C.S.; Biffinger, J.C.; Fitzgerald, L.A.; Ringeisen, B.R.; Lieber, C.M. Nanoparticle Facilitated Extracellular Electron Transfer in Microbial Fuel Cells. *Nano Lett.* **2014**, *14*, 6737–6742. [CrossRef]
35. El-Naggar, M.Y.; Wanger, G.; Leung, K.M.; Yuzvinsky, T.D.; Southam, G.; Yang, J.; Lau, W.M.; Nealson, K.H.; Gorby, Y.A. Electrical transport along bacterial nanowires from Shewanella oneidensis MR-1. *Proc. Natl. Acad. Sci. USA* **2010**, *107*, 18127–18131. [CrossRef]
36. Viggi, C.C.; Rossetti, S.; Fazi, S.; Paiano, P.; Majone, M.; Aulenta, F. Magnetite Particles Triggering a Faster and More Robust Syntrophic Pathway of Methanogenic Propionate Degradation. *Environ. Sci. Technol.* **2014**, *48*, 7536–7543. [CrossRef]
37. Wang, J.; Wan, W. Effect of Fe2+ concentration on fermentative hydrogen production by mixed cultures. *Int. J. Hydrogen Energy* **2008**, *33*, 1215–1220. [CrossRef]
38. Frey, M. Hydrogenases: Hydroge—activating enzymes. *ChemBioChem* **2002**, *3*, 153–160. [CrossRef]
39. Shanmugam, S.; Hari, A.; Pandey, A.; Mathimani, T.; Felix, L.; Pugazhendhi, A. Comprehensive review on the application of inorganic and organic nanoparticles for enhancing biohydrogen production. *Fuel* **2020**, *270*, 117453. [CrossRef]
40. Nadeem, F.; Jiang, D.; Tahir, N.; Alam, M.; Zhang, Z.; Yi, W.; Chaoyang, L.; Zhang, Q. Defect engineering in SnO2 nanomaterials: Pathway to enhance the biohydrogen production from agricultural residue of corn stover. *Appl. Mater. Today* **2020**, *21*, 100850. [CrossRef]
41. Shanmugam, S.; Krishnaswamy, S.; Chandrababu, R.; Veerabagu, U.; Pugazhendhi, A.; Mathimani, T. Optimal immobilization of Trichoderma asperellum laccase on polymer coated Fe_3O_4@SiO_2 nanoparticles for enhanced biohydrogen production from delignified lignocellulosic biomass. *Fuel* **2020**, *273*, 117777. [CrossRef]
42. Braga, J.K.; Stancari, R.A.; Motteran, F.; Malavazi, I.; Varesche, M.B.A. Metals addition for enhanced hydrogen, acetic and butyric acids production from cellulosic substrates by Clostridium butyricum. *Biomass Bioenergy* **2021**, *150*, 105679. [CrossRef]
43. Bhatia, S.K.; Jagtap, S.S.; Bedekar, A.A.; Bhatia, R.K.; Rajendran, K.; Pugazhendhi, A.; Rao, C.V.; Atabani, A.; Kumar, G.; Yang, Y.-H. Renewable biohydrogen production from lignocellulosic biomass using fermentation and integration of systems with other energy generation technologies. *Sci. Total Environ.* **2020**, *765*, 144429. [CrossRef]
44. Wang, J.; Wan, W. Influence of Ni2+ concentration on biohydrogen production. *Bioresour. Technol.* **2008**, *99*, 8864–8868. [CrossRef]
45. Yang, G.; Wang, J. Improving mechanisms of biohydrogen production from grass using zero-valent iron nanoparticles. *Bioresour. Technol.* **2018**, *266*, 413–420. [CrossRef]
46. Dolly, S.; Pandey, A.; Pandey, B.K.; Gopal, R. Process parameter optimization and enhancement of photo-biohydrogen production by mixed culture of Rhodobacter sphaeroides NMBL-02 and Escherichia coli NMBL-04 using Fe-nanoparticle. *Int. J. Hydrogen Energy* **2015**, *40*, 16010–16020. [CrossRef]
47. Hsieh, P.-H.; Lai, Y.-C.; Chen, K.-Y.; Hung, C.-H. Explore the possible effect of TiO2 and magnetic hematite nanoparticle addition on biohydrogen production by Clostridium pasteurianum based on gene expression measurements. *Int. J. Hydrogen Energy* **2016**, *41*, 21685–21691. [CrossRef]
48. Yin, Y.; Wang, J. Enhanced biohydrogen production from macroalgae by zero-valent iron nanoparticles: Insights into microbial and metabolites distribution. *Bioresour. Technol.* **2019**, *282*, 110–117. [CrossRef] [PubMed]
49. Engliman, N.S.; Abdul, P.M.; Wu, S.-Y.; Jahim, J.M. Influence of iron (II) oxide nanoparticle on biohydrogen production in thermophilic mixed fermentation. *Int. J. Hydrogen Energy* **2017**, *42*, 27482–27493. [CrossRef]
50. Malik, S.N.; Pugalenthi, V.; Vaidya, A.N.; Ghosh, P.C.; Mudliar, S.N. Kinetics of nano-catalysed dark fermentative hydrogen production from distillery wastewater. *Energy Procedia* **2014**, *54*, 417–430. [CrossRef]
51. Elreedy, A.; Fujii, M.; Koyama, M.; Nakasaki, K.; Tawfik, A. Enhanced fermentative hydrogen production from industrial wastewater using mixed culture bacteria incorporated with iron, nickel, and zinc-based nanoparticles. *Water Res.* **2019**, *151*, 349–361. [CrossRef]
52. Lin, R.; Cheng, J.; Ding, L.; Song, W.; Liu, M.; Zhou, J.; Cen, K. Enhanced dark hydrogen fermentation by addition of ferric oxide nanoparticles using Enterobacter aerogenes. *Bioresour. Technol.* **2016**, *207*, 213–219. [CrossRef] [PubMed]
53. Zaidi, A.A.; RuiZhe, F.; Shi, Y.; Khan, S.Z.; Mushtaq, K. Nanoparticles augmentation on biogas yield from microalgal biomass anaerobic digestion. *Int. J. Hydrogen Energy* **2018**, *43*, 14202–14213. [CrossRef]
54. Wang, J.; Wan, W. The effect of substrate concentration on biohydrogen production by using kinetic models. *Sci. China Ser. B Chem.* **2008**, *51*, 1110–1117. [CrossRef]

55. Reddy, K.; Nasr, M.; Kumari, S.; Kumar, S.; Gupta, S.K.; Enitan, A.M.; Bux, F. Biohydrogen production from sugarcane bagasse hydrolysate: Effects of pH, S/X, Fe^{2+}, and magnetite nanoparticles. *Environ. Sci. Pollut. Res.* **2017**, *24*, 8790–8804. [CrossRef]
56. Han, H.; Cui, M.; Wei, L.; Yang, H.; Shen, J. Enhancement effect of hematite nanoparticles on fermentative hydrogen production. *Bioresour. Technol.* **2011**, *102*, 7903–7909. [CrossRef]
57. Mullai, P.; Yogeswari, M.; Sridevi, K. Optimisation and enhancement of biohydrogen production using nickel nanoparticles–A novel approach. *Bioresour. Technol.* **2013**, *141*, 212–219. [CrossRef]
58. Taherdanak, M.; Zilouei, H.; Karimi, K. The effects of Fe0 and Ni0 nanoparticles versus Fe2+ and Ni2+ ions on dark hydrogen fermentation. *Int. J. Hydrogen Energy* **2016**, *41*, 167–173. [CrossRef]
59. Sun, Y.; Wang, Y.; Yang, G.; Sun, Z. Optimization of biohydrogen production using acid pretreated corn stover hydrolysate followed by nickel nanoparticle addition. *Int. J. Energy Res.* **2020**, *44*, 1843–1857. [CrossRef]
60. Zhang, J.; Zhao, W.; Yang, J.; Li, Z.; Zhang, J.; Zang, L. Comparison of mesophilic and thermophilic dark fermentation with nickel ferrite nanoparticles supplementation for biohydrogen production. *Bioresour. Technol.* **2021**, *329*, 124853. [CrossRef]
61. Zhao, W.; Zhang, Y.; Du, B.; Wei, D.; Wei, Q.; Zhao, Y. Enhancement effect of silver nanoparticles on fermentative biohydrogen production using mixed bacteria. *Bioresour. Technol.* **2013**, *142*, 240–245. [CrossRef] [PubMed]
62. Mohanraj, S.; Anbalagan, K.; Rajaguru, P.; Pugalenthi, V. Effects of phytogenic copper nanoparticles on fermentative hydrogen production by Enterobacter cloacae and Clostridium acetobutylicum. *Int. J. Hydrogen Energy* **2016**, *41*, 10639–10645. [CrossRef]
63. Mohanraj, S.; Anbalagan, K.; Kodhaiyolii, S.; Pugalenthi, V. Comparative evaluation of fermentative hydrogen production using Enterobacter cloacae and mixed culture: Effect of Pd (II) ion and phytogenic palladium nanoparticles. *J. Biotechnol.* **2014**, *192*, 87–95. [CrossRef] [PubMed]
64. Zhang, Y.; Shen, J. Enhancement effect of gold nanoparticles on biohydrogen production from artificial wastewater. *Int. J. Hydrogen Energy* **2007**, *32*, 17–23. [CrossRef]
65. Beckers, L.; Hiligsmann, S.; Lambert, S.D.; Heinrichs, B.; Thonart, P. Improving effect of metal and oxide nanoparticles encapsulated in porous silica on fermentative biohydrogen production by Clostridium butyricum. *Bioresour. Technol.* **2013**, *133*, 109–117. [CrossRef] [PubMed]
66. Ameen, F.; Alsamhary, K.; Alabdullatif, J.A.; ALNadhari, S. A review on metal-based nanoparticles and their toxicity to beneficial soil bacteria and fungi. *Ecotoxicol. Environ. Saf.* **2021**, *213*, 112027. [CrossRef]
67. Pu, Y.; Laratte, B.; Marks, R.S.; Ionescu, R.E. Impact of copper nanoparticles on porcine neutrophils: Ultrasensitive characterization factor combining chemiluminescence information and USEtox assessment model. *Mater. Today Commun.* **2017**, *11*, 68–75. [CrossRef]
68. El-Dalatony, M.M.; Zheng, Y.; Ji, M.-K.; Li, X.; Salama, E.-S. Metabolic pathways for microalgal biohydrogen production: Current progress and future prospectives. *Bioresour. Technol.* **2020**, *318*, 124253. [CrossRef]
69. Sun, Y.; Yang, G.; Zhang, L.; Sun, Z. Fischer-Trospch synthesis using iron-based catalyst in a microchannel reactor: Hybrid lump kinetic with ANNs/RSM. *Chem. Eng. Process. Process. Intensif.* **2017**, *122*, 181–189. [CrossRef]
70. Rambabu, K.; Show, P.-L.; Bharath, G.; Banat, F.; Naushad, M.; Chang, J.-S. Enhanced biohydrogen production from date seeds by Clostridium thermocellum ATCC 27405. *Int. J. Hydrogen Energy* **2020**, *45*, 22271–22280. [CrossRef]
71. Ulhiza, T.A.; Puad, N.I.M.; Azmi, A.S. Optimization of culture conditions for biohydrogen production from sago wastewater by Enterobacter aerogenes using Response Surface Methodology. *Int. J. Hydrogen Energy* **2018**, *43*, 22148–22158. [CrossRef]
72. Zainal, B.S.; Zinatizadeh, A.A.; Chyuan, O.H.; Mohd, N.S.; Ibrahim, S. Effects of process, operational and environmental variables on biohydrogen production using palm oil mill effluent (POME). *Int. J. Hydrogen Energy* **2018**, *43*, 10637–10644. [CrossRef]
73. Usman, M.; Kavitha, S.; Kannah, Y.; Yogalakshmi, K.; Sivashanmugam, P.; Bhatnagar, A.; Kumar, G. A critical review on limitations and enhancement strategies associated with biohydrogen production. *Int. J. Hydrogen Energy* **2021**, *46*, 31.
74. Sethupathy, A.; Kumar, P.S.; Sivashanmugam, P.; Arun, C.; Banu, J.R.; Ashokkumar, M. Evaluation of biohydrogen production potential of fragmented sugar industry biosludge using ultrasonication coupled with egtazic acid. *Int. J. Hydrogen Energy* **2021**, *46*, 1705–1714. [CrossRef]
75. Mirza, S.S.; Qazi, J.I.; Liang, Y.; Chen, S. Growth characteristics and photofermentative biohydrogen production potential of purple non sulfur bacteria from sugar cane bagasse. *Fuel* **2019**, *255*, 115805. [CrossRef]
76. Urbaniec, K.; Bakker, R.R. Biomass residues as raw material for dark hydrogen fermentation–A review. *Int. J. Hydrogen Energy* **2015**, *40*, 3648–3658. [CrossRef]
77. Hu, J.; Nagarajan, D.; Zhang, Q.; Chang, J.-S.; Lee, D.-J. Heterotrophic cultivation of microalgae for pigment production: A review. *Biotechnol. Adv.* **2018**, *36*, 54–67. [CrossRef]
78. Sun, Y.; Mang, J.-P.; Yang, G.; Li, Z.-H. Study on the spectra of spruce lignin with chlorine dioxide oxidation. *Spectrochim. Acta Part A Mol. Spectrosc.* **2007**, *27*, 1551–1554.
79. Tan, M.; Ma, L.; Rehman, M.S.U.; Ahmed, M.A.; Sajid, M.; Xu, X.; Sun, Y.; Cui, P.; Xu, J. Screening of acidic and alkaline pretreatments for walnut shell and corn stover biorefining using two way heterogeneity evaluation. *Renew. Energy* **2019**, *132*, 950–958. [CrossRef]
80. Meghana, M.; Shastri, Y. Sustainable valorization of sugar industry waste: Status, opportunities, and challenges. *Bioresour. Technol.* **2020**, *303*, 122929. [CrossRef]
81. Shuler, M.L. *Bioprocess Enzineering: Basic Concepts*; Prentice-Hall: New York, NY, USA, 2017; pp. 412–420.

82. Zhao, X.; Xing, D.; Qi, N.; Zhao, Y.; Hu, X.; Ren, N. Deeply mechanism analysis of hydrogen production enhancement of Ethanoligenens harbinense by Fe^{2+} and Mg^{2+}: Monitoring at growth and transcription levels. *Int. J. Hydrogen Energy* **2017**, *42*, 19695–19700. [CrossRef]
83. Palomo-Briones, R.; Razo-Flores, E.; Bernet, N.; Trably, E. Dark-fermentative biohydrogen pathways and microbial networks in continuous stirred tank reactors: Novel insights on their control. *Appl. Energy* **2017**, *198*, 77–87. [CrossRef]
84. Wimonsong, P.; Llorca, J.; Nitisoravut, R. Catalytic activity and characterization of Fe–Zn–Mg–Al hydrotalcites in biohydrogen production. *Int. J. Hydrogen Energy* **2013**, *38*, 10284–10292. [CrossRef]
85. Woodward, J.; Orr, M.; Cordray, K.; Greenbaum, E. Enzymatic production of biohydrogen. *Nature* **2000**, *405*, 1014–1015. [CrossRef]
86. Ergal, İ.; Gräf, O.; Hasibar, B.; Steiner, M.; Vukotić, S.; Bochmann, G.; Fuchs, W.; Simon, K.-M.R. Biohydrogen production beyond the Thauer limit by precision design of artificial microbial consortia. *Commun. Biol.* **2020**, *3*, 1–12. [CrossRef]
87. Wang, S.; Tang, H.; Peng, F.; Yu, X.; Su, H.; Xu, P.; Tan, T. Metabolite-based mutualism enhances hydrogen production in a two-species microbial consortium. *Commun. Biol.* **2019**, *2*, 1–11. [CrossRef]
88. Lu, Y.; Zhao, H.; Zhang, C.; Xing, X.-H. Insights into the global regulation of anaerobic metabolism for improved biohydrogen production. *Bioresour. Technol.* **2016**, *200*, 35–41. [CrossRef] [PubMed]
89. Banu, J.R.; Ginni, G.; Kavitha, S.; Kannah, R.Y.; Kumar, S.A.; Bhatia, S.K.; Kumar, G. Integrated biorefinery routes of biohydrogen: Possible utilization of acidogenic fermentative effluent. *Bioresour. Technol.* **2021**, *319*, 124241. [CrossRef] [PubMed]
90. Wang, L. *Sustainable Bioenergy Production*; CRC Press: Boca Raton, FL, USA, 2019.
91. Banu, J.R.; Kavitha, S.; Kannah, R.Y.; Bhosale, R.R.; Kumar, G. Industrial wastewater to biohydrogen: Possibilities towards successful biorefinery route. *Bioresour. Technol.* **2020**, *298*, 122378. [CrossRef] [PubMed]
92. Nawaz, M.Z.; Bilal, M.; Tariq, A.; Iqbal, H.M.; Alghamdi, H.A.; Cheng, H. Bio-purification of sugar industry wastewater and production of high-value industrial products with a zero-waste concept. *Crit. Rev. Food Sci. Nutr.* **2020**, 1–18. [CrossRef] [PubMed]
93. Liu, D.; Sun, Y.; Li, Y.; Lu, Y. Perturbation of formate pathway and NADH pathway acting on the biohydrogen production. *Sci. Rep.* **2017**, *7*, 1–8.
94. Gutekunst, K.; Hoffmann, D.; Westernströer, U.; Schulz, R.; Garbe-Schönberg, D.; Appel, J. In-vivo turnover frequency of the cyanobacterial NiFe-hydrogenase during photohydrogen production outperforms in-vitro systems. *Sci. Rep.* **2018**, *8*, 1–10.
95. Mangayil, R.; Karp, M.; Lamminmäki, U.; Santala, V. Recombinant antibodies for specific detection of clostridial [Fe-Fe] hydrogenases. *Sci. Rep.* **2016**, *6*, 1–9. [CrossRef]
96. Oladokun, O.; Ahmad, A.; Abdullah, T.A.T.; Nyakuma, B.B.; Kamaroddin, M.F.A.; Nor, S.H.M. Biohydrogen production from Imperata cylindrica bio-oil using non-stoichiometric and thermodynamic model. *Int. J. Hydrogen Energy* **2017**, *42*, 9011–9023. [CrossRef]
97. Show, K.-Y.; Lee, D.-J.; Chang, J.-S. Bioreactor and process design for biohydrogen production. *Bioresour. Technol.* **2011**, *102*, 8524–8533. [CrossRef] [PubMed]
98. Wang, Y.; Yang, G.; He, J.; Sun, G.; Sun, Z.; Sun, Y. Preparation of biochar catalyst from black liquor by spray drying and fluidized bed carbonation for biodiesel synthesis. *Process. Saf. Environ. Prot.* **2020**, *141*, 333–343. [CrossRef]
99. Park, J.-H.; Kim, D.-H.; Kim, H.-S.; Wells, G.F.; Park, H.-D. Granular activated carbon supplementation alters the metabolic flux of Clostridium butyricum for enhanced biohydrogen production. *Bioresour. Technol.* **2019**, *281*, 318–325. [CrossRef] [PubMed]
100. Jamali, N.S.; Jahim, J.M.; Isahak, W.N.R.W.; Abdul, P.M. Particle size variations of activated carbon on biofilm formation in thermophilic biohydrogen production from palm oil mill effluent. *Energy Convers. Manag.* **2017**, *141*, 354–366. [CrossRef]

energies

MDPI

Article

Optimization of a Mixed Refrigerant Based H_2 Liquefaction Pre-Cooling Process and Estimate of Liquefaction Performance with Varying Ambient Temperature

Steven Jackson * and Eivind Brodal

IAP, UiT-Norges Arktiske Universitetet, 9037 Tromsø, Norway; eivind.brodal@uit.no
* Correspondence: steve.jackson@uit.no

Abstract: Hydrogen used as an energy carrier can provide an important route to the decarbonization of energy supplies, but realizing this opportunity will require both significantly increased production and transportation capacity. One route to increased transportation capacity is the shipping of liquid hydrogen, but this requires an energy-intensive liquefaction step. Recent study work has shown that the energy required in this process can be reduced through the implementation of new and improved process designs, but since all low-temperature processes are affected by the available heat-sink temperature, local ambient conditions will also have an impact. The objective of this work is to identify how the energy consumption associated with hydrogen liquefaction varies with heat-sink temperature through the optimization of design parameters for a next-generation mixed refrigerant based hydrogen liquefaction process. The results show that energy consumption increases by around 20% across the cooling temperature range 5 to 50 °C. Considering just the range 20 to 30 °C, there is a 5% increase, illustrating the significant impact ambient temperature can have on energy consumption. The implications of this work are that the modelling of different liquified hydrogen based energy supply chains should take the impact of ambient temperature into account.

Keywords: hydrogen; liquefaction; optimization; ambient temperature; mixed refrigerant

Citation: Jackson, S.; Brodal, E. Optimization of a Mixed Refrigerant Based H_2 Liquefaction Pre-Cooling Process and Estimate of Liquefaction Performance with Varying Ambient Temperature. *Energies* **2021**, *14*, 6090. https://doi.org/10.3390/en14196090

Academic Editor: Bahman Shabani

Received: 2 September 2021
Accepted: 21 September 2021
Published: 24 September 2021

1. Introduction

Hydrogen used as a fuel, as an energy source for industrial processes or for generating electrical power can provide an important route to the decarbonization of energy supplies and the integration of renewable energy systems. The study of Acar and Dincer [1], for example, identifies that hydrogen can play "eight significant roles" in the green energy transition. Recent studies have also made the case that achieving a transition to carbon-free energy in the EU is impossible without a large increase in hydrogen production [2], and energy system modeling has found that "hydrogen and synfuels add up to between 20% and 50% of [EU] energy demand in transport in 2050" [3]. This positive view of the role that hydrogen could play in future low-carbon development is also reflected in political intent via the EU hydrogen strategy [4].

Reflecting this political support, research related to hydrogen energy has increased over recent years [5]. Important research topics include energy demand and supply modelling [2,3], the novel integration of renewable energy sources such as solar power [6], the development of enhanced electrolysis based production methods [7], the development of new applications such as use a reductant in steel manufacturing [8] and the assessment of alternative sources such as methanol [9].

In all envisaged future hydrogen-based economies, a significant increase in the transportation capacity for hydrogen is required. It is possible to transport hydrogen as a compressed gas or as a liquid at low temperature, in pipelines as a gas. The optimum transportation strategy will depending on both transportation capacity and the distance [10]. When the distance is significant it is reasonable to expect that shipping of hydrogen will be favored, and while researchers such as Ishimoto et al. [11] have studied the economics of shipping liquefied hydrogen, some commercial steps have also been made, with the world's first liquid hydrogen carrier ship launched in Japan in 2019 [12].

If hydrogen is transported at large-scale as a liquid, a key part of the supply chain will be the liquefaction process, which is very energy intensive. The specific energy consumption (SEC) of the most efficient currently operating large-scale hydrogen liquefaction (LHL) plants lies in the range 13 to 15 kWh/kg [13], which is much higher that even the most efficient LNG processes, which have a SEC of around 240 kWh/tonne. Because of this, there is significant interest in the development of new and improved LHL technologies that can help reduced SEC.

Research topics relating to improved LHL technologies include the integration of renewable energy sources, such as solar energy [14] and geothermal energy [15]; the use of mixed refrigerants (MR) for pre-cooling [16–18]; and the use of helium in the cryogenic cooling and liquefaction part of the process [19,20]. Other research has focused on the impact of the conversion of ortho-hydrogen to para-hydrogen on the liquefaction process [21,22] and the relative performance of different heat exchanger types [17,22,23]. The suggested efficiency of the proposed concepts for LHL studied lie in the range 5 to 8 kWh/kg [13], which represents a substantial motivation for the implementation of these technologies in the next generation of LHL plants.

The proposed use of a MR in the pre-cooling part of LHL processes represents a close parallel to the use of MR in the design of some of the largest and most efficient natural gas liquefaction processes and because of this, represents one of the most promising near-future improvements to LHL design. LNG plants based on the use of MRs include the Snøhvit plant located at Melkøya in northern Norway, which uses a cascade of three MR loops and is claimed to be the most efficient LNG plant in the world [24]. While the efficiency achieved by the Snøhvit LNG plant is due, in part, to its advanced design, the plant also benefits from its cold-climate location and subsequent access to a lower temperature heat sink than most other LNG plants.

The study of Rian and Ertesvåg [25] looked at the impact of ambient temperature on the Snøhvit LNG plant, finding that a reduction in the available heat sink temperature from 20 to 4 °C gives a reduction in exergy destruction of 10.9%. A small number of other studies have also considered the impact of ambient temperature on the performance of other types of LNG process [26–29] providing similar results. The study of Park et al., for example, finds that specific power consumption of single MR process increases by between 16% and 42% over the temperature range 10 to 25 °C, varying with the approach used in process optimization. This significance of this variation in energy consumption with ambient temperature is not only relevant to the design of LNG plant itself, it is large enough to affect the whole energy supply chain. For example, the study of Jackson et al. [29] finds that the CO_2 emissions for a power plant supplied by gas from an LNG plant located in northern Norway will be between 0.8 and 1.3% lower than if it were supplied by the same design of LNG plant located in the Middle East. It is therefore logical to expect that the performance of LHL plants using MR pre-cooling and the performance of energy supply chains based on LHL will be significantly affected by the ambient temperature at the liquefaction plant location.

Given the close parallel between MR based LNG processes and MR pre-cooled LHL processes and given the demonstrated impact of ambient temperature the performance of LNG processes, ambient temperature can be expected to have a significant impact on the performance of the type of LHL processes likely to be used in the near future. Although several studies have been made into the performance of MR pre-cooled type LHL process and studies have looked at the impact of ambient temperature on LNG process performance, no studies quantifying this impact of ambient temperature on LHL processes are currently found in the literature. The aim of this study is, therefore, to generate a set of data illustrating the impact of ambient temperature on the performance of MR pre-cooled type LHL process.

2. Materials and Methods

2.1. Selection of the Modeling Basis

Although only a handful different types of liquefaction process are used in current operating LHL plants, a wide range of improved processes have been proposed. In the present study a comprehensive review of the various improved liquefaction technologies is outside the scope of work. Instead, a single, representative, improved process was selected to be used as the basis for the present study. The details of the selection process are described below.

Because the results from this study are intended to support further research in future low-carbon energy supply and, specifically, how different supply chain configurations affect efficiency, the improved concepts of most relevance are those technologies likely to be used in the near future. Taking the techno-economic analysis of Cardella et al. [30] as a basis, the improved technology that fits best with the aim of the study is the use of a mixed refrigerant (MR) for pre-cooling of the hydrogen feed stream.

Most current, and much improved, hydrogen liquefaction processes are based on the division of the overall process into two parts: a pre-cooling step and a cryogenic-cooling step. In conventional LHL plant designs, the pre-cooling stage often uses liquid nitrogen (LIN) as a refrigerant, whereas the cryogenic-cooling step uses either helium in a Brayton cycle, or hydrogen in a Claude cycle [18]. In the cryogenic step, the hydrogen feed is generally cooled from below around −90 °C to the final liquefaction temperature. Although the break-point temperature between the pre-cooling and the cryogenic step, T_p, is potentially an optimization variable, the present study assumes that the impact of ambient temperature on operating parameters in the cryogenic step is small and, therefore, that T_p can be fixed.

Typical of the concepts for improved energy consumption using MRs is the process studied in the work of Skaugen et al. [17], which is based on a Claude cycle in cryogenic-cooling step and a MR in the pre-cooling step. In this process the pre-cooling step and the portion of the cryogenic step that operates above T_p are not integrated. This allows the present study to consider the optimization of the pre-cooling process independently from the operation of the cryogenic-cooling process. In addition, because the details of the composition and operating conditions for the proposed MR cycle are clearly set-out in the work of Skaugen et al. [17], the present study uses the work of Skaugen as the basis for model development and validation.

Although the operating parameters in the cryogenic-cooling step are assumed fixed in the present study (i.e., they are not affected by ambient temperature), the energy consumption of the cryogenic-cooling cycle compressor is still affected by the exit temperature that the inter and after-coolers, T_c, are designed to operate with, which would normally be set relative to the ambient temperature of the seawater, or air, used as the heat-sink. Because of this, modelling of the performance of the cryogenic cycle compressor as it varies with T_c does form part of the present study.

Another important factor in the design and optimization of hydrogen liquefaction processes is the conversion of ortho to para hydrogen. This process releases a significant quantity of heat, affecting both the process design and the selection of optimum operating parameters. The conversion of the ortho isomer during liquefaction is typically promoted using a catalyst. The effectiveness of the catalyst and the residence time in the heat exchangers affects the approach to the equilibrium concentration and, subsequently, the temperature profile in the heat exchangers. However, across the range of temperatures experienced in the pre-cooling process, the equilibrium concentration of para hydrogen varies by less than 5% [21]. Moreover, as in the study of Skaugen et al. [17]—which is a reference case for this study—catalytic conversion is assumed after the pre-cooling process. This will result in a low approach to the equilibrium conversion in the pre-cooling process and, therefore, in this study the modelling of the conversion of ortho to para hydrogen is set outside the scope of work.

2.2. Process Model Development

As described above, the process model used in this study consists of two separate parts: a model of the MR pre-cooling step, and a model of the cryogenic-cooling step cycle compressor. The development of these two models is described below. A block diagram showing the relationship between the cryogenic-cooling step and the pre-cooling step is also presented in Appendix A.

Figure 1 illustrates the process flow scheme used for the MR pre-cooling process, which is based on the flow scheme used in the reference study of Skaugen et al. [17]. The main equipment items shown in Figure 1 are a compressor (comprising RC-1 and RC-2), two process coolers (PC-1 and PC-2), a MR separator (VV-1), a pump (PP-1), and the main heat exchanger (HX-1). The MR compressor comprises two stages (RC-1 and 2), both with after-cooling (PC-1 and 2) to T_c. Any liquids condensed liquids after the first stage are separated in VV-1. Liquids separated in this way are pumped (PP-1) to the compressor discharge pressure—bypassing the second stage of compression (RC-2)—and mixed with the vapor stream entering the main heat exchanger (HX-1). The main heat exchanger is modelled as a multi-stream type heat exchanger with two hot streams: H_2 and high-pressure MR, and one cold stream: low-pressure MR. The low-pressure MR stream exiting the main heat exchanger returns to the MR compressor. Hydrogen leaving HX-1 is cooled to T_p.

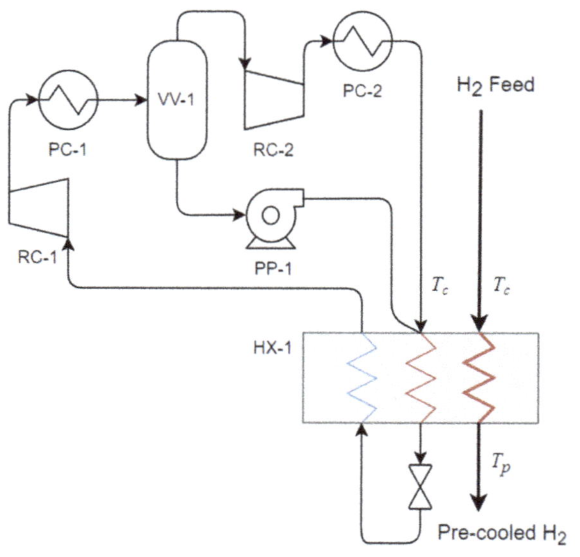

Figure 1. Flow diagram for the MR H_2 pre-cooling process.

To allow the calculation of process energy consumption a simplified model of the process presented in Figure 1 was developed in MATLAB [31] with the TREND software package [32] used to calculate thermo-physical properties. Table 1 presents the set of fixed modelling parameters, MP, used in the model of the MR pre-cooling process. In general, the parameters in Table 1 were selected to reflect those used in the reference study [17].

Table 1. Summary of MR process fixed modelling parameters.

Parameter	Value	Units
Hydrogen Feed Pressure	20	bara
H_2 pre-cooling temperature, T_p	−159	°C
Compressor/ pump efficiency	85	% *
HX-1 pressure-loss (hot streams)	0.5	bar
HX-1 pressure-loss (cold streams)	0.1	bar
PC-1 and 2 pressure-loss	0.5	bar

* Isentropic efficiency.

For simplicity, the pressure-loss in the main heat exchanger was scaled linearly with temperature and the two MR streams were assumed to be mixed before entering the heat exchanger and the combined MR stream enters the main heat exchanger at the H_2 feed temperature.

The temperature profiles for the combined hot streams and the cold stream in HX-1—the hot and cold composite curves—were estimated by splitting the heat exchanger into n equally sized temperature intervals, each sized $(T_{in} - T_{out})/n$ and stream enthalpies calculated for each temperature point ($n + 1$, total). Then the heat exchanger duty was also split into n equally sized intervals (Q_{HX-1}/n), and the hot and cold composite temperatures, T_{HC} and T_{CC}, interpolated at each point ($n + 1$, total) using linear interpolation of the temperature-enthalpy data. Finally, the temperature approach was calculated for each point, $\Delta T = T_{HC} - T_{CC}$. In both cases, n was set to 50 to give a high degree of accuracy to the calculations.

Figure 2 provides a sketch of the cycle compressor (comprising H2-1 to 4) for the cryogenic-cooling step which forms the basis of the present study. The stream LP H_2 represents the low-pressure hydrogen stream returning from the liquefaction process. This stream is compressed in two compressor stages (H2-1 and H2-2) before blending with medium-pressure hydrogen. The combined stream is then compressed in two further compressor stages (H2-3 and H2-4) before being passed-back to a multi-stream heat exchanger (HX-2), which cools the HP stream down to T_p. The compressor inter-stage pressures are calculated assuming equal stage pressure ratios.

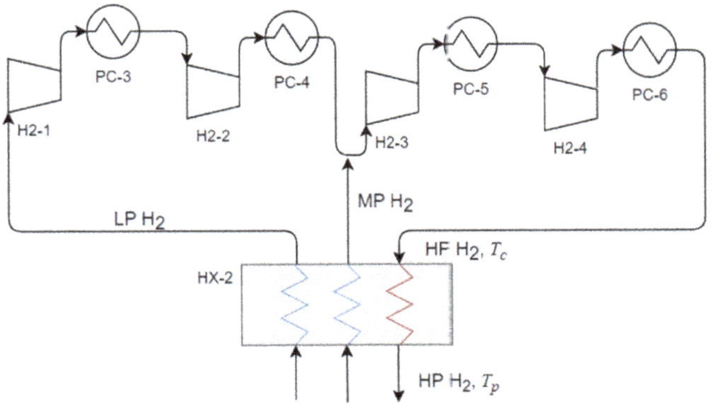

Figure 2. Flow diagram of the cryogenic-cooling step H_2 cycle compressor.

The model of the cryogen-cooling step compressor shown in Figure 2 was also developed in MATLAB using the same basis as the MR process model. Table 2 presents the fixed modelling parameters used in the study performance of this compressor, which are based on the reference model [17]. The outlet temperature of the four after-coolers (PC-3 to 6) were assumed equal to T_c and the inlet temperature of the LP and MP streams to the compressor was assumed to have a 2 °C approach to T_c in all cases.

Table 2. Summary of fixed modelling parameters for the cryogenic-cooling cycle compressor.

Parameter	Value	Units
LP H2 Feed Pressure	1.1	bara
LP H2 flowrate	51.5	tpd
MP H2 feed pressure	8.0	bara
MP H2 flowrate	1121.5	tpd
HP H2 return pressure	29.8	bara
PC-3 to 6 pressure-loss	0.5	bar

Isentropic efficiency.

In addition to the cycle compressor, the reference study describes several turbo-expanders within the cryogenic cooling step. These produce 2.8 MW of shaft power, which is assumed in the reference study to be recovered as electrical energy with as efficiency of 80% [17]. Assuming, as before, that the parameters in the cryogenic process remain constant with varying T_c, this recovered energy equates to a specific energy production for the expanders, SEC_{Ex}, of approximately 0.43 kWh/kg, which is a constant value for all cases studied in this work.

Where operating parameters were not available in the reference study, they have been inferred from the data that is presented there. Because of this, it cannot be claimed that there is any direct equivalence between the results presented here and the reference model.

2.3. MR Pre-Cooling Model Validation

An important aspect of successful optimization is the minimization of temperature differences in HX-1, and since the targeted minimum approach temperature is only 1 K, the accuracy of the property predications used in the process model is very important. In the TREND software package, several properties methods are available; to select the basis that is most appropriate for the present work, three of these were compared against results from the reference study: Peng Robinson (PR), Soave-Redlich-Kwong (SRK) and the TREND Helmholtz free energy model. Tables 3 and 4 present the parameters used in the validation work. The results of the validation work were used to select the properties method used in the later optimization work.

Table 3. Validation case MR composition.

Component	Mole Fraction
Nitrogen	0.101
Methane	0.324
Ethane	0.274
Propane	0.031
n-Butane	0.270

Table 4. Validation case MR modelling parameters.

Parameter	Value	Units
Hydrogen Feed Flow	125	tpd
MR feed temperature	12	°C
MR return temperature	−1.0	°C
MR feed pressure	35	bara
MR return pressure	4.25	bara

2.4. Optimization Problem Defenition

The objective of the optimization study was to minimize the energy consumption of the MR pre-cooling process whilst satisfying a minimum temperature approach constraint. The objective function was formulated as described in Equation (1):

$$\min\{\mathrm{SEC_{MR}}\}, \text{ such that } \begin{cases} \mathrm{lb}_i < \mathrm{OP}_i < \mathrm{lb}_i \\ \Delta T_{\mathrm{min}} - \Delta T_{\mathrm{acc}} > 0 \\ \dot{m}_{\mathrm{MR}} > 0 \end{cases}. \tag{1}$$

In Equation (1), $\mathrm{SEC_{MR}}$ is the specific energy consumption of the MR process, OP_i are the set of i optimization parameters (see Table 5), lb_i and ub_i are a set of lower and upper bounds for each parameter, ΔT_{min} is the minimum approach temperature in HX-1 ($\Delta T_{\mathrm{min}} = \min\{\Delta T_n\}$), ΔT_{acc} is the minimum acceptable approach temperature in HX-1 and $\dot{m}_{\mathrm{MR}}$ is the mass flowrate of the MR. $\mathrm{SEC_{MR}}$ was calculated from the sum the compression stage energy consumptions, W_{MR}, which are, in turn, a function of OP_i, MP_i (see Table 1) and T_c is described by Equation (2):

$$\mathrm{SEC_{MR}} = \sum W_{\mathrm{MR}}(\mathrm{OP}_i, \mathrm{MP}_i, T_c) / \dot{m}_{\mathrm{H2}}. \tag{2}$$

Table 5. Summary of Optimization Parameters with Initial ($\mathrm{OP}_{i,0}$) and Constraint Values.

Parameter	Description	$\mathrm{lb}_i < \mathrm{OP}_{i,0} < \mathrm{ub}_i$
OP_1	MR mole fraction N2	0.05 < 0.11 < 0.25
OP_2	Mole fraction CH4	0.20 < 0.32 < 0.50
OP_3	MR mole fraction C2	0.15 < 0.27 < 0.50
OP_4	MR mole fraction C3	0.00 < 0.03 < 0.10
OP_5	RC-1, Pin (bara)	2.00 < 4.25 < 6.00

In Equation (2), $\dot{m}_{\mathrm{H2}}$ is the mass flowrate of hydrogen in the pre-cooling process.

The set of optimization parameters, OP_i, used in the study are summarized in Table 5 along with the initial values used ($\mathrm{OP}_{i,0}$) and initial values of the boundary constraints (lb_i and ub_i).

Although the ultimate purpose of the boundary constraints shown in Table 5 was to limit the optimization process to physically meaningful solutions—e.g., component mole fractions greater than zero—the initial boundary constraints were also used to limit the search area around the likely optimum values. This was done to reduce optimization time. The initial values of lb and ub shown in Table 5 were set based on results from the reference case, but where the optimization solution was found close to the initial limits, the bounds were extended to ensure that the overall optimum solution was not missed.

In addition to the optimization parameters listed in Table 5, the MR compressor inter-stage pressure, MR compressor discharge pressure and HX-1 warm-end approach temperature could be considered as optimization parameters. However, in this work these have been excluded to limit complexity. The MR compressor discharge pressure is, therefore, fixed at the value used in the reference study, the MR inter-stage pressure set in each case to maintain equal stage pressure ratios, and the HX-1 warm-end approach set to 5 °C. The MR mole fraction for butane is also not identified as an optimization parameter because it is calculated from the sum of the other components.

2.5. Optimization Algorithm

In a phase of initial testing the *Fmincon* (FMC) algorithm with the SQP option was found to provide fast and generally accurate optimization results, although in some cases local minima were found. In all subsequent cases, FMC was used with the solution tolerance set to 0.001 kWh/kg and all other options left as default.

To help identify the global minimum solutions for each T_c, the boundary constraints shown in Table 5 were evaluated in a manual, stepwise, process: after the initial results had been gathered, new initial guesses were specified when the original initial guess was found to be a long way from the solution. When a stable set of bounds enclosing the global solution had been found, the *MultiStart*, MS, and *GlobalSearch*, GS, algorithms were used to help test the quality of the results. In both cases the MS and GS runs were again based on the FMC algorithm with the parameters as before.

The quality each optimization result was assessed qualitatively using the results from other T_c cases. The basis of this assessment was the assumption that a simple, monotonic, relationship was likely between each of the optimization parameters and T_c. In addition to this assessment, the temperature profiles in HX-1 for each case were reviewed qualitatively to determine if ΔT_{acc} was consistently approached throughout the heat exchanger.

2.6. Performance Variation with Cooling Temperature

Performance variation with cooling temperature was studied for the MR pre-cooling process by finding the optimum operating parameters, OP_i, for each cooling temperature, T_c case. The fixed modelling parameters shown in Table 2 were used as the basis in all cases. The cooling temperature range studied was 5 to 50 °C.

In the model developed for the cryogenic-cooling step, process parameters were not optimized: flowrates and pressure levels in the cryogenic cycle were held constant at the values shown in Table 3. The variation of the energy consumption of the cryogenic cycle compressor with T_c was modelled using the more simplistic assumption that, since the composite cooling curves in HX-2 are straight and parallel, a constant warm-end approach temperature exists across the range of cooling temperatures studied. The energy consumption of the cryogenic cycle compressor was calculated using the same basis as that of the MR pre-cooling process. A 2 °C warm-end approach temperature was assumed across the cooling temperature range 5 to 50 °C.

The overall SEC for the hydrogen liquefaction process was calculated as the sum of the energy consumption for the MR pre-cooling step, SEC_{MR}, and the cryogenic-cooling step, SEC_{CY}, which was—in turn—calculated as the sum of the cycle compressor stage energy consumptions minus the energy recovered in the cryogenic-cooling step expanders as described in Equations (3) and (4):

$$SEC = SEC_{MR} + SEC_{CY} \tag{3}$$

$$SEC_{CY} = \sum W_{H2}(MP_{H2}, T_c)/\dot{m}_{H2} - SEC_{Ex} \tag{4}$$

In Equation (3), W_{H2} is the energy consumption of the cycle compressors shown in Figure 2, and in Equation (4), MP_{H2} is the set of fixed modelling parameters for the cryogenic-cooling cycle compressor (see Table 2).

To provide an independent means of reviewing the trends shown in the results, the SEC for an ideal process that cooled the hydrogen from T_c to a final temperature of −259 °C was also calculated. This ideal energy consumption, SEC_{ID}, was then used to calculate a second law efficiency, $\eta_{id} = SEC/\ SEC_{ID}$, for the overall process. The method used to calculate SEC_{ID} was to summate the ideal Carnot cycle energy consumption for a set of very small temperature steps along temperature–enthalpy data for hydrogen as explained previously by Jackson et al. [29].

3. Results and Discussion

3.1. Process Modelling and Validation

Table 6 shows the results from the model validation work. In addition to the results from the reference study, three sets of results are presented in Table 6: Case A uses the TREND implementation of the Peng Robinson (PR) equation of state; Case B the TREND/SRK equation of state; and Case C the TREND/ Helmholtz free energy properties method.

Table 6. Summary of modelling parameters for the model validation work.

	Reference	Case A	Case B	Case C	
Properties method	-	PR	SRK	Hel.	
MP supply pressure	35.0	35.0	35.0	35.0	bara
MR return pressure	4.25	3.0 **	4.25	4.25	bara
MR return temp.	112	112.8	112.3	109.6	°C
MR mass flowrate	1600 *	1395	1703	1709	tpd
HX-1 min. approach	1.00	1.05	0.49	0.51	°C
HX-1 duty	12.6	11.2	13.2	12.9	kW

* Given only as an approximate value in the reference case. ** Adjusted to give a positive value for min. approach.

Of the three cases compared in Table 6, Case C—using the TREND/ Helmholtz free energy properties method—is considered to represent the closest match to the reference case, but since Case B also offers good agreement and significantly reduced calculation time, SRK is selected as the basis for further work.

Figure 3 presents the composite temperature profile data for Case B and C in Table 6. The results show that, although the shape of the curves differs between the two cases, the results from both cases show a very good fit between the warm and cold curves throughout the heat exchanger. These results, therefore, add confidence to the validation work and the selection of Case B as the modelling basis.

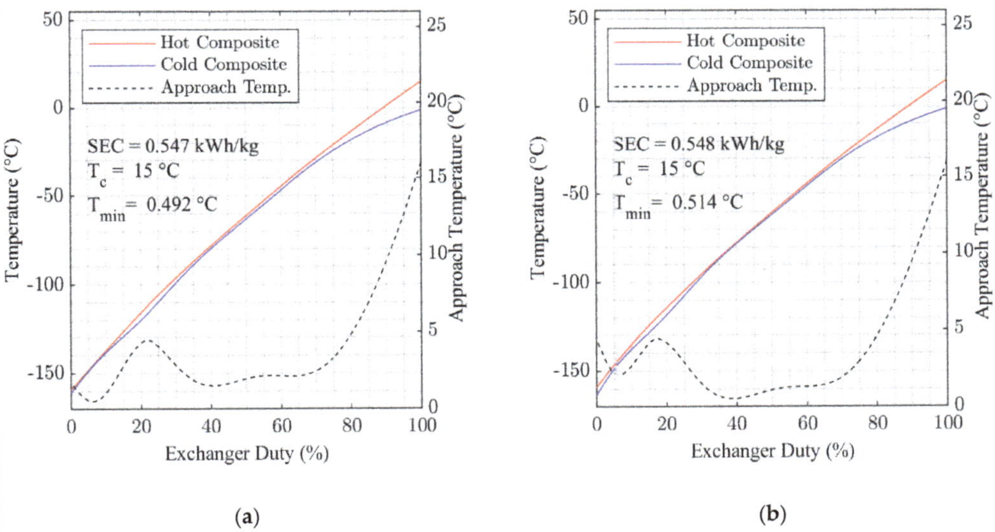

Figure 3. Composite Curves and Key Performance Parameters for HX-1, MR Pre-cooling Process: (**a**) Case B; (**b**) Case C.

While Table 6 and Figure 3 show that the selection of a good modelling basis is important to the determination of the optimum operating parameters for this process, no claim is made here that the modelling basis selected is the one that is most accurate for the modelling of this process, just that it provided a good match with the reference case in the validation work presented.

A limitation of the present study is that the heat generated during ortho-para hydrogen conversion is omitted form the model. This is a simplification that limits the extent to which this modelling work reflects the performance of a hydrogen liquefaction process operating in the real world. The main claim made here regarding the modelling basis is that it provides provide a consistent basis to study performance across the operating cases considered. The implication of this for further work is that the study of the variation in energy consumption with cooling temperature made here is valid and can provide some insight into how the performance real hydrogen liquefaction processes can be expected to vary when designed for utility cooling at different temperatures.

3.2. Performance Variation with Cooling Temperature

Figures 4 and 5 show how the five optimization parameters vary with T_c, and Figures 6 and 7 provide two examples of the optimized cooling curves resulting from these runs. In Figures 4 and 5 all of the data collected over the final set of optimization runs (two using GS and two using MS) are presented as points and the overall optimum datasets are connected by dotted lines.

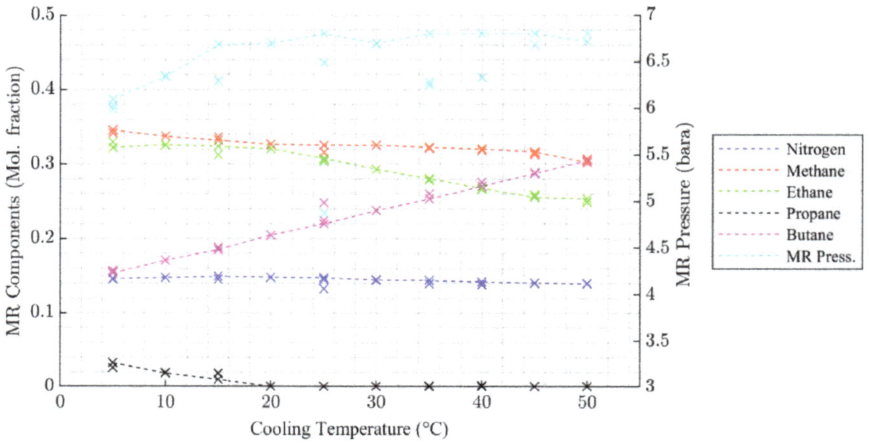

Figure 4. Variation in OP for the MR pre-cooling step with cooling temperature.

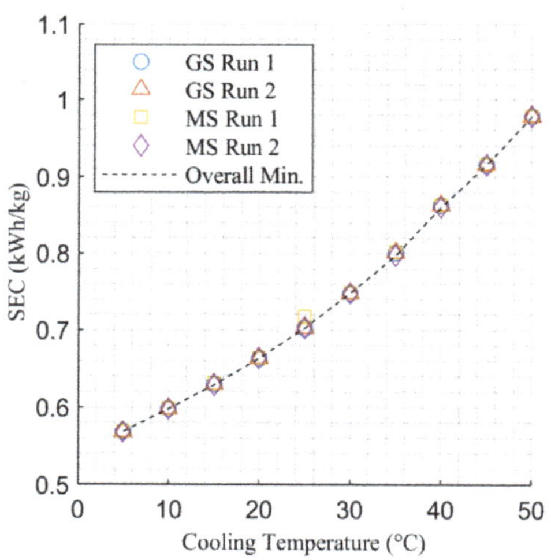

Figure 5. Variation in SEC for the MR pre-cooling step with cooling temperature.

The results presented in Figure 4 for MR composition show quite clear trends with the component mole fraction of each component a monotonic function of cooling temperature in the majority of cases. The impact on butane is largest, which is due to the steadily increasing heat duty at the warm end of HX-1 as the cooling temperature increases. The impact on the optimum nitrogen content in the MR is affected least by cooling temperature, reflecting the relatively static conditions at the cold end of HX-1.

The data presented in Figure 4 that represents optimum MR pressure solutions is less consistent with a slight upward trend visible across the range of cooling temperatures considered. This indicates that the optimum combination of MR composition and MR operating pressure is more difficult to determine and that the overall minimum may not have been found in all cases. However, Figure 5 shows a very consistent trend in how the SEC for the MR pre-cooling process varies with T_c, which provides confidence that a solution close to the overall minimum was found in all cases.

Figures 6 and 7 present the hot and cold composite cooling curves for the overall minimum SEC solutions found for T_c = 5 °C and T_c = 50 °C. Generally, the results in Figures 6 and 7 show that the optimization algorithm has found a good fit for the cooling curves, with the 2 °C pinch temperature approached in multiple locations within HX-1 in both cases. The cooling curves for each of the temperature points studied between T_c = 5 °C and T_c = 50 °C are presented in Figures A2–A9, which are contained in the Appendix A.

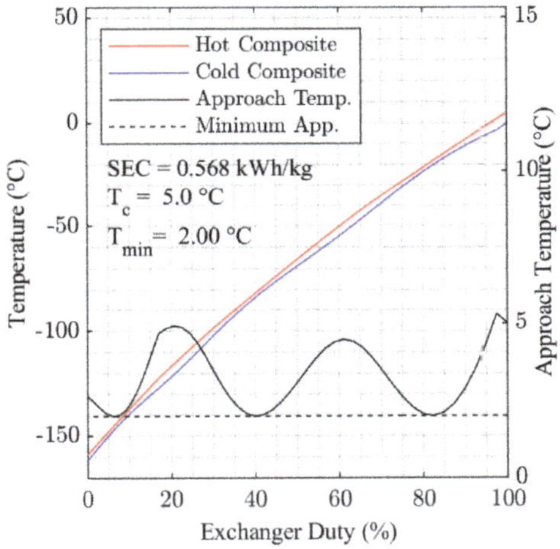

Figure 6. Composite curves in HX-1 and key performance parameters, 5 °C cooling temperature case.

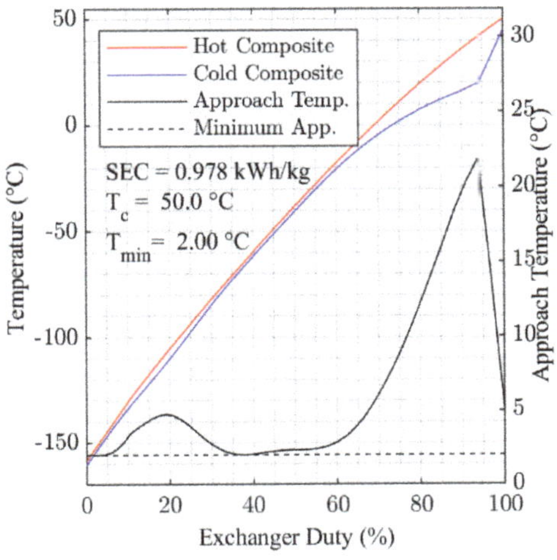

Figure 7. Composite curves in HX-1 and key performance parameters, 50 °C cooling temperature case.

Comparing the variation in minimum approach temperature data presented in Figure 6 for T_c = 5 °C with that presented in Figure 7 for T_c = 50 °C, it can also be observed that the optimization process has found a set of parameters that better minimize the temperature approach in HX-1 for the T_c = 5 °C case. Looking at the T_c = 50 °C case, we see that it becomes more difficult to maintain a close approach at the warm end of the heat exchanger suggesting that SEC could be reduced further through the addition of heavier components to the MR.

Figure 8 presents the SEC for the pre-cooling step, the cryogenic-cooling step, and the overall process. Figure 9 presents the same data in terms of the % change relative to the 25 °C case. Moreover, presented in Figure 9 are the corresponding second law efficiencies expressed as a percentage.

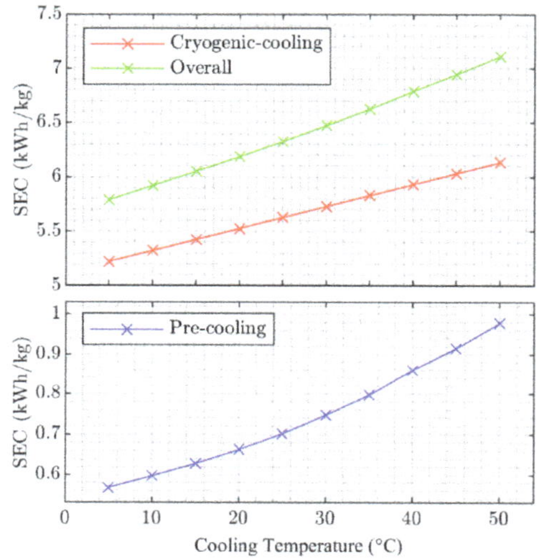

Figure 8. Variation in SEC with cooling temperature for the pre-cooling, cryogenic-cooling and overall cooling processes.

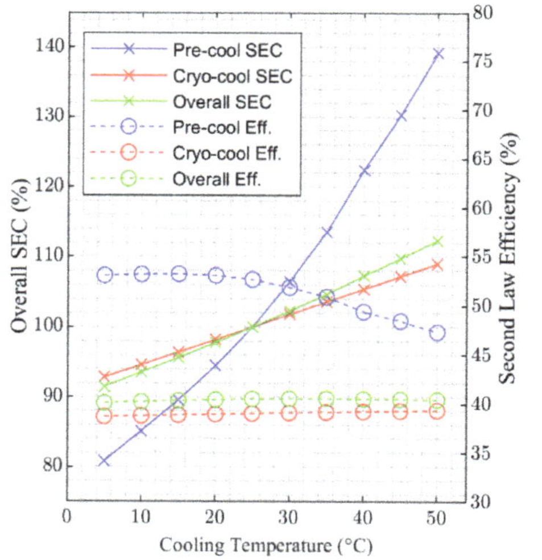

Figure 9. Percentage variation in overall SEC and second law efficiency for the pre-cooling, cryogenic-cooling and overall cooling processes.

Figure 8 shows that the contribution of the pre-cooling process to overall SEC across the range of cooling temperatures investigated is approximately 10%. In addition, Figure 8 shows non-linear variation in SEC with cooling temperature for the pre-cooling part of the overall process that contrasts with the linear relationship between SEC and cooling temperature for the cryogenic process. This non-linear relationship for the pre-cooling process reflects the fact that lower cooling temperatures both reduced cooling duty and increase efficiency, whereas the close to linear impact on the cryogenic process is a result of only reduced increased efficiency. Further insight into this is provided by the results presented in Figure 9.

The results presented in Figure 9 show that energy consumption for the overall liquefaction process increases by around 20% across the cooling temperature range 5 to 50 °C and 5% over the range 20 to 30 °C. For the pre-cooling process the increase is close to 80% over the full temperature range. Figure 9 also shows that while the second law efficiency of the cryogenic-cooling process increases slightly across the range of temperatures considered, the efficiency of the pre-cooling process drops above 25 °C. The cause of this drop in efficiency as the cooling temperature increases can be seen in Figures 6 and 7, which show that the mean temperature approach for the higher temperature cases is higher than that of the lower temperature cases. It is this reduced level of optimization as cooling temperature increases above 25 °C that accentuates the non-linear behavior notable in Figure 8.

The implication of the results presented in Figure 9 is the same as discussed earlier: that design changes in the MR process could help to improve performance for the cases where the cooling temperature is higher than 25 °C. Both the addition of heavier components to the MR mixture could provide a more optimized design or the division of the MR loop into additional pressure levels. Both of these design alternatives could form the basis of further study work.

4. Conclusions

A model for a hydrogen liquefaction process has been developed and validated against results from an independent study. Although the validation process highlighted the significant impact that different properties models can have on model predictions, the validation results also indicate that the present model is suitable for the study of the impact of ambient temperature on process performance.

A set of optimization parameters were selected, and an optimization method developed that was shown to be suitable for the study of process performance across a range of process cooling temperatures through the consistency of the results obtained. The MR studied is limited to a mixture of five components. It is indicated in the results presented that the addition of heavier components could be used to improve efficiency for cooling temperatures above 25 °C, although the available gains would be small.

The results of the optimization work show that the specific energy consumption, SEC, of the MR pre-cooling process increases by around 80%, from approximately 0.57 to 1.0 kWh/kg, across the cooling temperature range 5 to 50 °C. These results, combined with the calculated process performance for the cryogenic-cooling step (not optimized here), show that total energy consumption for the hydrogen liquefaction process increases by around 20%, from 5.8 to 7.1 kWh/kg, across the same temperature range. Considering just the range 20 to 30 °C, there is a 5% increase, which illustrates the significant impact ambient temperature can have on energy consumption.

The variation in energy consumption with cooling temperature implies a significant benefit for liquefaction processes operating in low ambient temperature locations, especially given that the hydrogen liquefaction process represents a very energy intensive step in the supply of liquid hydrogen. The aim of further work is to combine these results into a larger system model that considers the impact of ambient temperature on the supply of low-carbon energy from natural gas.

Author Contributions: Conceptualization, S.J.; methodology, S.J. and E.B.; validation, S.J.; formal analysis, S.J.; investigation, S.J.; data curation, S.J.; writing—original draft preparation, S.J.; writing—review and editing, S.J. and E.B.; supervision, E.B. Both authors have read and agreed to the published version of the manuscript.

Funding: This research received no external funding.

Institutional Review Board Statement: Not applicable.

Informed Consent Statement: Not applicable.

Data Availability Statement: Not applicable.

Conflicts of Interest: The authors declare no conflict of interest.

Appendix A

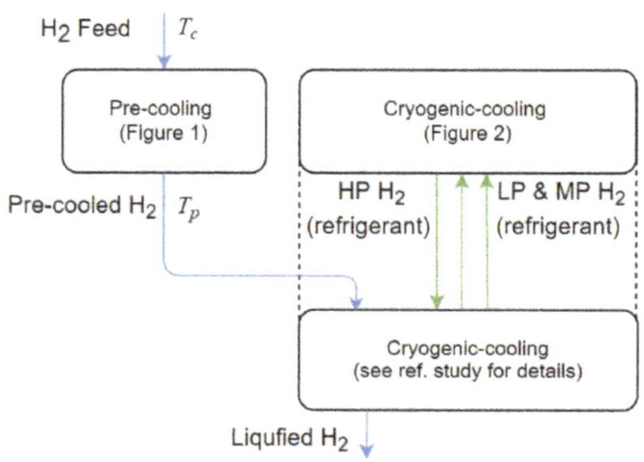

Figure A1. Block flow diagram of the overall liquefaction process.

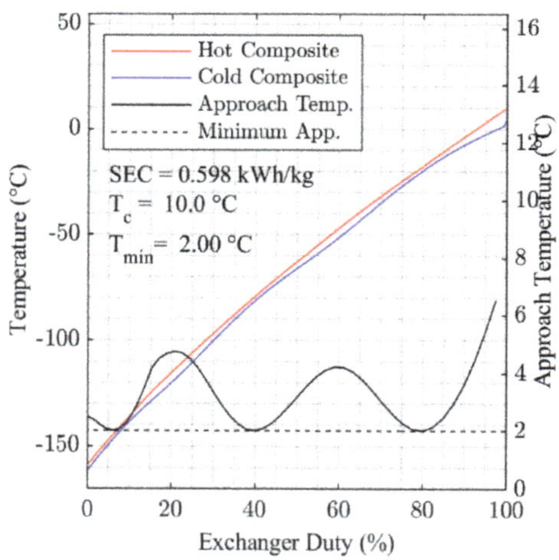

Figure A2. Optimized composite cooling curves for 10 °C cooling temperature.

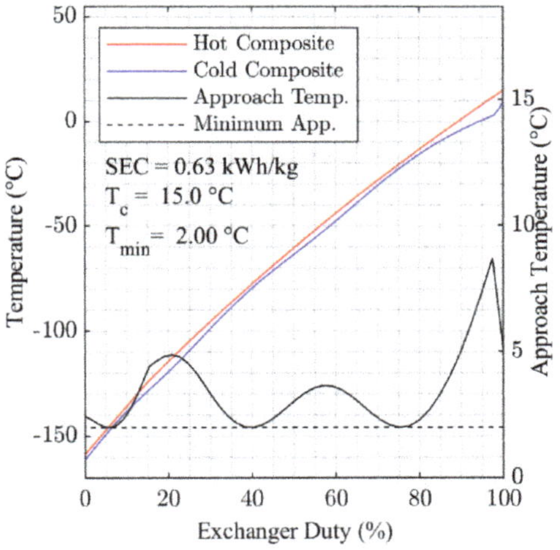

Figure A3. Optimized composite cooling curves for 15 °C cooling temperature.

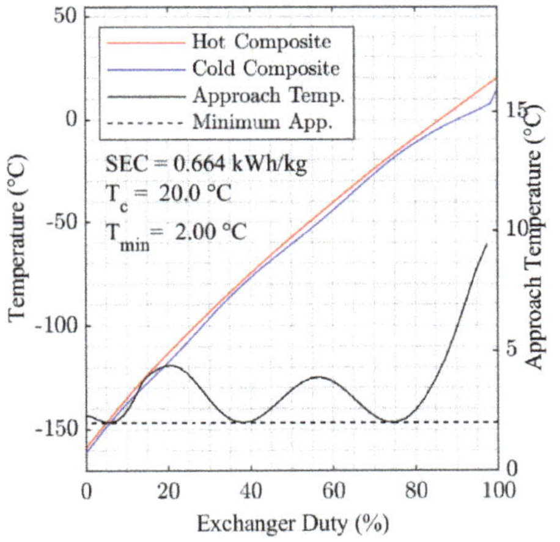

Figure A4. Optimized composite cooling curves for 20 °C cooling temperature.

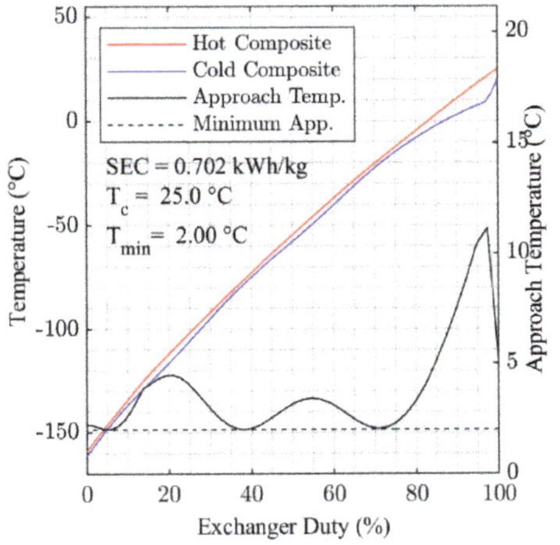

Figure A5. Optimized composite cooling curves for 25 °C cooling temperature.

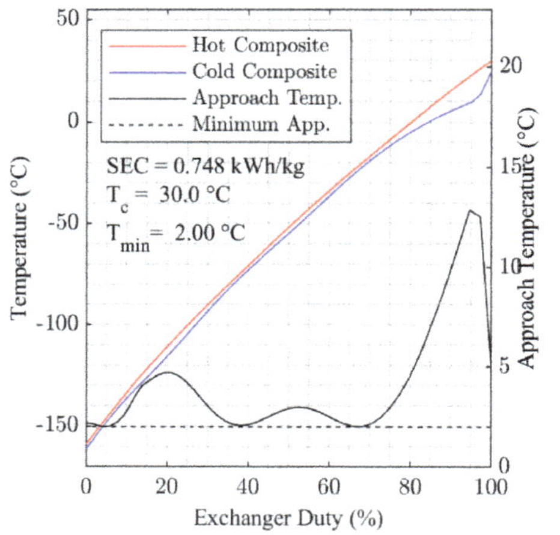

Figure A6. Optimized composite cooling curves for 30 °C cooling temperature.

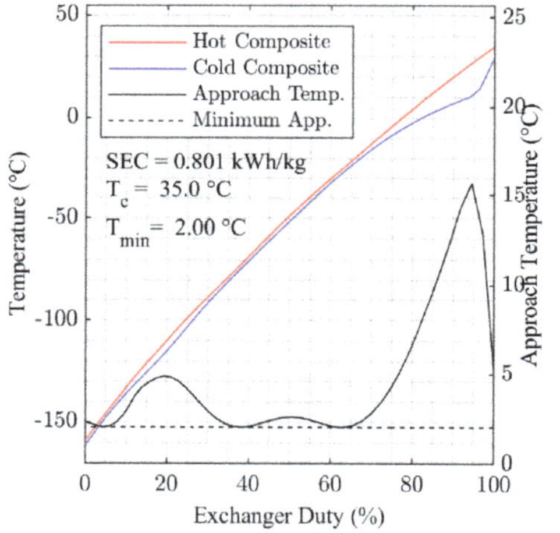

Figure A7. Optimized composite cooling curves for 35 °C cooling temperature.

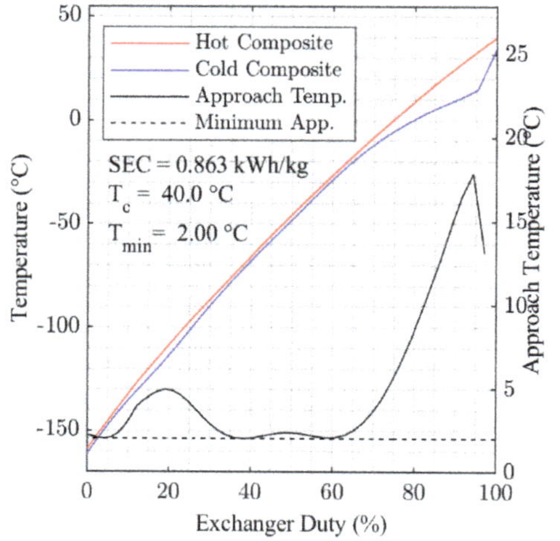

Figure A8. Optimized composite cooling curves for 40 °C cooling temperature.

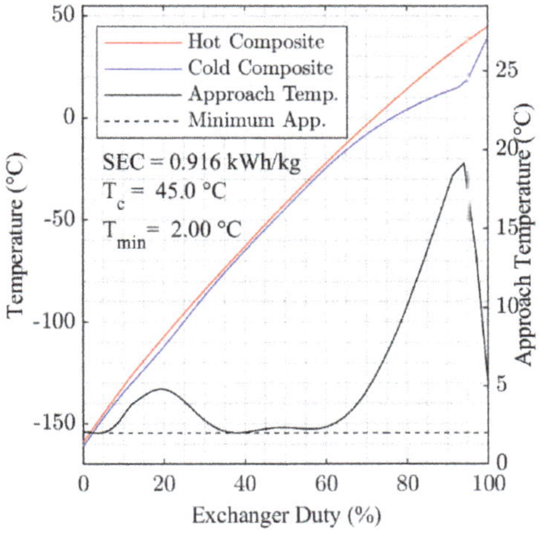

Figure A9. Optimized composite cooling curves for 45 °C cooling temperature.

References

1. Acar, C.; Dincer, I. Review and evaluation of hydrogen production options for better environment. *J. Clean. Prod.* **2019**, *218*, 835–849. [CrossRef]
2. Cell, F.; Undertaking, H.J. Hydrogen Roadmap Europe—A Sustainable Pathway for the European Energy Transition. *Fuel Cells Hydrogen Jt. Undert.* **2019**. [CrossRef]
3. Moya, J.I.; Dalius, W. Hydrogen Use in EU Decarbonisation Scenarios. Available online: https://ec.europa.eu/jrc/sites/default/files/final_insights_into_hydrogen_use_public_version.pdf (accessed on 1 September 2021).
4. The European Comission. *A Hydrogen Strategy for a Climate-Neutral Europe*; The European Comission: Brussels, Belgium, 2020.
5. Kang, J.-N.; Wei, Y.-M.; Liu, L.-C.; Han, R.; Yu, B.-Y.; Wang, J.-W. Energy systems for climate change mitigation: A systematic review. *Appl. Energy* **2020**, *263*, 114602. [CrossRef]
6. Yuksel, Y.E.; Ozturk, M.; Dincer, I. Energetic and exergetic assessments of a novel solar power tower based multigeneration system with hydrogen production and liquefaction. *Int. J. Hydrogen Energy* **2019**, *44*, 13071–13084. [CrossRef]
7. Sher, F.; Al-Shara, N.K.; Iqbal, S.Z.; Jahan, Z.; Chen, G.Z. Enhancing hydrogen production from steam electrolysis in molten hydroxides via selection of non-precious metal electrodes. *Int. J. Hydrogen Energy* **2020**, *45*, 28260–28271. [CrossRef]
8. Karakaya, E.; Nuur, C.; Assbring, L. Potential transitions in the iron and steel industry in Sweden: Towards a hydrogen-based future? *J. Clean. Prod.* **2018**, *195*, 651–663. [CrossRef]
9. Pethaiah, S.S.; Sadasivuni, K.K.; Jayakumar, A.; Ponnamma, D.; Tiwary, C.S.; Sasikumar, G. Methanol Electrolysis for Hydrogen Production Using Polymer Electrolyte Membrane: A Mini-Review. *Energies* **2020**, *13*, 5879. [CrossRef]
10. Yang, C.; Ogden, J. Determining the lowest-cost hydrogen delivery mode. *Int. J. Hydrogen Energy* **2007**, *32*, 268–286. [CrossRef]
11. Ishimoto, Y.; Voldsund, M.; Nekså, P.; Roussanaly, S.; Berstad, D.; Gardarsdottir, S.O. Large-scale production and transport of hydrogen from Norway to Europe and Japan: Value chain analysis and comparison of liquid hydrogen and ammonia as energy carriers. *Int. J. Hydrogen Energy* **2020**, *45*, 32865–32883. [CrossRef]
12. Hoshi, M. World's First Liquid Hydrogen Carrier Ship Launches in Japan. Available online: https://asia.nikkei.com/Business/Energy/World-s-first-liquid-hydrogen-carrier-ship-launches-in-Japan (accessed on 15 April 2021).
13. Aasadnia, M.; Mehrpooya, M. Large-scale liquid hydrogen production methods and approaches: A review. *Appl. Energy* **2018**, *212*, 57–83. [CrossRef]
14. Ghorbani, B.; Mehrpooya, M.; Aasadnia, M.; Niasar, M.S. Hydrogen liquefaction process using solar energy and organic Rankine cycle power system. *J. Clean. Prod.* **2019**, *235*, 1465–1482. [CrossRef]
15. Yilmaz, C. Optimum energy evaluation and life cycle cost assessment of a hydrogen liquefaction system assisted by geothermal energy. *Int. J. Hydrogen Energy* **2019**, *45*, 3558–3568. [CrossRef]
16. Ansarinasab, H.; Mehrpooya, M.; Mohammadi, A. Advanced exergy and exergoeconomic analyses of a hydrogen liquefaction plant equipped with mixed refrigerant system. *J. Clean. Prod.* **2017**, *144*, 248–259. [CrossRef]
17. Skaugen, G.; Berstad, D.; Wilhelmsen, Ø. Comparing exergy losses and evaluating the potential of catalyst-filled plate-fin and spiral-wound heat exchangers in a large-scale Claude hydrogen liquefaction process. *Int. J. Hydrogen Energy* **2020**, *45*, 6663–6679. [CrossRef]
18. Cardella, U.; Decker, L.; Sundberg, J.; Klein, H. Process optimization for large-scale hydrogen liquefaction. *Int. J. Hydrogen Energy* **2017**, *42*, 12339–12354. [CrossRef]
19. Yuksel, Y.E.; Ozturk, M.; Dincer, I. Analysis and assessment of a novel hydrogen liquefaction process. *Int. J. Hydrogen Energy* **2017**, *42*, 11429–11438. [CrossRef]

20. Chang, H.M.; Ryu, K.N.; Baik, J.H. Thermodynamic design of hydrogen liquefaction systems with helium or neon Brayton refrigerator. *Cryogenics* **2018**, *91*, 68–76. [CrossRef]
21. Donaubauer, P.J.; Cardella, U.; Decker, L.; Klein, H. Kinetics and Heat Exchanger Design for Catalytic Ortho-Para Hydrogen Conversion during Liquefaction. *Chem. Eng. Technol.* **2019**, *42*, 669–679. [CrossRef]
22. Wilhelmsen, O.; Berstad, D.; Aasen, A.; Neksa, P.; Skaugen, G. Reducing the exergy destruction in the cryogenic heat exchangers of hydrogen liquefaction processes. *Int. J. Hydrogen Energy* **2018**, *43*, 5033–5047. [CrossRef]
23. Skaugen, G.; Wilhelmsen, O. Comparing the Performance of Plate-Fin and Spiral Wound Heat Exchangers in the Cryogenic Part of the Hydrogen Liquefaction Process. In *Proceedings of the 15th Cryogenics 2019 Iir International Conference*; Prague, Czech Republic, 8–11 April 2019, Chrz, V., Haberstroh, C., Herzog, R., Kaiser, Z., Klier, J., Kralik, T., Lansky, M., Mericka, P., Schustr, P., Srnka, A., et al., Eds.; Refrigeration Science and Technology; International Institute of Refrigeration: Paris, France, 2019; pp. 318–324.
24. Bauer, H.C. Mixed fluid cascade, experience and outlook. In Proceedings of the AIChE Spring Meeting and Global Congress on Process Safety, Houston, TX, USA, 1–4 April 2012.
25. Rian, A.B.; Ertesvåg, I.S. Exergy Evaluation of the Arctic Snøhvit Liquefied Natural Gas Processing Plant in Northern Norway—Significance of Ambient Temperature. *Energy Fuels* **2012**, *26*, 1259–1267. [CrossRef]
26. Xu, X.; Liu, J.; Jiang, C.; Cao, L. The correlation between mixed refrigerant composition and ambient conditions in the PRICO LNG process. *Appl. Energy* **2013**, *102*, 1127–1136. [CrossRef]
27. Castillo, L.; Dahouk Majzoub, M.; Di Scipio, S.; Dorao, C.A. Conceptual analysis of the precooling stage for LNG processes. *Energy Convers. Manag.* **2013**, *66*, 41–47. [CrossRef]
28. Park, K.; Won, W.; Shin, D. Effects of varying the ambient temperature on the performance of a single mixed refrigerant liquefaction process. *J. Nat. Gas Sci. Eng.* **2016**, *34*, 958–968. [CrossRef]
29. Jackson, S.; Eiksund, O.; Brodal, E. Impact of Ambient Temperature on LNG Liquefaction Process Performance: Energy Efficiency and CO 2 Emissions in Cold Climates. *Ind. Eng. Chem. Res.* **2017**, *56*, 3388–3398. [CrossRef]
30. Cardella, U.; Decker, L.; Klein, H. Roadmap to economically viable hydrogen liquefaction. *Int. J. Hydrogen Energy* **2017**, *42*, 13329–13338. [CrossRef]
31. The MathWorks, Inc. *MATLAB*; The MathWorks, Inc.: Natick, MA, USA, 2018.
32. Span, R.; Eckermann, T.; Herrig, S.; Hielscher, S.; Jäger, A.; Thol, M. *TREND. Thermodynamic Reference and Engineering Data 5.0*; Lehrstuhl für Thermodynamik, Ruhr-Universität Bochum: Bochum, Germany, 2020.

energies MDPI

Article

Parametric Study for Thermal and Catalytic Methane Pyrolysis for Hydrogen Production: Techno-Economic and Scenario Analysis

Seunghyun Cheon [1,†], Manhee Byun [1,†], Dongjun Lim [1], Hyunjun Lee [1] and Hankwon Lim [1,2,*]

[1] School of Energy and Chemical Engineering, Ulsan National Institute of Science and Technology (UNIST), 50 UNIST-gil, Eonyang-eup, Ulju-gun, Ulsan 44919, Korea; tsi03075@unist.ac.kr (S.C.); qusaksgml@unist.ac.kr (M.B.); dongjun1993@unist.ac.kr (D.L.); owl12@unist.ac.kr (H.L.)
[2] Department of Energy Engineering, Ulsan National Institute of Science and Technology (UNIST), 50 UNIST-gil, Eonyang-eup, Ulju-gun, Ulsan 44919, Korea
* Correspondence: hklim@unist.ac.kr; Tel.: +82-52-217-2935
† These authors contributed equally to this paper.

Abstract: As many countries have tried to construct a hydrogen (H_2) society to escape the conventional energy paradigm by using fossil fuels, methane pyrolysis (MP) has received a lot of attention owing to its ability to produce H_2 with no CO_2 emission. In this study, a techno-economic analysis including a process simulation, itemized cost estimation, and sensitivity and scenario analysis was conducted for the system of thermal-based and catalyst-based MP (TMP-S1 and CMP-S2), and the system with the additional H_2 production processes of carbon (C) gasification and water–gas shift (WGS) reaction (TMPG-S3 and CMPG-S4). Based on the technical performance expressed by H_2 and C production rate, the ratio of H_2 combusted to supply the heat required and the ratio of reactants for the gasifier (C, Air, and water (H_2O)), unit H_2 production costs of USD 2.14, 3.66, 3.53, and 3.82 kgH_2^{-1} from TMP-S1, CMP-S2, TMPG-S3, and CMPG-S4, respectively, were obtained at 40% H_2 combusted and a reactants ratio for C-Air-H_2O of 1:1:2. Moreover, trends of unit H_2 production cost were obtained and key economic parameters of the MP reactor, reactant, and C selling price were represented by sensitivity analysis. In particular, economic competitiveness compared with commercialized H_2 production methods was reported in the scenario analysis for the H_2 production scale and C selling price.

Keywords: thermal methane pyrolysis; catalytic methane pyrolysis; H_2 production; process simulation; economic analysis; unit H_2 production cost

Citation: Cheon, S.; Byun, M.; Lim, D.; Lee, H.; Lim, H. Parametric Study for Thermal and Catalytic Methane Pyrolysis for Hydrogen Production: Techno-Economic and Scenario Analysis. *Energies* **2021**, *14*, 6102. https://doi.org/10.3390/en14196102

Academic Editor: Dmitri A. Bulushev

Received: 13 August 2021
Accepted: 16 September 2021
Published: 24 September 2021

1. Introduction

Many countries have tried to accomplish a successful transition of an energy system to hydrogen (H_2) based on various political strategies such as 'The National Hydrogen Strategy' (2020) in Germany [1], 'EU Hydrogen Strategy' (2020) in the EU [2], 'Basic Hydrogen Strategy' (2017)', 'Strategic Energy Plan' (2018), and 'The Strategic Road Map for Hydrogen and Fuel Cells' (2019) in Japan [3–5], 'Hydrogen in a Low-carbon Economy' (2018) in the UK [6], 'H_2@Scale' (2021) in USA [7], and 'National Hydrogen Roadmap' (2018) in Australia [8]. These active approaches to an H_2-based energy system come from the diverse advantages of H_2 as a clean energy carrier: it can be utilized in various energy sectors and easily combined in already constructed infrastructure, and, even though its volumetric energy density is relatively low, it shows a very high energy density of 120–142 MJ kg^{-1} in the compressed state [9–15]. For the conventional production of H_2, energy-intensive processes such as reforming, partial oxidation, and auto-thermal reforming of carbon-based fuels such as methane (CH_4) and hydrocarbon have been mainly used. However, these conventional methods, including steam methane reforming (SMR), have led to negative environmental effects due to large emissions of carbon dioxide (CO_2) [16–19]. Due to the environmental issues of conventional methods, a lot of recent research has suggested water electrolysis (WE) powered by renewable energy as an alternative eco-friendly solution to produce H_2 because there is no CO_2 emission in the procedure [20–22]. However, there are still technical and economic challenges to be immediately utilized [23–25]; most H_2 production still depends on conventional methods with additional processes to reduce CO_2 such as SMR with CO_2 capture and storage (SMR with CCS) [26–28].

To overcome the limitations of current H_2 production methods, the concept of thermal methane pyrolysis (TMP), where H_2 and carbon (C) are directly produced in the gas phase (Equation (1)), has been paid attention as an alternative, novel H_2 production method owing to several technical, economic, and environmental benefits as follows: (a) there is no oxygen (O_2) in the reaction leading to no CO_2 emissions or additional separation process, theoretically; (b) the process can be relatively simplified and lower energy is required than other methods such as reforming or partial oxidation; (c) reactant of the process, methane, is abundant and cheap leading to a cost effective operation of the process; (d) C products can be marketed because they are usually used as raw materials in various valuable materials such as rubber, tires, and pigments, etc.; (e) the separation of C is much easier than the separation of CO_2; (f) it requires a lower amount of heat compared to SMR (Equation (2a–c)) and WE (Equation (3)), which are the most common and novel H_2 production methods (Equations (1)–(3)) [29–38].

$$CH_4 \rightarrow 2H_2 + C\,(s) \qquad \Delta H^{\circ} = 75\ \text{kJ mol}^{-1} \tag{1}$$
$$CH_4 + H_2O \rightarrow 3H_2 + CO \qquad \Delta H^{\circ} = 206\ \text{kJ mol}^{-1} \tag{2a}$$
$$CO + H_2O \rightarrow H_2 + CO_2 \qquad \Delta H^{\circ} = -41\ \text{kJ mol}^{-1} \tag{2b}$$
$$CH_4 + 2H_2O \rightarrow 4H_2 + CO_2 \qquad \Delta H^{\circ} = 165\ \text{kJ mol}^{-1} \tag{2c}$$
$$H_2O \rightarrow H_2 + \tfrac{1}{2}O_2 \qquad \Delta H^{\circ} = 285\ \text{kJ mol}^{-1} \tag{3}$$

Because of its endothermicity and strong C–H bonding, TMP is usually operated at over 1373 K to obtain reasonable yields of H_2 and C leading to cost ineffectiveness and a large amount of energy being required [39,40]. These problems of having to use high temperature can be reduced by catalytic methane pyrolysis (CMP) where various types of catalysts (non-supported, metal supported, metal oxide supported, and carbonaceous, etc.) are adopted. Among the various catalysts used in CMP, the metal-based catalyst has very critical systematic limitations such as high toxicity of metal and rapid deactivation of the catalyst due to encapsulation of the active metal sites with C product [41]. Thus, carbon-based CMP has a lot of attention owing to the properties of carbon catalysts such as lower cost, higher stability, temperature resistance, and their ability to be safely stored due to their non-toxicity.

Based on the benefits of the concept of MP, many kinds of research have been conducted: Nishii et al. [42] carried out MP with different carbon-based catalysts (activated carbon, carbon black, mesoporous carbon, and carbon nanofiber) and found that all of these catalysts continued to maintain a CH_4 conversion of about 17% for longer than 600 min by catalyzing the produced C. It was reported that the produced C covered the catalyst surface, resulting in a specific surface area of 10 $m^2\ g^{-1}$ and an intensity of D-Raman peak and G-Raman peak (I_d/I_g) from 1.5 to 1.57 irrespective of the original structures of C. Tezel et al. [43] designed an experiment using CMP with a calcium silicate-based Ni–Fe catalyst with different Fe loading by using the co-impregnation method. It was revealed that the addition of Fe can delay the deactivation of the Ni catalyst and an increase in the CH_4 flow rate can decrease the initial reactant conversion and lifetime of the catalyst. It was reported that the highest methane conversion of 69% is obtained at 973 K with the catalyst that has the highest Fe addition. Quan et al. [44] investigated the optimization of a fluidized bed reactor (FLBR) for CMP using 40 wt% Fe/Al_2O_3 catalyst, and catalyst activity and stability were investigated after optimization in terms of the catalyst bulk density, bed height, and particle size, etc. It was reported that the reaction conditions of 12 L $(g_{cat}\ h)^{-1}$ feed dilution of 20% H_2-CH_4, and CO_2-regeneration of deactivated catalysts are the best conditions for MP. Patzschke et al. [45] investigated promising catalysts for particle suspension in molten

NaBr-KBr and reported that mixed Co–Mn catalysts can be optimal candidates for methane pyrolysis in molten salts owing to their fast kinetics and stability. The authors reported that increasing the ratio of molar Co–Mn from 0 to 2 improved the conversion of CH_4 from 4.8% to 10.4% at 1273 K for the smallest catalyst particle size range, which shows that closeness between the catalytic surface and the gas phase can improve conversions. Karaismailoglu et al. [46] investigated the effect of the doping of yttria on a nickel catalyst synthesized by the sol–gel citrate method and reported CH_4 conversion of 50% with this type of catalyst. It was reported that the addition of Yttria can improve the stability and activity of catalysts at elevated temperatures and that a lower nickel ratio in the catalyst reduces the formation of carbon. Not only experimental studies but also systematic approaches using process simulations and works for economic feasibility have been reported. Chen et al. [47] designed the vacuum promoted methane decomposition with carbon separation (VPMDCS), which include a reactor of MP continuously generating H_2 and a C separation reactor converting carbon into CO. It was reported that VPMDCS showed CH_4 conversion of 99.2% and produced high-purity H_2 and CO with concentrations of both 99.6%. By economic analysis, the unit hydrogen cost of EUR 5.4 kg^{-1} was reported. Riley et al. [48] simulated two concepts of CMP that used H_2 combustion and CH_4 combustion by Aspen Plus® comparing CO_2 emissions and H_2 production cost. It was revealed that the quality of produced C and its selling price are major factors in H_2 selling price, and H_2 production cost in the capacity of 216 ton d^{-1} is less than USD 3.25 $kgH_2{}^{-1}$ without considering the sale of C. Perez et al. [49] designed an MP process using a quartz bubble column including molten gallium, which is used for catalyst and heat transfer agents, with a porous plate distributor. The authors found that a maximum CH_4 conversion of 91% was achieved at a reactor temperature of 1392 K where gallium occupied 43% of the total reactor volume with a residence time for a bubble of 0.5 s. Additionally, by techno-economic analysis, it was concluded that a molten metal system can be competitive with SMR if a CO_2 tax of EUR 50 ton^{-1} is imposed and produced C is marketed. Kerscher et al. [50] designed two concepts of MP using electron beam plasma, which was generated from renewable electricity. The techno-economic assessment reported that levelized costs of H_2 for the electron beam plasma method ranged from 2.55 to 5.00 € $kgH_2{}^{-1}$, and CO_2 emission ranged from 1.9 to 6.4 $kgCO_2$ eq. $kgH_2{}^{-1}$ from a carbon footprint assessment, which shows a high potential for reducing life cycle emissions. Zhang et al. [51] investigated the CO_2 mitigation costs of CMP and the integrated power generation process in a fuel cell comparing a combined-cycle gas turbine power plant system with and without CCS. It was revealed that CMP shows low life cycle emissions per unit of electricity output of 0.13 tCO_2 eq MWh^{-1} but shows a high levelized cost of electricity of EUR 177 MWh^{-1}, concluding that it has high potential when assumed that produced C can be sold at current prices. Timmerberg et al. [52] assessed the levelized hydrogen production costs and life cycle greenhouse gas (GHG) emissions from MP in three systems where molten metal, plasma, and thermal gas reactors were used. It was reported that the plasma-based system using electricity from renewable sources shows the lowest emissions of 43 gCO_2 MJ^{-1}, and the molten metal and thermal gas system shows relatively higher GHG emissions due to the additional combustion and natural gas supply chain.

Even though many types of research have been conducted on the concepts of TMP and CMP, very few studies revealing both technical and economic viability of those technologies are reported, to the best of our knowledge. Therefore, in this study, a preliminary techno-economic parametric study is conducted to comprehensively investigate the feasibility of the concept of methane pyrolysis (MP). Firstly, a process simulation using Aspen Plus® for various MP processes, namely TMP and CMP, and with additional carbon gasification (TMPG and CMPG) are performed with detailed reaction kinetics under various technical parameters of reaction temperature, ratio of fuel combusted, and ratio of reactants for gasifier (C-Air-H_2O) (Figure 1). Based on the technical performance from the process simulation, yields of H_2 and C, and the amount of fuel required to supply heat to the MP reactor and gasification unit are obtained, and then, economic feasibility in terms of unit H_2

production cost is reported. In addition, to suggest future economic guidelines of this novel concept when this is commercialized, sensitivity and scenario analysis regarding various H_2 production scales and different C selling price scenarios are conducted revealing the cost competitiveness compared to the conventional H_2 production methods of SMR and SMR with CCS.

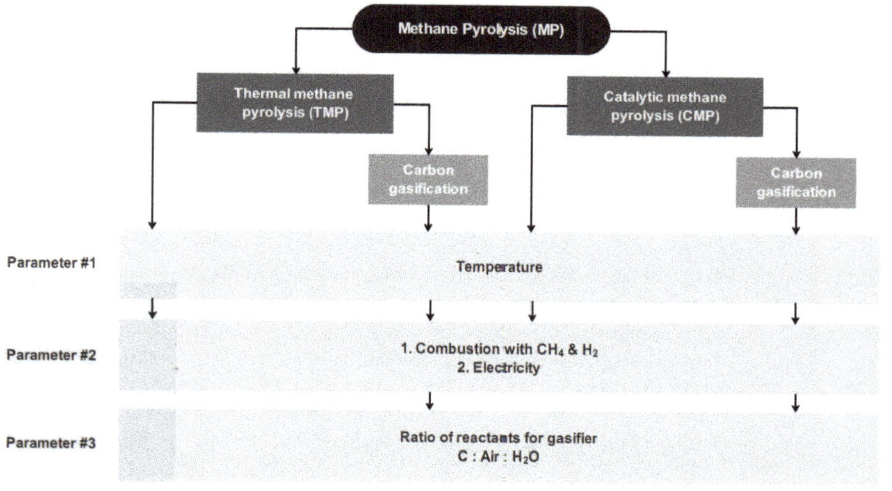

Figure 1. Schematic diagram of techno-economic parametric study for investigated systems for methane pyrolysis (MP).

2. Methods

2.1. Process Simulation

In this study, four systems for MP, classified as TMP-S1, CMP-S2, TMPG-S3, and CMPG-S4, were simulated in Aspen Plus® (Aspen Technology, Inc., Bedford, MA, USA) with detailed reactor validation based on kinetics reported by Keipi et al. [53] for TMP and Kim et al. [54] for CMP. As a result, Figure 2 shows the closeness of methane conversion between experimental and simulated methods at each investigated operating conditions validating proper insertion of reported kinetics to Aspen Plus®.

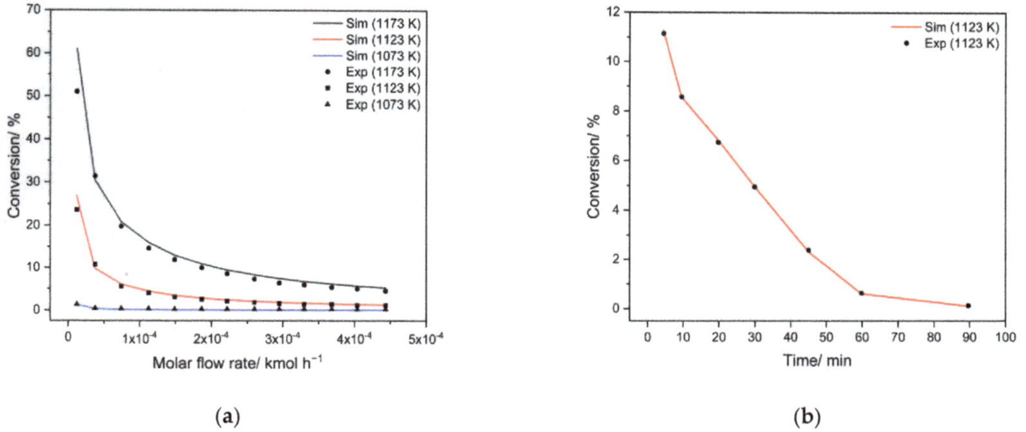

Figure 2. Results of kinetic validation for (**a**) thermal methane pyrolysis (TMP) and (**b**) catalytic methane pyrolysis (CMP) reactions.

For all systems, CH_4 entered the validated reactor in different temperature ranges of 1073–1373 K for TMP-S1 and TMPG-S3, and 1023–1173 K for CMP-S2 and CMPG-S4, then, product stream containing remained CH_4 and produced H_2 and C passed through the units of cyclone and pressure swing adsorption (PSA) for separating solid C and H_2, respectively (Figure 3). We assumed the pressure drop of cyclone as 0.01 bar and number of cyclones as only one and assumed separation efficiency of PSA as 100%. Especially for CMP-S2 and CMPG-S4, purified C entered the gasification unit to produce additional H_2 and carbon monoxide (CO) with different ratios of C, air, and water (H_2O) (1:1:1, 1:1:2, 1:1:3, 1:2:1, and 1:3:1), and water-gas shift (WGS) (Equation (2b)) reactor was followed to convert the produced CO to H_2. Additionally, for the heat supply system, various heat supply scenarios were assumed and classified as 100% electricity-based and different ratios of H_2 combusted (0%–100% matched with 100%–0% CH_4 combusted). Based on the result of the process simulation, material balance was obtained at the temperature of 1273 K for TMP-S1 and TMPG-S3 and 1173 K for CMP-S2 and CMPG-S4, with the ratio of H_2 combusted of 40%, and the ratio of reactants for the gasifier of 1:1:2 (C-Air-H_2O) (Table 1).

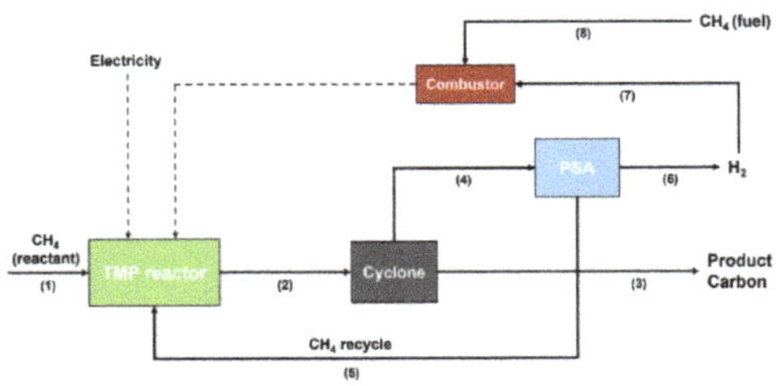

(**a**)

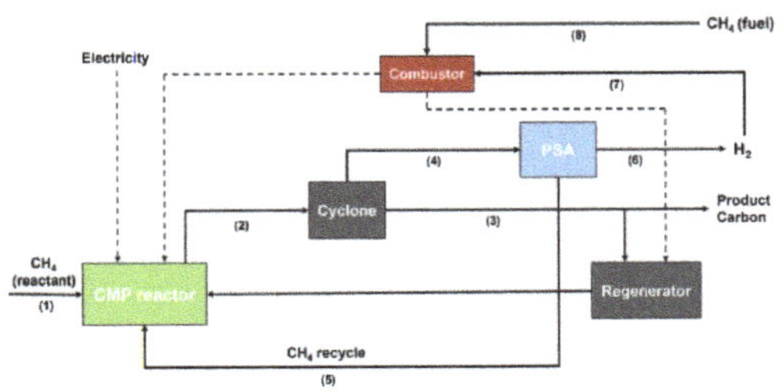

(**b**)

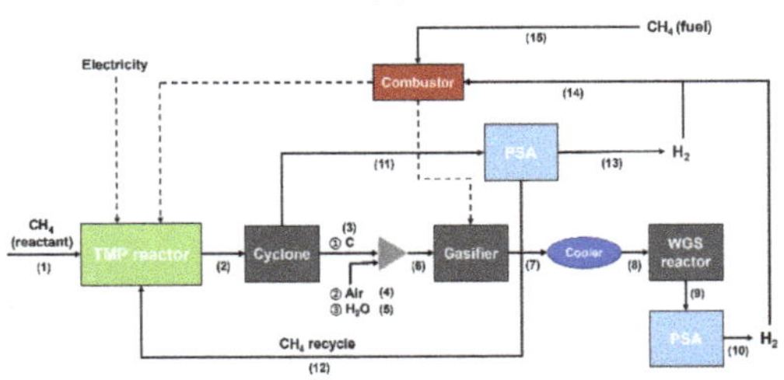

(**c**)

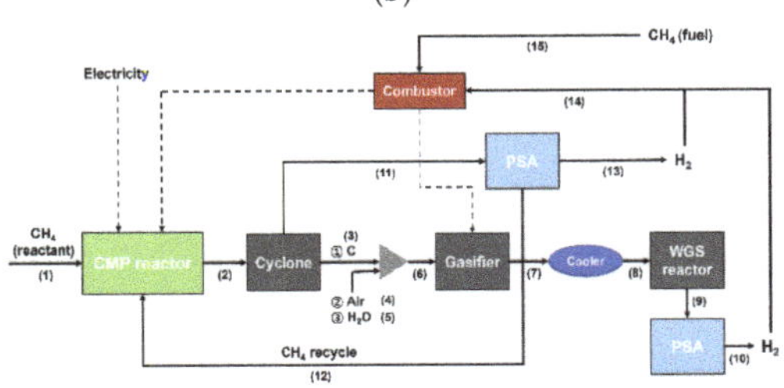

(**d**)

Figure 3. Block flow diagrams for methane pyrolysis (MP) systems of (**a**) thermal methane pyrolysis (TMP-S1), (**b**) catalytic methane pyrolysis (CMP-S2), and systems with additional gasification and WGS reaction of (**c**) TMPG-S3 and (**d**) CMPG-S4.

Table 1. Material balance for methane pyrolysis (MP) systems of (**a**) thermal methane pyrolysis (TMP-S1), (**b**) catalytic methane pyrolysis (CMP-S2), and systems with additional gasification and WGS reaction of (**c**) TMPG-S3 and (**d**) CMPG-S4.

(**a**) TMP-S1	(1)	(2)	(3)	(4)	(5)	(6)	(7)	(8)
Temperature/K	298	1273	1273	1273	1273	1273	1273	298
Pressure/bar	1.00	1.00	1.00	0.99	0.99	0.99	0.99	1.00
Molar flow/kmol h^{-1}	1.00	3.02	1.00	2.02	0.02	2.00	0.27	0.12
Mole fraction								
CH_4	1.00	0.01	0	0.01	1.00	0	0	1.00
C	0	0.33	1.00	0	0	0	0	0
H_2	0	0.66	0	0.99	0	1.00	1.00	0

Table 1. *Cont.*

(**b**) CMP-S2	(1)	(2)	(3)	(4)	(5)	(6)	(7)	(8)
Temperature/K	298	1173	1173	1173	1173	1173	1173	298
Pressure/bar	1.00	1.00	1.00	0.99	0.99	0.99	0.99	1.00
Molar flow/kmol h^{-1}	1.00	5.27	0.93	4.34	2.42	1.85	0.24	0.11
Mole fraction								
CH_4	1.00	0.48	0	0.57	1.00	0	0	1.00
C	0	0.18	1.00	0	0	0	0	0
H_2	0	0.35	0	0.43	0	1.00	1.00	0
(**c**) TMPG-S3	(1)	(2)	(3)	(4)	(5)	(6)	(7)	(8)
Temperature/K	298	1273	1273	298	298	341	973	623
Pressure/bar	1.00	1.00	1.00	1.00	1.00	1.00	1.00	1.00
Molar flow/kmol h^{-1}	1.00	3.02	1.00	1.00	2.00	4.00	3.78	3.78
Mole fraction								
CH_4	1.00	0.01	0	0	0	0	0	0
C	0	0.33	1.00	0	0	0.25	0	0
H_2	0	0.66	0	0	0	0	0.30	0.30
O_2	0	0	0	0.21	0	0.05	0	0
H_2O	0	0	0	0	1.00	0.50	0.23	0.23
N_2	0	0	0	0.79	0	0.20	0.21	0.21
CO_2	0	0	0	0	0	0	0.15	0.15
CO	0	0	0	0	0	0	0.12	0.12
(**c**) TMPG-S3	(9)	(10)	(11)	(12)	(13)	(14)	(15)	
Temperature/K	623	623	1273	1273	1273	1273	298	
Pressure/bar	1.00	1.00	0.99	0.99	0.99	0.99	1.00	
Molar flow/kmol h^{-1}	3.78	1.38	2.02	0.02	2.00	0.45	0.20	
Mole fraction								
CH_4	0	0	0.01	1.00	0	0	1.00	
C	0	0	0	0	0	0	0	
H_2	0.36	1.00	0.99	0	1.00	1.00	0	
O_2	0	0	0	0	0	0	0	
H_2O	0.16	0	0	0	0	0	0	
N_2	0.21	0	0	0	0	0	0	
CO_2	0.21	0	0	0	0	0	0	
CO	0.05	0	0	0	0	0	0	
(**d**) CMPG-S4	(1)	(2)	(3)	(4)	(5)	(6)	(7)	(8)
Temperature/K	298	1173	1173	298	298	337	923	623
Pressure/bar	1.00	1.00	1.00	1.00	1.00	1.00	1.00	1.00
Molar flow/kmol h^{-1}	1.00	5.08	0.92	0.92	1.84	3.68	3.45	3.45
Mole fraction								
CH_4	1.00	0.46	0	0	0	0	0	0
C	0	0.18	1.00	0	0	0.25	0	0
H_2	0	0.36	0	0	0	0	0.30	0.30
O_2	0	0	0	0.21	0	0.05	0	0
H_2O	0	0	0	0	1.00	0.50	0.22	0.22
N_2	0	0	0	0.79	0	0.20	0.21	0.21
CO_2	0	0	0	0	0	0	0.16	0.16
CO	0	0	0	0	0	0	0.10	0.10
(**d**) CMPG-S4	(9)	(10)	(11)	(12)	(13)	(14)	(15)	
Temperature/K	623	623	1173	1173	1173	1173	298	
Pressure/bar	1.00	1.00	0.99	0.99	0.99	0.99	1.00	
Molar flow/kmol h^{-1}	3.45	1.29	4.16	2.24	1.84	0.40	0.18	
Mole fraction								
CH_4	0	0	0.56	1.00	0	0	1.00	
C	0	0	0	0	0	0	0	
H_2	0.37	1.00	0.44	0	1.00	1.00	0	
O_2	0	0	0	0	0	0	0	
H_2O	0.15	0	0	0	0	0	0	
N_2	0.21	0	0	0	0	0	0	
CO_2	0.23	0	0	0	0	0	0	
CO	0.03	0	0	0	0	0	0	

2.2. Itemized Cost Estimation

To investigate the economic feasibility of each MP system, itemized cost estimation proposed by Turton et al. [55] was conducted considering various parameters of reaction temperature, types of fuel combusted and its ratio, and the ratio of reactant for gasifier. In this method, the unit H_2 production cost (USD $kgH_2{}^{-1}$) is obtained from the sum of the total cost (USD y^{-1}) divided by the total amount of H_2 produced (kg y^{-1}). In this study, the total cost is defined as annualized capital cost (USD y^{-1}), which is estimated from original capital cost (USD) by applying capital recovery factor (CRF) as shown in Equation (4), and operating cost (USD y^{-1}). Table 2 shows the list of these.

$$\mathrm{CRF} = \frac{i(1+i)^N}{(1+i)^N - 1} \tag{4}$$

where i is a discount rate and N is an economic analysis period.

In addition, to properly estimate each capital cost, the six-tenth rule (Equation (5)) and concept of chemical engineering plant cost index (CEPCI) (Equation (6)) are applied to consider economics of scale and effects of inflation.

$$C_2 = C_1\left(\frac{S_2}{S_1}\right)^{0.6} \tag{5}$$

where C is an equipment cost (USD) and S is a scale of the certain chemical process.

$$C_2 = C_1\left(\frac{I_2}{I_1}\right) \tag{6}$$

where C is an equipment cost (USD) and I is a CEPCI.

Table 2. List of economic parameters and assumptions used in itemized cost estimation.

Economic Parameters	
Capital Cost	
MP reactor [56]	EUR 2740 k
WGS reactor [56]	EUR 59 k
Regenerator [57]	USD 12,112,138
Catalyst [58]	USD 1.138 kg^{-1}
Gasifier [56]	EUR 211 k
PSA [59]	$\frac{\text{CEPCI}}{392.6} \times 1{,}510{,}000 \times \left(\frac{\text{inlet flow rate}}{500}\right)^{0.6}$ (USD)
Cyclone [58]	USD 31,400
Supplement	20% of (Capital cost-Supplement) (USD)
Operating Cost	
CH_4 [60]	USD 0.005 MJ_{LHV}^{-1}
Catalyst operating cost	Assumed as 10% loss per month
Water [61]	USD 12 ton^{-1}
Electricity [32]	USD 56 MWh^{-1}
Labor [62]	USD 11 hr^{-1}
PSA operating cost [59]	6.11 × 100 × (inlet flow rate except H_2) (USD)
C selling price [63]	EUR 500 ton^{-1}
Maintenance [64]	1% of (Capital cost-Supplement) (USD y^{-1})
Other cost [64]	2% of (Capital cost-Supplement) (USD y^{-1})
Economic Assumptions	
CEPCI (2021)	655.9
i	0.045
Exchange rate	USD 1 = EUR 0.85
N	20 years for MP reactor 10 years for WGS reactor, regenerator, PSA, PSA operating cost, cyclone, and supplement 1 year for catalyst
Stream factor	0.95

2.3. Sensitivity and Scenario Analysis

Key economic parameters of each system of MP and future unit H_2 production cost with varied C selling price, which can be advantages for economic feasibility, were quantitatively investigated by sensitivity and scenario analysis, respectively. For sensitivity analysis, with varied capital and operating cost in a range of ±20% with other parameters fixed, the degree of variation in unit H_2 production cost for each MP system was investigated. In addition, trends of unit H_2 production cost for each scenario according to different H_2 production scales and C selling prices were obtained and compared to conventional H_2 production methods of SMR and SMR with CCS.

3. Results and Discussion

3.1. H_2 and C Production Rates

Based on the result of the process simulation, H_2 and C production rates for each MP system with a feed CH_4 rate of 1 kmol h^{-1} were obtained at the different operating temperatures of 1073–1373 K for TMP-S1 and TMPG-S3 and 1023–1173 K for CMP-S2 and CMPG-S4, with the ratio of H_2 combusted of 0%–100%, and the ratio of reactants composed of C, Air, and H_2O (Figure 4).

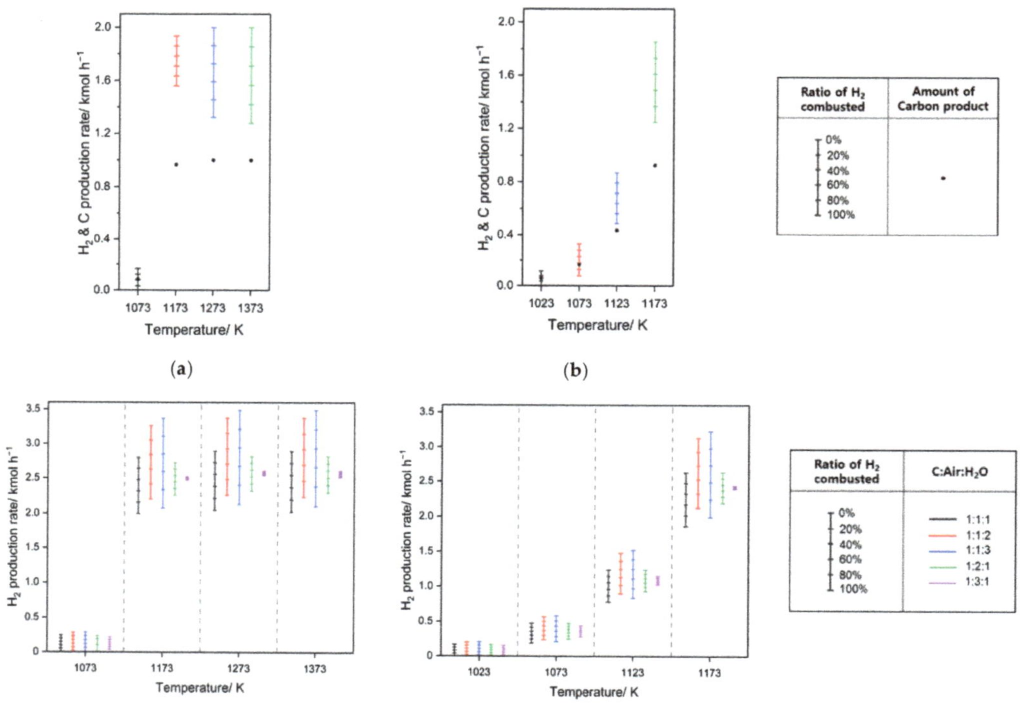

(a) (b) (c) (d)

Figure 4. Technical performance for methane pyrolysis (MP) systems of (**a**) thermal methane pyrolysis (TMP-S1), (**b**) catalytic methane pyrolysis (CMP-S2), and systems with additional gasification and WGS reaction of (**c**) TMPG-S3 and (**d**) CMPG-S4.

For TMP-S1 (Figure 4a), net H_2 production rates of 0–0.17, 1.56–1.93, 1.32–2.00, and 1.28–2.00 kmol h^{-1} and C production rates of 0.08, 0.97, 1.00, and 1.00 kmol h^{-1} were obtained at a temperature of 1073 K, 1173 K, 1273 K, and 1373 K, respectively. As reaction temperature increased, the range of net H_2 production rate was highly dependent on the ratio of H_2 combusted due to the large amount of heat required and the different thermodynamic properties of each fuel, and produced C was maximized from 1173 K, not even the maximum investigated temperature.

For CMP-S2 (Figure 4b), 0.00–0.12, 0.08–0.33, 0.49–0.87, and 1.25–1.85 kmol h^{-1} for net H_2 production rates and 0.06, 0.17, 0.43, and 0.93 kmol h^{-1} for C production rates were obtained at temperatures of 1023 K, 1073 K, 1123 K and 1173 K, respectively. Even though an increasing trend of H_2 and C production rate and a high dependence of the ratio of H_2 combusted was shown, which are similar to the results from TMP-S1, no theoretical maximum amounts of H_2 and C (2 and 1 kmol h^{-1}, respectively) were produced.

For TMPG-S3 (Figure 4c), even with the various ratios of C, Air, and H_2O for the gasification unit, very low H_2 production rates of 0.00–0.29 kmol h^{-1} were obtained at 1073 K. That poor technical performance, lower than the theoretical H_2 production rate of 2 kmol h^{-1} for the previous system of TMP-S1, was raised by higher reaction temperatures of MP, leading to the improved technical performance of 2.00–3.49 kmol h^{-1} H_2 production rates at 1173–1373 K. For the ratio of reactants in the gasifier, H_2 production rates at 1373 K dramatically increased 2.02–2.89 kmol h^{-1} for a 1:1:1 ratio to rates of 2.10–3.49 kmol h^{-1} for a ratio of 1:1:3, proving the importance of H_2O in the additional H_2 production processes of C gasification and the WGS reactor. Compared to the effect of H_2O on net H_2 production rates, an opposite effect of air was shown with decreased maximum H_2 production rates of 2.81 and 2.61 kmol h^{-1} for the ratios of 1:2:1 and 1:3:1, respectively, down from 2.89 kmol h^{-1} for a 1:1:1 ratio, thereby identifying its disadvantage in technical performance.

For CMPG-S4 (Figure 4d), lower H_2 production rates of 0.00–0.20, 0.19–0.58, and 0.78–1.52 kmol h^{-1} were obtained than those from TMPG-S3 in a range of similar investigated temperatures (1073–1273 K). Even though the technical performance was improved at a higher reaction temperature of 1173 K showing H_2 production rates of 1.86–3.23 kmol h^{-1}, this is still lower than those from TMPG-S3. For the effect of air and H_2O on H_2 production rates, similar trends to those for TMPG-S3 of increased rates from 1.86–2.63 kmol h^{-1} (1:1:1) to 1.99–3.23 kmol h^{-1} (1:3:1) and decreased rates from 2.19–2.63 kmol h^{-1} (1:2:1) to 2.39–2.44 kmol h^{-1} (1:3:1).

As a result, through the trends of H_2 and C production rates obtained from the process simulation, the detailed effects of temperature, the ratio of H_2 combusted, and the ratio of reactants entering the gasifier were confirmed.

3.2. Parametric Study—Fuel Consumption

To investigate the amount of fuel combusted to cover the total heat required in each system, the required amount of fuel (H_2 and CH_4) was obtained (Figures 5 and 6).

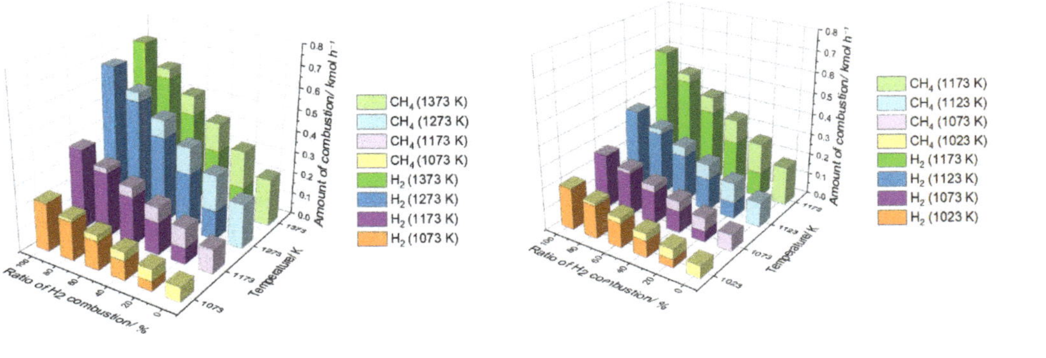

(a) (b)

Figure 5. Amount of fuel (CH_4 and H_2) consumption for methane pyrolysis (MP) systems of (**a**) thermal methane pyrolysis (TMP-S1) and (**b**) catalytic methane pyrolysis (CMP-S2) in temperature of 1073–1373 K and 1023–1173 K.

Figure 5 shows the required amount of fuel in TMP-S1 and CMP-S2 according to temperature and the ratio of H_2 combustion (0%, 20%, 40%, 60%, 80%, and 100%). From the process simulation, the total amount of heat required of 3425, 5739, 10,380, and 11,079 cal s^{-1} for TMP-S1 were needed at 1073 K, 1173 K, 1273 K, and 1373 K, respectively, and that of 3016, 3941, 5868, and 9246 cal s^{-1} for CMP-S2 were needed at 1023 K, 1073 K, 1123 K, and

1173 K, respectively, showing the higher amount of heat is needed for TMP-S1 than CMP-S2 due to its higher reaction temperature and the technical benefit of the catalyst-based MP.

For TMP-S1, the amounts of CH_4 consumption were estimated as 0.07, 0.11, 0.21, and 0.22 kmol h^{-1} and the amounts of H_2 consumption of 0.22, 0.37, 0.67, and 0.72 kmol h^{-1} were obtained when CH_4 and H_2 covered the total amount of heat required, respectively, showing an increasing trend as temperature increased. Similarly, for CMP-S2, 0.06–0.18 kmol h^{-1} of CH_4 consumption and 0.20–0.60 kmol h^{-1} of H_2 consumption were estimated when each type of fuel totally covered the required heat.

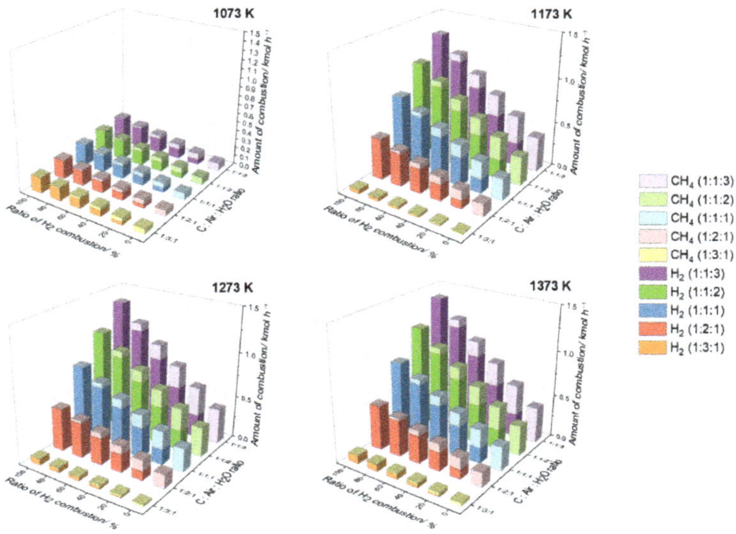

(a)

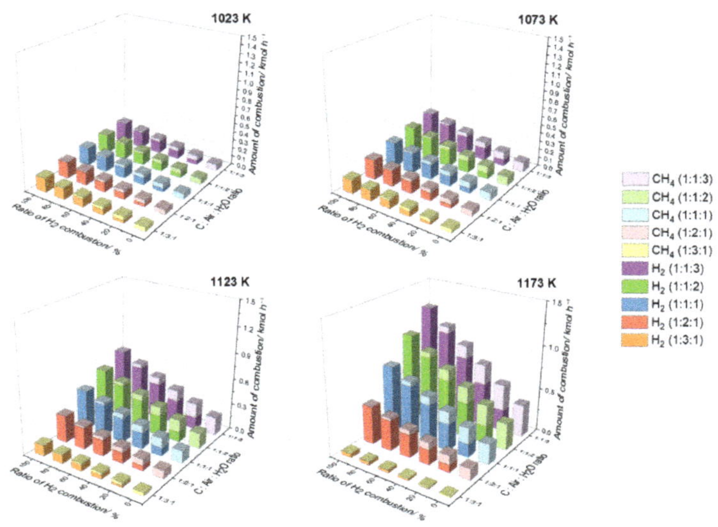

(b)

Figure 6. Amount of fuel (CH_4 and H_2) consumption for methane pyrolysis (MP) systems of thermal and catalytic methane pyrolysis with gasification and WGS reaction ((**a**) TMPG-S3 and (**b**) CMPG-S4) in temperature of 1073–1373 K and 1023–1173 K.

In Figure 6, the amount of required fuel to supply heat for TMPG-S3 and CMPG-S4 is shown. In TMPG-S3 and CMPG-S4, much larger amounts of heat of 2691–4330, 725–19852, 1047–20,821, and 1464–21,245 cal s^{-1} for TMPG-S3 and 2503–3643, 2442–5737, 1925–10,514, and 745–18,954 cal s^{-1} for CMPG-S4 were obtained as the temperature increased, showing much greater increase than previous systems due to the additional endothermic process of C gasification. For both systems in particular, the ratio of reactants of 1:1:3 showed the highest amount of heat required compared with other ratios due to its high reaction extent, represented by high H_2 and C production rates in Figure 4. In addition, the very high impact of H_2O in the amount of heat required was confirmed again with trends of the required fuel at different ratios of reactants for the gasifier.

For TMPG-S3, the amount of CH_4 fuel required of 0.01–0.42 kmol h^{-1} when it covers total heat required can be replaced by the amount of H_2 combusted of 0.05–1.39 kmol h^{-1}; for CMPG-S4, the H_2 consumption range of 0.05–1.24 kmol h^{-1} was estimated to replace the amount of CH_4 required of 0.01–0.37 kmol h^{-1}.

3.3. Itemized Cost Estimation

Based on the results from the process simulation, the itemized cost estimation for each MP system using only CH_4 and H_2 as fuel was conducted to investigate unit H_2 production cost at the temperature of 1273 K for TMP-S1 and TMPG-S3 and 1173 K for CMP-S2 and CMPG-S4, with the ratio of H_2 combusted of 40%, and the ratio of reactants for the gasifier of 1:1:2 (C-Air-H_2O) (Table 3).

Table 3. Results of itemized cost estimation for methane pyrolysis (MP) systems of (**a**) thermal methane pyrolysis (TMP-S1), (**b**) catalytic methane pyrolysis (CMP-S2), and systems with additional gasification and WGS reaction of (**c**) TMPG-S3 and (**d**) CMPG-S4.

	(a) TMP-S1		(b) CMP-S2		(c) TMPG-S3		(d) CMPG-S4	
Items	Annualized Cost/USD y^{-1}	Unit H_2 Production Cost/USD kgH_2^{-1}	Annualized Cost/USD y^{-1}	Unit H_2 Production Cost/USD kgH_2^{-1}	Annualized Cost/USD y^{-1}	Unit H_2 Production Cost/USD kgH_2^{-1}	Annualized Cost/USD y^{-1}	Unit H_2 Production Cost/ USD kgH_2^{-1}
1. Capital cost								
MP reactor	37,139	1.29	35,590	1.33	50,967	1.05	48,840	1.08
WGS reactor	-	-	-	-	1804	0.04	1729	0.04
Regenerator	-	-	19,233	0.72	-	-	-	-
Catalyst	-	-	6	0.00	-	-	6	0.00
Gasifier	-	-	-	-	6452	0.13	6183	0.14
PSA	11,323	0.39	17,924	0.67	21,320	0.44	25,460	0.56
Cyclone	224	0.01	214	0.01	224	0.00	213	0.00
Supplement	8833	0.31	11,665	0.44	13,819	0.28	13,854	0.31
2. Operating cost								
Reactant	33,704	1.17	33,704	1.26	33,704	0.69	33,704	0.74
Catalyst operating cost	-	-	6	0.00	-	-	6	0.00
Water	-	-	-	-	2448	0.05	2251	0.05
Fuel	4150	0.14	3696	0.14	6848	0.14	6189	0.14
Labor	7524	0.26	7524	0.28	7524	0.15	7524	0.17
PSA operating cost	2	0.00	183	0.01	178	0.00	341	0.01
Maintenance	11,489	0.40	15,173	0.57	17,975	0.37	18,021	0.40
Other costs	5745	0.20	7587	0.28	8988	0.18	9011	0.20
3. C selling price	−58,700	−2.04	−54,352	−2.03	-	-	-	-
4. Total cost	61,431	2.14	98,152	3.66	172,250	3.53	173,332	3.82

For TMP-S1, unit H_2 production cost of USD 2.14 kgH_2^{-1} was estimated considering the capital cost of the MP reactor, PSA, cyclone, and supplement, and the operating cost of reactant, fuel, labor, PSA operating cost, maintenance, and other costs. In this estimation, the costs of the MP reactor and reactant account for 31% and 28% of the production cost with no consideration of the C selling price (USD 4.17 kgH_2^{-1}), respectively, showing its high importance in the economic feasibility. For CMP-S2, unit H_2 production cost of USD 3.66 kgH_2^{-1} was estimated with additional items related to a catalyst such as the cost of the regenerator and catalyst, and its operating cost. Among economic parameters, it is clear that the costs of the MP reactor and reactant are the most influential economic factors, showing high ratios of 23% and 22% of the production cost without considering the C selling price (USD 5.69 kgH_2^{-1}). In both TMP-S1 and CMP-S2, where units of the gasifier and WGS reactor were not constructed in the process simulation, the cost of the MP reactor and reactant and the C selling price have a very high economic impact on H_2 production.

For TMPG-S3 and CMPG-S4, additional economic parameters related to the gasification of C and the WGS reaction means that the costs of the WGS reactor, gasifier, and water were considered and compared to both TMP-S1 and CMP-S2. For both systems, the relatively increased unit H_2 production costs of USD 3.53 and 3.82 kgH_2^{-1} for TMPG-S3 and CMPG-S4, respectively, were estimated compared to those of USD 2.14 and 3.66 kgH_2^{-1} for TMP-S1 and CMP-S2, respectively. In addition, similar to TMP-S1 and CMP-S2, costs of the MP reactor and reactant were found to be the most influential economic parameters showing portions of 30% and 19%, and 28% and 19% for TMPG-S3 and CMPG-S4, respectively.

Our results indicated that the selling of C can be a very effective way to obtain economic competitiveness through the concept of MP, and showed the importance of the cost of the MP reactor and reactant for economic feasibility.

3.4. Parametric Study—Economic Aspects

To investigate the effects of the important parameters of reaction temperature, the ratio of H_2 combusted to supply the heat required in the process, and the ratio of reactants composed of C, Air, and H_2O for the gasifier on economic feasibility, a comprehensive parametric study revealing trends of unit H_2 production cost was conducted (Figure 7).

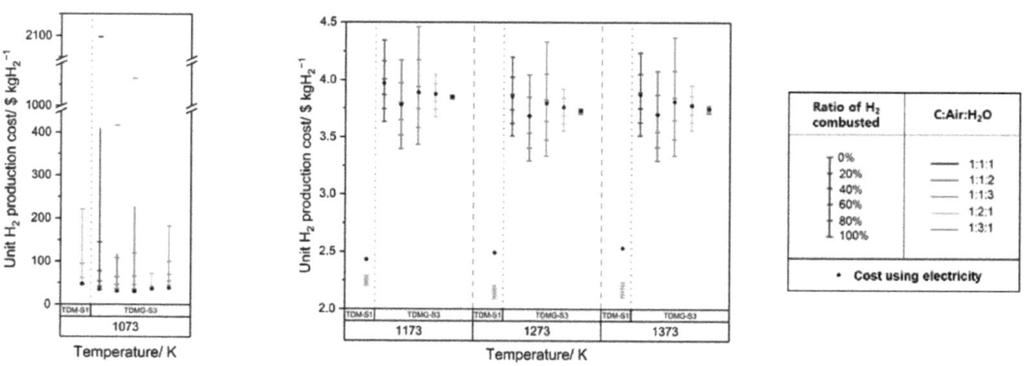

(a) (b)

Figure 7. *Cont.*

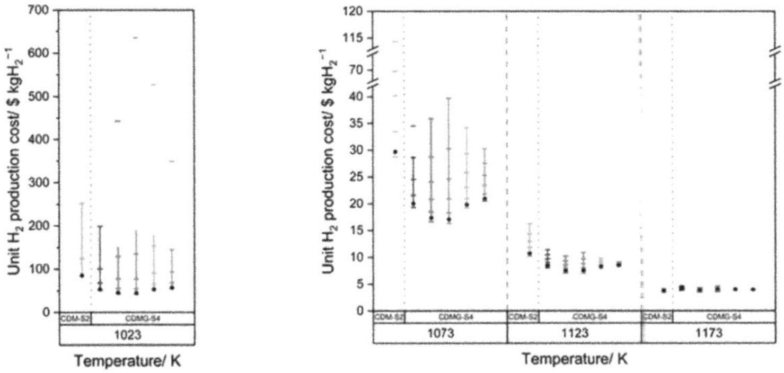

(c) (d)

Figure 7. Unit H_2 production cost for methane pyrolysis (MP) systems of (**a**) thermal methane pyrolysis (TMP-S1), (**b**) catalytic methane pyrolysis (CMP-S2), and the systems with additional gasification and WGS reaction of (**c**) TMPG-S3 and (**d**) CMPG-S4.

In the case of low temperature use where 1073 K for thermal-based systems (TMP-S1 and TMPG-S3) and 1023 K for catalyst-based systems (CMP-S2 and CMPG-S4) were considered, there were no economic benefits in either system. The minimum unit H_2 production costs of USD 46.09 kgH_2^{-1} (TMP-S1) and USD 29.48–38.06 kgH_2^{-1} (TMPG-S3) for thermal-based systems and USD 83.40 kgH_2^{-1} (CMP-S2) and USD 43.11–55.56 kgH_2^{-1} (CMPG-S4) for catalyst-based systems were reported. In addition, for the case of using electricity as the heat source, the cost was slightly higher but almost the same as the lowest cost using fuel combustion for both the thermal-based and catalyst-based system. Therefore, it is advantageous to use electricity as a heat source at low temperatures, but it still seems it would be difficult to gain economic benefits because of the high production costs.

In cases of high temperature, where 1173–1373 K for thermal-based systems and 1073–1173 K for catalyst-based systems were studied, cost reductions as temperature increased were shown. For TMP-S1, as temperature increased unit H_2 production costs of USD 2.20–2.29, 2.09–2.21, and 2.10–2.23 kgH_2^{-1} and a cost reduction of 5.03% for each minimum cost were reported, which are much cheaper than the minimum costs for TMPG-S3 of USD 3.40, 3.29, and 3.30 kgH_2^{-1}, proving the economic weakness of adopting the additional H_2 production processes of C gasification and WGS reaction. Interestingly, the lowest hydrogen production cost was found at temperatures of 1273K, even though it was not the highest temperature. In addition, for the case of using electricity, the cost for TMP-S1 is much higher at USD 2.43, 2.49, and 2.53 kgH_2^{-1} than in the combustion case as temperature increased. However, in TMPG-S3, it shows values of USD 3.78–3.97, 3.69–3.86, and 3.70–3.87 kgH_2^{-1} that are close to the average for combustion; thus, electricity can be beneficial as a heat source depending on the ratio of H_2 combusted.

Compared to the trend of unit H_2 production cost for the thermal-based system, dramatic cost reductions in CMP-S2 were obtained for the catalyst-based system showing decreased unit H_2 production costs of USD 28.77–114.35, 10.17–16.28, 3.43–4.16 kgH_2^{-1} due to its technical improvement as temperature increased. In addition, except at the temperature of 1173 K, the economic benefit of the processes for C gasification and WGS reaction was reported showing lower minimum unit H_2 production cost ranges of USD 16.34–20.52 and 7.02–8.42 kgH_2^{-1} for CMPG-S4 than those of USD 28.77 and 10.17 kgH_2^{-1} for CMP-S2 at 1073 K and 1123 K, respectively. In the case of electricity as the heat source, CMP-S2 shows costs of USD 29.68, 10.69, and 3.81 kgH_2^{-1} and CMPG-S4 shows costs of USD 17.10–20.95, 7.55–8.55, and 3.93–4.20 kgH_2^{-1} as temperature increased, showing slightly higher costs than the lowest cost using fuel combustion.

From these results, very critical effects, especially for the catalyst-based MP process, of temperature on economic feasibility and the need for the proper adoption of additional H_2 production processes can be revealed.

3.5. Sensitivity and Scenario Analysis

To investigate key economic parameters in each system and the possibility of the commercialization of each system, sensitivity and scenario analyses were conducted. In this study, one economic parameter was varied in the range of ±20% with the remaining parameters fixed, and variations of unit H_2 production cost for each system were obtained with key factors showing high variation remarked. Figure 8a,b reveal the economic importance of C selling price in both TMP-S1 and CMP-S2 showing very high variations of 19.1% and 11.1%, respectively. Costs of the MP reactor and reactant were also figured out as the next influential parameters showing variations of 12.1% and 10.9%, and 7.2% and 6.9% for TMPG-S3 and CMPG-S4, respectively. For TMPG-S3 and CMPG-S4, where all C is combusted leading to no profits from the selling of C, the same economic parameters of costs of the MP reactor and reactant were reported as the most influential factors with variations of 6.0% and 3.9%, and 5.6% and 3.9%, respectively (Figure 8c,d).

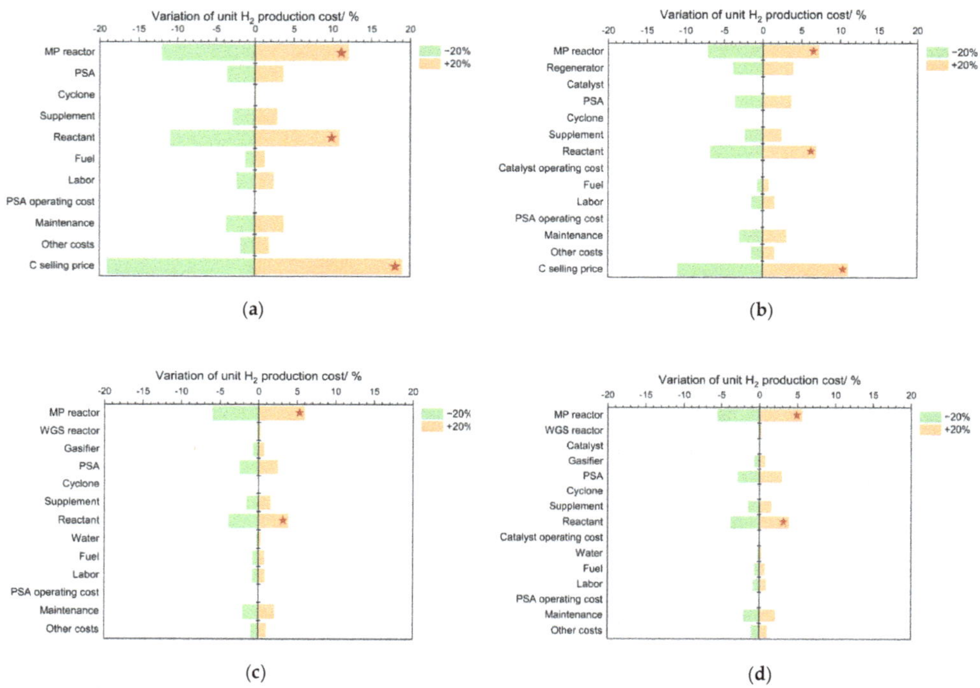

Figure 8. Results of sensitivity analysis for methane pyrolysis (MP) systems of (**a**) thermal methane pyrolysis (TMP-S1), (**b**) catalytic methane pyrolysis (CMP-S2), and the systems with additional gasification and WGS reaction of (**c**) TMPG-S3 and (**d**) CMPG-S4.

On the other hand, the scale of certain chemical engineering processes is also known to be a very influential factor to determine their economic feasibility [65]. Therefore, a scenario analysis for H_2 production scale, which can lead to a cost reduction in capital cost as reported [66], and the different C selling price in the system of TMP-S1 and CMP-S2, which is a very crucial economic factor as reported by the sensitivity analysis, to reflect a pessimistic price fluctuation and low product quality of C, was conducted and compared to conventional H_2 production methods of SMR (USD 0.94–1.78 kgH_2^{-1}) and SMR with CCS (USD 1.45–2.38 kgH_2^{-1}) [67] (Figure 9). As shown in Figure 9, there was a clear cost reduction as the H_2 production scale increased due to the economics of scale. For CMPG-S4, unit H_2 production cost decreased from USD 3.82 to 1.99 kgH_2^{-1} (−47.9%) proving it can compete in price with SMR+CCS but not with SMR. Similarly, larger cost reductions of 51.6% and 60.7% were obtained for TMP-S1 and CMP-S2, respectively, with no C selling price assumed (represented by USD 0 ton^{-1}), which are still not enough to compete with conventional H_2 production methods of SMR.

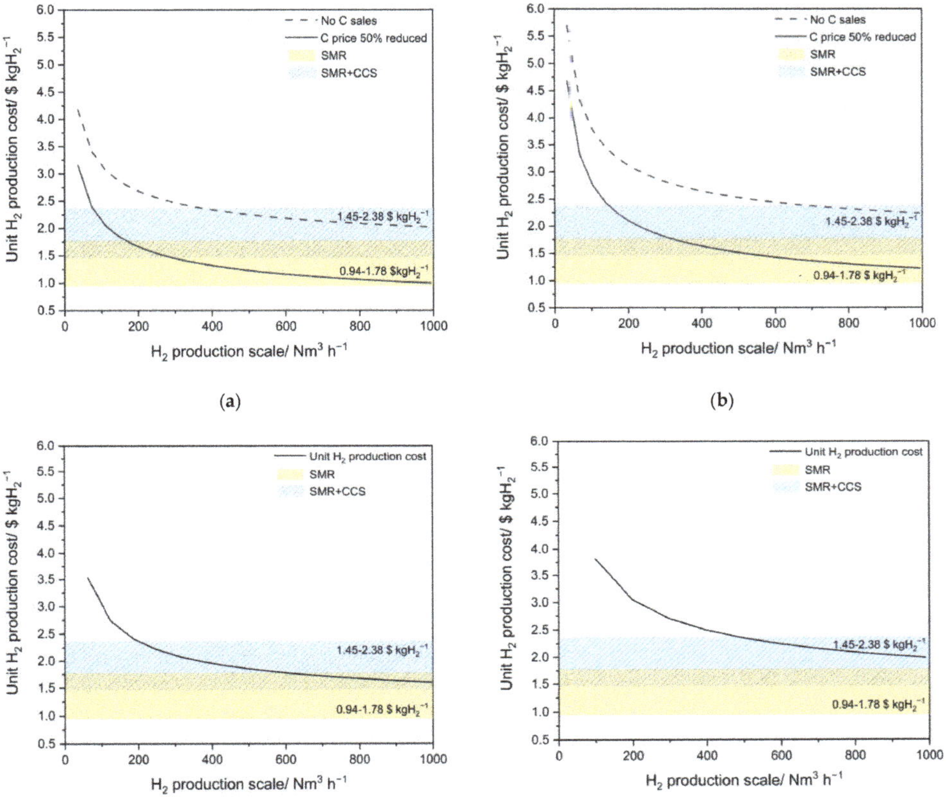

(a) (b) (c) (d)

Figure 9. Results of scenario analysis up to 1000 $Nm^3\ h^{-1}$ for methane pyrolysis (MP) systems of (**a**) thermal methane pyrolysis (TMP-S1), (**b**) catalytic methane pyrolysis (CMP-S2), and the systems with additional gasification and WGS reaction of (**c**) TMPG-S3 and (**d**) CMPG-S4 with comparison to systems of steam methane reforming (SMR) of USD 0.94–1.78 $kgH_2{}^{-1}$ and SMR with carbon capture and storage (CCS) of USD 1.45–2.38 $kgH_2{}^{-1}$.

However, for TMP-S1 and CMP-S2, with a C selling price of EUR 250 ton^{-1} (−50% of the assumed C selling price) and for TMPG-S3, with unit H_2 production costs decreased from USD 3.16 to 1.00 $kgH_2{}^{-1}$, USD 4.68 to 1.22 $kgH_2{}^{-1}$, and USD 3.53 to 1.60 $kgH_2{}^{-1}$ showing reductions of 68.3%, 73.9% and 54.8%, respectively, proving their economic competitiveness with conventional commercialized H_2 production methods of SMR and SMR with CCS.

In short, the key economic parameters of C selling price and costs of MP reactor and reactant were calculated and the future economic competitiveness for all systems with high H_2 production scales and C selling prices, even pessimistic prices, was confirmed.

4. Conclusions

The promising alternative clean concept of H_2 production of methane pyrolysis (MP) was technically and economically investigated by preliminary techno-economic analysis consisting of a process simulation using Aspen Plus®, itemized cost estimation, and sensitivity and scenario analysis for various parameters such as temperature, the ratio of fuel (CH_4 and H_2 produced from MP) combusted, and the ratio of reactants for C gasification

(C, Air, and H_2O). To investigate various H_2 production scenarios, thermal and catalytic methane pyrolysis (TMP-S1 and CMP-S2) and systems with additional H_2 production processes composed of C gasification and WGS reaction (TMPG-S3 and CMPG-S4) were considered in this study. The results of the process simulation indicated that reaction temperature is the most influential process variable for determining the technical performance of all investigated systems, and catalyst-based MP (CMP-S2 and CMPG-S4) showed much lower net H_2 and C production rates than thermal-based systems (TMP-S1 and TMPG-S3), where theoretical maximum yields were obtained, due to different kinetics used in the simulation. In addition, the amount of H_2O added in TMPG-S3 and CMPG-S4 was reported as the most important factor to increase the amount of H_2 produced. For an aspect of fuel consumption estimated from the amount of heat required for each system, trends similar to the net H_2 and C production were obtained showing the amounts of heat of 3425–11,079 cal s^{-1} for TMP-S1, 3016–9246 cal s^{-1} for CMP-S2, 725–21,245 cal s^{-1} for TMPG-S3, and 744–18,945 cal s^{-1} for CMPG-S4 matched with the required fuel amounts of 0.068–0.219, 0.060–0.183, 0.014–0.420, and 0.015–0.375 kmol h^{-1} and 0.223–0.723, 0.197–0.603, 0.047–1.386, 0.049–1.236 kmol h^{-1} for CH_4 and H_2, respectively. From the itemized cost estimation for each system at 1273 K for TMP-S1 and TMPG-S3 and 1173 K for CMP-S2 and CMPG-S4, with 40% H_2 combusted, and the ratio of 1:1:2 for C-Air-H_2O, unit H_2 production costs of USD 2.14, 3.66, 3.53, and 3.82 $gH_2{}^{-1}$ for each system, respectively, were obtained showing a very high portion of the costs of the MP reactor and reactant, and the economic benefits of the carbon (C) selling price. To investigate the effects of each parameter of temperature, the ratio of fuel combusted, and ratios of C, Air, and H_2O on economic feasibility, a parametric study was conducted proving the economic benefits of high temperature and the additional H_2 production process of C gasification and WGS reaction for CMPG-S4 but not for thermal-based MP process. The importance of the costs of the MP reactor and reactant and the C selling price for economic feasibility was calculated again by sensitivity analysis, where the variation of ±20% was assumed, with variations of unit H_2 production cost of 19.1% and 11.1% for C selling price in TMP-S1 and CMP-S2, respectively, and 12.1%–5.6% and 10.9%–3.9% for costs of the MP reactor and reactant, respectively, in all investigated systems. In addition, the effects of H_2 production scale for all systems and the C selling price for TMP-S1 and CMP-S2 on the unit H_2 production costs were investigated suggesting that all systems can compete with SMR with CCS, and especially TMP-S1 and CMP-S2, even with the pessimistic 50% reduced C selling price; for TMPG-S3, economic competitiveness with the commercialized H_2 production method of SMR can be achieved when an H_2 production scale larger than 1000 Nm^3 h^{-1} is assumed.

Conclusively, the techno-economic feasibility of MP processes, classified as the four systems of TMP-S1, CMP-S2, TMPG-S3, and CMPG-S4, was investigated with detailed H_2 and C production rates, the amount of fuel required to supply heat in each system, the trends of unit H_2 production cost according to temperature, the ratio of H_2 combusted, and the ratio of reactants used in the C gasifier, and revealed key economic parameters of the costs of the MP reactor and reactant and the C selling price. Although several techno-economic enhancements such as scale-up should be researched further to accomplish the economic competitiveness of MP compared to SMR, the environmental benefits of MP are clearly shown in this study based on its theoretical reaction stoichiometry and trends of its consumption of fuels. Based on the results, the potential of both thermal and catalytic MP for promising H_2 production is clearly presented.

Author Contributions: Conceptualization, H.L. (Hankwon Lim) and S.C.; methodology, S.C. and M.B.; validation, M.B.; investigation, D.L. and H.L. (Hyunjun Lee); writing-original draft preparation, S.C. and M.B.; writing-review and editing, S.C. and M.B.; supervision, H.L. (Hankwon Lim); funding acquisition, H.L. (Hankwon Lim). All authors have read and agreed to the published version of the manuscript.

Funding: This work was supported by the Korea Institute of Energy Technology Evaluation and Planning (KETEP) and the Ministry of Trade, Industry and Energy (MOTIE) of the Republic of Korea

(No. 20203020040010) and supported by the 2021 Research Fund (1.210103.01) of UNIST (Ulsan National Institute of Science and Technology).

Institutional Review Board Statement: Not applicable.

Informed Consent Statement: Not applicable.

Data Availability Statement: Not applicable.

Conflicts of Interest: The authors declare no conflict of interest.

Nomenclature

CO_2	Carbon dioxide
CO	Carbon monoxide
CCS	Carbon capture and storage
CMP	Catalytic methane pyrolysis
CMPG	Catalytic methane pyrolysis with carbon gasification
CRF	Capital recovery factor
CEPCI	Chemical engineering plant cost index
MP	Methane pyrolysis
FLBR	Fluidized bed reactor
GHG	Greenhouse gas
H_2	Hydrogen
CH_4	Methane
PSA	Pressure swing adsorption
SMR	Steam methane reforming
SMR with CCS	Steam methane reforming with carbon capture and storage
TMP	Thermal methane pyrolysis
TMPG	Thermal methane pyrolysis with carbon gasification
VPMDCS	Vacuum promote methane decomposition with carbon separation
H_2O	Water
WE	Water electrolysis
WGS	Water-gas shift

References

1. Federal Ministry for Economic Affairs and Energy. Available online: https://www.bmwi.de/Redaktion/EN/Publikationen/Energie/the-national-hydrogen-strategy.html (accessed on 20 July 2021).
2. European Commission. Available online: https://ec.europa.eu/energy/topics/energy-system-integration/hydrogen_en#eu-hydrogen-strategy (accessed on 20 July 2021).
3. Ministry of Economy, Trade and Industry. 2017. Available online: https://www.meti.go.jp/english/press/2017/1226_003.html (accessed on 20 July 2021).
4. Ministry of Economy, Trade and Industry. 2018. Available online: https://www.meti.go.jp/english/press/2018/0703_002.html (accessed on 20 July 2021).
5. Ministry of Economy, Trade and Industry. 2019. Available online: https://www.meti.go.jp/english/press/2019/0312_002.html (accessed on 20 July 2021).
6. Climate Change Committee. Available online: https://www.theccc.org.uk/publication/hydrogen-in-a-low-carbon-economy/ (accessed on 20 July 2021).
7. U.S. Department of Energy. Available online: https://www.energy.gov/eere/fuelcells/h2scale (accessed on 20 July 2021).
8. The Commonwealth Scientific and Industrial Research Organisation (CSIRO). Available online: https://www.csiro.au/en/work-with-us/services/consultancy-strategic-advice-services/csiro-futures/futures-reports/hydrogen-roadmap (accessed on 20 July 2021).
9. Nikolaidis, P.; Poullikkas, A. A comparative overview of hydrogen production processes. *Renew. Sustain. Energy Rev.* **2017**, *67*, 597–611. [CrossRef]
10. Mazloomi, K.; Gomes, C. Hydrogen as an energy carrier: Prospects and challenges. *Renew. Sustain. Energy Rev.* **2012**, *16*, 3024–3033. [CrossRef]
11. Dawood, F.; Anda, M.; Shafiullah, G. Hydrogen production for energy: An overview. *Int. J. Hydrog. Energy* **2020**, *45*, 3847–3869. [CrossRef]
12. Staffell, I.; Scamman, D.; Abad, A.V.; Balcombe, P.; Dodds, P.E.; Ekins, P.; Shah, N.; Ward, K.R. The role of hydrogen and fuel cells in the global energy system. *Energy Environ. Sci.* **2019**, *12*, 463–491. [CrossRef]

13. Acar, C.; Dincer, I. Review and evaluation of hydrogen production options for better environment. *J. Clean. Prod.* **2019**, *218*, 835–849. [CrossRef]
14. Reuß, M.; Grube, T.; Robinius, M.; Preuster, P.; Wasserscheid, P.; Stolten, D. Seasonal storage and alternative carriers: A flexible hydrogen supply chain model. *Appl. Energy* **2017**, *200*, 290–302. [CrossRef]
15. Tuomi, S.; Santasalo-Aarnio, A.; Kanninen, P.; Kallio, T. Hydrogen production by methanol–Water solution electrolysis with an alkaline membrane cell. *J. Power Sources* **2013**, *229*, 32–35. [CrossRef]
16. Lee, B.; Heo, J.; Kim, S.; Sung, C.; Moon, C.; Moon, S.; Lim, H. Economic feasibility studies of high pressure PEM water electrolysis for distributed H2 refueling stations. *Energy Convers. Manag.* **2018**, *162*, 139–144. [CrossRef]
17. Ishaq, H.; Dincer, I. Comparative assessment of renewable energy-based hydrogen production methods. *Renew. Sustain. Energy Rev.* **2020**, *135*, 110192. [CrossRef]
18. Parkinson, B.; Balcombe, P.; Speirs, J.F.; Hawkes, A.D.; Hellgardt, K. Levelized cost of CO_2 mitigation from hydrogen production routes. *Energy Environ. Sci.* **2018**, *12*, 19–40. [CrossRef]
19. Navarro, R.M.; Sánchez-Sánchez, M.C.; Alvarez-Galvan, M.C.; Del Valle, F.; Fierro, J.L.G. Hydrogen production from renewable sources: Biomass and photocatalytic opportunities. *Energy Environ. Sci.* **2008**, *2*, 35–54. [CrossRef]
20. Nguyen, T.; Abdin, Z.; Holm, T.; Mérida, W. Grid-connected hydrogen production via large-scale water electrolysis. *Energy Convers. Manag.* **2019**, *200*, 112108. [CrossRef]
21. Zhang, C.; Greenblatt, J.B.; Wei, M.; Eichman, J.; Saxena, S.; Muratori, M.; Guerra, O.J. Flexible grid-based electrolysis hydrogen production for fuel cell vehicles reduces costs and greenhouse gas emissions. *Appl. Energy* **2020**, *278*, 115651. [CrossRef]
22. Christopher, K.; Dimitrios, R. A review on exergy comparison of hydrogen production methods from renewable energy sources. *Energy Environ. Sci.* **2012**, *5*, 6640–6651. [CrossRef]
23. Vincent, I.; Bessarabov, D. Low cost hydrogen production by anion exchange membrane electrolysis: A review. *Renew. Sustain. Energy Rev.* **2018**, *81*, 1690–1704. [CrossRef]
24. Hosseini, S.E.; Wahid, M.A. Hydrogen production from renewable and sustainable energy resources: Promising green energy carrier for clean development. *Renew. Sustain. Energy Rev.* **2016**, *57*, 850–866. [CrossRef]
25. Stojić, D.L.; Marčeta, M.P.; Sovilj, S.P.; Miljanić, S. Hydrogen generation from water electrolysis—Possibilities of energy saving. *J. Power Sources* **2003**, *118*, 315–319. [CrossRef]
26. Boot-Handford, M.E.; Abanades, J.C.; Anthony, E.J.; Blunt, M.; Brandani, S.; Mac Dowell, N.; Fernández, J.R.; Ferrari, M.-C.; Gross, R.; Hallett, J.; et al. Carbon capture and storage update. *Energy Environ. Sci.* **2013**, *7*, 130–189. [CrossRef]
27. Wilberforce, T.; Baroutaji, A.; Soudan, B.; Al-Alami, A.H.; Olabi, A.G. Outlook of carbon capture technology and challenges. *Sci. Total Environ.* **2018**, *657*, 56–72. [CrossRef]
28. Bui, M.; Adjiman, C.S.; Bardow, A.; Anthony, E.J.; Boston, A.; Brown, S.; Fennell, P.S.; Fuss, S.; Galindo, A.; Hackett, L.A.; et al. Carbon capture and storage (CCS): The way forward. *Energy Environ. Sci.* **2018**, *11*, 1062–1176. [CrossRef]
29. Younessi-Sinaki, M.; Matida, E.A.; Hamdullahpur, F. Kinetic model of homogeneous thermal decomposition of methane and ethane. *Int. J. Hydrog. Energy* **2009**, *34*, 3710–3716. [CrossRef]
30. Kang, D.; Rahimi, N.; Gordon, M.J.; Metiu, H.; McFarland, E.W. Catalytic methane pyrolysis in molten $MnCl_2$-KCl. *Appl. Catal. B Environ.* **2019**, *254*, 659–666. [CrossRef]
31. Parfenov, V.E.; Nikitchenko, N.V.; Pimenov, A.A.; Kuz'Min, A.E.; Kulikova, M.V.; Chupichev, O.B.; Maksimov, A.L. Methane Pyrolysis for Hydrogen Production: Specific Features of Using Molten Metals. *Russ. J. Appl. Chem.* **2020**, *93*, 625–632. [CrossRef]
32. Bhaskar, A.; Assadi, M.; Somehsaraei, H.N. Can methane pyrolysis based hydrogen production lead to the decarbonisation of iron and steel industry? *Energy Convers. Manag. X* **2021**, *10*, 100079. [CrossRef]
33. Fincke, J.R.; Anderson, R.P.; Hyde, T.A.; Detering, B.A. Plasma Pyrolysis of Methane to Hydrogen and Carbon Black. *Ind. Eng. Chem. Res.* **2002**, *41*, 1425–1435. [CrossRef]
34. Wang, I.-W.; Kutteri, D.A.; Gao, B.; Tian, H.; Hu, J. Methane Pyrolysis for Carbon Nanotubes and COx-Free H2 over Transition-Metal Catalysts. *Energy Fuels* **2018**, *33*, 197–205. [CrossRef]
35. Rahimi, N.; Kang, D.; Gelinas, J.; Menon, A.; Gordon, M.J.; Metiu, H.; McFarland, E.W. Solid carbon production and recovery from high temperature methane pyrolysis in bubble columns containing molten metals and molten salts. *Carbon* **2019**, *151*, 181–191. [CrossRef]
36. Farmer, T.C.; McFarland, E.W.; Doherty, M.F. Membrane bubble column reactor model for the production of hydrogen by methane pyrolysis. *Int. J. Hydrog. Energy* **2019**, *44*, 14721–14731. [CrossRef]
37. Abbas, H.F.; Daud, W.W. Hydrogen production by methane decomposition: A review. *Int. J. Hydrog. Energy* **2010**, *35*, 1160–1190. [CrossRef]
38. Muradov, N. Low to near-zero CO_2 production of hydrogen from fossil fuels: Status and perspectives. *Int. J. Hydrog. Energy* **2017**, *42*, 14058–14088. [CrossRef]
39. Hu, C.; Shen, H.; Zhang, S.; Li, H. Methane pyrolysis in preparation of pyrolytic carbon: Thermodynamic and kinetic analysis by density functional theory. *Chin. J. Aeronaut.* **2019**, *33*, 1064–1073. [CrossRef]
40. Guéret, C.; Daroux, M.; Billaud, F. Methane pyrolysis: Thermodynamics. *Chem. Eng. Sci.* **1997**, *52*, 815–827. [CrossRef]
41. Sánchez-Bastardo, N.; Schlögl, R.; Ruland, H. Methane Pyrolysis for CO_2-Free H_2 Production: A Green Process to Overcome Renewable Energies Unsteadiness. *Chem. Ing. Tech.* **2020**, *92*, 1596–1609. [CrossRef]

42. Nishii, H.; Miyamoto, D.; Umeda, Y.; Hamaguchi, H.; Suzuki, M.; Tanimoto, T.; Harigai, T.; Takikawa, H.; Suda, Y. Catalytic activity of several carbons with different structures for methane decomposition and by-produced carbons. *Appl. Surf. Sci.* **2018**, *473*, 291–297. [CrossRef]
43. Tezel, E.; Figen, H.E.; Baykara, S.Z. Hydrogen production by methane decomposition using bimetallic Ni–Fe catalysts. *Int. J. Hydrog. Energy* **2019**, *44*, 9930–9940. [CrossRef]
44. Qian, J.X.; Enakonda, L.R.; Wang, W.J.; Gary, D.; Del-Gallo, P.; Basset, J.-M.; Bin Liu, D.; Zhou, L. Optimization of a fluidized bed reactor for methane decomposition over Fe/Al2O3 catalysts: Activity and regeneration studies. *Int. J. Hydrog. Energy* **2019**, *44*, 31700–31711. [CrossRef]
45. Patzschke, C.F.; Parkinson, B.; Willis, J.J.; Nandi, P.; Love, A.M.; Raman, S.; Hellgardt, K. Co-Mn catalysts for H_2 production via methane pyrolysis in molten salts. *Chem. Eng. J.* **2021**, *414*, 128730. [CrossRef]
46. Karaismailoglu, M.; Figen, H.E.; Baykara, S.Z. Hydrogen production by catalytic methane decomposition over yttria doped nickel based catalysts. *Int. J. Hydrog. Energy* **2019**, *44*, 9922–9929. [CrossRef]
47. Chen, Z.; Zhang, R.; Xia, G.; Wu, Y.; Li, H.; Sun, Z.; Sun, Z. Vacuum promoted methane decomposition for hydrogen production with carbon separation: Parameter optimization and economic assessment. *Energy* **2021**, *222*, 119953. [CrossRef]
48. Riley, J.; Atallah, C.; Siriwardane, R.; Stevens, R. Technoeconomic analysis for hydrogen and carbon Co-Production via catalytic pyrolysis of methane. *Int. J. Hydrog. Energy* **2021**, *46*, 20338–20358. [CrossRef]
49. Pérez, B.J.L.; Jiménez, J.A.M.; Bhardwaj, R.; Goetheer, E.; Annaland, M.V.S.; Gallucci, F. Methane pyrolysis in a molten gallium bubble column reactor for sustainable hydrogen production: Proof of concept & techno-economic assessment. *Int. J. Hydrog. Energy* **2020**, *46*, 4917–4935. [CrossRef]
50. Kerscher, F.; Stary, A.; Gleis, S.; Ulrich, A.; Klein, H.; Spliethoff, H. Low-carbon hydrogen production via electron beam plasma methane pyrolysis: Techno-economic analysis and carbon footprint assessment. *Int. J. Hydrog. Energy* **2021**, *46*, 19897–19912. [CrossRef]
51. Zhang, X.; Kätelhön, A.; Sorda, G.; Helmin, M.; Rose, M.; Bardow, A.; Madlener, R.; Palkovits, R.; Mitsos, A. CO_2 mitigation costs of catalytic methane decomposition. *Energy* **2018**, *151*, 826–838. [CrossRef]
52. Timmerberg, S.; Kaltschmitt, M.; Finkbeiner, M. Hydrog. and hydrogen-derived fuels through methane decomposition of natural gas—GHG emissions and costs. *Energy Convers. Manag. X* **2020**, *7*, 100043. [CrossRef]
53. Keipi, T.; Li, T.; Løvås, T.; Tolvanen, H.; Konttinen, J. Methane thermal decomposition in regenerative heat exchanger reactor: Experimental and modeling study. *Energy* **2017**, *135*, 823–832. [CrossRef]
54. Kim, M.H.; Lee, E.K.; Jun, J.H.; Kong, S.J.; Han, G.Y.; Lee, B.K.; Lee, T.-J.; Yoon, K.J. Hydrog. production by catalytic decomposition of methane over activated carbons: Kinetic study. *Int. J. Hydrog. Energy* **2003**, *29*, 187–193. [CrossRef]
55. Turton, R.; Bailie, R.C.; Whiting, W.B.; Shaeiwitz, J.A.; Bhattacharyya, D. *Analysis, Synthesis, and Design of Chemical Processes*, 4th ed.; Pearson Education: Upper Saddle River, NJ, UAS, 2013; pp. 157–226.
56. Keipi, T.; Tolvanen, H.; Konttinen, J. Economic analysis of hydrogen production by methane thermal decomposition: Comparison to competing technologies. *Energy Convers. Manag.* **2018**, *159*, 264–273. [CrossRef]
57. Vasalos, I.A.; Lappas, A.A.; Kopalidou, E.P.; Kalogiannis, K.G. Biomass catalytic pyrolysis: Process design and economic analysis. *Wiley Interdiscip. Rev. Energy Environ.* **2016**, *5*, 370–383. [CrossRef]
58. Liu, L.; Qian, H.; Mu, L.; Wu, J.; Feng, X.; Lu, X.; Zhu, J. Techno-economic analysis of biomass processing with dual outputs of energy and activated carbon. *Bioresour. Technol.* **2020**, *319*, 124108. [CrossRef]
59. Hoffman, Z. Simulation and Economic Evaluation of Coal Gasification with SETS Reforming Process for Power Production. Master's Thesis, Louisiana State University (LSU), Baton Rouge, LA, USA, 2005. Available online: https://digitalcommons.lsu.edu/gradschool_theses/2269 (accessed on 21 July 2021).
60. Cruellas, A.; Bakker, J.; Annaland, M.V.S.; Medrano, J.; Gallucci, F. Techno-economic analysis of oxidative coupling of methane: Current state of the art and future perspectives. *Energy Convers. Manag.* **2019**, *198*, 111789. [CrossRef]
61. Parkinson, B.; Tabatabaei, M.; Upham, D.C.; Ballinger, B.; Greig, C.; Smart, S.; McFarland, E. Hydrog. production using methane: Techno-economics of decarbonizing fuels and chemicals. *Int. J. Hydrog. Energy* **2018**, *43*, 2540–2555. [CrossRef]
62. Chen, J.; Tyagi, R.D.; Li, J.; Zhang, X.; Drogui, P.; Sun, F. Economic assessment of biodiesel production from wastewater sludge. *Bioresour. Technol.* **2018**, *253*, 41–48. [CrossRef]
63. Keipi, T.I.-M.; Hankalin, V.; Nummelin, J.; Raiko, R. Techno-economic analysis of four concepts for thermal decomposition of methane: Reduction of CO_2 emissions in natural gas combustion. *Energy Convers. Manag.* **2016**, *110*, 1–12. [CrossRef]
64. Gim, B.-J.; Kim, J.-W.; Ko, H.-M. Economic evaluation of domestic low-temperature water electrolysis hydrogen production. *Trans. Korean Hydrog. New Energy Soc.* **2011**, *22*, 559–567. [CrossRef]
65. Dolan, M.; Beath, A.; Hla, S.; Way, J.; Abu El Hawa, H. An experimental and techno-economic assessment of solar reforming for H_2 production. *Int. J. Hydrog. Energy* **2016**, *41*, 14583–14595. [CrossRef]
66. Gim, B.; Yoon, W.L. Analysis of the economy of scale and estimation of the future hydrogen production costs at on-site hydrogen refueling stations in Korea. *Int. J. Hydrog. Energy* **2012**, *37*, 19138–19145. [CrossRef]
67. International Energy Agency (IEA). Available online: https://www.iea.org/data-and-statistics/charts/hydrogen-production-costs-using-natural-gas-in-selected-regions-2018-2 (accessed on 25 July 2021).

energies

Review

Modern Trends in Design of Catalysts for Transformation of Biofuels into Syngas and Hydrogen: From Fundamental Bases to Performance in Real Feeds

Vladislav Sadykov *, Mikhail Simonov, Nikita Eremeev and Natalia Mezentseva

Department of Heterogeneous Catalysis, Boreskov Institute of Catalysis, 630090 Novosibirsk, Russia; smike@catalysis.ru (M.S.); yeremeev21@catalysis.ru (N.E.); mnv@catalysis.ru (N.M.)
* Correspondence: sadykov@catalysis.ru; Tel.: +7-913-903-8312

Abstract: This review considers problems related to design of efficient structured catalysts for natural gas and biofuels transformation into syngas. Their active components are comprised of fluorite, perovskite and spinel oxides or their nanocomposites (both bulk and supported on high surface area Mg-doped alumina or $MgAl_2O_4$) promoted by platinum group metals, nickel and their alloys. A complex of modern structural, spectroscopic and kinetic methods was applied to elucidate atomic-scale factors controlling their performance and stability to coking, such as dispersion of metals/alloys, strong metal-support interaction and oxygen mobility/reactivity as dependent upon their composition and synthesis procedures. Monolithic catalysts comprised of optimized active components loaded on structured substrates with a high thermal conductivity demonstrated high activity and stability to coking in processes of natural gas and biofuels reforming into syngas. A pilot-scale axial reactor equipped with the internal heat exchanger and such catalysts allowed to efficiently convert into syngas the mixture of natural gas, air and liquid biofuels in the autothermal reforming mode at low (~50–100 °C) inlet temperatures and GHSV up to 40,000 h^{-1}.

Keywords: biofuel reforming; structured catalysts; nanocomposite active components; design; mechanism; performance; coking stability

Citation: Sadykov, V.; Simonov, M.; Eremeev, N.; Mezentseva, N. Modern Trends in Design of Catalysts for Transformation of Biofuels into Syngas and Hydrogen: From Fundamental Bases to Performance in Real Feeds. *Energies* **2021**, *14*, 6334. https://doi.org/10.3390/en14196334

Academic Editor: Byong-Hun Jeon

Received: 1 September 2021
Accepted: 1 October 2021
Published: 4 October 2021

1. Introduction

Production of syngas by methods alternative to steam reforming of methane now attracts a lot of attention due to both environmental and commercial reasons [1,2]. Conversion of oxygenates obtained from biomass and dry reforming of natural gas appear to be most promising [1–3]. In dry reforming of biogas or natural gas containing CH_4 + CO_2 these greenhouse gases are transformed into syngas with H_2/CO ratio ~1, which is a suitable feed for Fischer–Tropsch and oxygenates synthesis. Oxygenates obtained from biomass and glycerol- byproduct of biodiesel production are considered as attractive alternatives to fossil fuels for syngas production [1,4,5].

Efficient catalysts of such processes are usually based upon supported noble (Pt, Rh, Ru) or transition (mainly Ni) metals [1–12]. The main problem of these processes is coking of catalysts leading to their deactivation. Even though noble metals are much more stable to coking, their high price makes their broad-scale application impossible. Hence, great efforts were devoted to design of Ni-based catalysts stable to coking. Next approaches were found to be successful.

1. Instead of traditional supports (SiO_2, alumina, zeolites, etc.) use mixed oxides of rare earth and transition metals with variable oxidation states of cations/oxygen stoichiometry. As the result, such oxides with fluorite [13–29], perovskite [30–67] and spinel [68–76] structures as well as their nanocomposites [35,72,73] have sufficient amount of reactive surface/bulk oxygen species characterized also by a high mobility providing their fast migration to metal particles, where they react with activated

fuel molecules, thus preventing coking [71–73]. These oxides are the most promising supports for catalysts of hydrocarbons or oxygenates reforming to syngas without coking [11–77]. Their oxygen mobility can be tuned by changing their composition as well as synthesis procedures [13,32,35].

2. In the case of bimetallic Ni-containing nanoalloys with Ru, Co, Fe, etc. coking is also much smaller than for pure Ni [8]. This is explained by dilution of the alloy surface layers by added atoms, thus preventing clustering of Ni atoms responsible for graphitic carbon nucleation. Moreover, guest metals suppress migration of carbon atoms into the bulk of alloy particle, thus preventing nucleation of carbon fibers at the metal/support interface [8]. While for traditional as well as fluorite-like supports nanoalloys were mainly loaded via impregnation route, for perovskites and spinels both Ni and other metal cations can be incorporated into the mixed oxide lattice during synthesis. Subsequent reduction generates nanocomposites comprised of segregated metal alloy nanoparticles strongly interacting with oxide matrix, which improves coking resistance and stability to sintering [12,54–61]. However, even for Ce–Zr–O fluorites application of such modern method as solvothermal one-pot synthesis in supercritical alcohols [22–24] allowed to provide incorporation of Ni cations into the mixed oxide lattice with the same Ni nanoparticles exsolution in reducing conditions.
3. Suppression of traditional supports (alumina, etc.) acidity by doping with basic cations such as Mg allows to minimize effect of side reactions (especially for such biofuels as ethanol) leading to coking [12,70,75,78–81].
4. Since for perovskite and fluorite oxides their specific surface area is usually not high, to enhance performance and thermal stability of catalysts on their bases they can be supported on $MgAl_2O_4$ [28,78,81] or Mg-doped alumina [82].

For any practical application catalysts for transformation of biofuels into syngas are to be supported as thin layers on heat-conducting monolithic substrates, which allows to minimize or even avoid heat and mass transfer limitations typical for granulated catalysts beds [77–86].

This review is devoted to analysis of these trends in design of catalysts for transformation of biofuels into syngas based on results of our research in frames of international collaboration in last 10 years compared with those published in literature. The most important new aspect of this review is detailed description of oxygen mobility in catalysts of fuels transformation into syngas comprised of mixed oxides with fluorite, perovskite and spinel structures promoted with Ni and Pt-group metals. It is based on application of unique techniques of oxygen isotope heteroexchange of these catalysts with $^{18}O_2$ or $C^{18}O_2$ in the gas phase in flow installations including experiments in the temperature-programmed mode as well as in the steady-state of catalytic reactions (Steady-State Isotope Transients Kinetic Analysis, SSITKA) [18,35,68,69,71,72,87]. Even though it is well known that a high oxygen mobility and reactivity in these catalysts allows to prevent coking in the reactions of biofuels transformation into syngas by a fast transfer of oxygen species to the metal-support interface, where they interact with activated fuels fragments transforming them into syngas, only in our works such strict characteristics of oxygen mobility as oxygen self-diffusion coefficients were systematically estimated using sophisticated software for isotope exchange kinetics data analysis. Atomic-scale features controlling oxygen mobility in these catalysts were elucidated using modern structural and spectroscopic methods, while their surface oxygen bonding strength was estimated by pulse microcalorimetry, which provided foundations for optimization of their compositions and synthesis procedures [78,82,87].

2. Synthesis of Active Components

The method of synthesis should provide a high dispersion of complex oxides along with spatial uniformity of elements distribution in their particles. A lot of methods including co-precipitation, solvothermal method, sol-gel method, Pechini polymeric precursor method, microemulsions, sonochemical method, microwave-assisted self-combustion and ultrasonic spray pyrolysis were used for synthesis of oxides [30,33,35,36,87]. Among advanced methods, synthesis in flow regimes (including that in supercritical conditions) characterized by continuous generation of nanoparticles appears to be very promising [20–26,87]. Note that single-phase complex oxides (such as ceria-zirconia mixed oxides, etc.) including cations differing by charge and size and, hence, inherent acidity, could not be prepared by traditional precipitation with alkaline solutions added to a mixed metal salts solution [87].

2.1. Pechini Method

Ester polymeric precursors (Pechini) method [88] is based upon using citric acid and ethylene diamine as chelating agents. With ethylene glycol solution single-phase nanocrystalline doped cerium–zirconium oxides [13,18,89], perovskites [31,32,34,35] and spinel oxides [69] were obtained possessing a high spatial uniformity of cations distribution. In the case of aqueous solutions such spatial uniformity was not obtained [62–65]. For preparation of perovskite-fluorite nanocomposites Pechini method was further modified. It was made by adding fluorite oxide nanopowder into the polymeric precursor solution containing cations of perovskite followed by ultrasonic treatment and evaporation. After polymeric matrix decomposition and calcinations under air this provides nanocomposites with a high specific surface area and developed interphases between perovskite and fluorite domains [31].

2.2. Synthesis in Supercritical Alcohols

Complex $Ce_{1-x}Zr_xO_{2-\delta}$ oxides were synthesized in supercritical ethanol and isopropanol using $Zr(OBu)_4$, $ZrOCl_2$ and $Ce(NO_3)_3 \cdot 6H_2O$ solutions in isopropanol at 400–480 °C and pressure 120–140 atm [22–26]. Single-phase samples with uniform spatial distribution of cations were obtained only with solutions containing acetylacetone (AA) with AA/Zr molar ratio 2. This method allowed also to obtain single-phase samples of $Ce_{1-x}Zr_xO_{2-\delta}$ doped with Ti and Nb cations [21,22] as well as to promote them with Ni cations in so-called one-pot route of synthesis [21–24]. In reducing conditions Ni cations are exsolved from the fluorite lattice providing small Ni clusters strongly interacting with support, which helps to suppress coking and sintering.

2.3. Mesoporous Nanocomposites

Specific surface area of perovskites prepared via Pechini method is in the range of 10–15 m^2/g, which is too small for their good performance as active components of structured catalysts. To deal with this problem perovskites were loaded on Mg-doped γ-Al_2O_3 [90] or mesoporous $MgAl_2O_4$ prepared by self-assembly method induced by evaporation (EISA) with copolymer Pluronic P123 [78].

Even though for (Ru + Ni)-promoted doped $MnCr_2O_4$ spinels specific surface areas were reasonably high (~100 m^2/g), to improve their sintering resistance these active components were loaded on Mg-doped γ-Al_2O_3 as well [90].

3. Characterization of Nanocomposite Materials

3.1. Structural Features

For doped ceria and Ce–Zr–O oxides, a complex of modern techniques was applied for studies of their structure. This includes high resolution transmission electron microscopy with elemental mapping, diffraction studies using X-ray synchrotron radiation and neutron diffraction, wide-angle X-ray scattering (WAXS), infrared and Raman spectroscopies. This allowed to elucidate effects of samples chemical composition and preparation procedures on their phase homogeneity, spatial cations distribution in particles, types and concentrations of defects and features of local coordination environment of Ce and Zr cations [73,77,89–98]. For $Ce_{0.5}Zr_{0.5}O_{2-y}$ composition having the highest oxygen mobility, doping with La, Gd, Pr, Sm cations (thus producing $Ln_x(Ce_{0.5}Zr_{0.5})_{1-x}O_{2-y}$ oxides with $x = 0.1 \div 0.3$) stabilizes the pseudo-cubic structure in humid environment and reduces domain sizes. For these samples the effect of domain boundaries on the oxygen mobility is significant [18,94,95,98].

Prepared in optimized (with addition of AA complexing agent) supercritical conditions $Ce_{0.5}Zr_{0.5}O_{2-y}$ samples have a cubic structure with the crystallite size of ~5.5 nm. Doping by Ti and Nb cations increases oxygen deficiency due to generation of Ce^{3+} cations [21,22]. According to TEM data, nickel oxide particles supported by impregnation (5 wt.%) have sizes from 20 to 40 nm, while for one-pot route they are smaller (~10 nm).

Perovskites of $LnFe_{0.7-x}Ru_xNi_{0.3}O_{3-\delta}$ (Ln = La, Pr, Sm; x = 0–0.1) composition prepared by Pechini method are single-phase rhombohedral samples. Their reduction produces nanocomposites comprised of Ni–Fe–(Ru) nanoalloys and LnO_x situated in the surface layers of remaining Ln–Fe–O particles [32,34,35].

Freshly prepared Ru/($La_{0.8}Pr_{0.2}Mn_{0.2}Cr_{0.8}O_3$ + 10 wt.% NiO + 10 wt.% YSZ) nanocomposite mainly consists of the perovskite phase, with Ni and Ru cations being mainly dissolved in its surface layers. YSZ disorders perovskite structure and hampers sintering due to interfaces between its nanoparticles and perovskite domains [32].

Perovskite + fluorite (P+F) nanocomposites prepared by optimized procedures [31,35] are characterized by the developed interphase, a higher specific area as compared to the mechanical mixture of P+F phases. Cations redistribution between perovskite and fluorites nanodomains helps to improve oxygen mobility.

Oxides with a spinel structure based on $MnCr_2O_4$ prepared by Pechini method [69,90] have 2–40% of admixture phase with corundum structure due to segregation of $(Mn,Cr)_2O_3$ oxide during annealing in air. Doping with Fe and Zn cations as well as supporting Ru + Ni decrease the content of this admixture due to spinel structure stabilization.

After supporting up to 10 wt.% of spinel, fluorite or perovskite oxides on $MgAl_2O_4$ or 10 wt.% Mg-doped γ-Al_2O_3 followed by supporting Ru + Ni by impregnation when required, epitaxial layers of these oxides are formed along with incorporation of rare-earth and transition metal cations into the surface layers of these supports [78,90]. In reducing conditions Ni–Ru alloy nanoparticles are formed strongly interacting with layers of rare-earth or transition metal oxides on the surface of these high surface area supports [78,90].

3.2. Surface Properties

The most detailed characteristics of the surface of catalysts were obtained with the help of X-ray Photoelectron Spectroscopy (XPS) and Fourier-transformed Infra-red Spectroscopy of adsorbed CO (FTIRS of adsorbed CO). While the first method gives information about the charge state of ions on the surface (as judged by their binding energies (BE) in XPS spectra) as well as their surface concentrations, the second one allows to estimate the number of coordinatively unsaturated sites (metal atoms, cations) as well as their charges reflected in intensities as well as frequencies of carbonyl absorption bands in FTIRS spectra [13,94–97]. Secondary Ions Mass Spectrometry (SIMS) allows to estimate variation of the content of cations along the depth of the surface layer sputtered by the beam of argon ions [13].

For doped Ce–Zr oxides the surface was found to be enriched by large Pr, Ce and La cations as revealed by XPS and SIMS [94,95]. This implies domain boundaries enrichment by the same cations, which could affect their transport properties.

The surface layer of $MnCr_2O_4$ spinel is enriched by Mn as judged by XPS data, which is explained by segregation of Mn^{2+} cations on the surface of spinel obtained by Pechini method, where decomposition of polymeric precursor under contact with air initially occurs in rather reducing conditions [69]. Apparently even after complete oxidation of all organic residues and transformation of charge state of Mn surface cations mainly to 3+ state with an admixture of 4+ state, they remain on the surface as revealed by its enrichment by Mn. This helps to provide a high mobility and reactivity of the surface oxygen having a lower bonding strength with Mn cations than with Cr cations, thus ensuring a high coking resistance of these catalysts in fuels reforming [68,69,71].

For Pt/Ln–Ce–Zr–O catalysts (Ln = La, Pr, Gd) pretreated in O_2 platinum was found to be present in three oxidation states: 0, 2+ and 4+ (XPS binding energies BE equal to

71, 72 and 75 eV, respectively), the content of Pt cations being the highest in the case of Pr-doped samples [13,95,97]. FTIRS of adsorbed CO also revealed several states of Pt on the surface reflected in bands of linear carbonyls Pt^0-CO (νCO 2046–2084 cm^{-1}), Pt^+-CO (νCO 2125–2140 cm^{-1}) and Pt^{2+}-CO (νCO 2170–2180 cm^{-1}) [13,94–98]. FTIRS spectra did not contain bands which could be assigned to Pt^{4+}-CO carbonyls, since these cations are able to oxidize CO even at liquid nitrogen temperature, being reduced to 2+ and 1+ states. For Pt/La–Ce–Zr–O sample the highest concentration of coordinatively unsaturated Pt^{2+} cations was revealed by these methods, which can be explained by their stabilization with strongly basic La cations. Hence, strong metal-support interaction for these catalysts results in stabilization of Pt^{n+} cations on the surface.

For Ni/Ce–Zr–O samples, mainly FTIRS bands of terminal carbonyls Ni-CO at νCO~2105 cm^{-1} are observed [89]. This is explained by decoration of Ni nanoparticles surface by Ce–Zr–O fragments due to strong metal-support interaction, thus hampering appearance of neighboring Ni atoms able to stabilize bridging carbonyls. In a similar way, FTIRS bands corresponding to terminal Ni/Ru carbonyls were mainly observed for Ni + Ru loaded oxides, where alloys are formed after reduction [78,90,91]. Hence, these effects of dilution and decoration are vital to prevent coking, since carbon nucleation on Ni particles requires ensembles of the same surface atoms >6 or stepped faces having coordinatively unsaturated atoms [8].

3.3. Oxygen Species: Bonding Strength and Mobility

The bonding strength of surface oxygen species is determined by such methods as pulsed microcalorimetry and temperature-programmed desorption (TPD) of O_2. Experimental data are compared with calculations for model structures of surface sites using the semiempirical interacting bonds method [35,99–104]. The oxygen mobility was estimated using such methods as oxygen isotope heteroexchange with $^{18}O_2$ or $C^{18}O_2$ [24,68,69,105–107], steady state isotopic transient kinetic analysis (SSITKA) [108,109] and modeling of reforming processes kinetic relaxations [19,104,108].

For catalysts based on fluorite, perovskite and spinel oxides containing transition and rare-earth cations with variable charges, the bonding strength of surface oxygen species depends on their stoichiometry. In the initial oxidized state, it is ~150–200 kJ/mol O_2, corresponding to M–O forms of oxygen [35,99,100]. After removing more than one monolayer of oxygen its binding strength increases to ~400 kJ/mol for spinels ($MnCrO_x$), ~500 kJ/mol for perovskites (PrFeOx, LaPrMnCrOx) and ~650 kJ/mol for LnCeZrO fluorites, respectively, corresponding to M_2O bridging forms of the surface oxygen [99–103]. In the stationary state of these catalysts in the reactions of fuels reforming, only bridging forms of oxygen are present on the surface being regenerated by CO_2 or H_2O pulses forming also CO and H_2 as products [101–103]. For PrCeZrO fluorite and $MnCrO_x$ spinel layers supported on mesoporous $MgAl_2O_4$ the coverage by reactive oxygen species decreases while their bonding strength increases due to interaction with support [104].

Diffusion coefficients of oxygen for these catalysts are high enough (up to 10^{-12} cm^2/s at 700 °C, Table 1, Figure 1) to provide fast oxygen migration to the metal–support interface, which is required to transform activated fuel fragments into syngas.

Even though for stoichiometric spinel $MnCr_2O_4$ the oxygen diffusion in the bulk is not too fast, it is at least close to that of La-doped ceria-zirconia, which allows to suggest usage of much less expensive spinel for the catalysts design. Oxygen diffusion coefficients along domain boundaries in Pt-supported fluorites exceed by 1–3 order of magnitude those for the bulk diffusion (Table 1). The amount of oxygen involved into this fast diffusion channel decreases when using oxygen exchange with $C^{18}O_2$ or SSITKA instead of heteroexchange with $^{18}O_2$, apparently reflecting also its lower amount (oxygen storage capacity) in real reaction conditions of biofuels reforming.

Table 1. Oxygen self-diffusion coefficients in the bulk (D_{bulk}) and along grain boundaries ($D_{interface}$) at 700 °C [19,24,68,69,104–109].

Sample, Type of Exchange Molecule	D_{bulk}, cm^2/s	$D_{interfaces}$, cm^2/s
$Pt/Pr_{0.3}Ce_{0.35}Zr_{0.35}O_{2-\delta}$, $^{18}O_2$	4×10^{-14}	$>3.3 \times 10^{-11}$
$Pt/Pr_{0.3}Ce_{0.35}Zr_{0.35}O_{2-\delta}$, $C^{18}O_2$	-	$>2 \times 10^{-12}$
$Pt/La_{0.3}Ce_{0.35}Zr_{0.35}O_{2-\delta}$, $^{18}O_2$	4×10^{-15}	$5 \times 10^{-13} \div 7 \times 10^{-13}$
$LaNiPt/Pr_{0.15}Sm_{0.15}Ce_{0.35}Zr_{0.35}O_{2-\delta}$, $^{18}O_2$	3×10^{-14}	$>2.5 \times 10^{-11}$
$LaNiPt/Pr_{0.15}Sm_{0.15}Ce_{0.35}Zr_{0.35}O_{2-\delta}$, $C^{18}O_2$	-	$>5 \times 10^{-12}$
$Co_{1.8}Mn_{1.2}O_4$, $^{18}O_2$	8×10^{-13}	-
$Ni_{0.33}Co_{1.33}Mn_{1.33}O_4$, $^{18}O_2$	1.5×10^{-12}	-
$Ni_{0.6}Co_{1.2}Mn_{1.2}O_4$, $^{18}O_2$	10^{-11}	-
$PrNi_{0.5}Co_{0.5}O_3$–$Ce_{0.9}Y_{0.1}O_{2-\delta}$-nanocomposite, $C^{18}O_2$	$10^{-11} \div 10^{-9}$	$10^{-8} \div 10^{-7}$
$Ce_{0.65}Pr_{0.25}Y_{0.1}O_{2-\delta}$, $C^{18}O_2$	10^{-8}	-
$Ni/Pr_{0.2}Ce_{0.4}Zr_{0.4}O_{2-\delta}$, $C^{18}O_2$	4.8×10^{-12}	-
$Ni_{0.5}Cu_{0.5}O/Nd_{5.5}WO_{11.25-\delta}$ nanocomposite, $C^{18}O_2$	Fast 2.2×10^{-11} Slow ~10^{-13}	-
5 wt.% $Ni/Ce_{0.75}Zr_{0.25}O_2$, $C^{18}O_2$	1.3×10^{-14}	-
2 wt.% Ni+2 wt.% $Ru/MnCr_2O_4$, $C^{18}O_2$	2.6×10^{-15}	-

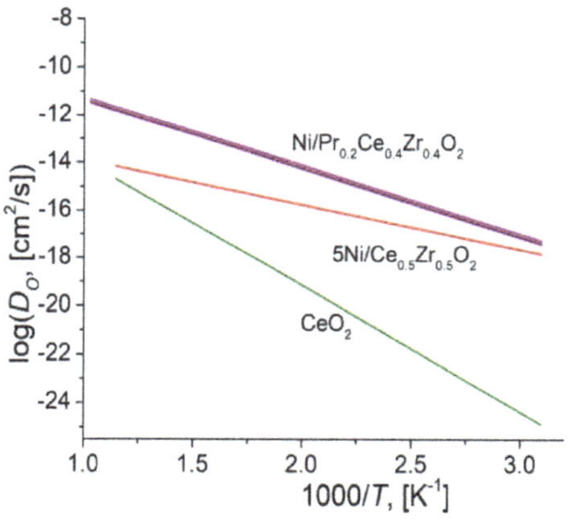

Figure 1. Temperature dependencies of oxygen self-diffusion coefficients D_O for ceria and ceria-zirconia based samples prepared via modified Pechini route.

For $PrNi_{0.5}Co_{0.5}O_3$–$Ce_{0.9}Y_{0.1}O_{2-\delta}$ nanocomposite the method of isotopic heteroexchange of oxygen with $C^{18}O_2$ also demonstrated coexistence of fast and slow channels of oxygen diffusion [106]. Fast migrations go through perovskite–fluorite interfaces as well as via $Ce_{0.65}Pr_{0.25}Y_{0.1}O_{2-\delta}$ nanodomains (Table 1), while diffusion through perovskite domains is slow.

4. Catalytic Properties

4.1. Catalysts Based on Cerium–Zirconium Mixed Oxides

In methane dry reforming (MDR) Ni-supported biphasic ceria-zirconia sample prepared in supercritical ethanol without adding AA complexion was very fast completely deactivated in MDR due to coking [20]. In contrary, a high and stable performance of catalysts with Ni supported by impregnation on single-phase $Ce_{0.5}Zr_{0.5}O_2$ oxide prepared either by modified Pechini route or in supercritical alcohols with addition of AA (*vide supra*) was maintained even at 600 °C [20–24,89]. At 700 °C, the effective first-order rate constants for these catalysts were in the range of 1–4 s^{-1} cm^{-2}, being close to values for Pt or Ru- supported Pr (Pr+Sm)–doped ceria-zirconia (Figure 2) and exceeding by an order of magnitude k values for Ni supported on these fluorites (~0.2 s^{-1} cm^{-2} at 700 °C) [87]. This stresses importance of the spatial homogeneity of ceria-zirconia mixed oxide for ensuring a high oxygen mobility and, hence, stability to coking, which is usually not taken into account.

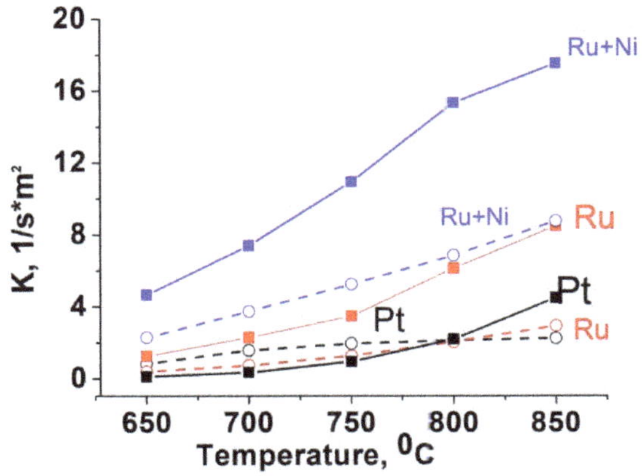

Figure 2. Specific rate constants temperature dependence for methane dry reforming on metal -supported SmPrCeZrO (filled symbols) and PrCeZrO (empty symbols) catalysts. Feed 7% CH_4 +7% CO_2 in He.

Mixed $Ce_{0.75}Zr_{0.25}O_2$ oxides doped with Ti, Nb and Ti+Nb (Table 2) were prepared in supercritical isopropanol with addition of AA. 5 wt.% Ni were supported either by impregnation (I) or via one-pot synthesis from mixed solutions in supercritical conditions (O) [24]. For samples with Ni added in one-pot route its surface content estimated as Ni/Ce+Zr atomic ratio from X-ray photoelectron spectroscopy data was twice as low in comparison with impregnated samples. This indicates Ni incorporation into the bulk of fluorite particles for one-pot samples reflected in the increase of oxygen diffusion coefficients (Table 2) due to generation of additional oxygen vacancies. Such disordering also leads to stronger sintering of samples during calcinations step, so specific surface area for one-pot samples was twice as low [24]. However, both reagents' conversions and reaction rate related to the surface concentration of nickel atoms estimated by hydrogen chemisorption (TOF, turn-over frequency) of one-pot samples were quite close to those of impregnated samples of the same chemical composition (Table 2), which indicates on a higher dispersion of Ni on the surface of one-pot samples. TOF values strongly depend on the support composition, and the catalyst with titanium and niobium co-doped ceria-zirconia support prepared by impregnation is three times more active than that with the unmodified ceria-zirconia support, apparently due to optimized interaction of Ni with more disordered doped ceria-zirconia.

Table 2. Kinetic parameters of MDR and oxygen mobility for catalysts prepared in supercritical conditions [24].

Sample [1]	k_{eff} (700 °C), s^{-1}	TOF, s^{-1}	D_O (700 °C), 10^{-15} cm^2/s
5 wt.% Ni/$Ce_{0.75}Zr_{0.25}O_2$-I	46	2.9	7.1
5 wt.% Ni/$Ce_{0.75}Zr_{0.25}O_2$-O	51	2.4	13
5 wt.% Ni/$Ce_{0.75}Ti_{0.1}Zr_{0.15}O_2$-I	38	3.4	3.2
5 wt.% Ni/$Ce_{0.75}Ti_{0.05}Nb_{0.05}Zr_{0.15}O_2$-I	64	9.0	1.7
5 wt.% Ni/$Ce_{0.75}Ti_{0.05}Nb_{0.05}Zr_{0.15}O_2$-O	38	2.6	7.8

[1] I—impregnated sample, O—one-pot sample.

Oxygen diffusion coefficients (D_o) estimated by the isotope exchange method, of the order of 10^{-15}–10^{-14} cm^2/s (Table 2), are quite close to those for $Ni/Ce_{0.5}Zr_{0.5}O_2$ with single-phase oxide support prepared via modified Pechini route (Figure 1). They are high enough for efficient oxygen transport to the metal-support interface and contribute to a high value of catalytic activity and stability against catalyst coking.

For catalysts based on $SmPrCeZrO_2$ oxide support, in MDR the rate constant was higher for supported Ru (~7 s^{-1} m^{-2} at 850 °C) than for Pt (Figure 2) [87]. Ni-supported catalyst is much less active (k~0.2 s^{-1} cm^{-2} at 700 °C), which is explained by coking of the catalyst. For supported $LaNiO_3$ activity is higher by an order of magnitude (k~1.8 s^{-1} cm^{-2} at 700 °C). This is explained by decoration of Ni nanoparticles by LaO_xCO_3 species preventing coking. Ru+Ni-supported catalyst demonstrates the highest activity (Figure 2) since NiRu clusters are not coked [87]. For Ru co-supported with $LaNiO_3$, specific activity is significantly lower being identical with that for Ru alone supported on this fluorite. According to CO chemisorption data the metal surface area for this catalyst (~0.05 m^2/g) is much lower than that for Ru + Ni–loaded sample (0.5 m^2/g). Hence, the rate constant related to the metal surface is even higher for Ru+$LaNiO_3$–loaded catalyst due to stronger metal–support interaction/interface.

For $Pt/Ln_x(Ce_{0.5}Zr_{0.5})_{1-x}O_{2-y}$ catalysts specific rate of hydrogen production in methane partial oxidation into syngas (POM) in diluted feed (Figure 3) is the highest for La-doped catalysts due to a higher content of Pt^{n+} cations transformed into clustered metal species in reaction media [13,18,72,94,95].

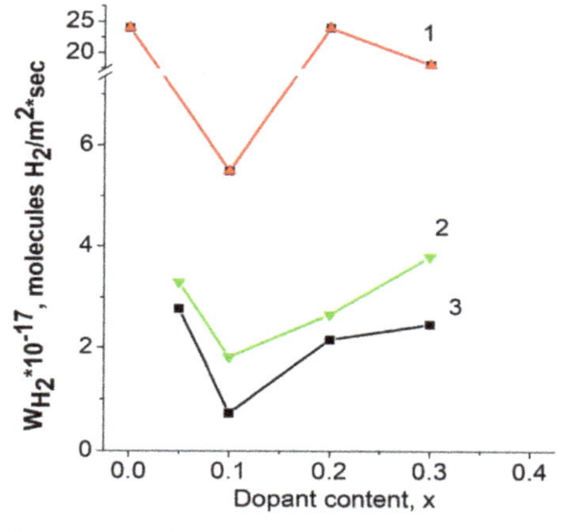

Figure 3. Specific rate of H_2 production in partial oxidation of methane for $Pt/La_x(Ce_{0.5}Zr_{0.5})_{1-x}O_{2-y}$ O_{2-y} (1), $Pt/Pr_x(Ce_{0.5}Zr_{0.5})_{1-x}O_{2-y}$ (2) and $Pt/Gd_x(Ce_{0.5}Zr_{0.5})_{1-x}O_{2-y}$ (3) catalysts. Feed 1% CH_4 + 0.5% O_2 in He, 770 °C, contact time 5 ms.

In the autothermal reforming of acetone (Figure 4) [95], specific catalytic activity of Pt-supported doped ceria-zirconia catalysts correlates with the oxygen mobility required to prevent coking, being the highest for $Pt/Pr_{0.3}Ce_{0.35}Zr_{0.35}O_2$ catalyst [18].

While testing catalysts in real concentrated feeds in POM, methane steam reforming and MDR, the temperature gradient along the length of reactor equipped with the catalyst fraction or granules caused by exothermicity or endothermicity of these reactions emerges, which complicates data analysis. To avoid this problem, catalytic layers were supported onto the inner walls of single channels cut from corundum honeycomb monoliths, which allowed to avoid any temperature gradients and estimate rate constants (Table 3) [109–112]. Among $Pt/Ln_x(Ce_{0.5}Zr_{0.5})_{1-x}O_{2-y}$ catalysts the highest activity was revealed for Pr-doped catalyst possessing the highest oxygen mobility, which stresses importance of this characteristic.

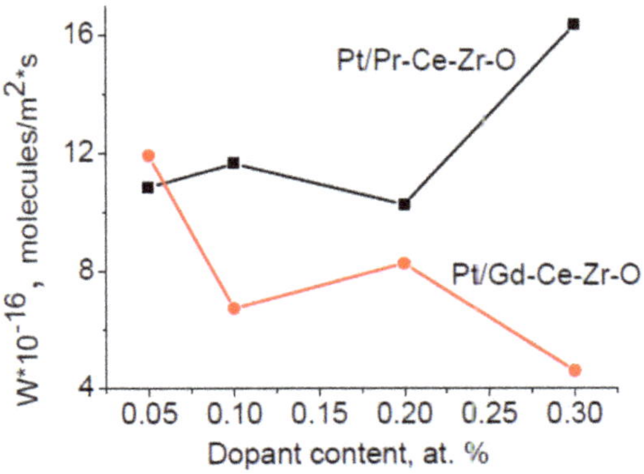

Figure 4. Specific rate of H_2 production in autothermal reforming of acetone for $Pt/Pr_x(Ce_{0.5}Zr_{0.5})_{1-x}O_{2-y}$ and $Pt/Gd_x\ (Ce_{0.5}Zr_{0.5})_{1-x}O_{2-y}$ catalysts.

Table 3. Effective first-order rate constants (s^{-1}) of CH_4 reforming at 700 °C on separate corundum channels with supported active components [111,112].

Catalyst Composition	Partial Oxidation [1]	Steam Reforming [2]	Dry Reforming [3]
$LaNiO_x/Ce_{0.2}Zr_{0.8}O_2$	80		
$LaNiPt/Ce_{0.2}Zr_{0.8}O_2$ (0.4 wt.% Pt)	63	16	62
0.4 wt.% $Pt/Ce_{0.2}Zr_{0.8}O_2$	40	20	5
0.4 wt.% Pt+2.8 $La/Ce_{0.2}Zr_{0.8}O_2$	94	44	44
1.8 wt.% $Pt/Ce_{0.2}Zr_{0.8}O_2$	40	4	1
1.4 wt.% $Pt/Pr_{0.05}(Ce_{0.5}Zr_{0.5})_{0.95}O$		40	
1.4 wt.% $Pt/Pr_{0.3}Ce_{0.35}Zr_{0.35}O_2$	80	30	30
1.4 wt.% $Pt/Gd_{0.3}\ Ce_{0.35}Zr_{0.35}O_2$	60		8
1.4 wt.% $Pt/La_{0.3}\ Ce_{0.35}Zr_{0.35}O_2$	30		4

[1] Feed 7% CH_4 + 3.5% O_2, N_2 balance; [2] Feed 7% CH_4 + 21% H_2O, N_2 balance; [3] Feed 7% CH_4 + 7% CO_2, N_2 balance.

To accelerate preparation of ceria-zirconia based catalysts, an automated workstation was used allowing wet impregnation of La-doped γ-Al_2O_3 by mixed solutions, so series of samples with supported CeZrO layers doped by Pr or Sm and promoted by Pt, Ru, Cu, Cu + Ni were made [113]. The ruthenium-promoted $Ru/Ce_{0.4}Zr_{0.4}Sm_{0.2}\ O_{2-y}/La$-$\gamma$-$Al_2O_3$ catalyst demonstrated the best activity and coking stability in ethanol steam reforming, which was explained by a high mobility and reactivity of oxygen in this sample.

4.2. Catalysts Based on Perovskite Oxides

4.2.1. Reactions of Methane Reforming

A lot of ABO_3 perovskites with partial substitution of La in A sublattice for alkaline-earth (Ca, Sr) or rare-earth (Ce, Pr) cations as well as containing in B sublattice, along with Ni, such metals as Fe, Co, Rh and Ru, were studied in these reactions [30–48,50,51,53–60] as well as in diesel fuel reforming [49,61]. Main attention was paid to the effect of perovskite initial composition on segregated in reaction conditions Ni (or Ni alloys) dispersion, interaction of a metal component with remaining oxide matrix and ability of the latter to activate oxidants and provide oxygen species diffusion to the metal-support interface. Clearly, formation of Ni nanoalloys with Co, Fe, Ru, Rh for reduced perovskites doped in B sublattice prevents coking. For Fe-containing perovskites [28,32,34,39,57] incomplete reduction of Fe cations in reaction conditions helps to stabilize a part of perovskite Ln-Fe-O matrix strongly interacting with segregated nanoalloys. In a similar way, substitution of La in A sublattice for Ce and Pr cations able to change their charge state increases oxygen mobility and reactivity in remaining perovskite matrix, thus improving resistance to coking [32,34,58].

A high and stable performance in MDR without any coking was demonstrated also for $LaMnO_3$-based perovskite with *in-situ* exsolved Ni nanoparticles strongly interacting with perovskite surface layers [114].

Detailed review on methane dry reforming over perovskite derived catalysts is presented in [39], where basically the same trends in ensuring high and stable performance related to oxygen mobility and strong interaction of metal nanoparticles with remaining oxide matrix are analyzed.

4.2.2. Ethanol Reforming on Bulk Perovskites

In reactions of oxygenates reforming a high activity and stability of catalysts based on perovskites precursors was shown as well [32,63–67,108–110,115,116]. The same factors—exsolution of Ni-containing nanoalloys from perovskite lattice in reducing conditions, their stabilization by strong interaction with remaining matrix and oxygen transfer to metal-support interface determine their resistance to coking and activity. It has been shown that even Ni/La_2O_3 catalysts obtained from $LaNiO_3$ have a higher stability to coking than catalysts of the same composition prepared by traditional impregnation.

In our work [32], $LnFe_{0.7-x}M_xNi_{0.3}O_3$ perovskites (Ln = La, Pr, M = Mn, Ru, x = 0–0.3) as precursors of robust catalysts were prepared by modified Pechini route. Before testing in EtOH steam reforming in concentrated feed 10% EtOH+40% H_2O + N_2 at contact time 70 ms they were either reduced by H_2 at 850 °C or pretreated in the reaction feed. The highest activity was shown by Pr and Ru-containing samples, with specific rate constants at 800 °C varying from 4 to 8 $s^{-1}m^{-2}$, and their performance was stable due to presence of only a trace admixture of ethylene in products, which is usually responsible for coking in this reaction.

4.2.3. Perovskite-Fluorite Nanocomposites

Since perovskite + fluorite nanocomposites possess a higher oxygen mobility compared to separate phases [106], a lot of research was devoted to their design not only as cathodes of solid oxide fuel cells (SOFC) and materials for oxygen separation membranes [35,87], but also as active components of structured catalysts for fuels reforming and SOFC anodes operating in the internal reforming mode [31,32,35,72,73,77,87,117–120]. As fluorites doped ceria or zirconia were employed. Different methods of synthesis were used for preparation, the most efficient was modified Pechini method when prepared nanocrystalline fluorite was dispersed in a polymeric precursor of perovskite followed by its decomposition and calcinations [31], while (100 − x) wt.% $LnFeNi_{0.3}$ + x wt.% $Ce_{0.9}Gd_{0.1}O_{2-\delta}$ (GDC) (x = 5–50 wt.%) nanocomposites were prepared from polymeric precursors containing all cations. Their testing in MDR (feed 10% CH_4 +10% CO_2 in He, contact time 15 ms) revealed that at 800 °C specific first-order rate constant k ($s^{-1}m^{-2}$) goes through the maximum at x = 10 wt.% (k = 5.5), its values being nearly the same at x = 0 (k = 2.5) and 20% (k = 2.0). Such trend is apparently explained by the positive effect of perovskite structure disordering at a low content of fluorite dopant followed by subsequent surface blocking by Ce and Gd cations due to their segregation as a result of their bigger sizes.

Nanocomposite (10 wt.% Ni + 2 wt.% Ru)/($La_{0.8}Pr_{0.2}Mn_{0.2}Cr_{0.8}O_3$ + 10 wt.% YSZ) [72,73,77] showed a high and stable performance in MDR. Its specific activity is close to that of (Ru + Ni)/SmPrCeZrO catalyst (Figure 5a), and it provides a high methane conversion into syngas at high temperatures even at short contact times (Figure 5b).

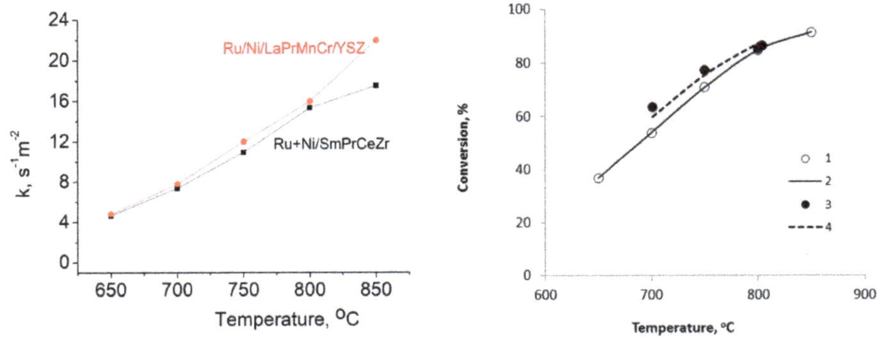

(a) (b)

Figure 5. (**a**) Temperature dependence of specific rate constants for methane dry reforming on fractions of (Ru+Ni)/SmPrCeZrO and (Ru+Ni)/LaPrMnCr/YSZ catalysts. Feed 7% CH_4 + 7% CO_2, contact time 15 ms; (**b**) Temperature dependence of CH_4 conversion in CH_4 dry reforming on fraction of (Ru+Ni)/(LaPrMnCr+YSZ) catalyst (1, 2; feed 7% CH_4 + 7% CO_2 in He, contact time 15 ms) and on the stack of microchannel plates with this active component (3, 4; feed 20% CH_4 + 20% CO_2 in Ar, contact time 0.4 s). 1,3-experimental data, 2,4-fitting.

4.3. Catalysts Based on Spinels

$Co_{1.8}Mn_{1.2}O_4$, $Ni_{0.33}Co_{1.33}Mn_{1.33}O_4$ and $Ni_{0.6}Co_{1.2}Mn_{1.2}O_4$ catalysts were prepared by thermal decomposition of nitrates and studied in ethanol steam reforming reaction. The highest activity was found for $Ni_{0.6}Co_{1.2}Mn_{1.2}O_4$ catalyst, which is explained by a high content of mixed Ni-Co clusters segregated at the surface in reaction conditions along with the highest oxygen mobility (Table 1) preventing coking [68].

Mixed manganese-chromium oxides $Mn_xCr_{3-x}O_4$ (x = 0.3–2.7) prepared by modified Pechini route and promoted by 2 wt.% Ni + 2 wt.% Ru were studied in steam reforming of ethanol [69]. The density of surface metal sites was estimated by CO pulse chemisorption. The highest conversion of ethanol and yield of syngas were obtained for stoichiometric $MnCr_2O_4$ composition. Turnover frequencies (TOF, s^{-1}) at 500 °C vary in the row (Ru+Ni)/$Mn_{0.3}Cr_{2.7}O_4$ (0.88) < (Ru+Ni)/Mn_2CrO_4 (1.13) < (Ru+Ni)/$Mn_{2.7}Cr_{0.3}O_4$ (1.36) < (Ru+Ni)/$MnCr_2O_4$ (3.28). In this series of catalysts, the highest oxygen diffusion coefficient was found for (Ru+Ni)/$Mn_{2.7}Cr_{0.3}O_4$ sample with the excess of manganese. This implies that namely metal-support interaction provides the highest TOF for (Ru+Ni) $MnCr_2O_4$ catalyst, while sufficient oxygen mobility (Table 1) ensures coking stability. Even in concentrated feed 10% C_2H_5OH + 40% H_2O in N_2 at short contact time 70 ms concentration of byproduct CH_4 was only 2% at 700 °C, while only trace admixtures of usual byproducts such as ethylene and acetaldehyde were observed [69]. This means that such known intermediates as ethoxy complexes and acetaldehyde are rapidly transformed into syngas due to a high concentration of reactive oxygen forms and metal sites [104,121].

Testing (Ru + Ni)/$MnCr_2O_4$ catalyst in the autothermal reforming of glycerol (feed 10.9% $C_3H_8O_3$ + 9.5% O_2 + 44.5% H_2O + 35.1% N_2, contact time ~40 ms) also revealed its high efficiency and stable performance [73]. Thus, already at 750 °C complete conversion of glycerol was achieved with the content of main byproducts CH_4 and C_2H_4 less than 2% and syngas yield approaching equilibrium.

4.4. Catalysts Based on High Surface Area Supports

Since specific surface area of perovskites, fluorites and spinel is lower than that of traditional supports such as γ-alumina, silica, zeolites, etc., a lot of research was devoted to design of catalysts for fuels reforming where active components—metals or their combination with reactive oxides are loaded on high surface area supports [12,28,39,72,75,77–80,90,122–133]. Since acidity of supports is well known to be responsible for coking, the best results for activity and stability of Ni-loaded catalysts in fuels reforming, as expected, were obtained in the case of Mg-doped alumina [72,90,129,131,133] or $MgAl_2O_4$ [12,28,70,75,78–80,123,124,126,128,129].

In our studies [72,73,77,90,104,129], both Mg–γ-Al_2O_3 supports prepared by supporting Mg on γ-Al_2O_3 as well as mesoporous Mg–γ-Al_2O_3 and $MgAl_2O_4$ supports prepared by EISA method with Pluronic P123 were used. Ni was supported either by impregnation or during one-pot synthesis. Nanocomposite active components comprised of mixed oxides with perovskite, fluorite and spinel structures described above were supported by impregnation and then promoted by Ni+Ru. Catalysts were tested in dry reforming of methane and steam/autothermal reforming of ethanol as fractions as well as layers supported on small plates of heat-conducting substrates. Even for catalysts containing only Ni prepared by one-pot synthesis a high and stable performance in ethanol steam reforming was demonstrated due to a high dispersion of metal and acidity suppression [129]. The most promising active components comprised of mesoporous $MgAl_2O_4$ with supported $PrNi_{0.9}Ru_{0.1}O_3$, $MnCr_2O_4$ or $Ce_{0.35}Zr_{0.35}Pr_{0.3}O_3$ promoted with Ni+Ru demonstrated a high efficiency and resistance to coking in dry reforming of natural gas and autothermal reforming of such fuels as ethanol and ethyl acetate [78]. In a similar way, Ni/$CeZrO_2$/$MgAl_2O_4$ catalyst revealed a high activity and coking stability in tri-reforming of methane due to a small size of Ni nanoparticles and moderate basicity of support [28]. In steam reforming of methane [80] a high activity of Rh-Ni/$MgAl_2O_4$ washcoated FeCrAlloy honeycomb monolith was observed and explained by a high active metal dispersion as well as absence of heat and mass transfer limitations.

5. Mechanisms of Main Reactions

5.1. Partial Oxidation and Dry Reforming of Methane

Mechanism of these reactions was studied by using such methods as SSITKA, kinetic transients and pulse techniques (pulse microcalorimetry, pulse studies in flow conditions including those carried out in vacuum systems called TAP) [19,72,82,87,101,102,108,109,111,112,134–137]. For majority of efficient catalysts based on mixed oxides with a high mobility and storage capacity of reactive oxygen species with supported Pt group metals and/or Ni mechanism can be described by so called bifunctional type, where molecules of oxidants are activated on support vacancies producing oxygen species (and CO in the case of CO_2), while methane is activated on metal sites by C-H bond rupture (rate-limiting stage). These steps are conjugated by rapid transfer of surface oxygen species to the metal sites where they interact with CH_x fragment transforming them into CO and H_2. A typical feature of such redox scheme in the case of MDR is the same degree of methane conversion and syngas selectivity in pulses containing only CH_4 or CH_4 + CO_2, as well as identical CO_2 conversion into CO in mixed and CO_2—containing pulses. Simplified scheme of methane dry reforming and sequence of elementary steps successfully applied for modeling of transient over Ni-Ru-$Sm_{0.15}Pr_{0.15}Ce_{0.35}Zr_{0.35}O_2$ catalyst [72] are shown below:

$CH_4 + 2s \rightarrow CH_{3s} + H_s$ $CH_{3s} + s \rightarrow CH_{2s} + H_s$ $CH_{2s} + s \rightarrow CH_s + H_s$ $CH_s + s \rightarrow C_s + H_s$	Methane activation and decomposition
$O_{ls} + s \Leftrightarrow O_s + V_{Os}$ $C_s + O_s \rightarrow CO_s + s$ $CO_s \Leftrightarrow CO + s$	CO formation from CH_4
$H_s + H_s \Leftrightarrow H_{2s} + s$ $H_{2s} \Leftrightarrow H_2 + s$	H_2 formation
$O_{ls} + s \Leftrightarrow O_s + V_{Os}$ $H_{sm} + O_s \Leftrightarrow OHs + s$ $OH_s + H_s \Leftrightarrow H_2O_s + s$ $H_2O_s \Leftrightarrow H_2O + s$	Support reduction
$CO_2 + V_{os} \Leftrightarrow CO + O_{ls}$	Support oxidation and CO formation from CO_2

Here $\rightarrow$—denotes irreversible steps; $\Leftrightarrow$—reversible steps; s—surface sites; O_{ls}—lattice oxygen atoms; V_{Os}—lattice oxygen vacancy, sm—surface metal sites

For catalysts comprised of Pt/doped CeZrO oxides direct route of CH_4 partial oxidation into syngas, which generates CO and H_2 even in the presence of oxygen, was reliably demonstrated in our studies. It is explained by stabilization of Pt cations due to metal-support interaction and fast migration of oxygen species activated on support to Pt. Moreover, Pt cations are less efficient in oxidation of CO and H_2, while being more efficient in C-H bond activation in CH_4.

Basic scheme of methane partial oxidation [109] is presented here:

(1) $O_2 + 2Pt \Leftrightarrow 2PtO$
(2) $H_2O + z \Leftrightarrow H_2 + zO$
(3) $zO + Pt \Leftrightarrow PtO + z$ spillover
(4) $O_{bulk} + z \Leftrightarrow zO + V_O$ (bulk) diffusion
(5) $CH_4 + PtO \rightarrow CO + 2H_2 + Pt$
(6) $CH_4 + 4PtO \rightarrow CO_2 + 2H_2O + 4Pt$

(7) $CO + PtO \rightarrow CO_2 + Pt$
(8) $H_2 + PtO \rightarrow H_2O + Pt$
(9) $Pt + CO + H_2O \Leftrightarrow CO_2 + H_2 + Pt$

Clearly this mechanism is impossible in the case of such supported metals as Ru, Pd, Ni, Co, etc., which can only combust methane, CO and H_2 in the oxidized state. For these catalysts indirect scheme of methane partial oxidation is realized, in which all oxygen is consumed for methane combustion in the inlet part of the catalytic layer, while syngas is generated via steam and dry reforming of methane in the main part of catalytic layer where O_2 is absent in the gas phase and supported metals are in the reduced state [82].

These unique features of Pt cations were also reflected in specificity of MDR relaxation after contact of oxidized Pt/PrCeZrO catalyst with reaction feed, where both CH_4 and CO_2 conversions decline with time-on-stream apparently caused by the catalyst progressing reduction [19,136,137]. Mathematical modeling using scheme of methane dry reforming mechanism given below allowed to describe such transients taking into account a high efficiency of Pt cations in CH_4 activation, while CO_2 transformation occurs via carbonates adsorbed on Pt^{n+}–Pr^{4+}–O oxidized sites:

(1) $CO_2 + [PtO] \leftrightarrow [PtCO_3]$
(2) $CH_4 + [PtCO_3] \rightarrow 2\,CO + 2\,H_2 + [PtO]$
(3) $CH_4 + [PtO] \rightarrow CO + 2\,H_2 + [Pt]$
(4) $[Pt] + [O_s] \leftrightarrow [PtO] + [V_s]$
(5) $CO_2 + [V_s] \rightarrow CO + [O_s]$
(6) $H_2 + [PtO] \rightarrow H_2O + [Pt]$

Here [PtO] and [Pt] denote the oxidized and vacant Pt-centers, $[PtCO_3]$ is the carbonate complex, $[O_s]$ and $[V_s]$ are the oxidized and vacant sites inside the lattice layer of Pt/PrSmCeZrO complex oxide composite,

Note that for catalysts with all other supported transition and precious metals in oxidized state catalytic activity in MDR is negligible and begins to increase to the steady-state value in the process of catalysts reduction by reaction mixture at sufficiently high temperatures.

5.2. Reactions of Ethanol Transformation into Syngas

For steady-state of catalysts based on oxides with high oxygen mobility and reactivity in ethanol reforming reactions pulse studies revealed that ethanol conversion and products selectivities in pulses containing reaction mixture or only ethanol are practically the same [103,104,121]. This proves realization of step-wise redox mechanism of these reactions. FTIRS studies combined with SSITKA identified main intermediates of ethanol conversion into syngas, such as ethoxy species and acetaldehyde and estimated rate constants of their transformation, while acetates were shown to be spectators. The rate-limiting step is the cleavage of C–C bond in ethoxy species fixed on rare-earth or transition metal cations due to incorporation of terminal oxygen species located on neighboring Me sites as supported by DFT calculations for the (001) face of $MnCr_2O_4$ spinel doped by Ru [104]. Kinetic scheme of the reaction of ethanol partial oxidation used in analysis of SSITKA data is as follows:

(1) $O_2 + 2[Z] \rightarrow 2\,[ZO]$
(2) $C_2H_5OH + [ZO] \rightarrow C_2H_4O + H_2O + [Z]$
(3) $C_2H_4O + (2{+}n)[ZO] \rightarrow CO + CO_2 + nH_2O + (2{-}n)H_2 + (2{+}n)[Z]$, $n = 0–2$
(4) $CO + [ZO] \leftrightarrow CO_2 + [Z]$
(5) $H_2 + [ZO] \leftrightarrow H_2O + [Z]$

where [ZO] and [Z] correspond to oxidized and reduced site of catalysts surface, respectively, without any differentiation of the nature of active sites related to different transition metal cations (Mn, Cr) and metal atoms (Ni, Ru) at this level of analysis similar to that for catalysts based on bulk $MnCr_2O_4$ [104].

6. Development of Structured Catalysts

For design of such catalysts substrates made of metals, alloys and cermets were used [72,73,77,82,84,86,87,90,91,98,117]. Detonation spraying [138] was applied to cover thin foils, microchannel plates or gauzes by protective nonporous layer of alumina or zirconia. Known procedures of stacking and winding were applied to produce monolithic substrates from these constituents. Foam substrates comprised of NiAl alloys, ceramics, etc. were prepared via polyurethane foam duplication followed by required mechanical and thermal treatment [73,86]. These substrates were covered by catalytic layers using suspensions of active components in isopropanol. Structured catalysts were tested in pilot reactors in processes of natural gas, liquid fuels and biofuels reforming/autothermal reforming under realistic conditions. For best catalysts on substrates with a high thermal conductivity a high syngas yield was provided even at short contact times due to more uniform temperature profile along the catalysts' length [73,82,84]. In the autothermal reforming mode it helps to transfer heat generated in the inlet part of monolith due to exothermal combustion of a part of fuel into the following parts where O_2 in the gas phase is absent, so endothermic reactions of steam and dry reforming occur. For strongly endothermic reaction of MDR this positive effect of a high thermal conductivity of substrate is demonstrated by the identical temperature dependence of methane conversion in the layer of catalyst fraction for diluted feed and for the stack of microchannel plates with the same active component even in more concentrated feed (Figure 5b). In dry reforming of real natural gas (NG) containing up to 6% of C_2–C_4 alkanes also high and stable performance of structured catalyst was demonstrated for concentrated feed (Figure 6).

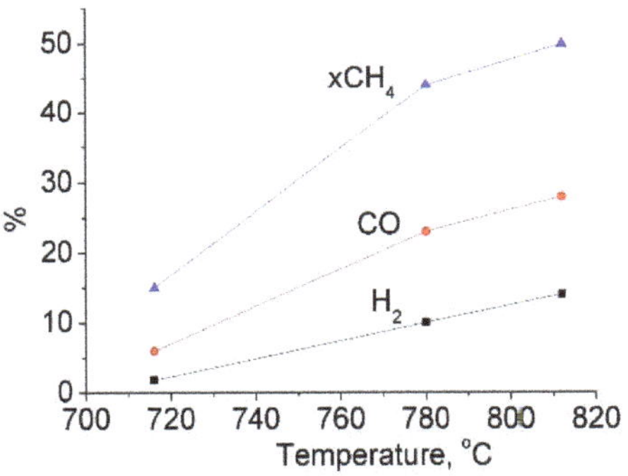

Figure 6. Temperature dependence of CH_4 conversion and CO/H_2 concentrations in NG dry reforming on the package of 5 Fechraloy microchannel plates with 1% NiO+1% Ru /SmPrCeZrO active component. Feed 50% CO_2 +40% NG +N_2, contact time 0.1 s.

This advantage of structured catalysts on heat-conducting metal/cermet substrates allowed to carry out efficient transformation of a lot of fuels including gasoline and diesel into syngas via partial oxidation and steam/autothermal reforming [82–87,117,139–142]. Reforming of biofuels such as acetone, ethyl acetate and glycerol is known to be accompanied by gas-phase reactions yielding ethylene, which is easily transformed into coke on catalysts. However, structured catalysts with nanocomposite active components demonstrated high and stable performance in the autothermal reforming of such biofuels as ethanol, acetone, ethyl acetate (Figure 7) and glycerol (Figure 8) with fuels content up to 25% and O_2 content in the mixture with steam up to 20% [73,82,86,87,117]. Note that Figures 7 and 8 show a high yield of syngas achieved at very short contact times, which is provided by a high efficiency of active components supported on mesoporous $MgAl_2O_4$ or Mg-doped –γ-Al_2O_3. Even anisole (content in the feed up to 10%), sunflower oil (content up to 0.7%) [82,117] and turpentine oil (commercial bio-oil mainly containing $C_{10}H_{16}$ and $C_{10}H_{18}O$ components, such as α-pinen, α-terpineol, etc., content in the feed up to 6% [90]) (Figure 9) were successfully converted into syngas in the autothermal reforming on developed structured catalysts. Another important problem in design of efficient syngas generators, especially for the small-scale application, is related to the heat management, since endothermic reactions of fuels steam and dry reforming require extensive preheat of inlet streams. This can be dealt with by using the exit stream for preheating the inlet feed in specially designed heat exchangers conjugated with catalytic reactors. Another aspect of energy efficiency problem solution is bound with possibility to conjugate exothermal partial oxidation of methane into syngas with endothermal processes of biofuels steam/dry reforming. These problems were solved in design of a radial-type reactor equipped with the heat exchanger described in detail in our previous work [82,117]. Here a cylindrical stack of catalytic microchannel washers was wrapped by gauze sheets with supported active components as well as with microspherical catalysts loaded between gauzes. The feed enters the central part of the stack of washers and flows in the radial direction. The reformed gas is collected into a plenum around the catalyst arrangement and exited from a single pipe, while inlet feed is heated by passing through the heat exchanger situated around the reactor as an outer shell. Such design made it possible to efficiently carry out partial oxidation of a mixture of natural gas and liquid fuel (ethanol, ethyl acetate, turpentine oil) to syngas at high flow rates (up to 40,000 h^{-1}) with preheating the mixture up to 50–100 °C at the reactor inlet (Figures 10–12).

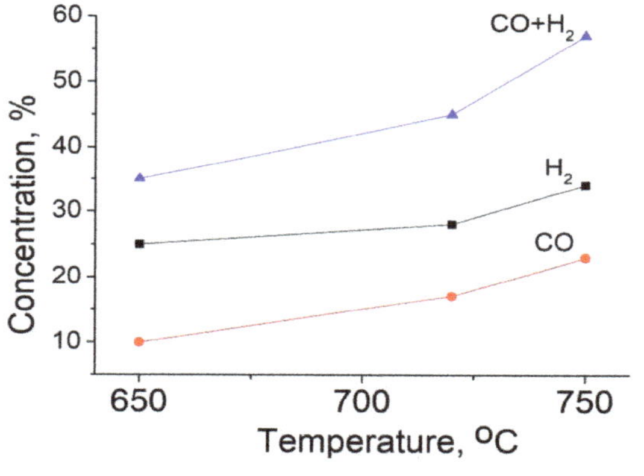

Figure 7. Temperature dependence of product concentrations in ATR of ethylacetate over the monolithic honeycomb catalyst comprised of FeCrAl gauzes loaded with NiRu/CeZrPrO on mesoporous $MgAl_2O_4$. Feed 30% EtAc + 60% H_2O + 8% O_2, N_2—balance, contact time 0.1 s.

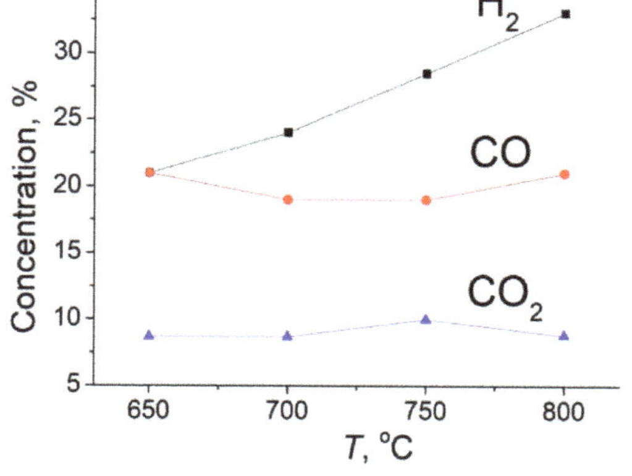

Figure 8. Temperature dependence of products concentration in the process of glycerol autothermal reforming on microchannel CrAlO substrate with supported (Ni + Ru)/$MnCr_2O_4$/10% MgO–γ-Al_2O_3 active component. Feed 15% O_2 + 22% glycerol + 22% H_2O + N_2, contact time 0.06 s.

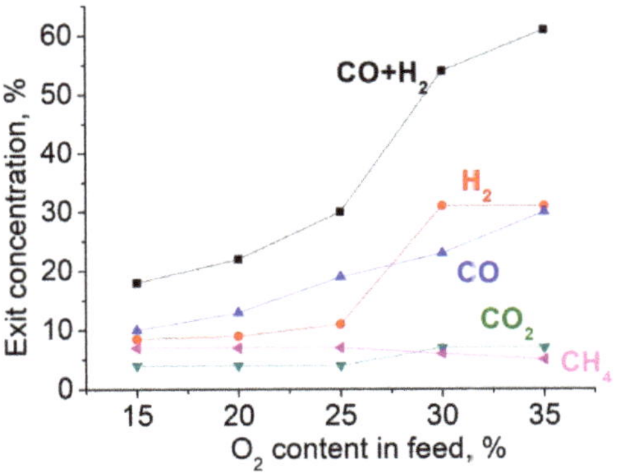

Figure 9. Dependence of products concentrations on oxygen content in the feed in the process of turpentine oil autothermal reforming on 3 4% LaNiPt/La–Ce–Zr–O/Fechraloy thin foil honeycomb at 800 °C. Feed 7% turpentine + 40% H_2O+O_2+N_2, contact time 0.5 s.

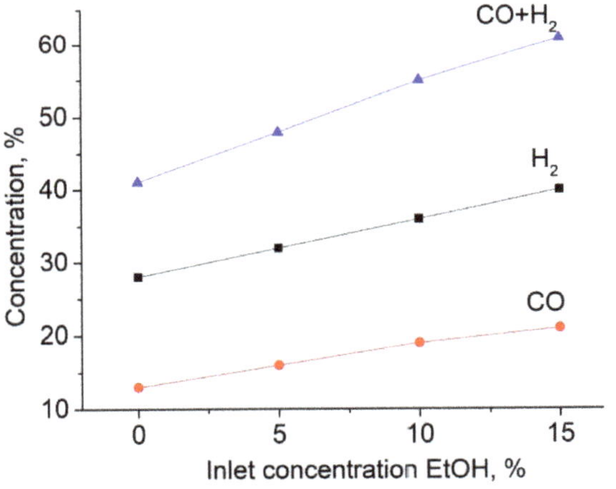

Figure 10. Dependence of products concentration on ethanol content in feed 24% NG + 16–25% O_2 + N_2 + EtOH in the process of autothermal reforming in the radial reactor with the heat exchanger. Flow rate 2.5 m^3/h, LaNi(Pt)O_3/La$_{0.1}$Ce$_{0.45}$Zr$_{0.45}$O$_2$ active component.

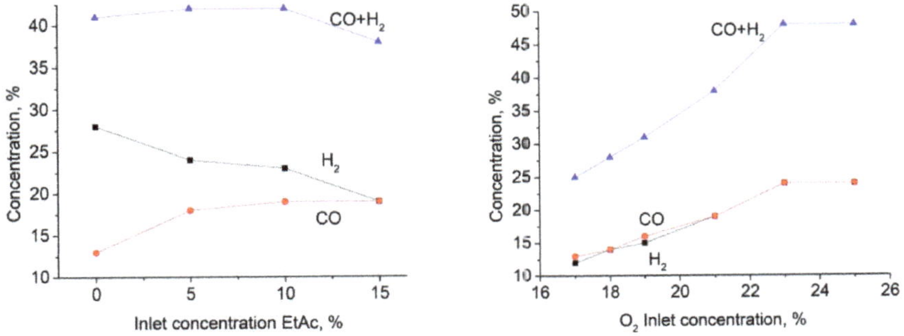

(**a**) (**b**)

Figure 11. Dependence of products concentration on ethyl acetate (**a**) or O_2 (**b**) content in the autothermal reforming of the mixture of natural gas and ethyl acetate in radial reactor with the heat exchanger and LaNi(Pt)O_3/La$_{0.1}$Ce$_{0.45}$Zr$_{0.45}$O$_2$ active component. Feed 20% NG + 18% O_2 +N_2 +EtAc (**a**) or 10% NG + O_2 + N_2 + 15% EtAc (**b**), flow rate 2.5 m^3/h.

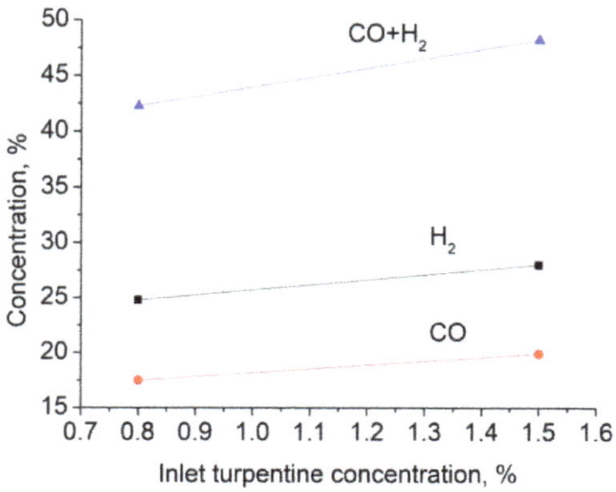

Figure 12. Dependence of products concentration on turpentine content in feed 20% NG +21% O_2 +N_2 +turpentine oil in the process of autothermal reforming in the radial reactor with the heat exchanger. Flow rate 2.5 m^3/h, $LaNi(Pt)O_3/La_{0.1}Ce_{0.45}Zr_{0.45}O_2$ active component.

7. Conclusions

Pechini method and synthesis in supercritical alcohols allowed to provide uniformity of the spatial distribution of elements in nanodomains of fluorite, perovskite and spinel oxides required for controlling their oxygen mobility and reactivity. Promoting these oxides and their nanocomposites of optimized composition by platinum group metals and nickel allowed to create effective and coking-resistant catalysts for biogas and biofuels transformation into syngas. Due to developed metal-support interface and strong metal–support interactions bifunctional scheme of reaction mechanism is realized with activation of oxidants (O_2, CO_2, H_2O) on the oxide support sites, while fuel molecules are activated on metal centers. Fast diffusion of surface oxygen species to the metal–oxide interface provides conjugation of these steps resulting in efficient syngas generation and coking suppression. Preparation of mesoporous supports comprised of $MgAl_2O_4$ or Mg-doped alumina with enhanced basicity allowed to design core-shell systems with a high working area of supported catalytic nanocomposite layers, increase their activity, thermal and coking stability and decrease content of rare-earth elements. Structured heat-conducting substrates loaded with optimized active components provide efficient heat- and mass-transfer in reactors for biofuels transformation into syngas. Pilot reactors with internal heat exchangers permit efficient operation in the autothermal mode on the mixture of natural gas, air and real biofuels such as ethanol, glycerol and turpentine oil with the inlet temperature 50–100°C and GHSV up to 40,000 h^{-1}.

Author Contributions: Writing—original draft preparation, V.S., N.E., M.S. and N.M.; writing—review and editing, V.S. All authors have read and agreed to the published version of the manuscript.

Funding: Design and studies of structured catalysts were carried out in the framework of the budget project of the Boreskov Institute of Catalysis, Siberian Branch, Russian Academy of Sciences (project AAAA-A21-121011390007-7). The authors acknowledge support from the Russian Science Foundation according to project no. 18-73-10167 for the synthesis of catalysts in supercritical environment fluids and their characterization.

Institutional Review Board Statement: Not applicable.

Informed Consent Statement: Not applicable.

Acknowledgments: Authors are glad to acknowledge efficient international collaboration in frames of INTAS and FP7 Projects OCMOL and BIOGO in this area of research.

Conflicts of Interest: The authors declare no conflict of interest.

References

1. Bepari, S.; Kuila, D. Steam reforming of methanol ethanol and glycerol over nickel-based catalysts—A review. *Int. J. Hydrogen Energy* **2020**, *45*, 18090–18113. [CrossRef]
2. Chattahathan, S.A.; Adhikari, A.; Abdoulmoumine, N. A review on current status of hydrogen production from bio-oil. *Renew. Sustain. Energy Rev.* **2012**, *16*, 2366–2372. [CrossRef]
3. Aziz, M.A.A.; Setiabudi, H.D.; Teh, L.P.; Annuar, N.H.R.; Jalil, A.A. A review of heterogeneous catalysts for syngas production via dry reforming. *J. Taiwan Inst. Chem. Eng.* **2019**, *101*, 139–158. [CrossRef]
4. Chen, J.; Sun, J.; Wang, Y. Catalysts for steam reforming of bio-oil: A review. *Ind. Eng. Chem. Res.* **2016**, *56*, 4627–4637. [CrossRef]
5. Dou, B.; Song, Y.; Wang, C.; Chen, H.; Xu, Y. Hydrogen production from catalytic steam reforming of biodiesel byproduct glycerol: Issues and challenges. *Renew. Sustain. Energy Rev.* **2014**, *30*, 950–960. [CrossRef]
6. Anil, C.; Modak, J.M.; Madras, G. Syngas production via CO_2 reforming of methane over noble metal (Ru, Pt, and Pd) doped $LaAlO_3$ perovskite catalyst. *Mol. Catal.* **2020**, *484*, 110805. [CrossRef]
7. Pakhare, D.; Spivey, J. A review of dry (CO_2) reforming of methane over noble metal catalysts. *Chem. Soc. Rev.* **2014**, *43*, 7813–7837. [CrossRef] [PubMed]
8. Yentekakis, I.V.; Panagiotopoulou, P.; Artemakis, G. A review of recent efforts to promote dry reforming of methane (DRM) to syngas production via bimetallic catalyst formulations. *Appl. Catal. B Environ.* **2021**, *296*, 120210. [CrossRef]
9. Bae, J.; Lee, S.; Kim, S.; Oh, J.; Choi, S.; Bae, M.; Kang, I.; Katikaneni, S.P. Liquid fuel processing for hydrogen production: A review. *Int. J. Hydrogen Energy* **2016**, *41*, 19990–20022. [CrossRef]
10. Pham, T.T.P.; Ro, K.S.; Chen, L.; Mahajan, D.; Siang, T.J.; Ashik, U.P.M.; Hayashi, J.; Minh, D.H.; Vo, D.-V.N. Microwave-assisted dry reforming of methane for syngas production: A review. *Environ. Chem. Lett.* **2020**, *18*, 1987–2019. [CrossRef]
11. Gao, X.; Ashok, J.; Kawi, S. Smart designs of anti-coking and anti-sintering Ni-based catalysts for dry reforming of methane: A recent review. *Reactions* **2020**, *1*, 162–194. [CrossRef]
12. Cho, E.; Lee, Y.-H.; Kim, H.; Jang, E.J.; Kwak, J.H.; Lee, K.; Ko, C.H.; Yoon, W.L. Ni catalysts for dry methane reforming prepared by A-site exsolution on mesoporous defect spinel magnesium aluminate. *Appl. Catal. A Gen.* **2020**, *602*, 117694. [CrossRef]
13. Sadykov, V.A.; Kuznetsova, T.G.; Alikina, G.M.; Frolova, Y.V.; Lukashevich, A.I.; Muzykantov, V.S.; Rogov, V.A.; Batuev, L.C.; Kriventsov, V.V.; Kochubei, D.I.; et al. Ceria-based fluorite-like oxide solid solutions promoted by precious metals as catalysts of methane transformation into syngas. In *New Topics in Catalysis Research*; McReynolds, D.K., Ed.; Nova Science Publishers: Hauppauge, NY, USA, 2007; pp. 97–196.
14. Charisiou, N.D.; Iordanidis, A.; Polychronopoulou, K.; Yentekakis, I.V.; Goula, M.A. Studying the stability of Ni supported on modified with CeO_2 alumina catalysts for the biogas dry reforming reaction. *Mater. Today Proc.* **2018**, *5*, 27607–27616. [CrossRef]
15. Faria, E.C.; Neto, R.C.R.; Colman, R.C.; Noronha, F.B. Hydrogen production through CO_2 reforming of methane over $Ni/CeZrO_2/Al_2O_3$ catalysts. *Catal. Today* **2014**, *228*, 138–144. [CrossRef]
16. Grabchenko, M.; Pantaleo, G.; Puleo, F.; Kharlamova, T.S.; Zaikovskii, V.I.; Vodyankina, O.; Liotta, L.F. Design of Ni based catalysts supported over binary La-Ce oxides: Influence of La/Ce ratio on the catalytic performances in DRM. *Catal. Today* **2021**. [CrossRef]
17. Vagia, E.C.; Lemonidou, A.A. Investigations on the properties of ceria–zirconia-supported Ni and Rh catalysts and their performance in acetic acid steam reforming. *J. Catal.* **2010**, *269*, 388–396. [CrossRef]
18. Sadykov, V.A.; Kuznetsova, T.G.; Frolova, Y.V.; Alikina, G.M.; Lukashevich, A.I.; Rogov, V.A.; Muzykantov, V.S.; Pinaeva, L.G.; Sadovskaya, E.M.; Ivanova, Y.A.; et al. Fuel-rich methane combustion: Role of the Pt dispersion and oxygen mobility in a fluorite-like complex oxide support. *Catal. Today* **2006**, *117*, 475–483. [CrossRef]
19. Mirodatos, C.; van Veen, A.C.; Pokrovskaya, S.A.; Chumakova, N.A.; Sazonova, N.N.; Sadykov, V.A. Modeling of transient studies on the reaction kinetics over catalysts with lattice oxygen mobility: Dry reforming of CH_4 over a Pt/PrCeZrO catalyst. *Chem. Eng. J.* **2018**, *343*, 530–543. [CrossRef]
20. Smirnova, M.Y.; Pavlova, S.N.; Krieger, T.A.; Bespalko, Y.N.; Anikeev, V.I.; Chesalov, Y.A.; Kaichev, V.V.; Mezentseva, N.V.; Sadykov, V.A. The synthesis of $Ce_{1-x}Zr_xO_2$ oxides in supercritical alcohols and catalysts for carbon dioxide reforming of methane on their basis. *Russ. J. Phys. Chem. B* **2017**, *11*, 1–10. [CrossRef]
21. Simonov, M.; Bespalko, Y.; Smal, E.; Valeev, K.; Fedorova, V.; Krieger, T.; Sadykov, V. Nickel-containing ceria-zirconia doped with Ti and Nb. Effect of support composition and preparation method on catalytic activity in methane dry reforming. *Nanomaterials* **2020**, *10*, 1281. [CrossRef] [PubMed]
22. Bespalko, Y.; Smal, E.; Simonov, M.; Valeev, K.; Fedorova, V.; Krieger, T.; Cherepanova, S.; Ishchenko, A.; Rogov, V.; Sadykov, V. Novel $Ni/Ce(Ti)ZrO_2$ catalysts for methane dry reforming prepared in supercritical alcohol media. *Energies* **2020**, *13*, 3365. [CrossRef]
23. Pavlova, S.; Smirnova, M.; Bobin, A.; Cherepanova, S.; Kaichev, V.; Ishchenko, A.; Selivanova, A.; Rogov, V.; Roger, A.-C.; Sadykov, V. Structural, textural, and catalytic properties of $Ni\text{-}Ce_xZr_{1-x}O_2$ catalysts for methane dry reforming prepared by continuous synthesis in supercritical isopropanol. *Energies* **2020**, *13*, 3728. [CrossRef]
24. Fedorova, V.; Simonov, M.; Valeev, K.; Bespalko, Y.; Smal, E.; Eremeev, N.; Sadovskaya, E.; Krieger, T.; Ishchenko, A.; Sadykov, V. Kinetic regularities of methane dry reforming reaction on nickel-containing modified ceria-zirconia. *Energies* **2021**, *14*, 2973. [CrossRef]

25. Auxéméry, A.; Frias, B.B.; Smal, E.; Dziadek, K.; Philippot, G.; Legutko, P.; Simonov, M.; Thomas, S.; Adamski, A.; Sadykov, V.; et al. Continuous supercritical solvothermal preparation of nanostructured ceria-zirconia as supports for dry methane reforming catalysts. *J. Supercrit. Fluids* **2020**, *162*, 104855. [CrossRef]
26. Smirnova, M.Y.; Bobin, A.S.; Pavlova, S.N.; Ishchenko, A.V.; Selivanova, A.V.; Kaichev, V.V.; Cherepanova, S.V.; Krieger, T.A.; Arapova, M.V.; Roger, A.-C.; et al. Methane dry reforming over Ni catalysts supported on Ce–Zr oxides prepared by a route involving supercritical fluids. *Open Chem.* **2017**, *15*, 412–425. [CrossRef]
27. Kambolis, A.; Matralis, H.; Trovarelli, A.; Papadopoulou, C.H. Ni/CeO_2-ZrO_2 catalysts for the dry reforming of methane. *Appl. Catal. A Gen.* **2010**, *377*, 16–26. [CrossRef]
28. Lino, A.V.P.; Rodella, C.B.; Assaf, E.M.; Assaf, J.M. Methane tri-reforming for synthesis gas production using Ni/$CeZrO_2$/$MgAl_2O_4$ catalysts: Effect of Zr/Ce molar ratio. *Int. J. Hydrogen Energy* **2020**, *45*, 8418–8432. [CrossRef]
29. Shoynkhorova, T.B.; Simonov, P.A.; Potemkin, D.I.; Snytnikov, P.V.; Belyaev, V.D.; Ishchenko, A.V.; Svintsitskiy, D.A.; Sobyanin, V.A. Highly dispersed Rh-, Pt-, Ru/$Ce_{0.75}Zr_{0.25}O_{2-\delta}$ catalysts prepared by sorption-hydrolytic deposition for diesel fuel reforming to syngas. *Appl. Catal. B Environ.* **2018**, *237*, 237–244. [CrossRef]
30. Shahnazi, A.; Firoozi, S. Improving the catalytic performance of $LaNiO_3$ perovskite by manganese substitution via ultrasonic spray pyrolysis for dry reforming of methane. *J. CO_2 Util.* **2021**, *45*, 101455. [CrossRef]
31. Naurzkulova, S.M.; Arapova, M.V.; Ishchenko, A.V.; Krieger, T.A.; Saraev, A.A.; Kaichev, V.V.; Rogov, V.A.; Krasnov, A.V.; Massalimova, B.K.; Sadykov, V.A. Ni–Ru-containing mixed oxide-based composites as precursors for ethanol steam reforming catalysts: Effect of the synthesis methods on the structural and catalytic properties. *Open Chem.* **2021**, *19*, 696–708. [CrossRef]
32. Sadykov, V.A.; Pavlova, S.N.; Alikina, G.M.; Sazonova, N.N.; Mezentseva, N.V.; Arapova, M.V.; Rogov, V.A.; Krieger, T.A.; Ishhenko, A.V.; Gulyaev, R.V.; et al. Perovskite-based catalysts for transformation of natural gas and oxygenates into syngas. In *Perovskite: Crystallography, Chemistry and Catalytic Performance*; Zhang, J., Li, H., Eds.; Nova Science Publishers: Hauppauge, NY, USA, 2013; pp. 1–58.
33. Barros, B.S.; Melo, D.M.A.; Libs, S.; Kiennemann, A. CO_2 reforming of methane over La_2NiO_4/α-Al_2O_3 prepared by microwave assisted self-combustion method. *Appl. Catal. A Gen.* **2010**, *378*, 69–75. [CrossRef]
34. Pavlova, S.; Kapokova, L.; Bunina, R.; Alikina, G.; Sazonova, N.; Krieger, T.; Ishchenko, A.; Rogov, V.; Gulyaev, R.; Sadykov, V.; et al. Syngas production by CO_2 reforming of methane using LnFeNi(Ru)O_3 perovskites as precursors of robust catalysts. *Catal. Sci. Technol.* **2012**, *2*, 2099–2108. [CrossRef]
35. Sadykov, V.A.; Pavlova, S.N.; Kharlamova, T.S.; Muzykantov, V.S.; Uvarov, N.F.; Okhlupin, Y.S.; Ishchenko, A.V.; Bobin, A.S.; Mezentseva, N.V.; Alikina, G.M.; et al. Perovskites and their nanocomposites with fluorite-like oxides as materials for solid oxide fuel cells cathodes and oxygen-conducting membranes: Mobility and reactivity of the surface/bulk oxygen as a key factor of their performance. In *Perovskites: Structure, Properties and Uses*; Borowski, M., Ed.; Nova Science Publishers: Hauppauge, NY, USA, 2010; pp. 67–178.
36. Goldwasser, M.R.; Rivas, M.E.; Pietri, E.; Pérez-Zurita, M.J.; Cubeiro, M.L.; Grivobal-Constant, A.; Leclercq, G. Perovskites as catalysts precursors: Synthesis and characterization. *J. Mol. Catal. A Chem.* **2005**, *228*, 325–331. [CrossRef]
37. Tanaka, H.; Misono, M. Advances in designing perovskite catalysts. *Curr. Opin. Solid State Mater. Sci.* **2001**, *5*, 381–387. [CrossRef]
38. Goldwasser, M.R.; Rivas, M.E.; Pietri, E.; Pérez-Zurita, M.J.; Cubeiro, M.L.; Gingembre, L.; Leclercq, L.; Leclercq, G. Perovskites as catalysts precursors: CO_2 reforming of CH_4 on $Ln_{1-x}Ca_xRu_{0.8}Ni_{0.2}O_3$ (Ln = La, Sm, Nd). *Appl. Catal. A Gen.* **2003**, *255*, 45–57. [CrossRef]
39. Bhattar, D.; Abedin, M.A.; Kanitkar, S.; Spivey, J.J. A review on dry reforming of methane over perovskite derived catalysts. *Catal. Today* **2021**, *365*, 2–23. [CrossRef]
40. Mawdsley, J.R.; Krause, T.R. Rare earth-first-row transition metal perovskites as catalysts for the autothermal reforming of hydrocarbon fuels to generate hydrogen. *Appl. Catal. A Gen.* **2008**, *334*, 311–320. [CrossRef]
41. Batiot-Dupeyrat, C.; Gallego, G.A.S.; Mondragon, F.; Barrault, J.; Tatibouët, J.-M. CO_2 reforming of methane over $LaNiO_3$ as precursor material. *Catal. Today* **2005**, *107–108*, 474–480. [CrossRef]
42. Pereniguez, R.; Gonzalez-DelaCruz, V.M.; Holgado, J.P.; Caballero, A. Synthesis and characterization of a $LaNiO_3$ perovskite as precursor for methane reforming reactions catalysts. *Appl. Catal. B Environ.* **2010**, *93*, 346–353. [CrossRef]
43. Moradi, G.R.; Rahmanzade, M.; Sharifni, S. Kinetic investigation of CO_2 reforming of CH_4 over La–Ni based perovskite. *Chem. Eng. J.* **2010**, *162*, 787–791. [CrossRef]
44. Nam, J.W.; Chae, H.; Lee, S.H.; Jung, H.; Lee, K.-Y. Methane dry reforming over well-dispersed Ni catalyst prepared from perovskite-type mixed oxides. *Stud. Surf. Sci. Catal.* **1998**, *119*, 843–848.
45. Valderrama, G.; Goldwasser, M.R.; Urbina de Navarro, C.; Tatibouët, J.M.; Barrault, J.; Batiot-Dupeyrat, C.; Martınez, F. Dry reforming of methane over Ni perovskite type oxides. *Catal. Today* **2005**, *107–108*, 785–791. [CrossRef]
46. Rynkowski, J.; Samulkiewicz, P.; Ladavos, A.K.; Pomonis, P.J. Catalytic performance of reduced $La_{2-x}Sr_xNiO_4$ perovskite-like oxides for CO_2 reforming of CH_4. *Appl. Catal. A Gen.* **2004**, *263*, 1–9. [CrossRef]
47. Chettapongsaphan, C.; Charojrochkul, S.; Assabumrungrat, S.; Laosiripojana, N. Catalytic H_2O and CO_2 reforming of CH_4 over perovskite-based $La_{0.8}Sr_{0.2}Cr_{0.9}Ni_{0.1}O_3$: Effects of pre-treatment and co-reactant/CH_4 on its reforming characteristics. *Appl. Catal. A Gen.* **2010**, *386*, 194–200. [CrossRef]
48. Valderrama, G.; Kiennemann, A.; Goldwasser, M.R. La-Sr-Ni-Co-O based perovskite-type solid solutions as catalyst precursors in the CO_2 reforming of methane. *J. Power Sources* **2010**, *195*, 1765–1771. [CrossRef]

49. Mota, N.; Álvarez-Galván, M.C.; Al-Zahrani, S.M.; Navarro, R.M.; Fierro, J.L.G. Diesel fuel reforming over catalysts derived from $LaCo_{1-x}Ru_xO_3$ perovskites with high Ru loading. *Int. J. Hydrogen Energy* **2012**, *37*, 7056–7066. [CrossRef]
50. Gallego, G.S.; Batiot-Dupeyrat, C.; Barrault, J.; Florez, E.; Mondragon, F. Dry reforming of methane over $LaNi_{1-y}B_yO_{3\pm\delta}$ (B = Mg, Co) perovskites used as catalyst precursor. *Appl. Catal. A Gen.* **2008**, *334*, 251–258. [CrossRef]
51. Valderrama, G.; Kienneman, A.; Goldwasser, M.R. Dry reforming of CH_4 over solid solutions of $LaNi_{1-x}Co_xO_3$. *Catal. Today* **2008**, *133*, 142–148. [CrossRef]
52. Goldwasser, M.R.; Rivas, M.E.; Lugo, M.L.; Pietri, E.; Perez-Zurita, J.; Cubeiro, M.L.; Griboval-Constant, A.; Leclercq, G. Combined methane reforming in presence of CO_2 and O_2 over $LaFe_{1-x}Co_xO_3$ mixed-oxide perovskites as catalysts precursors. *Catal. Today* **2005**, *107–108*, 106–113. [CrossRef]
53. Bian, Z.; Wang, Z.; Jiang, B.; Hongmanorom, P.; Zhong, W.; Kawi, S. A review on perovskite catalysts for reforming of methane to hydrogen production. *Renew. Sustain. Energy Rev.* **2020**, *134*, 110291. [CrossRef]
54. Provendier, H.; Petit, C.; Estournes, C.; Kiennemann, A. Steam reforming of methane on $LaNi_xFe_{1-x}O_3$ ($0\leq x\leq 1$) perovskites. Reactivity and characterization after test. *C. R. Acad. Sci.—Ser. IIC—Chem.* **2001**, *4*, 57–66.
55. Provendier, H.; Petit, C.; Estournes, C.; Libs, S.; Kiennemann, A. Stabilization of active nickel catalysts in partial oxidation of methane to synthesis gas by iron addition. *Appl. Catal. A Gen.* **1999**, *180*, 163–173. [CrossRef]
56. Kapokova, L.; Pavlova, S.; Bunina, R.; Alikina, G.; Krieger, T.; Ishchenko, A.; Rogov, V.; Sadykov, V. Dry reforming of methane over $LnFe_{0.7}Ni_{0.3}O_{3-\delta}$ perovskites: Influence of Ln nature. *Catal. Today* **2011**, *164*, 227–233. [CrossRef]
57. Kim, W.Y.; Jang, J.S.; Ra, E.C.; Kim, K.Y.; Kim, E.H.; Lee, J.S. Reduced perovskite $LaNiO_3$ catalysts modified with Co and Mn for low coke formation in dry reforming of methane. *Appl. Catal. A Gen.* **2019**, *575*, 198–203. [CrossRef]
58. Gallego, G.S.; Marín, J.G.; Batiot-Dupeyrat, C.; Barrault, J.; Mondragon, F. Influence of Pr and Ce in dry methane reforming catalysts produced from $La_{1-x}A_xNiO_{3-\delta}$ perovskites. *Appl. Catal. A Gen.* **2009**, *369*, 97–103. [CrossRef]
59. Rivas, M.E.; Fierro, J.L.G.; Goldwasser, M.R.; Pietri, E.; Pérez-Zurita, M.J.; Griboval-Constant, A.; Lecler, G. Structural features and performance of $LaNi_{1-x}Rh_xO_3$ system for the dry reforming of methane. *Appl. Catal. A Gen.* **2008**, *344*, 10–19. [CrossRef]
60. Araujo, G.C.; Lima, S.M.; Assaf, J.M.; Pena, M.A.; Fierro, J.L.G.; Rangel, M.D.C. Catalytic evaluation of perovskite-type oxide $LaNi_{1-x}Ru_xO_3$ in methane dry reforming. *Catal. Today* **2008**, *133–135*, 129–135. [CrossRef]
61. Mota, N.; Alvarez-Galvána, M.C.; Navarro, R.M.; Al-Zahrani, S.M.; Goguet, A.; Daly, H.; Zhang, W.; Trunschke, A.; Schlögl, R.; Fierro, J.L.G. Insights on the role of Ru substitution in the properties of $LaCoO_3$-based oxides as catalysts precursors for the oxidative reforming of diesel fuel. *Appl. Catal. B Environ.* **2012**, *113–114*, 271–280. [CrossRef]
62. Chen, H.; Yu, H.; Peng, F.; Yang, G.; Wang, H.; Yang, J.; Tang, Y. Autothermal reforming of ethanol for hydrogen production over perovskite $LaNiO_3$. *Chem. Eng. J.* **2010**, *160*, 333–339. [CrossRef]
63. Liu, J.Y.; Lee, C.C.; Wang, C.H.; Yeh, C.T.; Wang, C.B. Application of nickel–lanthanum composite oxide on the steam reforming of ethanol to produce hydrogen. *Int. J. Hydrogen Energy* **2010**, *35*, 4069–4075. [CrossRef]
64. De Lima, S.M.; da Silva, A.M.; da Costa, L.O.O.; Assaf, J.M.; Jacobs, G.; Davis, B.H.; Mattos, L.V.; Noronha, F.B. Evaluation of the performance of Ni/La_2O_3 catalyst prepared from $LaNiO_3$ perovskite-type oxides for the production of hydrogen through steam reforming and oxidative steam reforming of ethanol. *Appl. Catal. A Gen.* **2010**, *377*, 181–190. [CrossRef]
65. Chen, S.Q.; Wang, H.; Liu, Y. Perovskite La–St–Fe–O (St = Ca, Sr) supported nickel catalysts for steam reforming of ethanol: The effect of the A site substitution. *Int. J. Hydrogen Energy* **2009**, *34*, 7995–8005. [CrossRef]
66. Chen, S.Q.; Li, Y.D.; Liu, Y.; Bai, X. Regenerable and durable catalyst for hydrogen production from ethanol steam reforming. *Int. J. Hydrogen Energy* **2011**, *36*, 5849–5856. [CrossRef]
67. De Lima, S.M.; da Silva, A.M.; da Costa, L.O.O.; Assaf, J.M.; Mattos, L.V.; Sarkari, R.; Venugopale, A.; Noronha, F.B. Hydrogen production through oxidative steam reforming of ethanol over Ni-based catalysts derived from $La_{1-x}Ce_xNiO_3$ perovskite-type oxides. *Appl. Catal. B Environ.* **2012**, *121–122*, 1–9. [CrossRef]
68. Sadovskaya, E.M.; Frolov, D.D.; Goncharov, V.B.; Fedorova, A.A.; Morozov, I.V.; Klyushin, A.Y.; Vinogradov, A.S.; Smal, E.A.; Sadykov, V.A. Mixed spinel-type Ni-Co-Mn oxides: Synthesis, structure and catalytic properties. *Catal. Sustain. Energy* **2016**, *3*, 25–31. [CrossRef]
69. Smal, E.A.; Simonov, M.N.; Mezentseva, N.V.; Krieger, T.A.; Larina, T.V.; Saraev, A.A.; Glazneva, T.S.; Ishchenko, A.V.; Rogov, V.A.; Eremeev, N.F.; et al. Spinel-type $Mn_xCr_{3-x}O_4$-based catalysts for ethanol steam reforming. *Appl. Catal. B Environ.* **2021**, *283*, 119656. [CrossRef]
70. Medeiros, R.L.B.A.; Macedo, H.P.; Melo, V.R.M.; Oliveira, A.A.S.; Barros, J.M.F.; Melo, M.A.F.; Melo, D.M.A. Ni supported on Fe-doped $MgAl_2O_4$ for dry reforming of methane: Use of factorial design to optimize H_2 yield. *Int. J. Hydrogen Energy* **2016**, *41*, 14047–14057. [CrossRef]
71. Pinaeva, L.G.; Sadovskaya, E.M.; Smal, E.A.; Bobin, A.S.; Sadykov, V.A. Mechanism and kinetics of ethanol oxidation over Ru(Pt) supported onto Mn-Cr-O mixed oxides. *Chem. Eng. J.* **2018**, *333*, 101–110. [CrossRef]
72. Sadykov, V.A.; Bobrova, L.N.; Mezentseva, N.V.; Pavlova, S.N.; Fedorova, Y.E.; Bobin, A.S.; Vostrikov, Z.Y.; Glazneva, T.S.; Smirnova, M.; Sazonova, N.N.; et al. Methane dry reforming on nanocomposite catalysts: Design, kinetics and mechanism. In *Small-Scale Gas to Liquid Fuel Synthesis*; Kannelopoulos, N., Ed.; CRC Press, Taylor & Francis Group: Boca Raton, FL, USA, 2015; pp. 315–376.

73. Sadykov, V.; Mezentseva, N.; Simonov, M.; Smal, E.; Arapova, M.; Pavlova, S.; Fedorova, Y.; Chub, O.; Bobrova, L.; Kuzmin, V.; et al. Structured nanocomposite catalysts of biofuels transformation into syngas and hydrogen: Design and performance. *Int. J. Hydrogen Energy* **2015**, *40*, 7511–7522. [CrossRef]
74. Korneeva, E.V.; Kardash, T.Y.; Rogov, V.A.; Smal, E.A.; Sadykov, V.A. Catalytic steam reforming of ethanol over W-, V-, or Nb–modified Ni-Al-O hydrotalcite-type precursors. *Catal. Sustain. Energy* **2017**, *4*, 17–24. [CrossRef]
75. Hadian, N.; Rezaei, M. Combination of dry reforming and partial oxidation of methane over Ni catalysts supported on nanocrystalline $MgAl_2O_4$. *Fuel* **2013**, *113*, 571–579. [CrossRef]
76. Bian, Z.; Das, S.; Wai, M.H.; Hongmanorom, P.; Kawi, S. A review on bimetallic nickel-based catalysts for CO_2 reforming of methane. *ChemPhysChem* **2017**, *18*, 3117–3134. [CrossRef] [PubMed]
77. Sadykov, V.; Mezentseva, N.; Alikina, G.; Bunina, R.; Pelipenko, V.; Lukashevich, A.; Vostrikov, Z.; Rogov, V.; Krieger, T.; Ishchenko, A.; et al. Nanocomposite catalysts for steam reforming of methane and biofuels: Design and performance. In *Advances in Nanocomposites—Synthesis, Characterization and Industrial Applications*; Reddy, B., Ed.; INTECH: Vienna, Austria, 2011; pp. 909–946.
78. Sadykov, V.; Pavlova, S.; Fedorova, J.; Bobin, A.; Fedorova, V.; Simonov, M.; Ishchenko, A.; Krieger, T.; Melgunov, M.; Glazneva, T.; et al. Structured catalysts with mesoporous nanocomposite active components loaded on heat-conducting substrates for transformation of biogas/biofuels into syngas. *Catal. Today* **2021**, *379*, 166–180. [CrossRef]
79. Wysocka, J.H.I.; Rogala, A. Catalytic activity of nickel and ruthenium–nickel catalysts supported on SiO_2, ZrO_2, Al_2O_3, and $MgAl_2O_4$ in a dry reforming process. *Catalysts* **2019**, *9*, 540. [CrossRef]
80. Katheria, S.; Deo, G.; Kunzru, D. Rh-Ni/$MgAl_2O_4$ catalyst for steam reforming of methane: Effect of Rh doping, calcination temperature and its application on metal monoliths. *Appl. Catal. A Gen.* **2019**, *570*, 308–318. [CrossRef]
81. Lino, A.V.P.; Assaf, E.M.; Assaf, J.M. Adjusting process variables in methane tri-reforming to achieve suitable syngas quality and low coke deposition. *Energy Fuels* **2020**, *34*, 16522–16531. [CrossRef]
82. Sadykov, V.; Bobrova, L.; Pavlova, S.; Simagina, V.; Makarshin, L.; Parmon, V.; Ross, J.R.H.; Van Veen, A.C.; Mirodatos, C. *Syngas Generation from Hydrocarbons and Oxygenates with Structured Catalysts*; Nova Science Publishers: Hauppauge, NY, USA, 2012; p. 140.
83. Deutschman, O. Catalytic reforming of logistic fuels at high-temperatures. *Catalysis* **2012**, *24*, 48–82.
84. Bobrova, L.N.; Sadykov, V.A.; Mezentseva, N.V.; Pelipenko, V.V.; Vernikovskaya, N.V.; Klenov, O.P.; Smorygo, O.L. Catalytic performance of structured packages coated with perovskite-based nanocomposite in the methane steam reforming reaction. *Int. J. Hydrogen Energy* **2016**, *41*, 4632–4645. [CrossRef]
85. Shoynkhorova, T.B.; Rogozhnikov, V.N.; Ruban, N.V.; Shilov, V.A.; Potemkin, D.I.; Simonov, P.A.; Belyaev, V.D.; Snytnikov, P.V.; Sobyanin, V.A. Composite Rh/$Zr_{0.25}Ce_{0.75}O_{2-\delta}$-$\gamma$-$Al_2O_3$/Fecralloy wire mesh honeycomb module for natural gas, LPG and diesel catalytic conversion to syngas. *Int. J. Hydrogen Energy* **2019**, *44*, 9941–9948. [CrossRef]
86. Smorygo, O.L.; Sadykov, V.A.; Bobrova, L.N. *Open Cell Foams as Substrates for Design of Structured Catalysts, Solid Oxide Fuel Cells and Supported Asymmetric Membrane*; Nova Science Publishers: Hauppauge, NY, USA, 2016; p. 207.
87. Sadykov, V.A. (Ed.) *Advanced Nanomaterials for Catalysis and Energy. Synthesis, Characterization and Applications*, 1st ed.; Elsevier: Amsterdam, The Netherlands, 2019; p. 567.
88. Pechini, M.P. Method of Preparing Lead and Alkaline Earth Titanates and Niobates and Coating Method Using the Same to Form a Capacitor. US Patent No. 3330697, 1967. Available online: https://patents.google.com/patent/US3330697A/ (accessed on 6 August 2021).
89. Sadykov, V.A.; Simonov, M.N.; Mezentseva, N.V.; Pavlova, S.N.; Fedorova, Y.E.; Bobin, A.S.; Bespalko, Y.N.; Ishchenko, A.V.; Krieger, T.A.; Glazneva, T.S.; et al. Ni-loaded nanocrystalline ceria-zirconia solid solutions prepared via modified Pechini route as stable to coking catalysts of CH_4 dry reforming. *Open Chem.* **2016**, *14*, 363–376. [CrossRef]
90. Sadykov, V.; Pavlova, S.; Smal, E.; Arapova, M.; Simonov, M.; Mezentseva, N.; Rogov, V.; Glazneva, T.; Lukashevich, A.; Roger, A.-C.; et al. Structured catalysts for biofuels transformation into syngas with active components based on perovskite and spinel oxides supported on Mg-doped alumina. *Catal. Today* **2017**, *293–294*, 176–185. [CrossRef]
91. Sadykov, V.; Mezentseva, N.; Fedorova, Y.; Lukashevich, A.; Pelipenko, V.; Kuzmin, V.; Simonov, M.; Ishchenko, A.; Vostrikov, Z.; Bobrova, L.; et al. Structured catalysts for steam/autothermal reforming of biofuels on heat-conducting substrates: Design and performance. *Catal. Today* **2015**, *251*, 19–27. [CrossRef]
92. Shmakov, A.N.; Cherepanova, S.V.; Zyuzin, D.A.; Fedorova, Y.E.; Bobrikov, I.A.; Roger, A.-C.; Adamski, A.; Sadykov, V.A. The crystal structure of compositionally homogeneous mixed ceria-zirconia oxides by high resolution X-ray and neutron diffraction methods. *Open Chem.* **2017**, *15*, 438–445. [CrossRef]
93. Kol'ko, V.P.; Zyuzin, D.A.; Sadykov, V.A.; Kriventsov, V.V.; Moroz, E.M. Structure of the mixed oxides $Me_x(Ce_{0.5}Zr_{0.5})_{1-x}O_y$ (Me = Gd, Pr). *Glass Phys. Chem.* **2007**, *33*, 335–339. [CrossRef]
94. Sadykov, V.A.; Kriventsov, V.V.; Moroz, E.M.; Borchert, Y.V.; Zyuzin, D.A.; Kol'ko, V.P.; Kuznetsova, T.G.; Ivanov, V.P.; Boronin, A.I.; Mezentseva, N.V.; et al. Ceria-zirconia nanoparticles doped with La or Gd: Effect of the doping cation on the real structure. *Solid State Phenom.* **2007**, *128*, 81–88. [CrossRef]

95. Sadykov, V.A.; Mezentseva, N.V.; Alikina, G.M.; Lukashevich, A.I.; Borchert, Y.V.; Kuznetsova, T.G.; Ivanov, V.P.; Trukhan, S.N.; Paukshtis, E.A.; Muzykantov, V.S.; et al. Pt-supported nanocrystalline ceria-zirconia doped with La, Pr or Gd: Factors controlling syngas generation in partial oxidation/autothermal reforming of methane or oxygenates. *Solid State Phenom.* **2007**, *128*, 239–248. [CrossRef]
96. Sadykov, V.; Mezentseva, N.; Alikina, G.; Lukashevich, A.; Muzykantov, V.; Bunina, R.; Boronin, A.; Pazhetnov, E.; Paukshtis, E.; Kriventsov, V.; et al. Doped nanocrystalline Pt-promoted ceria-zirconia as anode catalysts for IT-SOFC: Synthesis and properties. *Mater. Res. Soc. Symp. Proc.* **2007**, *1023*, 26–31. [CrossRef]
97. Kuznetsova, T.G.; Sadykov, V.A. Specific features of the defect structure of metastable nanodisperse ceria, zirconia, and related materials. *Kinet. Catal.* **2008**, *49*, 840–858. [CrossRef]
98. Cortés Corberán, V.; Rives, V.; Mezentseva, N.V.; Sadykov, V.A.; Martínez-Tamayo, E. Nanostructured metal oxide catalyst. In *Nanotechnology for Sustainable Manufacturing*; Rickerby, D., Ed.; CRC Press: Boca Raton, FL, USA, 2014; pp. 153–198.
99. Bulgakov, N.N.; Sadykov, V.A.; Lunin, V.V.; Kemnitz, E. Lattice defects and oxygen absorption/migration in ceria/ceria-zirconia solid solutions: Analysis by semiempirical interacting bonds method. *React. Kinet. Catal. Lett.* **2002**, *76*, 103–110. [CrossRef]
100. Bulgakov, N.N.; Sadykov, V.A.; Lunin, V.V.; Kemnitz, E. Surface energies and heats of oxygen adsorption in ceria/ceria-zirconia solid solutions: Analysis by semiempirical interacting bonds method. *React. Kinet. Catal. Lett.* **2002**, *76*, 111–116. [CrossRef]
101. Sadykov, V.; Rogov, V.; Ermakova, E.; Arendarsky, D.; Mezentseva, N.; Alikina, G.; Sazonova, N.; Bobin, A.; Pavlova, S.; Schuurman, Y.; et al. Mechanism of CH_4 dry reforming by pulse microcalorimetry: Metal nanoparticles on perovskite/fluorite supports with high oxygen mobility. *Thermochim. Acta* **2013**, *567*, 27–34. [CrossRef]
102. Simonov, M.N.; Rogov, V.A.; Smirnova, M.Y.; Sadykov, V.A. Pulse microcalorimetry study of methane dry reforming reaction on Ni/ceria-zirconia catalyst. *Catalysts* **2017**, *7*, 268. [CrossRef]
103. Simonov, M.N.; Sadykov, V.A.; Rogov, V.A.; Bobin, A.S.; Sadovskaya, E.M.; Mezentseva, N.V.; Ishchenko, A.V.; Krieger, T.A.; Roger, A.-C.; van Veen, A.C. Ethanol selective oxidation into syngas over Pt-promoted fluorite-like oxide: SSITKA and pulse microcalorimetry study. *Catal. Today* **2016**, *277*, 157–163. [CrossRef]
104. Sadykov, V.A.; Eremeev, N.F.; Sadovskaya, E.M.; Chesalov, Y.A.; Pavlova, S.N.; Rogov, V.A.; Simonov, M.N.; Bobin, A.S.; Glazneva, T.S.; Smal, E.A.; et al. Detailed mechanism of ethanol transformation into syngas on catalysts based on mesoporous $MgAl_2O_4$ support loaded with Ru+Ni/(PrCeZrO or $MnCr_2O_4$) active components. *Top. Catal.* **2020**, *63*, 166–177. [CrossRef]
105. Sadykov, V.; Sadovskaya, E.; Bobin, A.; Kharlamova, T.; Uvarov, N.; Ulikhin, A.; Argirusis, C.; Sourkouni, G.; Stathopoulos, V. Temperature-programmed $C^{18}O_2$ SSITKA for powders of fast oxide-ion conductors: Estimation of oxygen self-diffusion coefficients. *Solid State Ionics* **2015**, *271*, 69–72. [CrossRef]
106. Sadykov, V.; Eremeev, N.; Sadovskaya, E.; Bobin, A.; Ishchenko, A.; Pelipenko, V.; Muzykantov, V.; Krieger, T.; Amanbaeva, D. Oxygen mobility and surface reactivity of $PrNi_{1-x}Co_xO_{3-\delta}$ perovskites and their nanocomposites with $Ce_{0.9}Y_{0.1}O_{2-\delta}$ by temperature-programmed isotope exchange experiments. *Solid State Ionics* **2015**, *273*, 35–40. [CrossRef]
107. Sadovskaya, E.M.; Ivanova, Y.A.; Pinaeva, L.G.; Grasso, G.; Kuznetsova, T.G.; van Veen, A.; Sadykov, V.A.; Mirodatos, C. Kinetics of oxygen exchange over CeO_2-ZrO_2 fluorite-based catalysts. *J. Phys. Chem. A* **2007**, *111*, 4498–4505. [CrossRef]
108. Pinaeva, L.G.; Sadovskaya, E.M.; Ivanova, Y.A.; Kuznetsova, T.G.; Prosvirin, I.P.; Sadykov, V.A.; Schuurman, Y.; van Veen, A.C.; Mirodatos, C. Water gas shift and partial oxidation of CH_4 over CeO_2–ZrO_2(–La_2O_3) and Pt/CeO_2–ZrO_2(–La_2O_3): Performance under transient conditions. *Chem. Eng. J.* **2014**, *257*, 281–291. [CrossRef]
109. Sadykov, V.A.; Sazonova, N.N.; Bobin, A.S.; Muzykantov, V.S.; Gubanova, E.L.; Alikina, G.M.; Lukashevich, A.I.; Rogov, V.A.; Ermakova, E.N.; Sadovskaya, E.M.; et al. Partial oxidation of methane on Pt-supported lanthanide doped ceria-zirconia oxides: Effect of the surface/lattice oxygen mobility on catalytic performance. *Catal. Today* **2011**, *169*, 125–137. [CrossRef]
110. Reshetnikov, S.I.; Lukashevich, A.I.; Alikina, G.M.; Sadykov, V.A. Effect of oxygen mobility in solid catalyst on transient regimes of catalytic reaction of methane partial oxidation at short contact times. *Catal. Lett.* **2006**, *110*, 235–242. [CrossRef]
111. Pavlova, S.N.; Sazonova, N.N.; Sadykov, V.A.; Alikina, G.M.; Lukashevich, A.I.; Gubanova, E.L.; Bunina, R.V. Study of synthesis gas production over structured catalysts based on $LaNi(Pt)O_x$- and Pt(LaPt)- CeO_2-ZrO_2 supported on corundum. *Stud. Surf. Sci. Catal.* **2007**, *167*, 343–348.
112. Sazonova, N.N.; Sadykov, V.A.; Bobin, A.S.; Pokrovskaya, S.A.; Gubanova, E.L.; Mirodatos, C. Dry reforming of methane over fluorite-like mixed oxides promoted by Pt. *React. Kinet. Catal. Lett.* **2009**, *98*, 35–41. [CrossRef]
113. Yaseneva, P.; Pavlova, S.; Sadykov, V.; Alikina, G.; Lukashevich, A.; Rogov, V.; Belochapkine, S.; Ross, J. Combinatorial approach to the preparation and characterization of catalysts for biomass steam reforming into syngas. *Catal. Today* **2008**, *137*, 23–28. [CrossRef]
114. Wei, T.; Jia, L.; Zheng, H.; Chi, B.; Pu, J.; Li, J. $LaMnO_3$-based perovskite with *in-situ* exsolved Ni nanoparticles: A highly active, performance stable and coking resistant catalyst for CO_2 dry reforming of CH_4. *Appl. Catal. A.* **2018**, *564*, 199–207. [CrossRef]
115. Arapova, M.V.; Pavlova, S.N.; Rogov, V.A.; Krieger, T.A.; Ishchenko, A.V.; Roger, A.-C. Ni(Co)-containing catalysts based on perovskite-like ferrites for steam reforming of ethanol. *Catal. Sustain. Energy* **2014**, *1*, 10–20. [CrossRef]
116. Marinho, A.L.A.; Rabelo-Neto, R.C.; Noronha, F.B.; Mattos, L.V. Steam reforming of ethanol over Ni-based catalysts obtained from $LaNiO_3$ and $LaNiO_3$/$CeSiO_2$ perovskite-type oxides for the production of hydrogen. *Appl. Catal. A Gen.* **2016**, *520*, 53–64. [CrossRef]
117. Sadykov, V.; Sobyanin, V.; Mezentseva, N.; Alikina, G.; Vostrikov, Z.; Fedorova, Y.; Pelipenko, V.; Usoltsev, V.; Tikhov, S.; Salanov, A.; et al. Transformation of CH_4 and liquid fuels into syngas on monolithic catalysts. *Fuel* **2010**, *89*, 1230–1240. [CrossRef]

118. Frolova-Borchert, Y.V.; Sadykov, V.A.; Alikina, G.M.; Lukashevich, A.I.; Moroz, E.M.; Kochubey, D.I.; Kriventsov, V.V.; Zaikovskii, V.I.; Zyryanov, V.V.; Uvarov, N.F. Nanocomposites comprised of doped cerium dioxide and lanthanum manganite for syngas production. *Solid State Ionics* **2006**, *177*, 2533–2538. [CrossRef]
119. Sadykov, V.; Mezentseva, N.; Alikina, G.; Bunina, R.; Rogov, V.; Krieger, T.; Belochapkine, S.; Ross, J. Composite catalytic materials for steam reforming of methane and oxygenates: Combinatorial synthesis, characterization and performance. *Catal. Today* **2009**, *145*, 127–137. [CrossRef]
120. Sadykov, V.; Mezentseva, N.; Alikina, G.; Bunina, R.; Pelipenko, V.; Lukashevich, A.; Tikhov, S.; Usoltsev, V.; Vostrikov, Z.; Bobrenok, O.; et al. Nanocomposite catalysts for internal steam reforming of methane and biofuels in solid oxide fuel cells: Design and performance. *Catal Today.* **2009**, *146*, 132–140. [CrossRef]
121. Sadykov, V.A.; Chub, O.V.; Chesalov, Y.A.; Mezentseva, N.V.; Pavlova, S.N.; Arapova, M.V.; Rogov, V.A.; Simonov, M.N.; Roger, A.-C.; Parkhomenko, K.V.; et al. Mechanism of ethanol steam reforming over Pt/(Ni+Ru)-promoted oxides by FTIRS in situ. *Top. Catal.* **2016**, *59*, 1332–1342. [CrossRef]
122. Marinho, A.L.A.; Toniolo, F.S.; Noronha, F.B.; Epron, F.; Duprez, D.; Bion, N. Highly active and stable Ni dispersed on mesoporous CeO_2-Al_2O_3 catalysts for production of syngas by dry reforming of methane. *Appl. Catal. B Environ.* **2021**, *281*, 119459. [CrossRef]
123. Guilhaume, N.; Bianchi, D.; Wandawa, R.A.; Yin, W.; Yves Schuurman, Y. Study of CO_2 and H_2O adsorption competition in the combined dry/steam reforming of biogas. *Catalysis Today* **2021**, *375*, 282–289. [CrossRef]
124. Di Michele, A.; Dell'Angelo, A.; Tripodi, A.; Bahadori, E.; Sanchez, F.; Motta, D.; Dimitratos, N.; Rossetti, I.; Ramis, G. Steam reforming of ethanol over Ni/$MgAl_2O_4$ catalysts. *Int. J. Hydrogen Energy* **2019**, *44*, 952–964. [CrossRef]
125. Zhang, S.; Ying, M.; Yu, J.; Zhan, W.; Wang, L.; Guo, Y.; Guo, Y. $Ni_xAl_1O_{2-\delta}$ mesoporous catalysts for dry reforming of methane: The special role of $NiAl_2O_4$ spinel phase and its reaction mechanism. *Appl. Catal. B Environ.* **2021**, *291*, 120074. [CrossRef]
126. Lino, A.V.P.; Assaf, E.M.; Assaf, J.M. X-ZrO_2 addition (X= Ce, La, Y and Sm) on Ni/$MgAl_2O_4$ applied to methane tri-reforming for syngas production. *J. CO_2 Util.* **2019**, *33*, 273–283. [CrossRef]
127. Mao, X.; Foucher, A.C.; Stach, E.A.; Gorte, R.J. Changes in Ni-NiO equilibrium due to $LaFeO_3$ and the effect on dry reforming of CH_4. *J. Catal.* **2020**, *381*, 561–569. [CrossRef]
128. Messaoudia, H.; Thomas, S.; Djaidja, A.; Slyemia, S.; Barama, A. Study of La_xNiO_y and La_xNiO_y/$MgAl_2O_4$ catalysts in dry reforming of methane. *J. CO_2 Util.* **2018**, *24*, 40–49. [CrossRef]
129. Pavlova, S.; Fedorova, Y.; Ishchenko, A.; Melgunov, M.; Melgunova, E.; Simonov, M.; Rogov, V.; Krieger, T.; Sadykov, V.; Roger, A.-C. Ni-containing catalysts based on ordered mesoporous MgO–Al_2O_3 for methane dry reforming. *Catal. Sustain. Energy* **2018**, *5*, 59–66. [CrossRef]
130. Jabbour, K.; El Hassan, N.; Davidson, A.; Casaleb, S.; Massiani, P. Factors affecting the long-term stability of mesoporous nickel-based catalysts in combined steam and dry reforming of methane. *Catal. Sci. Technol.* **2016**, *6*, 4616–4631. [CrossRef]
131. Xu, L.; Song, H.; Chou, L. Carbon dioxide reforming of methane over ordered mesoporous NiO–MgO–Al_2O_3 composite oxides. *Appl. Catal. B Environ.* **2011**, *108–109*, 177–190. [CrossRef]
132. Xu, L.; Song, H.; Chou, L. Ordered mesoporous MgO-Al_2O_3 composite oxides supported Ni based catalysts for CO_2 reforming of CH_4: Effects of basic modifier and mesopore structure. *Int. J. Hydrogen Energy* **2013**, *38*, 7307–7325. [CrossRef]
133. Jabbour, K.; Massiani, P.; Davidson, A.; Casale, S.; El Hassan, N. Ordered mesoporous "one-pot" synthesized Ni-Mg(Ca)-Al_2O_3 as effective and remarkably stable catalysts for combined steam and dry reforming of methane (CSDRM). *Appl. Catal. B Environ.* **2017**, *201*, 527–554. [CrossRef]
134. Bobin, A.S.; Sadykov, V.A.; Rogov, V.A.; Mezentseva, N.V.; Alikina, G.M.; Sadovskaya, E.M.; Glazneva, T.S.; Sazonova, N.N.; Smirnova, M.Y.; Veniaminov, S.A.; et al. Mechanism of CH_4 dry reforming on nanocrystalline doped ceria–zirconia with supported Pt, Ru, Ni, and Ni–Ru. *Top. Catal.* **2013**, *56*, 958–968. [CrossRef]
135. Sadykov, V.A.; Tikhov, S.F.; Isupova, L.A. *Heterogeneous Catalytic Redox Reactions*; De Gruyter: Berlin, Germany; Boston, MA, USA, 2020.
136. Sadykov, V.A.; Gubanova, E.L.; Sazonova, N.N.; Pokrovskaya, S.A.; Chumakova, N.A.; Mezentseva, N.V.; Bobin, A.S.; Gulyaev, R.V.; Ishchenko, A.V.; Krieger, T.A.; et al. Dry reforming of methane over Pt/PrCeZrO catalyst: Kinetic and mechanistic features by transient studies and their modeling. *Catal. Today* **2011**, *171*, 140–149. [CrossRef]
137. Sadykov, V.; Mezentseva, N.; Muzykantov, V.; Efremov, D.; Gubanova, E.; Sazonova, N.; Bobin, A.; Paukshtis, E.; Ishchenko, A.; Voronin, V.; et al. Real structure-oxygen mobility relationship in nanocrystalline doped ceria-zirconia fluorite-like solid solutions promoted by Pt. *MRS Online Proc. Libr.* **2008**, *1122*, 1122-O05-03. [CrossRef]
138. Ulianitsky, V.; Shtertser, A.; Sadykov, V.; Smurov, I. Development of catalytic converters using detonation spraying. *Mater. Manuf. Process.* **2016**, *31*, 1433–1438. [CrossRef]
139. Rogozhnikov, V.N.; Kuzin, N.A.; Snytnikov, P.V.; Potemkin, D.I.; Shoynkhorova, T.B.; Simonov, P.A.; Shilov, V.A.; Ruban, N.V.; Kulikov, A.V.; Sobyanin, V.A. Design, scale-up, and operation of a Rh/$Ce_{0.75}Zr_{0.25}O_{2-\delta}$-η-$Al_2O_3$/FeCrAl alloy wire mesh honeycomb catalytic module in diesel autothermal reforming. *Chem. Eng. J.* **2019**, *374*, 511–519. [CrossRef]
140. Zazhigalov, S.V.; Rogozhnikov, V.N.; Snytnikov, P.V.; Potemkin, D.I.; Simonov, P.A.; Shilov, V.A.; Ruban, N.V.; Kulikov, A.V.; Zagoruiko, A.N.; Sobyanin, V.A. Simulation of diesel autothermal reforming over Rh/$Ce_{0.75}Zr_{0.25}O_{2-\delta}$-η-$Al_2O_3$/FeCrAl wire mesh honeycomb catalytic module. *Chem. Eng. Process.* **2020**, *150*, 107876. [CrossRef]

141. Potemkin, D.I.; Rogozhnikov, V.N.; Ruban, N.V.; Shilov, V.A.; Simonov, P.A.; Shashkov, M.V.; Sobyanin, V.A.; Snytnikov, P.V. Comparative study of gasoline, diesel and biodiesel autothermal reforming over Rh-based FeCrAl-supported composite catalyst. *Int. J. Hydrogen Energy* **2020**, *45*, 26197–26205. [CrossRef]
142. Shilov, V.A.; Rogozhnikov, V.N.; Zazhigalov, S.V.; Potemkin, D.I.; Belyaev, V.D.; Shashkov, M.V.; Zagoruiko, A.N.; Sobyanin, V.A.; Snytnikov, P.V. Operation of $Rh/Ce_{0.75}Zr_{0.25}O_{2-\delta}$-η-$Al_2O_3$/FeCrAl wire mesh honeycomb catalytic modules in diesel steam and autothermal reforming. *Int. J. Hydrogen Energy* **2021**, in press. [CrossRef]

energies

Review

Recent Achievements in Microalgal Photobiological Hydrogen Production

Eleftherios Touloupakis [1], Cecilia Faraloni [2], Ana Margarita Silva Benavides [3,4] and Giuseppe Torzillo [2,3,*]

1 Istituto di Ricerca sugli Ecosistemi Terrestri, CNR, Via Madonna del Piano 10, 50019 Sesto Fiorentino, Italy; eleftherios.touloupakis@cnr.it
2 Istituto per la Bioeconomia, CNR, Via Madonna del Piano 10, 50019 Sesto Fiorentino, Italy; cecilia.faraloni@ibe.cnr.it
3 Centro de Investigacion en Ciencias Del Mar y Limnología, Universidad de Costa Rica, San José 2060, Costa Rica; ana.silva@ucr.ac.cr
4 Escuela de Biologia, Universidad de Costa Rica, San José 2060, Costa Rica
* Correspondence: giuseppe.torzillo@cnr.it

Citation: Touloupakis, E.; Faraloni, C.; Silva Benavides, A.M.; Torzillo, G. Recent Achievements in Microalgal Photobiological Hydrogen Production. *Energies* **2021**, *14*, 7170. https://doi.org/10.3390/en14217170

Academic Editor: Wei-Hsin Chen

Received: 6 October 2021
Accepted: 27 October 2021
Published: 1 November 2021

Abstract: It is well known that over the last 60 years the trend of long-lived greenhouse gas emissions have shown a strong acceleration. There is an increasing concern and a mounting opposition by public opinion to continue with the use of fossil energy. Western countries are presently involved in a so-called energy transition with the objective of abandoning fossil energy for renewable sources. In this connection, hydrogen can play a central role. One of the sustainable ways to produce hydrogen is the use of microalgae which possess two important natural catalysts: photosystem II and hydrogenase, used to split water and to combine protons and electrons to generate gaseous hydrogen, respectively. For about 20 years of study on photobiological hydrogen production, our scientific hopes were based on the application of the sulfur protocol, which indisputably represented a very important advancement in the field of hydrogen production biotechnology. However, as reported in this review, there is increasing evidence that this strategy is not economically viable. Therefore, a change of paradigm for the photobiological production of hydrogen based on microalgae seems mandatory. This review points out that an increasing number of microalgal strains other than *Chlamydomonas reinhardtii* are being tested and are able to produce sustainable amount of hydrogen without nutrient starvation and to fulfill this goal including the application of co-cultures.

Keywords: biohydrogen; microalgae; *Chlamydomonas reinhardtii*; *Chlorella* sp.; photobioreactors; light conversion efficiency

1. Introduction

Microalgae are capable of converting light energy into chemical energy. Biofuels such as biodiesel, biohydrogen, and bioethanol can be derived from microalgae [1]. Photosynthesis in microalgae is coupled to the splitting of water and the evolution of oxygen (O_2). This process is catalyzed by the membrane-bound multi-protein complex photosystem II (PSII) [2].

It has has been known since 1942, when Gaffron and co-workers noticed that under anaerobic conditions *Scenedesmus obliquus* cells can transiently produce hydrogen (H_2) upon illumination when deprived of oxygen [3]. In microalgae, hydrogenase enzyme catalyzes H_2 production in a light-dependent process [4]. Upon illumination, after a dark incubation period, due to light-driven electron transport from ferredoxin to hydrogenase, H_2 production is observed. H_2 production in microalgae can be divided into direct or indirect processes [5]. A direct process occurs when electrons (e^-) from water splitting are transferred via PSII and ferredoxin to hydrogenase. An indirect process occurs when e^- are derived from the metabolism of carbohydrates, previously accumulated during the (light) aerobic phase, and then utilized for H_2 production via both a photo-fermentation process involving photosystem I (PSI) and in a process in the dark, involving the enzyme pyruvate:ferredoxin oxidoreductase (PFR). In *Chlamydomonas reinhardtii* (hereafter *C. reinhardtii*), PFR enzyme catalyzes the reduction of ferredoxin (Fdx) and the transfer of e^- to hydrogenase in a similar pathway to that utilized by bacteria (Figure 1) [6]. In *C. reinhardtii*, up to 92% of the final H_2 output comes from the direct photolysis coupled to the water oxidation operated by PSII [7]. Contribution of dark fermentation to the overall H_2 output is considered negligible (about 4%) in *C. reinhardtii*, but it can be significant in other microalgae such as *Chlorella*, as recently shown [8]. Microalgal hydrogenase enzymes are inactivated by the presence of molecular oxygen, and their expression is induced under anaerobic conditions.

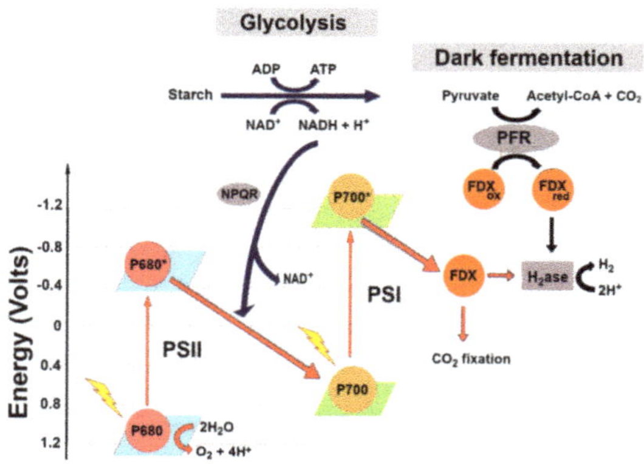

Figure 1. Metabolic hydrogen production pathways used by *Chlamydomonas reinhartii*.FDX: ferredoxin; H_2ase: hydrogenase; NPQR: NADPH−plastoquinone oxidoreductase; PFR: pyruvate:ferredoxin oxidoreductase; PSI: photosystem I; PSII: photosystem II.

In recent years, energy-related H_2 demands have prompted scientists to develop methods that greatly enhance the H_2-evolving ability of microalgae. The most promising approach has been the so-called "two-stage process" of photosynthesis (stage 1) and H_2 production (stage 2) [9]. In this process, there is a separation of the reactions of oxygen and hydrogen production. This bypasses the sensitivity of the hydrogenase enzyme to oxygen. Under such conditions, it was possible to produce significant volumes of H_2 by *C. reinhardtii* in a sustained process.

Several microalgae species have been studied for H_2 production, especially *C. reinhardtii*, *Chlorella vulgaris*, and *Chlorella pyrenoidosa* [10–13]. Among them, *C. reinhardtii* is a model microorganism widely recognized as an H_2 producer, presenting a hydrogenase with an enzymatic activity 10 to 100 times higher than other species [14]. H_2 production requires many optimization steps in order to reach a sustainable process [8,14–17]. Some of these parameters include choosing a proper microalgae strain and selecting appropriate culture conditions (growth media, light, pH, temperature, chlorophyll concentration) and proper photobioreactor (PBR) designs [18–20].

Many works have reported improved H_2 production in many microalgal strains by using sulfur, phosphorus, or nitrogen-depleted media [12,21–23]. In such culture conditions, microalgae sustain H_2 production only for some days since macro/micronutrient depletion in the culture compromises cell viability. This is the major drawback in microalgal H_2 production processes carried out by nutrient deprivation. Microalgae-based H_2 production requires anaerobic conditions due to the sensitivity of hydrogenase to O_2 [24]. O_2 sensitivity of hydrogenase is a major issue for H_2 production; therefore, there are many studies on oxygen suppression in order to improve H_2 production yield. Genetic and metabolic engineering of microalgae [25,26], nutrient stresses [27,28], light conditions optimization [29], and elimination of competing pathways for electrons [30] are examples of strategies used to improve H_2 evolution in microalgae.

This review provides an overview of the most relevant achievements in the photobiological production of H_2 by microalgae, and proposes a change of paradigm for the future research in the field.

2. Genetic Modification

Krishna et al. reported that sustained H_2 production is achieved by altering the ratio between PSI and PSII [31]. In this work, a *C. reinhardtii* C3 mutant with a modified PSI/PSII ratio (0.33) produced H_2 with a rate of 3 mL H_2/L/d for 42 days. Chen et al. identified a *C. reinhardtii* mutant strain hpm91 lacking proton gradient regulation 5, with 30-fold H_2 production yield compared to wild type (WT) [32]. Characterization of the hpm91 strain revealed an increased reactive-oxygen-species-scavenging capacity. This translates into an enhanced stability of PSII complex and increased H_2 production yield. Steinbeck et al. investigated the capacity of *C. reinhardtii pgr5* and *pgr5 pgrl1* double mutant to produce H_2 [33]. The *pgr* mutants showed four times higher maximal enhanced H_2 production rate (7 mL/L/h) than the WT. Pinto et al. studied a *Chlamydomonas* mutant with reduced rubisco levels, activity, and stability [34]. This mutant was used to reduce carbon fixation by Calvin cycle activity, which is the main competitor for the reducing power required by the hydrogenase. In this work, the rubisco mutant presented 15 times higher H_2 production than the WT. Eilenberg et al. studied the in vivo H_2 production efficiency of a *C. reinhardtii* strain Fd-HydA containing ferredoxin fused to HydA. H_2 production rate was 4.5 times higher than that of the native HydA in vivo [35]. Torzillo et al. showed that the in vivo H_2 production of the *C. reinhardtii* mutant strain L159I-N230Y was up to 5-fold higher (16 nmoles $H_2/\mu g_{chl}$/h) than that of *C. reinhartii* CC 124 [36,37]. Batyrova et al. developed a genetically modified *C. reinhardtii* strain that activates photosynthesis in a cyclical manner. In this strain, the low O_2 production benefits H_2 production [38]. In comparison with the WT, this genetically modified strain presented higher H_2 production levels. Kosourov et al. showed that a truncated light antenna *C. reinhardtii* mutant could produce six times more H_2 compared to the WT strain [39]. Xu et al. introduced a catalase gene from *Synechococcus elongatus* PCC7942 and an *Escherichia coli* pyruvate oxidase gene, both driven by a HSP70A/RBCS2 promoter, into the chloroplast of *C. reinhardtii* [40]. Under low light, these microalgal cells consumed more O_2 than WT, resulting in a lower O_2 content and increased H_2 production [40]. Kruse et al. used the *Chlamydomonas* strain Stm6, which has a modified respiratory metabolism and large starch reserves compared with the WT [41]. *Chlamydomonas* strain Stm6 presented 5–13 times increased H_2 production rate (540 mL $H_2/L_{culture}$) compared to the WT [41]. Later, Volgusheva et al. obtained similar results by using the *Chlamydomonas* Stm6 mutant [42]. They attained an anaerobic condition much faster in the Stm6 strain than in the WT. This was a result of the higher respiration rate and lower initial O_2 production rate. H_2 production was four times higher in the Stm6 strain compared to the WT. Oey and co-workers reported the knock-down of the LHCMB 1, 2, and 3 proteins in the *C. reinhardtii* strain Stm6Glc4 [43]. The produced *C. reinhardtii* mutant exhibited increased light-to-H_2 and biomass conversion efficiencies of 180% and 165%, respectively. Wu et al. introduced a leghemoglobin gene (lba) into chloroplasts of *C. reinhardtii*. The genetically modified *Chlamydomonas* with lba gene consumed O_2 faster than WT, thus improving H_2 production [44]. Noone et al. introduced the clostridial hydrogenase gene into *C. reinhardtii* that contains insertionally inactivated hydrogenase genes. The presence of the more O_2-tolerant clostridial hydrogenase led to more sustained H_2 production [45].

Nowadays, the primary current challenge of such a process is the development of an oxygen-resistant hydrogenase. However, other bottlenecks may also be of significant importance, such as the oxygen sensitivity of hydrogenases. In this case, a number of other scientific and engineering issues are very likely to arise. They may include: (1) maximizing photosynthetic light-conversion efficiency (LCE); finding the proper redox potential balance in the organism to facilitate H_2 production; (2) preventing the effect of the buildup of high relative H_2 partial pressure restricting the process by feedback inhibition; (3) addressing inefficient metabolic processes such as unneeded ATP generation during H_2 production in microalgae; (4) examining issues associated with the generation of destructive, active-oxygen species; and (5) minimizing the production of alternative, carbon-containing products that drain usable reducing power from the system. Recently, an increased H_2 output was attained by bioengineering photosynthesis [46].

In the following paragraphs, some of the most recent strategies used for sustained photobiological H_2 production by microalgae are summarized.

3. O_2 Removal

The use of inert gas (such as N_2 or Ar) is another type of strategy to remove the O_2 in microalgal cultures [47,48]. Alternatively, O_2 scavengers can be employed to remove the O_2 in order to induce anaerobiosis in the culture. Paramesh and Chandrasekhar screened three O_2 scavengers individually in order to improve H_2 production in *Chlorococcum minutum* [49]. In the presence of all three O_2 scavengers, efficient H_2 generation was found. They found that sodium sulfite was the best one for enhancement of H_2 production. Nagy et al. showed that the simultaneous addition of glucose, glucose oxidase, and ascorbate to the *C. reinhardtii* culture resulted in reduced O_2 content in the headspace and tenfold-increased H_2 production [30]. Su et al. created an O_2-consuming sandwich-like layer by using tannic acid, polydopamine, and laccase, in order to generate anaerobiosis around the *Chlorella pyrenoidosa* cells [50]. This layer enabled the encapsulated cell to switch from O_2 production to H_2 production. Márquez-Reyes et al. found that the chemical reducing agent cysteine induced anaerobic H_2 production in cultures of *Chlamydomonas gloeopara* and *Scenedesmus obliquus* cultures [51]. In the presence of cysteine, H_2 production was 5 times higher compared to the sulfur-starvation protocol. Chen and coworkers found that *C. reinhardtii* produce H_2 at a rate of 0.44 μmol H_2/h/mg_{chl} per month by using a chemoenzymatic cascade system (CEC). The CEC system contained four components: glucose oxidase, catalase, glucose, and $Mg(OH)_2$. In this CEC, they combined O_2 consumption, cell aggregation, and pH maintenance to activate hydrogenase [52]. Nagy et al. showed that the application of an iron-based O_2 absorbent ($O20_{TM}$) in *C. reinhardtii* cultures, in which the activation of the Calvin–Benson–Bassham cycle in the light was prevented, presented a H_2 production yield of 2.58 mL/L/h, to which corresponded a mean LCE (light to H_2) of 0.27% [30].

4. Co-Cultures

Another approach to create an anaerobic environment is the addition of living aerobic bacteria to the microalgae cultures (co-culture) [53]. Many works have proven the possibility of increasing H_2 production by co-culturing microalgae and bacteria [28,54–57]. The main advantage of co-culturing microalgae with heterotrophic bacteria is the efficient removal of the O_2 from the growth media. Simultaneously, the CO_2 released during bacterial fermentation of an organic substrate can support microalgae growth. Moreover, many metabolites can be exchanged between these microorganisms, such as carbon, nitrogen, phosphorous, and sulfur sources, and vitamins [58,59]. The presence of bacteria inside the microalgal culture enhances starch accumulation [60]. Different *Chlamydomonas* WT co-cultures incubated in sulfur depleted TAP medium employing *Pseudomonas* sp. or *Bradirizhobium japonicum* have achieved high H_2 production rates (165–170 mL H_2/L) [54,60]. Fakhimi et al. evaluated H_2 production by *C. reinhardtii* in co-cultures with different bacteria strains [28]. They found that co-culturing *Pseudomonas* spp. with *Chlamydomonas* significantly improved microalgal H_2 production. Interestingly, the integration of the photobiological and the fermentative H_2 production in *Chlamydomonas* and *Escherichia coli* co-cultures resulted in H_2 production 60% higher than the sum of the respective monocultures [28]. *Chlamydomonas* co-cultures with *Pseudomonas* sp. and *Bradyrhizobium japonicum* (not H_2-producing bacteria) in sulfur-depleted TAP medium improved H_2 production by 22.7 times and 32.3 times compared to the pure microalgal cultures, respectively [57]. Furthermore, the production of H_2 by *C. reinhardtii* in nutrient-replete cultures is strongly limited by the O_2 release, unless it is performed under very low light irradiance (lower than 20 μmol photons/m^2/s), but it may become feasible under higher light irradiance by using different consortia, which allow the maintenance of anaerobiosis conditions, thus creating an opportunity to use full medium and much higher light irradiance, enhancing the H_2 output. Finally, it must be pointed out that large-scale production of H_2 with *Chlamydomonas* and other microalgae will be necessarily carried out not with axenic cultures, but rather a microalgae–bacteria consortium, therefore understanding the complex interplay between microalgae H_2 producers and bacteria is important for the economic exploitation of an industrial H_2 production process.

5. Immobilization

Microalgae immobilization can increase the H_2 production yield [61]. The main reason for their higher performance is that such experiments are usually carried out by using a much higher chlorophyll concentration compared to liquid cultures. The process of switching between oxygenic photosynthesis (aerobiosis) and H_2 production (anaerobiosis) can be facilitated by using cell immobilization systems [62]. One of the most used materials for microalgae encapsulation is calcium alginate [63]. Immobilization in calcium alginate matrix allows high cell density and protection from mechanical stress and contamination and is easy to scale-up [64]. Immobilization of microalgal cells could increase their LCE [62]. Using *Calothrix*, *Anabaena*, and *Chlamydomonas* cells immobilized on thin calcium alginate films gave an LCE of 2.5% of the photosynthetically active radiation [48,65]. *C. reinhardtii* immobilized on calcium alginate films in nutrient-depleted cultures (-P,-S) presented a H_2 production rate of 12.5 μmol/mgChl/h [66]. Ruiz-Marin et al. proposed immobilization of *Chlorella vulgaris* and *Scenedesmus obliquus* cells in calcium alginate for the production of H_2 [11]. These microalgae were grown in urban wastewater under sulfur starvation and blue or purple light conditions. The maximum H_2 production obtained under red light was 204.8 mL H_2/L/d for *Scenedesmus obliquus* and 39.1 mL H_2/L/d for *Chlorella vulgaris* [11]. Maswanna et al. studied H_2 production by *Tetraspora* sp. CU2551 cells immobilized in a 4% *w/v* calcium alginate matrix in their recent work. They obtained a maximum H_2 production rate of 182 ± 20 nmol/mg of cell dry weight/h [67].

Our group recently tested the capability of immobilized *Chlorella vulgaris* (BEIJ G-120 strain) cells in a calcium alginate (3%*w/v*) gel matrix to produce H_2 in a direct light-driven process under continuous illumination. Calcium alginate beads were stable, showing minimal cell leakage, and they measured 4.69 ± 0.02 mm in diameter and 54.01 ± 0.03 μL in volume, carrying 145.5 ± 8.9 μg of microalgal cells (biomass dry weight/bead) (Figure 2). Immobilized cells retained their viability for more than 30 days (Figure 3A). Immobilized *Chlorella* cells were capable of generating H_2 without nutrient deprivation with a maximum rate of 162 mL/L (Figure 3B). Anaerobiosis was maintained by the presence of glucose and the high respiration rate of the strain.

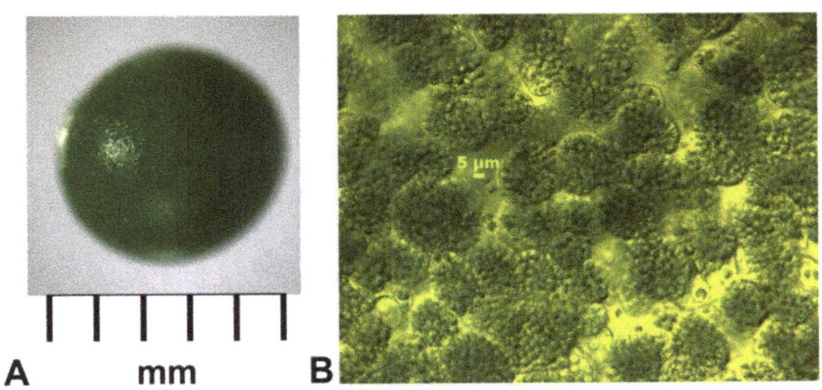

Figure 2. *Chlorella vulgaris* (BEIJ G-120 strain) immobilized in calcium alginate beads. (**A**) Calcium-alginate bead; (**B**) *Chlorella vulgaris* cells inside the calcium-alginate beads.

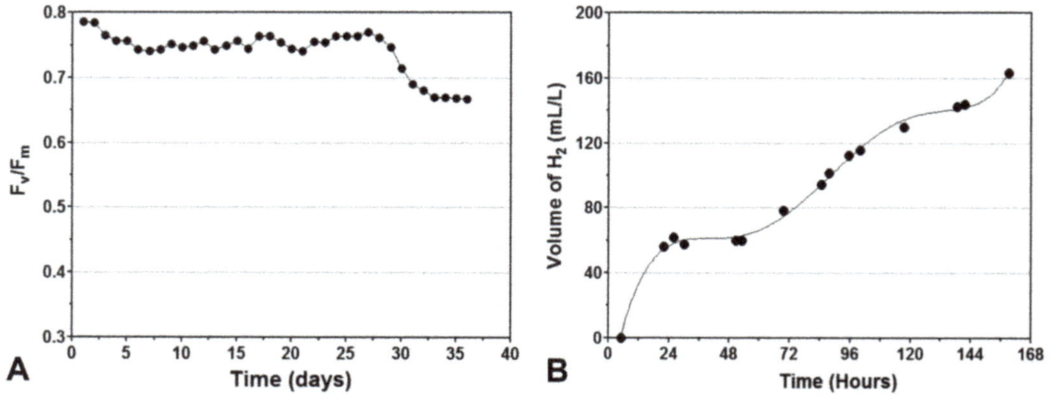

Figure 3. (**A**) F_v/F_m of immobilized *Chlorella vulgaris* cells as a function of time; (**B**) H_2 production of the immobilized *Chlorella vulgaris* cells as a function of time.

6. Hydrogen Production without Nutrient Starvation

The H_2 production protocol by Melis and coworkers based on sulfur starvation greatly improved light-driven, algal H_2 production, and particularly the possibility for researchers to study the process [68]. However, in the recent years it has become clear that it is not adequate for an industrial development of the process since it requires one to eliminate sulfur residues. Moreover, the severe reduction of PSII activity caused by the sulfur deprivation greatly reduces the H_2 production and thus the viability of the process. Awareness of these limits has prompted several workers to eliminate the sulfur deprivation phase by selecting strains with high respiration-to-photosynthesis ratios.

Liu et al. presented a work on H_2 production of *Chlorella pyrenoidosa* using $NaHCO_3$ as a carbon source and N′-(3,4-Dichlorophenyl)-N,N-dimethylurea (DCMU) [69]. In this work, *Chlorella pyrenoidosa* cells showed an overall H_2 production of 93.86 mL/L. In a recent work, Li et al. constructed a transgenic *C. reinhardtii* strain (amiRNA-D1) with a heat-inducible expression system targeting D1 gene (*psbA*). After a heat-shock, the transgenic *C. reinhardtii* strain presented a 73% decrease of *psbA* gene expression and a 60% increase of H_2 content compared to the WT strain [70]. Ben-Zvi et al. explored the in vivo H_2 production of HydA–SOD fusion phenotype in *C. reinhardtii* and found that expression of an active hydrogenase superoxide dismutase fusion protein resulted in sustained H_2 production with a rate of 20 mL H_2/L/d for 8 days [71]. Hwang et al. showed that the over-expression of the hydrogenase gene in *Chlorella vulgaris* resulted in H_2 production under aerobic conditions with continuous illumination using CO_2 as the sole source of carbon [72]. Under 5% O_2 and 10% CO_2, *Chlorella vulgaris* strains YSL01 and YSL16 produced 1.9 mL H_2/h and 1.2 mL H_2/h in 3 and 4 days, respectively. In another of their works, this group studied and compared the photosynthetic activities of *C. reinhardtii* and *Chlorella sorokiniana* with different acetate/Cl^- ratios [73]. They found that maintaining acetate/Cl^- ratios greater than 60–100 led to continuous O_2 depletion. Using fermenter effluents, at an acetate/Cl^- ratio of 150, *Chlorella sorokiniana* and *C. reinhardtii* presented an H_2 production rate of 0.25–0.33 mmol/L/min and 0.20–0.38 mmol/L/min, respectively. Kosourov et al. demonstrated sustained H_2 production by *C. reinhardtii* by shifting the culture light conditions from continuous illumination to a set of light pulses interrupted by longer dark phases [29]. In a recent work, Sirawattanamongkol et al. demonstrated that *Chlorella* sp. strain KLS Sc59 was able to produce up to 750 mL H_2/L in the presence of reducing agents such as ethanol and dithionite [74]. H_2 production rates in various microalgae strains are summarized in Table 1.

Table 1. Comparison of H_2 production rates in various microalgae strains.

Microalgal Species	Growth Mode	H_2 Production	References
C. reinhardtii	TAP	25 μmol/mg_{Chl}/h	[29]
Chlorella sp. AARL G014	TAP-S	0.49 mmol/mg_{Chl}/h	[75]
C. reinhardtii CC-503	TAP co-culture	255 mmol/mg_{Chl}	[55]
Chlorella vulgaris strains YSL01	BBM-$EDTANa_2$	1.9 mL/L	[72]
Chlorella lewinii KU201	TAP-S	13.03 mL/L	[18]
Chlorella sp. IOAC707S	TAP-NaCl	38.00 mL/L	[38]
Chlorella sorokiniana KU204	TAP-P	69.00 mL/L	[18]
Chlorella protothecoides	TAP-NS	82.50 mL/L	[76]
Chlorella sorokiniana KU204	TAP-S	89.64 mL/L	[18]
Chlorella pyrenoidosa	TCP + DCMU	93.86 mL/L	[69]
C. reinhardtii Stm6	TAP-S	540 mL/L	[41]
C. reinhardtii C3	TAP	3.0 mL/L/d	[31]
C. reinhardtii (HS-14)	TAP	20 mL/L/d	[71]
Immobilized *Chlorella vulgaris*	Artificial wastewater-S	39.1 mL/L/d	[11]
Chlorella vulgaris MACC360	TAP co-culture	56.0 mL/L/d	[77]
Immobilized *Scenedesmus obliquus*	Artificial wastewater-S	204.8 mL/L/d	[11]
Chlorella salina Mt	TAP-S	0.5 mL/L/h	[78]
C. reinhardtii CC124	TAP-S	0.6 mL/L/h	[79]
C. reinhardtii CC-124	TAP-S	3.3 mL/L/h	[80]
C. reinhardtii pgr5/pgrl1	TAP-S	7.0 mL/L/h	[33]
C. reinhardtii L159I-N230Y	TAP-S	11.1 mL/L/h	[37]
Chlorella vulgaris BEIJ (G-120)	HM + glucose	5.0 mL/L/h.	[8]
Immobilized Chlorella vulgaris NIER-10003	MA-S + glucose	238 mL/L/h	[81]
Chlorella sorokiniana	150 of acetate/Cl^- ratio	0.33 mmol/L/min	[73]
C. reinhardtii	150 of acetate/Cl^- ratio	0.38 mmol/L/min	[73]
C. reinhardtii (ΨH1)	TAP	3.6 mL/L/h	[46]
C. reinhardtti	HSM + O_2 absorbent	2.58 mL/L/h	[30]

In our recent work, we reported H_2 production by *Chlorella vulgaris* (strain BEIJ G-120) without the use of nutrient deprivation [8]. This *Chlorella* strain presents two main properties: high respiration rate and high light compensation point. By exploiting these two properties, it was possible to efficiently consume the photosynthetically produced O_2, thus maintaining anaerobiosis, even under light conditions. In this work, *Chlorella* cells presented a maximum H_2 production rate of 12 mL/L/h and an average rate of 4.98 mL/L/h. The strain was capable of producing H_2 in the dark as well, by fermentation of glucose. The excessive accumulation of byproducts of the fermentation (e.g., acetate, formate, lactate, ethanol) may inhibit H_2 production. However, the possibility of also producing H_2 in the dark by microalgae is desirable for the development of the process under natural light/dark cycle. On the other hand, some of the byproducts of dark fermentation, such as acetate, can be used as substrate for mixotrophic grown during the following light phase.

7. Theoretical Limit for Biological Hydrogen Production

Only a small fraction of the total solar light radiation (>1,100,000 EJ per year) can potentially be transformed into H_2 energy using the process of photosynthesis, according to the following general equation:

$$2H_2O + 8 \text{ photons} = O_2\uparrow + 4e^- + 4H^+ = 2H_2\uparrow$$

Step 1 indicates the total incident radiation received at the surface of the culture (100%) (Figure 4). It follows that:

1. Approximately 10% is lost by reflection and scattering (90% of initial remaining).
2. Approximately 55% of radiation is not available to drive photosynthesis since it falls outside of the photosynthetically active radiation (400–700 nm) and thus is not utilized

by photosynthetic pigments. As a result, the total amount of available light drops to 41%.

3. About 20.4% of the radiation is lost as heat [82].
4. Assuming as quantum requirement that 8 photons are required to produce 2 mol of H_2, and considering that 1 mol of H_2 is 286 KJ, and the mean energy for charge separation at PSII and PSI is 173.5 KJ/mol, it follows that the efficiency of the process will be the following: (286 KJ/mol × 2)/(173.5 KJ/mol × 8) × 100 = 41.2%, with a corresponding loss of energy of 59%. Consequently, the theoretical LCE for H_2 production, attainable by direct biophotolysis is about 13.4% of incident solar light [83].
5. With a LCE of about 10%, assuming that approximately 20% of the energy can be lost for cell maintenance, it might be possible to produce about 600,000 m^3/ha/y of H_2 in sunny areas.

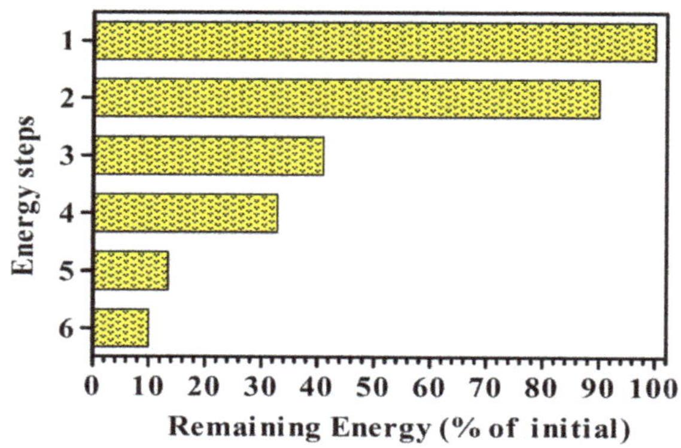

Figure 4. The energy losses of the incident solar light during the different steps of the photobiological H_2 production process.

8. Chlorophyll Fluorescence Measurements as a Tool for Monitoring Changes of Photochemical Efficiency during the Hydrogen Production Process

Chlorophyll fluorescence is a fast and non-invasive tool for monitoring residual photosynthetic activity during the H_2 production process through changes in the maximum quantum yield of PSII (F_v/F_m) and the effective quantum yield ($\Delta F/F'_m$) [38]. In particular, it has been observed that in *C. reinhardtii*, at the occurrence of anaerobiosis, the value of $\Delta F/F'_m$ rapidly declined, and this drop could be ascribed to the state 1 to state 2 transition, controlled by the redox state of plasto-quinone (PQ)-pool [36,84–86]. This mechanism regulates the migration of the light-harvesting complex (LHC) from PSII (state 1) to PSI (state 2), and it is induced under high level of PQ-pool reduction and excess of light energy. The start of the H_2 production induces a partial oxidation of the photosynthetic electron chains, comprising PQ-pool, with a partial recovery of $\Delta F/F'_m$. In *C. reinhardtii*, migration of LHC can involve up to 80% of the total LHC. The redox potential of the cells represents another important parameter related to the cell physiology under anaerobiosis, as it is the result of a balance between starch degradation, the capacity of PSII to perform photosynthesis, and the ability of cell to dissipate electrons from PQ-pool. Indeed, after establishing anaerobiosis, the value of the redox potential changes from a positive initial value to a very low value (about –550mV in *C. reinhardtii*). The changes of the values of the redox potential lag behind the changes in the yield and are less rapid that the chlorophyll fluorescence changes. In Figure 5, an example of the typical kinetics of chlorophyll fluorescence yield and redox potential in the different phases of the H_2 production process is reported.

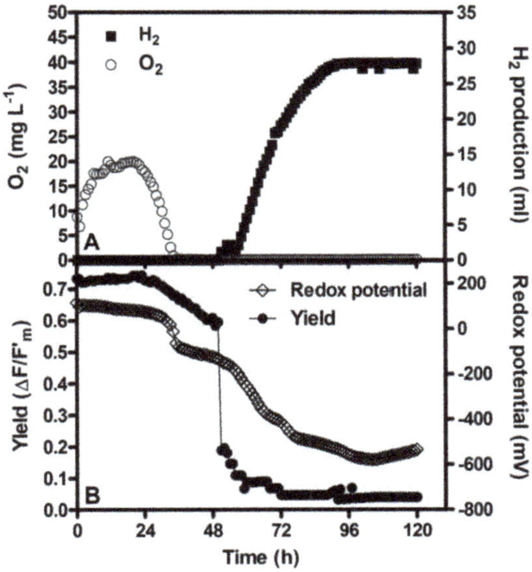

Figure 5. *C. reinhardtii* kinetics of chlorophyll fluorescence and redox potential during induction of H_2 production under sulfur deprivation. (**A**) Time courses in dissolved oxygen (empty circle) and output of hydrogen (H_2) (filled square). (**B**) The time courses in the effective quantum yield of PSII ($\Delta F/F'_m$) (filled circle) and the redox potential (Eh) (empty diamond) in *C. reinhardtii* under sulfur deprivation with 70 μmol photons/m^2/s, supplied on both sides of the reactor.

Other important information on the changes of the photosynthetic efficiency can be provided by the chlorophyll fluorescence rise kinetics (OJIP curve), strictly reflecting the progressive reduction of the photosynthetic electron transport chain [60,87], which can indicate and quantify the reduction of electrons transport for each step [88]. The most evident change occurs at the J-step level, indicating the reduced transfer of electrons further than Q_A, measured by V_J parameter, and thus, an accumulation of reduced Q_A^- [88]. An example of the changes of the shape of the OJIP curve during the occurrence of anaerobiosis in *C. reinhardtii* is reported in Figure 6.

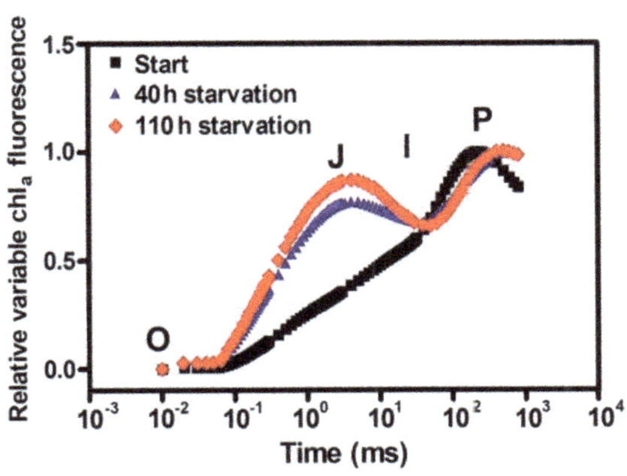

Figure 6. Effect of sulfur deprivation on the chlorophyll a fluorescence transient in *C. reinhardtii* cultures. Start (dark squares); 40 h of sulfur deprivation (blue triangles); 110 h of sulfur deprivation (red diamonds). Relative variable fluorescence ($V_t = (F_t - F_0)/(F_m - F_0)$) [89].

In *Chlorella sorokiniana,* nutrient starvation conditions reduce photosynthetic activity and induce anaerobiosis and H_2 production, indicated by the decrease of both F_v/F_m and $\Delta F/F'_m$. Similarly, to what was observed in *C. reinhardtii*, the maintenance of a residual PSII activity provides electron to hydrogenase enzyme [90]. The same behavior was observed in *Chlorella vulgaris* without nutrient starvation with a strain able to reach anaerobiosis in complete medium, showing a decline of F_v/F_m and $\Delta F/F'_m$ within 24 h [8].

In conclusion, the use of fluorescence measurements to monitor changes in photosynthetic activity can help us to better understand the physiological status of microalgae during the H_2 production process, making it easier to interfere in the cell metabolism or enhance the production process. Moreover, the application of chlorophyll fluorescence helps in selecting strains more resistant to the stress imposed by anaerobic conditions, and with higher potential H_2 output.

9. Photobiological Hydrogen Production in Outdoor Photobioreactors

Until now, H_2 production experiments using *C. reinhardtii* have been carried out mostly under laboratory conditions. Mean LCE in sulfur-deprived laboratory cultures grown in well-mixed PBR has hardly surpassed 1% (light to H_2). The necessity to downregulate the PSII activity to the level of the respiration is considered the main reason for such a low efficiency. As a matter of fact, the LCE strongly increased when it was possible to use microalgal strains with high respiration-to-photosynthesis ratio. This was the case of the *Chlorella* strain G-120, which averaged 3.2%, over the 8-day period [8].

The utilization of solar energy is mandatory for the economical scale-up of the H_2 production process. However, under solar light, the light energy received by microalgae cells exceeds their ability of light conversion into valuable biomass. This leads to either energy dissipation as heat or to photodamage and cell death, which strongly reduce the LCE. To reduce the "saturation effect", a number of PBR designs have been proposed [91]. Torzillo and coworkers reported H_2 production of about 21% of that attained under laboratory in an outdoor 50 L tubular PBR using *C. reinhardtii* under sulfur deprivation (Figure 7) [92]. The PBR consisted of ten glass tubes (2.0 m length and 4.85 cm internal diameter) placed horizontally and connected by polyvinylchloride U-bends (Figure 7). The PBR was placed in a stainless-steel container with temperature-controlled water. A polyvinyl chloride pump allowed the culture to circulate.

Figure 7. The tubular photobioreactor (working volume 50 L) used for the outdoor H_2 production experiments. The photobioreactor was equipped with probes for measurement and control of pH and temperature. Culture speed can be adjusted to reach the desired turbulence.

The low performance of the culture was explained by the rapid PSII inactivation by the high light irradiation, which during the experiments reached as much as 1850 μmol photons/m^2/s in the middle of day. In order to avoid the problem of light saturation, Giannelli and Torzillo, 2012 [79] proposed a 110 L PBR in which the culture tubes were immersed in water with light-scattering silica nanoparticles. The PBR contained 64 glass tubes (length 2.0 m, internal diameter 27.5 mm) arranged on an 8 × 8 square pitch cell and connected by polyvinylchloride U-bends. The PBR was immersed in a rectangular parallelepiped tank made of isotactic polypropylene, except for the opposite square faces, which were made of transparent Plexiglas. The culture was circulated with a peristaltic pump. The light scattering promoted by nanoparticles permitted a homogeneous distribution of light on the surface of the PBR (Figure 8). Solar light was collected by two sun-tracking mirrors, which delivered light to the opposite faces of the reactors through two light ducts (Figure 8c).

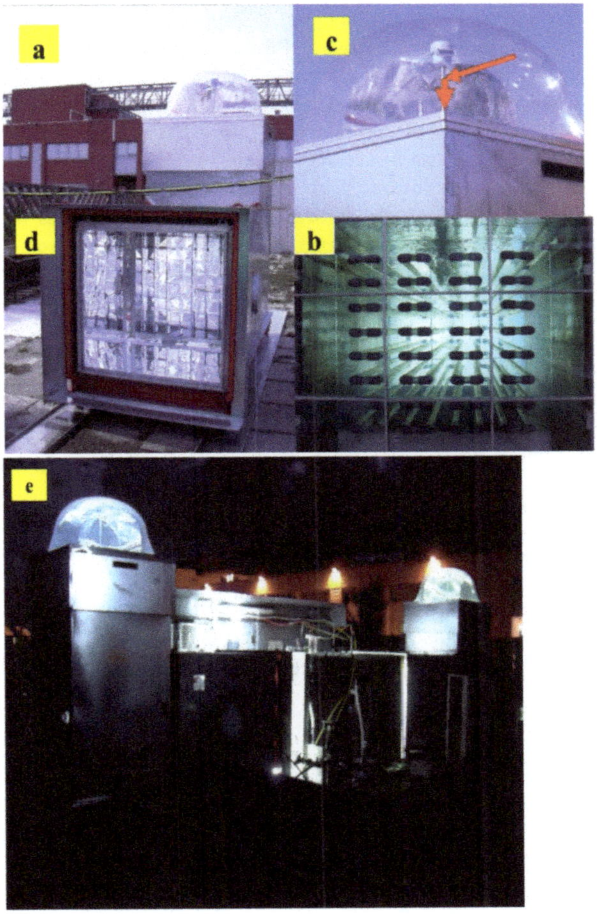

Figure 8. The 110 L photobioreactor utilized for the hydrogen production experiments. (**a**) General view of the photobioreactor; (**b**) frontal view of the photobioreactor; (**c**) sun-tracking mirrors collecting light; (**d**) light ducts delivering light to the photobioreactor from the opposite sides; (**e**) nocturnal view of the photobioreactor with two light bulbs (1000 W each) placed in the light duct to provide illumination to the culture during the night.

The total amount of H_2 collected reached 3.5 L, which was almost 2-fold higher than that obtained with the 50 L tubular PBR per unit of volume of reactor. Furthermore, in the scaled-up PBR, the LCE of the process increased from 0.055% in the 50 L horizontal PBR to 0.213% in the 110 L PBR, which was much closer to that attained in the laboratory with sulfur-deprived *C. reinhardtii* cultures. This still-low efficiency was the result of a number of factors: (i) the necessity to carry out the process according to the two-step protocol (sulfur starvation), which strongly reduces the contribution of PSII; (ii) the need to attain sulfur starvation by culture dilution, which normally yields lower H_2 output; and (iii) the longer mixing time of tubular PBRs, which delays reaching H_2-saturation levels in the cultures.

10. Optimal Photobioreactor Design for the Hydrogen Production

The experience acquired from experiments with different indoor and outdoor PBRs has provided some guidelines useful for the optimal design of future PBRs for H_2 production [93,94]. An in-depth discussion on the influence of the PBR design on the H_2 output can be found in a recent book on the subject of H_2 production [94].

The development of an economically feasible PBR system is the most important factor for successful H_2 production. Closed PBRs (such as flat-panel and tubular) are mandatory for H_2 production. Tubular PBRs contain one or more glass tubes arranged in various configurations and orientations. Flat-panel PBRs consist of one or more transparent panels containing the culture, which is circulated between the panels by a pump. Flat-panel PBRs probably meet most of the above-mentioned requirements for H_2 production. They present high area-to-volume ratio and good biomass productivity, and shorter mixing time compared to tubular PBRs, which reduces the risk of H_2 oversaturation in the reactor. A drawback is the difficulty of controlling the temperature, and the high power consumption for mixing. Although the setup costs of closed PBRs are high, they provide several advantages, such as optimal growth, minimizing the risk of culture contamination, and reduced water and CO_2 consumption.

The International Energy Agency established the commercial cost target for H_2 production at 0.3 USD/kg. According to James et al., with 10% LCE, the cost could be 2.99 USD per gallon of gasoline equivalent [95]. The authors estimated that by using robust microalgae strains presenting 1.5% LCE, the cost of H_2 would be 8.44 USD per gallon of gasoline equivalent. Greater costs for PBR construction are materials, manufacturing, and personnel costs. H_2 production prices should be more promising than market prices. Show et al. showed that the costs of fabrication materials and chemical nutrients are the main expenses (84% of the total cost) for PBR development [96]. Recycling metabolic products of PBRs (such as organic acids) and/or considering potentially cheaper nutrient sources are possible ways to reduce the chemical cost of nutrients necessary for microalgal growth [97]. Finally, production cost based on direct bio-photolysis, were estimated to be, 18.45 $/kgH$_2$, for the Netherlands, which is expected to drop significantly in the future (potential cost of 3.10 $/kgH$_2$) [98].

11. Concluding Remarks

This review provides evidence that photobiological H_2 production by microalgae and cyanobacteria might be a viable option. The discovery of the sulfur starvation method has allowed maintenance of H_2 production for several days, and thus, it has represented an opportunity to study the process beyond scientific curiosity.

Nowadays, however, a substantial change of paradigm in photobiological H_2 production is necessary. The opportunity to improve the economic feasibility of the process could come from employing strains which do not need sulfur starvation. These strains feature a high respiration-to-photosynthesis ratio and a higher level of saturation irradiance compared to the *C. reinhardtii* strains currently available. These characteristics are usually found in microalgal strains with reduced antenna size, which is a very important biotechnological condition to allow penetration of the light deep through the culture layers [99–102]. Important achievements could be expected from microalgal cultures growing both mixotrophically and heterotrophically in PBRs and fermenter, respectively, in the presence of glucose. A *Chlorella* strain with such characteristics was recently studied by us under laboratory conditions, but its better H_2 performance needs to be proved under solar light in PBRs. Of course, the use of an expensive source such as glucose as a respiratory substrate to maintain anaerobiosis, and thus the functioning of the hydrogenase, halves the efficiency of the process and strongly reduces its sustainability. Therefore, it will be important to consider the potential of much cheaper sources of organic substrates such as wastewater from sugar factories, baker's yeast, and breweries. The O_2 consumption, through respiration of organic substrates, produces H_2 with high purity (close to 98%), which strongly reduces the investment cost for H_2 purification. The use of molasses, which is very rich in glucose and sucrose, could represent an option.

In conclusion, until a hydrogenase resistant to oxygen is discovered, the selection of strains with higher resistance to oxygen and/or with high respiration-to-photosynthesis ratio represent nowadays the only realistic possibility for the success of photobiological H_2 production. This research should proceed in parallel with efforts to engineer organisms with O_2-resistant hydrogenases. Success in either direction will lead to expected improvements in technologies to: (1) increase effective conversion efficiency of photosynthesis; (2) reduce or possibly eliminate competing pathways, such as CO_2 fixation; (3) increase starch biosynthesis.

Author Contributions: Writing—original draft preparation, E.T., G.T., C.F. and A.M.S.B.; writing—review and editing, E.T. and G.T. All authors have read and agreed to the published version of the manuscript.

Funding: This work received no funds by any research agency.

Institutional Review Board Statement: Not applicable.

Informed Consent Statement: Not applicable.

Data Availability Statement: The data that support the findings of this study are available from the corresponding author upon reasonable request.

Conflicts of Interest: The authors declare no conflict of interest.

References

1. Moshood, T.D.; Nawanir, G.; Mahmud, F. Microalgae biofuels production: A systematic review on socioeconomic prospects of microalgae biofuels and policy implications. *Environ. Chall.* **2021**, *5*, 100207. [CrossRef]
2. Masojídek, J.; Torzillo, G.; Koblízek, M. Photosynthesis in microalgae. In *Handbook of Microalgal Culture: Applied Phycology and Biotechnology*, 2nd ed.; John Wiley & Sons: Chichester, UK, 2013; pp. 21–36.
3. Gaffron, H.; Rubin, J. Fermentative and photochemical production of hydrogen in algae. *J. Gen. Physiol.* **1942**, *26*, 219–240. [CrossRef]
4. Martinez-Burgos, W.J.; de Souza Candeo, E.; Pedroni Medeiros, A.B.; Cesar de Carvalho, J.; Oliveira de Andrade Tanobe, V.; Soccol, C.R.; Sydney, E.B. Hydrogen: Current advances and patented technologies of its renewable production. *J. Clean. Prod.* **2021**, *286*, 124970. [CrossRef]
5. Wang, Y.; Yang, H.; Zhang, X.; Han, F.; Tu, W.; Yang, W. Microalgal hydrogen production. *Small Methods* **2020**, *4*, 1900514. [CrossRef]
6. Dębowski, M.; Dudek, M.; Zieliński, M.; Nowicka, A.; Kazimierowicz, J. Microalgal hydrogen production in relation to other biomass-based technologies—A review. *Energies* **2021**, *14*, 6025. [CrossRef]
7. Kosourov, S.; Nagy, V.; Shevela, D.; Jokel, M.; Messinger, J.; Allahverdiyeva, Y. Water oxidation by photosystem II is the primary source of electron for sustained H_2 photoproduction in nutrient-replete green algae. *Proc. Natl. Acad. Sci. USA* **2020**, *117*, 29629–29636. [CrossRef] [PubMed]
8. Touloupakis, E.; Faraloni, C.; Silva Benavides, A.M.; Masojídek, J.; Torzillo, G. Sustained photobiological hydrogen production by *Chlorella vulgaris* without nutrient starvation. *Int. J. Hydrog. Energy* **2021**, *46*, 3684–3694. [CrossRef]
9. Melis, A. Photosynthetic H_2 metabolism in *Chlamydomonas reinhardtii* (unicellular green algae). *Planta* **2007**, *226*, 1075–1086. [CrossRef]
10. Liu, J.Z.; Ge, Y.M.; Sun, J.Y.; Chen, P.; Addy, M.; Huo, S.H.; Li, K.; Cheng, P.F.; Ruan, R. Exogenic glucose as an electron donor for algal hydrogenases to promote hydrogen photoproduction by *Chlorella pyrenoidosa*. *Bioresour. Technol.* **2019**, *289*, 121762. [CrossRef] [PubMed]
11. Ruiz-Marin, A.; Canedo-López, Y.; Chávez-Fuentes, P. Biohydrogen production by *Chlorella vulgaris* and *Scenedesmus obliquus* immobilized cultivated in artificial wastewater under different light quality. *AMB Express* **2020**, *10*, 191. [CrossRef]
12. Li, L.; Litao, Z.; Jianguo, L. Proteomic analysis of hydrogen production in *Chlorella pyrenoidosa* under nitrogen deprivation. *Algal. Res.* **2021**, *53*, 102143. [CrossRef]
13. Grechanik, V.; Naidov, I.; Bolshakov, M.; Tsygankov, A. Photoautotrophic hydrogen production by nitrogen-deprived *Chlamydomonas reinhardtii* cultures. *Int. J. Hydrog. Energy* **2021**, *46*, 3565–3575. [CrossRef]

14. Amaro, H.M.; Esquível, M.G.; Pinto, T.S.; Malcata, F.X. Hydrogen Production by Microalgae. In *Natural and Artificial Photosynthesis: Solar Power as an Energy Source*, 1st ed.; Razeghifard, R., Ed.; John Wiley and Sons, Inc.: Hoboken, NJ, USA, 2013; pp. 231–241.
15. Oliveira, F.; Araujo, A.P.C.; Romao, B.B.; Cardoso, V.L.; Ferreira, J.S.; Batista, F.R.X. Hydrogen photo-production using *Chlorella* sp. through sulfur-deprived and hybrid system strategy. *Chem. Eng. Trans.* **2015**, *43*, 301–306.
16. Jimenez-Llanos, J.; Ramirez-Carmona, M.; Rendon-Castrillon, L.; Ocampo-Lopez, C. Sustainable biohydrogen production by *Chlorella* sp. microalgae: A review. *Int. J. Hydrog. Energy* **2020**, *45*, 8310–8328. [CrossRef]
17. Nagarajan, D.; Dong, C.D.; Chen, C.Y.; Lee, D.J.; Chang, J.S. Biohydrogen production from microalgae-Major bottlenecks and future research perspectives. *Biotechnol. J.* **2021**, *16*, e2000124. [CrossRef] [PubMed]
18. Pongpadung, P.; Liu, J.; Yokthongwattana, K.; Techapinyawat, S.; Juntawong, N. Screening for hydrogen producing strains of green microalgae in phosphorus or sulfur deprived medium under nitrogen limitation. *Sci. Asia* **2015**, *41*, 97. [CrossRef]
19. Rashid, N.; Lee, K.; Han, J.I.; Gross, M. Hydrogen production by immobilized *Chlorella vulgaris*: Optimizing pH, carbon source and light. *Bioproc. Biosyst. Eng.* **2013**, *36*, 867–872. [CrossRef]
20. Alalayah, W.M.; Alhamed, Y.A.; Al-Zahrani, A.; Edris, G. Influence of culture parameters on biological hydrogen production using green algae *Chlorella vulgaris*. *Rev. Chim.* **2015**, *66*, 788–791.
21. Melis, A. Green alga hydrogen production: Progress, challenges and prospects. *Int. J. Hydrog. Energy* **2002**, *27*, 1217–1228. [CrossRef]
22. Tsygankov, A.A.; Kosourov, S.N.; Tolstygina, I.V.; Ghirardi, M.L.; Seibert, M. Hydrogen production by sulfur-deprived *Chlamydomonas reinhardtii* under photoautotrophic conditions. *Int. J. Hydrog. Energy* **2006**, *31*, 1574–1584. [CrossRef]
23. Batyrova, K.; Gavrisheva, A.; Ivanova, E.; Liu, J.G.; Tsygankov, A. Sustainable hydrogen photoproduction by phosphorus-deprived marine green microalgae *Chlorella* sp. *Int. J. Mol. Sci.* **2015**, *16*, 2705–2716. [CrossRef]
24. Rashid, N.; Lee, K.; Mahmood, Q. Bio-hydrogen production by *Chlorella vulgaris* under diverse photoperiods. *Bioresour. Technol.* **2011**, *102*, 2101–2104. [CrossRef]
25. Khetkorn, W.; Rastogi, R.P.; Incharoensakdi, A.; Lindblad, P.; Madamwar, D.; Pandey, A.; Larroche, C. Microalgal hydrogen production—A review. *Bioresour. Technol.* **2017**, *243*, 1194–1206. [CrossRef]
26. Kosourov, S.; Böhm, M.; Senger, M.; Berggren, G.; Stensjö, K.; Mamedov, F.; Lindblad, P.; Allahverdiyeva, Y. Photosynthetic hydrogen production: Novel protocols, promising engineering approaches and application of semi-synthetic hydrogenases. *Physiol. Plant* **2021**, in press. [CrossRef] [PubMed]
27. Gonzalez-Ballester, D.; Jurado-Oller, J.L.; Fernandez, E. Relevance of nutrient media composition for hydrogen production in *Chlamydomonas*. *Photosynth. Res.* **2015**, *125*, 395–406. [CrossRef]
28. Fakhimi, N.; Dubini, A.; Tavakoli, O.; González-Ballester, D. Acetic acid is key for synergetic hydrogen production in *Chlamydomonas*-bacteria co-cultures. *Bioresour. Technol.* **2019**, *289*, 121648. [CrossRef] [PubMed]
29. Kosourov, S.; Jokel, M.; Aro, E.M.; Allahverdiyeva, Y. A new approach for sustained and efficient H_2 photoproduction by: *Chlamydomonas reinhardtii*. *Energy Environ. Sci.* **2018**, *11*, 1431–1436. [CrossRef]
30. Nagy, V.; Podmaniczki, A.; Vidal-Meireles, A.; Tengölics, R.; Kovács, L.; Rákhely, G.; Scoma, A.; Tóth, S.Z. Water-splitting-based, sustainable and efficient H_2 production in green algae as achieved by substrate limitation of the Calvin–Benson–Bassham cycle. *Biotechnol. Biofuels* **2018**, *11*, 69. [CrossRef]
31. Krishna, P.S.; Styring, S.; Mamedov, F. Photosystem ratio imbalance promotes direct sustainable H_2 production in *Chlamydomonas reinhardtii*. *Green Chem.* **2019**, *21*, 4683–4690. [CrossRef]
32. Chen, M.; Zhang, J.; Zhao, L.; Xing, J.; Peng, L.; Kuang, T.; Rochaix, J.D.; Huang, F. Loss of algal Proton Gradient Regulation 5 increases reactive oxygen species scavenging and H_2 evolution. *J. Integr. Plant Biol.* **2016**, *58*, 943–946. [CrossRef]
33. Steinbeck, J.; Nikolova, D.; Weingarten, R.; Johnson, X.; Richaud, P.; Peltier, G.; Hermann, M.; Magneschi, L.; Hippler, M. Deletion of proton gradient regulation 5 (PGR5) and PGR5-Like 1 (PGRL1) proteins promote sustainable light-driven hydrogen production in *Chlamydomonas reinhardtii* due to increased PSII activity under sulfur deprivation. *Front. Plant Sci.* **2015**, *6*, 1–11. [CrossRef]
34. Pinto, T.S.; Malcata, F.X.; Arrabaça, J.D.; Silva, J.M.; Spreitzer, R.J.; Esquivel, M.G. Rubisco mutants of *Chlamydomonas reinhardtii* enhance photosynthetic hydrogen production. *Appl. Microbiol. Biotechnol.* **2013**, *97*, 5635–5643. [CrossRef]
35. Eilenberg, H.; Weiner, I.; Ben-Zvi, O.; Pundak, C.; Marmari, A.; Liran, O.; Wecker, M.S.; Milrad, Y.; Yacoby, I. The dual effect of a ferredoxin-hydrogenase fusion protein in vivo: Successful divergence of the photosynthetic electron flux towards hydrogen production and elevated oxygen tolerance. *Biotechnol. Biofuels* **2016**, *9*, 182. [CrossRef] [PubMed]
36. Torzillo, G.; Scoma, A.; Faraloni, C.; Ena, A.; Johanningmeier, U. Increased hydrogen photoproduction by means of a sulfur-deprived *Chlamydomonas reinhardtii* D1 protein mutant. *Int. J. Hydrog. Energy* **2009**, *34*, 4529–4536. [CrossRef]
37. Scoma, A.; Krawietz, D.; Faraloni, C.; Giannelli, L.; Happe, T.; Torzillo, G. Sustained H_2 production in a *Chlamydomonas reinhardtii* D1 protein mutant. *J. Biotechnol.* **2012**, *157*, 613–619. [CrossRef]
38. Batyrova, K.; Hallenbeck, P.C. Hydrogen Production by a *Chlamydomonas reinhardtii* strain with inducible expression of photosystem II. *Int. J. Mol. Sci.* **2017**, *18*, 647. [CrossRef] [PubMed]
39. Kosourov, S.N.; Ghirardi, M.L.; Seibert, M. A truncated antenna mutant of *Chlamydomonas reinhardtii* can produce more hydrogen than the parental strain. *Int. J. Hydrog. Energy* **2011**, *36*, 2044–2048. [CrossRef]
40. Xu, F.Q.; Ma, W.M.; Zhu, X.G. Introducing pyruvate oxidase into the chloroplast of *Chlamydomonas reinhardtii* increases oxygen consumption and promotes hydrogen production. *Int. J. Hydrog. Energy* **2011**, *36*, 10648. [CrossRef]

41. Kruse, O.; Rupprecht, J.; Bader, K.P.; Thomas-Hall, S.; Schenk, P.M.; Finazzi, G.; Hankamer, B. Improved photobiological H_2 production in engineered green algal cells. *J. Biol. Chem.* **2005**, *280*, 34170–34177. [CrossRef]
42. Volgusheva, A.; Styring, S.; Mamedov, F. Increased photosystem II stability promotes H_2 production in sulfur-deprived *Chlamydomonas reinhardtii. Proc. Natl. Acad. Sci. USA* **2013**, *110*, 7223. [CrossRef] [PubMed]
43. Oey, M.; Ross, I.L.; Stephens, E.; Steinbeck, J.; Wolf, J.; Radzun, K.A.; Kugler, J.; Ringsmuth, A.K.; Kruse, O.; Hankamer, B. RNAi knock-down of LHCBM1, 2 and 3 increases photosynthetic H_2 production efficiency of the green alga *Chlamydomonas reinhardtii. PLoS ONE* **2013**, *8*, e61375. [CrossRef]
44. Wu, S.X.; Yan, G.Y.; Xu, L.L.; Wang, Q.X.; Liu, X.L. Improvement of hydrogen production with expression of lba gene in chloroplast of *Chlamydomonas reinhardtii. Int. J. Hydrog. Energy* **2010**, *35*, 13419. [CrossRef]
45. Noone, S.; Ratcliff, K.; Davis, R.; Subramanian, V.; Meuser, J.; Posewitz, M.C.; Ghirardi, M.L. Expression of a clostridial [FeFe]-hydrogenase in *Chlamydomonas reinhardtii* prolongs photo-production of hydrogen from water splitting. *Algal. Res.* **2017**, *22*, 116–121. [CrossRef]
46. Kanygin, A.; Milrad, Y.; Thummala, C.; Reischneider, K.; Baker, P.; Pini, M.; Jacoby, I.; Redding, K.E. Rewriting photosynthesis, I-hydrogenase chimera that makes H_2 in vivo. *Energy Environ. Sci.* **2020**, *13*, 2903. [CrossRef]
47. Touloupakis, E.; Silva Benavides, A.M.; Cicchi, N.; Torzillo, G. Growth and hydrogen production of outdoor cultures of *Synechocystis* PCC 6803. *Algal. Res.* **2016**, *16*, 78–85. [CrossRef]
48. Kosourov, S.; Murukesan, G.; Seibert, M.; Allahverdiyeva, Y. Evaluation of light energy to H_2 energy conversion efficiency in thin films of cyanobacteria and green alga under photoautotrophic conditions. *Algal. Res.* **2017**, *28*, 253–263. [CrossRef]
49. Paramesh, K.; Chandrasekhar, T. Improvement of photobiological hydrogen production in *Chlorococcum minutum* using various oxygen scavengers. *Int. J. Hydrog. Energy* **2020**, *45*, 7641–7646. [CrossRef]
50. Su, D.; Qi, J.; Liu, X.; Wang, L.; Zhang, H.; Xie, H.; Huang, X. Enzyme-modulated anaerobic encapsulation of *Chlorella* cells allows switching from O_2 to H_2 production. *Angew. Chem. Int. Ed. Engl.* **2019**, *58*, 3992–3995. [CrossRef] [PubMed]
51. Márquez-Reyes, L.A.; Sánchez-Saavedra, M.P.; Valdez-Vazquez, I. Improvement of hydrogen production by reduction of the photosynthetic oxygen in microalgae cultures of *Chlamydomonas gloeopara* and *Scenedesmus obliquus. Int. J. Hydrog. Energy* **2015**, *40*, 7291–7300. [CrossRef]
52. Chen, J.; Li, J.; Li, Q.; Wang, S.; Wang, L.; Liu, H.; Fan, C. Engineering a chemoenzymatic cascade for sustainable photobiological hydrogen production with green algae. *Energy Environ. Sci.* **2020**, *13*, 2064. [CrossRef]
53. Scognamiglio, V.; Giardi, M.T.; Zappi, D.; Touloupakis, E.; Antonacci, A. Photoautotrophs–bacteria co-cultures: Advances, challenges and applications. *Materials* **2021**, *14*, 3027. [CrossRef] [PubMed]
54. Ban, S.; Lin, W.; Wu, F.; Luo, J. Algal-bacterial cooperation improves algal photolysis-mediated hydrogen production. *Bioresour. Technol.* **2018**, *251*, 350–357. [CrossRef] [PubMed]
55. He, J.; Xi, L.; Sun, X.; Ge, B.; Liu, D.; Han, Z.; Pu, X.; Huang, F. Enhanced hydrogen production through co-cultivation of *Chlamydomonas reinhardtii* CC-503 and a facultative autotrophic sulfide-oxidizing bacterium under sulfurated conditions. *Int. J. Hydrog. Energy* **2018**, *43*, 15005–15013. [CrossRef]
56. Fakhimi, N.; Tavakoli, O.; Marashi, S.-A.; Moghimi, H.; Mehrnia, M.R.; Dubini, A.; González-Ballester, D. Acetic acid uptake rate controls H_2 production in *Chlamydomonas*-bacteria co-cultures. *Algal. Res.* **2019**, *42*, 101605. [CrossRef]
57. Fakhimi, N.; Gonzalez-Ballester, D.; Fernández, E.; Galván, A.; Dubini, A. Algae-bacteria consortia as a strategy to enhance H_2 production. *Cells* **2020**, *9*, 1353. [CrossRef]
58. Hom, E.; Aiyar, P.; Schaeme, D.; Mittag, M.; Sasso, S. A chemical perspective on microalgal-microbial interactions. *Trends Plant Sci.* **2015**, *20*, 689–693. [CrossRef] [PubMed]
59. Fuentes, J.L.; Nores, I.G.; Cuaresma, M.; Montero, Z.; Del Valle, M.G.; Vílchez, C. Impact of microalgae-bacteria interactions on the production of algal biomass and associated compounds. *Mar. Drugs* **2016**, *14*, 100. [CrossRef]
60. Xu, L.; Li, D.; Wang, Q.; Wu, S. Improved hydrogen production and biomass through the co-cultivation of *Chlamydomonas reinhardtii* and *Bradyrhizobium japonicum. Int. J. Hydrog. Energy* **2016**, *41*, 9276–9283. [CrossRef]
61. Khosravitabar, F. Microalgal biohydrogen photoproduction: Scaling up challenges and the ways forward. *J. Appl. Phycol.* **2020**, *32*, 277–289. [CrossRef]
62. Kosourov, S.N.; He, M.; Allahverdiyeva, Y.; Seibert, M. Immobilization of microalgae as a tool for efficient light utilization in H_2 production and other biotechnology applications. In *Microalgal Hydrogen Production*; Royal Society of Chemistry: London, UK, 2018; pp. 355–384.
63. Touloupakis, E.; Rontogiannis, G.; Silva Benavides, A.M.; Cicchi, B.; Ghanotakis, D.F.; Torzillo, G. Hydrogen production by immobilized *Synechocystis* sp. PCC 6803. *Int. J. Hydrog. Energy* **2016**, *41*, 15181–15186. [CrossRef]
64. Tsygankov, A.; Kosourov, S. Microbial BioEnergy: Hydrogen production, advances in photosynthesis and respiration. In *Immobilization of Photosynthetic Microorganisms for Efficient Hydrogen Production*; Zannoni, D., De Philippis, R., Eds.; Springer Science & Business Media: Dordrecht, The Netherlands, 2014; pp. 321–347.
65. Laurinavichene, T.V.; Fedorov, A.S.; Ghirardi, M.L.; Seibert, M.; Tsygankov, A.A. Demonstration of sustained hydrogen photoproduction by immobilized, sulfur-deprived *Chlamydomonas reinhardtii* cells. *Int. J. Hydrog. Energy* **2006**, *31*, 659–667. [CrossRef]
66. Kosourov, S.N.; Seibert, M. Hydrogen photoproduction by nutrient-deprived *Chlamydomonas reinhardtii* cells immobilized within thin alginate films under aerobic and anaerobic conditions. *Biotechnol. Bioeng.* **2009**, *102*, 50–58. [CrossRef] [PubMed]

67. Maswanna, T.; Lindblad, P.; Maneeruttanarungroj, C. Improved biohydrogen production by immobilized cells of the green alga *Tetraspora* sp. CU2551 incubated under aerobic condition. *J. Appl. Phycol.* **2020**, *32*, 2937–2945. [CrossRef]
68. Melis, A.; Zhang, L.; Forestier, M.; Ghirardi, M.L.; Seibert, M. Sustained photobiological hydrogen gas production upon reversible inactivation of oxygen evolution in the green alga *Chlamydomonas reinhardtii*. *Plant Physiol.* **2000**, *122*, 127–136. [CrossRef]
69. Liu, J.-Z.; Ge, Y.-M.; Xia, S.-Y.; Sun, J.-Y.; Mu, J. Photoautotrophic hydrogen production by *Chlorella pyrenoidosa* without sulfur deprivation. *Int. J. Hydrog. Energy* **2016**, *41*, 8427–8432. [CrossRef]
70. Li, H.; Liu, Y.; Wang, Y.; Chen, M.; Zhuang, X.; Wang, C.; Wang, J.; Hu, Z. Improved photobio-H_2 production regulated by artificial miRNA targeting psbA in green microalga *Chlamydomonas reinhardtii*. *Biotechnol. Biofuels* **2018**, *11*, 36. [CrossRef]
71. Ben-Zvi, O.; Dafni, E.; Feldman, Y.; Yacoby, I. Re-routing photosynthetic energy for continuous hydrogen production in vivo. *Biotechnol. Biofuels* **2019**, *12*, 266. [CrossRef]
72. Hwang, J.H.; Kim, H.C.; Choi, J.A.; Abou-Shanab, R.A.; Dempsey, B.A.; Regan, J.M.; Kim, J.R.; Song, H.; Nam, I.H.; Kim, S.N.; et al. Photoautotrophic hydrogen production by eukaryotic microalgae under aerobic conditions. *Nat. Commun.* **2014**, *5*, 3234. [CrossRef]
73. Hwang, J.-H.; Lee, M.; Kang, E.H.; Lee, W.H. Renewable algal photo H_2 production without S control using acetate enriched fermenter effluents. *Int. J. Hydrog. Energy* **2021**, *46*, 1740–1751. [CrossRef]
74. Sirawattanamongkol, T.; Maswanna, T.; Maneeruttanarungroj, C. A newly isolated green alga *Chlorella* sp. KLSc59: Potential for biohydrogen production. *J. Appl. Phycol.* **2020**, *32*, 2927–2936. [CrossRef]
75. Duangjan, K.; Nakkhuntho, W.; Pekkoh, J.; Pumas, C. Comparision of hydrogen production in microalgae under autotrophic mixotrophic media. *Bot. Lith.* **2017**, *23*, 169–177.
76. Zhang, L.; He, M.; Liu, J.; Li, L. Role of the mitochondrial alternative oxidase pathway in hydrogen photoproduction in *Chlorella protothecoides*. *Planta* **2015**, *241*, 1005–1014. [CrossRef]
77. Shetty, P.; Boboescu, I.Z.; Pap, B.; Wirth, R.; Kovacs, K.L.; Bíro, T.; Futo, Z.; White, R.A.; Maroti, G. Exploitation of algal-bacterial consortia in combined biohydrogen generation and wastewater treatment. *Front. Energy Res.* **2019**, 7. [CrossRef]
78. Chader, S.; Hacene, H.; Agathos, S.N. Study of hydrogen production by three strains of *Chlorella* isolated from the soil in the Algerian Sahara. *Int. J. Hydrog. Energy* **2009**, *34*, 4941–4946. [CrossRef]
79. Giannelli, L.; Torzillo, G. Hydrogen production with the microalga *Chlamydomonas reinhardtii* grown in a compact tubular photobioreactor immersed in a scattering light nanoparticle suspension. *Int. J. Hydrog. Energy* **2012**, *37*, 16951–16961. [CrossRef]
80. Faraloni, C.; Ena, A.; Pintucci, C.; Torzillo, G. Enhanced hydrogen production by means of sulfur-deprived *Chlamydomonas reinhardtii* cultures grown in pretreated olive mill wastewater. *Int. J. Hydrog. Energy* **2011**, *36*, 5920–5931. [CrossRef]
81. Song, W.; Rashid, N.; Choi, W.; Lee, K. Biohydrogen production by immobilized *Chlorella* sp. using cycles of oxygenic photosynthesis and anaerobiosis. *Bioresour. Technol.* **2011**, *102*, 8676–8681. [CrossRef]
82. Zhu, X.-G.; Long, S.P.; Ort, D.R. What is the maximum efficiency with which photosynthesis can convert solar energy into biomass? *Curr. Opin. Biotechnol.* **2008**, *19*, 153–159. [CrossRef]
83. Melis, A. Solar energy conversion efficiencies in photosynthesis: Minimizing the chlorophyll antennae to maximize efficiency. *Plant Sci.* **2009**, *177*, 272–280. [CrossRef]
84. Finazzi, G.; Barbagallo, R.P.; Bergo, E.; Barbato, R.; Forti, G. Photoinhibition of *Chlamydomonas reinhardtii* in State 1 and State 2: Damages to the photosynthetic apparatus under linear and cyclic electron flow. *J. Biol. Chem.* **2001**, *276*, 22251. [CrossRef]
85. Antal, T.K.; Krendeleva, T.E.; Laurinavichene, T.V.; Makarova, W.; Ghirardi, M.L.; Rubin, A.B.; Tsygankov, A.A.; Seibert, M. The dependence of algal H_2 production on photosystem II and O_2 consumption activities in sulphur-deprived *Chlamydomonas reinhardtii* cells. *Biochem. Biophys. Acta* **2003**, *1607*, 153–160.
86. Burlacot, A.; Sawyer, A.; Cuiné, S.; Auroy-Tarrago, P.; Blangy, S.; Happe, T.; Peltiera, G. Flavodiiron-mediated O_2 photoreduction links H_2 production with CO_2 fixation during the anaerobic induction of photosynthesis. *Plant Physiol.* **2018**, *177*, 1639–1649. [CrossRef]
87. Antal, T.K.; Kukarskikh, G.P.; Volgusheva, A.A.; Krendeleva, T.E.; Tyystjärvi, E.; Rubin, A.B. Hydrogen photoproduction by immobilized S-deprived *Chlamydomonas reinhardtii*: Effect of light intensity and spectrum, and initial medium pH. *Algal. Res.* **2016**, *17*, 38–45. [CrossRef]
88. Strasser, R.; Srivastava, A.; Govindjee. Polyphasic chlorophyll a fluorescence transient in plants and cyanobacteria. *Photochem. Photobiol.* **1995**, *61*, 32–42. [CrossRef]
89. Faraloni, C.; Torzillo, G. Xanthophyll cycle induction by anaerobic conditions under low light in *Chlamydomonas reinhardtii*. *J. Appl. Phycol.* **2013**, *25*, 1457–1471. [CrossRef]
90. Pongpadung, P.; Zhang, L.; Sathasivam, R.; Yokthongwattana, K.; Juntawon, N.; Liu, J. Stimulation of hydrogen photoproduction in *Chlorella sorokiniana* subjected to simultaneous nitrogen limitation and sulfur- and/or phosphorus-deprivation. *J. Pure Appl. Microbiol.* **2018**, *12*, 1719–1727. [CrossRef]
91. Torzillo, G.; Chini Zittelli, G. Tubular photobioreactors. Products and biorefinery design. In *Algal Biorefineries*; Prokop, A., Bajpai, R.K., Zappi, M.E., Eds.; Springer International Publishing: Cham, Switzerland, 2015; Volume 2.
92. Scoma, A.; Giannelli, L.; Faraloni, C.; Torzillo, G. Outdoor H_2 production in a 50-L tubular photobioreactor by means of a sulfur-deprived culture of the microalga *Chlamydomonas reinhardtii*. *J. Biotechnol.* **2012**, *157*, 620–627. [CrossRef] [PubMed]

93. Lindblad, P.; Fuente, D.; Borbe, F.; Cicchi, B.; Conejero, J.A.; Couto, N.; Čelešnik, H.; Diano, M.; Dolinar, M.; Esposito, S.; et al. CyanoFactory, a European consortium to develop technologies needed to advance cyanobacteria as chassis for production of chemicals and fuels. *Algal. Res.* **2019**, *41*, 101510. [CrossRef]
94. Siebert, M.; Torzillo, G. *Microalgal Hydrogen Production: Achievements and Perspectives*; The Royal Society of Chemistry: Cambridge, UK, 2018.
95. James, B.D.; Baum, G.N.; Perez, J.; Baum, K.N. *Technoeconomic Analysis of Photoelectrochemical (PEC) Hydrogen Production*; (US DOE Contract no. GS-10F-009J); Directed Technologies Inc.: Arlington, VA, USA, 2009.
96. Show, K.-Y.; Yan, Y.; Zong, C.; Guo, N.; Chang, J.-S.; Lee, D.-J. State of the art and challenges of biohydrogen from microalgae. *Bioresour. Technol.* **2019**, *289*, 121747. [CrossRef]
97. Silva Benavides, A.M.; Campos Rudin, M.; Villalobos, N.; Touloupakis, E.; Torzillo, G. Growth and hydrogen production by three *Chlamydomonas* strains cultivated in a commercial fertilizer. *Int. J. Hydrog. Energy* **2019**, *44*, 9849–9855. [CrossRef]
98. Frowijn, L.S.F.; van Sark, W.G.J.H.M. Analysis of photon-driven solar-to-hydrogen production methods in the Netherlands, *Sustain. Energy Technol. Assess.* **2021**, *48*, 101631.
99. Perrine, Z.; Negi, S.; Sayre, R. Optimization of photosynthetic light energy utilization by microalgae. *Algal. Res.* **2012**, *1*, 134–142. [CrossRef]
100. Cazzaniga, S.; Dall'Osto, L.; Szaub, J.; Ballottari, M.; Purton, S.; Bassi, R. Domestication of the green alga *Chlorella sorokiniana*: Reduction of antenna size improves light-use efficiency in a photobioreactor. *Biotechnol. Biofuels* **2014**, *7*, 157–170. [CrossRef]
101. Shin, W.-S.; Lee, B.; Jeong, B.-R.; Chang, Y.K. Truncated light-harvesting chlorophyll antenna size in *Chlorella vulgaris* improves biomass productivity. *J. Appl. Phycol.* **2016**, *28*, 3193–3202. [CrossRef]
102. Hu, G.-R.; Fan, Y.; Zhen, Y.-L.; Xu, F.; Zhang, L.; Li, F.-L. Photoprotection capacity of microalgae improved by regulating the antenna size of high-harvesting complexes. *J. Appl. Phycol.* **2020**, *32*, 1027–1039. [CrossRef]

energies MDPI

Article

CFD Simulation of Hydrogen Generation and Methane Combustion Inside a Water Splitting Membrane Reactor

Te Zhao [1], Chusheng Chen [2] and Hong Ye [1,*]

[1] Department of Thermal Science and Energy Engineering, University of Science and Technology of China, Hefei 230027, China; alamode@mail.ustc.edu.cn
[2] Collaborative Innovation Center of Chemistry for Energy Materials, Laboratory of Materials for Energy Conversion, Department of Materials Science and Engineering, University of Science and Technology of China, Hefei 230026, China; ccsm@ustc.edu.cn
* Correspondence: hye@ustc.edu.cn

Abstract: Hydrogen production from water splitting remains difficult due to the low equilibrium constant (e.g., $Kp \approx 2 \times 10^{-8}$ at 900 °C). The coupling of methane combustion with water splitting in an oxygen transport membrane reactor can shift the water splitting equilibrium toward dissociation by instantaneously removing O_2 from the product, enabling the continuous process of water splitting and continuous generation of hydrogen, and the heat required for water splitting can be largely compensated for by methane combustion. In this work, a CFD simulation model for the coupled membrane reactor was developed and validated. The effects of the sweep gas flow rate, methane content and inlet temperature on the reactor performance were investigated. It was found that coupling of methane combustion with water splitting could significantly improve the hydrogen generation capacity of the membrane reactor. Under certain conditions, the average hydrogen yield with methane combustion could increase threefold compared to methods that used no coupling of combustion. The methane conversion decreases while the hydrogen yield increases with the increase in sweep gas flow rate or methane content. Excessive methane is required to ensure the hydrogen yield of the reactor. Increasing the inlet temperature can increase the membrane temperature, methane conversion, oxygen permeation rate and hydrogen yield.

Keywords: oxygen transport membrane reactor; water splitting; methane combustion; CFD simulation; hydrogen yield

Citation: Zhao, T.; Chen, C.; Ye, H. CFD Simulation of Hydrogen Generation and Methane Combustion Inside a Water Splitting Membrane Reactor. *Energies* **2021**, *14*, 7175. https://doi.org/10.3390/en14217175

Academic Editors: Mahesh Suryawanshi and Bahman Shabani

Received: 20 September 2021
Accepted: 27 October 2021
Published: 1 November 2021

1. Introduction

As a clean, highly efficient and sustainable energy carrier, hydrogen is considered one of the most promising forms of alternative energy to conventional fossil fuels [1,2]. Under the increasingly severe situation of energy and environmental issues, the development of affordable, clean, and efficient hydrogen production technology is particularly important for hydrogen utilization [3]. Currently, hydrogen production methods can be broadly classified into three major categories based on the nature of their chemical processes and/or energy inputs: thermochemical, electrochemical, and biological methods [4]. Among these, thermochemical water splitting has attracted considerable attention because water is considered an ideal source due to its clean, abundant, and renewable characteristics [1,5]. However, the water splitting reaction for hydrogen generation is a thermodynamically limited reaction. The efficient hydrogen production from water ($2H_2O \rightleftarrows H_2+O_2$) remains difficult due to the low equilibrium constant, e.g., $Kp \approx 2 \times 10^{-8}$ at 900 °C, and only low equilibrium concentrations of $P_{O2} = 4.6 \times 10^{-6}$ bar and $P_{H2} = 9.2 \times 10^{-6}$ bar are achieved [6]. Even at high temperatures, only a small amount of hydrogen can be obtained, e.g., the generated hydrogen concentration is only 0.1% at 1600 °C.

Recently, a technique of the oxygen transport membrane (OTM) reactor was developed for hydrogen production via water splitting. The OTMs are made of mixed ionic and electronic conducting materials that can simultaneously conduct electrons and oxygen ions. Thus, the membrane has a 100% selectivity to oxygen while it is impermeable to other gases; only oxygen can permeate through the membrane [7]. In a water splitting membrane reactor, one side of the membrane (defined as the feed side) is exposed to the water vapor, where water splits into oxygen and hydrogen at an elevated temperature. Inert gases (such as N_2, He or Ar) or reactive gases (such as reducing gases: CO, syngas, or CH_4) are introduced into the other side (defined as the sweep side) to take away or react with the permeated oxygen to form a low oxygen partial pressure on this side. Driven by the oxygen partial pressure difference between the two sides, oxygen can continuously transport from the feed side to the sweep side. Through the instantaneous removal of water decomposition products O_2, the equilibrium of the water splitting reaction will be broken and shifted toward the decomposition into oxygen and hydrogen. Then, the continuous production of a substantial quantity of hydrogen can be achieved [2]. It is feasible to use the OTM reactor for water splitting to produce hydrogen. Park et al. [8] reported an experiment of using an OTM reactor based on the $La_{0.7}Sr_{0.3}Cu_{0.2}Fe_{0.8}O_{3-\delta}$ (LSCuF−7328) membrane for hydrogen generation with water vapor fed into the feed side and coal gas (CO/CO_2) fed into the sweep side. The coal gas consumed the permeated oxygen, thus improving the oxygen permeation rate and hydrogen generation (e.g., a hydrogen yield of 4.7 $cm^3/min \cdot cm^2$ at 900 °C could be achieved). Zhu et al. [1] systematically investigated the behavior of water splitting in a $La_{0.9}Ca_{0.1}FeO_{3-\delta}$ (LCF−91) OTM reactor under different reducing atmospheres (i.e., CO, H_2/CO, and CH_4). The results show that the LCF-91 membrane exhibits a favorable oxygen permeability and hydrogen production rates under reducing atmospheres (i.e., 6.17×10^{-8}, 5.23×10^{-8} and 3.90×10^{-8} $mol/cm^2 \cdot s$ under CO, H_2/CO and CH_4, respectively). If the water splitting reaction is fast enough, then the oxygen permeation process will be the controlling step of the hydrogen production. The studies of Park et al. [8] on the $La_{0.7}Sr_{0.3}Cu_{0.2}Fe_{0.8}O_{3-\delta}$ (LSCuF−7328) membrane reactor, Hong et al. [9] and Habib et al. [10] on the $La_{0.1}Sr_{0.9}Co_{0.9}Fe_{0.1}O_{3-\delta}$ (LSCoF−1991) membrane reactor, Ben-Mansour et al. [11] on the $Ba_{0.5}Sr_{0.5}Co_{0.8}Fe_{0.2}O_x$ (BSCoF−5582) membrane reactor, Jiang et al. [12] on the $BaCo_xFe_yZr_{1-x-y}O_{3-\delta}$ (BCoFZ) membrane reactor, Lee et al. [13] on $La_{0.6}Sr_{0.4}Ti_{0.2}Fe_{0.8}O_{3-\delta}$ (LSTF−6428) membrane reactor, and Zhu et al. [1] on $La_{0.9}Ca_{0.1}FeO_{3-\delta}$ (LCF−91) membrane reactor all show that, compared with inert gas, the use of reducing/reacting gas as a sweep gas can improve the oxygen permeability of the membrane reactor, thus leading to a higher hydrogen yield.

In recent years, researchers have proposed the concept of coupling water splitting with the partial oxidation of methane (POM) reaction, allowing the two reactions to proceed simultaneously in one apparatus [1,14–18]. The feed side is fed with water vapor while the sweep side is fed with methane. At 800–900 °C, the water splitting first occurs on the feed side; the product O_2 permeates from the water splitting side to the sweep side through the OTM to provide the oxygen required for the POM reaction. A valuable advantage of this membrane reactor is that it can produce hydrogen and syngas simultaneously. In addition, POM is a slightly exothermic reaction, the heat released by the POM reaction can partially compensate for the heat required for water splitting. However, at a high temperature such as 800 °C, the enthalpy change for the water splitting is approximately +248 kJ/mol, and is only −23 kJ/mol for the POM reaction. Therefore, a large amount of heat needs to be provided to the reactor from the outside to facilitate the proceeding of water splitting. For this reason, if the methane combustion reaction is coupled with water splitting in an OTM reactor, more heat can be provided for the water splitting by the complete combustion of methane to further improve the hydrogen yield.

In this work, the $La_{0.7}Sr_{0.3}Cu_{0.2}Fe_{0.8}O_{3-\delta}$ (LSCuF−7328), with a high oxygen permeability and appreciable stability, was selected as the membrane material, a CFD model for water splitting coupled with methane combustion in the LSCuF−7328 membrane reactor was developed and validated. The endothermic effect of the water splitting reaction was taken into account by adding an energy source term on the feed side. The contribution of this work is that the effect of the coupling of methane combustion on the water splitting reactor performance is analyzed and compared with the reactor without combustion. Furthermore, the effects of the sweep gas flow rate, the methane content and the inlet temperature on the reactor performance, such as the membrane temperature distribution, the oxygen permeation rate, the methane conversion and the hydrogen production rate were investigated.

2. Model

2.1. Descriptions of the Membrane Reactor

Figure 1 shows the schematic diagram of the 2D axisymmetric representation of the water splitting membrane reactor coupled with methane combustion, with x representing the axial direction and r representing the radial direction. The geometry consists of two concentric horizontal tubes with the outer tube made of quartz and the inner tube made of the LSCuF−7328 oxygen transport membrane. The reactor is divided into two zones by the membrane: a feed side and a sweep side. High-temperature water vapor in the mixture with N_2 is fed into the feed side, while CH_4 in mixture with CO_2 as the sweep gas is introduced into the sweep side (i.e., the sweep gas is a mixture of CH_4 and CO_2). Both N_2 and CO_2 are used as the carrier gases. At an elevated temperature, oxygen and hydrogen are produced by a water splitting reaction on the feed side. Driven by the oxygen partial pressure differences across the membrane, oxygen permeates through the membrane from the feed side to the sweep side and then reacts with CH_4. The permeated oxygen is consumed rapidly, the equilibrium of the water splitting reaction continues to move in the direction of generating hydrogen and oxygen, and then hydrogen is generated continuously.

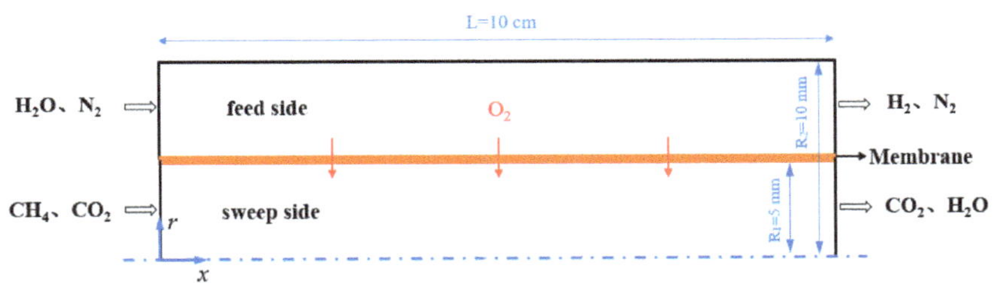

Figure 1. Schematic diagram of 2D axisymmetric representation of the coupled membrane reactor.

2.2. Governing Equations

Due to the low velocity of the inlet gases, the flow is laminar and in a steady state at both the feed and sweep sides. The outer quartz tube of the reactor is assumed to be isothermal and impermeable. The permeation process of oxygen through the OTM is achieved by adding sink and source terms in the areas adjacent to the membrane surface [19]. That is, oxygen is allowed to disappear from the feed side through the sink term and reappear on the sweep side through the source term. The mathematic model describing the processes of the flow, heat and mass transfer, as well as reaction of the two-dimensional axisymmetric ($x-$ and $r-$coordinates) reactor geometry, includes the following governing conservation equations [20]:

Continuity equation:

$$\frac{\partial}{\partial x}(\rho u) + \frac{1}{r}\frac{\partial}{\partial r}(r\rho v) = S_i \tag{1}$$

Momentum conservation equations:

x-momentum equation:

$$\frac{\partial}{\partial x}(\rho u u) + \frac{1}{r}\frac{\partial}{\partial r}(r\rho v u) = -\frac{\partial p}{\partial x} + \frac{\partial}{\partial x}\left(\mu\frac{\partial u}{\partial x}\right) + \frac{1}{r}\frac{\partial}{\partial r}\left(r\mu\frac{\partial u}{\partial r}\right) \tag{2}$$

r-momentum equation:

$$\frac{\partial}{\partial x}(\rho u v) + \frac{1}{r}\frac{\partial}{\partial r}(r\rho v v) = -\frac{\partial p}{\partial r} + \frac{\partial}{\partial x}\left(\mu\frac{\partial v}{\partial x}\right) + \frac{1}{r}\frac{\partial}{\partial r}\left(r\mu\frac{\partial v}{\partial r}\right) - \mu\frac{v}{r^2} \quad (3)$$

Energy conservation equation:

$$\frac{\partial}{\partial x}\left(\rho c_p u T\right) + \frac{1}{r}\frac{\partial}{\partial r}\left(r\rho c_p v T\right) = \frac{\partial}{\partial x}\left(k\frac{\partial T}{\partial x}\right) + \frac{1}{r}\frac{\partial}{\partial r}\left(rk\frac{\partial T}{\partial r}\right) + S_h \quad (4)$$

Species balance equation:

$$\frac{\partial}{\partial x}(\rho u Y_i) + \frac{1}{r}\frac{\partial}{\partial r}(r\rho v Y_i) = \frac{\partial}{\partial x}\left(\rho D_{i,m}\frac{\partial Y_i}{\partial x}\right) + \frac{1}{r}\frac{\partial}{\partial r}\left(r\rho D_{i,m}\frac{\partial Y_i}{\partial r}\right) + S_i + R_i \quad (5)$$

where u and v are the velocity components of the fluid in the x- and r-directions, respectively, p is the local pressure, T is the local temperature, ρ is the density of the fluid, μ is the dynamic viscosity of the fluid, k is the effective thermal conductivity, S_h is the energy source term due to chemical reaction, R_i is the production rate of species i (as the chemical reaction and calculation of R_I requires the chemical kinetic mechanism discussed below), and Y_i is the mass fraction of species i. The diffusion coefficient $D_{i,m}$ is determined by specifying the binary mass diffusion coefficient of the component i in the component j. The corresponding diffusion coefficient in the mixture can be calculated as follows [21]:

$$D_{i,m} = \frac{1 - X_i}{\sum_{j,j\neq i}\left(\frac{X_i}{D_{i,j}}\right)} \quad (6)$$

where X_i is the mole fraction of the species i, $D_{i,j}$ is binary mass diffusion coefficient which can be calculated using the Chapman–Enskog formula on the basis of kinetic theory [22].

Since the research on the kinetic mechanism of the water splitting reaction in a membrane reactor is still insufficient, a mixture of H_2 and O_2 is adopted to substitute the water vapor fed to the feed side. A mass source term and an energy source term are added to simulate the water splitting and oxygen permeation process. The mass of hydrogen gradually increases on the feed side while the mass of oxygen gradually decreases on the feed side and increases on the sweep side; this process is simulated through the mass source or the sink term, i.e., S_i (kg/m^3·s). The value of S_i is equal to zero except when $i = O_2$ or $i = H_2$. The expression of S_i is given as follows:

$$S_i = \begin{cases} +\frac{J_{H_2}\cdot A_{cell}\cdot MW_{H_2}}{V_{cell}}, & \text{at feed side} \\ -\frac{J_{O_2}\cdot A_{cell}\cdot MW_{O_2}}{V_{cell}}, & \text{at feed side} \\ +\frac{J_{O_2}\cdot A_{cell}\cdot MW_{O_2}}{V_{cell}}, & \text{at sweep side} \end{cases} \quad (7)$$

where J_{O_2} and J_{H_2} are the oxygen permeation rate (mol/m^2·s) and hydrogen production rate (mol/m^2·s), respectively; A_{cell} and V_{cell} are the area (m^2) and volume (m^3) of the cells, respectively, and MW_{O_2} and MW_{O_2} are the molecular weight of oxygen and hydrogen (kg/mol), respectively. The oxygen permeation model for the LSCuF−7328 membrane that combines surface exchange on the feed and sweep sides, and the bulk diffusion in terms of the oxygen partial pressures, is given as follows [23]:

$$J_{O_2} = \frac{\frac{k_r}{k_f}\left(P_2^{-0.5} - P_1^{-0.5}\right)}{\frac{1}{k_f P_1^{0.5}} + \frac{2L}{D_V} + \frac{1}{k_f P_2^{0.5}}} \quad (8)$$

where P_1 and P_2 are the oxygen partial pressures at the feed and sweep side, respectively, and L is the membrane thickness. The values of k_f, k_r, and D_V are the forward and reverse surface exchange rate constant and the oxygen vacancy bulk diffusion coefficient, respectively, with the following expressions [7]:

$$k_f = 6.23 \times 10^4 \exp(-27{,}300/T)\ (\text{m}/\text{atm}^{0.5}\,\text{s}) \tag{9}$$

$$k_r = 2.91 \times 10^8 \exp(-29{,}000/T)\ (\text{mol}/\text{m}^2\,\text{s}) \tag{10}$$

$$D_V = 1.58 \times 10^{-6} \exp(-8851.7/T)\ (\text{m}^2/\text{s}) \tag{11}$$

The research results of Wang et al. [24] and Lee at al. [13] both revealed that, based on the water splitting reaction and oxygen permeation process in a membrane reactor, the hydrogen production rate is twice the oxygen permeation rate:

$$J_{H_2} = 2\,J_{O_2} = \frac{2\cdot\frac{k_r}{k_f}\left(P_2^{-0.5} - P_1^{-0.5}\right)}{\frac{1}{k_f P_1^{0.5}} + \frac{2L}{D_V} + \frac{1}{k_f P_2^{0.5}}} \tag{12}$$

The endothermic process of water splitting is simulated through the energy source term on the feed side, which is given as follows:

$$S_h = 2\cdot\Delta H\cdot J_{O_2}\cdot\frac{A_{cell}}{V_{cell}} \tag{13}$$

where S_h is the energy source term (W/m^3), representing the heat absorption rate of water. ΔH is the enthalpy of formation of the water vapor (kJ/mol), and its value can be fitted via the HSC software, which is given as follows:

$$\Delta H = -\,243.33 + 1.30T^{0.5} - 0.16T + 0.0087T^{1.5} - 0.00028T^2 + 4.54T^{2.5} - 2.93T^3 \tag{14}$$

In this work, the discrete ordinates (DO) model is used to solve the radiative transfer equation (RTE) for modeling the solid and gas mixture radiation heat transfer. The RTE in the direction $\vec{s}$ is given as follows:

$$\frac{dI\left(\vec{r},\vec{s}\right)}{ds} + (a+\sigma_s)I\left(\vec{r},\vec{s}\right) = an^2\frac{\sigma T^4}{\pi} + \frac{\sigma_s}{4\pi}\int_0^{4\pi} I\left(\vec{r},\vec{s}'\right)\Phi\left(\vec{s},\vec{s}'\right)s\Omega' \tag{15}$$

where I is the radiation intensity, and r and s are the position and path length, respectively. The values of a and σ_s are absorption and scattering coefficients, respectively; n is the refractive index. The DO model converts the RTE into an equation for radiation intensity in the spatial coordinates (x,y,z) [21]. The weighted sum of the gray gases model (WSGGM) is used to evaluate the absorption coefficient of the gases.

The laminar, finite-rate model is adopted to calculate the methane combustion reaction. The reaction rates calculated from the Arrhenius rate expressions appear as source terms in the species balance equation. In this simulation, a single-step kinetic reaction mechanism is used, and the kinetic equations are given as follows [25]:

$$CH_4 + 2O_2 \rightarrow CO_2 + 2H_2O \tag{16}$$

$$R = k[CH_4]^{n_{CH_4}}[O_2]^{n_{O_2}} \tag{17}$$

$$k = AT^{\beta}\exp\left(-\frac{E_a}{RT}\right) \tag{18}$$

where R is the Arrhenius reaction rate; k is the reaction rate constant; A is the frequency factor and $A = 2.119 \times 10^{11}$; E_a is the activation energy and $E_a = 2.027 \times 10^8$ J/kmol, and n_{CH_4} and n_{O_2} take the value of 0.2 and 1.3, respectively.

The methane conversion can be calculated as follows:

$$X_{CH_4} = \frac{F_{CH_4,\ inlet} - F_{CH_4,\ outlet}}{F_{CH_4,\ inlet}} \tag{19}$$

where X_{CH_4} is the methane conversion, and $F_{CH_4,\ inlet}$ and $F_{CH_4,\ outlet}$ are the inlet and outlet flow rates of methane.

2.3. Geometry and Boundary Conditions

The internal radii of the quartz tube and the membrane tube are set as 10 and 5 mm, respectively. The tube length is 10 cm. The membrane thickness is 1 mm. The density and the thermal conductivity of the membrane are taken as 6000 kg/m^3 and 4 W/(m·K), respectively [7]. The emissivity of the membrane and the quartz tube walls are set to 0.8 [7]. These parameters are summarized in Table 1.

Table 1. Composition, dimensions, and physical parameters of membrane reactor.

Oxygen transport membrane	$La_{0.7}Sr_{0.3}Cu_{0.2}Fe_{0.8}O_{3-\delta}$ (LSCuF−7328)
Effective length of membrane reactor	10 cm
Internal radii of quartz tube	10 cm
Internal radii of membrane tube	5 cm
Thickness of membrane	1 cm
Density of membrane	6000 kg/m^3
Thermal conductivity of membrane	4 W/(m·K)
Emissivity of membrane and quartz tube	0.8

The geometry is divided into two flow zones, one for the hydrogen generation side, the feed zone, and the other side, the sweep zone. The two zones are separated by a wall serving as the membrane. Each zone has inlet and outlet boundaries. The mass flow inlet conditions are specified at the inlet boundary of the feed and sweep zones, while the pressure outlet conditions are specified for the two outlet cross-sections. A mixture gas of H_2, O_2 and N_2 with a fixed molar ratio of 0.4/0.2/0.4 is supplied to the feed side, and the mass flow rate of the feed gas is fixed at 5×10^{-6} kg/s. While CH_4 in the mixture with CO_2 is introduced to the sweep side with a varied composition and flow rate. The wall of the the quartz tube is assumed to be adiabatic. The pressures of the inlet gas streams are assumed to be 1 atm for the two sides. Table 2 gives the variation range of the boundary condition parameters in the subsequent parametric studies.

Table 2. Variation range of boundary condition parameters in parametric studies.

Sweep Gas Flow Rate (kg/s)	CH_4/CO_2 Mass Ratio	Inlet Temperature (K)
-	-	1053
1×10^{-7}	-	1063
2×10^{-7}	0.05/0.95~0.4/0.6	1073
3×10^{-7}	-	1083
-	-	1093

The effect of change in the sweep gas flow rate is investigated while keeping the mass ratio of CH_4/CO_2 at 0.2/0.8 and the inlet temperature of the two sides at 1073 K. In a similar fashion, the effect of change in the CH_4/CO_2 mass ratio is investigated while keeping the sweep gas flow rate at 1×10^{-7} kg/s and the inlet temperature at 1073 K. Lastly, the effect of the change in the inlet temperature is investigated while keeping the sweep gas flow rate at 1×10^{-7} kg/s and the CH_4/CO_2 mass ratio at 0.2/0.8. The sweep gas flow rate of 1×10^{-7} kg/s, the CH_4/CO_2 mass ratio of 0.2/0.8, and the inlet temperature of 1073 K are the base case values.

2.4. Solution Procedures

The commercial CFD software FLUENT is used to simulate the coupling process of hydrogen generation and methane combustion in the water splitting membrane reactor. Due to the symmetry nature of the reactor around its axis, the geometry is modelled as a two-dimensional axisymmetric unit. The segregated solver is used for solving the governing equations, while a steady state is considered. The Reynolds numbers in the feed side and the sweep side are approximately 12.37 and 2.90, respectively. Therefore, the laminar flow is considered. The energy model is used to solve the energy equation. The discrete ordinate model is considered for radiation. The incompressible ideal gas assumption is made for all the gases. The laminar finite-rate chemical kinetics model is used to simulate the volumetric reactions. The SIMPLE Pressure–Velocity coupling scheme is chosen. The Second Order Upwind discretization scheme is used for pressure, momentum, energy, species, and the discrete ordinate model. The convergence criteria for all the species, continuity, momentum, and energy residuals are set as 10^{-9}.

Two numerical models are utilized in the present work. The first model is the water splitting reaction to produce hydrogen and oxygen permeation through the OTM using the set of Equations (7)–(14). These equations are defined in the FLUENT via a series of user-defined functions (UDFs) written in C++ language; the code is then compiled and connected to the FLUENT software. The second model in the present work is the chemical kinetics model of methane combustion (Equations (16)–(18)), which can be simulated by using the laminar finite-rate model in the FLUENT and be set directly in FLUENT panel using the default data from the FLUENT database. There are heat and mass transfers between the feed and sweep sides, which couple the two models. This coupling is achieved by adding the mass sink/source term in the continuity equation, as well as the species balance equation and the energy source in the energy conservation equation. In order to ensure the accuracy of the numerical solution, a grid independence test is carried out. The profile of the mesh distribution and refinement is given in Figure 2. As shown in the Figure 2, a structured mesh is used with 29,336 quadrilateral cells. Grid refinement is performed on the regions near the outer wall and the regions adjacent to each side of the membrane where the gradient is relatively high. In addition, the numerical model is validated with the available experimental data in the literature, as presented in the next section.

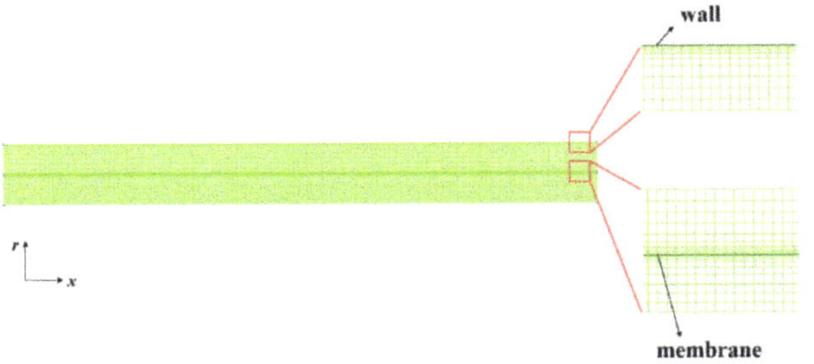

Figure 2. Profile of mesh distribution and refinement.

3. Validation of the Model

Due to the lack of experimental data fitted to the flow configuration described in the present work, a second model with the geometry used by Park et al. [8] was established to represent the experimental setup for the validation of the water splitting model. The simulation results of the second model were compared with the the experimental data by Park et al. The schematic diagram of the second model is given in Figure 3. As shown in this figure, a LSCuF−7328 membrane disk was affixed to one end of an Al_2O_3 tube, then they were placed inside a larger Al_2O_3 tube with one side of the membrane (defined as the feed side) exposed to the humidified nitrogen (N_2/H_2O = 0.51/0.49) and the other side (defined as the sweep side) exposed to a mixture of hydrogen and helium (H_2/He = 0.8/0.2) [26]. The membrane thickness is 0.05 mm and the diameter of the membrane disk is 12 mm.

The flow rate for both the feed and sweep gases was 150 mL/min [27]. The pressures of the inlet gas streams were 1 atm for the two sides. The numerical calculations were carried out with the reported experimental conditions, and a mixture of H_2 and O_2 was used to substitute the water vapor ($N_2/H_2/O_2 = 0.51/0.327/0.163$) in the simulation. The comparison of the H_2 production rate between the simulation results and experimental data is shown in Figure 4. It can be seen that the hydrogen production is enhanced as the temperature increases. This is because increasing the operating temperature can shift the equilibrium of the water splitting reaction to generate oxygen and hydrogen, and also lead to a reduction in the membrane resistance to oxygen permeation, thereby the amount of hydrogen generated increases. The simulation results are in good agreement with the experimental data. When the temperature is 1093 K, the maximum deviation between the simulation results of the hydrogen production rate and the experimental data is approximately 12%, which suggests that the simulation model is appropriate for the numerical investigation of the hydrogen generation process via water splitting in an oxygen transport membrane reactor.

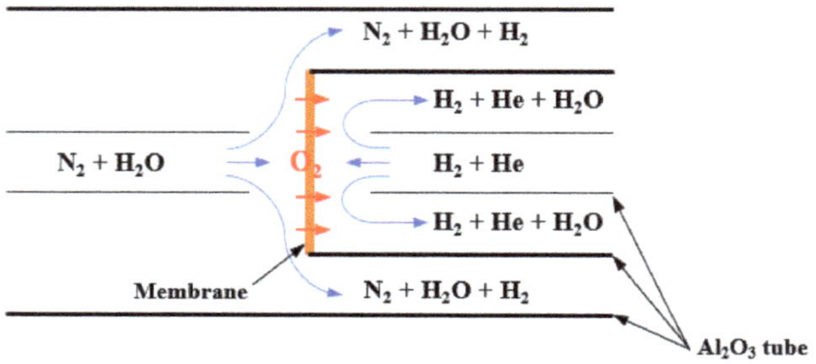

Figure 3. Schematic diagram of the second model used for validation.

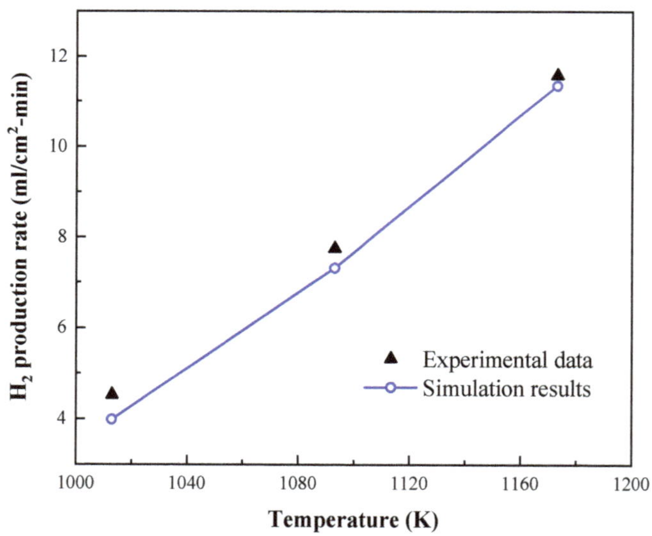

Figure 4. Comparison of H_2 production rate between simulation results and experimental data.

4. Results and Discussion

4.1. Comparison of Non−Reactive and Reactive Flow on Sweep Side

In this section, the results in terms of the temperature distribution and oxygen permeation rate under the non−reactive flow and reactive flow on the sweep side are compared. For the case of the non−reactive flow on the sweep side, CH_4 is considered as an inert gas with no combustion reaction taking place on this side, the process only involves the water splitting and oxygen permeation. For the case of the reactive flow on the sweep side, all of the water splitting, oxygen permeation, and methane combustion occur in the membrane reactor. The two cases are under the same operating conditions taking the base case values of the boundary condition parameters described in Section 2.3.

The temperature contours of the membrane reactor for the two cases are depicted in Figure 5. It is found that when there is no methane combustion reaction on the sweep side in the radial direction, the closer to the OTM, the lower the temperature is. Additionally, the closer to the OTM, the larger the radial temperature gradient is on the feed side. Along the axis, the temperature gradually decreases from the inlet to the outlet. This is because the sweep gas continually takes away the oxygen across the membrane generated by the water splitting. The oxygen on the feed side is then continuously consumed, the water splitting equilibrium continues to move toward the direction of generating oxygen and hydrogen. Water splitting is a highly endothermic process, and the reaction mainly occurs in the area on the feed side adjacent to the membrane; therefore, the closer to the OTM, the lower the temperature is and the larger the radial temperature gradient is on the feed side. Due to the continuous endothermic reaction, the temperature gradually decreases along the length of the tube. When there is a methane combustion on the sweep side, in the radial direction, the closer to the OTM, the lower the temperature is due to the endothermic effect of the water splitting. The closer to the rotation axis, the higher the temperature is. Due to the exothermic methane combustion process, the temperature on the sweep side in the axial direction gradually rises and peaks at approximately 1/5 of the tube length. Then, the temperature gradually decreases due to the continuous permeation of low-temperature oxygen.

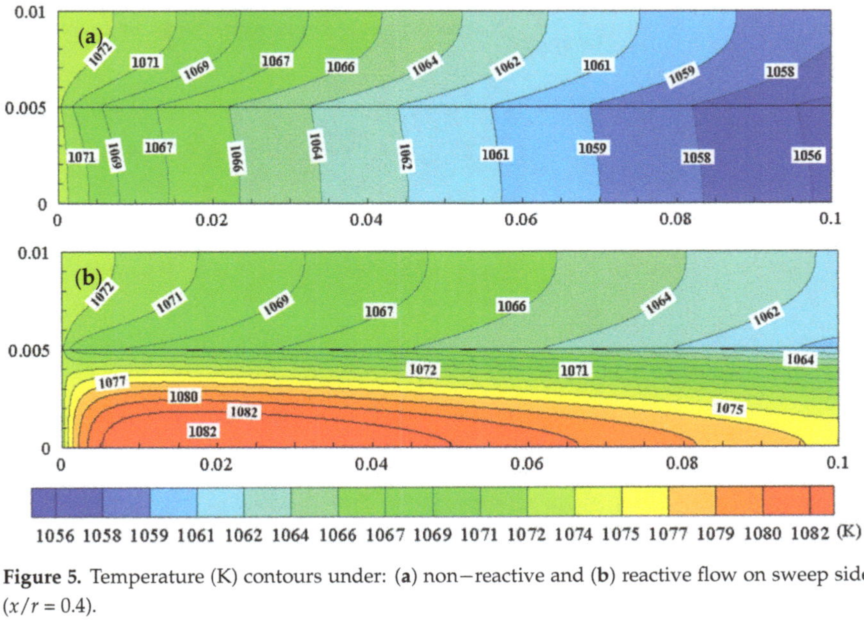

Figure 5. Temperature (K) contours under: (**a**) non−reactive and (**b**) reactive flow on sweep side ($x/r = 0.4$).

The flow pattern and velocity contour in the x−direction is provided in Figure 6. The contours of species concentrations (mass fraction of H_2 and CH_4) are given in Figure 7. It can be seen that the closer to the wall or the membrane, the larger the velocity gradient is. In the feed side, the mass fraction of H_2 increases along the length of the tube due to water splitting, and the closer to the membrane, the higher the H_2 concentration is. (Because a mixture of H_2 and O_2 is adopted to substitute the water vapor in this simulation, the concentration of H_2 is not zero at the inlet of feed side.) In the sweep side, the mass fraction of CH_4 decreases along the length of the tube due to the combustion reaction with O_2, and the closer it is to the membrane, the lower the CH_4 concentration.

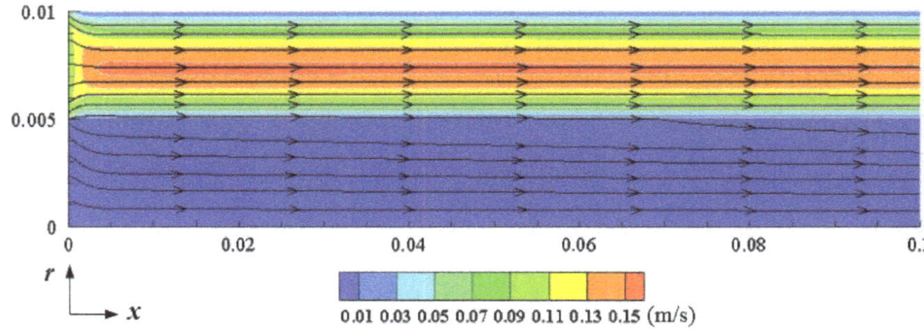

Figure 6. Flow pattern and velocity contour in $x-$direction under reactive flow (x/r = 0.4).

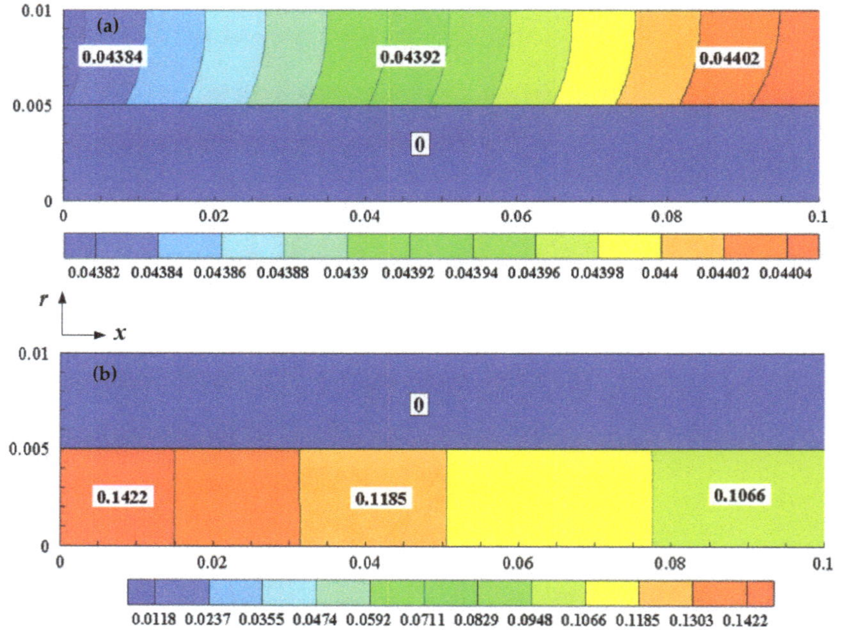

Figure 7. Contours of species concentrations under reactive flow: (**a**) mass fraction of H_2; (**b**) mass fraction of CH_4; (x/r = 0.4).

Figure 8 shows the distribution of the oxygen permeation rate for the two cases. It can be seen that when there is no combustion, the oxygen permeation rate along the length of the tube gradually decreases from 1.41×10^{-8} to 0.5×10^{-8} mol/cm^2·s. The maximum permeation occurs at the inlet of the sweep side. This is because there is no oxygen in the sweep side inlet that provides the maximum oxygen partial pressure difference, the driving force for the permeation. As the oxygen gradually permeates into the sweep side, the difference in the oxygen partial pressure between the two sides gradually decreases; therefore, the oxygen permeation rate decreases along the tube length. When there is a methane combustion reaction, the oxygen permeation rate decreases gradually along the length of the tube. A maximum value of 2.99×10^{-8} mol/cm^2·s is obtained at the inlet, and the value at the outlet is about 2.07×10^{-8} mol/cm^2·s, which is significantly higher than in the case of no methane combustion. On the one hand, methane consumes part of the oxygen during the combustion process, which reduces the oxygen partial pressure on the sweep side, thereby increasing the oxygen partial pressure difference between the two sides. On the other hand, the heat released by the methane combustion increases the temperature of the reactor, thereby reducing the oxygen permeation resistance of the membrane; therefore, the oxygen permeation rate is higher compared to the rate without combustion. Clearly, hydrogen generation is directly related to the oxygen permeation rate. Hence, the improvement in the oxygen permeation rate will directly enhance the hydrogen generation. The averaged hydrogen production rate without the methane combustion is 1.86×10^{-8} mol/cm^2·s, and it is 5.63×10^{-8} mol/cm^2·s with combustion, which is a rate approximately three times higher than that of the former. Therefore, the coupling of methane combustion and water splitting reaction can significantly improve the hydrogen production capacity of the membrane reactor.

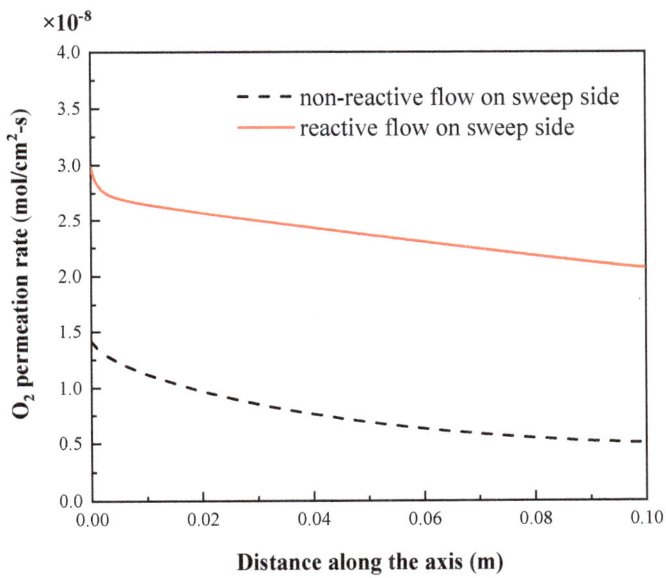

Figure 8. Distribution of O_2 permeation rate under non−reactive and reactive flow on sweep side.

4.2. Effects of Sweep Gas Flow Rate and Fuel Composition

In this section, the effects of the sweep gas flow rate and fuel composition (i.e., methane content) are analyzed. The sweep gas (fuel) is a mixture of CH_4 and CO_2, where CO_2 is used as a carrier gas to adjust the temperature.

Figure 9 shows the distributions of the membrane temperature and oxygen permeation rate along the axis at different sweep gas flow rates. It is found that the temperature of the OTM gradually decreases with the increase in the fuel flow rate. Along the tube length, the oxygen permeation rate first increases near the inlet and then decreases with the increase in the sweep gas flow rate. This is because, at the inlet, the amount of CH_4 increases with the increasing sweep gas flow rate. Then, the amount of O_2 consumed by CH_4 increases, and thus the oxygen partial pressure difference between the two sides of the membrane increases. Therefore, the oxygen permeation flux first increases. However, as the oxygen permeating across the membrane increases, the equilibrium of the water splitting reaction accelerates toward the direction of generating oxygen and hydrogen. The reaction absorbs more heat and the membrane temperature decreases further. It takes 2 mol of H_2O to produce 1 mol of O_2, while only 0.5 mol of CH_4 is required to consume 1 mol of oxygen. From a stoichiometric point of view, burning 1 mol of CH_4 can promote the decomposition of 4 mol of H_2O. However, according to the enthalpy changes in the water splitting and methane combustion reactions, e.g., at 1073 K, the enthalpy changes of these two reactions are 248.2 and −801.7 kJ/mol, respectively. It is found that the heat released by the combustion of 1 mol of methane is less than that absorbed by the decomposition of 4 mol of water vapor. Hence, the temperature in the reactor keeps decreasing with the proceeding of the reactions. After the inlet section, the influence of temperature on the oxygen permeation process is dominant, hence the oxygen permeation rate gradually decreases with the increase in the sweep gas flow rate thereafter.

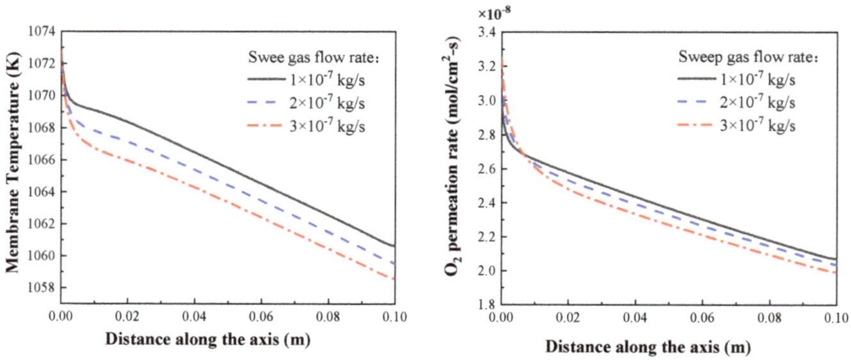

(a) (b)

Figure 9. Effects of sweep gas flow rate on: (**a**) membrane temperature; (**b**) oxygen permeation rate.

Figure 10 presents the distribution of the membrane temperature and oxygen permeation rate along the axis at different methane mass fractions. When the methane content is low, such as 6% by mass, the membrane temperature reaches a minimum near the outlet and then increases slightly. At a low methane content, the methane is completely consumed with oxygen somewhere near the outlet. Due to the endothermic effect of water splitting on the other side, the membrane temperature drops to the lowest. Then, driven by the oxygen partial pressure difference between the two sides, the oxygen continues to permeate into the sweep side. At this time, the oxygen partial pressure in the sweep side begins to increase, and the partial pressure difference gradually decreases, and thus the oxygen permeation rate will decrease accordingly. Hence, the water splitting reaction starts to slow down, and the temperature rises slightly thereafter. When the methane content is high, due to the excessive amount of the gas in the fuel, methane cannot be exhausted in the reactor, hence there is no minimum peak in the membrane temperature. With the continuing increase in the methane content, the membrane temperature changes slightly. At a high methane content, the oxygen permeation rate also increases slightly with the increase in the methane mass fraction.

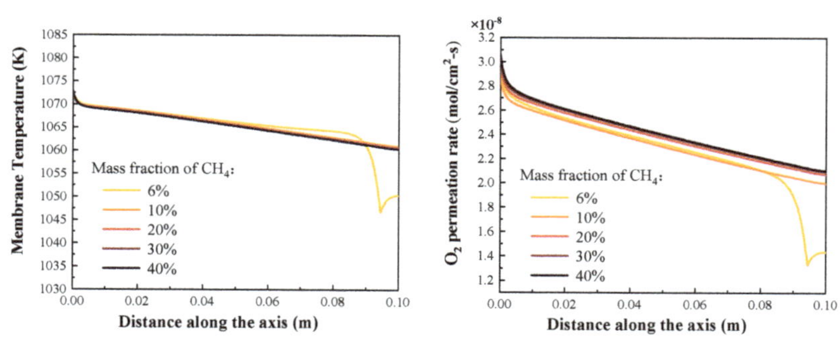

(a) (b)

Figure 10. Effects of methane content on: (**a**) membrane temperature; (**b**) oxygen permeation rate.

The change of methane conversion under different sweep gas flow rates and methane contents is presented in Figure 11. The methane conversion decreases with the increase in the sweep gas flow rate or methane content. At a sweep gas flow rate of 1×10^{-7} kg/s and methane mass fraction of less than 8.5%, the methane is completely consumed and the conversion is 100% due to the insufficient methane content. When the mass fraction exceeds 9%, the methane conversion begins to decrease due to the limited oxygen permeability of the membrane. When the mass fraction reaches 40%, the methane conversion decreases to 17.3%. When the sweep gas flow rate increases to 2×10^{-7} and 3×10^{-7} kg/s, and the methane content increases from 5% to 40%, the methane conversion decreases from 62.4% to 8.4% and from 41.2% to 5.4%, respectively. Since the methane conversion depends on the temperature and the amount of oxygen available for combustion, according to the above analysis, the membrane temperature decreases and the average oxygen permeation rate only has a slight increase. Hence, the methane conversion also decreases accordingly with the increase in the sweep gas flow rate. As the methane content increases, the oxygen permeation rate increases slightly, yet the increased amount of oxygen is not enough to consume the increased methane; therefore, the methane conversion also decreases.

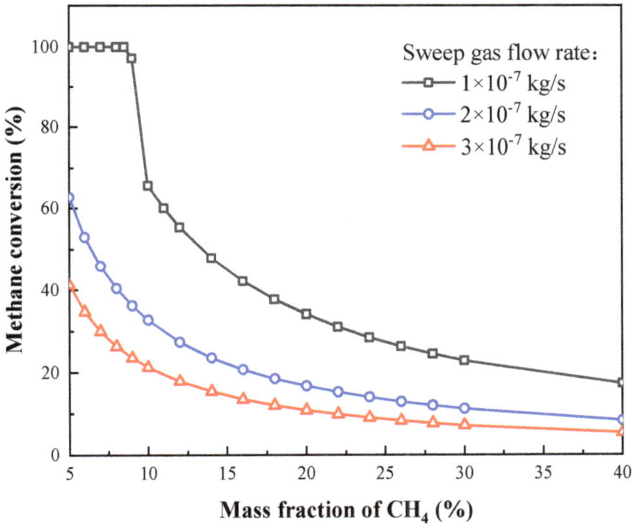

Figure 11. Methane conversion at different sweep gas flow rate and methane content.

Figure 12 shows the changes in the hydrogen production rate of the membrane reactor under different sweep gas flow rates and methane contents. It is found that the hydrogen production rate increases with the increase in the fuel flow rate and methane content. At a sweep gas flow rate of 1×10^{-7} kg/s and a methane mass fraction of less than 8.5%, the hydrogen production rate has a slight increase with the increase in methane content. When the methane content increases from 5% to 8.5%, the hydrogen production rate only increases from 4.63×10^{-8} to 4.80×10^{-8} mol/cm^2·s. In this case, the amount of methane is insufficient, and the membrane temperature and oxygen permeation rate both show a low value, and thus the hydrogen production rate will be a very low value. As the mass fraction of methane increases from 8.5% to 9%, the hydrogen yield increases rapidly from 4.80×10^{-8} to 5.40×10^{-8} mol/cm^2·s. As the methane content reaches 9%, it can be seen from Figure 11 that the methane content is similarly excessive; there is combustion reaction on the sweep side in the entire tube. Both the membrane temperature and oxygen permeation rate increase, hence the hydrogen production rate increases rapidly. After that, when the methane content increases to 40%, the hydrogen production rate only increases from 5.40×10^{-8} to 5.75×10^{-8} mol/cm^2·s. That is to say, when the methane is excessive, the hydrogen production rate increases slowly with the increase in methane content. This is because the membrane temperature is almost unchanged at this time, and the oxygen permeation rate also increases slightly; therefore, the hydrogen production rate has a slight increase. When the methane is excessive, the hydrogen production rate also increases slightly with the increase in the sweep gas flow rate. When the flow rate increases from 1×10^{-7} kg/s to 2×10^{-7} and 3×10^{-7} kg/s and the methane content is 40%, the hydrogen production rate only increases from 5.75×10^{-8} to 5.79×10^{-8} and 5.82×10^{-8} mol/cm^2·s, respectively. This is because at these flow rates, methane is excessive even when the mass fraction is 5%. Therefore, in order to ensure the hydrogen yield of the membrane reactor, the methane needs to be excessive. Additionally, due to the limited oxygen permeation capacity of the membrane reactor, when the methane is excessive, with continuous increase in the methane content or sweep gas flow rate, the hydrogen production rate only has a slight increase.

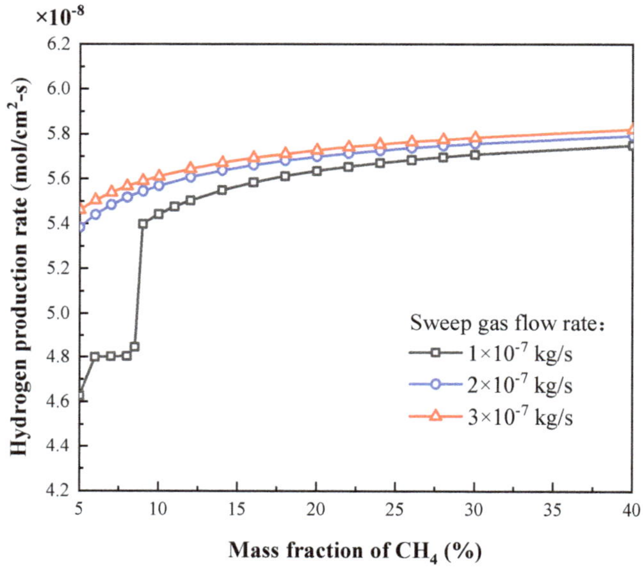

Figure 12. Hydrogen production rate at different sweep gas flow rates and methane contents.

4.3. Effects of Inlet Temperature

In this section, the effects of the inlet gas temperature on the membrane temperature, the oxygen permeation rate, the hydrogen production rate, and the methane conversion are investigated. The inlet temperature of the sweep gas is the same as that of the feed gas.

Figure 13 shows the distribution of temperature along the axis at different inlet temperatures. It is found that as the reaction continues, the temperature of the OTM gradually decreases along the length of the tube. As the inlet temperature increases, the reaction rate of methane combustion increases, and the heat released by combustion also increases, hence the temperature of the membrane also increases. It is also found that the temperature difference between the reactor inlet and outlet increases as the inlet temperature increases. This is because the water splitting reaction is enhanced with the increase in the inlet temperature, and the heat absorbed by the reaction increases accordingly, hence the temperature of the membrane decreases more along the length of the tube.

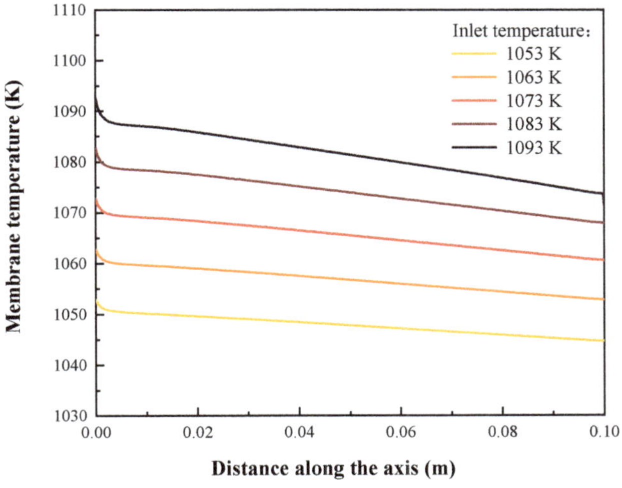

Figure 13. Effect of inlet temperature on membrane temperature.

Figure 14 shows the distribution of the oxygen permeation rate along the axis at different inlet temperatures. It can be seen that, as the inlet temperature increases, the oxygen permeation rate gradually increases due to the decrease in the oxygen permeation resistance. Along the axis, due to the decrease in the membrane temperature, the oxygen permeation rate gradually decreases. Additionally, the higher the inlet temperature is, the larger the difference in the oxygen permeation rate between the inlet and the outlet of the reactor. Figure 15 shows the hydrogen production rate, as well as the methane conversion rate, at different inlet temperatures. Since the temperature directly affects the oxygen permeation process, increasing the inlet temperature can consume more oxygen permeated from the feed side and enhance the water splitting reaction, and thus effectively increasing the hydrogen production rate. When the inlet temperature increases from 1053 to 1093 K, the hydrogen production rate increases from 3.61×10^{-8} to 8.23×10^{-8} mol/cm^2·s, a 1.28-fold increase. In addition, the methane conversion also increases with the increase in the inlet temperature. When the inlet temperature increases from 1053 to 1093 K, the methane conversion increases from 21.5% to 50.8%.

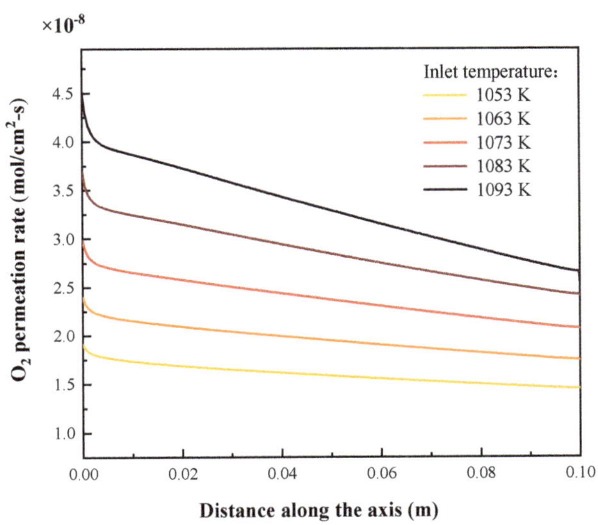

Figure 14. Effect of inlet temperature on oxygen permeation rate.

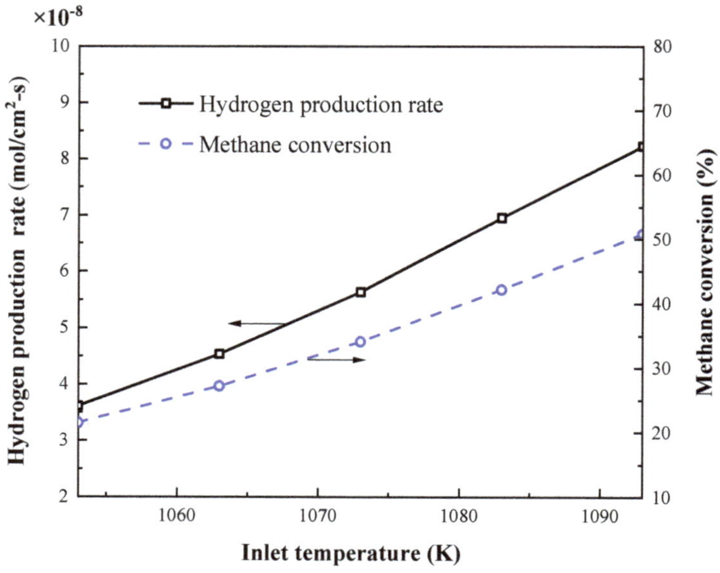

Figure 15. Effects of inlet temperature on hydrogen production rate and methane conversion.

5. Conclusions

A CFD simulation model for the coupled membrane reactor was developed by replacing the water vapor with a mixture of hydrogen and oxygen on the feed side, adding the energy source term on this side and adding a mass source term on both sides of the membrane, respectively, and the validity of the model was verified. The effects of the sweep gas flow rate, methane content and inlet temperature on the performance of membrane reactor were investigated when coupled with the methane combustion reaction. It is found that the coupling of methane combustion and water splitting can significantly improve the hydrogen generation capacity of the membrane reactor. Under certain conditions, the average hydrogen yield with methane combustion can be increased to three times, compared to the hydrogen yield without the methane combustion reaction. The methane conversion decreases with the increase in the sweep gas flow rate and methane content, while the trend of the hydrogen yield is opposite. In order to ensure the hydrogen yield of the membrane reactor, the methane needs to be excessive. Due to the limited oxygen permeability of the membrane reactor, when the methane is excessive, the hydrogen yield only increases slightly with the increase in the methane content or sweep gas flow rate. Increasing the inlet gas temperature can increase the membrane temperature, methane conversion, oxygen permeation rate and hydrogen yield.

Author Contributions: Conceptualization, T.Z. and C.C.; methodology, T.Z.; software, T.Z.; validation, T.Z.; investigation, T.Z.; writing—original draft preparation, T.Z.; writing—review and editing, T.Z. and H.Y.; visualization, T.Z.; supervision, H.Y. and C.C. All authors have read and agreed to the published version of the manuscript.

Funding: This research was funded by the National Natural Science Foundation of China, grant number U1832113.

Institutional Review Board Statement: Not applicable.

Informed Consent Statement: Not applicable.

Data Availability Statement: The data that support the findings of this study are available from the corresponding author upon reasonable request.

Conflicts of Interest: The authors declare no conflict of interest.

References

1. Zhu, X.; Shi, C.Z.; Li, K.Z.; Zhai, K.; Wang, H.; Wei, Y.G.; Tian, D.; Zeng, C.H. Water splitting for hydrogen generation over lanthanum-calcium-iron perovskite-type membrane driven by reducing atmosphere. *Int. J. Hydrogen Energy* **2017**, *42*, 19776–19787. [CrossRef]
2. Li, W.P.; Zhu, X.F.; Cao, Z.W.; Wang, W.P.; Yang, W.S. Mixed ionic-electronic conducting (MIEC) membranes for hydrogen production from water splitting. *Int. J. Hydrogen Energy* **2015**, *40*, 3452–3461. [CrossRef]
3. Demirci, U.B.; Miele, P. Overview of the relative greenness of the main hydrogen production processes. *J. Clean Prod.* **2013**, *52*, 1–10. [CrossRef]
4. Chaubey, R.; Sahu, S.; James, O.O.; Maity, S. A review on development of industrial processes and emerging techniques for production of hydrogen from renewable and sustainable sources. *Renew. Sust. Energ. Rev.* **2013**, *23*, 443–462. [CrossRef]
5. Fang, W.; Steinbach, F.; Cao, Z.W.; Zhu, X.F.; Feldhoff, A. A highly efficient sandwich-like symmetrical dual-phase oxygen-transporting membrane reactor for hydrogen production by water splitting *Angew. Chem. Int. Ed.* **2016**, *55*, 8648–8651. [CrossRef] [PubMed]
6. Wei, Y.Y.; Yang, W.S.; Caro, J.; Wang, H.H. Dense ceramic oxygen permeable membranes and catalytic membrane reactors. *Chem. Eng. J.* **2013**, *220*, 185–203. [CrossRef]
7. Nemitallah, M.A.; Habib, M.A.; Salaudeen, S.A.; Mansir, I. Hydrogen production, oxygen separation and syngas oxy-combustion inside a water splitting membrane reactor. *Renew. Energ.* **2017**, *113*, 221–234. [CrossRef]
8. Park, C.Y.; Lee, T.H.; Dorris, S.E.; Balachandran, U. Hydrogen production from fossil and renewable sources using an oxygen transport membrane. *Int. J. Hydrogen Energy* **2010**, *35*, 4103–4110. [CrossRef]
9. Hong, J.; Kirchen, P.; Ghoniem, A.F. Numerical simulation of ion transport membrane reactors: Oxygen permeation and transport and fuel conversion. *J. Membr. Sci.* **2012**, *407*, 71–85. [CrossRef]
10. Habib, M.A.; Ahmed, P.; Ben-Mansour, R.; Badr, H.M.; Kirchen, P.; Ghoniem, A.F. Modeling of a combined ion transport and porous membrane reactor for oxy-combustion. *J. Membr. Sci.* **2013**, *446*, 230–243. [CrossRef]
11. Ben-Mansour, R.; Habib, M.A.; Badr, H.M.; Nemitallah, M.A. Characteristics of oxy-fuel combustion in an oxygen transport reactor. *Energ. Fuel.* **2012**, *26*, 4599–4606. [CrossRef]
12. Jiang, H.Q.; Liang, F.Y.; Czuprat, O.; Efimov, K.; Feldhoff, A.; Schirrmeister, S.; Schiestel, T.; Wang, H.H.; Caro, J. Hydrogen production by water dissociation in surface-modified $BaCo_xFe_yZr_{1-x-y}O_{3-\delta}$ hollow-fiber membrane reactor with improved oxygen permeation. *Chem. Eur. J.* **2010**, *16*, 7898–7903. [CrossRef]
13. Lee, K.J.; Choe, Y.J.; Lee, J.S.; Hwang, H.J. Fabrication of a microtubular $La_{0.6}Sr_{0.4}Ti_{0.2}Fe_{0.8}O_{3-\delta}$ membrane by electrophoretic deposition for hydrogen production. *Adv. Mater. Sci. Eng.* **2015**, *2015*, 505989. [CrossRef]
14. Jiang, H.Q.; Wang, H.H.; Werth, S.; Schiestel, T.; Caro, J. Simultaneous production of hydrogen and synthesis gas by combining water splitting with partial oxidation of methane in a hollow-fiber membrane reactor. *Angew. Chem. Int. Ed.* **2008**, *47*, 9341–9344. [CrossRef]
15. Jiang, H.Q.; Wang, H.H.; Liang, F.Y.; Werth, S.; Schirrmeister, S.; Schiestel, T.; Caro, J. Improved water dissociation and nitrous oxide decomposition by in situ oxygen removal in perovskite catalytic membrane reactor. *Catal. Today* **2010**, *156*, 187–190. [CrossRef]
16. Wu, X.Y.; Ghoniem, A.F.; Uddi, M. Enhancing co-production of H_2 and syngas via water splitting and POM on surface-modified oxygen permeable membranes. *AIChE J.* **2016**, *62*, 4427–4435. [CrossRef]
17. Liang, W.Y.; Zhou, H.Y.; Caro, J.; Jiang, H.Q. Methane conversion to syngas and hydrogen in a dual phase $Ce_{0.8}Sm_{0.2}O_{2-\delta}$-$Sr_2Fe_{1.5}Mo_{0.5}O_{5+\delta}$ membrane reactor with improved stability. *Int. J. Hydrogen Energy* **2018**, *43*, 14478–14485. [CrossRef]
18. Markov, A.A.; Merkulov, O.V.; Patrakeev, M.V.; Leonidov, I.A. Hydrogen and synthesis gas co-production on oxygen membranes of mixed conductor: Scale-sensitive features of the process. *Int. J. Hydrogen Energy* **2019**, *44*, 26807–26815. [CrossRef]
19. Habib, M.A.; Salaudeen, S.A.; Nemitallah, M.A.; Ben-Mansour, R.; Mokheimer, E.M. Numerical investigation of syngas oxy-combustion inside a LSCF-6428 oxygen transport membrane reactor. *Energy* **2016**, *96*, 654–665. [CrossRef]
20. Coltrin, M.E.; Glarborg, P. *Chemically Reacting Flow: Theory and Practice*; Wiley-Interscience: Hoboken, NJ, USA, 2003.
21. Fluent, I. *FLUENT 6.3 User's Guide*; Fluent Inc.: Lebanon, NH, USA, 2006.
22. McGee, H.A. *Molecular Engineering*; McGraw-Hill: New York, NY, USA, 1991.
23. Rui, Z.B.; Zhang, K.; Li, Y.; Lin, Y. Simulation of methane conversion to syngas in a membrane reactor: Part I A model including product oxidation. *Int. J. Hydrogen Energy* **2008**, *33*, 2246–2253. [CrossRef]
24. Wang, H.; Gopalan, S.; Pal, U.B. Hydrogen generation and separation using $Gd_{0.2}Ce_{0.8}O_{1.9\delta}-Gd_{0.08}Sr_{0.88}Ti_{0.95}Al_{0.053}O_{3\pm\delta}$ mixed ionic and electronic conducting membranes. *Electrochim. Acta* **2011**, *56*, 6989–6996. [CrossRef]
25. Mancini, N.D.; Mitsos, A. Ion transport membrane reactors for oxy-combustion-Part II: Analysis and comparison of alternatives. *Energy* **2011**, *36*, 4721–4739. [CrossRef]
26. Song, S.J.; Lee, T.; Wachsman, E.; Chen, L.; Dorris, S.; Balachandran, U. Defect structure and transport properties of Ni–$SrCeO_{3-\delta}$ cermet for hydrogen separation membrane. *J. Electrochem. Soc.* **2005**, *152*, 125–129. [CrossRef]
27. Lee, T.; Dorris, S.; Balachandran, U. Thin film preparation and hydrogen pumping characteristics of $BaCe_{0.8}Y_{0.2}O_{3-\delta}$. *Solid State Ion.* **2005**, *176*, 1479–1484. [CrossRef]

energies

MDPI

Article

Economic Dispatch Model of Nuclear High-Temperature Reactor with Hydrogen Cogeneration in Electricity Market

James Richards [1,2,*], Cristian Rabiti [1], Hiroyuki Sato [3], Xing L. Yan [3] and Nolan Anderson [1]

1 Nuclear Science and Technology Directorate, Idaho National Laboratory, Idaho Falls, ID 83402, USA; c.rabiti@usnc.com (C.R.); nolan.anderson@inl.gov (N.A.)
2 Nuclear Engineering Department, University of Idaho, Idaho Falls, ID 83402, USA
3 Sector of Fast Reactor and Advanced Reactor Research and Development, Japan Atomic Energy Agency, 2-4, Shirakata, Tokai 319-1195, Japan; sato.hiroyuki09@jaea.go.jp (H.S.); yan.xing@jaea.go.jp (X.L.Y.)
* Correspondence: rich2057@vandals.uidaho.edu

Abstract: Hydrogen produced without carbon emissions could be a useful fuel as nations look to decarbonize their electricity, transport, and industry sectors. Using the iodine–sulfur (IS) cycle coupled with a nuclear heat source is one method for producing hydrogen without the use of fossil fuels. An economic dispatch model was developed for a nuclear-driven IS system to determine hydrogen sale prices that would make such a system profitable. The system studied is the HTTR-GT/H_2, a design for power and hydrogen cogeneration at the Japan Atomic Energy Agency's High Temperature Engineering Test Reactor. This study focuses on the development of the economic model and the role that input data plays in the final calculated values. Using a historical price duration curve shows that the levelized cost of hydrogen (LCOH) or breakeven sale price of hydrogen would need to be 98.1 JPY/m^3 or greater. Synthetic time histories were also used and found the LCOH to be 67.5 JPY/m^3. The price duration input was found to have a significant effect on the LCOH. As such, great care should be used in these economic dispatch analyses to select reasonable input assumptions.

Keywords: hydrogen; nuclear; economic dispatch; integrated energy systems; iodine–sulfur cycle; stochastic optimization

Citation: Richards, J.; Rabiti, C.; Sato, H.; Yan, X.L.; Anderson, N. Economic Dispatch Model of Nuclear High-Temperature Reactor with Hydrogen Cogeneration in Electricity Market. *Energies* **2021**, *14*, 8289. https://doi.org/10.3390/en14248289

Academic Editor: Muhammad Aziz

Received: 1 September 2021
Accepted: 31 October 2021
Published: 9 December 2021

1. Introduction

In recent years, research on hydrogen use, production methods, and economics has increased as countries have begun attempting to reduce their carbon footprints. As a power source, hydrogen offers flexible electricity generation, with the potential to serve as the load following or peaking power units. Hydrogen could also be used to shift electricity demand to off-peak hours, acting as a large-scale demand response or energy storage medium. Producing hydrogen via nuclear power and using it as a flexible load resource is being investigated by numerous organizations [1–3]. Several of these nuclear hydrogen configurations are also currently in development.

These nuclear integrated energy systems (IES) could provide economic benefits to nuclear power plants (NPPs). Competing with cheap fossil resources and declining renewable energy costs has left NPPs at an economic disadvantage [4]. Hydrogen production allows NPPs to diversify their revenue streams and has potential to increase NPP profitability [5].

In Japan, fossil fuel import requirements have led to high electricity prices and investigations into methods of producing electricity cheaply and locally [6]. Nuclear power could be advantageous in decreasing our dependence on fossil fuels, since uranium is much more energy dense, requires less frequent imports, and can be stored onsite for future use. Furthermore, following their initial installation, sources of renewable energy do not require any additional imports. Combining these technologies in a way that also reduces carbon emissions while maintaining low electricity prices is important for the future of Japan's electricity system. An IES that enables NPPs to sell a secondary commodity instead of losing money on electricity sales could help boost overall system profitability.

Besides addressing cost and security concerns, hydrogen produced via nuclear energy could help in meeting the greenhouse gas reduction goals set by Japan's Ministry of Economy, Trade, and Industry (METI). METI has also set cost-reduction goals for hydrogen produced via low- or zero-emission sources [7]. With sufficient infrastructure, this clean hydrogen could be used to aid in decarbonizing Japan's industry or transport sections. Currently in Japan, hydrogen is sold at a wholesale price of ~100 JPY/Nm^3. METI's goal is to reduce this price to 30 JPY/Nm^3 by 2030, and to 20 JPY/Nm^3 by approximately 2050 [7].

Government and research entities in Japan have also achieved expertise in nuclear high-temperature gas cooled reactors (HTGRs) and the applications thereof. The operating high-temperature engineering test reactor (HTTR) has aided in acquiring HTGR experimental and operational experience. The HTTR is a 30-MWt, helium-cooled reactor that uses graphite moderated prismatic fuel assemblies. The outlet temperature is 950 °C—high enough to integrate different process applications (e.g., hydrogen production) for testing purposes [8].

The iodine–sulfur (IS) cycle for hydrogen production appears to be a strong candidate for pairing with an HTGR [9]. The IS cycle utilizes a Bunsen reaction to convert water, I_2, and SO_2 into HI and H_2SO_4. The HI is then split up into its hydrogen and iodine components. A side reaction converts the H_2SO_4 into SO_2, water, and oxygen, thus completing the cycle. The reactions are listed in Table 1.

Table 1. IS cycle reactions.

Stage	Reaction
Bunsen Reaction	$I_2 + SO_2 + 2H_2O \rightarrow 2HI + H_2SO_4$
H_2SO_4 Decomposition	$2H_2SO_4 \rightarrow 2SO_2 + 2H_2O + O_2$
HI Decomposition	$2HI \rightarrow I_2 + H_2$
Net Inputs/Outputs	$2H_2O \rightarrow 2H_2 + O_2$

Several difficulties have inhibited the deployment of IS cycles, such as heat input and material requirements. This cycle requires high quality heat at upwards of 800 °C for the H_2SO_4 decomposition reaction, meaning that coupling with the current fleet of light-water reactors is difficult because they output steam at approximately 300 °C [10]. Additionally, material challenges associated with catalyst, reactant and container interactions or highly corrosive environments require special materials, such as Hastelloy C-276 [11], zirconium alloys [12], or special design features that isolate highly acidic environments from metals to avoid acidic oxidation.

Because of the unique positioning with an operating high temperature reactor, the Japan Atomic Energy Agency (JAEA) has emphasized the development of the IS cycle for hydrogen production [11], going so far as to design an HTTR and IS cycle cogeneration facility known as the HTTR-GT/H_2.

The HTTR-GT/H_2 is a design for coupling the HTTR with an IS cycle in order to demonstrate hydrogen–HTGR coupling capabilities. The process diagramed in [12] adds an intermediate heat exchange system to the HTTR in order to send heat to the nuclear-IS. A turbine for generating electricity is also planned. Thus, the demonstration could entail the choice of whether to dispatch and sell hydrogen or electricity, depending on regional electricity prices, hydrogen agreements, or other economic incentives.

While the technical development of the HTTR-GT/H_2 has been detailed in previous studies, this report focuses on developing a techno-economic model to flexibly dispatch the HTTR-GT/H_2 for electricity and/or hydrogen cogeneration. The goal is to investigate the potential impacts of different input assumptions or real-world conditions on the profitability of such a system. This work seeks to improve our understanding of the assumptions necessary for eventually making investment decisions pertaining to commercial hydrogen systems.

The HTTR-GT/H_2 system was chosen for this economic model due to its simple design and the availability of process modeling data. Compared to commercial-scale systems, the HTTR-GT/H_2 is relatively small, both in terms of nuclear plant size and hydrogen production. The small size means that electricity price feedback to the operation changes of the HTTR-GT/H_2 would be minimal. The HTTR-GT/H_2 has undergone detailed process modeling and has developed operation modes. Knowing the operating conditions for both the electricity sale and hydrogen sale modes makes the economic dispatch easier to model, and the smaller nature of this system helps further simplify the problem, since the system would participate in fewer electricity markets and have less of an impact on the electricity and hydrogen markets at large. This makes the impact of certain inputted data (e.g., electricity price data) more readily apparent. These effects and assumptions should be known prior to expanding this modeling methodology to larger, commercial systems as part of a broader study.

The HTTR-GT/H_2 dispatch model acts as a price-taker model. Electricity is sold when regional electricity prices exceed the HTTR's operating costs and hydrogen is produced when the electricity price falls below HTTR-GT/H_2 electricity production costs. The hydrogen is produced via the co-located IS cycle, as detailed in [13]. The price-taker assumption means that the model does not have any feedback between the changing load from the nuclear-IS cycle and grid electricity prices. This assumption is generally made for small generators and loads, such as the HTTR-GT/H_2.

The goal of the dispatch model is to determine the price at which the system can sell hydrogen while breaking even economically. This price, also known as the levelized cost of hydrogen (LCOH), is the point at which sufficient money is made to justify building the hydrogen facility and dispatching energy to hydrogen production instead of selling only electricity.

The dispatch model was developed using the Risk Analysis Virtual Environment (RAVEN) model, developed at Idaho National Laboratory [14]. Two RAVEN plugins, the Holistic Energy Resource Optimization Network (HERON) and the Tool for Economic Analysis (TEAL)—also developed at Idaho National Laboratory—were used for creating the dispatch algorithm and tracking the economic parameters within the model [15].

2. HTTR-GT/H_2 Dispatch Model Methodology

The HTTR-GT/H_2 dispatch model was developed to generate insights into the optimal dispatch of nuclear IES and how different factors can affect that dispatch. As such, this model demonstrates how the HTTR-GT/H_2 might best be dispatched in response to fluctuating hourly electricity prices throughout the year. The model also allows for the investigation of different input assumptions and their effects on the stochastic optimization of decisions to dispatch hydrogen or electricity.

2.1. Modeling Framework

The RAVEN framework is a multi-purpose optimization, data analysis, and uncertainty quantification code. It can be used in conjunction with the HERON plugin to develop economic dispatch models.

HERON creates a two-loop dispatch algorithm that incorporates RAVEN's optimization and synthetic time history generation abilities. The general structure of the stochastic dispatch model is given in Figure 1. The outer loop optimizes some grid parameter(s) (e.g., generator capacity), while the inner loop samples synthetic time histories, performs the economic dispatch, and tracks discounted cash flows via the TEAL plugin.

The HERON plugin was used to build the HTTR-GT/H_2 dispatch model in RAVEN. HERON simplified the creation of this model, which might otherwise have proven complicated for typical RAVEN users. HERON enables users to quickly develop inputs based on technology prices, commodities such as electricity or hydrogen, and the hydrogen and electricity markets. HERON then translates these user-friendly inputs into RAVEN scripts that utilize RAVEN's sampling, data transfer, and stochastic optimization capabilities to perform the dispatch.

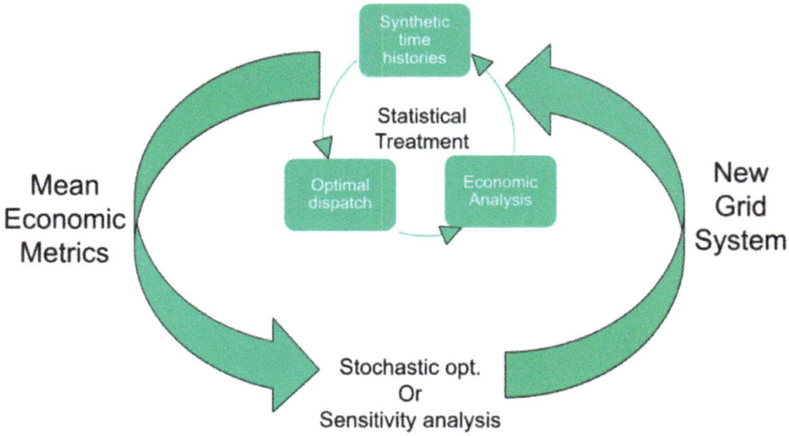

Figure 1. General schematic of the HERON dispatch model workflow [15].

Figure 2 shows the decision process for the HTTR-GT/H_2 dispatch model. This specific dispatch model utilizes an inner loop to perform the hydrogen/electricity (e- in Figure 2) dispatch and an outer loop to track hydrogen prices.

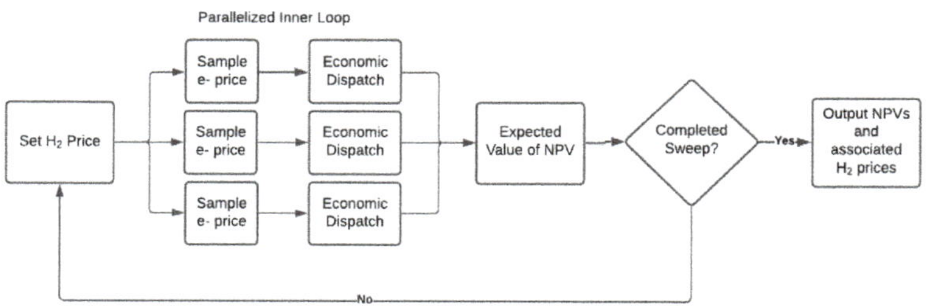

Figure 2. Algorithm used in the HTTR-GT/H_2 price-taker dispatch model. The model can run a different inner loop for each stochastic time history in order to generate an expected net present value.

2.2. Synthetic Time History Production

For the dispatch model to perform the stochastic optimization, a method of producing synthetic data was required. An autoregressive moving average (ARMA) model coupled with a Fourier series detrending model, known as a FARMA model, was used to produce synthetic electricity-price time histories. RAVEN's capabilities allow users to train a FARMA on historical, time-dependent datasets, then sample that model to create synthetic time histories. The FARMA model takes an input time history and uses a Fourier series decomposition to extract the trends that occur due to temporal variations. For example, in a load profile, there are variations that occur due to seasonal weather variations or weekly use patterns. These trends are not noise and should be quantified and extracted prior to the ARMA producing synthetic noise. A Fourier fast transform was used to extract the time scales of these correlations. This process is detailed in depth in [16]. The Fourier detrending equation, used on the time scales found from the fast Fourier transform, is given in Equation (1), as taken from [16]:

$$x_t = y_t - \sum_{i=1}^{p} [a_m \sin(2\pi f_m t) + b_m \cos(2\pi f_m t)] \tag{1}$$

After the Fourier detrending, the ARMA statistically quantifies the noise and allows for stochastic reproduction in future samples. The Fourier detrend pulls out all the strong, time-dependent trends in the dataset, leaving the noise. An ARMA algorithm can be used to model that noise. Equation (2) describes the ARMA process:

$$x_t = \sum_{i=1}^{p} \phi_i x_{t-1} + \alpha_t + \sum_{j=1}^{q} \theta_j \alpha_{j-1} \quad (2)$$

where x is the output vector for a given dimension n, the input vectors, θ and ϕ, are n by n matrices, and α is the error term. The variables p and q are the autoregressive and moving average terms, respectively. When parameter p is zero, only the moving average is used. When q is zero, the process is exclusively autoregressive.

The dispatch model can then sample the FARMA model and produce large numbers of synthetic time histories for stochastic optimization purposes. As a price-taker model, its dispatch is based on economic decisions dependent on the electricity prices from trained FARMA.

Historical electricity prices for the HTTR operating region (i.e., the Tokyo region of Japan's electricity system) were used for training the FARMA model. The HTTR-GT/H_2 was assumed ineligible for the non-fossil or baseload markets due to its status as a small-scale test reactor. Larger commercial reactors could likely participate in the spot and intraday markets in addition to the baseload and non-fossil markets. The input data were separated into 30 min increments covering a 1-year period. Prices reflect 2018 historical prices.

RAVEN's advanced clustering methods were leveraged to improve the accuracy of the synthetic price data [17]. While Fourier detrending is useful for capturing seasonal effects, the clustering takes it a step further by isolating those segments with major differences.

RAVEN clustered the data set into representative four-day periods. Each representative four-day period is known as a cluster, and each cluster was trained as an individual FARMA. By training these individual four-day clusters—as opposed to a single FARMA—over the year, RAVEN can achieve improved accuracy by further isolating the effects of long-term seasonal trends. Each specific four-day window, or segment, is then assigned to the cluster that best represents it. The clustering algorithm offers improved accuracy compared to overtraining the FARMA over the entire year.

Figure 3 shows each cluster and the time at which it occurred in the year. Each panel shows the four-day periods that are similar to each other and are thus representable by a single FARMA model. Note that the four-day periods in the shoulder months (usually in the spring and fall) tend to be similar. Additionally, the summer or winter peaks may have only a few four-day segments in their cluster. This is a feature of the clustering algorithm: by training a different FARMA for each representative window, the peak price events will not impact the production of synthetic data for the more typical shoulder months.

In Figure 4, a complete synthetic time history is plotted against the original data. This synthetic history reflects a possible time history of electricity prices that is statistically similar to the original input data. The FARMA can be sampled many times over to produce a broad range of synthetic time histories statistically similar to the input price profile.

For each model run, the FARMA was sampled 100 times to reduce the modeling uncertainties in the input electricity prices. The price duration curve (PDC) is shown in Figure 5.

The historical PDC is largely identical to the average synthetic PDC, except when comparing the 100 or so highest electricity price hours.

This dispatch model will be used to investigate the impact of the PDC discrepancy found in the 100 or so highest electricity price hours. Therefore, the model was run in two modes: one using 100 synthetic price histories and returning the expected breakeven sale price of hydrogen, and the other using the historical PDC to determine the expected breakeven sale price of hydrogen.

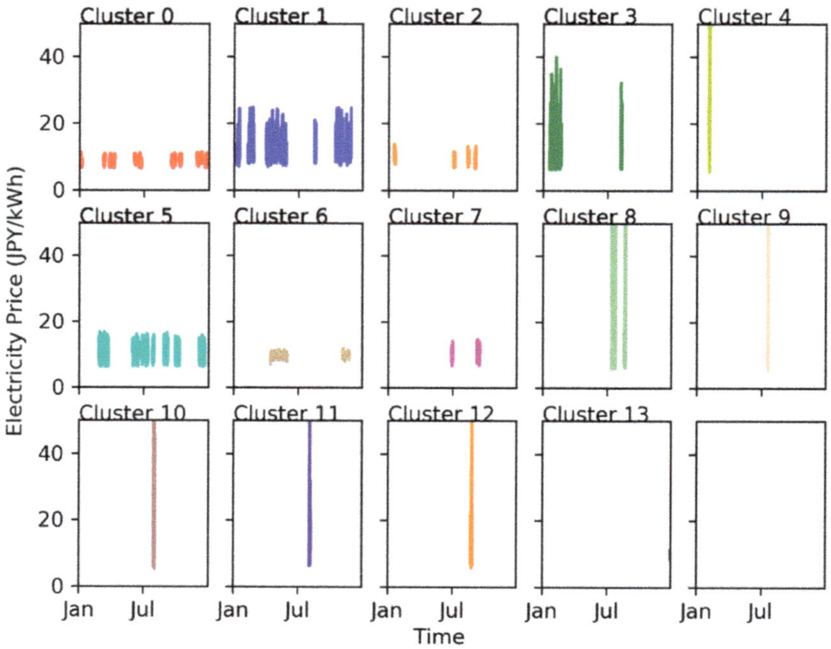

Figure 3. Four-day segments plotted by cluster, as produced by RAVEN when training the FARMA.

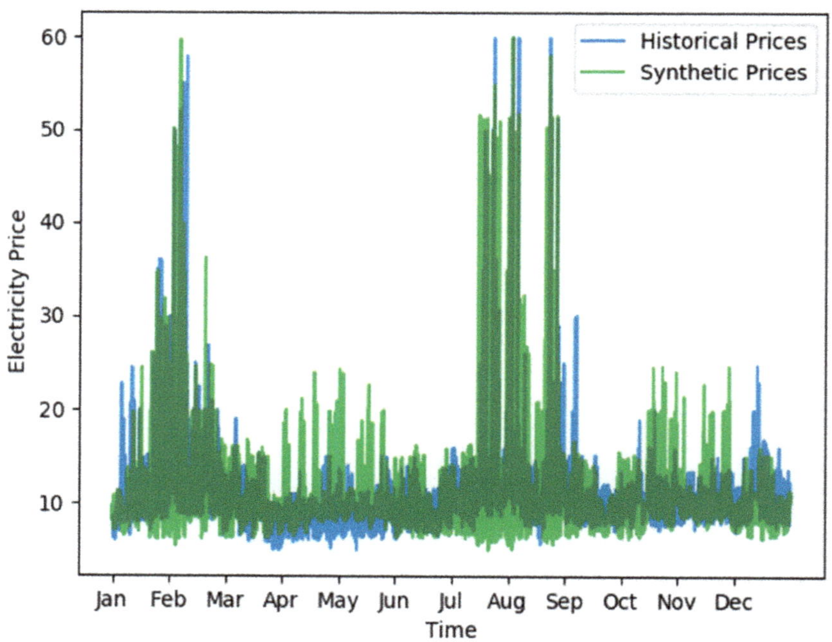

Figure 4. Historical 2018 Tokyo region electricity prices plotted against the synthetic time history produced by sampling the RAVEN FARMA.

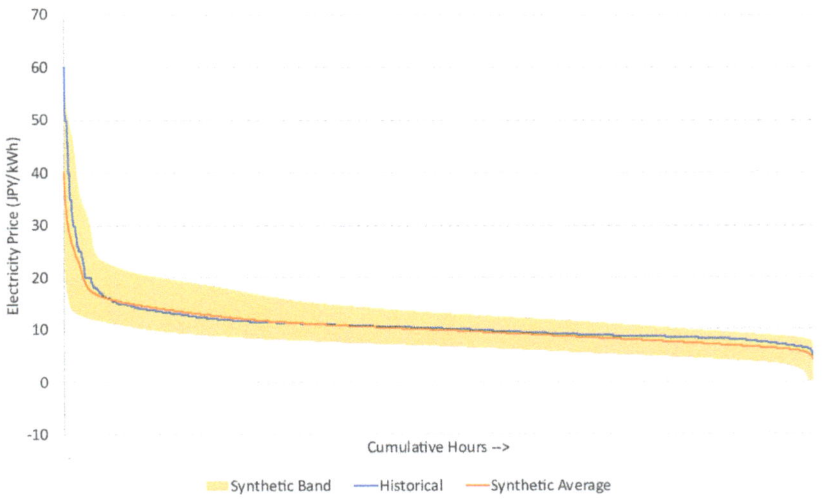

Figure 5. PDC comparison between the synthetic and historical data.

2.3. *HTTR-GT/H_2 Dispatch Dynamics*

After sampling the FARMA to produce a synthetic dataset, the model dispatches either electricity or hydrogen depending on the electricity price at each time step. When the electricity price exceeds the cost of producing electricity, electricity is produced and sold. When the electricity price falls below the cost of producing electricity, hydrogen is produced and sold. In this manner, the model chooses the most economically advantageous commodity to produce and sell during each hour.

The dispatch algorithm assumes that the HTTR-GT/H_2 operates in one of two modes, as put forth by Yan et al. [13]. These modes are the electricity and hydrogen mode in Table 2. The dispatch algorithm decides between dispatching electricity in accordance with electricity operation mode or dispatching hydrogen in accordance with hydrogen mode. The system is assumed to be flexible enough to switch back and forth between modes within the 30-min time steps allotted. Note that these configurations are based on proposed test designs for the HTTR-GT/H_2, which may not necessarily utilize the HTTR's entire 30-MWt heat for hydrogen/electricity production.

Table 2. HTTR-GT/H_2 operational modes (data reproduced from [13]).

Parameters	Electricity Mode	Hydrogen Mode
Reactor outlet temp, °C	850	950
Reactor power, MWt	30	30
Power generation, MWe	0.6	0.3
Hydrogen production, Nm^3/h	0	29.5

The IS cycle design developed by JAEA is sized to provide 29.5 Nm^3/h of hydrogen, as per the design in [13].

The amount of hydrogen delivered during each hour (m_{H_2}) is represented in Equation (3). When the price of electricity (P_{elec}) falls below the electricity production cost (C_{elec}), the system dispatches hydrogen in accordance with the previously defined operation modes. When electricity price is higher than the cost of producing electricity, hydrogen is not produced; instead, the power is used to make electricity.

$$\dot{m}_{H_2} = \begin{cases} 29.5\ \frac{\text{Nm}^3}{\text{h}}, & P_{elec} < C_{elec} \\ 0\ \frac{\text{Nm}^3}{\text{h}}, & P_{elec} \geq C_{elec} \end{cases} \quad (3a)$$

$$e = \begin{cases} 0.3\ \text{MW}, & P_{elec} < C_{elec} \\ 0.6\ \text{MW}, & P_{elec} \geq C_{elec} \end{cases} \quad (3b)$$

2.4. Cash Flow Analysis

Once the dispatch is complete, the model collects economic data to produce a system net present value (NPV). These cash flows include the capital cost, operating and maintenance costs of the IS cycle, and an assumed hydrogen storage cost. Revenue comes from the sale of hydrogen and electricity.

The NPVs in this report represent a differential NPV, shown in Equation (4). NPV_{ref} is the NPV of the HTTR-GT/H_2 when only electricity is sold and no hydrogen process has been built. NPV_{ref} serves as a baseline against which NPV_{cogen} is compared. When ΔNPV is positive, the cogeneration system is more profitable than only selling electricity. When ΔNPV is negative, the system would be more profitable focusing on electricity and not building the IS unit. Thus, when ΔNPV is 0, the profitability of the cogeneration system equals that of only generating electricity. This is the breakeven point, at which the hydrogen price represents the LCOH for this system.

$$\Delta NPV = NPV_{Cogen} - NPV_{ref} \quad (4)$$

Using ΔNPV means that only cash flows that differ between reference and cogeneration cases need to be tracked. Expenditures such as fixed HTTR costs and capital investments associated with the nuclear reactor can be disregarded, as they are equivalent in both the reference case and cogeneration cases. The limitation of this method is that ΔNPV only reflects the nuclear-IS profitability relative to the reference case rather than determining its absolute profitability. More information on the economics of the HTTR-GT/H_2 are required before an analysis of total system profitability can be conducted.

Equation (5) gives the mathematical basis for disregarding equivalent cash flows that appear in both the reference and cogeneration NPVs.

$$NPV_{ref} = CF_{e-sales} - CF_{nuc,FOM} - CF_{nuc,marginal} - CF_{nuc,CAPEX} \quad (5a)$$

$$\begin{aligned} NPV_{Cogen} = \ & CF_{H_2-sales} + CF_{e-sales,\ cogen} - CF_{nuc,FOM} - CF_{nuc,marginal} \\ & - CF_{nuc,CAPEX} - CF_{IS,CAPEX} - CF_{IS,FOM} - CF_{IS,marginal} \end{aligned} \quad (5b)$$

$$\Delta NPV = (CF_{H_2-sales} + CF_{e-sales,\ cogen} - CF_{IS,CAPEX} - CF_{IS,FOM} - CF_{IS,marginal}) - CF_{e-sales} \quad (5c)$$

The NPVs are calculated by summing the discounted cash flows associated with each case. Equation (6) details the NPV calculation. For this analysis, the discount rate, r, is 8%.

$$NPV = \sum_{yr=0}^{lifetime} \frac{CF_{total,yr}}{(1+r)^{yr}} \quad (6)$$

The cash flows accounted for in the cogeneration case are (1) cost of electricity generation from HTTR-GT/H_2, (2) IS capital and operating cost, (3) hydrogen storage, (4) revenue from electricity sale, and (5) revenue from hydrogen sale. Only the cost of electricity generation and revenue from electricity sale are tabulated in the reference case. The simulation is run for 1 year and used for every year of the project's 30-year lifetime.

The output for this model is the breakeven cost of hydrogen. Hydrogen prices exceeding the LCOH would make building the nuclear-IS system and strategically dispatching hydrogen more profitable than just selling electricity. Prices below the LCOH mean that the system would lose money relative to only selling electricity. The model allows for investigating the uncertainty that certain model inputs (e.g., electricity price data) impose on the LCOH.

To find the LCOH, the hydrogen price was varied, and the point at which ΔNPV equaled zero was found. This can either be achieved via optimization or by sweeping the solution space on a grid and locating the zero point. For this analysis, the grid sweep was used since the only variable being perturbed was the hydrogen price.

2.5. Economic Parameters

The cost of hydrogen production from the nuclear-IS system is given in Table 3, as estimated by JAEA in [18]. The capital cost is driven by the capacity of the IS cycle. For example, the provided capital cost of 3.4 JPY/m^3 was multiplied by the IS cycle capacity of 29.5 m^3/h and the 8760 h in the year. The loss of chemicals during operation of the IS was treated as a variable operating cost.

Table 3. Cost breakdown of hydrogen production by nuclear-IS system. Note that, for this analysis, the capital cost is taken on a capacity basis (i.e., Nm^3 of capacity). The table uses data from [18].

Parameter	Cost
Capital Cost	3.4 JPY/m^3
Operational Cost	2.0 JPY/m^3
Loss of Chemicals	0.2 JPY/m^3
Nuclear Heat	19.9 JPY/m^3

The dispatch model also assumes a hydrogen storage cost for a tank sized to hold 4 h of production from the IS cycle. Storage flexing and hydrogen overproduction is not included in this analysis. The storage acts as a simple addition to the capital cost. A price of $600/kg was used [5], equivalent to 5326.5 JPY/Nm^3 at an exchange rate of 106 JPY = 1 USD.

3. Results

Two scenarios were run: dispatch using synthetic price histories and dispatch using the historical PDC. For the synthetic case, each dispatch instance was run with 100 synthetically generated electricity price time histories to produce a more stochastic optimization. The historical case used the 2018 historical electricity prices as inputs. The outer loop varied hydrogen prices from 0 to 120 JPY/Nm^3.

A sample 8 h dispatch window is shown in Figure 6. The amount of revenue that the system would generate during each hour is calculated for hydrogen and electricity sales while operating in hydrogen production mode and electricity production mode, respectively, as shown in Figure 6a. Hydrogen or electricity is then produced, depending on which opportunity cost is greater (see Figure 6b).

(a)

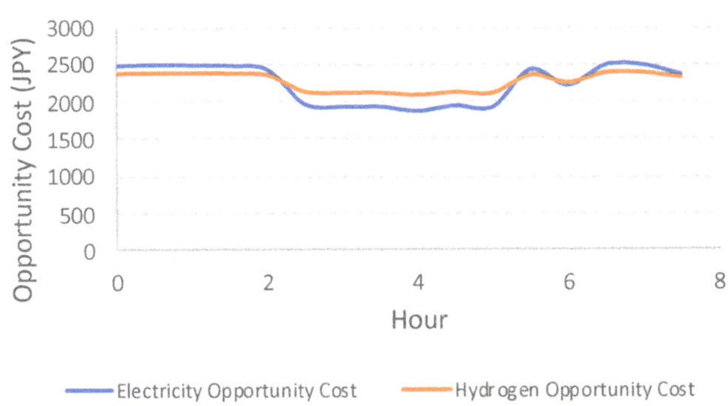

Figure 6. *Cont.*

(b)

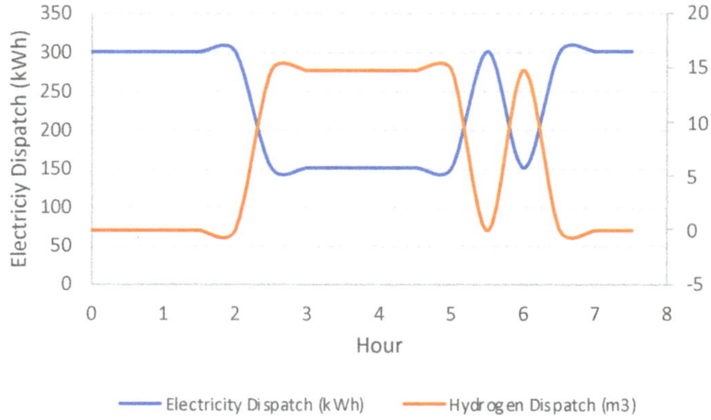

Figure 6. Example of dispatch logic over an 8 h period. (**a**) The opportunity cost for producing hydrogen or electricity. (**b**) Hydrogen or electricity modes dispatched in accordance with higher opportunity cost. This strategy ensures that electricity is sold only when profitable.

3.1. Stochastic Optimization of LCOH

The stochastic optimization case performed economic dispatch on 100 different synthetic price time histories generated by sampling the trained electricity-price FARMA. The individual economic parameters were gathered for each of these runs, and the model returned the expected ΔNPV.

Figure 7 shows the relationship between hydrogen price and ΔNPV. Breakeven LCOH occurs at 67.5 JPY/m^3, when the ΔNPV is zero. Hydrogen prices were evaluated in increments of 10 JPY/m^3 (from 20 to 120 JPY/m^3), with higher resolution around the breakeven price of hydrogen.

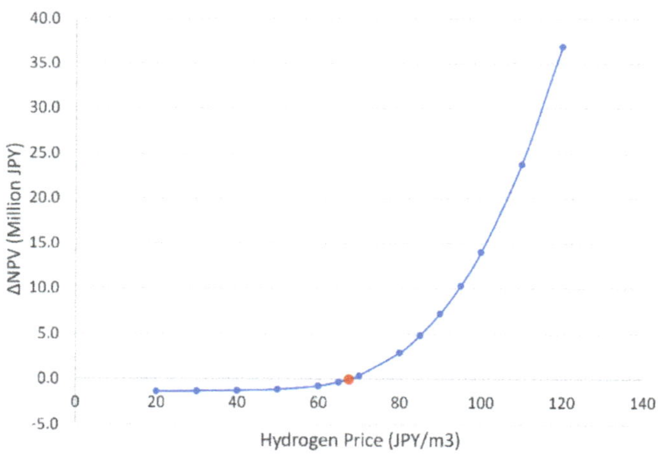

Figure 7. ΔNPV for various hydrogen prices using the synthetic PDC as inputs. The red dot represents the breakeven LCOH.

Hydrogen prices above and below the LCOH offer insight into the system dynamics. With the IS cycle dispatched while hydrogen prices are less than the LCOH, too few hydrogen-producing hours exist to recover the capital expenditure incurred from building the IS unit. With hydrogen prices greater than the LCOH, hydrogen sale becomes economically advantageous in ample time, ultimately recovering—even exceeding—the capital cost.

Figure 8 shows the number of hours per year during which the IS cycle dispatches hydrogen. At 40 JPY/m^3 or less, the hydrogen price is so low that the IS unit is never economically advantageous to dispatch. An LCOH of 67.5 JPY/m^3 equates to 431 expected hours of hydrogen production per year. Price increases result in boosting the number of hours in which hydrogen production is economically advantageous. At a high enough hydrogen price, the system would choose to dispatch hydrogen exclusively.

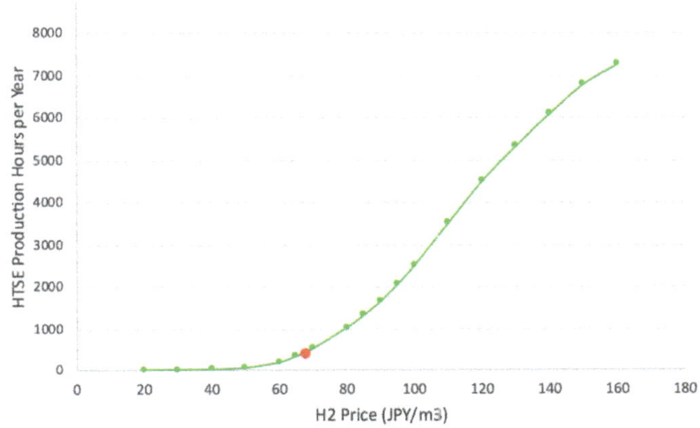

Figure 8. Utilization rate of the IS unit, plotted against the hydrogen price in the stochastic optimization scenario. As the hydrogen price rises, hydrogen deployment becomes increasingly more economically advantageous than electricity sale. Thus, the number of hydrogen production hours increases. The red dot represents the breakeven LCOH.

Table 4 summarizes the expected parameters for dispatch at the LCOH of 67.5 JPY/m^3 using synthetic price inputs.

Table 4. Expected dispatch values for the system at a levelized hydrogen cost of 67.5 JPY/m^3.

Parameter	Expected Value (per Year)
Hydrogen Produced	127,020.4 m^3
Electricity Produced	5126.6 MWh
Hours of Hydrogen Production	431.2 h

3.2. LCOH with Historical Price Duration Curve

Comparing the stochastically optimized LCOH to one optimized using historical PDC data is useful for understanding the implications of the distribution tails on cost. The stochastic optimization case outputs the expected LCOH under a wide range of possible synthetic PDC. LCOH optimization using the historical dataset gives an example of the LCOH found on a PDC that is slightly skewed to higher prices.

As with the stochastic case, Figure 9 shows that hydrogen is not dispatched at low hydrogen prices and that there is a range in which a small amount of energy is dispatched for hydrogen production despite the inability to recover the IS capital cost. In this case, when hydrogen prices fall below approximately 80 JPY/m^3, the system does not dispatch hydrogen. At 80–98 JPY/m^3, a small amount of hydrogen is dispatched. At 98.1 JPY/m^3, the ΔNPV equals zero, thus representing the LCOH.

Figure 10 shows the IS unit's utilization rate plotted against the hydrogen price. The utilization rate is zero hours when the hydrogen price is low. At an LCOH of 98.1 JPY/m^3, the IS unit produces hydrogen for 637 h. Figure 10 shows the utilization rate of the IS under different hydrogen price conditions.

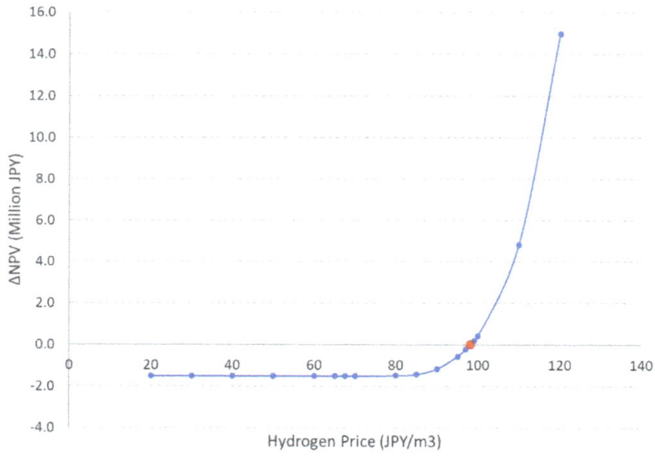

Figure 9. ΔNPV for various hydrogen prices using the historical PDC as input. The red dot represents the breakeven LCOH.

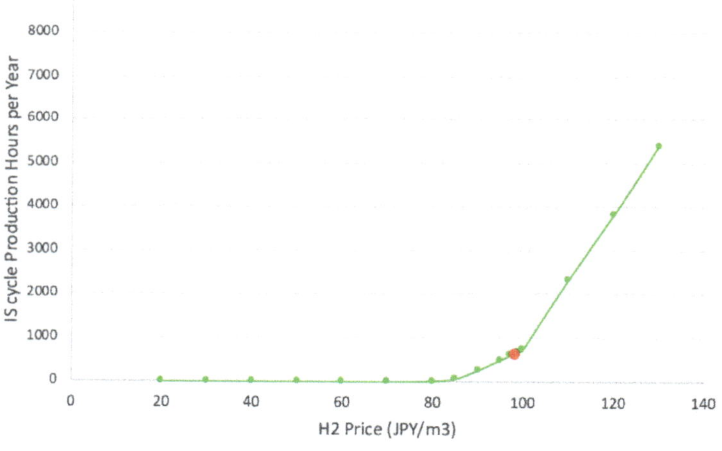

Figure 10. Utilization rate of the IS unit plotted against the hydrogen price. As the hydrogen price rises, hydrogen deployment becomes increasingly more economically advantageous than electricity sale, so the number of hydrogen production hours increases. The red dot represents the breakeven LCOH.

Table 5 summarizes the dispatch parameters found at an LCOH of 98.1 JPY/m^3 using historical price inputs.

Table 5. Dispatch values for the system at a levelized cost of hydrogen of 98.1 JPY/m^3.

Parameter	Expected Value (per Year)
Hydrogen Produced	18,791.5 m^3
Electricity Produced	5064.9 MWh
Hours of Hydrogen Production	637.0 h

4. Discussion

The reported LCOH values should not be relied on as a basis for making investment decisions. Rather, they help us understand the implications of different inputs so that when economic competitiveness is evaluated, the correct breadth of input data can be applied.

The effects of price distribution can be viewed by comparing the stochastic optimization case to the historical PDC case. The hydrogen dispatch is driven by two factors: the hours having the lowest electricity prices and the price of hydrogen. By raising the price of hydrogen, selling hydrogen becomes more profitable during more hours. Lower electricity prices and more incidences of low electricity prices also make hydrogen more economically advantageous than electricity.

The lowest-priced hours of the electricity price distribution are what dictate system profitability, since the capacity is fixed and the hydrogen price varied. The stochastic optimization case uses synthetic price histories in an attempt to produce the expected LCOH. On average, the synthetic histories showed lower electricity prices at the tail than did the historical price distribution. This led to a lower LCOH than in the historical case.

This lowest-priced hour distribution phenomenon is illustrated in Table 6. The lowest 500 h of electricity prices from the year are averaged and compared with the LCOHs for several different synthetic histories. The average electricity price over the year is also provided. The LCOH shares a stronger correlation with the bottom-hour average than with the total yearly price average. By way of comparison, the LCOH at the stochastic mean and historical points are \$0.64/m^3 and \$0.78/m^3, respectively at an exchange rate of 106 JPY/USD.

Table 6. Impacts of the distribution tail on LCOH.

Scenario	Average Electricity Price, Cheapest 500 h (JPY/kWh)	Average Yearly Electricity Price (JPY/kWh)	LCOH (JPY/m^3)
Minimum	2.14	6.91	38.1
Minus 1 Standard Deviation	3.16	7.73	45.0
Mean	5.92	10.56	68.2
Historical [1]	6.80	11.09	83.2
Plus 1 Standard Deviation	7.87	13.40	88.0
Maximum	8.43	14.84	94.0

[1] The historical scenario is the only scenario not produced by sampling the ARMA model.

Synthetic data produced using the ARMA method outputted cheaper bottom-500 h price averages, as well as overall average prices that were lower than the historical averages. This meant that the distribution of PDCs was slightly more favorable to hydrogen dispatch than the historical PDC. As such, the LCOH was lower in the stochastic case than in the historical case.

This analysis demonstrates that careful consideration should be taken when applying PDCs to this type of economic dispatch problem. The breakeven price of hydrogen highly depends on the PDC input. Stochastic optimization helps reduce uncertainty, but care should still be taken to produce PDCs that are meaningful with regard to the chosen timeframe of analysis. For example, using a 2020 PDC to predict the 2030 LCOH would be inappropriate. A projection of 2030 prices would be acceptable, but the best practice would be to use a host of projected possibilities to produce an expected LCOH.

The results from this study also show that lower overall electricity prices and more incidences of low prices would provide greater economic incentives for hydrogen production. This means that NPPs in locations with depressed electricity prices due to factors such as zero- or negative-bid renewable energies, mild climates, or low electricity demand could provide hydrogen at a lower price yet still break even or potentially turn a profit.

Several other pathways exist for reducing the LCOH. Reducing capital expenditure would depress the LCOH. The effects and sizes of potential storage options could be explored in more detail. Additional cashflows generated by the NPP's ability to participate in other areas of the electricity market would lower the ΔNPV and thus the LCOH, as well. Before investment decisions are made, each of these sensitivities should be investigated to better understand their feedback.

5. Conclusions

This analysis explored the economics of dispatching a nuclear-IS cogeneration unit. The results demonstrate the economic potential of such a system when compared to only selling electricity. These results highly depend on input assumptions, specifically the magnitude and distribution of electricity prices. Historical input electricity price data model runs show that a 98.1 JPY/m^3 LCOH was needed to break even under the specified conditions. Model runs with synthetically produced electricity price data for stochastic optimization found that an LCOH of 67.5 JPY/m^3 was required.

The LCOH in this report should not be taken as a final value for the HTTR-GT/H_2's profitability, but as an exploration of the impacts of input assumptions on the final answer. Special care should be taken in this type of dispatch analysis to produce a host of meaningful electricity price time histories that represent possibilities for the evaluation years. In this regard, the FARMA approach shows great potential.

This study also serves as another indicator that dispatching hydrogen and electricity could be more economically advantageous than just selling electricity under the right conditions. Much of the nuclear hydrogen production and dispatch work focuses on light water reactors and U.S. electricity markets while focusing on electrolysis hydrogen production technology. This study performs the economic dispatch on a unique reactor, hydrogen production system, and electricity market and shows the breakeven price of hydrogen. Performing this analysis at different locations and with different technologies is important for understanding the economic competitiveness of producing hydrogen from nuclear energy.

Efforts to further this research could include running a larger stochastic optimization case aimed at optimizing the size of the IS unit on a commercial reactor or at optimizing different sensitivities (e.g., capital cost).

Author Contributions: Conceptualization, J.R. and C.R.; methodology, J.R. and C.R.; software, J.R.; validation, J.R. and C.R.; formal analysis, J.R.; investigation, J.R.; resources, C.R.; data curation, J.R.; writing—original draft preparation, J.R.; writing—review and editing, J.R., C.R., N.A., H.S. and X.L.Y.; visualization, J.R.; supervision, C.R., H.S. and N.A.; project administration, C.R., H.S. and N.A.; funding acquisition, C.R., H.S., X.L.Y. All authors have read and agreed to the published version of the manuscript.

Funding: Prepared for the U.S. Department of Energy Office of Nuclear Energy under DOE Idaho Operations Office Contract DE-AC07-05ID14517.

Data Availability Statement: Restrictions apply to the availability of these data. The data were obtained from the Japan Atomic Energy Agency and are available from the authors with the permission of JAEA and INL.

Acknowledgments: This manuscript has been authored by a contractor of the US Government for the U.S. Department of Energy, Office of Nuclear Energy (DOE-NE), under DOE-NE Idaho Operations Office contract DEAC0705ID14517.

Conflicts of Interest: The authors declare no conflict of interest.

List of Acronyms

Abbreviation	Meaning
ARMA	Auto-Regressive Moving Average
FARMA	Fourier Series Detrending ARMA
HERON	Holistic Energy Resource Optimization Network
HTGR	High Temperature Gas-cooled Reactor
HTTR	High Temperature Engineering Test Reactor
HTTR-GT/H_2	High Temperature Engineering Test Reactor with Gas Turbine and IS Cycle
IES	Integrated Energy Systems
IS	Iodine–Sulfur Cycle

Abbreviation	Meaning
JAEA	Japan Atomic Energy Agency
JPY	Japanese Yen
LCOH	Levelized Cost of Hydrogen
METI	Japan Ministry of Economy, Trade, and Industry
NPP	Nuclear Power Plant
NPV	Net Present Value
PDC	Price Duration Curve
RAVEN	Risk Analysis Virtual Enviro rment
TEAL	Tool for Economic Analysis
USD	United States Dollar

References

1. Bragg-Sitton, S.M.; Boardman, R.; Rabiti, C.; O'Brien, J. Reimagining future energy systems: Overview of the US program to maximize energy utilization via integrated nuclear-renewable energy systems. *Int. J. Energy Res.* **2020**, *44*, 8156–8169. [CrossRef]
2. Juárez-Martínez, L.C.; Espinosa-Paredes, G.; Vázquez-Rodríguez, A.; Romero-Paredes, H. Energy optimization of a Sulfur–Iodine thermochemical nuclear hydrogen production cycle. *Nucl. Eng. Technol.* **2020**, *53*, 2066–2073. [CrossRef]
3. Prawira, A.Y.; Kim, P.S.; Yim, M.-S. Techno-Economic Analysis of Hydrogen Production Using Nuclear Power Plant Electricity Generation in Korea. *Fuel* **2020**, *30390*, 24000.
4. Szilard, R.; Sharpe, P.; Kee, E.; Davis, E.; Grecheck, G. *Economic and Market Challenges Facing the U.S. Nuclear Commercial Fleet—Cost and Revenue Study*; Idaho National Lab. (INL): Idaho Falls, ID, USA, 2017. Available online: www.inl.gov (accessed on 22 January 2020).
5. Frick, K.L.; Talbot, P.W.; Wendt, D.S.; Boardman, R.D.; Rabiti, C.; Bragg-Sitton, S.M.; Hawkins, T. *Evaluation of Hydrogen Production Feasibility for a Light Water Reactor in the Midwest*; Idaho National Lab. (INL): Idaho Falls, ID, USA, 2019. Available online: http://www.inl.gov (accessed on 7 September 2020).
6. JEPIC. The Electric Power Industry in Japan. 2019. Available online: http://www.aect.net/library/electricity-101/ (accessed on 15 September 2020).
7. Hydrogen and Fuel Cell Strategy Council. *The Strategic Road Map for Hydrogen and Fuel Cells—Industry-Academia-Government Action Plan to Realize a 'Hydrogen Society'—Hydrogen and Fuel Cell Strategy Council*; Tokyo, Japan, 2019. Available online: https://www.meti.go.jp/english/press/2019/pdf/0312_002b.pdf (accessed on 15 September 2020).
8. Yan, X.L. Status of HTTR and Technology Developments for Near Term Deployment of Nuclear Process Heat Applications in Japan. In Proceedings of the Nuclear High Temperature Heat for Industrial Processes, Vienna, Austria, 19 September 2017; Available online: http://inis.iaea.org/Search/search.aspx?orig_q=RN:48078629 (accessed on 10 September 2020).
9. Kasahara, S.; Imai, Y.; Suzuki, K.; Iwatsuki, J.; Terada, A.; Yan, X.L. Conceptual design of the iodine–sulfur process flowsheet with more than 50% thermal efficiency for hydrogen production. *Nucl. Eng. Des.* **2018**, *329*, 213–222. [CrossRef]
10. Schultz, K.; Brown, L.; Besenbruch, G.; Hamilton, C. *Large-Scale Production of Hydrogen by Nuclear Energy for The Hydrogen Economy*; General Atomics: Oakland, CA, USA, 2003. [CrossRef]
11. Noguchi, H.; Takegami, H.; Kamiji, Y.; Tanaka, N.; Iwatsuki, J.; Kasahara, S.; Kubo, S. R&D status of hydrogen production test using IS process test facility made of industrial structural material in JAEA. *Int. J. Hydrog. Energy* **2019**, *44*, 12583–12592. [CrossRef]
12. Vitart, X.; Carles, P.; Anzieu, P. A general survey of the potential and the main issues associated with the sulfureiodine thermochemical cycle for hydrogen production using nuclear heat. *Prog. Nucl. Energy* **2008**, *50*, 402–410. [CrossRef]
13. Yan, X.L.; Sato, H.; Sumita, J.; Nomoto, Y.; Horii, S.; Imai, Y.; Suyama, K. Design of HTTR-GT/H_2 test plant. *Nucl. Eng. Des.* **2018**, *329*, 223–233. [CrossRef]
14. Alfonsi, A.; Rabiti, C.; Mandelli, D.; Cogliati, J.; Wang, C.; Talbot, P.W.; Maljovec, D.P. *RAVEN User Guide*; Idaho National Lab. (INL): Idaho Falls, ID, USA. [CrossRef]
15. Talbot, P.W.; Gairola, A.; Prateek, P.; Alfonsi, A.; Rabiti, C.; Boardman, R D. *HERON as a Tool for LWR Market Interaction in a Deregulated Market*; Idaho National Lab. (INL): Idaho Falls, ID, USA, 2020. [CrossRef]
16. Talbot, P.W.; Rabiti, C.; Alfonsi, A.; Krome, C.; Kunz, M.R.; Epiney, A.; Mandelli, D. Correlated synthetic time series generation for energy system simulations using Fourier and ARMA signal processing. *Int. J. Energy Res.* **2020**, *44*, 8144–8155. [CrossRef]
17. Stauff, N.; Maronati, G.; Ponciroli, R.; Ganda, F.; Kim, T.; Taiwo, T.; Kim, S. *Nuclear Fuel Cycle and Supply Chain Daily Market Analysis Capability and Results*; Argonne National Lab. (ANL): Argonne, IL, USA, 2019. [CrossRef]
18. Iwatsuki, J.; Kasahara, S.; Kubo, S.; Inagaki, Y.; Kunitomi, K.; Ogawa, M. *Economic Evaluation of HTGR IS Process Hydrogen Production System*; JAEA-Review 2014-037; Japan Atomic Energy Agency: Ibaraki, Japan, 2014.

energies

MDPI

Article

An Intelligent Site Selection Model for Hydrogen Refueling Stations Based on Fuzzy Comprehensive Evaluation and Artificial Neural Network—A Case Study of Shanghai

Yan Zhou [1,2], Xunpeng Qin [1,*], Chenglong Li [1,3] and Jun Zhou [4]

1 Hubei Collaborative Innovation Center for Automotive Components Technology, Hubei Key Laboratory of Advanced Technology for Automotive Components, School of Automotive Engineering, Wuhan University of Technology, Wuhan 430070, China; 974248850@whut.edu.cn (Y.Z.); lichenglong@whut.edu.cn (C.L.)
2 School of Mechanical and Electrical Engineering, Wuhan Business University, Wuhan 430056, China
3 Automobile Technology and Service College, Wuhan City Polytechnic, Wuhan, 430064, China
4 China Automotive Technology and Research Center (Wuhan), Wuhan 430056, China; zhoujun@catarc.ac.cn
* Correspondence: qxp915@hotmail.com

Abstract: With the gradual popularization of hydrogen fuel cell vehicles (HFCVs), the construction and planning of hydrogen refueling stations (HRSs) are increasingly important. Taking operational HRSs in China's coastal and major cities as examples, we consider the main factors affecting the site selection of HRSs in China from the three aspects of economy, technology and society to establish a site selection evaluation system for hydrogen refueling stations and determine the weight of each index through the analytic hierarchy process (AHP). Then, combined with fuzzy comprehensive evaluation (FCE) method and artificial neural network model (ANN), FCE method is used to evaluate HRS in operation in China's coastal areas and major cities, and we used the resulting data obtained from the comprehensive evaluation as the training data to train the neural network. So, an intelligent site selection model for HRSs based on fuzzy comprehensive evaluation and artificial neural network model (FCE-ANN) is proposed. The planned HRSs in Shanghai are evaluated, and an optimal site selection of the HRS is obtained. The results show that the optimal HRSs site selected by the FCE-ANN model is consistent with the site selection obtained by the FCE method, and the accuracy of the FCE-ANN model is verified. The findings of this study may provide some guidelines for policy makers in planning the hydrogen refueling stations.

Keywords: hydrogen refueling station; evaluation index system; analytic hierarchy process; fuzzy comprehensive evaluation; artificial neural network

Citation: Zhou, Y.; Qin, X.; Li, C.; Zhou, J. An Intelligent Site Selection Model for Hydrogen Refueling Stations Based on Fuzzy Comprehensive Evaluation and Artificial Neural Network—A Case Study of Shanghai. *Energies* **2022**, *15*, 1098. https://doi.org/10.3390/en15031098

Academic Editor: Attilio Converti

Received: 2[illegible] November 2021
Accepted: 28 January 2022
Published: 1 February 2022

1. Introduction

Hydrogen energy is known as the cleanest secondary energy, and hydrogen fuel cell vehicles (HFCVs) have attracted much attention due to their near-zero emission and pollution-free advantages [1,2]. Hydrogen fuel cell vehicles can be used efficiently and conveniently, and rationally set HRSs play a key role [3,4]. The site selection of HRSs involves the influence of various factors such as technology, economy and society. For a long time, establishing scientific and reasonable site selection criteria has been a problem academic circles are striving to solve [5,6]. In the past five years, various countries have begun to build HRSs on a large scale to promote the development of the HFCVs industry [1,7–9]. According to statistics, the number of HRSs in the world has exceeded 900, of which more than 300 are in China. Also, there were more than 500 HRSs in operation in the world by the end of 2021 [10,11]. Figure 1 shows the operation, construction and planning of HRSs in China and around the world. As can be seen from Figure 1, compared with the world, China has a relatively small proportion in normal operation and a large proportion in planning. As the largest country in the production and sales of new energy vehicles, relevant institutions and enterprises highly value site selection [12]. However, the HRS layout and planning are not simple. There are many factors and obstacles to consider, such as high construction costs, immature hydrogen transportation technology, long payback periods, and a shortage of HFCVs [5,13–15]. These obstacles lead to the actual operation of HRSs not being positive. Compared with gas stations, HRSs have greater safety risks; the recent HRS explosion in Norway confirmed this [16]. Especially in a city, and faced with a relatively dense population and complex and specific road environments, factories, and other building facilities, the construction of hydrogen refueling stations is more difficult [17,18]. The motivation of this research is to seek to establish an intelligent decision-making model for the site selection of HRSs, which specifically guides the location and construction of HRSs, so as to promote the rapid development of the hydrogen fuel cell vehicle industry. The relevant results are verified in Shanghai, China.

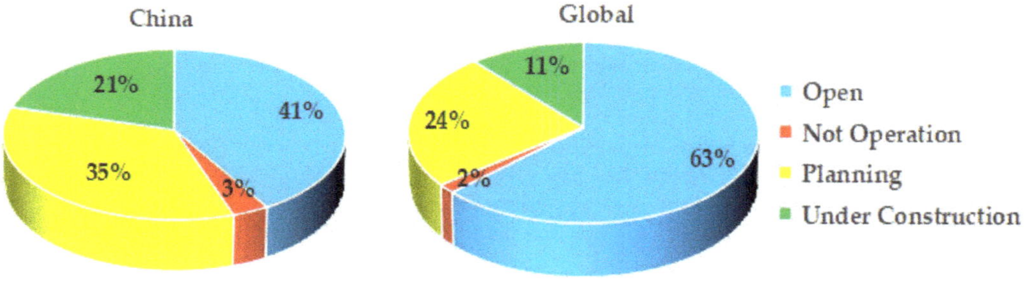

Figure 1. Structure diagram of HRSs in China and Global by the end of 2021 [10,11].

It has been found according to previous studies that many scholars have begun to carry out related research on HRSs. In order to solve different problems, the research themes are also different. For example, on the issue of factors affecting the construction of HRSs, Xu, C. et al, focusing on the obstacles in the development of hydrogen refueling stations, a modified fuzzy DEMATEL approach was adopted to establish a comprehensive barrier list for the construction of HRSs from four perspectives: technology, economy, society and politics [5]. Bai, W. et al. aimed at how to choose an appropriate financing model in the stage of investment and construction of a HRS, established a financing model decision-making index system, and determined the priority of each factor based on Fuzzy AHP [13]. The above research results provide a reference for the sustainable development of HRSs. In the study of the HRS site selection model, Zhou, J. et al. established a two-stage site selection optimization model based on geographic information big data to provide technical support for the layout of HRSs. In this study, ten types of geographic information were selected. The layout of the HRS is planned based on the elements and six evaluation indicators, and Shandong, China is selected as a case to demonstrate the model, which provides a reference and basis for the development of hydrogen energy strategy in Shandong Province [1]. Sun et al. proposed a HRS location model based on hydrogen life cycle cost and consumer hydrogen cost. The optimization goal is to minimize the total cost of HRS construction and operation [19], and establish a city HRS location and capacity calculation. Using the multi-dimensional particle swarm algorithm to solve the model [6,20,21], there are still a large number of studies focusing on the development of hydrogen energy vehicles and the construction of HRSs under the background of reducing energy consumption and carbon emissions, And the hydrogen safety research for risk assessment, etc. [22–26]. From the above research questions, it can be found that there are few studies on the evaluation and prediction of the location of HRSs. The cost and location of the HRSs are important in the process of site selection and planning. The sustainable development of HRSs plays an important role, so it must be meaningful to use the comprehensive evaluation results of the already operating HRSs as a reference in the planning process of the site selection of HRSs.

Artificial neural network (ANN) has the ability to learn to predict behaviors and patterns from a limited set of correct data to make statistics and predict future developments [27,28]. There are many types of ANN algorithms, such as BP, RBF, MLP, GMDH, GA etc. Among them, the BP algorithm has high self-learning and adaptive capabilities, and can automatically extract "reasonable rules" between input and output data through learning [29]. The generalization ability of RBF is better than that of BP in many aspects; however, when solving problems with the same accuracy requirements, the structure of BP network is simpler than that of RBF network [30,31]. The biggest feature of the GMDH network is that the input quantity taken is automatically determined during the training process, while neural networks such as BP require a lot of prior knowledge to determine the network structure, but the problem of time-consuming nonlinear modeling of the GMDH algorithm is unavoidable [32,33]. MLP is a generalization of single-layer perceptron, which can solve nonlinear problems that single-layer perceptron cannot solve [34]. Roshani, G. H. et al. using an ANN model, the volume fraction of gas, oil and water in three-phase flow independent of flow regime was predicted with high accuracy. The ANN model was developed in MATLAB software using a multilayer perceptron (MLP) neural network [35].

Amir Sattari, M. et al. used the ANN method in the prediction of flow pattern type and gas volume percentage to realize the identification of the flow state and high-precision prediction of the porosity of each flow state [36]. The above research shows that ANN model has obvious advantages in statistics and prediction. At the same time, the training and learning process of neural network requires a large amount of data. For more accurate planning of HRSs, it is undoubtedly a very effective learning data to select the evaluation data of the operational HRSs as the learning and training data of the neural network. The evaluation of the completed HRSs is a complex multi-criteria decision-making process, "AHP, FCE, Expert System, and fuzzy multi-criteria decision-making (FMCDM)" have become the preferred method for solving this type of problems [37–42]. The AHP can classify and deal with many influential indicators, it has the characteristics of systematization and modeling, and it can determine the weight of each indicator in the evaluation system [43,44]. Some factors are difficult to quantify or quantitative cost is higher, and the quantitative value is subject to subjective factors of evaluators. Factors such as policies can only be described by fuzzy qualitative language, and it is difficult to use classical mathematical models to accurately calculate and determine the optimal scheme from selected schemes. The FCE method can solve these problems well [45–47]. However, the calculation process of FCE is more complicated, and implementation is easily influenced by subjective factors [48]. Using fuzzy control theory to quantify unquantifiable factors, and using ANN to analyze experimental data, error is small, there is high precision, and complex forecasting problems have good generalization ability [35,36]. Therefore, in order to supplement the previous research on the site selection and planning of HRSs, there is a lack of research on the comprehensive evaluation of the already built hydrogen refueling stations as a reference decision-making standard. This paper will establish an evaluation system for HRSs planning through AHP, and use fuzzy comprehensive evaluation and artificial neural network model (FCE-ANN) to evaluate and predict the site selection of HRSs under planning.

Based on the overview and analysis of the above researches, the main findings are as follows:

1. At present, there are few studies on the evaluation of the site selection of HRSs. The existing studies tend to be macro-development plans, and the evaluation results of specific HRSs cannot be given.

2. The existing HRSs site selection planning model lacks the comprehensive evaluation of the operational HRSs as a reference decision-making standard.

3. The layout of HRSs is affected by many factors, most of which are complex and not easy to be measured specifically. According to the literature survey, few people have specialized in training neural networks based on the evaluation data of operational HRSs.

On the basis of the above analysis, we can position and clarify this paper's contribution:

1. According to the existing literature and expert consultation, determine the evaluation factors of China's HRS site selection planning, and establish a FCE index system. Use AHP to determine the weight of each evaluation index. Establish an expert decision-making committee to conduct a comprehensive evaluation of the operational HRSs in China, and obtain FCE results.

2. An intelligent site selection model based on FCE-ANN model is proposed. The ANN has good adaptive ability and can give an objective evaluation to the multi-index comprehensive evaluation problem, which is very beneficial for weakening the human factors in the weights.

3. Selected the evaluation result data of 50 operational HRSs in coastal and major cities in China as the training data of the neural network. 8 HRSs under planning in Shanghai were selected for site selection and model validation.

4. The results of this study are intended to provide some guidance and suggestions for policy makers of HRS planning; at the same time, the method of this study is intended to provide some reference and thinking for relevant researchers in conducting complex site selection decision-making research.

The rest of the paper is organized as follows: Section 2 introduces the basis for the selection of elements and indices in the evaluation system of HRSs, and establishes a comprehensive evaluation index system for HRSs. Section 3 describes the three-stage key steps and core algorithms of the FCE-ANN model for HRSs in detail, and gives the research framework of this study. Section 4 takes China's coastal and major cities as the research area to conduct an empirical study to verify the FCE-ANN model. In addition, a discussion and in-depth analysis of the results; Section 5 summarizes and prospects this study.

2. Evaluation Index System for Site Selection of HRSs

HRSs construction and planning are affected by many factors. Although few people directly study the influencing factors of HRSs location planning, potential influencing factors can be found from other aspects. On the basis of the existing literature and research, and our expert consulting and workshop discussion, we further subdivided influence factors of HRSs planning. The influencing factors selected in the study are divided into three categories: economic, technical, and social factors. Under each factor, there are three specific sub-factors, as shown in Table 1.

Table 1. Factors affecting HRSs planning in China.

Categories	Factors	References
Economic factors	Construction scale	[5,49,50]
	Investment strength	[5,13,51]
	Operational costs	[13,16,49,52]
Technological factors	Hydrogen production technology	[16,17,50]
	Hydrogen storage technology	[16,51]
	Transportation technology	[5,16,53,54]
Social factors	The population density	[25,55,56]
	Environmental factors	[14,52,57]
	Social identity	[58] and EC

EC: expert consultation.

2.1. Factors Affecting HRSs Planning

2.1.1. Economic Factors U_1

(1) Construction scale U_{11}

HRSs planning and construction are generally approved and managed by national governmental departments; the construction party needs to submit a series of qualification documents before construction and operation qualification can be obtained [5]. Fixed HRSs also need to apply for planning and construction permits, but skid-mounted stations do not need this procedure [59]. Therefore, the construction scale is also a factor affecting HRSs planning. The larger the construction scale is, the more complicated the approval procedure is, and the corresponding cost would far exceed the infrastructure construction cost [50].

(2) Investment strength U_{12}

The main difficulty affecting HRS development is the high construction cost. Although local governments pay increasing attention to the construction of HRSs, and planning and supporting policies are issued one after another, the high construction cost requires investors to have certain financial strength. All HRSs in China are in the form of external high-pressure hydrogen storage. The initial cost of constructing an HRS larger than 200 kg in China exceeds CNY 10 million [5]. In addition, if debugging, engineering-design, and management costs, and other factors are taken into account, the superposition of these factors directly leads to almost all completed HRSs in China hardly being profitable [13].

(3) Operational costs U_{13}

The cost of operation and maintenance is an important part of the life-cycle cost of a HRS [51]. On the one hand, the construction and development of HRSs in China is still in the initial stage. Compared with the mature operation and maintenance technologies of HRSs in the United States and Japan, China lacks maintenance experience, resulting in relatively high maintenance costs of HRSs [59]. Besides the initial equipment investment of HRSs construction, the amount of the continuous daily fault-free hydrogenation of hydrogenation power consumption, and the cost of employing personnel should also be considered as influencing factors of operating HRSs costs [49]. In other words, operation and maintenance costs are an important factor to be considered in the planning and construction of HRSs.

2.1.2. Technological Factors U_2

(1) Hydrogen production technology U_{21}

Hydrogen production from natural gas, coal gasification, industrial byproducts, oil refineries, and water electrolysis are the main sources of industrial hydrogen [13,16,50]. However, because of technical barriers, China still relies on imports for key components in most hydrogen production facilities [15]. Expensive equipment makes it difficult to reduce the cost, and the incomplete domestic hydrogen production technology is the main factor affecting the construction of HRSs.

(2) Hydrogen storage technology U_{22}

Mature hydrogen storage technology and unbreakable key links are key factors affecting the construction of HRSs in China [13,16]. Hydrogen can be stored in three ways: high-pressure gaseous hydrogen storage, low-temperature liquid hydrogen storage, and metal alloy hydrogen storage [51]. High-pressure gaseous hydrogen storage and low-temperature liquid hydrogen storage are widely used at present. Low-temperature liquid hydrogen storage has a high density, but it is difficult to liquefy hydrogen, which leads to high cost and the requirement for container heat insulation [51,60]. The advantages and disadvantages of different hydrogen storage technologies are compared in Table 2, which reflects the immature hydrogen storage technology in China [60].

(3) Transportation technology U_{23}

There are three common hydrogen transport modes: tube trailers, pipeline, and liquid trucks [5]. China mainly uses long-tube trailer transport, and its long tube trailer transport equipment industry is mature. According to transport industry standard JT/T 617.1-2018 Dangerous Goods Road Transport Rules of the People's Republic of China, long-tube trailers are used for transport, and transport and storage pressure is 20 MPa. Due to cost limitations, this method is suitable for short-distance hydrogen transportation—the economic transportation radius is about 200 km. Low-pressure pipeline transportation is suitable for large-scale point-to-point transportation [54]. Due to the existence of hydrogen embrittlement between hydrogen and some metals, pipe materials have special requirements, so the investment cost is high [53]. In China, large-scale low-pressure pipeline transportation does not exist. The cost of a hydrogen pipeline is more than twice that of a natural-gas pipeline, and hydrogen density is much less than that of natural gas, resulting in a higher transportation cost of hydrogen than that of natural gas. Liquid hydrogen tank transport is widely used in other countries, but it is only used in the aerospace and military fields in China, and is not yet commercialized [5]. To sum up, the current mainstream hydrogen transportation technology needs to be improved, and hydrogen transportation technology is the main factor affecting the construction of HRSs in China [59]. Table 3 summarizes and analyzes current mainstream modes of transportation.

Table 2. Advantages and disadvantages of hydrogen storage technologies [60].

Hydrogen Storage Technology	Theoretical Hydrogen Storage	Advantages	Disadvantages
High-pressure gaseous hydrogen storage	1–2 wt %	Low cost, fast charging and discharging speed, can be carried out at room temperature	Low hydrogen storage capacity, needs high-pressure vessels, high energy consumption, high transportation costs, poor safety
Low-temperature liquid hydrogen storage	>10 wt %	High volume energy density, high liquid purity	Liquefaction has high energy consumption, and demanding storage and maintenance requirements
Metal alloy storage hydrogen	1–8 wt %	Large hydrogen storage capacity, high safety, good stability, good operability	Hydrogen storage performance is poor, easy to powder, transportation is not convenient
Organic liquid hydrogen storage	5–10 wt %	Large amount of hydrogen storage, safe and convenient transportation, can be recycled, low energy consumption	High cost of catalytic hydrogenation and dehydrogenation unit, complex technical operation, and low efficiency in dehydrogenation

Transportation technology U_{23}.

Table 3. Comparison of scale, cost, and energy consumption of three modes of hydrogen transport [59].

Hydrogen Transport Mode		Tube Trailer	Pipeline	Liquid Truck
Cost	Scale	Cost not affected by scale	Large scale, low cost	Scale is large, and cost is greatly reduced
	Distance	Short distance, low cost	Short distance, low cost	Distance is long, cost rises slightly
Energy consumption	Scale	Unit energy consumption of the three methods has nothing to do with hydrogen transport scale		
	Distance	Long distance, high cost	Minimal energy consumption	Minimal impact

2.1.3. Social Factors U_3

(1) The population density U_{31}

HRSs built in urban areas are located in densely populated areas, and are far more dangerous than conventional gas stations are because of the energy produced by hydrogen explosions. Hydrogen as an energy source is widely used in transportation, and safety is greatly important, which is one of the reasons why the hydrogen industry is developing relatively slowly [25,55] In June 2019, the explosion of a hydrogenation station in Sanvika, Norway, also impacted China's hydrogen and fuel-cell vehicle industry, with negative market sentiment [55].

(2) Environmental factors U_{32}

Hydrogen is lighter than natural gas, has a wider explosive limit, spreads faster, and is likely to cause more dangerous accidents [45]. High-pressure hydrogen embrittlement is common, and high pressure demands more stringent safety spaces in the surrounding environment [51]. Hydrogenation station construction must consider whether there are dangerous sources or crowded places around, such as schools and hospitals [5,52]. Therefore, in the process of the expert evaluation of built hydrogenation stations in China, the surrounding environment of all hydrogenation stations is first objectively evaluated by experts.

(3) Social identity factors U_{33}

New-energy vehicles are the future development trend of automobiles and are strongly supported by the state [57]. The performance of the car is similar to that of traditional fuel cars, but battery life and the cost of late replacement make most people balk. Due to a lack of an adequate understanding of hydrogen energy, it has been regarded as a dangerous source by the public for a long time, so the public's recognition of hydrogen fuel vehicles and hydrogenation stations affects its development [46,59].

To summarize, a dual hierarchical evaluation index system for the site selection of HRSs was constructed, as shown in Figure 2.

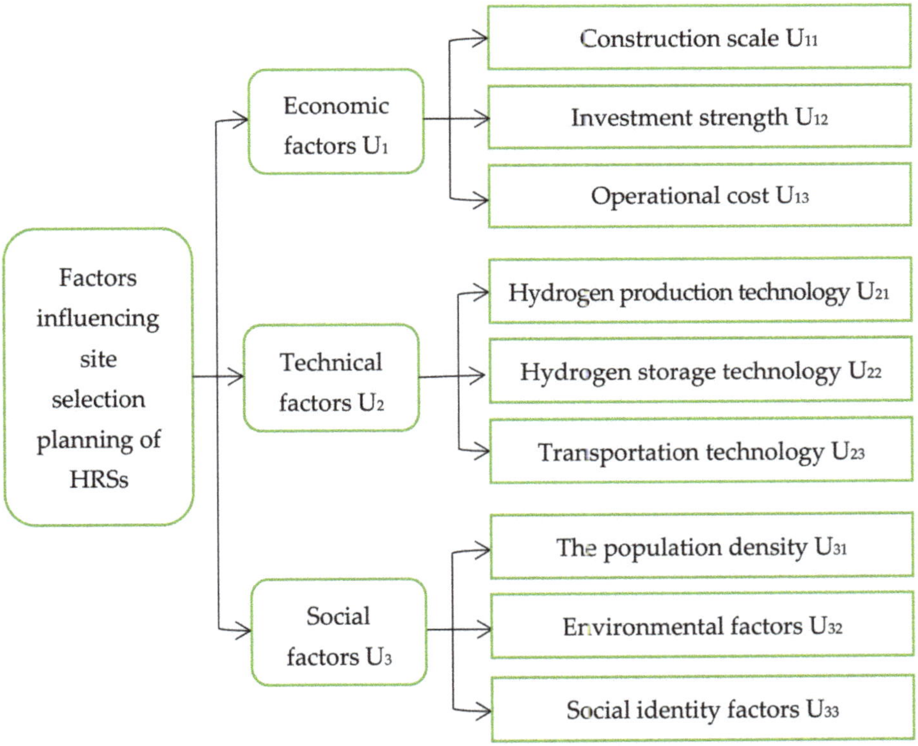

Figure 2. Evaluation index system of HRSs site selection.

3. Model

On the basis of the above evaluation index system of HRSs site selection, the process of the intelligent location model of HRSs based on FCE-ANN will be introduced in detail. The proposed FCE-ANN intelligent site selection decision-making process was divided into three stages. The first is described in detail using the AHP method, namely, a questionnaire survey to invite expert consultation to determine the HRSs site selection evaluation index system of the weight of each index value of the calculation process. Second, Experts and scholars in the hydrogen energy industry are invited to form an expert evaluation group to conduct a FCE of HRSs. Lastly, expert evaluation levels and evaluation results were taken as the data of the training ANN, the ANN model was used for comparative decisions, and the feasibility of the intelligent site selection decision method was obtained. The details for each phase are described in the following sections. This study thus includes three modules: AHP method, FCE method, and ANN model. The research framework is shown in Figure 3.

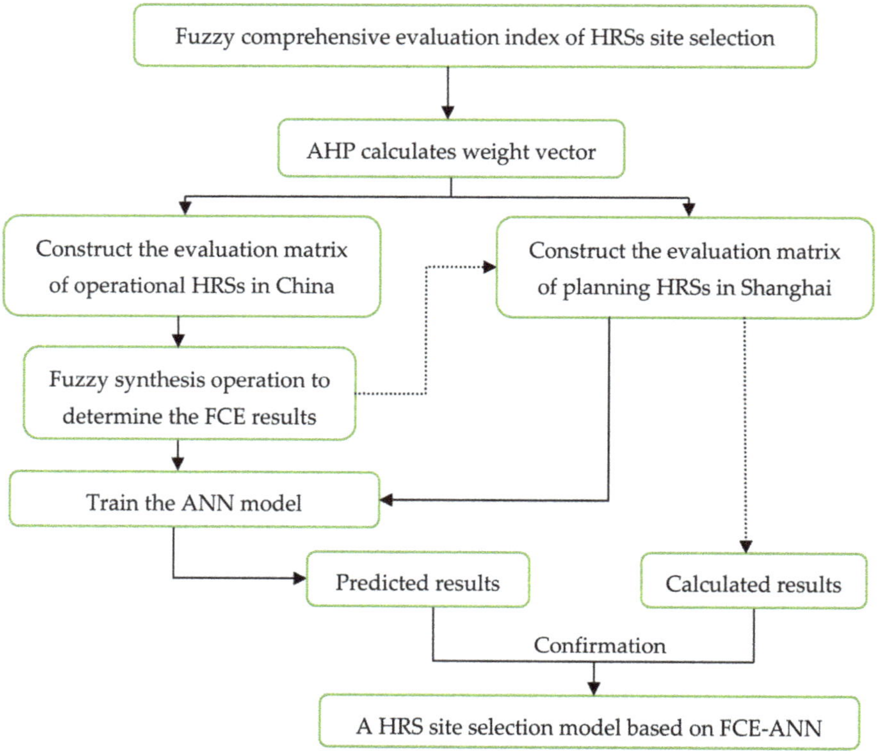

Figure 3. Research framework proposed in this study.

3.1. Analytic Hierarchy Process (AHP)

The evaluation process of HRS site selection is a multi-index decision-making process, and different indices have different influences, so they should be accordingly weighed [13]. Methods to determine the weight value are usually AHP, the entropy weight method (EWM) and the Criteria Importance Though Intercrieria Correlation (CRITIC)method [37,42,57]. Researching the literature showed that AHP is usually used for criteria weighting [61]. It is rated, and the weight is calculated by using the relative size information of the value between two pairs of comparison. The entropy method calculates the weight of the amount of information of the data entropy value. The CRITIC method mainly uses data fluctuations or the correlation between data to calculate a weight [13,42,44,62].

Because the site selection of HRSs involves many indices, these indices lack accurate data descriptions. Through hierarchical analysis, the importance of the two indices in a multilevel system can be compared. Therefore, an expert questionnaire was designed in this study; 20 respondents from the hydrogen energy industry, scientific research institutions, and doctoral students in universities were selected to fill in the questionnaire and conduct detailed consultations to conduct a graded evaluation of our evaluation indicators, using AHP method to determine the most practical and reliable coefficients of each index.

The specific steps of AHP are as follows:

(1) Project weight calculation

An evaluation index for classification was set up for the calculation of the average values of experts for indices at all levels; it was used as the initial weight, and Table 4 Saaty 1–9 scale method was referenced to quantify [42]; the judgment matrix was constructed.

Table 4. Scales and meanings of the judgment matrix [42].

Number	Importance Level	a_{ij}
1	i and j are equally important	1
2	i is slightly more important than j	3
3	i is obviously more important than j	5
4	i is more strongly important than j	7
5	i is much more important than j	9
6	i is a little less important than j	1/3
7	i is significantly less important than j	1/5
8	i is strongly less than j	1/7
9	i is much less than j	1/9

The product root method is used to calculate the geometric mean value of each row of judgment matrix $\overline{w_i}$:

$$\overline{w_i} = \left(\prod_{j=1}^{n} a_{ij}\right)^{\frac{1}{n}} \tag{1}$$

Among them, a_{ij} is the relative importance of the index i and the index j compared in pairs, and n is the number of indices. Feature vectors w_i, namely, the weight coefficients of each index, are obtained by normalizing the geometric mean values of each line [42,61].

(2) Consistency test of the judgment matrix

Calculate the maximal eigenvalue of the judgment matrix:

$$\lambda_{\max} = \frac{1}{n}\sum_{i=1}^{n} \frac{\left(\sum_{j=1}^{n} a_{ij}w_j\right)}{w_i} \tag{2}$$

Calculating consistency indicators (CI). The specific formula is:

$$CI = \frac{\lambda_{\max} - 1}{n-1} \tag{3}$$

Calculating consistency ratio (CR, where RI is the average random consistency index [47]:

$$CR = \frac{CI}{RI} \tag{4}$$

When $CR < 0.10$, the consistency of the judgment matrix is considered to be acceptable; otherwise, the judgment matrix should be appropriately modified [63,64]. The consistency of the judgment matrix in this paper was tested in MATLAB R2018a, and the consistency of the judgment matrix was less than 0.1, which was acceptable.

On the basis of expert judgments and statistical data, the constructed judgment matrices are shown in Tables 5–8.

Table 5. Judgment matrix U–U_i.

U	U_1	U_2	U_3	Weights w_i
U_1	1	3/2	5/3	0.4289
U_2	2/3	1	5/2	0.3747
U_3	3/5	2/5	1	0.1964

Note: CI = 0.0368, CR = 0.0634, r = 3.0735.

Table 6. Judgment matrix U_1–U_{1j}.

U_1	U_{11}	U_{12}	U_{13}	Weights w_{1j}
U_{11}	1	5/2	2	0.5257
U_{12}	2/5	1	3/4	0.2058
U_{13}	1/2	4/3	1	0.2685

Note: CI = 0.0231, CR = 0.0398, λ = 3.0005.

Table 7. Judgment matrix U_2–U_{2j}.

U_2	U_{21}	U_{22}	U_{23}	Weights w_{2j}
U_{21}	1	3/4	3/2	0.3415
U_{22}	4/3	1	4/3	0.3978
U_{23}	2/3	3/4	1	0.2607

Note: CI = 0.0091, CR = 0.0158, λ = 3.0183.

Table 8. Judgment matrix U_3–U_{3j}.

U_3	U_{31}	U_{32}	U_{33}	Weights w_{3j}
U_{31}	1	2/3	2/5	0.2000
U_{32}	3/2	1	3/5	0.3000
U_{33}	5/2	5/3	1	0.5000

Note: CI = 0.0024, CR = 0.0381, λ = 3.0000.

(3) Calculate the combined weight of each index

The combined weight of each index is calculated according to the weights of the criterion layer and the sub-criteria layer [63]. Taking the combined weight as an example, $A_{11} = w_1 \cdot w_{11}$, and other combined weights are analogous, so the combined weight A of the evaluation index system is:

$$A = (0.2255, 0.0883, 0.1152, 0.1280, 0.1491, 0.0977, 0.0393, 0.0589, 0.0982)$$

The combined weight not only takes into account the weight distribution of the secondary indicators in the indicators of this level, but also takes into account its overall evaluation. The weight distribution in the index system, the larger the combined weight, the more important the secondary index is in the whole evaluation index system [61].

3.2. Principle of Fuzzy Comprehensive Evaluation(FCE) Method

Fuzzy comprehensive evaluation (FCE) can reasonably combine these attributes and factors with items affected by multiple attributes or their overall strengths or weaknesses [43,48]. Therefore, the FCE method based on a fuzzy set evaluates subordinate-level status evaluation from multiple indicators [41]. The evaluation interval is, on the one hand, divided into changes, so that the evaluation criteria and the fuzziness of the influencing factors can reflect the level of the considered objects; on the other hand, the experience of people can be given full play in the evaluation, so that evaluation results are more objective and in line with the actual situation [46,63], FCE can achieve the combination of qualitative and quantitative aspects, expand the amount of information, thereby improving the amount of credible evaluation conclusions [43].

The steps of establishing the FCE model of HRSs site selection are as follows:

(1) A set of indices affecting the site selection of HRSs was established:

$$U = \{u_1, u_2, \cdots u_n\} \tag{5}$$

There are 9 indices in U: In order to facilitate the following calculations, redefine u_1 construction scale; u_2 investment strength, u_3 operational costs, u_4 hydrogen production technology, u_5 hydrogen storage technology, u_6 transportation technology, u_7 the population density, u_8 environmental factors, and u_9 social identity.

(2) The evaluation set of influencing factors of hydrogenation station siting was established:

$$V = \{v_1, v_2, \cdots, v_m\} \tag{6}$$

Correctly determining the membership function is the basis of applying fuzzy theory to properly and quantitatively describe a fuzzy concept, and the key to solving various practical problems by using the fuzzy method. The corresponding membership function is selected according to different evaluation factors. Determine the weight of each evaluation factor:

$$A = \{a_1, a_2, \cdots, a_i\} \tag{7}$$

a_i is the weight of the first factor, $\sum_{i=1}^{n} a_i = 1$.

(3) Evaluate each factor u_i to determine the degree to which it belongs to comment, and the evaluation result is written as r_i:

$$r_i = \{r_{i1}, r_{i2}, \cdots, r_{im}\} \tag{8}$$

In this paper, 10 relatively authoritative experts were selected to evaluate all indices, and the expert evaluation level was converted into the specific quantitative value of each index of each HRS according to the assigned value. Due to the large number of experts, the average value was taken as the final evaluation result of a factor. Thus, fuzzy relation R from U to V could be determined by all single factor evaluation results [44,48].

R is the fuzzy comprehensive evaluation matrix.

$$R = \begin{bmatrix} r_{11} & r_{12} & \cdots & r_{1m} \\ r_{21} & r_{22} & \cdots & r_{2m} \\ \cdots & \cdots & \cdots & \cdots \\ r_{n1} & r_{n2} & \cdots & r_{nm} \end{bmatrix} \tag{9}$$

(4) Calculate the value of B:

$$B = A \bullet R \tag{10}$$

This can be used as fuzzy comprehensive evaluation vector after the normalization of B obtained by the operation, where A is the index weight. The FCE result of HRSs is then obtained [44,48].

(5) Composition operator selection:

A basic step of the FCE of HRSs site selection is to determine the evaluation grade.

$$V = \{v_1, v_2, \cdots, v_m\} = \{\text{“very good”}, \text{“good”}, \text{“general”}, \text{“poor”}, \text{“very poor”}\}.$$

The FCE can synthesize and average the functions of various influencing factors according to the weight, which has obvious advantages in the comprehensive evaluation treatment of various influencing factors. In the calculation process, two types of fuzzy operators are often used, namely, principal factor prominence and weighted average [45]. The salient feature of the main factor operator is to reduce the interference of other data, while the salient feature of the weighted average operator is to avoid information loss. This can take into account the functions of various influencing factors as a whole. In view of the large number of factor sets in this study, in order to avoid information loss and make the collected data information play its role to the maximum, the weighted average fuzzy operator synthesis model is adopted [48,63].

3.3. *Artificial Neural Networks (ANN) Theory*

An artificial neural network (ANN) is an empirical model that mimics the function of a human neural network. There is often a highly nonlinear mapping relationship between its input and output, and it is generally difficult to write its expression, so it is called "black box" [28].The characteristic of ANN is that it can store information or knowledge distribution in a large number of neurons or the whole system. It has the potential of self-learning and self-organization [65]. In addition, it has strong fault tolerance and can deal with noisy or incomplete data [27,66]. Therefore, ANN provides a powerful tool in solving complex nonlinear multi-index comprehensive evaluation problems. The BP (Back propagation) neural network model is a multi-layer feedforward neural network trained according to the error back propagation algorithm, which is the most widely used neural network. So in this study, the back propagation algorithm (BP algorithm) is used, and the BP neural network model has the advantages of fast calculation speed and efficient solution. It can better simulate the process of comprehensive evaluation by evaluation experts [28,37]. In order to simulate the process of comprehensive evaluation by experts, considering the input, hidden and output layers, a BP neural network is designed. The comprehensive evaluation results of the 9 evaluation indices of 50 HRSs in coastal and major cities in China were taken as the input of the ANN, and the FCE results was obtained by fuzzy logic calculation are used as the output to train the neural network. The architecture is shown in Figure 4.

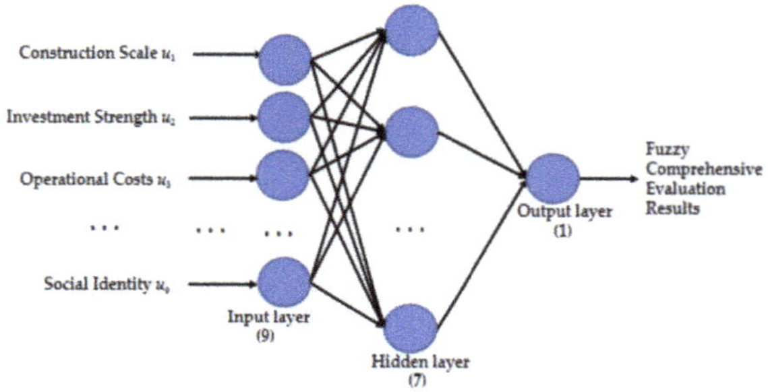

Figure 4. Architecture of artificial neural network for HRSs site selection.

ANN is an intelligence-based processing structure consisting of interconnected processing units called neurons. It includes input nodes, output nodes and one or more layers of hidden node. The input layer to the output layer needs to be processed by the hidden layer. The neurons of each hidden layer basically perform two tasks: (1) the weighted sum of all process inputs, and (2) the weighted sum of the nonlinear transformation of the neuron transfer function produces the output of each neuron. Common transfer functions include hyperbolic tangent S-type (Tansig), S-type (Logsig), positive linear (Poslin), and purelin linear (Purelin) transfer functions, as shown in Formulas (11)–(14) [28,67]. The output layer is used for use error estimates to predict outcomes. Initially in the back-propagation algorithm, the input is propagated to the hidden layer, which propagates the sensitivity back to reduce errors. At the end of the process, it updates the weights and biases. The performance index of the neural network is generally expressed by mean square error (*MSE*), mean relative error (*MRE*) and mean absolute error (*MAE*), which are used to judge the errors between network output values and target values, as shown in Formulas (15)–(17) [35,36].

$$\tan sig(n) = \frac{e^n - e^{-n}}{e^n + e^{-n}} \tag{11}$$

$$\log sig(n) = \frac{1}{1 + e^{-n}} \tag{12}$$

$$poslin(n) = \begin{cases} n, n > 0 \\ 0, n \leq 0 \end{cases} \tag{13}$$

$$purelin(n) = n \tag{14}$$

$$MSE = \frac{1}{n}\sum\nolimits_{t=1}^{n} (y - y')^2 \tag{15}$$

$$MAE = \frac{1}{n}\sum_{t=1}^{n} (y - y') \quad (16)$$

$$MRE\% = \frac{100}{n}\sum_{t=1}^{n} \left| \frac{y - y'}{y} \right| \quad (17)$$

where n represents the number of training sets, y represents the output result of the target, and y' represents the output result of the neural network.

The basic idea of establishing an intelligent location decision model of HRSs was as follows.

Determine the evaluation index set, and the number of indices determines the input node of the neural network. If the number of specific evaluation indices is 9, the number of input nodes of neural network is 9.

Each index in the index set is the input factor of the neural network. According to the characteristics of each index, the input factor is graded, the input value of the neural network is determined by the expert evaluation method, and the expected output value of the artificial neural network is obtained by the FCE method. After grading each factor, each score value is normalized into the score value in the (0,1) domain to meet the requirements of the BP neural network range [28,66].

4. Prediction of HRSs Site Selection Based on FCE-ANN

In this study, 50 HRSs already in operation in coastal and major cities of China and 8 HRSs that are being planned in Shanghai were selected as analytical examples. According to the established evaluation index system for site selection of HRSs, the FCE-ANN model is verified by an example through the steps of model establishment, data analysis and calculation, and result comparison and analysis.

4.1. Overview of HRSs in Operation in China

At present, coastal and major cities are the main developed areas for HRSs construction in China. The specific distribution is shown in Figure 5 [10], highlighted by the red dots. It can be seen that Guangdong is the province with the largest number of HRSs currently in operation. In order to better reflect the construction of HRSs in China, this paper selects 50 HRS from the provinces marked in the figure as research objects.

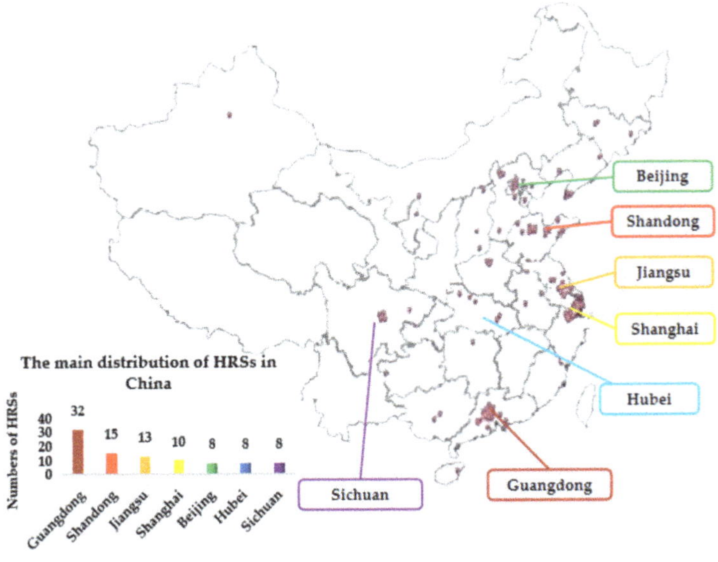

Figure 5. The Main Distribution of HRSs in China [10].

The collected information showed that the hydrogen storage, transportation, and hydrogenation of operational HRSs in China are mostly the same, and they were all constructed and are maintained at a relatively low cost. However, there are relatively few HRSs built with high costs and relatively advanced technologies such as hydrogen production and pipeline transportation. Among the 50 selected HRSs, only Shanghai YiLanShunGong has adopted hydrogen production and pipeline transportation in the industrial zone.

4.2. Fuzzy Comprehensive Evaluation of Site Selection of HRSs

In this study, 10 experts were invited to form an expert evaluation group, of which 3 were from the hydrogen energy industry with technical backgrounds such as hydrogen production and hydrogen storage for more than 5 years; 5 were from Chinese universities and other research institutions (including 2 doctoral students), specializing in HRSs research, and very concerned about the construction and planning of HRSs in China; the other two are government officials, familiar with the planning policies, population density and economic level of HRSs in various regions. With our organization, the expert evaluation group received as much research information as possible from 50 HRSs in China one week in advance. The formal evaluation is in a conference room. First, 10 experts interpret and discuss the information of 50 HRSs according to their respective areas of expertise. Secondly, experts conduct single-index evaluation of the 9 evaluation indices of each HRS. Finally, collect the evaluation grades of various evaluation indices of all HRSs by experts, and assign the quantitative results to each index of each HRS according to the membership function, and finally take the average value as the final evaluation result of a certain index, and the evaluation levels and corresponding assignments are shown in Table 9.

Table 9. Corresponding values of evaluation levels.

Evaluation Levels	Very Good	Good	General	Poor	Very Poor
Corresponding values	1	0.8	0.5	0.2	0

Among these indices, the evaluation requirements of each index are different. For example, when the cost of hydrogen is relatively low, the hydrogen cost level of the hydrogenation station is relatively high, both in construction cost and operation and maintenance cost. Among technical factors, the more advanced the technology of each index is, the higher the corresponding evaluation level is. For example, when the hydrogen production technology is more advanced, the hydrogen production technology of the HRS has a relatively high level, and the level of hydrogen storage technology is the same as that of transportation technology. Among the social influencing factors, the higher the population density, the more target users for the hydrogenation station, so the level of the population density index of the hydrogenation station is relatively high. However, for environmental factors, the less environmental damage caused, the higher the level. The higher the safety index is, the higher the corresponding evaluation level.

The values of 50 HRSs based on the evaluation levels of various indices of operating HRSs were collected and sorted here. The comprehensive evaluation results and neural network training data of HRSs are shown in Appendix A (Table A1) after the results had been arranged in order.

The above AHP calculation confirms that the combined weight A of each index in the sub-criteria layer to the target layer in the evaluation system is:

$$A = (A1, A2, A3) = (0.2255, 0.0883, 0.1152; 0.128, 0.1491, 0.0977; 0.0393, 0.0589, 0.0982)$$

The 50 HRSs were divided into 5 groups with 10 HRSs in each group. The 10 former HRSs were taken as examples.

The comprehensive results of economic factors are:

$$B1 = A1 \cdot R1 = (0.2255, 0.0883, 0.1152) \cdot \begin{bmatrix} 0.98 & 0.98 & 0.98 & 0.93 & 0.98 & 0.93 & 0.91 & 0.91 & 0.94 & 0.93 \\ 0.96 & 0.96 & 0.98 & 0.98 & 0.97 & 0.98 & 0.93 & 0.96 & 0.80 & 0.98 \\ 0.98 & 0.98 & 0.88 & 0.91 & 0.97 & 0.93 & 0.93 & 0.96 & 0.96 & 0.93 \end{bmatrix}$$
$$= (0.4187, 0.4187, 0.4089, 0.4011, 0.4184, 0.4034, 0.3945, 0.4006, 0.3932, 0.4034)$$

The comprehensive results of technical factors are:

$$B2 = A2 \cdot R2 = (0.128, 0.1491, 0.0977) \cdot \begin{bmatrix} 0.93 & 0.93 & 0.93 & 0.93 & 0.93 & 0.93 & 0.91 & 0.96 & 0.93 & 0.91 \\ 0.98 & 0.93 & 0.98 & 0.98 & 0.91 & 0.93 & 0.93 & 0.91 & 0.92 & 0.93 \\ 1.00 & 0.93 & 0.93 & 0.93 & 0.88 & 0.88 & 0.88 & 0.88 & 0.96 & 0.88 \end{bmatrix}$$
$$= (0.3629, 0.3486, 0.3560, 0.3560, 0.3407, 0.343437, 0.3411, 0.3445, 0.3500, 0.3411)$$

The comprehensive results of social factors are:

$$B3 = A3 \cdot R3 = (0.0393, 0.0589, 0.0982) \cdot \begin{bmatrix} 0.71 & 0.93 & 0.88 & 0.98 & 0.83 & 0.94 & 0.76 & 0.94 & 0.89 & 0.96 \\ 0.91 & 0.91 & 0.91 & 0.88 & 0.87 & 0.88 & 0.88 & 0.91 & 0.86 & 0.88 \\ 0.91 & 0.91 & 0.88 & 0.88 & 0.90 & 0.88 & 0.88 & 0.88 & 0.90 & 0.88 \end{bmatrix}$$
$$= (0.1709, 0.1795, 0.1746, 0.1768, 0.1722, 0.1752, 0.1681, 0.1770, 0.1740, 0.1760)$$

So, the overall results of the top 10 are:

$$B = A \cdot R = [A1, A2, A3] \cdot \begin{bmatrix} R1 \\ R2 \\ R3 \end{bmatrix} = (0.9524, 0.9467, 0.9395, 0.9339, 0.9313, 0.9223, 0.9037, 0.9221, 0.9172, 0.9205)$$

Similarly, the fuzzy comprehensive evaluation results of the other 40 HRSs are calculated in the same way. Please refer to Appendix A (Table A1) for the specific results.

Comprehensive evaluation results of the above 50 HRSs were carried out in accordance with the comprehensive evaluation results reorder in Appendix A (Table A1). The last column sorting result shows that YiLanShunGong had the highest evaluation levels in the 50 HRSs, and the site of the construction scale, operational cost, hydrogen source, and the impact on the society were the most reasonable.

4.3. ANN Training and Prediction

4.3.1. ANN Construction and Training

In this study, a BP neural network was modeled and simulated in the MATLAB software. The BP (Back propagation) neural network model is a multi-layer feedforward neural network trained according to the error back propagation algorithm, which is the most widely used neural network. The backpropagation algorithm is slow to converge and sometimes leads to overfitting. To address these issues, Levenberg–Marquardt (LM) backpropagation was developed to achieve fast convergence without overfitting [35]. By reduction in error response of the network became more smother and it also reduces the problem of overfitting. LM back propagation algorithm uses the conjugate gradient technique to reduce the sum of squares at each titration. The FCE results of the 9 evaluation indices of 50 HRSs in major cities in China were taken as the input of the ANN, and the ranking of FCE results was obtained by fuzzy logic calculation are used as expected output layer for constructing and training the neural network [27].

The number of hidden-layer neurons affects the performance and accuracy of a neural network. If the number of hidden layer neurons is too small, the data fitting of the neural network may not be carried out well. If the number of neurons is too large, the neural network may perform well in training, but poorly in the prediction results of unknown samples. Therefore, it is very important to determine the appropriate number of neurons. In order to ensure the best model performance, different neural networks with from 3 to 10 neurons using Tansig transfer function were tested, and the neural network model with the minimal *MAE*, *MRE*, *MSE* was selected [35,36]. Test results are shown in Table 10.

Table 10. Influence of number of hidden-layer neurons on neural network using Tansig transfer function.

Number of Neurons	3	4	5	6	7	8	9	10
MAE	0.2098	0.0428	0.0522	0.065	0.0289	0.0522	0.0836	0.0835
MRE	0.8352	0.1657	0.2246	0.2529	0.1129	0.2266	0.3176	0.3283
MSE (1×10^{-7})	24	2.42	3.65	5.07	0.463	9.95	5.67	9.11

As shown in Table 10, when the number of neurons is 7, the *MAE*, *MRE*, *MSE* reach the minimums, and the results are 0.0289, 0.1129 and 0.463×10^{-7} espectively. Therefore, the number of neurons in this case model is 7. In addition to the number of neurons in the hidden layer affecting the performance of the neural network, different transfer functions in the hidden layer also affect the performance of the neural network. Figure 6 shows the *MAE*, *MRE*, *MSE* of different transfer functions in the hidden layer. It can be seen that the Purelin transfer function of this neural network structure has the best effect. The number of epochs is set 2000 so that the neural network can fully iterate. The three-layer BP neural network can approximate any nonlinear continuous function with arbitrary precision. For the small amount of data in this paper, the number of hidden layers is set to 1. The finally relevant parameters of this ANN model are shown in Table 11.

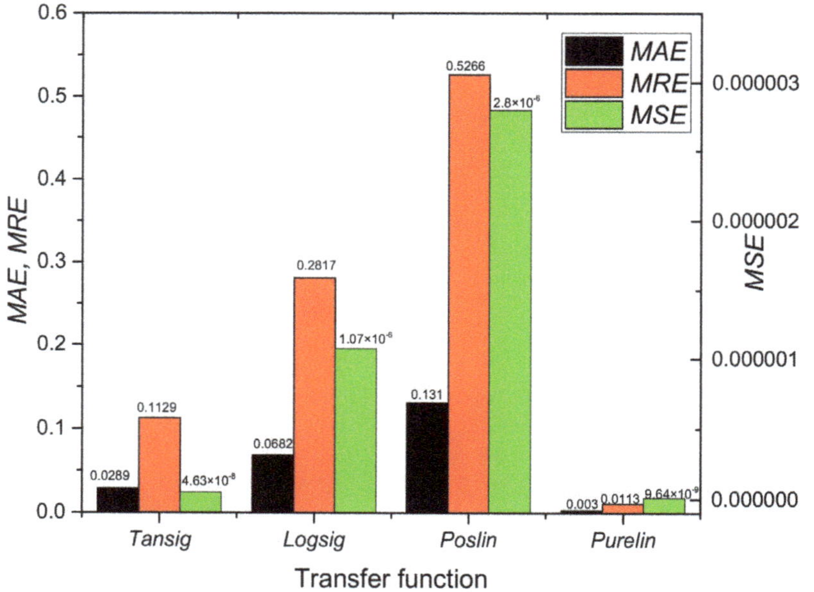

Figure 6. The *MAE*, *MRE*, *MSE* of different transfer functions in the hidden layer.

Table 11. Parameters of ANN model.

Neuron Transfer Function	Algorithm	Epochs	Training Target Error
Purelin	LM	2000	1×10^{-12}
Input nodes	Output nodes	Number of hidden neurons	Number of hidden layers
9	1	7	1

The neural network code was constructed in MATLAB R2018a, and the data in Appendix A (Table A1) were substituted into the neural network for training. The results of expert evaluation were taken as the output to obtain the structure of the neural network. The correlation coefficient of neural network is shown in Figure 7. The samples number for training, validation, and testing data are 34 (about 70%), 8 (about 15%), and 8 (about 15%), respectively. The training set of a neural network learns the dataset of samples and constructs the required model by training the sample data. The function of the validation set is to stop the training of the neural network model in time when the performance of the model fails to continue to improve or change during training. Using ANN, the process of training is prone to overfitting and underfitting. The test set does not participate in the training and mainly tests the accuracy of the training model and can detect whether the model produces overfitting. If the results of the test set are too large, an overfitting may occur. The Figure 7 shows that the regression coefficients of the training, validation, and test sets of the model were all close to 1, indicating that the ANN model was well-trained and there is no phenomenon of underfitting and overfitting. So the ANN model could be used for the site selection prediction of HRSs.

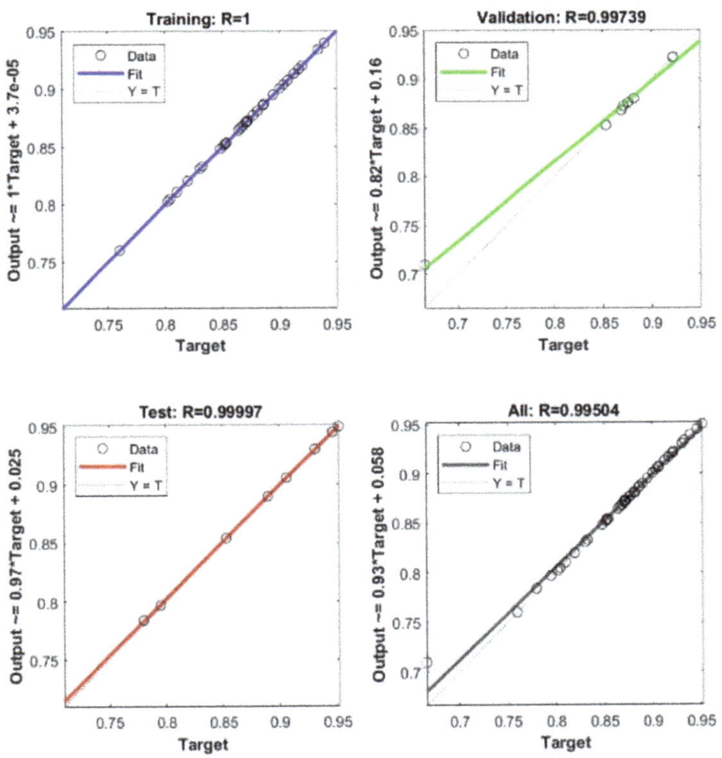

Figure 7. Correlation coefficients between model objectives and ANN prediction based on training, validation, and test sets, and the whole dataset.

4.3.2. Location Prediction of HRSs

After training the ANN model, 8 HRSs that are being planned in Shanghai are selected for prediction, and the accuracy of the model is evaluated. The FCE results and ranking of the 8 HRSs are shown in Appendix A (Table A2). The scores of each index of {D1,D2,D3,D4,D5,D6,D7,D8} of the 8 HRSs were input into the ANN model for prediction and evaluation. The result of its operation is as follows:

$$YP = (0.9682, 0.9493, 0.9473, 0.94516, 0.9319, 0.8767, 0.8322, 0.8138)$$

A comparison of the predicted and evaluation results is shown in Table 12.

Table 12 shows that the expert evaluation results obtained by FCE were very close to the prediction result of the ANN model, the ANN prediction results are the same as the comprehensive evaluation. The comprehensive ranking of the eight HRSs was consistent. It shows that the ANN model we constructed can predict the location of the HRS well, and further shows that there is no underfitting and overfitting phenomenon. Among the two evaluation methods, the ShuiChan Road HRS was the best site selection result, which shows that HRSs site selection can be decided quickly and accurately through ANN model. According to the comprehensive evaluation results of experts, the values of the top four HRSs are relatively close, and the results were discussed and analyzed. According to the actual survey data, the planned daily hydrogen refueling capacity of these four HRSs is 500kg and the refueling pressure is 35Mpa. Except for ShangChai HRS, which is a fixed station, the other three HRSs are Skid-mounted stations. The skid-mounted station has a certain advantage in terms of construction scale and operation and maintenance cost, while the fixed station invests heavily. Therefore, the four HRSs have little difference in economic factors. Similarly, in terms of technical factors, experts generally believe that the ShuiChan Road HRS is located on the main road in BaoShan District, and the transportation is more convenient, which makes the cost of hydrogen relatively low. At the same time, the surrounding population density is relatively large, the demand for HRSs is relatively large, and the public acceptance is relatively high. The two HRSs planned in the lower-ranked JiaDing District are oil-hydrogen joint construction stations, which are far inferior to other HRSs in terms of major economic factors. Moreover, Jiading District has a low population density and is far from the core area [68]. The transportation cost is high, and the hydrogen storage technology in the station is far less than that of the independent HRS, so they rank low. After the above discussion and actual investigation, it is worthy of affirmation that the comprehensive score of the HRS of the ShuiChan road is the highest, which also proves the objectivity of the expert evaluation and the accuracy of the ANN.

Table 12. Comparison table of predicted sample results.

Planned HRSs	FCE Results	ANN Results
ShuiChan Road HRS D1	0.9590	0.9682
ShangChai HRS D2	0.9497	0.9493
BaoShan Northern Suburb HRS D3	0.9480	0.9473
Military road HRSD4	0.9421	0.9416
WenXiang Road HRS D5	0.9329	0.9319
BaoPan HRS(JiaDing) D6	0.8782	0.8767
GaoJia HRS(JiaDing) D7	0.8323	0.8322
ShangHai LinGang HRS D8	0.8140	0.8138

5. Conclusions

The evaluation index system of HRSs site selection is a complex nonlinear system, which needs to comprehensively consider economic, technical, and social factors. Therefore, it is particularly important to establish an efficient and accurate mathematical model to select the construction sites of hydrogen stations. Therefore, this paper provides a new site selection model for HRSs based on FCE-ANN, which evaluates the operational HRSs in China through the AHP and the FCE method, and uses the obtained data to train the ANN model. The input of the ANN model is the evaluation index results for site selection of HRSs, and the output is the FCE results of the site selection of HRSs. The trained ANN model is used to predict the site selection planning of Shanghai HRSs, and it is concluded that the ShuiChan Road HRS is the best site selection in Shanghai planning. The FCE-ANN model prediction results were compared with the results obtained by the FCE method. The results of the two methods were consistent, clearly showing that a FCE-ANN can objectively and accurately evaluate alternatives to reduce the influence of human factors, and that a FCE- ANN model can quickly and accurately solve HRSs site selection decision-making problems. It can also provide reference for scholars to study other site selection decision-making issues. According to the evaluation system of HRSs site selection established in this paper, a comprehensive understanding of HRSs can be obtained, and on the basis of the evaluation results, it can provide reference for government decision-makers in formulating macro-control policies. In the evaluation system, economic factors have the largest weight, which also reflects that in the process of investment and construction of hydrogen refueling stations, it is very important to evaluate the economic strength of the investor and constructor. Currently, for China, the hydrogen energy industry is in its infancy, and investment in construction a HRS has a relatively long return on investment period, so the government should also give maximum policy and financial support, so as to promote the healthy and sustainable development of the hydrogen energy supply chain industry. This study uses the FCE-ANN model to solve the complex multi-factor evaluation problem, but requires sufficient historical data for model training. However, there are not many hydrogen refueling stations in China, and the historical data and evaluation indicators available for reference are limited. The problem can be better solved in the future. The HRSs evaluation index system established in this paper is based on the current development stage of the hydrogen energy industry. In the future, with the development of the hydrogen energy industry, the evaluation index system of the HRS should be continuously improved and supplemented. It is suggested to explore the deep influencing factors from the perspective of the public's recognition and satisfaction of HRSs. At the same time, the influence of uncertain factors on HRSs in the future should be considered, and fuzzy mathematics and mathematical logic should be combined with ANN to solve problems such as uncertainty and fuzziness, so as to consider the development strategy of HRSs in the long run.

Author Contributions: Conceptualization, X.Q. and Y.Z.; methodology, Y.Z.; software, C.L.; validation, Y.Z., C.L. and J.Z.; formal analysis, Y.Z.; investigation, J.Z.; resources, J.Z.; data curation, Y.Z.; writing—original draft preparation, Y.Z.; writing—review and editing, X.Q.; visualization, Y.Z.; supervision, X.Q.; project administration, X.Q.; funding acquisition, X.Q. All authors have read and agreed to the published version of the manuscript.

Funding: This research was funded by Major Project of Technological Innovation in Hubei Province, grant number 2020BED010 and 2019AAA075.

Acknowledgments: Gratitude is expressed to all the experts and scholars who participated in the establishment of the evaluation system of this study through interviews and questionnaires; and gratitude is expressed to 10 experts and scholars who participated in the comprehensive evaluation of hydrogen refueling stations in this study. Sincere thanks also to Orange Group Database for providing the information of HRSs.

Conflicts of Interest: The authors declare no conflict of interest.

Appendix A

Table A1. Evaluation results of indices and ranking of 50 operational HRSs in China.

Operational HRSs	u_1	u_2	u_3	u_4	u_5	u_6	u_7	u_8	u_9	Results	Rank
YiLan Shun Gong HRS	0.98	0.96	0.98	0.93	0.98	1.00	0.71	0.91	0.91	0.9524	1
Nan Tong Bai Ying HRS	0.98	0.96	0.98	0.93	0.93	0.93	0.93	0.91	0.91	0.9467	2
Changshu Changjia HRS	0.98	0.98	0.88	0.93	0.98	0.93	0.88	0.91	0.88	0.9395	3
Yancheng Chuang Yong HRS	0.93	0.98	0.91	0.93	0.98	0.93	0.98	0.88	0.88	0.9339	4
Nanhai Songgang HRS	0.98	0.97	0.97	0.93	0.91	0.88	0.83	0.87	0.90	0.9313	5
Meijin Qingdao bridgehead HRS	0.93	0.98	0.93	0.93	0.93	0.88	0.94	0.88	0.88	0.9223	6
Zhongshan Sha Lang HRS	0.91	0.93	0.93	0.91	0.93	0.88	0.76	0.88	0.88	0.9221	7
Frodo Road HRS	0.91	0.96	0.96	0.96	0.91	0.88	0.94	0.91	0.88	0.9205	8
Yunfu Xinxing West Second Ring Road HRS *	0.94	0.8	0.96	0.93	0.92	0.96	0.89	0.86	0.90	0.9172	9

Table A1. *Cont.*

Operational HRSs	u_1	u_2	u_3	u_4	u_5	u_6	u_7	u_8	u_9	Results	Rank
Shenhua RuGao HRS	0.93	0.98	0.93	0.91	0.93	0.88	0.96	0.88	0.88	0.9133	10
Jinhong Yunan HRS	0.92	0.83	0.93	0.93	0.92	0.96	0.89	0.88	0.90	0.9131	11
Shunde Shunfeng HRS	0.93	0.93	0.93	0.87	0.90	0.92	0.81	0.88	0.91	0.9074	12
Lianxin Guangzhou Development Zone HRS	0.92	0.92	0.94	0.93	0.92	0.93	0.73	0.80	0.88	0.9063	13
Venus in the region of Pidu HRS	0.93	0.93	0.93	0.91	0.93	0.88	0.89	0.88	0.88	0.9037	14
Jiading River bridge HRS	0.91	0.93	0.91	0.91	0.88	0.91	0.71	0.88	0.88	0.9007	15
Beijing yongfeng HRS	0.91	0.94	0.91	0.91	0.88	0.94	0.76	0.88	0.88	0.8949	16
Pujiang gas HRS	0.91	0.93	0.91	0.91	0.88	0.86	0.71	0.88	0.88	0.8900	17
Shanghai electric drive HRS (JiaDing)	0.86	0.91	0.91	0.91	0.88	0.86	0.96	0.88	0.88	0.8868	18
Shiyan Dongfeng HRS	0.86	0.88	0.88	0.91	0.88	0.86	0.93	0.88	0.84	0.8858	19
Dongguan Energy Shatian HRS	0.90	0.95	0.96	0.82	0.89	0.84	0.72	0.88	0.86	0.8817	20
Tongji—New Source Dalian HRS	0.92	0.89	0.78	0.92	0.86	0.84	0.91	0.93	0.78	0.8813	21
Wuhan Development Zone Xiongzhong HRS	0.84	0.86	0.86	0.91	0.91	0.86	0.86	0.88	0.87	0.8774	22
Shanghai power HRS (FengXian)	0.84	0.91	0.91	0.91	0.86	0.86	0.91	0.88	0.92	0.8756	23
DongHui HRS	0.86	0.91	0.91	0.91	0.86	0.86	0.91	0.88	0.92	0.8728	24
Huangpu Knowledge City, Guangzhou HRS	0.86	0.88	0.88	0.91	0.88	0.86	0.81	0.88	0.86	0.8719	25
Yunfu Silao HRS	0.87	0.94	0.86	0.84	0.84	0.86	0.88	0.86	0.90	0.8717	26
Zhongshan Dayang motor HRS	0.83	0.88	0.88	0.91	0.88	0.86	0.82	0.88	0.84	0.8711	27
Shanghai Anting HRS	0.87	0.86	0.86	0.84	0.84	0.86	0.78	0.78	0.90	0.8708	28
Shenzhen Kaihaoda HRS	0.87	0.84	0.86	0.84	0.84	0.86	0.83	0.78	0.90	0.8687	29
Donghua harbor city HRS	0.86	0.86	0.88	0.91	0.88	0.86	0.88	0.88	0.84	0.8685	30
Nanhai Taoyuan HRS	0.8	0.81	0.88	0.93	0.91	0.88	0.83	0.87	0.90	0.8662	31
Changshu Toyota HRS	0.84	0.86	0.91	0.91	0.91	0.86	0.86	0.88	0.87	0.8645	32
Foshan Chancheng, Longsha Road HRS	0.78	0.77	0.86	0.93	0.90	0.88	0.87	0.87	0.88	0.8540	33
Yunfu Zhongtong HRS	0.92	0.61	0.81	0.93	0.86	0.84	0.91	0.88	0.78	0.8532	34
MingTian hydrogen HRS	0.87	0.61	0.91	0.93	0.86	0.84	0.91	0.88	0.81	0.8530	35
Geely Commercial Vehicles HRS	0.89	0.91	0.80	0.86	0.84	0.84	0.86	0.88	0.78	0.8528	36
Sanshui Guanghai Avenue West HRS	0.77	0.81	0.85	0.93	0.92	0.93	0.74	0.88	0.88	0.8528	37
Yunfu Luo Ding 1 HRS	0.84	0.86	0.88	0.91	0.88	0.86	0.91	0.88	0.84	0.8514	38
Jiashan Edelman HRS	0.87	0.84	0.86	0.86	0.89	0.84	0.98	0.91	0.86	0.8482	39
Guangzhou Huangpu East District, HRS	0.78	0.76	0.87	0.93	0.86	0.85	0.72	0.79	0.86	0.8328	40
Foshan, Gaoming District, Genghe Town HRS	0.78	0.73	0.78	0.93	0.90	0.79	0.90	0.88	0.84	0.8303	41
Foshan, Gaoming District, Tram HRS	0.75	0.81	0.84	0.88	0.83	0.78	0.90	0.88	0.84	0.8197	42
Nanhai Hanlan Jiujiang Longgao HRS	0.74	0.80	0.84	0.93	0.78	0.80	0.84	0.83	0.82	0.8102	43
Foshan Nanhai Guicheng HRS	0.74	0.80	0.83	0.90	0.78	0.80	0.82	0.83	0.82	0.8044	44
Foshan Foqi Foluo Road HRS	0.74	0.80	0.84	0.88	0.78	0.80	0.82	0.83	0.81	0.8020	45
Nanhai Ruihui HRS	0.71	0.72	0.78	0.88	0.86	0.80	0.84	0.83	0.82	0.7950	46
Guangdong Southwest Hekou HRS	0.73	0.71	0.75	0.87	0.82	0.78	0.84	0.83	0.76	0.7801	47
Foshan Hejiao HRS *	0.73	0.72	0.70	0.75	0.78	0.80	0.85	0.83	0.80	0.7601	48
Foshan Nanhai Zhangkeng HRS *	0.73	0.72	0.70	0.75	0.78	0.80	0.85	0.83	0.80	0.7601	49
Yunfu Zhaotong HRS *	0.63	0.61	0.65	0.80	0.60	0.50	0.89	0.86	0.70	0.6659	50

* Oil-hydrogen joint construction station.

Table A2. Evaluation results of indices and ranking of planning HRSs in Shanghai.

Planning HRSs	u_1	u_2	u_3	u_4	u_5	u_6	u_7	u_8	u_9	Results	Rank
Shui Chan Road HRS (Bao Shan) D1	0.96	0.96	0.96	0.96	0.96	0.96	0.93	0.96	0.96	0.9590	1
Shang Chai HRS (Yang Pu) D2	0.96	0.98	0.94	0.93	0.96	0.96	0.96	0.94	0.91	0.9497	2
Baoshan Northern Suburb HRS D3	0.93	0.98	0.96	0.93	0.96	0.96	0.90	0.96	0.95	0.9480	3
Military Road HRS (YangPu) D4	0.93	0.96	0.94	0.96	0.96	0.93	0.96	0.94	0.91	0.9421	4
Wen Xiang Road HRS (SongJiang) D5	0.93	0.93	0.93	0.93	0.93	0.93	0.85	0.98	0.96	0.9329	5
ShangHai LinGang HRS D6	0.90	0.90	0.78	0.93	0.86	0.80	0.80	0.93	0.96	0.8782	6
GaoJia HRS(JiaDing) D7	0.85	0.83	0.93	0.78	0.76	0.80	0.83	0.78	0.92	0.8323	7
BaoPan HRS(JiaDing) D8	0.83	0.80	0.92	0.73	0.75	0.79	0.83	0.76	0.92	0.8140	8

References

1. Zhou, J.; Wu, Y.; Tao, Y.; Gao, J.; Zhong, Z.; Xu, C. Geographic information big data-driven two-stage optimization model for location decision of hydrogen refueling stations: An empirical study in China. *Energy* **2021**, *225*, 120330. [CrossRef]
2. Bansal, S.; Zong, Y.; You, S.; Mihet-Popa, L.; Xiao, J. Technical and economic analysis of one-stop charging stations for battery and fuel cell EV with renewable energy sources. *Energies* **2020**, *13*, 2855. [CrossRef]
3. Apostolou, D.; Xydis, G. A literature review on hydrogen refuelling stations and infrastructure. Current status and future prospects. *Renew. Sustain. Energy Rev.* **2019**, *113*, 109292. [CrossRef]
4. Viktorsson, L.; Heinonen, J.; Skulason, J.; Unnthorsson, R. A Step towards the hydrogen economy—A life cycle cost analysis of a hydrogen refueling station. *Energies* **2017**, *10*, 763. [CrossRef]
5. Xu, C.; Wu, Y.; Dai, S. What are the critical barriers to the development of hydrogen refueling stations in China? A modified fuzzy DEMATEL approach. *Energy Policy* **2020**, *142*, 111495. [CrossRef]
6. Lopez Jaramillo, O.; Rinebold, J.; Kuby, M.; Kelley, S.; Ruddell, D.; Stotts, R.; Krafft, A.; Wentz, E. Hydrogen station location planning via geodesign in connecticut: Comparing optimization models and structured stakeholder collaboration. *Energies* **2021**, *14*, 7747. [CrossRef]
7. Khan, U.; Yamamoto, T.; Sato, H. Consumer preferences for hydrogen fuel cell vehicles in Japan. *Transp. Res. Part D Transp. Environ.* **2020**, *87*, 102542. [CrossRef]
8. Sandri, O.; Holdsworth, S.; Hayes, J.; Willand, N.; Moore, T. Hydrogen for all? Household energy vulnerability and the transition to hydrogen in Australia. *Energy Res. Soc. Sci.* **2021**, *79*, 102179. [CrossRef]
9. Lee, D.-Y.; Elgowainy, A.; Vijayagopal, R. Well-to-wheel environmental implications of fuel economy targets for hydrogen fuel cell electric buses in the United States. *Energy Policy* **2019**, *128*, 565–583. [CrossRef]
10. OrangGroup. Available online: http://www.china-orangegroup.com/ (accessed on 25 October 2021).
11. H2stations.org. Available online: https://www.h2stations.org/ (accessed on 25 October 2021).
12. Meng, X.; Gu, A.; Wu, X.; Zhou, L.; Zhou, J.; Liu, B.; Mao, Z. Status quo of China hydrogen strategy in the field of transportation and international comparisons. *Int. J. Hydrogen Energy* **2020**, *46*, 28887–28899. [CrossRef]
13. Bai, W.; Zhang, L. How to finance for establishing hydrogen refueling stations in China? An analysis based on Fuzzy AHP and PROMETHEE. *Int. J. Hydrogen Energy* **2020**, *45*, 34354–34370. [CrossRef]
14. Ren, X.; Dong, L.; Xu, D.; Hu, B. Challenges towards hydrogen economy in China. *Int. J. Hydrogen Energy* **2020**, *45*, 34326–34345. [CrossRef]
15. Song, P.; Sui, Y.; Shan, T.; Hou, J.; Wang, X. Assessment of hydrogen supply solutions for hydrogen fueling station: A Shanghai case study. *Int. J. Hydrogen Energy* **2020**, *45*, 32884–32898. [CrossRef]
16. Gye, H.-R.; Seo, S.-K.; Bach, Q.-V.; Ha, D.; Lee, C.-J. Quantitative risk assessment of an urban hydrogen refueling station. *Int. J. Hydrogen Energy* **2019**, *44*, 1288–1298. [CrossRef]
17. Li, Y.; Cui, F.; Li, L. An integrated optimization model for the location of hydrogen refueling stations. *Int. J. Hydrogen Energy* **2018**, *43*, 19636–19649. [CrossRef]
18. Cho, S.; Kim, J. Multi-site and multi-period optimization model for strategic planning of a renewable hydrogen energy network from biomass waste and energy crops. *Energy* **2019**, *185*, 527–540. [CrossRef]
19. He, C.; Sun, H.; Xu, Y.; Lv, S. Hydrogen refueling station siting of expressway based on the optimization of hydrogen life cycle cost. *Int. J. Hydrogen Energy* **2017**, *42*, 16313–16324. [CrossRef]
20. Sun, H.; He, C.; Wang, H.; Zhang, Y.; Lv, S.; Xu, Y. Hydrogen station siting optimization based on multi-source hydrogen supply and life cycle cost. *Int. J. Hydrogen Energy* **2017**, *42*, 23952–23965. [CrossRef]
21. Zheng, Q.; Lv, H.; Zhou, W.; Zhang, C. Research on multi-period hydrogen refueling station location model in Jiading district. *World Electr. Veh. J.* **2021**, *12*, 146. [CrossRef]

22. Lin, R.-H.; Ye, Z.-Z.; Wu, B.-D. A review of hydrogen station location models. *Int. J. Hydrogen Energy* **2020**, *45*, 20176–20183. [CrossRef]
23. Kurtz, J.; Bradley, T.; Winkler, E.; Gearhart, C. Predicting demand for hydrogen station fueling. *Int. J. Hydrogen Energy* **2020**, *45*, 32298–32310. [CrossRef]
24. Kim, H.; Eom, M.; Kim, B.-I. Development of strategic hydrogen refueling station deployment plan for Korea. *Int. J. Hydrogen Energy* **2020**, *45*, 19900–19911. [CrossRef]
25. Chang, X.; Ma, T.; Wu, R. Impact of urban development on residents' public transportation travel energy consumption in China: An analysis of hydrogen fuel cell vehicles alternatives. *Int. J. Hydrogen Energy* **2019**, *44*, 16015–16027. [CrossRef]
26. Liang, Y.; Pan, X.; Zhang, C.; Xie, B.; Liu, S. The simulation and analysis of leakage and explosion at a renewable hydrogen refuelling station. *Int. J. Hydrogen Energy* **2019**, *44*, 22608–22619. [CrossRef]
27. Xu, A.; Li, R.; Chang, H.; Xu, Y.; Li, X.; Lin, G.; Zhao, Y. Artificial neural network (ANN) modeling for the prediction of odor emission rates from landfill working surface. *Waste Manag.* **2022**, *138*, 158–171. [CrossRef] [PubMed]
28. Geetha, A.; Santhakumar, J.; Sundaram, K.M.; Usha, S.; Thentral, T.M.T.; Boopathi, C.S.; Ramya, R.; Sathyamurthy, R. Prediction of hourly solar radiation in Tamil Nadu using ANN model with different learning algorithms. *Energy Rep.* **2022**, *8*, 664–671. [CrossRef]
29. Sang, B. Application of genetic algorithm and BP neural network in supply chain finance under information sharing. *J. Comput. Appl. Math.* **2021**, *384*, 113170. [CrossRef]
30. Liu, Q.; Sun, P.; Fu, X.; Zhang, J.; Yang, H.; Gao, H.; Li, Y. Comparative analysis of BP neural network and RBF neural network in seismic performance evaluation of pier columns. *Mech. Syst. Signal Processing* **2020**, *141*, 106707. [CrossRef]
31. Liu, J.; Zhang, J.; Zhang, X. Semi-discretized numerical solution for time fractional convection–diffusion equation by RBF-FD. *Appl. Math. Lett.* **2022**, *128*, 107880. [CrossRef]
32. He, Y.; Meng, Z.; Xu, H.; Zou, Y. A dynamic model of evaluating differential automatic method for solving plane problems based on BP neural network algorithm. *Phys. A Stat. Mech. Its Appl.* **2020**, *556*, 124845. [CrossRef]
33. Mulashani, A.K.; Shen, C.; Nkurlu, B.M.; Mkono, C.N.; Kawamala, M. Enhanced group method of data handling (GMDH) for permeability prediction based on the modified Levenberg Marquardt technique from well log data. *Energy* **2022**, *239*, 121915. [CrossRef]
34. Huang, B.; Chen, W.; Lin, C.-L.; Juang, C.-F.; Wang, J. MLP-BP: A novel framework for cuffless blood pressure measurement with PPG and ECG signals based on MLP-Mixer neural networks. *Biomed. Signal Processing Control.* **2022**, *73*, 103404. [CrossRef]
35. Roshani, G.H.; Nazemi, E.; Roshani, M.M. Flow regime independent volume fraction estimation in three-phase flows using dual-energy broad beam technique and artificial neural network. *Neural Comput. Appl.* **2016**, *28*, 1265–1274. [CrossRef]
36. Amir Sattari, M.; Hossein Roshani, G.; Hanus, R.; Nazemi, E. Applicability of time-domain feature extraction methods and artificial intelligence in two-phase flow meters based on gamma-ray absorption technique. *Measurement* **2021**, *168*, 108474. [CrossRef]
37. Yariyan, P.; Zabihi, H.; Wolf, I.D.; Karami, M.; Amiriyan, S. Earthquake risk assessment using an integrated fuzzy analytic hierarchy process with artificial neural networks based on GIS: A case study of Sanandaj in Iran. *Int. J. Disaster Risk Reduct.* **2020**, *50*, 101705. [CrossRef]
38. De Campos Souza, P.V. Fuzzy neural networks and neuro-fuzzy networks: A review the main techniques and applications used in the literature. *Appl. Soft Comput.* **2020**, *92*, 106275. [CrossRef]
39. Rahmati, O.; Panahi, M.; Ghiasi, S.S.; Deo, R.C.; Tiefenbacher, J.P.; Pradhan, B.; Jahani, A.; Goshtasb, H.; Kornejady, A.; Shahabi, H.; et al. Hybridized neural fuzzy ensembles for dust source modeling and prediction. *Atmos. Environ.* **2020**, *224*, 117320. [CrossRef]
40. Wang, M.; Niu, D. Research on project post-evaluation of wind power based on improved ANP and fuzzy comprehensive evaluation model of trapezoid subordinate function improved by interval number. *Renew. Energy* **2019**, *132*, 255–265. [CrossRef]
41. Liang, D.; Dai, Z.; Wang, M. Assessing customer satisfaction of O2O takeaway based on online reviews by integrating fuzzy comprehensive evaluation with AHP and probabilistic linguistic term sets. *Appl. Soft Comput.* **2021**, *98*, 106847. [CrossRef]
42. Wang, H.; Cheng, M.; Zhang, S.; Fan, J.; Feng, H.; Zhang, F.; Wang, X.; Sun, L.; Xiang, Y. Optimization of irrigation amount and fertilization rate of drip-fertigated potato based on analytic hierarchy process and fuzzy comprehensive evaluation methods. *Agric. Water Manag.* **2021**, *256*, 107130. [CrossRef]
43. Zhang, H.; He, X.; Mitri, H. Fuzzy comprehensive evaluation of virtual reality mine safety training system. *Saf. Sci.* **2019**, *120*, 341–351. [CrossRef]
44. Qin, G.; Zhang, M.; Yan, Q.; Xu, C.; Kammen, D.M. Comprehensive evaluation of regional energy internet using a fuzzy analytic hierarchy process based on cloud model: A case in China. *Energy* **2021**, *228*, 120569. [CrossRef]
45. Zhou, Z.; Zhang, X.; Dong, W. Fuzzy comprehensive evaluation for safety guarantee system of reclaimed water quality. *Procedia Environ. Sci.* **2013**, *18*, 227–235. [CrossRef]
46. Nie, B.-S.; Zhao, P.-F.; Guo, J.-H.; Niu, P.-P.; Wang, G. Fuzzy comprehensive evaluation of coal mine safety investment structure based on the M(1,2,3) model. *Energy Procedia* **2012**, *16*, 592–597. [CrossRef]
47. He, Z.; Li, M.; Cai, Z.; Zhao, R.; Hong, T.; Yang, Z.; Zhang, Z. Optimal irrigation and fertilizer amounts based on multi-level fuzzy comprehensive evaluation of yield, growth and fruit quality on cherry tomato. *Agric. Water Manag.* **2021**, *243*, 106360. [CrossRef]

48. Su, X.; Tao, L.; Liu, H.; Wang, L.; Suo, M. Real-time hierarchical risk assessment for UAVs based on recurrent fusion autoencoder and dynamic FCE: A hybrid framework. *Appl. Soft Comput.* **2021**, *106*, 107286. [CrossRef]
49. Ajanovic, A.; Haas, R. Economic prospects and policy framework for hydrogen as fuel in the transport sector. *Energy Policy* **2018**, *123*, 280–288. [CrossRef]
50. Gökçek, M.; Kale, C. Techno-economical evaluation of a hydrogen refuelling station powered by Wind-PV hybrid power system: A case study for İzmir-Çeşme. *Int. J. Hydrogen Energy* **2018**, *43*, 10615–10625. [CrossRef]
51. Moradi, R.; Groth, K.M. Hydrogen storage and delivery: Review of the state of the art technologies and risk and reliability analysis. *Int. J. Hydrogen Energy* **2019**, *44*, 12254–12269. [CrossRef]
52. Blazquez-Diaz, C. Techno-economic modelling and analysis of hydrogen fuelling stations. *Int. J. Hydrogen Energy* **2019**, *44*, 495–510. [CrossRef]
53. Stehlík, K.; Tkáč, M.; Bouzek, K. Recent advances in hydrogen technologies in the Czech Republic. *Int. J. Hydrogen Energy* **2019**, *44*, 19055–19060. [CrossRef]
54. Minutillo, M.; Perna, A.; Forcina, A.; di Micco, S.; Jannelli, E. Analyzing the levelized cost of hydrogen in refueling stations with on-site hydrogen production via water electrolysis in the Italian scenario. *Int. J. Hydrogen Energy* **2021**, *46*, 13667–13677. [CrossRef]
55. Tsunemi, K.; Kihara, T.; Kato, E.; Kawamoto, A.; Saburi, T. Quantitative risk assessment of the interior of a hydrogen refueling station considering safety barrier systems. *Int. J. Hydrogen Energy* **2019**, *44*, 23522–23531. [CrossRef]
56. Yoo, B.-H.; Wilailak, S.; Bae, S.-H.; Gye, H.-R.; Lee, C.-J. Comparative risk assessment of liquefied and gaseous hydrogen refueling stations. *Int. J. Hydrogen Energy* **2021**, *46*, 35511–35524. [CrossRef]
57. Mayer, T.; Semmel, M.; Guerrero Morales, M.A.; Schmidt, K.M.; Bauer, A.; Wind, J. Techno-economic evaluation of hydrogen refueling stations with liquid or gaseous stored hydrogen. *Int. J. Hydrogen Energy* **2019**, *44*, 25809–25833. [CrossRef]
58. Nie, X.; Wei, X.; Xia, Q.; Zhou, M. Customers' purchase intention for hydrogen vehicle and industrial agglomeration: Evidence from Jiangsu Province, China. *Int. J. Hydrogen Energy* **2021**, *46*, 18011–18019. [CrossRef]
59. Li, Z.; Wang, W.; Ye, M.; Liang, X. The impact of hydrogen refueling station subsidy strategy on China's hydrogen fuel cell vehicle market diffusion. *Int. J. Hydrogen Energy* **2021**, *46*, 18453–18465. [CrossRef]
60. Hassan, I.A.; Ramadan, H.S.; Saleh, M.A.; Hissel, D. Hydrogen storage technologies for stationary and mobile applications: Review, analysis and perspectives. *Renew. Sustain. Energy Rev.* **2021**, *149*, 111311. [CrossRef]
61. Coffey, L.; Claudio, D. In defense of group fuzzy AHP: A comparison of group fuzzy AHP and group AHP with confidence intervals. *Expert Syst. Appl.* **2021**, *178*, 114970. [CrossRef]
62. Ghosh, A.; Ghorui, N.; Mondal, S.P.; Kumari, S.; Mondal, B.K.; Das, A.; Gupta, M.S. Application of hexagonal fuzzy MCDM methodology for site selection of electric vehicle charging station. *Mathematics* **2021**, *9*, 393. [CrossRef]
63. Pu, H.; Luo, K.; Zhang, S. Risk assessment model for different foodstuff drying methods via AHP-FCE method: A case study of "coal-burning" fluorosis area of Yunan and Guizhou Province, China. *Food Chem.* **2018**, *263*, 74–80. [CrossRef] [PubMed]
64. Zhong, C.; Yang, Q.; Liang, J.; Ma, H. Fuzzy comprehensive evaluation with AHP and entropy methods and health risk assessment of groundwater in Yinchuan Basin, northwest China. *Environ. Res.* **2022**, *204*, 111956. [CrossRef]
65. Asghar, A.B.; Farooq, S.; Khurram, M.S.; Jaffery, M.H.; Ejsmont, K. Estimation of the solid circulation rate in circulating fluidized bed system using adaptive neuro-fuzzy algorithm. *Energies* **2021**, *15*, 211. [CrossRef]
66. Li, R.; Xu, A.; Zhao, Y.; Chang, H.; Li, X.; Lin, G. Genetic algorithm (GA)—Artificial neural network (ANN) modeling for the emission rates of toxic volatile organic compounds (VOCs) emitted from landfill working surface. *J. Environ. Manag.* **2022**, *305*, 114433. [CrossRef]
67. Hui, L.; Yunfei, H.; En, Z.; Kai, S. Prediction of safety objective of an enterprise using fuzzy neural network. *Procedia Eng.* **2012**, *43*, 162–167. [CrossRef]
68. Chen, Q.; Gu, Y.; Tang, Z.; Wang, D.; Wu, Q. Optimal design and techno-economic assessment of low-carbon hydrogen supply pathways for a refueling station located in Shanghai. *Energy* **2021**, *237*, 121584. [CrossRef]

energies

MDPI

Article

Hydrogen Production by Water Electrolysis with Low Power and High Efficiency Based on Pre-Magnetic Polarization

Ke Li [1,*], Heng Zhang [1], Xiaoyu Zheng [1], Chang Liu [2] and Qianding Chen [1]

[1] School of Information Engineering, Southwest University of Science and Technology, Mianyang 621000, China; zhswust123@163.com (H.Z.); zxyswust@163.com (X.Z.); cqdmj@163.com (Q.C.)
[2] School of Environment and Resource, Southwest University of Science and Technology, Mianyang 621000, China; liuc@swust.edu.cn
* Correspondence: tulip110000@163.com

Abstract: In this paper, a method of efficient hydrogen production using low-power electrolysis based on pre-magnetic polarization was proposed in order to improve the rate of hydrogen production by water electrolysis, with reduced energy consumption, molecular polarity, and stress–strain characteristics of distilled water under the condition of a pre-magnetic field. By constructing a microphysical model of hydrogen proton energy-level transition and a macroscopic mathematical model corresponding to magnetization vector-polarization hydrogen proton concentration in the pre-magnetic field, the ionic conductivity, electrolyte current density, interelectrode voltage, and hydrogen production efficiency under a varying magnetic field were qualitatively and quantitatively analyzed. In addition, an adjustable pre-magnetic polarization hydrolyzing hydrogen production test platform was set up to verify the effectiveness of the proposed method. The repeated test results, within a magnetic field strength range of 0–10,000 GS, showed that the conductivity of distilled water after pre-magnetic polarization treatment increased by 2–3 times, the electrolytic current density of the PEM (Proton Exchange Membrane) increased with increasing magnetic field strength, the voltage between the poles continuously decreased, and the hydrogen production rate was significantly improved. When the magnetic field strength reached 10,000 GS, the rate of hydrogen production by the electrolysis of distilled water increased by 15–20% within a certain period of time.

Keywords: pre-magnetic polarization; hydrogen production rate; electrical conductivity; current density

Citation: Li, K.; Zhang, H.; Zheng, X.; Liu, C.; Chen, Q. Hydrogen Production by Water Electrolysis with Low Power and High Efficiency Based on Pre-Magnetic Polarization. *Energies* **2022**, *15*, 1878. https://doi.org/10.3390/en15051878

Academic Editor: George Avgouropoulos

Received: 17 February 2022
Accepted: 2 March 2022
Published: 3 March 2022

1. Introduction

To cope with the challenges of climate security and to promote the low-carbon, green, and sustainable development of the global economy, China, the European Union (EU), Japan, South Korea, and several other major economies have successively announced their own carbon neutrality goals; thus, the carbon reduction and decarbonization process of the energy industry will be further accelerated. As an efficient and undefiled ideal secondary energy, hydrogen has the advantages of no pollution, recyclability, and high calorific value per unit mass. It can directly convert chemical energy into electric energy through fuel cells and will be an important support when building a clean-energy society in the future [1,2]. At present, the mainstream hydrogen production technology mainly includes fossil energy hydrogen production (represented by coal and natural gas), industrial by-product hydrogen production (represented by the chlor-alkali industry and coke oven tail gas), and multi-energy complementary hydroelectricity hydrogen production (represented by clean energy such as water and light). Among these, hydrogen production by water electrolysis has been widely noted by the majority of scientific research institutions and scholars for its advantages of having rich raw materials, as well as being green, clean, and low-carbon [3,4].

However, the production of hydrogen by water electrolysis accounts for less than 10% of the total production of industrial hydrogen in China at present [5]. The main problems faced by the development of hydrogen production by water electrolysis technology at present are its high energy consumption and low energy-exchange efficiency. At the 2020 global hydrogen energy summit, Professor Zhang Jiujun, academician of the Academy of Science of the Royal Society of Canada, pointed out that most of the present hydrogen production methods using water electrolysis have an energy efficiency of about 50%, and the consumption of electric energy is relatively high. Professor Zhang mentioned that improving the efficiency and reducing the cost of hydrogen production has become a hot issue that urgently needs to be solved. Therefore, a large number of scientific research institutions and scholars have begun to study the changes in various physicochemical characteristics in water electrolysis, and adopted various auxiliary methods such as super gravity [6], ultrasonic [7], and magnetic fields [8], attempting to mitigate the above problems by enhancing the performance of electrolytic cells [9].

Some scholars have studied the influence of an external magnetic field in the process of hydrogen production using the electrolysis of water and have obtained good results. For instance, according to the research conducted by T-Iida et al. [10], after placing the whole electrolyzer in the magnetic field, the electrolytic cell voltage significantly decreased due to the reduction in mass transfer loss and electrolytic resistance. In addition, the relationship between electrolytic cell characteristics and magnetic flux intensity was obtained. X. He [11] established a magnetohydrodynamics (MHD) model of electrolyte under the action of magnetic field, and designed the experiment to prove that adding a certain size of magnetic field could improve the electrolytic efficiency of the electrolyzer. S. Jing et al. [12] developed a system that was capable of producing a repetitive, pulsed, high magnetic field, and improved the performance of an Nd–Fe–B electrocatalyst in an alkaline water electrolyzer. H.M.et al. [13] observed the magnetic field's effect on electrode bubble coverage using a transparent electrode, and found that the magnetic field's effects became more evident with increasing working current density. M.F. Kaya et al. [14] displayed the positive effects of different magnetic fields on the performance of a PEM electrolyzer under a variable water flow. In addition, at a higher limit current diffusion value, the introduction of a magnetic field can improve the mass transfer inside the electrolytic cell by increasing the convection of the electrolyte, thereby improving the performance of the electrolytic cell [15–17].

The influence of the applied magnetic field on the process of hydrogen production by electrolysis has been analyzed in detail in the above literature, and some achievements have been achieved. However, this research focuses on analyzing the effect of a magnetic field on electrode material and bubble rate during electrolysis. There is a lack of research on the development of and dynamic variation in the hydrogen proton micropolarization process in electrolytes, and the effect of circulating water velocity on hydrogen production rate. In addition, the external magnetic field relies on electricity for generation, which increases the electrolytic energy consumption. Therefore, this paper studies the molecular polarity and stress–strain characteristics of distilled water in a pre-magnetic field, proposes an efficient hydrogen production method using low-power water electrolysis based on pre-magnetic polarization, and demonstrates this method through experiments. This study can not only supplement and improve the existing hydrogen production using the water electrolysis system, but also provides a new idea for efficient hydrogen production using low-power electrolysis.

2. Theoretical Analyses

The effect of the magnetic field on water molecules in the pre-magnetic polarization treatment of distilled water is schematically shown in Figure 1.

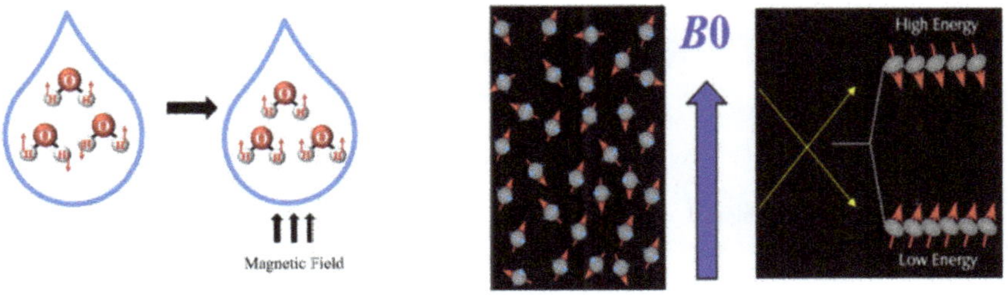

(a) (b)

Figure 1. Effect of magnetic field on water molecules: (**a**) Reorientation of proton spin under magnetic field effect; (**b**) sketch of energy-level splitting of hydrogen proton.

From the microscopic perspective, according to the two different spin directions of hydrogen protons, water molecules simultaneously have paramagnetic and diamagnetic behaviors [18,19]. Due to the diamagnetic behavior of water, when an external magnetic field is applied, the positive and negative charges will be subjected to the action of forces in the opposite direction. Then, the number of intermolecular hydrogen bonds in the solution will decrease due to structural damage, and the opening of macromolecular water clusters will accelerate the diffusion of particles in water [20]. The Lorentz force is formulated by:

$$F = qvB \tag{1}$$

When the intensity of the pre-polarized magnetic field increases, the Lorentz force acted on particles in water will increase. The Lorentz force will act on particles in water to increase their internal energy and, as a result, the particles in water will collide more violently, and the probability of collision ionization will increase. In addition, due to the increase in the free stroke of protons, the kinetic energy possessed by the protons during collision will be larger. Hence, they can provide greater energy to overcome the potential barrier, so as to break H–O bonds and increase OH-concentration in solution. Meanwhile, the alternating action of a magnetic force on the water system can cause a macro-ordered vibration of water particles. The energy-level fracture of hydrogen protons occurs in the polarization field, forming two energy states with different proton numbers [21]. In this way, the hydrogen protons can maintain higher activity, which is conducive to the realization of efficient and rapid hydrogen production by electrolysis.

In the actual electrolysis hydrogen production process, the optimal method is to generate the same amount of hydrogen by consuming less electric energy. Therefore, the conversion efficiency of the electrolyzer is defined as:

$$\eta = \frac{QP}{IV} \times 100\% \tag{2}$$

where Q represents the volume flow rate of the evolved hydrogen in L/S; P represents the pressure of the evolved hydrogen in Pa; I represents the current passing through the electrolyzer in A; and V represents the inter-electrode voltage of the electrolyzer in V.

It can be seen from (2) that the more hydrogen is produced by the electrolysis of electric energy per unit power, the higher the conversion efficiency of the electrolyzer. According to Faraday's laws of electrolysis, under ideal conditions, the current effect of water electrolysis is 100%. Thus, (2) is deformed according to the following formulas:

$$\begin{cases} Q = 22.4 \times \frac{1}{nF} \\ \eta = \frac{QP}{IV} \times 100\% = 22.4 \times \frac{1}{nF} \times \frac{P}{IV} \times 100\% = \frac{P}{8614V} \times 100\% \end{cases} \tag{3}$$

where n represents the stoichiometric number of electrons consumed in the hydrogen evolution reaction (n = 2) and F represents the Faraday constant (96,485 C/mol).

There are only two variables, P and V, left in (3), and the gas pressure P is generally constant. When using (3) to quantitatively analyze the energy-conversion efficiency of the electrolyzer, only the variable V needs to be considered, and the energy conversion efficiency of the electrolyzer can only be improved by reducing the actual inter-electrode voltage of the electrolyzer. The essential reason for this is that, during electrolysis hydrogen production, the mass transfer process of the diffusion layer of water treated by magnetic polarization is strengthened, hence speeding up the speed of electrolysis hydrogen production.

No additional magnetic field is applied in the process of electrolytic hydrogen production after pre-magnetic polarization. According to the Stokes–Einstein formula and Nernst–Einstein equation, the following relations can be obtained:

$$\begin{cases} D = K_B T / 6\pi\eta r \\ \sigma = nq^2 D / K_B T \end{cases} \tag{4}$$

where D represents the diffusion coefficient in m^2/s; K_B represents the Boltzmann constant in J/K; T represents the temperature in K; η represents the viscosity in Pa·s; r represents the diffusion radius in m; σ represents the ionic conductivity in S/m; n represents the number of current carriers per unit volume; and q represents the quantity of electric charge in C.

The research shows that the viscosity of water solution will decrease after magnetic polarization [9]. According to the Raman spectrum analysis, B. Zheng et al. concluded that, after magnetic polarization, the hydrogen bond structure of water solution was destroyed, the outer shell layer of the hydrated ions fell off [20], and the hydrated ion radius was small. After magnetic polarization, when temperature and other conditions remained unchanged, the diffusion coefficient D of particles in water increased, the conductivity of the water solution also increased, and the resistivity ρ decreased; without changing the structure of the electrolyzer and the magnitude of the current, the decrease in ρ could effectively reduce the resistance over-potential, thereby reducing the actual inter-electrode voltage of the electrolyzer.

As shown in Figure 2, there are differences between pre-magnetic polarization and the conventional electrolysis hydrogen production process. Specifically, compared with traditional hydrogen production by water electrolysis, hydrogen production by water electrolysis based on pre-magnetic polarization applies a variable magnetic field to the distilled water before electrolysis. This could polarize the water molecules before entering the electrolyzer, so the protons could obtain the excitation energy provided by the magnetic field in advance. Then, when the protons enter the electrolyzer for electrolysis, they would only need a smaller electric field to become free electrons, participate in electric conduction, overcome the potential barrier, and break the hydrogen–oxygen (H–O) bond, thereby producing hydrogen.

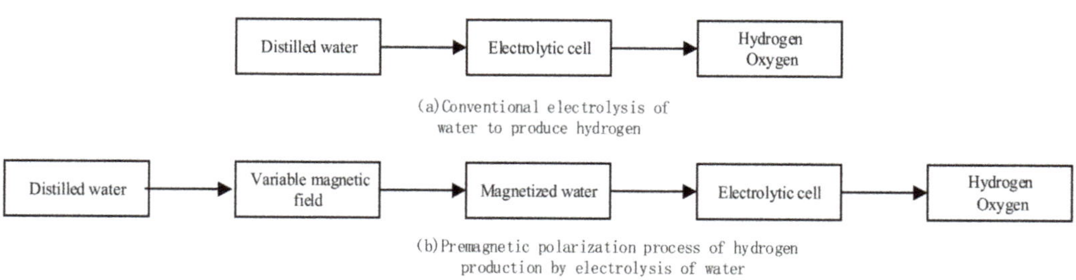

Figure 2. Comparison of conventional hydrogen production from water electrolysis and pre-magnetic electrolysis.

In summary, hydrogen production by water electrolysis under pre-magnetic polarization conditions has the following main advantages. Firstly, it can reduce the energy consumption of hydrogen production by electrolysis. To be more specific, there is a macroscopic orderly vibration of the magnetic force in the water system. The hydrogen protons undergo energy-level transitions in the polarized field, and the activity is enhanced. Therefore, hydrogen can be generated with less energy during electrolysis. Secondly, it can improve energy conversion efficiency. Specifically, after pre-magnetic polarization treatment, the physical properties of water have changed. Its diffusion and mass transfer process are enhanced during electrolysis, and the hydrogen production rate is accelerated under the same conditions.

3. Experimental

3.1. Experimental Equipment

To verify the feasibility of hydrogen production by water electrolysis under the condition of pre-magnetic polarization, a set of experimental equipment for hydrogen production by water electrolysis, based on strong pre-magnetic polarization, was designed (Figure 3). The equipment used an Nd–Fe–B permanent magnet to form a pre-polarization magnetic field, and the electrolyte was distilled water with an initial conductivity $\sigma \approx 0.4$ μs/cm. In particular, it should be noted that there are materials costs for the polarization loss rate of the permanent magnet, which is difficult to assess from the short running time of the experimental process. An intelligent peristaltic pump was used to control the flow rate and magnetic polarization time of distilled water, and the magnetic field intensity was calibrated by a Tesla meter. The change in conductivity after electrolyte magnetization was measured by a conductivity meter. Meanwhile, the PEM electrolyzer was used to produce hydrogen, and the inter-electrode voltage was measured by a high-precision bench-type digital multimeter. The changing process of hydrogen production was recorded by a high-speed camera. The relation diagram of the actual experimental equipment is shown in Figure 4.

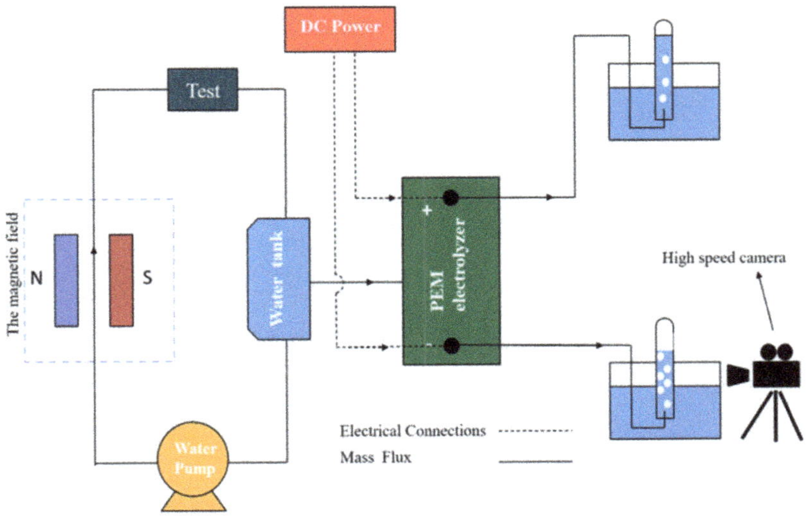

Figure 3. Schematic diagram of low-power electrolytic hydrogen production experimental device based on strong magnetic pre-polarization.

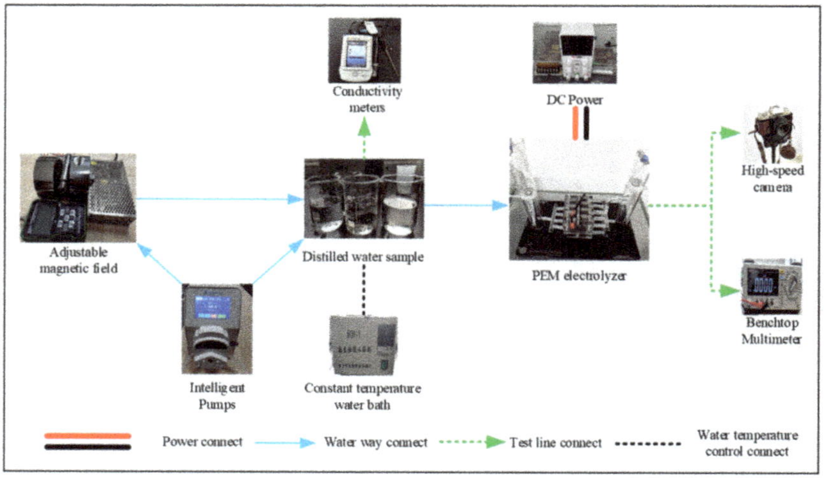

Figure 4. Schematic diagram of low-power electrolytic hydrogen production experimental device based on strong magnetic pre-polarization.

3.2. Experimental Process

The experimental process is shown in Figure 5, and is divided into two main stages. In the first stage of pre-magnetic polarization, the distilled water was divided into 50 samples, with 350 mL in each sample. Then, an adjustable pre-magnetic polarization field was set up using permanent magnets, and each sample of distilled water was controlled by the intelligent peristaltic pump to cut the magnetic line of force at a flow rate of 500 mL/min and in a perpendicular direction to the magnetic line of force. The magnetic field gradients were set as 500 GS, 1000 GS, 1500 GS, 2000 GS, 2500 GS, and 10,000 GS. After pre-magnetic polarization treatment, the changes in the properties of the distilled water, such as viscosity and hydrogen bonds, were analyzed. In addition, the conductivity of the distilled water was recorded every 1 min over 40 min during the pre-magnetic polarization. The curve chart of conductivity with the pre-magnetic polarization time was generated according to the data. After pre-magnetic polarization, the samples were statically placed, and the conductivity was measured every 1 min to observe the change trend. For the second stage of electrolysis hydrogen production, the magnetized distilled water samples were put into the PEM electrolyzer to produce hydrogen. After the samples underwent 35 min pre-magnetic polarization treatment under different magnetic field intensities, the data on the changes in their hydrogen production in the first 100 s were recorded by the high-speed camera. Finally, the experimental data were compared, the time-varying curve chart of hydrogen production from distilled water under different magnetic fields was drawn, and the changing trend was analyzed.

To minimize the data error, uncertainty (Uc) is introduced to measure the data validity, which is expressed as:

$$Uc = \sqrt{\frac{\sum_{i=1}^{n} (s_i - s_a)^2}{(n-1)n}} \quad (5)$$

where s_i represents the data of ith experiment; s_a represent the arithmetic mean of the data in repeated experiments; and n represent the number of experiments.

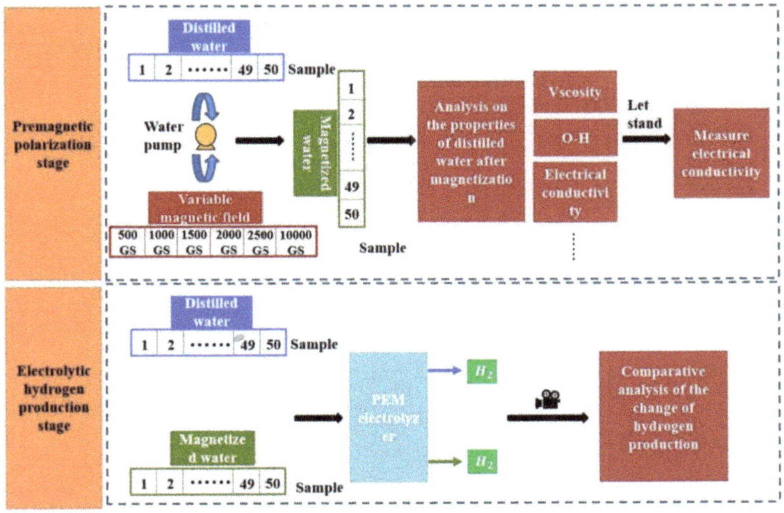

Figure 5. Experimental process diagram.

4. Results and Discussion

4.1. Conductivity Change Inmagnetic Polarized Water

Figure 6 shows the changing curve of water conductivity with the magnetic polarization time after pre-magnetic polarization in different magnetic fields at 20 ± 0.1 °C. It can clearly be seen that the conductivity fluctuates and increases with increasing magnetic polarization time. The movement of particles in water molecules is irregular. After applying a magnetic field, the Lorentz force bends and breaks the hydrogen bonds between water molecules, resulting in the separation of the molecules. However, the energy generated by each cyclic polarization is small, which provides water molecules with enough time for self-repairing. By increasing the time for cyclic pre-magnetic polarization, the increase in conductivity gradually slows down.

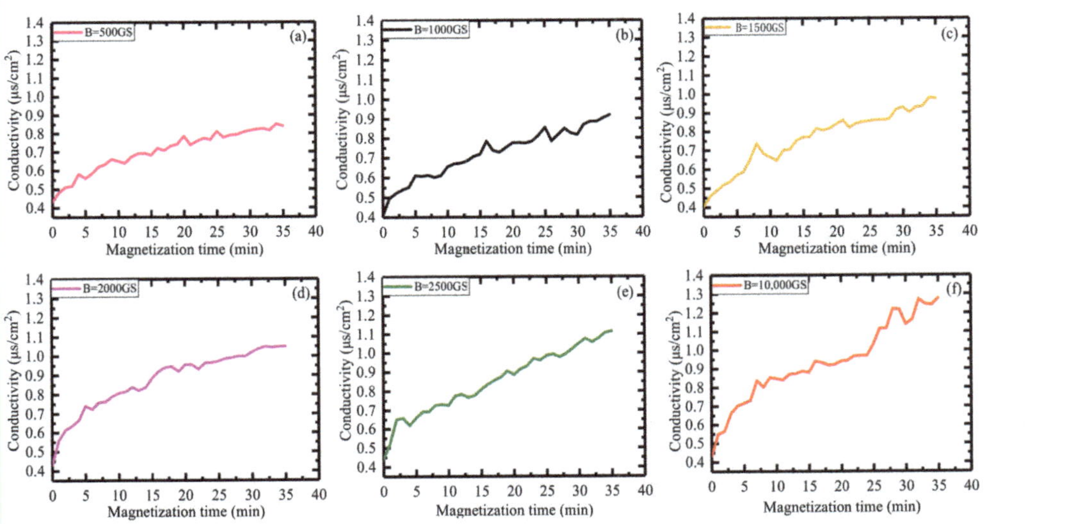

Figure 6. Change in conductivity of distilled water with magnetization time under different magnetic fields. Magnetic induction intensity: (**a**) 500 GS; (**b**) 1000 GS; (**c**) 1500 GS; (**d**) 2000 GS; (**e**) 2500 GS; (**f**) 10,000 GS.

Figure 7 is generated by the above data, processed with the non-linear curve fitting function Allometric of Origin software. It can be seen from the figure that the conductivity of distilled water obtained the fastest increases in different magnetic fields within 0–10 min, and 30 min later, the increase rate slowed down with the increase in magnetic field intensity and magnetization time.

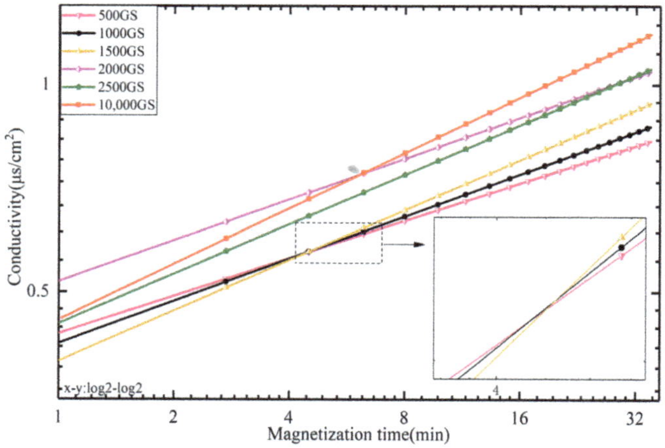

Figure 7. Fitting curve of conductivity of distilled water with magnetization time under different magnetic fields.

As shown in Table 1, it can be seen that the error between experimental data and expected value is less than 1%.

Table 1. The uncertainty of conductivity in these experiments.

	Magnetization Time (min)							
	0	5	10	15	20	25	30	35
Magnetic Field Intensity (GS)	Uncertainty of Conductivity Change (%)							
500	0.61	0.46	0.58	0.91	0.79	0.77	0.63	0.56
1000	0.86	0.38	0.78	0.92	0.85	0.72	0.64	0.68
1500	0.59	0.63	0.52	0.89	0.42	0.49	0.69	0.62
2000	0.67	0.71	0.59	0.57	0.75	0.66	0.87	0.52
2500	0.44	0.39	0.49	0.63	0.87	0.94	0.74	0.76
10,000	0.72	0.74	0.87	0.65	0.56	0.89	0.92	0.56

4.2. Change in Current Density and Actual Inter-Electrode Voltage in Magnetic Polarized Water in PEM Electrolyzer

The quantified distilled water, treated by polarization in different magnetic fields, was put into the PEM electrolyzer. When the control power supply provides a constant current of 0.7 A, the fitted changing curve of the actual inter-electrode voltage of the electrolyzer is shown as curve (a) in Figure 8. It can be observed that, with continuous increases in the magnetic field's intensity, the actual inter-electrode voltage shows a decreasing trend, and the decrease rate continuously slows. The fitted changing curve of the electrolyzer electrode plate's current density when controlling the DC power supply to provide an 11 V constant voltage is shown as curve (b) in Figure 8. The figure shows that, with the increase in the magnetic field's intensity, the actual inter-electrode voltage shows an increase trend, and the increase rate continuously slows. The experimental phenomena show that, with pre-magnetic polarization treatment, the properties of distilled water are changed. The ionic diffusion coefficient in water increased, the resistivity of the electrolyte decreased, and the resistance potential was effectively reduced, thereby reducing the actual inter-electrode voltage of the electrolyzer.

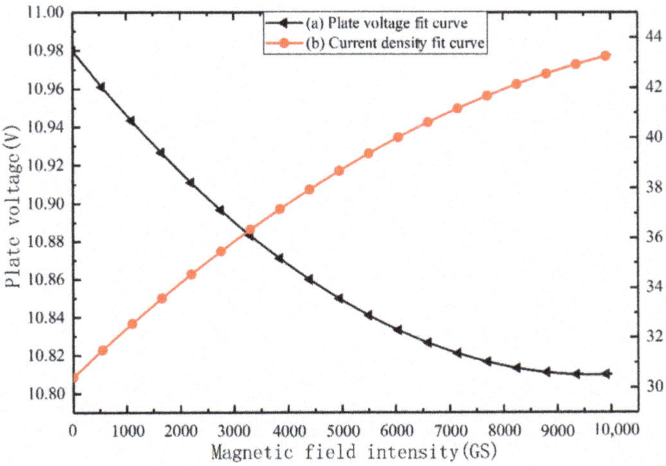

Figure 8. Fitting curves of inter-pole voltage and plate current density in PEM electrolyzer for hydrogen production by electrolysis of distilled water after different magnetic field treatments.

4.3. Hydrogen Production Rate and Energy Consumption

A total of 350 mL distilled water was taken, either without magnetic polarization treatment or with different magnetic field magnetic polarization treatments, and the sample was passed into the PEM electrolytic cell for hydrogen production experiment. Meanwhile, in different time periods, the changes in the hydrogen production of each sample were recorded by a high-speed camera in different time periods. Due to the electrolyzer's defects, only the changes in hydrogen production within 100 s were observed. In order to reduce the experimental error, the research group conducted a large number of repeated experiments. Figure 9 shows the time-variant hydrogen production of electrolyzing distilled water, which pre-polarization in different magnetic fields. It can be seen that the hydrogen production rate in the PEM electrolyzer significantly increased with pre-magnetic polarization treatment.

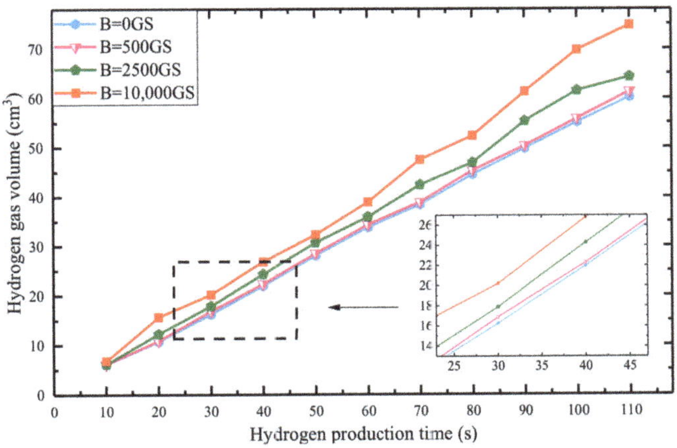

Figure 9. Variation in hydrogen production from distilled water with time after pre-polarization of different magnetic fields.

As shown in Table 2, it can be seen that the error between experimental data and expected value is less than 3%.

Table 2. The uncertainty of hydrogen production rate in these experiments.

	Hydrogen Production Time (s)							
	10	20	40	60	80	90	100	110
Magnetic Field Intensity (GS)	Uncertainty of Hydrogen Volume (%)							
500	1.63	1.78	2.02	1.97	2.14	1.53	2.34	1.63
1000	1.52	1.38	2.21	2.03	1.69	2.54	1.14	1.83
1500	1.49	2.31	1.39	2.19	2.63	1.84	1.58	2.93
2000	1.69	1.47	1.76	2.63	3.04	2.93	2.79	1.73
2500	1.75	1.39	2.64	1.98	1.67	1.64	2.38	2.89
10,000	1.32	2.12	1.96	2.07	2.37	1.79	2.25	2.34

As shown in Table 3, the electric energy consumption needed to produce the same amount of hydrogen in the two experiments was calculated. This shows that, for the same amount of hydrogen production, the electric energy consumed by distilled water pretreated in 10,000 GS magnetic field was significantly reduced.

Table 3. Comparison of electric energy consumption needed to produce the same amount of hydrogen from distilled water without pre-magnetic polarization and with 10,000 gs magnetic polarization.

	0 GS	10,000 GS
Hydrogen Volume (cm^3)	Power Consumption (kW·h)	
10	0.04273	0.03148
20	0.08323	0.06091
40	0.15579	0.13242
60	0.23058	0.19961

As shown in Figure 10, the changes in hydrogen production between the two experiments were compared at different times, which indicates that the hydrogen production rate by distilled water pre-polarized by 10,000 GS magnetic field increased by 15–20%. At the same time, the energy consumption of hydrogen production by pre-magnetically polarized water was reduced, and its hydrogen production rate was improved. This verified the theoretical analysis in this work that, under the applied magnetic field, Lorentz force has a significant impact on the charge transfer in water electrolysis, and the relevant properties of water have been changed, making it easier to electrolyze.

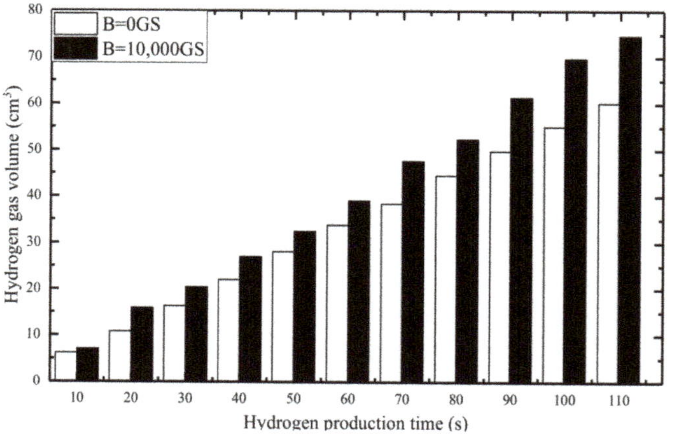

Figure 10. Comparison of hydrogen production of distilled water without pre-polarization and after 10,000 GS magnetic field polarization.

5. Conclusions

Combined with the pre-magnetic polarization mechanism of hydrogen protons in electrolytes under a static magnetic field, this paper explored the molecular polarity and stress–strain characteristics of distilled water under pre-magnetic polarization. Additionally, a new hydrogen production method based on low-power and high-efficiency electrolysis under pre-magnetic polarization was proposed. By establishing the microscopic physical model of the energy-level transition of hydrogen protons in the magnetic field environment and the macroscopic mathematical model corresponding to the magnetization vector and the concentration of polarized hydrogen protons, both qualitative and quantitative analyses were carried out to study the ionic conductivity, electrolyte current density, interelectrode voltage, and hydrogen production under varying magnetic fields. The feasibility of the proposed method was verified by a water electrolysis hydrogen production experiment using a self-designed adjustable pre-magnetic polarization water electrolysis hydrogen production test platform. The test results show that the ionic conductivity of distilled water after pre-magnetic polarization treatment increased by 2–3 times. With the increasing magnetic field strength, the current density of PEM electrolysis continuously increases, while the voltage between electrodes continuously decreases and the hydrogen production rate significantly increases. When the magnetic field strength reaches 10,000 GS, the hydrogen production rate of electrolytic distilled water increases by 15–20% within a certain period of time.

The research described in this paper provides a theoretical basis for the parameter selection of hydrogen production using pulse current electrolysis. In future studies, a multiphysics numerical model for magnetized water will be developed, and pulse current techniques will be employed to further study.

Author Contributions: Conceptualization, K.L.; methodology, K.L. and H.Z.; validation, K.L., H.Z. and X.Z.; formal analysis, K.L.; investigation, H.Z. and X.Z.; resources, K.L. and C.L.; data curation, H.Z. and X.Z.; writing—original draft preparation, K.L. and H.Z.; writing—review and editing, K.L., C.L. and Q.C.; supervision, C.L.; project administration, Q.C.; funding acquisition, K.L. All authors have read and agreed to the published version of the manuscript.

Funding: This research was funded by National Natural Science Foundation of China, grant number 61901400; Sichuan Science and Technology Program, grant number 2021YFG0253.

Data Availability Statement: The data presented in this study are available on request from the corresponding author. The data are not publicly available due to future patent protection.

Conflicts of Interest: The authors declare no conflict of interest.

References

1. Zhan, S.Q.; Huang, Y.J.; Wang, J.F. Experimental study on bubble growth characteristics on horizontal microcelectrode surface in vertical magnetic field. *J. Cent. South Univ. Sci. Technol.* **2021**, *52*, 249–258. [CrossRef]
2. Mao, Z.Q.; Mao, Z.M.; Yu, H. *Hydrogen Production Process and Technology*; Chemical Industry Press: Beijing, China, 2018.
3. Wang, M.G.; Wang, Z.; Guo, Z.C. Water electrolysis enhanced by super gravity field for hydrogen producting. *Int. J. Hydrogen Energy* **2010**, *35*, 3198–3205. [CrossRef]
4. Lin, M.Y.; Hourng, L.W. Ultrasonic wave field effects on hydrogen production by water electrolysis. *J. Chin. Inst. Eng.* **2014**, *237*, 1080–1089. [CrossRef]
5. Taqieddin, A.; Nazari, R.; Rajic, L. Review: Physicochemical hydrodynamics of gas bubbles in two phase electrochemical systems. *J. Electrochem. Soc.* **2017**, *164*, E448–E459. [CrossRef] [PubMed]
6. Bidin, N.; Azni, S.R. The effect of magnetic and optic field in water electrolysis. *Int. J. Hydrogen Energy* **2017**, *42*, 16325–16332. [CrossRef]
7. Purnami; Hamidi, N.; Sasongko, M.N.; Widhiyanuriyawan, D.; Wardana, I.N.G. Strengthening external magnetic fields with activated carbon graphene for increasing hydrogen production in water electrolysis. *Int. J. Hydrogen Energy* **2020**, *45*, 19370–19380. [CrossRef]
8. Nieminen, J.O.; Voigt, J.; Hartwig, S. Improved contrast in ultra-low-field MRI with time-dependent bipolar prepolarizing fields: Theory and NMR demonstrations. *Metrol. Meas. Syst.* **2013**, *20*, 327–336. [CrossRef]
9. Chizhik, V.I.; Kupriyanov, P.A.; Mozzhukhin, G.V. NMR in magnetic field of the earth: Pre-polarization of nuclei with alternating magnetic field. *Appl. Magn. Reson.* **2014**, *45*, 641–651. [CrossRef]

10. Iida, T.; Matsushima, H.; Fukunaka, Y. Water electrolysis under a magnetic field. *J. Electrochem. Soc.* **2007**, *154*, E112–E115. [CrossRef]
11. He, X.K. *Research on the Ineluence of External Magnetic Field on the Electrolytic Efficiency of Oxyy-Hydrogen Generator*; Harbin Institute of Technology: Harbin, China, 2015.
12. Jing, S.; Hu, H.; Wang, S. Design of pulsed power supply for repetitive pulsed high magnetic field for water electrolysis. *Rev. Sci. Instrum.* **2021**, *92*, 114708. [CrossRef] [PubMed]
13. Matsushima, H.; Iida, T.; Fukunaka, Y. Gas bubble evolution on transparent electrode during water electrolysis in a magnetic field. *Electrochim. Acta* **2013**, *100*, 261–264. [CrossRef]
14. Kaya, M.F.; Demir, N.; Rees, N.V.; El-Kharouf, A. Improving PEM water electrolyser's performance by magnetic field application. *Appl. Energy* **2020**, *264*, 114721. [CrossRef]
15. Costa, C.M.; Merazzo, K.J.; Goncalves, R. Magnetically active lithium-ion batteries towards battery performance improvement. *Iscience* **2021**, *24*. [CrossRef] [PubMed]
16. Zhang, Y.X.; Fan, L.; Li, X.G. Effect of Alternating Magnetic Field on Electrochemical Behavior of 316L and TA2 in Simulated Seawater. *J. Mater. Eng. Perform.* **2021**, *30*, 9377–9389. [CrossRef]
17. Yin, Y.; Huang, G.; Tong, Y.; Liu, Y.; Zhang, L. Electricity production and electrochemical impedance modeling of microbial fuel cells under static magnetic field. *J. Power Sources* **2013**, *237*, 58–63. [CrossRef]
18. Lin, M.Y.; Hourng, L.W.; Kuo, C.W. The effect of magnetic force on hydrogen production efficiency in water electrolysis. *Int. J. Hydrogen Energy* **2012**, *37*, 1311–1320. [CrossRef]
19. Deng, B.; Pang, X.F. Static Magnetic Field Influence on Properties of Water. *J. Univ. Electron. Sci. Technol. China* **2008**, *37*, 959–962.
20. Zheng, B.M.; Chen, Y.G.; Qian, G.Y.; Gong, X.Z.; Wang, Z. Enhancement mechanism of high strength circulating magnetic field on electrodeposition of low concentration copper ion solution. *Min. Metall.* **2021**, *3*, 109–115. [CrossRef]
21. Pietra, A.; Gianni, M.; Zuliani, N.; Malabotti, S.; Taccani, R. Experimental Characterization of an Alkaline Electrolyser and a Compression System for Hydrogen Production and Storage. *Energies* **2021**, *14*, 5347. [CrossRef]

energies

Review

Law and Policy Review on Green Hydrogen Potential in ECOWAS Countries

Abdoulaye Ballo [1,*], Koffi Kouakou Valentin [2], Bruno Korgo [1,3], Kehinde Olufunso Ogunjobi [1], Solomon Nwabueze Agbo [4], Daouda Kone [2] and Moumini Savadogo [2]

1 West African Science Service Centre on Climate Change and Adapted Land Use (WASCAL), Competence Centre, Ouaga BP 9507, Burkina Faso; korgo.b@wascal.org (B.K.); ogunjobi.k@wascal.org (K.O.O.)
2 West African Science Service Centre on Climate Change and Adapted Land Use (WASCAL) Headquarters, Cantonments Accra PMB CT 504, Ghana; valenkof@yahoo.fr (K.K.V.); kone.d@wascal.org (D.K.); savadogo.m@wascal.org (M.S.)
3 Laboratoire des Energies Thermiques Renouvelables (LETRE), Université Joseph KI-ZERBO, Ouaga BP 13495, Burkina Faso
4 Forschungszentrum Jülich GmbH, Wilhelm-Johnen-Strasse, 52425 Jülich, Germany; s.agbo@fz-juelich.de
* Correspondence: balloabdoulaye09@yahoo.fr; Tel.: +223-75-38-20-54

Abstract: This paper aims to review existing energy-sector and hydrogen-energy-related legal, policy, and strategy documents in the ECOWAS region. To achieve this aim, current renewable-energy-related laws, acts of parliament, executive orders, presidential decrees, administrative orders, and memoranda were analyzed. The study shows that ECOWAS countries have strived to design consistent legal instruments regarding renewable energy in developing comprehensive legislation and bylaws to consolidate it and to encourage investments in renewable energy. Despite all these countries having a legislative basis for regulating renewable energy, there are still weaknesses that revolve around the law and policy regarding its possible application in green hydrogen production and use. The central conclusion of this review paper is that ECOWAS member states presently have no official hydrogen policies nor bylaws in place. The hydrogen rise presents a challenge and opportunity for members to play an important role in the fast-growing global hydrogen market. Therefore, these countries need to reform their regulatory frameworks and align their policies by introducing green hydrogen production in order to accomplish their green economy transition for the future and to boost the continent's sustainable development.

Keywords: renewable energy; green hydrogen; green economy; laws and policies; ECOWAS

Citation: Ballo, A.; Valentin, K.K.; Korgo, B.; Ogunjobi, K.O.; Agbo, S.N.; Kone, D.; Savadogo, M. Law and Policy Review on Green Hydrogen Potential in ECOWAS Countries. *Energies* **2022**, *15*, 2304. https://doi.org/10.3390/en15072304

Academic Editor: Ben McLellan

Received: 3 February 2022
Accepted: 8 March 2022
Published: 22 March 2022

1. Introduction

Energy is key to economic growth, and it is crucial for development. Access to affordable and clean energy is a key component to meet sustainable Development Goals for the Economic Community of West African States (ECOWAS) countries, whose vision is to increase the share of renewable energy in the overall electricity capacity to 48 percent by 2030 [1,2]. Over the last decade, ECOWAS members, which comprise 15 countries, experienced an increased population of over 334.6 million people, which represents one-third of Sub-Saharan Africa's total population [3,4]. ECOWAS countries are facing several challenges in terms of access to electricity by their population. Only 42% of the total population and 8% of rural residents have access to electricity [5,6]. The electricity access varies widely, from Niger with an electrification rate of just 13% in 2019 to 96% in Cabo Verde which has achieved nearly universal access. In some countries, such as Sierra Leone and Burkina Faso, the estimated share of rural populations with access to electricity may be as low as 1 to 4% [7]. This low access emanates from a combination of several factors, which may include socio-economic, technical, political, financial, and institutional policy framework barriers. ECOWAS members can meet their electricity demand through the promotion of green hydrogen because of their huge renewable energy potential. For instance, hydropower potential is sufficient to cover all the energy demands in the ECOWAS countries [8]. The region, throughout the year, also has huge solar energy potential with a very high radiation mean of 5 to 6 kWh/m^2 [9]. According to IRENA (2018), West Africa's solar potential could reach 20 GW by 2030 [10]. Despite the available renewable resource potential (solar, hydro, and wind), the ECOWAS countries' objective of transiting to a green economy using renewable energy resources still faces constraints that hinder its further economic development.

The transition to a green economy using renewable energy implies strong political engagement, important investment, the modernization of infrastructure systems, and a concerted effort from different scientific disciplines, which is a significant challenge. Therefore, in 2019, as a recommendation of COP21, many countries such as China, France, Japan, and South Korea set investment targets to deploy hydrogen energy technologies globally. In early 2021, more than 10 countries including Australia, Chile, Finland, Germany, Norway, Portugal, and Spain, plus the European Union (EU) developed and adopted hydrogen strategies and roadmaps. This was largely catalyzed by the United Nations Climate Change Conference in Glasgow (COP26) in November 2021 [11]. Nine further countries are expected to unveil their strategies in the near future [12]. The EU hydrogen strategy places particular emphasis on cooperation and complementarities with sub-Saharan African countries as part of its green hydrogen strategy and economic transition. According to Elkhan (2021), the deployment of green hydrogen will bridge the gap between the energy transition between Europe and Africa on the one hand, and climate and development goals on the other [13]. At the same time, ECOWAS countries have entered into global agreements and processes where the green economy is defined explicitly as a Sustainable Development Goal. These are, for example, the decision of the Paris Climate Agreement; the African Union's Comprehensive Strategy on Climate Change; the African Development Bank's Green Growth Framework; UN Sustainable Development Goals (SDG 7); Nations Framework Convention on Climate Change (UNFCCC) in order to move to a low-carbon economy, etc. Further, to facilitate this sustainable green economy transition in ECOWAS countries, appropriate research and development efforts that are transdisciplinary are necessary. Such efforts will be required to update the existing infrastructure and technologies-based fossil fuel and regulations framework concerning green energy policies.

ECOWAS countries have many existing law and policy framework conditions that may boost renewable energy development and other green energy options such as green hydrogen energy. The development of green hydrogen fits perfectly into the global approach towards reducing greenhouse gas emissions and protecting the environment and climate. Indeed, green hydrogen technologies release no greenhouse gases and make it possible to decarbonize several sectors of human activity (transport, combustion, industrial processes, production and storage of electricity) while offering an opportunity for economic growth for West African countries. Investing in green hydrogen energy in the ECOWAS region will be very helpful to achieve the ambitious goal of the European Union (EU), the ECOWAS, and the Union Economique et Monétaire Ouest Africaine (UEMOA) which includes [14]: (i) universal access to sustainable energy services by 2030, (ii) 31% of the share of renewable energy in the electricity mix by 2030 [15]. Hydrogen has the highest specific energy content among all conventional fuels and can be extracted from water. It has been projected that green hydrogen will play a great role in the future scenario of energy sectors [16]. Hydrogen utilization is free of toxic gas formation as well as CO_2 emission compared with other fossil fuels and the energy yield of hydrogen is about 122 kJ/g, which is 2.75 times greater than that of hydrocarbon fuels [17]. This will help decarbonize and help achieve greater green energy and economic transition that lay emphasis on clean energy technologies such as green hydrogen technology.

Additionally, in many ECOWAS countries, the status of renewable energy resources is not clearly defined, but rather incorporated within the competitive and regulatory frameworks of the electricity or energy sector. This approach poses several challenges for investors wishing to invest in renewable energy projects [18,19]. Based on the over-emphasized gaps in the insecure legislation in the renewable energy sector, there is an urgent need for ECOWAS member countries to clearly define a legal statutory policy for renewable energy resources with robust regulatory laws which will significantly contribute to attracting large-scale investment in the sector.

The research problem is, therefore, to investigate the gaps that exist in hydrogen energy policy and strategies that can meet sustainable development goals on green hydrogen deployment at the national and regional levels. The paper also attempts to examine the policy barriers to the development and the implementation of renewable energy and hydrogen technologies across ECOWAS member countries. The significance of this review is to make recommendations that will guide ECOWAS governments on green hydrogen policy and strategies for the transition to a green economy.

2. Methods

To review the current renewable energy focusing on hydrogen laws and regulation in the ECOWAS region, a review of relevant documentation was carried out based on scientific literature, articles in journals, acts, executive orders, presidential decrees, administrative orders on hydrogen-based energy, web pages, and research platforms. In addition, this paper reviews a large number of reports and information regarding hydrogen and renewable energy law and policy published by the ECREEE, the International Energy Agency (IEA), International Renewable Energy Agency (IRENA), and other reports from research and development institutions that are relevant to the renewable energy and hydrogen economy [20,21]. Comparisons of energy policies and legislation of countries were also carried out. The selected countries included Benin, Burkina Faso, Cabo Verde, Cote d'Ivoire, The Gambia, Ghana, Guinea, Guinea-Bissau, Liberia, Mali, Niger, Nigeria, Senegal, Sierra Leone, and Togo.

As a rule, this review paper recommends improving the performance of current renewable energy strategies by using new opportunities like green hydrogen. The paper will be useful in helping decision-makers and stakeholders to have a better overview of the concept of hydrogen energy to facilitate access to clean and affordable electricity by the population. In order for more detail, the different stages undertaken for this review are explained below:

Stage 1: Reviews the different policies and laws concerning renewable energy and hydrogen energy in the fifteen (15) ECOWAS countries for the promotion of hydrogen deployment.

Stage 2: Review of policies that support green hydrogen promotion and avoid barriers to hydrogen development.

Stage 3: Evaluates the general energy policies in all the ECOWAS countries, which will allow policymakers and investors to know more about each country's energy gaps and uncertainties, which needed to be addressed.

Stage 4: Establishes the significant role of government and policies in the development of renewable energy and the hydrogen economy. This will help respective governments to know their responsibilities and the importance of achieving their energy target.

Stage 5: Explains the significance of renewable energy and hydrogen energy concerning UN SDGs, and how the use of hydrogen-based energy can achieve their energy target by 2030.

Stage 6: Gives strategies that can be used to accomplish green hydrogen policy formulation in ECOWAS countries.

3. Status of Hydrogen Energy in ECOWAS

Currently, the West African Science Service Centre on Climate Change and Adapted Land Use (WASCAL) produces the Atlas of Green Hydrogen Generation Potential (H2Atlas) in fifteen (15) West African countries. The H2Atlas–Africa project is funded by the German Ministry of Education and Research BMBF's Energy and Hydrogen Technologies Department [22]. It is estimated that many ECOWAS countries have very good prerequisites to produce green hydrogen. The region's potential was estimated at up to 165,000 terawatt-hours of green hydrogen per year, which is about 1500 times Germany's estimated hydrogen demand for 2030. It was also estimated that in ECOWAS countries three-quarters of the land area is suitable for wind turbines. By using wind and solar energy, ECOWAS countries could produce up to 165,000 terawatt-hours of green hydrogen annually, 120,000 terawatt-hours of which could already be produced for less than 2.50 euros per kilogram compared with 7–10 euros/kg in Germany [23]. Green hydrogen offers a real opportunity to initiate development in the ECOWAS region that will make the region a global green hydrogen powerhouse, and will also help member states meet their energy needs and benefit economically from the export of green hydrogen. To support local capacity in this regard, WASCAL, in partnership with RWTH Aachen University and Forschungszentrum Jülich GmbH, took up the topic of hydrogen in its newly created Master's Programme in Energy and Green Hydrogen, which admits students from all ECOWAS member states [24].

3.1. Review of Hydrogen-Related Laws and Policies in ECOWAS

In the coming years, hydrogen, produced with renewable electricity, is projected to grow rapidly. Many ongoing and planned projects across the world point in this direction. Hydrogen from renewable power is already technically important and is rapidly becoming economically relevant for the global energy transition. For Africa to be part of this global movement towards green hydrogen, it is important to create legislative frameworks that facilitate hydrogen-based sector coupling. However, in common with many jurisdictions, the fifteen ECOWAS members do not yet have official hydrogen policies fully in place. The lack of a legislative framework creates many gaps and uncertainties, which need to be addressed before the hydrogen economy can start to flourish in the region. ECOWAS members need a dedicated new legislative framework for renewable hydrogen production which may facilitate its deployment and the related economic benefits. Although in many ECOWAS countries there exist specific laws that deal with renewable energy, the level of intervention and implementation remains a challenge and varies from country to country.

Benin. Law No. 2020-05 of 1 April 2020 guides the energy policy in Benin. It describes new-generation capacity regulations with the aims of promoting renewable energy sectors and increasing their share in the energy mix in order to improve the rate of national energy independence. The idea behind this law is for the government to promote all sources of renewable energies and, in particular, the promotion of technologies for the enhancement of local energy resources. This is based on in-depth knowledge of the real potential of renewable energies and on the existence of adequate human resources in setting up an environmental regulation policy conducive to its implementation. This law grants subsidies, tax advantages, or guarantees to companies, enterprises, and establishments that undertake to produce or promote the production of electricity from renewable energies. Therefore, the import, purchase, or acquisition of materials and equipment intended for the production and exploitation of electricity based on renewable energies and those intended for research and development in the field of renewable energies benefit from a total exemption of taxes except for road taxes, statistical tax, and community taxes (LOI No 2020-05, 2020). On these, an indicative program is adopted every five (05) years by the Council of Ministers in charge of energy. In addition, this law gives greater allocation to the development of energy efficiency and renewable energy but gives no explicit allocation for green hydrogen technology.

Burkina Faso: In Burkina Faso, Law No. 014-2017/AN of 20 April 2017 and Law no. 058-2017/AN of 20 December 2017 were established for renewable energy promotion and regulations. These laws fixed the conditions of eligibility to the exemption from tax on imports and sales of solar equipment and materials. The innovation under these laws is to achieve 50% renewable energy in the electricity mix by 2030 and increase the share in urban and rural electrification rates by 95% and 50% for 2030, respectively. For this, the share of renewable energies, excluding biomass for cooking needs, will reach 50% of the electricity mix in 2030 in terms of capacity. Therefore, the installed PV capacities would increase from 65 MW in 2015/16 to 205 MW in 2030 [25]. However, concerning hydrogen-based energy, Burkina Faso lacks a clear act or policy framework on hydrogen energy strategy.

Cape Verde: In Cape Verde the government established Decree-Law No. 14/2006 of 20 February and Law no. 39/2019 of 8 April, to guide the renewable energy policy framework. This law aimed to boost and target electricity production from renewable energy sources by 2025 and 2030, and set commitments for wind and solar energy use, promotion of energy efficiency, and combating energy losses. Furthermore, the government has decided on the inclusion of 50% renewable energy in the energy mix as a target for 2030 [26].

Côte d'Ivoire: In Côte d'Ivoire, Decree No. 2016-862 of 03 November 2016 was enacted to grant tax and customs benefits on renewable energy equipment, and projects that contribute to the improvement of energy efficiency. For the implementation of these laws, the Ivorian state has decided to double the rate of improvement of energy efficiency by 2030. By 2030, national objectives have been set to improve access to energy by the population, to improve the level of renewable energies, and to increase energy efficiency. To achieve this, indicators have been identified to assess the actions to be taken.. This is a 42% increase in the energy mix from renewable energies to meet electricity consumption, as well as a 26% increase for large and medium hydraulics, and 16% for various other renewable energy sources by 2030

The Gambia. The renewable energy sector in the Gambia is governed by the Bill/Act of 2013, which establishes a legal, economic, and institutional basis to promote the use of renewable energy resources and for connected matters. Furthermore, the Ministry of Energy has developed an energy strategy that facilitates the use of renewable energy resources for both power and non-power applications from import duty. The Act guides all renewable energy equipment that fulfills the eligibility to be exempted from value-added and any other retail tax for fifteen years from commissioning, and all proceeds from the sale of carbon emission credits shall be exempt from sales taxes [27]. Nevertheless, Gambia lacks a clear Decree/Act/Bill on renewable energy.

Ghana. Act 832 in Ghana provides for the development, management, and utilization of renewable energy sources for energy production in an efficient and environmentally sustainable manner. The Act was legislated in light of the policy direction and with the specific objective of accelerating the development and utilization of renewable energy technologies to achieve a 10% penetration of national electricity by 2020. Electricity generation capacity from renewables is projected to reach 1353.63 MW by 2030 which will contribute to the creation of 220,000 jobs, and carbon savings of about 11 million tonnes of CO_2 by 2030 [28]. Act 832 stipulates that fossil fuel-based wholesale electricity suppliers, fossil fuel producers, and any other companies that contribute to greenhouse gas emission shall invest in non-utility scale renewable energy to offset greenhouse gas emissions and mitigate the impact of climate change [29].

Guinea. Guinea has adopted law No 2014/30/1/6/1/2/N on the appropriate fiscal policy for the implementation of energy-saving measures for the promotion of renewable energies. For this, the state grants bonuses, subsidies, or loans at subsidized rates on renewable energy equipment by reducing customs duties to minimum rates and exemption from import VAT [30]. Nevertheless, the country lacks a clear Decree/Act on renewable energy.

Guinea Bissau. In Guinea Bissau, the legislative framework is composed of two laws that regulate the energy and electricity sector, namely Decree-Laws No. 2/2007 and 3/2007. Decree-Law No. 2/2007 defines the structure of the energy sector, its organization, and the provisions that apply to the different forms of energy. Decree-Law No. 3/2007 regulates the production, transport, distribution, import, and export of electrical energy within the country. However, no specific fiscal and investment frameworks or provisions exist for the energy sector currently. As the country disposes of several renewable energy resources such as solar, hydropower, and biomass, the exploitation of these resources will enable the country to reach its ambitious targets of installing more than 70 MW of renewable capacity by 2030. According to the government off-grid systems, with 80% penetration of renewable energy, this will contribute to meeting 100% of the demand by 2030 [31].

Liberia. On July 6, 2015, the Liberian government established a wholly autonomous Rural and Renewable Energy Agency (RREA). The RREA aims at addressing the challenges faced by the energy supply in rural areas. The RREA operates to ensure universal access to modern energy services in an affordable, sustainable, and environmentally- friendly manner in order to foster the economic, political, and social development of Liberia.

Mali. The development of renewable energies (RE) appears to be a priority in Malian public policy and strategic framework for growth and poverty reduction. Mali has several strategies, policies, acts, and regulations governing renewable energy. These include the following:

- The Malian Agency for the Development of Domestic Energy and Rural Electrification (AMADER), established by Law No. 03-006 of 21 May 2003. The vision is to implement the policy and regulatory framework of rural electrification and domestic energy.
- Mali Renewable Energy Agency (AER-Mali), established by Decree No 2015-0049/P-RM of 6 February 2015. AER is mandated for the popularization and promotion of clean energies in Mali. It works to enable an ecological transition and also to expand the use of renewable energies by facilitating its access to a large majority of the Malian population [32].
- Regulation, Electricity, and Water Commission (CREE), created by Law No 00-185/P-RM of 14 April. 2000 is responsible for regulating the electricity sector and the public drinking water service in urban centers. The Act mandates CREE to issue licenses to support the development of the public electricity and water service; defend the interests of users and the quality of public service; promote and organize a competition between operators [33].
- National Directorate of Energy (DNE), created by Law No 99- 0 1 3 IP-RM OF 01 April. 1999′s mission is to develop the elements of national energy policy. It vision is to increase the national electrification rate to 87% by 2030 (42% in 2017, source: DNE); and the share of renewable energies connected to the grid would be 977.4 MW or 52.5% by 2030 (including 538 MW of solar; 389 MW of hydroelectricity; 20 MW of wind; and 30 MW of bioelectricity [34]).
- National Energy Policy (PEN), adopted in 2006, with the aim to balance energy availability and national socio-economic development needs, foster synergy between the major energy sector stakeholders, and guide their interventions.
- National Strategy for Renewable Energy Development, adopted in 2006, aims to promote the widespread use of renewable energy technologies and equipment. It is intended to increase the share of renewable energy in national electricity generation and also to develop biofuel, and create better conditions to sustain renewable energy services. It further aims to search for sustainable and suitable financing mechanisms for renewable energy in Mali.

Niger. In Niger, the energy sector is governed by the Electricity Law No. 2016-05 of 17 May 2016, implemented by the Minister in charge of Energy. This law ensures the implementation of the entire energy policy framework and regulations, as well as increased financing for the country's energy and economic growth. The law established a fund that domiciles in an account at the Banque Centrale des États de l'Afrique de l'Ouest (BCEAO) for the financing of renewable energies and the control of electrical energy. These funds, for instance, consist in particular of state subsidies; the contribution of the carbon tax; and part of the cost of controlling renewable energy equipment. Therefore, due to the importance of renewable energy, the government of Niger is projecting an increase in the national electrification rate to 65% by 2030 (currently at 10%). This includes targets for 30% renewables in the national energy mix by 2030 (150 MW from grid solar; 100 MW from off-grid solar; 20 MW from wind) [35].

Nigeria. The Nigerian Electricity Regulatory Commission, through Act No. 6 of 2005, made the Regulations on Feed-In-Tariff for Renewable-Energy-Sourced Electricity to develop, promote, and harness the Renewable Energy (RE) resources of the country and incorporate all viable ones into the national energy mix. The commission is also intended to enhance the attainment of the national targets on renewable-energy-sourced electricity. Due to the restricted access to electricity supply in the largely off-grid rural areas of Nigeria, the Federal Government developed the Nigerian Renewable Energy and Energy Efficiency Policy (NREEEP) to facilitate its plan to utilize renewable energy as an alternative source of electricity for off-grid rural areas. The NREEEP therefore generally aims to set out a framework for action to address Nigeria's challenge of inclusive access to modern and clean energy resources, improved energy security, and climate objectives. Nigeria has several national policies and action plans on renewable energy and energy efficiency. These include:

- National Energy Policy (NEP) created in 2003 (revised 2006 and 2013). It covers all aspects of the energy sector, including renewable energy, energy efficiency, and rural electrification. It defines, among other aims, a national target for 75% electrification rate by 2020 and a reduction in electricity generation, transmission, and distribution losses from 15–40% in 2013 to less than 10% by 2020 [36].
- National Renewable Energy Master Plan (REMP) created in 2005 (revised 2012). The REMP was developed by the Energy Commission of Nigeria (ECN) in collaboration with the United Nations Development Programme (UNDP) and was later revised in 2012. The REMP sets out Nigeria's roadmap for increasing the national deployment of renewable energy and promoting sustainable development [37].
- Rural Electrification Strategy & Implementation Plan (RESIP) created in 2006 (revised 2014). It aims to expand electricity access in a cost-effective way, for both off-grid and on-grid electricity supply. One tool for achieving this is to focus subsidies on expanding access rather than on electricity consumption. The RESIP includes targets for 75% and 90% electricity access by 2020 and 2030, respectively, with at least a 10% renewable power share by 2025 [38].
- National Energy Master Plan (NEMP) created in 2007 (revised 2014). It covers all energy sources, energy consumption, capacity development, energy financing, energy databases, and the project cycle (planning, implementation, and monitoring and evaluation). The NEMP sets targets for a share of renewable energy (excluding large hydro) in the national energy sector of 10.6% in the long term (2016–2030) [39].
- National Biofuel Policy and Incentives created in 2007. This policy is aimed at creating a viable biofuels industry, reducing the nation's dependency on gasoline, and reducing pollution of the environment [40].
- Sustainable Energy for All (SEforAll) Action Agenda developed by the Federal Ministry of Power. The document provides useful information on energy access and energy efficiency as well as the renewable energy potential and market in Nigeria and relevant policies and barriers to be overcome [41].
- National Energy Efficiency Action Plan (NEEAP) created in 2016. It sets targets for energy savings and proposes actions for meeting the set targets. The NEEAP targeted 40% and 100% efficient lightening in households by 2020 and 2030, respectively; and efficient energy increase by 20% and 50% in the transport, power, and industrial sectors by 2020 and 2030, respectively [42].
- Intended Nationally Determined Contribution (INDC) approved in 2015. The INDC set conditional and unconditional objectives as 20% and 45%, respectively, and targeted 13 GW solar PV off-grid in the Nigeria electricity mix by 2030, with an annual increase in energy efficiency from 2% to reach 30% by 2030. [43].

Senegal. For the promotion and development of renewable energies, Senegal has put in place an appropriate legal framework, and sufficient incentive to allow the production in sufficient quantity, storage, and transport, as well as the marketing of these products throughout the national territory. Thus, Law No. 2010-21 on the Orientation of Renewable

Energies emphasizes the promotion of renewable energies by granting a total exemption on the acquisition of materials and equipment intended for production of renewable energy for domestic self-consumption, and also tax incentives for the acquisition of materials and equipment intended for production, operation, and research and development in the field of renewable energies (Loi No 2010-21, 2010).

Sierra Leone. In Sierra Leone, the electricity sector and water regulatory commission are governed by the National Electricity Act (2011). This was established on the basis of furthering renewable energy development to attract private investments, extending electricity to rural and remote areas, increasing the percentage contribution of solar energy to the total energy mix, improving healthcare, and enhancing other human services. In addition, the law commits the government to take numerous enabling measures to ensure that renewable energy becomes a significant part of its energy portfolio over the next fifteen years.

Togo. In Togo, Law No 2018-010 establishes the legal framework governing the equipment/materials, installations, and other necessary infrastructure for the production, storage, transport, distribution, marketing, and consumption of electricity based on renewable energy sources. For instance, Law No. 2018-010 in its Title II, articles 16 and 17 highlights this in the three (3) legal regimes for electricity production projects based on renewable energy sources. However, the system of declaration and freedom is applied to the activities of production of electrical energy based on renewable energy sources intended for the needs of clean consumption without injection into the national electricity grid. The Togolese government also grants tax and customs exemptions of up to 10 years for projects to build power plants and infrastructure based on renewable energy sources used for their own needs or the sale of electricity [44].

3.2. Policy Support for Hydrogen Energy

The deployment of hydrogen worldwide is facing several barriers through production, infrastructure, market, industry (see Table 1). The development of hydrogen requires policies in place to support its deployment. The lack of an existing hydrogen value chain represents one of the major obstacles to overcome for the development of the green hydrogen economy. Table 1 groups the types of barriers that hydrogen has to overcome in order to be the next energy paradigm into five blocks. The different elements (i.e., technological, infrastructure, industry, shipping, aviation) interact with each other, creating new and harder barriers for hydrogen deployment. As a preliminary example, one of the hydrogen problems is its electrolyzer [45], which is a technological barrier, but to avoid this barrier, investment is needed, and this investment has to be funded by an African investment bank. The lack of infrastructure for hydrogen production, storage, and transportation has been identified as the greatest challenge to market expansion. The electrolyzer and infrastructure barriers can easily be overturned if there are sufficient economic incentives for producers to research further in order to find a solution; these incentives will be only generated if a market for hydrogen exists, and if it is profitable for investors [22,45].

Table 1. Barriers and policies for hydrogen value chain [19].

Barriers	- Capital cost - Electricity cost - Lack of hydrogen market - Barriers to power market	- Limited existing infrastructure - Technical limitations of users - Lack of investment	- High cost - Lack of demand for green products - Global competition and carbon leakage	- High cost - Procurement of sustainable CO_2 - Policy focus on biofuels	- High cost - Technical barriers

Table 1. *Cont.*

	ELECTROLYSIS	INFRASTRUCTURE	INDUSTRY	AVIATION	SHIPPING
Policy Options	- Set capacity targets - Offer loans - Introduce feed-in premium - Allow participation in ancillary markets	- Collaborate on global trading of hydrogen - Identify priorities for conversion - Align blending targets - Provide financing	- Offer dedicated loans - Develop public procurement of green products - Phase out high emission technologies	- Set targets - Review policy focus - Expand emissions trading system	- Introduce fiscal incentives - Set targets for zero-emission vessels - Support infrastructure development

4. Results and Discussion

Table 2 shows general review of the energy policies in all the ECOWAS countries. It shows that in all ECOWAS countries there exist specific laws that deal with renewable energy. The level of intervention of each institution concerning renewable energy is well articulated in the legal text. Nevertheless, ECOWAS members currently have no official hydrogen policies in place. Such policy regulation is necessary for encouraging the generation of green energy.

Table 2. ECOWAS Energy Policy Evaluation.

	Energy Policy	Renewable Energy Policy	Hydrogen Energy Policy	Hydrogen Energy Initiative
Burkina Faso	Yes	Yes	No	No
Benin	Yes	Yes	No	No
Cape Verde	Yes	Yes	No	No
Cote d'Ivoire	Yes	Yes	No	Yes
Gambia	Yes	Yes	No	No
Ghana	Yes	Yes	No	No
Guinea	Yes	Yes	No	Yes
Guinea Bissau	Yes	Yes	No	No
Liberia	Yes	Yes	No	No
Mali	Yes	Yes	No	Yes
Niger	Yes	Yes	No	No
Nigeria	Yes	Yes	No	No
Senegal	Yes	Yes	No	No
Sierra Leone	Yes	Yes	No	No
Togo	Yes	Yes	No	No
ECOWAS	Yes	Yes	No	No

4.1. The Role of Effective Government Structures and Policies in the Development of Renewable Energy and Hydrogen Economy

The development of green hydrogen in ECOWAS countries requires a sincere partnership between governments and businesses, and national and international cooperation. This partnership will allow a good energy transition in the region towards a green hydrogen economy and will allow a fundamental modification of the economic relations of countries around the world. Significant and effective implementation of relevant supporting policies of governments in the development of the hydrogen economy will be necessary, at least until ECOWAS economies are ready to support the expectation of the population and the market forces. Government action can accelerate the development and deployment of green hydrogen. Governments must participate in research and development as well as investment in infrastructure. Investment here must be clearly in the realm of government, which will bring dividends. Policymakers at all levels, both regional and national, must legislate and adopt development strategies and encourage investment in energy-efficiency and renewable energy technologies. Renewable energy and green hydrogen targets must be paired with specific policy mechanisms designed to help meet national goals. In addition, practical and strategic investment in infrastructure must be addressed.

However, governments should know that one country's action alone will not bring enough about a hydrogen economy; therefore, there must be a collaboration between the country's governments and private organizations at all jurisdictions for establishing a long-term plan and the creation of a viable green hydrogen energy market. Therefore, the conceptions of governance will need to shift and adapt for developing public awareness and educating communities about the relevant safety and standards surrounding green hydrogen production and use. The role of effective government must be to restructure its regulation and law to provide much stronger economic incentives for the adoption of renewable electricity targets to outline sectorial development priorities. While developments in green hydrogen energy in the region are almost non-existent, green hydrogen implementation in the region will enter into commercial use as a storage medium to buffer intermittent renewable electricity supplies in the energy sector.

4.2. The Significance of Renewable Energy and Hydrogen Energy Concerning UN SDGs

The most significant effort to balance the increasing needs of energy in ECOWAS countries aligns with the UN Sustainable Development. ECOWAS regions are disproportionately underdeveloped due to extreme energy poverty and poor governance. The increased demography and resource requirements for energy responses coupled with the economic downturn and poor governance are likely to further constrain the available finance for energy investments and further heighten the risk of energy insecurity in ECOWAS member countries. Against this background, further solutions need to be found on the one hand by minimizing the use of non-renewable energy, and on the other hand by developing new global solutions using environmentally friendly sources and technologies. This would significantly contribute to the resilience and achievement of the UN Sustainable Development Goals (SDG 7) in the ECOWAS region. Building resilience requires a broader focus that addresses multiple dimensions of energy efficiency and securities. Therefore, renewable energy and green hydrogen production in the region have the potential to become key components. Scaling renewable energy for hydrogen production in the ECOWAS region can lead to significant opportunities for additional employment, especially for youth and women, which will allow countries to achieve the Sustainable Development Goal. Most of this employment will require skills, education, and training which need to be enhanced rapidly for the rapid deployment of the generated hydrogen [46]. Nevertheless, the region still lacks decision support on the decarbonization of the energy system through green hydrogen production, which was one of the key political goals of the Paris Agreement for the mid-21st century.

5. Policy Recommendation

ECOWAS countries have a vast potential for renewable energy generation at relatively low costs. Hydrogen production could leverage socio-economic opportunities in the region and provide local industries with green e-fuels and chemicals. It also opens new perspectives for exports and revenue-raising. However, a strong legal, regulatory, and institutional framework with a Framework Act is needed as a basis for attracting and encouraging private investment as well as providing financial incentives. To accomplish green hydrogen policy formulation in ECOWAS countries, governments would need to take up the following measures:

- At the regional level a clear vision of hydrogen energy policies needs to be put in place by members to achieve their energy targets. This can be done by combining efforts by identifying countries with a comparative advantage in renewable energy resources potential for green hydrogen production to support future energy needs for other countries. This collaboration and cooperation action will be to accelerate the development and deployment of green hydrogen and increase the amount of hydrogen gas that can be produced through electrolysis.

- Mapping the fifteen countries' existing infrastructure for future hydrogen storage and export and treating hydrogen as part of an integrated energy system.
- ECOWAS members must create a sub-regional partnership for hydrogen economy (SPHE) to foster intergovernmental cooperation on hydrogen. The objectives of the Sub-regional Partnership for Hydrogen Economy for ECOWAS (SPHE) must be to accelerate progress in hydrogen technologies and share information, resources, and infrastructure among member states. This partnership can also lead to exchanges in terms of initiatives, policies, and technology status, as well as on safety, regulations, codes, and standards. Its aims will also be to enhance public institutional finance to support research and development as well as investment in the technology.
- ECOWAS members should harmonize the regulatory framework by the implementation of a sound legal policy to attract large-scale investment and help hydrogen energy project developers develop investor confidence. Clear legal and regulatory renewable hydrogen production guidelines will contribute to stable and predictable energy production and distribution which attracts both local and international investors. It can provide a framework for the investors and can provide incentives to encourage greater investment.
- Close collaboration between ECOWAS countries and developed countries which have advanced in hydrogen technologies, as well as partnership between the public and private sector along the entire value chain. A consensus on the pivotal role of the private sector needs to be put in place for green hydrogen development and to encourage the flow of international investments. The consensus must include premium prices for green products, carbon taxes, quotas, standards, guarantees of origins, and tax incentives.
- Clear social and ecological criteria for the import of green hydrogen must be defined by the ECOWAS countries. Although the huge production possibilities for cheap green hydrogen have been emphasized by the government, the transport question and its costs need to be clearly defined.
- Priority should be given to the utilization of both the available renewable energy resources and the produced green hydrogen in ensuring that basic and affordable access to electricity is guaranteed for all the local people within ECOWAS. The option of export of green hydrogen or its derivatives should be considered if the former has been addressed.
- Strengths, weaknesses, opportunities, and threats (SWOT) of hydrogen deployment must be explored. These include:
  - ✓ develop green hydrogen economy with comprehensive legislation;
  - ✓ implement specific laws for safety and stability for utilization of hydrogen;
  - ✓ promote the use of hydrogen energy by stimulating public acceptance;
  - ✓ promote regulated hydrogen economy and green economy;
  - ✓ ncourage research, innovations, and development;
  - ✓ encourage investment and development in hydrogen technology;
  - ✓ encourage funding programs by the government;
  - ✓ encourage the government to develop strategies to increase competition between renewable energy resources and non-renewable energy; and
  - ✓ encourage the government to privilege public investments to financially support hydrogen economy projects.

Furthermore, these policy strategies will help decision-makers to enhance and guide green energy transition deployment in the ECOWAS countries and the integration of green hydrogen within this mix.

6. Conclusions

In ECOWAS countries, energy demand, which mainly relies on fossil fuels such as petrol and natural gas, as well as thermal and hydro, is becoming difficult to fulfill population demands and are also causing environmental damage. Therefore, these fossil fuels need to be replaced gradually by alternative renewable energy resources such as green hydrogen energy, solar energy, and wind energy. Green hydrogen is expected to play a significant role in a future climate-neutral economy, enabling zero carbon emission. It enables investment in sustainable energy infrastructure for generating green energy. Since the hydrogen-based energy sector has not yet taken off in the region, there is a need for policymakers to elaborate a clear and strong hydrogen energy legislation roadmap for future energy strategies. This legislation must take into account the country's level of ambition and how hydrogen will meet the social, economic, and environmental objectives to achieve the UN Sustainable Development Goals (SDG) in the ECOWAS region. In addition, national governments need to formulate policies and provide support framework around the following measures in order to achieve hydrogen deployment:

- continue financing Research and Development (R&D) to explore the potential application of hydrogen energy. The scientific research and development will require significant investment for the development of hydrogen energy technologies.
- elaborate on the cooperation among all stakeholders: energy actors, Research Development centers, industry partners, investors, international partners, and politicians. This cooperation should be made attractive by providing a win-win situation for all.
- implement specific laws that address safety and stability issues for the promotion of hydrogen.
- establish green hydrogen as a key player in the energy mix by identifying its contribution to national and regional energy strategies. Incorporate technical, political, and legal concepts to make this possible.
- increasing access to clean and affordable electricity should be considered a priority. This should be addressed by harnessing the huge renewable energy resources within the region. These resources should also be explored for green hydrogen production for both local use and for export.

It can be said that hydrogen has the potential to alter the ECOWAS regions' economic and political structure in ways more profound than any previous energy options. A clear regulatory, policy and legal framework that supports and enbles its deployment is needed to make this happen.

Author Contributions: Data curation: A.B. and K.K.V.; Writing—original A.B. and K.K.V.; Writing—draft preparation: A.B. and K.K.V.; Writing —review and editing: A.B. and K.K.V.; Funding acquisition: M.S., D.K. and S.N.A.; Project administration: K.O.O., B.K., D.K., M.S. and S.N.A.; Resources: A.B. and K.K.V. All authors have read and agreed to the published version of the manuscript.

Funding: This review article was sponsored by the West African Science Service Centre on Climate Change and Adapted Land Use (WASCAL).

Institutional Review Board Statement: Not applicable.

Informed Consent Statement: Not applicable.

Data Availability Statement: No new data were created or analyzed in this study. Data sharing from any achieve is not applicable to this article.

Acknowledgments: This review article is sponsored by the West African Science Service Centre on Climate Change and Adapted Land Use (WASCAL), an African partner to the H2Atlas-Afria project being implemented in partnership with Forschungszentrum Jülich GmbH, Germany. The H2Atlas–Africa is part of the Go-Green-Go African Initiative, funded by the German Federal Ministry of Education and Research (BMBF). Funding support from the BMBF is acknowledged. We also would like to thank WASCAL Competence Centre for providing facilities and assistance.

Conflicts of Interest: The authors declare no conflict of interest.

References

1. IRENA Planning and Prospects for Renewable Energy: West Africa 2013. 2013. Available online: http://www.irena.org (accessed on 1 February 2022).
2. Bazilian, M.; Nussbaumer, P.; Rogner, H.H.; Brew-Hammond, A.; Foster, V.; Pachauri, S.; Kammen, D.M. Energy access scenarios to 2030 for the power sector in sub-Saharan Africa. *Util. Policy* **2012**, *20*, 1–16. [CrossRef]
3. Bazyomo, S.D.; Lawin, E.A.; Ouedraogo, A. Seasonal Trends in Solar Radiation Available at the Earth's Surface and Implication of Future Annual Power Outputs Changes on the Photovoltaic Systems with One and Two Tracking Axes. *J. Climatol. Weather Forecast.* **2017**, *5*, 1–9. [CrossRef]
4. Akinyemi, O.; Efobi, U.; Osabuohien, E.; Alege, P. Regional Integration and Energy Sustainability in Africa: Exploring the Challenges and Prospects for ECOWAS. *African Dev. Rev.* **2019**, *31*, 517–528. [CrossRef]
5. Manneh, M.; Yaffa, S. Opportunities and Constraints of Scaling Up Electricity Access in the Gambia. *Am. Int. J. Bus. Manag.* **2021**, *4*, 71–79.
6. Aglina, M.K.; Agbejule, A.; Nyamuame, G.Y. Policy framework on energy access and key development indicators: ECOWAS interventions and the case of Ghana. *Energy Policy* **2016**, *97*, 332–342. [CrossRef]
7. World Bank. World Bank Global Electrification Database from 'Tracking SDG 7: The Energy Progress Report' led jointly by the custodian agencies: The International Energy Agency (IEA), the International Renewable Energy Agency (IRENA), the United Nations Statistics. Available online: https://data.worldbank.org/indicator/EG.ELC.ACCS.RU.ZS?end=2019&start=2002&view=chart (accessed on 1 February 2022).
8. Karaki, K. Understanding ECOWAS Energy Policy From national interests to regional markets and wider energy access? *Ecdpm.* 2017, 24. Available online: https://ecdpm.org/wp-content/uploads/ECOWAS-Energy-Background-Paper-PEDRO-Political-Economy-Dynamics-Regional-Organisations-Africa-ECDPM-2017.pdf (accessed on 1 February 2022).
9. Ramdé, E.W.; Azoumah, Y.; Brew-Hammond, A.; Rungundu, A.; Tapsoba, G. Site Ranking and Potential Assessment for Concentrating Solar Power in West Africa. *Nat. Resour.* **2013**, *4*, 146–153. [CrossRef]
10. IRENA. *IRENA Planning and Prospects for Renewable Power: West Africa*; International Renewable Energy Agency: Abu Dhabi, United Arab Emirates, 2018; Available online: https://www.irena.org/-/media/Files/IRENA/Agency/Publication/2018/Nov/IRENA_Planning_West_Africa_2018.pdf (accessed on 1 February 2022).
11. World Energy Council—Working Paper: Hydrogen on the Horizon: National Hydrogen Strategies. 2021. Available online: https://www.worldenergy.org/assets/downloads/Working_Paper_-_National_%0AHydrogen_Strategies_-_September_2021.pdf (accessed on 28 February 2022).
12. Threlfall, R.; Blicher-Hansen, J. National hydrogen strategies. *KPMG* **2021**, *14*, 1930.
13. Sadik-Zada, E.R. Political economy of green hydrogen rollout: A global perspective. *Sustainability* **2021**, *13*, 23. [CrossRef]
14. Akoutou, A.B.; Sohn, R.; Vogl, M.; Yeboah, D. *Understanding Regional Integration in West Africa-A Multi-Thematic and Comparative Analysis Content*; Center for European Integration Studies: Bonn, Germany, 2014.
15. ECREEE. *ECOWAS Centre for Renewable Energy and Energy Efficiency. ECOWAS Renewable Energy Policy*; ECREEE: Praia, Cape Verde, 2013. Available online: http://www.ecreee.org/page/ecowas-renewable-energy-policy-erep (accessed on 28 February 2022).
16. Axelsson, L.; Franzén, M.; Ostwald, M.; Berndes, G.; Lakshmi, G.; Ravindranath, N.H. Perspective: Jatropha cultivation in southern India: Assessing farmers' experiences. *Biofuels Bioprod. Biorefining* **2012**, *6*, 246–256. [CrossRef]
17. Hosseini, S.E.; Wahid, M.A. Hydrogen production from renewable and sustainable energy resources: Promising green energy carrier for clean development. *Renew. Sustain. Energy Rev.* **2016**, *57*, 850–866. [CrossRef]
18. Lu, Y.; Khan, Z.A.; Alvarez-Alvarado, M.S.; Zhang, Y.; Huang, Z.; Imran, M. A critical review of sustainable energy policies for the promotion of renewable energy sources. *Sustainability* **2020**, *12*, 1–30. [CrossRef]
19. Zhang, H. Prioritizing Access of Renewable Energy to the Grid in China: Regulatory Mechanisms and Challenges for Implementation. *Chin. J. Environ. Law* **2019**, *3*, 167–202. [CrossRef]
20. IRENA. Green Hydrogen: A Guide to Policy Making. 2020. Available online: https://www.irena.org/-/media/Files/IRENA/Agency/Publication/2020/Nov/IRENA_Green_hydrogen_policy_2020.pdf (accessed on 28 February 2022).
21. IRENA. *Global Renewables Outlook: Energy Transformation 2050*; International Renewable Energy Agency: Abu Dhabi, United Arab Emirates, 2020.
22. WASCAL Atlas of Green Hydrogen Generation Potential: H2Atlas. 2021. Available online: https://wascal.org/v3/wp-content/uploads/2020/08/H2-Atlas-project-flier.pdf (accessed on 29 December 2021).
23. Meza, E. Germany Eager to Partner with West African Countries on Green Hydrogen. Journalism for Energy Transition. 2021. Available online: https://www.cleanenergywire.org/news/germany-eager-partner-west-african-countries-green-hydrogen (accessed on 30 December 2021).
24. WASCAL. 60 West African Students Receive Full German Government Scholarship in International Master's in Green Hydrogen. 2021. Available online: https://wascal.org/60-west-african-students-receive-full-german-government-scholarship-in-international-masters-in-green-hydrogen/ (accessed on 30 December 2021).
25. MME. *Plan d 'Action National de l' Initiative Energie Durable Pour Tous «SE4ALL»*; Ministère des Mines et de l'Energie: Ouagadougou, Burkina Faso, 2015.

26. Nacional, R.I.O.; Ponto, D.O.; Renov, E.; Energ, N.; Em, T. Energias Renováveis e Eficiência Energética em São Tomé e Príncipe—Relatório Nacional do Ponto de Situação. 2020. Available online: https://dgrne.org/pt-pt/taxonomy/term/19 (accessed on 30 December 2021).
27. Renewable Energy Act. Renewable Energy Act 2013. 2013. Available online: https://www.mope.gm/download-file/f8e76bcc-d26e-11ea-94db-022a5fa1767e (accessed on 30 December 2021).
28. UNDP. *Ghana Renewable Energy Master Plan*; Energy Commission: Accra, Ghana, 2019.
29. Act. *Renewable Energy (Amendment) ACT, 2020, (Act 1045)*; PURC: Accra, Ghana, 2020; Volume 2020. Available online: https://www.purc.com.gh/categ/regulatory-framework/subcategories/renewable-energy-amendment-act-2020-act-1045 (accessed on 30 December 2021).
30. Decret D/2019/221/PRG/SGG, Code de l'Environnement de la Republique de Guinee. 2019. Available online: https://guilaw.com/code-lenvironnement-2019/ (accessed on 30 December 2021).
31. ALLER. *Energias Renováveis e Eficiência Energética na Guiné-Bissau—Relatório Nacional do Ponto de Situação*; UNIDO: Vienna, Austria, 2018.
32. JSTM. Agence des Energies Renouvelables du Mali: Atteindre un potentiel national de 38% d'ici 2030. *J. Sci. Tech. Mali* **2019**. Available online: https://www.jstm.org/agence-des-energies-renouvelables-du-mali-atteindre-un-potentiel-national-de-38-dici-2030/ (accessed on 28 February 2022).
33. Primature. CREE (Commission de Regulation de, l'Electricite et de l'Eau). 2000. Available online: https://primature.ml/cree-commission-de-regulation-de-lelectricite-et-de-leau/ (accessed on 28 February 2022).
34. MEE. *Prospectus d'Investissement de l'Energie Durable Pour Tous SEforALL du Mali*; Direction Nationale de l'Energie: Bamako, Mali. Available online: https://www.se4all-africa.org/fileadmin/uploads/se4all/Documents/Country_AAs/PI_SEforALL_MALI.pdf (accessed on 30 December 2021).
35. SEforALL Niger. Republique Du Niger Ministere de l'Energie Prospectus d'Investissement de l 'Energie Durable Pour Tous (Seforall) Du Niger. 2019. Available online: https://www.se4all-africa.org/fileadmin/uploads/se4all/Documents/Country_IPs/PI_SeforALL_Niger_FINAL__1_.pdf (accessed on 30 December 2021).
36. NEP. Federal Republic of Nigeria National Energy Policy the Presidency Energy Commission of Nigeria. Energy Commision of Nigeria: Abuja, Nigeria, 2003; pp. 1–89. Available online: http://rea.gov.ng/wp-content/uploads/2017/09/National_Energy_Policy_Nigeria.pdf (accessed on 28 February 2022).
37. Energy Commission of Nigeria. National Energy Master Plan. 2021. Available online: https://rise.esmap.org/data/files/library/nigeria/Clean%20Cooking/Supporting%20Documentation/Nigeria_National%20Energy%20Master%20Plan%20Draft.pdf (accessed on 30 December 2021).
38. RESIP. Rural Electrification Strategy and Implementation Plan. 2016. Available online: http://rea.gov.ng/wp-content/uploads/2017/09/RESIP.pdf (accessed on 30 December 2021).
39. NNPC. Federal Republic of Nigeria Official Gazette of the Nigerian Bio-fuel Policy and Incentives. 2007. Available online: https://www.lse.ac.uk/GranthamInstitute/wp-content/uploads/laws/1517.pdf (accessed on 30 December 2021).
40. Sterling, S.R.; Hulse, M. Sustainable Energy for All Action Agenda (Se4all), 2016. In *International Energy and Poverty: The Emerging Contours*; Routledge: London, UK, 2015; pp. 273–286. [CrossRef]
41. Robinson, F.A. National Energy Efficiency Action Plan (NEEAP), 2016 (2015–2030). *Phys. Bull.* **1976**, *27*, 374–375. [CrossRef]
42. NACOP. Sustainable Energy for All Action Agenda (SE4ALL-AA). 2016. Available online: https://www.seforall.org/sites/default/files/NIGERIA_SE4ALL_ACTION_AGENDA_FINAL.pdf (accessed on 30 December 2021).
43. Loi N 2010-21, Loi d'Orientations sur les Énergies Renouvelables. 2010, p. 6. Available online: https://www.crse.sn/sites/default/files/2018-11/LOI%202010-21%20Energies%20renouvelables.pdf (accessed on 30 December 2021).
44. Government of Togo. Loi N 2018-010 Relative à la Promotion de la Production de l'Electricite à Base des Sources d'Energies Renouvelables au Togo. 2018. Available online: http://www.arse.tg/wp-content/uploads/2018/10/LOI_ENERGIE_Renouvelable_Togo.pdf (accessed on 30 December 2021).
45. Pascuzzi, S.; Anifantis, A.S.; Blanco, I.; Mugnozza, G.S. Electrolyzer performance analysis of an integrated hydrogen power system for greenhouse heating a case study. *Sustainability* **2016**, *8*, 629. [CrossRef]
46. Bezdek, R.H. The hydrogen economy and jobs of the future. *Renew. Energy Environ. Sustain.* **2019**, *4*, 1. [CrossRef]

energies

Article

Influence of Low Inlet Pressure and Temperature on the Compressor Map Limits of Electrical Turbo Chargers for Airborne Fuel Cell Applications

Jonas Schröter, Daniel Frank, Valentin Radke, Christiane Bauer, Josef Kallo and Caroline Willich *

Institute for Energy Conversion and Storage, Ulm University, Albert-Einstein-Allee 47, 89081 Ulm, Germany; jonas.schroeter@uni-ulm.de (J.S.); daniel.frank@uni-ulm.de (D.F.); valentin.radke@uni-ulm.de (V.R.); christiane.bauer@uni-ulm.de (C.B.); josef.kallo@uni-ulm.de (J.K.)

* Correspondence: caroline.willich@uni-ulm.de

Abstract: For the optimal high-efficiency operation of a PEM fuel cell system, the temperature, pressure, humidity and mass flow of the supplied air must be tuned to the fuel cell stack requirements. Especially for aircraft applications, this requires a thorough understanding of the fuel cell air supply system behavior and how it changes when the ambient pressure is below 1 bar(a) during flight. This work investigates the influence of low inlet pressures and varying inlet temperatures on the compression map of an electrical turbo charger. This is especially relevant in airborne fuel cell application and not much literature can be found on that topic. Compressor limits are evaluated experimentally and theoretically. The theory of mass flow and speed correction is compared to experimental findings and found to be applicable for the surge and speed limit of the investigated turbo chargers as long as the compressor map is not limited by the power of the electric motor and inverter. Based on this, a prediction of the compressor map for altitudes up to 10,000 m is made with the help of a developed software tool.

Keywords: compressor map; low pressure; pressurized fuel cell

Citation: Schröter, J.; Frank, D.; Radke, V.; Bauer, C.; Kallo, J.; Willich, C. Influence of Low Inlet Pressure and Temperature on the Compressor Map Limits of Electrical Turbo Chargers for Airborne Fuel Cell Applications. *Energies* **2022**, *15*, 2896. https://doi.org/10.3390/en15082896

Academic Editors: Bahman Shabani and Mahesh Suryawanshi

Received: 21 March 2022
Accepted: 11 April 2022
Published: 15 April 2022

1. Introduction

Fuel cell-powered airplanes provide the potential to significantly reduce global greenhouse gas emissions in aviation [1,2]. The requirements related to system mass and volume-specific energy density, power and security are very demanding compared to conventional fields of fuel cell application [3–5].

One of the biggest differences compared to automotive applications is the greater variation in ambient conditions. While the pressure on the ground is around 1 bar(a), depending on the location and weather, ambient pressure during flight decreases significantly with higher flight altitudes. According to the International Standard Atmosphere [6], ambient pressure is decreased to 0.26 bar(a) at a height of 10 km, which is typical for long-range flights [7]. Temperature also decreases with altitude. For the mentioned altitude of 10 km, temperature decreases to −50 °C.

The performance of fuel cells increases with higher operation pressures. Lee et al. [2] simulated a pressurized air-cooled fuel cell stack and found an improvement in stack power by pressurizing the inlet air up to 5 atm. The lower air velocity resulting from higher air pressure mitigated the electrolyte dehydration. Pressurization also led to a more uniform current density distribution through reduced reactant depletion. Furthermore, the activation overpotential and the ohmic losses were reduced with increased pressure. These results are confirmed by other studies [4,5,8–10].

The changes in inlet conditions during flight affect the air supply to the fuel cell system. In order to ensure stable and powerful operation, an air supply system is needed that enables control of the inlet pressure of the stack. In an all-electric system an electrical turbo charger is typically used to adjust the pressure level [8,11–13], but it is necessary to take into account the varying performance of a given turbo charger depending on the inlet conditions [9,14].

Kurzke et al. [15–17] used the corrected mass flow and speed to predict compressor maps of conventional gas turbines for flight-relevant conditions from known lines in the compressor map, obtained at standard conditions. They stated that the analytical prediction of the pressure ratio is difficult and therefore used empirical data for their model. The predictions matched well with their measured compressor maps. Leufven et al. [18] also used the corrected mass flow and speed correction derived from dimensionless numbers to predict the compressor map for a gas turbine. Their model fit well with the measured data from their gas turbine test bench.

Not much attention has been paid thus far to the effects of ambient pressure and temperature changes on the limits due to power and the maximum speed of electric driven turbo chargers, as the relevance of this topic only recently arose with the growing interest in pressurized PEM fuel cells in recent years. The limiting factors for the compressor maps of gas turbines and electrical compressors are not necessarily the same, because the power sources for driving the compression are different. Gas turbines use the expansion of the exhaust gas in a turbine, while electric compressors use an electric motor. Therefore, there is a need to investigate the effects that limit the operation of electrical turbo compressors for low inlet pressures and temperatures, as this is relevant for airborne fuel cell applications.

This work investigates the shift of the compressor map and its limitations for electrically driven turbo chargers, aiming at an application in a fuel cell air supply in aviation. The focus lays on applying and confirming mass flow and speed correction in order to predict altitude dependent compressor map shifts. For this reason, the Rotrex EK10AA and the Fischer 150k fuel cell air compressor are used as representatives of commercially available electrical turbo chargers and are evaluated experimentally under simulated flight conditions in a climate chamber. The influences of pressure and temperature changes are tested separately as well as in combination, as occurs during real flight conditions. The experimental data is compared to analytical data obtained via a developed software tool which calculates corrected mass flow and corrected speed from compressor maps at ground conditions. This comparison shows in which cases the mass and speed correction terms, known from gas turbines, correctly describe the compressor map shift of electrical turbo chargers, and can be used for predictions of the compressor map changes for high altitudes. The results also show that for electrical compressors, whose performance is limited by the inverter power, the performance at high altitude is underestimated when using the mass and speed correction terms.

2. Operational Map of Electrical Turbo Compressors

2.1. Compressor Map Limits

The typical compressor map of an electrical turbo compressor shows the pressure ratio between in and outlet pressure over the air mass flow. This map is in principle limited by four relevant effects [19–25]: the surge limit, the choke limit, the speed limit and the power limit.

The surge line limits the compressor map at low mass flows and high pressure ratios. This limit is relevant for every turbo compressor. At the surge limit, a stall of the flow occurs which leads to an air oscillation and reversed flow that can cause severe damage to the compressor, especially in cases of air bearing.

The choke limit depends on the geometry of the compressor and is only relevant for some compressors. Once the air flowing through the blades and volute reaches the speed of sound, it cannot be further accelerated. Therefore, the maximum possible mass flow is reached in this point and the choke line limits the compressor map at higher mass flows.

The speed limit protects the compressor from structural damages by limiting the maximum impeller speed through the implementation of the inverter control. The speed limit therefore limits the compressor map at higher pressure ratios and at higher mass flows.

The power limit is only relevant for some electrical turbo chargers and depends on the manufacturer's technical configuration of the motor. If the power limit is reached, the inverter or the electrical motor prevents a further acceleration of the impeller. For some turbo charger/inverter combinations a minimum voltage is needed to reach a certain speed. Often a maximum current limit is implemented into the system for thermal reasons. As the inlet pressure decreases, the air mass flow also decreases, and therefore less power is required for compression [9]. This means that the power limit becomes less important at lower inlet pressures. In this case, the speed limit might become the relevant limit of the compressor map.

2.2. Compressor Map Prediction according to Corrected Mass Flow

The considered application for electrical turbo chargers in this work is the supply of air to a fuel cell in aviation. For this reason, the absolute compressor outlet pressure, which correlates to the fuel cell inlet pressure, is the variable of interest. The term absolute compressor map is used in this work to describe a map of the absolute outlet pressure of the turbo charger over the mass flow.

The prediction of the absolute compressor map for gas turbines in literature [15,17,18,20,21] is usually done with the help of a universal compressor map showing the pressure ratio over the air mass flow for defined standard conditions and the formula for mass flow correction Equation (5), assuming Mach similarity.

The Mach number in flow direction M_f is the ratio of flow velocity v to the speed of sound c.

$$M_f = \frac{v}{c} = \frac{v}{\sqrt{\kappa R_S T}} \rightarrow v = M_f \sqrt{\kappa R_S T} \tag{1}$$

where c is expressed through the adiabatic index for air $\kappa = 1.4$, the specific gas constant for air $R_s = 287 \frac{J}{kg \cdot K}$; and the air temperature.

The density of air ρ can be calculated from the ideal gas law with the pressure p; the gas constant R_s; and the air temperature T.

$$\rho = \frac{p}{R_S T} \tag{2}$$

Using the cross sectional area A, Equations (1) and (2) the air mass flow through the compressor can be expressed according to Equation (3).

$$\dot{m} = \rho A v = \frac{p}{R_s T} A \left(M_f \sqrt{\kappa R_s T} \right) \tag{3}$$

Comparing the air mass flow for two different inlet temperatures and pressures indexed "0" and "1" leads to

$$\frac{\dot{m}_1}{\dot{m}_0} = \frac{\frac{p_1}{R_S T_1} A \left(M_f \sqrt{\kappa R_S T_1} \right)}{\frac{p_0}{R_S T_0} A \left(M_f \sqrt{\kappa R_S T_0} \right)} = \frac{p_1 \sqrt{T_0}}{p_0 \sqrt{T_1}} \tag{4}$$

$$\dot{m}_1 = \dot{m}_0 \frac{p_1 \sqrt{T_0}}{p_0 \sqrt{T_1}} \tag{5}$$

Equation (5) is named mass flow correction and can be used to predict the compressor behavior for varying inlet conditions. The mass flow correction is combined with a speed correction for varying inlet temperatures, also derived from Mach similarity. For the speed correction, the radial Mach number is the relevant parameter [20]. In this case, the flow velocity v in Equation (1) can be expressed by the radius r times the rotation speed to get an expression for the radial Mach number M_r.

$$M_r = \frac{v}{c} = \frac{r\omega}{\sqrt{\kappa R_S T}} \tag{6}$$

Comparing two different inlet temperatures and rearranging Equation (6) to be similar to Equation (4) leads to the speed correction Equation (7),

$$\omega_1 = \frac{\omega_0 \cdot \sqrt{T_1}}{\sqrt{T_0}} \tag{7}$$

where ω_1 is the assigned speed for T_1, taking the speed ω_0 at temperature T_0 as the reference point. Varying inlet pressures do not affect the corrected speed.

3. Experimental Setup

To validate the developed equation, experimental tests were performed with an EK10AA turbo charger (Rotrex A/S, Ishøj, Denmark) [23] and an EMTC-150K AIR (FISCHER Fuel Cell Compressor AG, Herzogenbuchsee, Switzerland) [21] in a custom-made climate chamber [26]. The test bench setup, shown in Figure 1, was similar to the one described in [9]. Filtered ambient air was compressed with the specified electrical turbo charger. The heated air was cooled with an air-to-liquid intercooler before it passed the mass flow control valve. Sensors measured the compressors' static in- and outlet pressures (PT0517 and PT5504, ifm electronic GmbH, Essen, Germany) and temperatures (HFM5, Robert Bosch GmbH, Germany and TM4101 ifm electronic GmbH, Essen, Germany) as well as the air mass flow (HFM5, Robert Bosch GmbH, Gerlingen, Germany). Compressor speed and the state of the throttle valve were set manually in the control software. Each curve was recorded at its respective compressor speed. The maximum possible speed was 140 krpm for the Rotrex and 150 krpm for the Fischer compressor.

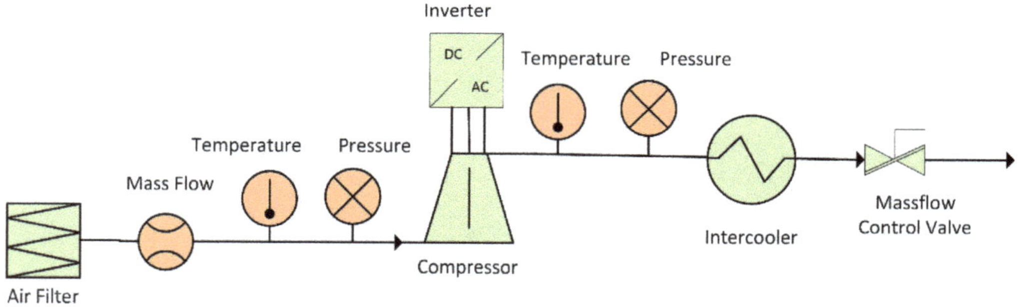

Figure 1. Schematic of the test bench setup, including air filter, compressor with inverter, intercooler, mass flow control valve and sensors.

According to Bernoulli's principle Equation (8), the dynamic pressure p_{dyn} calculated from mass flow and density was added to the measured static pressure p_{sta} to obtain the total pressure p_{tot}.

$$p_{tot} = p_{sta} + p_{dyn} = p_{stat} + \frac{1}{2} \cdot \frac{R_S \cdot T \cdot \dot{m}^2}{p_{sta} \cdot \pi^2 \cdot r^4} \tag{8}$$

Compressor outlet pressures were measured for inlet pressures of 940, 700 and 500 mbar(a) ± 20 mbar and inlet temperatures of −10, 5, 20 and 40 °C ė 4 °C. In a further experiment, variation of inlet pressure and temperature were combined according to Table 1 to simulate the operating conditions of real flight altitudes.

Table 1. Values used for simulated altitude, oriented towards the international standard atmosphere.

Altitude/m.	Pressure/mbar(a)	Temperature/°C
600	935	11
1000	900	8
2000	800	2
3000	700	−5
4000	580	−11

4. Results and Discussion

4.1. Pressure and Temperature Dependence of the Compressor Map

For the following considerations, the measurements with the Rotrex EK10AA turbo compressor were evaluated because its compressor map is not determined by the power limit. Figure 2 shows the measured outlet pressures over the air mass flow for varying compressor inlet pressures (940 mbar, 700 mbar and 500 mbar) and rotational speeds of 120 and 140 krpm. These speeds were chosen because they represent the upper, application-relevant part of the compressor map. The observed shifts in pressure and mass flow for constant speed were valid for lower speeds in the same way. Points close to the surge line were avoided and only tested for the combined pressure and temperature variations since it was not clear how much mechanical overstressing the compressor could handle.

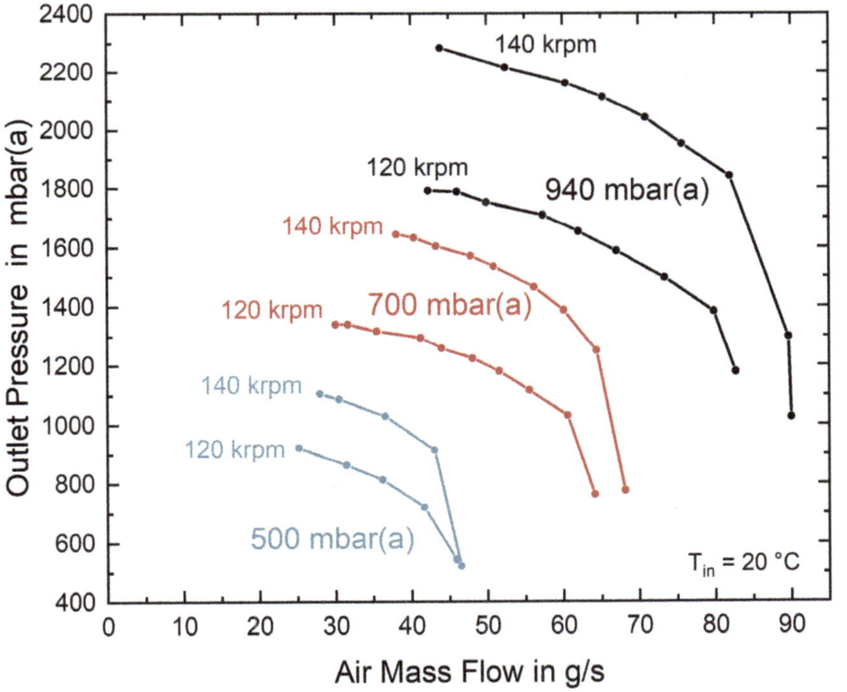

Figure 2. Absolute compressor map for 940, 700 and 500 mbar(a); inlet pressure and compressor speeds of 140 and 120 krpm; and 20 °C inlet temperature.

Figure 2 shows clearly that decreasing the inlet pressure led to a decrease in measured outlet pressure for the same air mass flow over the full compressor map, as was expected. For 140 krpm and an inlet pressure decrease from 940 mbar(a) to 700 mbar(a), the outlet pressure was reduced by more than 500 mbar for all mass flows. Further lowering to 500 mbar(a) inlet pressure led to an additional decrease of about 600 mbar for the same mass flow. The maximum air mass flow provided by the compressor also changed with inlet pressure. The maximum air mass flow reduced from 90 g/s for 940 mbar(a) inlet pressure to 69 g/s for 700 mbar(a) and 47 g/s for 500 mbar(a). A similar behavior of outlet pressure and mass flow reduction was observed for all compressor speeds. The absolute outlet pressure difference for varying inlet pressures was smaller for lower compressor speeds. For example, for 120 krpm and 50 g/s, the difference between 940 mbar(a) and 700 mbar(a) was 560 mbar, compared to 700 mbar for 940 mbar(a) inlet pressure. The reason for this was a lower compression ratio at lower speeds and a resulting lower absolute difference.

Figure 3 shows the outlet pressure curves for rotational speeds of 120 and 140 krpm for four different air inlet temperatures (−10 °C, 5 °C, 20 °C and 40 °C).

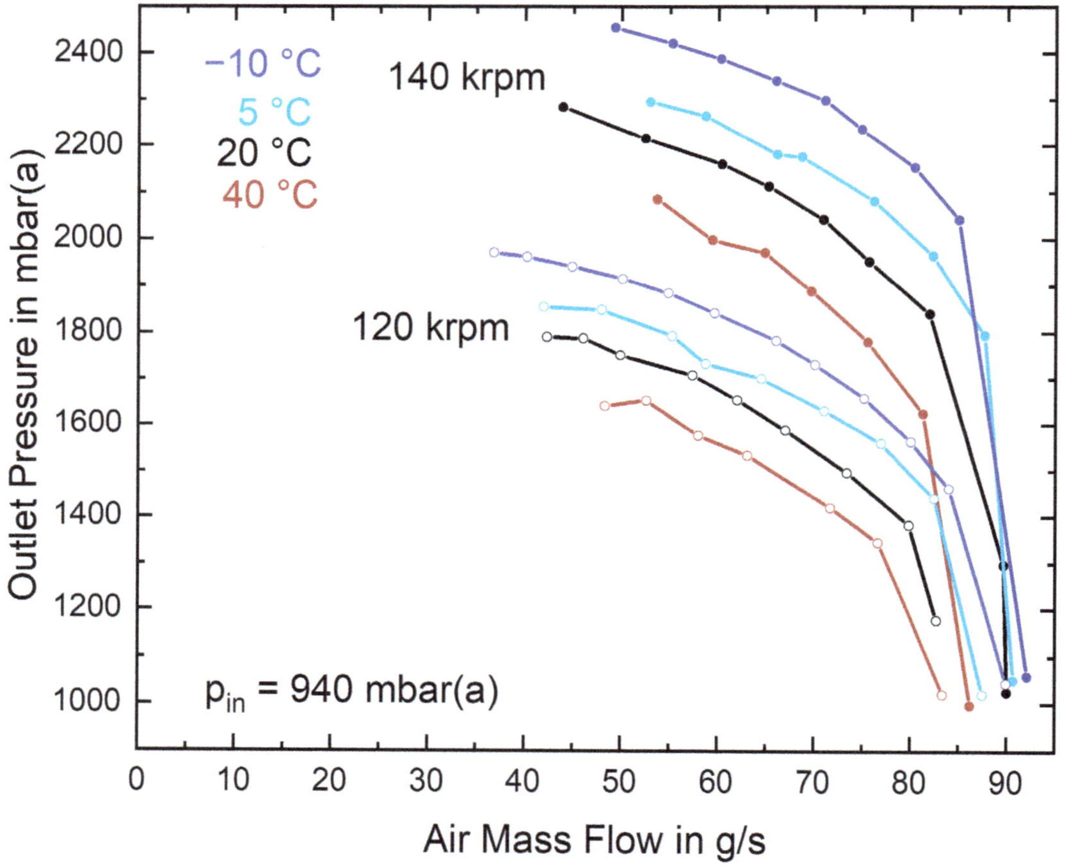

Figure 3. Absolute compressor map for varying inlet temperatures of −10 °C (dark blue), 5 °C (light blue), 20 °C (black) and 40 °C (red) and varying compressor speeds of 120 krpm (empty points) and 140 krpm (filled points).

From the measured curves, it can be seen that lower inlet temperatures enabled higher compressor outlet pressures. For the maximum compressor speed, a 50 K temperature decrease from 40 °C to −10 °C led to a higher outlet pressure of about 400 mbar for all mass flows. Furthermore, the maximum possible mass flow increased from 86 g/s to 92 g/s. These influences of the inlet temperature were also confirmed by the measurements for 5 °C and 20 °C. For lower compressor speeds, the difference in absolute outlet pressure again decreased because of the lower pressure ratios, as described for varying inlet pressures.

Figures 2 and 3 show that the inlet pressure and temperature both have a significant influence on the compressor outlet pressure and need to be considered for a fuel cell system in airplane applications.

4.2. Corrected Mass Flow

In order to compare the measured results to the theory of corrected mass flow calculation, the measurements were corrected with Equations (5) and (7) to 940 mbar(a) and 20 °C, as shown in Figure 4. For the speed correction required for varying inlet temperatures, it was necessary to use curves with different speeds than the measured ones. A linear interpolation between 100 and 140 krpm was used to correct the speeds for the 120 krpm curves and a linear interpolation between 120 and 140 krpm was used to correct the 140 krpm curves. In reality, an exponential behavior would be expected, but the error made by using the much simpler linear interpolation was estimated to be below 5%, and thus negligible compared to the measurement error. An interpolation is always necessary if the compressor limits are corrected at lower inlet temperatures. For the 40 °C curve, the corrected speed is 145 krpm which would not be possible with the measured compressor.

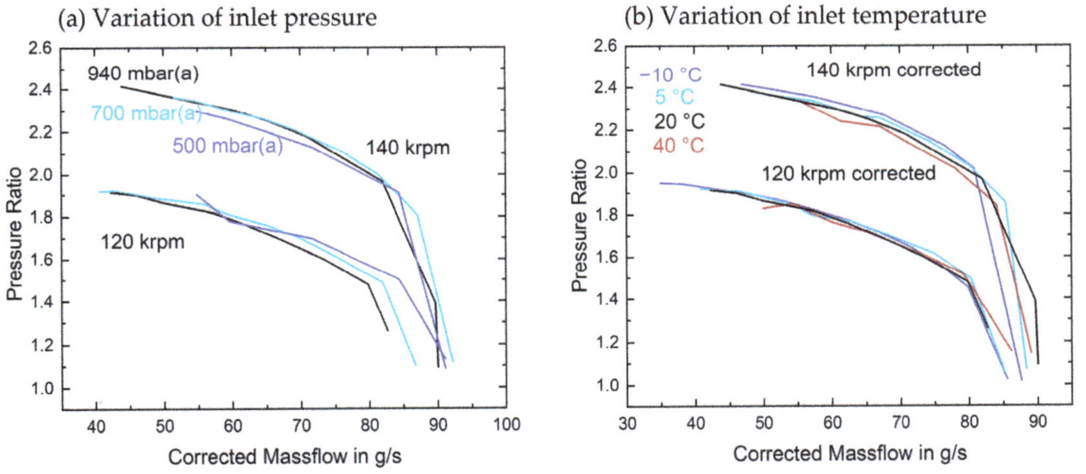

Figure 4. (**a**) Pressure ratio over corrected mass flow for measurements with 940 mbar(a) (black), 700 mbar(a) (light blue) and 500 mbar(a) (dark blue) inlet pressure at 20 °C, corrected to 20 °C and 940 mbar(a). (**b**) Pressure ratio over corrected mass flow for measurements with −10 °C (dark blue), 20 °C (black), 5 °C (light blue) and 40 °C (red) inlet temperature at 940 mbar(a), corrected to 20 °C and 940 mbar(a).

The resulting curves matched well within the range of the measurement error. Therefore, it can be concluded that Equations (5) and (7) correctly described the shift of the compressor map for the speed limit, as this is the relevant limit for the observed measurements.

4.3. Flight Conditions

During a real flight, pressure and temperature changes occur simultaneously. In order to assess the behavior for this combined change, tests were performed in the climate chamber emulating realistic flight conditions, as given in Table 1.

Figure 5 shows the compressor map using corrected mass flow and pressure ratio for different altitudes according to the standard atmosphere [6], as given in Table 1.

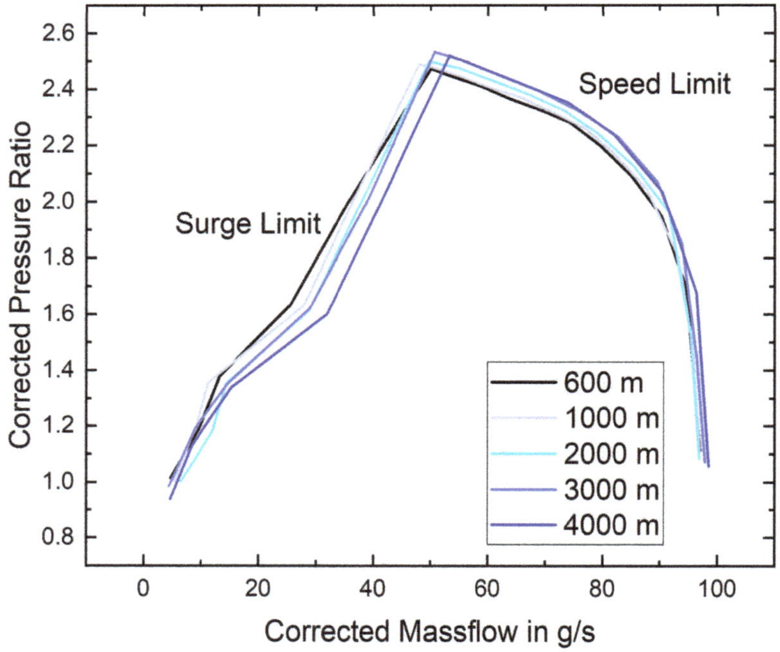

Figure 5. Pressure ratio over corrected air mass flow for measurements with simulated altitudes of 600, 1000, 2000, 3000 and 4000 m corrected to 0 m (15 °C, 1000 mbar(a)).

The correction terms, Equations (5) and (7), describe the behavior of the speed limit with varying altitudes well in the range of error, and it can be concluded that the correction terms are also valid for combined temperature and pressure changes with increasing altitude. Therefore, these equations can be used to calculate all relevant limits for the examined compressor. Of course, this is only valid if the speed, and not the power, limits the compressor map such as for the investigated Rotrex EK10AA.

4.4. Influence of Power Limit on the Compressor Map at Varying Inlet Pressures

For the second examined compressor, the Fischer 150k, the power limit, which is implemented in the inverter controller, limits part of the operation map of the compressor. In this case, the correction terms Equations (5) and (7) cannot be used to describe the operation range. Figure 6 shows the experimentally obtained values of the pressure ratio over the corrected air mass flow for different inlet pressures.

It can be observed that the compressor map at high inlet pressures was limited at higher air mass flows and pressures by the maximum power that the motor can provide. Once this power limit was reached, further acceleration of the impeller was prevented by the inverter. As mentioned previously, the power limit becomes less important with lower inlet pressures since the power depends on the air mass flow [9], which decreases with low inlet pressures. It can be seen from Figure 6 that at inlet pressures lower than 700 mbar(a), the compressor map of the investigated Fischer 150k was no longer determined by the motor power and the speed limit became the only relevant limit.

For the examined Fischer 150k, the power limit also depends on the pressure ratio since it varied between 4.5 and 5.3 kW during the measurements. Higher power was achieved at higher pressure ratios.

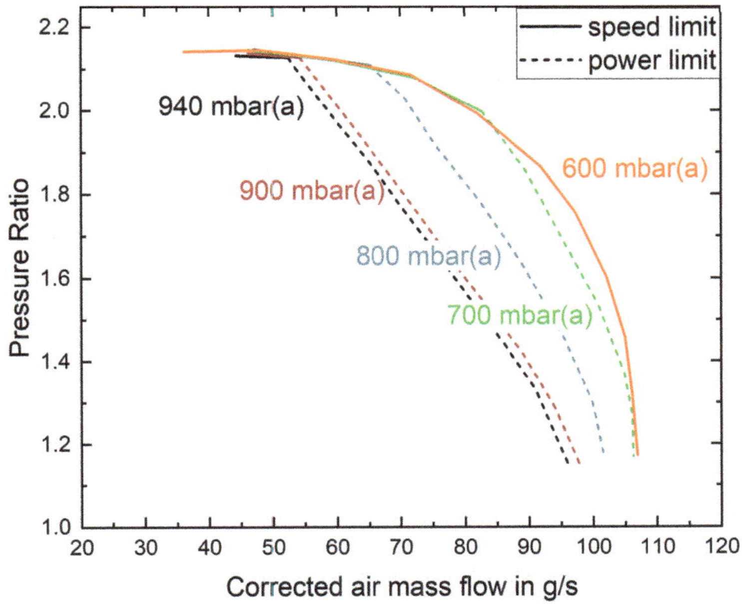

Figure 6. Corrected pressure ratios over corrected air mass flow for the measurements with the Fischer 150k compressor, for inlet pressures of 940, 900, 800, 700 and 600 mbar(a), corrected for 20 °C and 940 mbar(a). Dashed lines indicate the power limit.

4.5. Prediction of the Compressor Map

Using the mass and speed correction terms, the behavior of a compressor map without a power limit at even higher altitudes can be predicted. Based on Equations (5) and (7), a software tool (see Supplementary Material) was implemented in OriginLab to predict the changes of the absolute compressor map with altitude. The corresponding inlet pressure and temperature values are calculated according to the International Standard Atmosphere. The known surge and speed limits, often given in compressor data sheets, can be entered as data points and the mass and speed correction are used to calculate the new pressure and mass flow points for the desired altitude. For the speed correction, a linear fit through the origin is assumed. The error made by this assumption is less than 5% for the investigated compressors which is negligible compared to measurement errors. The resulting compressor map for the desired altitude is plotted automatically. The software tool is published together with this study.

Figure 7 shows the measured, as well as predicted values for the maximum reachable outlet pressure, the corresponding mass flow and the maximum possible air mass flow over altitude for the Rotrex EK10AA compressor, which is not limited by power.

It can be seen that the operation range regarding outlet pressure and air mass flow significantly decreases when going to higher altitudes. At and altitude of 5000 m, the point of maximum outlet pressure decreased from 2500 mbar(a) at 50 g/s to 1450 mbar(a) at 29 g/s compared to ground level. Going up to 10,000 m, which corresponds to −50 °C and 264 mbar(a), the maximum outlet pressure decreased further to 780 mbar(a) at 15 g/s. Furthermore, the maximum possible air mass flow decreased from 98 g/s at 0 m, to 55 g/s at 5000 m and to 29 g/s at 10,000 m. The measured points are in good agreement with the predicted points for the Rotrex EK10AA compressor.

The tool only works fine if the input operation map is not limited by the power, because in that case the speed limit is not determining the boundaries of the operational map. However, the maximum power can also be entered into the tool to calculate the power limit. To simplify the calculation, a constant efficiency of 60% [9] was assumed, which can also be changed if necessary. The calculation of the power limit is only relevant for two cases: 1. If the speed limit is known and used as input and a power limit is implemented afterwards. 2. If the input reference compressor map is taken at low pressures without the presence of the power limit, but the performance at higher inlet pressures for which the power limit is relevant, is of interest.

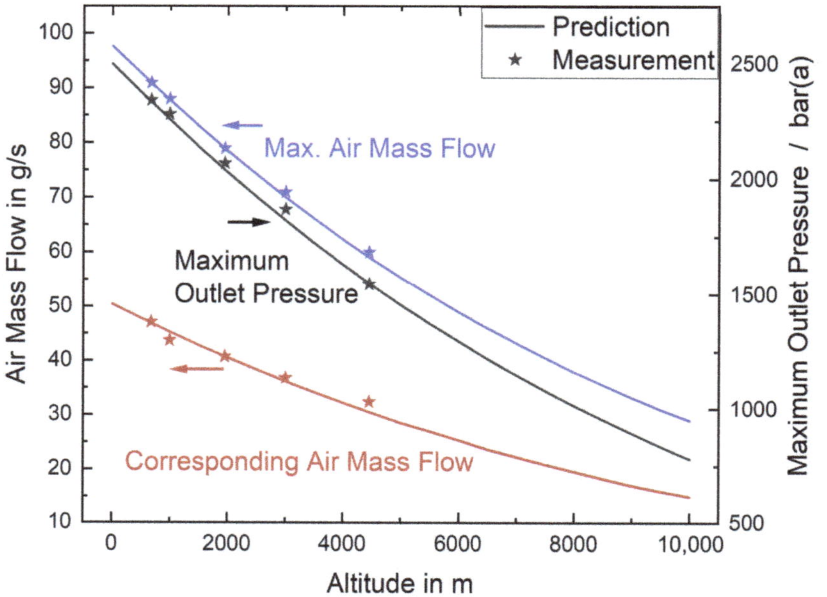

Figure 7. Predicted maximum pressure ratio with corresponding air mass flow and maximum air mass flow for the Rotrex EK10AA; measurement data are plotted in stars for comparison.

Figure 8 shows the measured compressor limit at 600 m as well as the measured and calculated limits for 4000 m. This change in altitude corresponds simultaneously to a change in pressure and temperature. Since the 600 m curve was measured closer to the surge limit than the 4000 m curve, the calculated curve for 4000 m contains points with a lower mass flow than the measured curve.

It can be seen that the prediction works fine in the range of error for the part of the compressor map that is limited by the speed limit. In contrast, the power limitation of the reference curve at 600 m leads to an underestimation of the compressor map for 4000 m because the power limit is used as input, although it becomes less relevant with increased altitude. As can be seen, the measured curve is only limited by the speed limit, and the correction only works for the parts of the compressor map that are not limited by the power.

If the surge and speed limit alone define the borders of the reference compressor map, the prediction according to the mass and speed correction, work reliably. This is shown by the good agreement of measured and calculated data presented in Figure 7, as well as for the very low mass flows in Figure 8. When using the mass and speed correction terms, only the available points of the reference compressor map are corrected. If points close to the surge line are missing, the tool cannot predict these points for other inlet conditions.

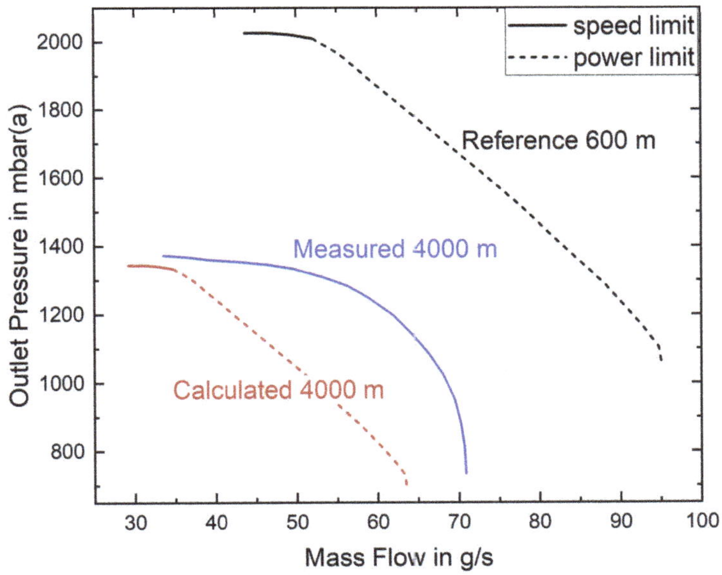

Figure 8. Measured compressor limit at 600 m and calculated corresponding limit at 4000 m, in comparison to the measured compressor limit for a 4000 m altitude for the Fischer 150k compressor. Speed limit is plotted in the solid line and the power limit is plotted in dashed lines.

5. Conclusions

This work examined the influence of varying inlet pressure and temperature resulting from varying flight altitudes on the compressor map of two electrical turbo chargers. Two commercially available turbo compressors, the Rotrex EK10AA and the Fischer 150k, are tested in a custom-made climate chamber at different ambient temperatures, pressures and combinations of both, that simulate flight altitudes. Decreasing the inlet pressure leads to a decrease in the measured outlet pressure for the same air mass flow over the full compressor map, while lower inlet temperatures enable higher compressor outlet pressures, as is to be expected. The results of the Rotrex EK10AA were compared to the theory of corrected mass flow and speed and confirmed the applicability of corrected mass flow and corrected speed relationships for the speed and the surge limit of the compressor at varying flight altitudes. This allows the selection of electrical turbo chargers for airborne fuel cell application with only one given compressor map, which is defined by the speed, choke and surge limit. Typically, manufacturers give the map for ground conditions. From this map the behavior of the compressor for other altitudes can be predicted.

Based on this, a software tool was developed to calculate the impact of altitude on the compressor map considering the combined change in pressure and temperature, and a prediction of the compressor map for up to 10,000 m altitude was given. The maximum possible outlet pressure and air mass flow range significantly decreased for elevated flight altitudes. The maximum outlet pressure and the maximum possible air mass flow for the investigated Rotrex EK10AA compressor dropped by 70% when going from sea level to 10,000 m.

However, for some commercially available compressors, like the examined Fischer 150k, the maximum power available from the motor and inverter limits the compressor map and corrected mass flow and speed can no longer be used for the prediction of the compressor map at low inlet pressures. This power limit becomes less relevant with increasing altitude, so using the mass and speed correction theory might underestimate the performance at high altitudes. In order to reliably predict the behavior at high altitudes the compressor map that is not limited by inverter or motor power has to be known.

The results gained in this study can be used to predict the operation range of a pressurized airborne fuel cell system. Further research is necessary to examine the effect of high altitude on the efficiency and power of the compressor and the corresponding fuel cell system.

Supplementary Materials: The following supporting information can be downloaded at: https://www.mdpi.com/article/10.3390/en15082896/s1, Video S1: Map prediction tool.

Author Contributions: Conceptualization, C.B. and C.W.; Data curation, Valentin Radke; Funding acquisition, C.B., J.K. and C.W.; Investigation, J.S., D.F. and V.R.; Methodology, J.S. and D.F.; Project administration, C.B.; Supervision, C.W.; Validation, J.S., D.F. and V.R.; Visualization, J.S.; Writing—original draft, J.S., D.F., V.R. and C.W.; Writing—review & editing, C.W. All authors have read and agreed to the published version of the manuscript.

Funding: This research was funded by the German Federal Ministry for Economic Affairs and Energy as part of the project HighV (20Y1701B) and the Federal Ministry for Transport and Digital Infrastructure as part of the project Go4Hy2 (03B10703A).

Institutional Review Board Statement: Not applicable.

Informed Consent Statement: Not applicable.

Data Availability Statement: Not applicable.

Acknowledgments: The authors want to thank Marcel Haag and Fatih Türk for their help in setting up the test bench and Florian Becker from DLR Hamburg for fruitful discussions.

Conflicts of Interest: The authors declare no conflict of interest.

Nomenclature

A	cross sectional area
c	speed of sound
$\dot{m}$	air mass flow
M_f	Mach number in flow direction
M_r	Mach number in radial direction
p	pressure
r	radius
R_s	specific gas constant for air
T	temperature
v	flow velocity
κ	adiabatic index for air
ρ	density of air
ω	rotation speed

References

1. EASA. European Aviation Environmental Report 2019. Available online: www.easa.europa.eu/eaer (accessed on 11 April 2022).
2. Lee, D.S.; Fahey, D.W.; Skowron, A.; Allen, M.R.; Burkhardt, U.; Chen, Q.; Doherty, S.J.; Freeman, S.; Forster, P.M.; Fuglestvedt, J.; et al. The contribution of global aviation to anthropogenic climate forcing for 2000 to 2018. *Atmos. Environ. (1994)* **2021**, *244*, 117834. [CrossRef] [PubMed]
3. Peters, R. *Brennstoffzellensysteme in der Luftfahrt*; Springer: Berlin/Heidelberg, Germany, 2015; ISBN 978-3-662-46797-8.
4. Flade, S.; Stephan, T.; Thalau, O.; Burberg, T.; Schirmer, J.; Kallo, J. Air Breathing PEM Fuel Cells in Aviation. *ECS Trans.* **2016**, *75*, 471–477. [CrossRef]
5. Werner, C.; Busemeyer, L.; Kallo, J. The impact of operating parameters and system architecture on the water management of a multifunctional PEMFC system. *Int. J. Hydrog. Energy* **2015**, *40*, 11595–11603. [CrossRef]
6. Heinkel, M. ICAO-Standardatmosphäre (ISA). Available online: www.dwd.de (accessed on 11 April 2022).
7. Joos/as. Lufthansa: Die Flotte des Kranichs. Available online: https://www.flugrevue.de/zivil/groesste-deutsche-fluggesellschaft-lufthansa-die-flotte/ (accessed on 11 April 2022).
8. Schröder, M.; Becker, F.; Kallo, J.; Gentner, C. Optimal operating conditions of PEM fuel cells in commercial aircraft. *Int. J. Hydrog. Energy* **2021**, *46*, 33218–33240. [CrossRef]

9. Schröter, J.; Graf, T.; Frank, D.; Bauer, C.; Kallo, J.; Willich, C. Influence of pressure losses on compressor performance in a pressurized fuel cell air supply system for airplane applications. *Int. J. Hydrog. Energy* **2021**, *46*, 21151–21159. [CrossRef]
10. Wahdame, B.; Candusso, D.; Kauffmann, J.-M. Study of gas pressure and flow rate influences on a 500W PEM fuel cell, thanks to the experimental design methodology. *J. Power Sources* **2006**, *156*, 92–99. [CrossRef]
11. Pischinger, S.; Schönfelder, C.; Bornscheuer, W.; Kindl, H.; Wiartalla, A. Integrated Air Supply and Humidification Concepts for Fuel Cell Systems. In *SAE Technical Paper Series, Proceedings of the SAE 2001 World Congress, Detroit, Michigan, 5–8 March 2001*; SAE International: Warrendale, PA, USA, 2001.
12. Campanari, S.; Manzolini, G.; Beretti, A.; Wollrab, U. Performance Assessment of Turbocharged Pem Fuel Cell Systems for Civil Aircraft Onboard Power Production. *J. Eng. Gas Turbines Power* **2008**, *130*, 021701. [CrossRef]
13. Garrett Motion Inc. Hydrogen Fuel Cell Vehicles to Accelerate Electrification in the Global Auto Industry. Available online: www.garrettmotion.com (accessed on 11 April 2022).
14. Fröhlich, P. Development of an oil free turbo compressor for mobile fuel cell applications–challenges and results. In Proceedings of the Konferenzband der ersten FC^3 Fuel Cell Conference, Chemnitz, Germany, 23–24 November 2019.
15. Kurzke, J.; Riegler, C. A New Compressor Map Scaling Procedure for Preliminary Conceptional Design of Gas Turbines. In *Volume 1: Aircraft Engine; Marine; Turbomachinery; Microturbines and Small Turbomachinery, Proceedings of the ASME Turbo Expo 2000: Power for Land, Sea, and Air, Munich, Germany, 8–11 May 2000*; American Society of Mechanical Engineers: New York, NY, USA, 2014; p. 05082000. ISBN 978-0-7918-7854-5.
16. Kurzke, J. Correlations Hidden in Compressor Maps. In Proceedings of the ASME 2011 Turbo Expo: Turbine Technical Conference and Exposition, Vancouver, BC, Canada, 6–10 June 2011.
17. Kurzke, J. Model Based Gas Turbine Parameter Corrections. In *Volume 1: Turbo Expo 2003, Proceedings of the ASME Turbo Expo 2003, Collocated with the 2003 International Joint Power Generation Conference, Atlanta, GA, USA, 16–19 June 2003*; American Society of Mechanical Engineers: New York, NY, USA, 2009; Volume 06162003, pp. 91–99. ISBN 0-7918-3684-3.
18. Leufven, O.; Eriksson, L. Engine Test Bench Turbo Mapping. In Proceedings of the SAE 2010 World Congress, Detroit, MI, USA, 13–15 April 2010; SAE International: Warrendale, PA, USA, 2010.
19. Garrett Motion Inc. Turbo Tech 103 | Expert: Compressor Mapping. Available online: www.garrettmotion.com (accessed on 11 April 2022).
20. Volker, Gümmer. Lecture Flight Propulsion 1 and Gas Turbines: Establishing Mach No. Similitude. Munich, Germany, 1 April 2021.
21. FISCHER Fuel Cell Compressor, AG. Data Sheet EMTC-150k. Available online: www.fischer-fuelcell-compressor.com (accessed on 11 April 2022).
22. Tiainen, J.; Jaatinen-Värri, A.; Grönman, A.; Fischer, T.; Backman, J. Loss development analysis of a micro-scale centrifugal compressor. *Energy Convers. Manag.* **2018**, *166*, 297–307. [CrossRef]
23. Rotrex. EK10AA Datasheet. Available online: www.rotrex.com (accessed on 11 April 2022).
24. Marelli, S.; Carraro, C.; Marmorato, G.; Zamboni, G.; Capobianco, M. Experimental analysis on the performance of a turbocharger compressor in the unstable operating region and close to the surge limit. *Exp. Therm. Fluid Sci.* **2014**, *53*, 154–160. [CrossRef]
25. Müller, D. Einfluss der Anordnung und des Betriebs von Abgasturboladern auf deren Pumpgrenze. Ph.D. Thesis, Gottfried Wilhelm Leibniz Universität, Hannover, Germany, 2008.
26. Schröter, J.; Steinbarth, D.; Bauer, C.; Reichmann, U.; Kallo, J.; Willich, C. Climate and Pressure Chamber for Simulation of Flight Conditions. Available online: https://oparu.uni-ulm.de/xmlui/bitstream/handle/123456789/38074/Climate_chamber.pdf?sequence=1&isAllowed=y (accessed on 11 April 2022).

energies

MDPI

Article

Thermoelectric Generator as the Waste Heat Recovery Unit of Proton Exchange Membrane Fuel Cell: A Numerical Study

Hossein Pourrahmani [1,*], Hamed Shakeri [2] and Jan Van herle [1]

[1] Group of Energy Materials, École Polytechnique Fédérale de Lausanne, 1951 Sion, Switzerland; jan.vanherle@epfl.ch

[2] School of Mechanical Engineering, Iran University of Science and Technology, Tehran 16846-13114, Iran; hamed.shakeri72@gmail.com

* Correspondence: hossein.pourrahmani@epfl.ch

Abstract: The proton exchange membrane fuel cell (PEMFC) is a prominent environmentally friendly alternative candidate to internal combustion engines in automotive applications. The recovery of the waste heat of light-duty diesel engines has been investigated recently, which is similarly relevant for PEMFCs. Thermoelectric generators (TEG) applied on the stack's walls have been already proposed and tested as a cooling method for small scale applications of the PEMFC. For the medium scale usages of the PEMFC stack, TEG technology may be further used to recover heat lost through the cooling water required for stack thermal management, which was the focus of the present study. Using an agglomerate model for the PEMFC and a computational fluid dynamic (CFD) thermal model for the TEG heat exchanger unit, the operation and performance of the PEMFC stack and heat recovery unit were simulated, respectively. After validation, results indicated that the transferred heat from the PEMFC to the cooling channel increased the temperature of the coolant from room temperature to 330.5 K at the current density of 0.8 A/cm^2. CFD analysis revealed that 37.7 W of the heated wasted by the PEMFC stack could be recovered by the currently available TEG material and geometry.

Keywords: proton exchange membrane (PEM); agglomerate model; waste heat recovery; thermoelectric generator (TEG); computational fluid dynamic (CFD)

Citation: Pourrahmani, H.; Shakeri, H.; Van herle, J. Thermoelectric Generator as the Waste Heat Recovery Unit of Proton Exchange Membrane Fuel Cell: A Numerical Study. *Energies* **2022**, *15*, 3018. https://doi.org/10.3390/en15093018

Academic Editors: Bahman Shabani and Mahesh Suryawanshi

Received: 15 March 2022
Accepted: 18 April 2022
Published: 20 April 2022

1. Introduction

Concentrated efforts have been made to reduce the usage of fossil fuels after ascertaining their adverse contributions to climate change and pollution [1]. The automotive industry still largely relies on internal combustion engines fueled by diesel [2]. However, fuel cells and batteries are becoming viable alternatives to diminish and also control the localization of pollutant emissions [3]. Herein, proton exchange membrane fuel cells (PEMFCs) are a prominent technology for the automotive industry that electrochemically converts hydrogen and oxygen into water, electricity, and heat [4].

In different industries, there exist considerable amounts of heat that can be recovered [5]. Figure 1 illustrates the share of waste heat energy in United Kingdom. Approximately half of the hydrogen utilized by the low-temperature PEMFC changes to electricity, while the rest mainly converts to heat, which should be removed to keep the operational temperature range around 65 °C to 85 °C [6]. This temperature range stems from constraints related to the electrolyte membrane, usually made of Nafion [7]. Operation at temperatures above or below leads to drying of the membrane or flooding, respectively [8]. The heat produced by PEMFCs can be removed through natural convection, water evaporation, or extra reactants, but mainly through the cooling system [9]. Thermal and water management, that is, the combined control of the spatial distribution of temperature and humidity in the stack, is required to guarantee high performance and durability for PEMFCs [10]. In this regard, a precise and accurate model is needed to characterize the performance of the PEMFC.

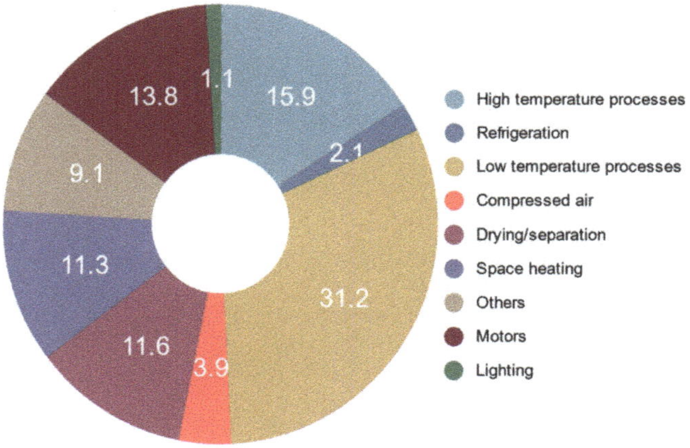

Figure 1. The percentage share of waste heat energy by different industries in United Kingdom [11].

Both experimental and numerical models are used to evaluate the output power density and voltage of the cell, hence the corresponding amount of waste heat [12]. The electrochemical performance of the catalyst layer (CL) can be simulated with varying levels of detail and accuracy in modeled phenomena and discretization, from zero to three-dimensional [13]. The corresponding physics encompasses electrochemistry coupled with multiphase liquid, gas, and solid phase transport in the reticulated porous and Pt-loaded carbon networks [14]. Among the proposed modeling approaches for the CL [15], the so-called agglomerate model proved to be precise, at the cost of higher complexity compared to the porous electrode, interface, and microscopic models.

The agglomerate model provides an improved representation of the CL by averaged microstructural parameters, in particular the electrocatalyst and carbon agglomerates covered by a thin layer of membrane. The reaction rate is further assumed to be homogeneous. Porous electrode models also consider the presence of porous agglomerates of ionomer-covered carbon particles in the CL. They, however, represent the system at the scale of the catalyst layer, while the agglomerate model accounts explicitly for the diffusion of products and reactants throughout the electrolyte film, which surrounds the electrocatalyst and carbon supports [16]. Simpler microscopic and single-pore models consider cylindrical Teflon-coated pores for gas diffusion, whereas interface models treat the CL as interface conditions between the gas diffusion layer (GDL) and membrane. In addition to the selection of the right model for the PEMFC, the most suitable strategy should be selected to recover the wasted heat and improve the thermal management [17–19].

Based on the size of the PEMFC stack, different cooling methods, such as air, passive, liquid, and phase-change cooling, can be used. The studies by Kwan [20–22] and Shen [23,24] explored thermoelectric generators (TEGs) implemented on the walls of the bipolar plates (BPs) to recover and convert the waste heat directly into electricity in small-scale applications. The performance is currently very modest: in the design developed by Kwan et al. [21], each kg of TEG material is expected to produce 0.15 W of electricity under optimized conditions.

TEGs generate electricity when subjected to a temperature gradient, and they are used in different applications such as temperature sensors, wearable energy harvesters, or for waste heat recovery [25]. Based on the trend described by He et al. [26] on the increase in the figure of merit for different materials, Shen et al. [24] evaluated the corresponding efficiency of different TEG units with different materials, i.e., considering different figures of merit, by changes in hot-side temperature. Figure 2 shows a similar trend for efficiency based on the results of Shen et al. [24] and Jouhara et al. [27] in 2016 concerning changes in the hot-side temperature and materials of TEG units in the constant cold-side temperature of 300 K. It should be noted that the data mentioned in Figure 2 were the highest possible values that could be obtained with recent TEG materials in 2016. For example, in a similar temperature range, the efficiency of Bi_2Te_3 varied between 2% and 9% in 2010, based on the values reported by Shen et al. [24].

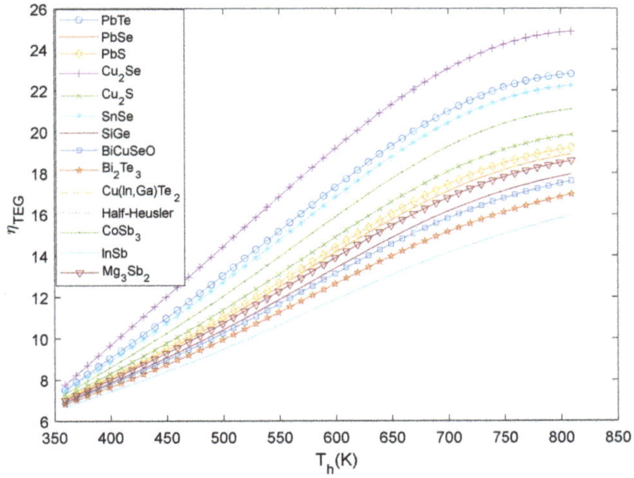

Figure 2. Changes in the efficiency of TEG materials considering the variation in TEG material and hot-side temperature in the constant cold-side temperature of 300 K.

Fernández-Yañez et al. [28] and Zorbas et al. [29] considered conventional TEG modules to recover the waste heat from the output exhaust gas of light-duty diesel engines. Bi_2Te_3 modules of 2.5 cm × 2.5 cm in size were installed at different positions on the exhaust pipe of a 96 kW engine car (1995 cm^3). The power and recovery efficiency for the TEG unit in the range of 300 W and 5.5% were achieved, respectively, with a cold-side temperature of 30 °C and at part-load engine conditions. Projections by Zorbas et al. [29] about potential progress in TEG technology suggest that a reduction in fuel consumption by approximately 20% may be reached in the future, compared to 5% with current technology [29]. From an economic standpoint, the cost of the technology considered by Zorbas et al. [29], with an estimated 5% reduction in fuel consumption using TEG technology, was in the range of 500 EUR for a 96 kW engine implementation with amortization expected within 2 to 3 years. Although there have been suggestions for using the thermoelectric generators for waste heat recovery at small scales, TEGs have not been used on a medium-scale level accounting for liquid cooling methods.

The main novelty of the current study was to investigate the performance of TEG modules for the recovery of waste heat from the PEMFC coolant channels using the liquid cooling method with medium-scale stacks of PEMFCs (see Figure 3). The expected recovery performance was assessed assuming standard TEG material performance in terms of temperature levels. An agglomerate PEMFC model and three-dimensional thermal modeling of the TEG-based recovery module were then used to simulate the performance of the current standard material technology and component geometries. The PEMFC model was validated at low and high current densities using the I-V characteristic curve. Furthermore, a 3-D CFD thermal model of the TEG unit was developed, and pressure drops compared with numerical data by Fernández-Yañez et al. [28]. Empirical relationships were used for calculations by post-processing of TEG performance corresponding to the thermal conditions. The waste heat from the PEMFC was computed by a modeling approach with agglomerate formalism for the CL. The recovery performance and corresponding spatial distribution of temperature along the TEG heat exchangers were simulated and discussed.

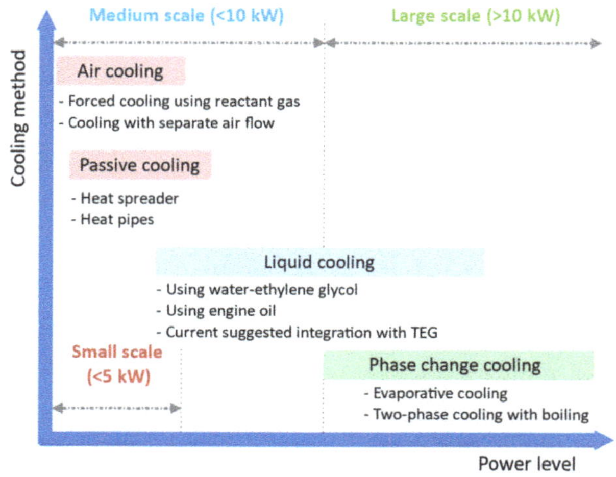

Figure 3. Different cooling methods based on the size of the PEMFC stack.

2. Problem Description and Governing Equations

The functional layers in a PEMFC are the membrane, the gas diffusion backing layers (either a single GDL or an assembly of GDL and microporous layers), bipolar plates (BPs), and CL. Studies have further shown the beneficial effect of inserting a microporous layer (MPL) between the GDL and CL to improve the effective liquid and gas-phase transport properties of the multilayered system for better water management characteristics [30]. In the absence of contaminants, hydrogen diffuses through the PEMFC anode side GDL and MPL until the CL to produce protons and electrons by electrochemical reaction (Figure 4). The membrane selectively transports protons to the cathode, where electrochemical reaction with oxygen produces water.

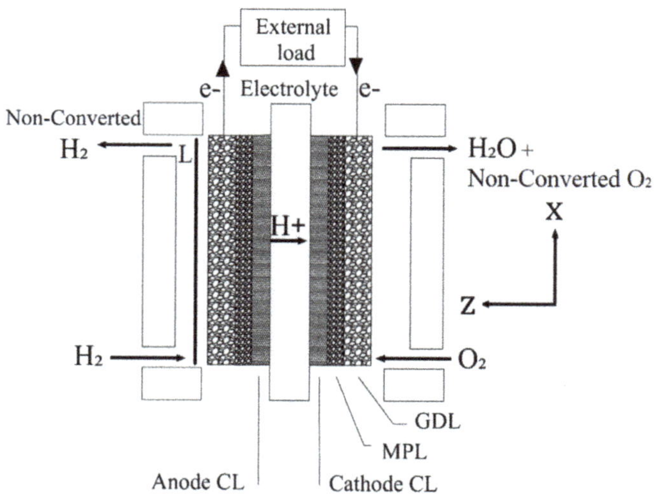

Figure 4. Schematic of PEMFC working principles and components.

Nafion is the common membrane material, which has a thickness ranging from 0.01 to 0.1 mm. Carbon-supported catalyst and ionomer porous composites are used for the CL with thicknesses spanning from 100 nm to 0.05 mm. GDL and MPL are porous carbon structures coated by polytetrafluoroethylene (PTFE), with the porosity of the MPL usually being lower to improve water management. In the present study, only variations and fluxes along the x direction were considered, as the corresponding gradients in composition and potentials were expected to be the largest. This assumption was considered acceptable to predict the output electrical power of the system with sufficient accuracy for the objective of the present study.

The PEMFC model was one-dimensional along the z-direction. It assumed incompressible flows, ideal gases, a laminar regime and single-gas phase transport in the voids. Additionally, the concentration of protons in the membrane was considered constant, that is, the potential gradient was the driving force for the flux of protons. The input flow was assumed to be pure hydrogen, that is, potential contamination of the cell due to trace elements in practice was neglected.

The geometry of the heat recovery module modeled in the current study corresponds to that analyzed by Fernández-Yañez [28]. The present study was indeed a feasibility investigation using a configuration for which data for validation were available. The module consisted of a TEG unit implemented in a water/water co-flow heat exchanger: the flow of cooling water exiting the PEMFC was the input for the water exchanger in the TEG unit (straight internal channels), and the cooling water from the environment to TEG (serpentine channel) acted as the cold source. Figure 5 shows the location of the suggested TEG unit to be integrated in the cooling channel of the PEMFC as a cooling method on medium-scale stacks given by Figure 3.

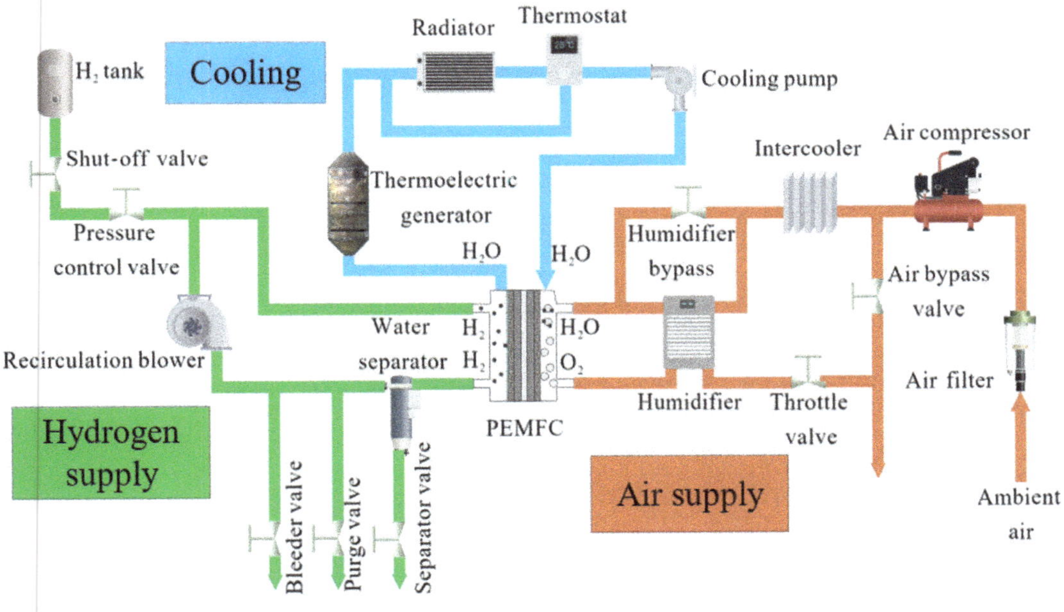

Figure 5. The suggested location of the TEG unit in the PEMFC's balance of plant.

TEGs were mounted on steel sheets below the cooling exchanger and above the water exchanger. The temperature difference between the hot and cold sides of TEG was the driving force for the generation of electricity. Fins were implemented in the water compartment for better heat transfer between the water and the steel layer. The locations and the geometry of these fins were adapted from Fernández-Yañez [28]. Figure 6 shows the schematic of the TEG unit.

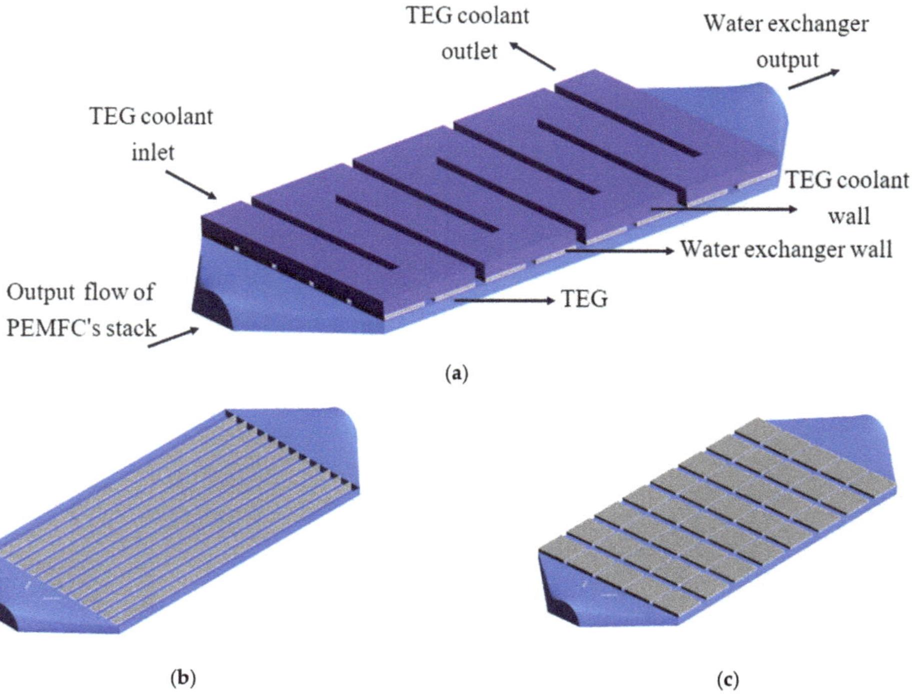

Figure 6. Schematic of the TEG unit for waste heat recovery: (**a**) assembled unit; (**b**) water channels on the hot side of the TEG; (**c**) arrangement of the 40 TEG modules in the assembly.

2.1. Membrane

The modeling of water and proton transport follows standard governing equations (see Equation (1)):

$$N_k = J_k + c_k u^{mix} \tag{1}$$

where u^{mix} ($m.s^{-1}$) is the mixture velocity in the z-direction in Figure 4, k refers to either water or protons, N_k ($mol.m^{-2}.s^{-1}$) is the molar flux due to electro-osmotic driving forces and convection, J_k ($mol.m^{-}.s^{-1}$) is the molar diffusion flux, and c_k ($mol.m^{-3}$) is the molar concentration. The molar flux of water accounts for back diffusion and electroosmotic drag:

$$J_{H_2O} = -D_{H_2O}\frac{\partial c_{H_2O}}{\partial x} + n_{drag}\frac{i_{protonic}}{F} \tag{2}$$

where $i_{protonic}$ ($A.m^{-2}$) is the protonic current density in the z-direction (Figure 4), n_{drag} is the dimensionless drag coefficient, c_{H_2O} ($mol.m^{-3}$) is the molar concentration of water, J_{H_2O} ($mol.m^{-2}.s^{-1}$) is the molar diffusion flux of water, F ($C.mol^{-1}$) is Faraday's constant, and D_{H_2O} ($m^2.s^{-1}$) is the diffusion coefficient. The electrolyte's resistance to transport is dependent upon the dimensionless water content in the membrane λ_m, and a semi-empirical relationship is used for the effective diffusion coefficient [31]:

$$n_{drag} = 2.5\frac{\lambda_m}{22} \tag{3}$$

$$\lambda_m = \frac{c_{H_2O}}{\frac{\rho^m_{dry}}{M^m} - bc_{H_2O}} \tag{4}$$

$$D_{H_2O} = D'\left[\exp\left(2416\cdot\left(\frac{1}{303} - \frac{1}{T}\right)\right)\right]\cdot\frac{1}{a(17.81 - 78.9a + 108a^2)}\cdot\lambda_m \quad (5)$$

where b is the dimensionless membrane extension coefficient in the z-direction (based on experiments, b = 0.0126 [31]), a is a dimensionless parameter called water activity, T (K) is temperature, M^m ($kg.mol^{-1}$) is the membrane molecular mass, ρ^m_{dry} ($kg.m^{-3}$) is the dry membrane density, and D' ($m^2.s^{-1}$) is the diffusion coefficient at the reference temperature. The total molar flux of water is consequently:

$$N_{H_2O} = J_{H_2O} + c_{H_2O}u^{mix} \quad (6)$$

Proton transport is computed according to Equations (7) and (8):

$$J_{H^+} = -\frac{F}{R_uT}D_{H^+}c_{H^+}\frac{\partial\phi_m}{\partial x} \quad (7)$$

$$N_{H^+} = J_{H^+} + c_{H^+}u^{mix} \quad (8)$$

where ϕ_m (V) is the electrostatic potential within the membrane, provided by Equation (9) with membrane conductivity Equation (10), c_{H^+} ($mol.m^{-3}$) is the molar concentration of protons, J_{H^+} ($mol.m^{-2}.s^{-1}$) is the molar diffusion flux of protons, R_u ($J.mol^{-1}.K^{-1}$) is the universal gas constant, and D_{H^+} is the proton diffusivity, assumed constant (4.5×10^{-9} $m^2.s^{-1}$)

$$\frac{\partial\phi_m}{\partial x} = -\frac{i_{protonic}}{\sigma_m} + \frac{F}{\sigma_m}c_{H^+}u^{mix} \quad (9)$$

$$\sigma_m = (0.005139\lambda_m - 0.00326)\exp\left[1268\left(\frac{1}{T_{ref}} - \frac{1}{T}\right)\right] \quad (10)$$

Here, σ_m is the dimensionless conductivity of the membrane.

The applied solution procedure and boundary conditions to solve the system of equations in the membrane follows the detailed description in Spiegel [31].

2.2. Catalyst Layer

The agglomerate CL model assumes the uniform distribution of the electrochemical reaction rate and aims at capturing the effect of the main microstructural feature of the CL, that is, an agglomerate type structure where the electrocatalyst is supported by a carbon agglomerate (see Figure 7).

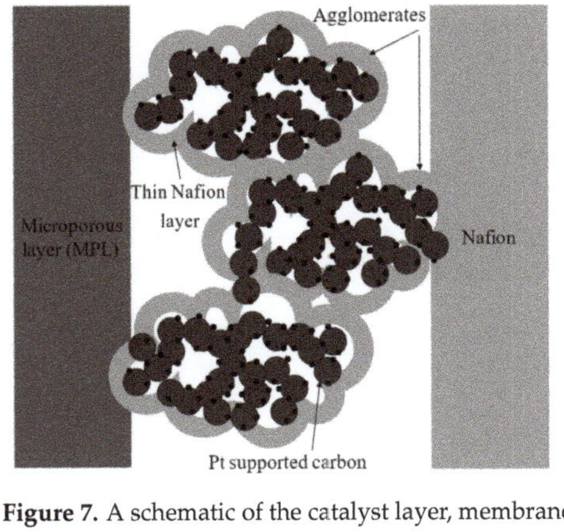

Figure 7. A schematic of the catalyst layer, membrane, and microporous layer indicating the agglomerates inside the catalyst layer.

The volumetric current density i_c, $(A.m^{-3})$ is calculated using Equation (11), considering uniform distribution of identical agglomerates in the cathode CL [32]:

$$i_c = 4F\frac{P_{O_2}}{H_{O_2,agg}}\left(\frac{1}{Ek'(1-\varepsilon_{CL})} + \frac{r_{agg}+\delta_i+\delta_w}{r_{agg}}\left(\frac{\delta_i}{a_{agg,i}D_{O_2,i}} + \frac{\delta_w}{a_{agg,w}D_{O_2,w}}\right)\right)^{-1} \quad (11)$$

where, P_{O_2} (Pa) is the partial pressure of oxygen, $H_{O_2,agg}$ $(Pa.m^3.mol^{-1})$ is Henry's constant for oxygen in the agglomerate, k' $(1.s^{-1})$ is the reaction rate parameter, $a_{agg,w}$ $(1.m^{-1})$ is the liquid water effective agglomerate surface area, and $a_{agg,\ i}$ $(1.m^{-1})$ is the ionomer effective agglomerate surface area. $D_{O_2,i}$ $(m^2.s^{-1})$ and $D_{O_2,w}$ $(m^2.s^{-1})$ are the oxygen diffusivity in the ionomer and liquid water, respectively, while δ_i (m) and δ_w (m) are the corresponding values for the thickness of ionomer film and the thickness of the liquid water film. r_{agg} (m) is the agglomerate radius, while ε_{CL} is the volume fraction of void in the catalyst layer:

$$\varepsilon_{CL} = 1 - L_{\frac{Pt}{C}} - L_i \quad (12)$$

Here, L_i and $L_{Pt/C}$ are the corresponding values of volume fractions of the ionomer phase and the Pt/C particles, as follows [32]:

$$L_{\frac{Pt}{C}} = \frac{m_{pt}}{t_{CL}}\left(\frac{1}{\rho_{pt}} + \frac{1-f}{f}\frac{1}{\rho_c}\right) \quad (13)$$

$$L_i = \frac{L_{\frac{Pt}{C}}}{r_{agg}^3(1-L_{i,agg})}\left[r_{agg}^3 L_{i,agg} + \left(\left(r_{agg+\delta_i}\right)^3 - r_{agg}^3\right)\right] \quad (14)$$

where t_{CL} (m) is the thickness of the CL, ρ_{pt} $(kg.m^{-3})$ is the density of platinum, and ρ_c $(kg.m^{-3})$ is that of carbon. m_{pt} $(mg.cm^{-2})$ is the platinum loading, while $L_{i,agg}$ is the volume fraction of the ionomer phase inside the agglomerate. Equation (15) also indicates the platinum loading divided by the sum of platinum loading and carbon loading (f):

$$f = \frac{m_{pt}}{m_{pt} + m_c} \quad (15)$$

Afterwards, the thickness of the ionomer film (δ_i) and the thickness of the liquid water film (δ_w) can be calculated [32]:

$$\delta_i = r_{agg}\left[\sqrt[3]{\frac{L_i(1-L_{i,agg})}{L_{\frac{Pt}{C}}} - L_{i,agg} + 1} - 1\right] \quad (16)$$

$$\delta_w = \sqrt[3]{\left(r_{agg}+\delta_i\right)^3 + \frac{3s\varepsilon}{4\pi N_v}} - \left(r_{agg}+\delta_i\right) \quad (17)$$

Here, s is the liquid water saturation, while N_v $(1.m^{-3})$ is the number of agglomerate particles per CL volume [32]:

$$N_v = \frac{3L_{\frac{Pt}{C}}}{4\pi r_{agg}^3(1-L_{i,agg})} \quad (18)$$

Equivalent governing equations are used for the cathode and anode, but only the former case is detailed hereafter. Assuming a first-order Tafel kinetic reaction yields, by solving the mass conservation equation in a spherical agglomerate, a relation for the effectiveness factor (Equation (19)) [33]. The effectiveness factor is the ratio of the average reaction rate in the agglomerate core to the reaction at the surface of the core.

$$E = \frac{1}{3\varphi^2}(3\varphi\coth(3\varphi) - 1) \tag{19}$$

Herein, φ is the Thiele modulus:

$$\varphi = \frac{r_{agg}}{3}\sqrt{\frac{k'}{L_{i,agg}^{1.5} D_{O_2}^{eff}}} \tag{20}$$

$D_{O_2}^{eff}$ ($m^2.s^{-1}$) is the effective oxygen diffusion coefficient in the catalyst layer [31].

$$k' = \frac{i_{0,ORR} a_{agg}}{4F(1 - \varepsilon_{CL})C_{O_2}^{ref}} \exp\left(-\frac{\alpha_c F}{R_u T}(\eta_{ORR})\right) \tag{21}$$

Here, $i_{0,ORR}$ ($A.m^{-2}$) is the reference exchange current density at cathode, $c_{O_2}^{ref}$ ($mol.m^{-3}$) is the reference molar concentration of oxygen, η_{ORR} (V) is the cathode overpotential, α_c is the symmetry cathodic charge transfer coefficient, and a_{agg} ($1.m^{-1}$) is the effective agglomerate surface area, as follows:

$$a_{agg} = \frac{3L_{\frac{Pt}{C}}\varepsilon_{CL}}{r_{agg}^3(1 - L_{i,agg})}(r_{agg} + \delta_i)^2 \tag{22}$$

The surface concentration can be assumed to be equal to the bulk concentration based on Fick's law, considering the low solubility of oxygen, steady-state conditions, and thinness of the film. The details of the derivation for this equation are available in refs. [31,33] for the two limiting cases $\varphi >> 1$ and $\varphi << 1$. The former indicates slow O_2 diffusion in the agglomerate compared to the ORR rate yielding $E \cong \frac{1}{\varphi}$, while the latter opposite situation results in $E \cong 1$.

Equation (11) holds for the cathode if the external mass transfer limitations can be neglected. These mass transfer limitations occur when the diffusion rate of the molecules is lower than the reaction rate.

The boundary conditions at the inlet of the gas flow channel are the temperature (353.15 K), mass flow rate, and the species mass fraction. At a current density of 1.0 $A.cm^{-2}$ (here referred to as the reference current density I_{ref}), Equations (23) and (24) provide the mass flow ($kg.s^{-1}$) of reactants at anode and cathode, respectively:

$$Q_a = \frac{\zeta_a M_{H_2}}{2FY_{H_2}} I_{ref} A_m \tag{23}$$

$$Q_c = \frac{\zeta_c M_{O_2}}{2FY_{O_2}} I_{ref} A_m \tag{24}$$

where $\zeta_a = 1.5$ is the stoichiometric ratio at the anode, while that of cathode is $\zeta_c = 2.0$. M_{H_2} (kg/mol) and M_{O_2} ($kg.mol^{-1}$) are the molecular weights of hydrogen and oxygen, respectively, while Y_k indicates the mass fraction of component k, and A_m (m^2) is the representative of the membrane's area. The inlet liquid water saturation is assumed to be zero, like the electrostatic potential at the anode terminal. The electrostatic potential at the cathode terminal is also equal to the cell's voltage (V_{cell}) at the operating temperature of 353.15 K. The overpotential terms to calculate the potential of the cell V_{cell} under polarization, which is based upon the Nernst equation (see Equation (25)), are computed from the above governing equations, as discussed in detail by Siegel et al. [34]:

$$V_{cell} = V_0 - V_{act} - V_{ohmic} - V_{conc} \tag{25}$$

Here, V_0 (V) is the Nernst potential, while V_{act} (V), V_{ohmic} (V), and V_{conc} (V) are the activation, ohmic, and concentration overpotentials, respectively.

The generated present study considers an upper bound limiting case, that is, all the output thermal power of the PEMFC system (Equation (26)) is assumed transferred to the coolant loop as an increase in fluid enthalpy, since the flux is constant.

$$P_{thermal} = -(V_{act} + V_{ohm} + V_{conc}).N_{cell}.i \quad (26)$$

2.3. Thermoelectric Generator (TEG)

ANSYS CFX module 19.2 has been used to solve the three-dimensional continuity, momentum, and energy equations for viscous, laminar, and incompressible flow of water in the hot- and cold-side heat exchangers, as well as heat transport in the solid domains (Figure 6). The CFD model, therefore, corresponds to a water–water heat exchanger. Figure 8 shows the details of the simulation procedure to analyze the performance of the TEG unit.

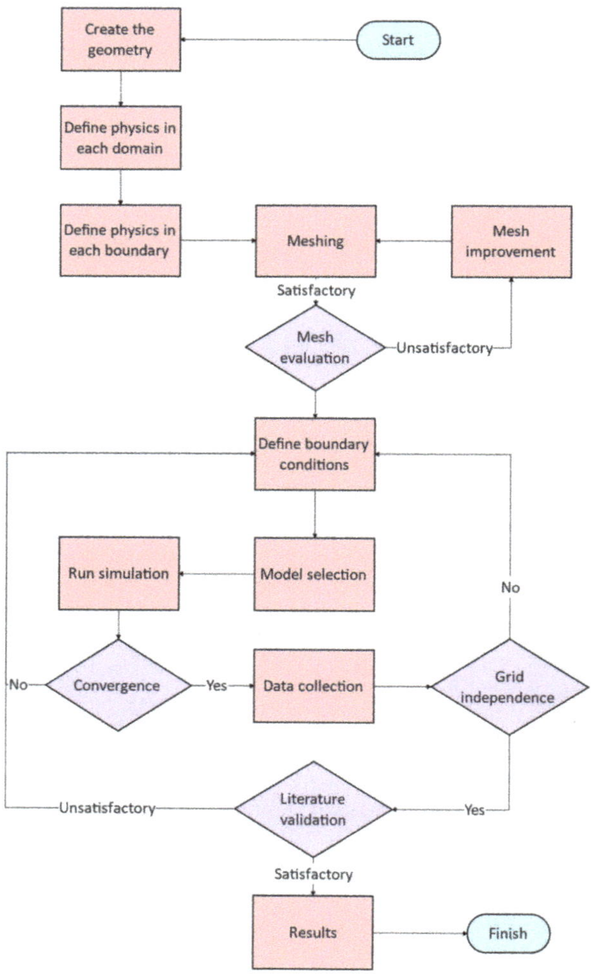

Figure 8. The flowchart of simulation procedure to perform the CFD analysis of the TEG unit.

Empirical relationships were derived to calculate during post-processing the TEG electrical performance under the thermal conditions simulated by the CFD exchanger model. In this regard, the possible additional resistance of the TEG unit, which may result in more energy consumption [35–37], was considered. The CFD simulations and TEG calculations were performed in an uncoupled manner. The local relationship between temperature gradient, potential, and current was computed using the freely available module TEG1-4199-5.3 [38], which shows close to linear current–voltage characteristics. For instance, the tabulated output voltage for Bi_2Te_3 was 6.7 V at a current, power, hot-side temperature, and cold-side temperature of 1.12 A, 7.5 W, 300 °C, and 30 °C, respectively. The corresponding open-circuit voltage of the module was 13.4 V under a heat flow flux of 9.5 $W.cm^{-2}$. It should be noted that the specific heat capacity of this TEG module was 156.1 ($J.kg^{-1}.K^{-1}$), while the other materials utilized in the TEG unit were aluminum and steel with specific heat capacities of 434 ($J.kg^{-1}.K^{-1}$) and 903 ($J.kg^{-1}.K^{-1}$), respectively. As the hot-side temperature was different in this study, a correlation was extrapolated based on the data in ref. [38]. Equation (27) is the fitted empirical expression used to calculate the TEG open voltage V_{TEG} (V) in the system, assuming a constant cold-side solid temperature of 30 °C.

$$V_{TEG} = -0.00004 \times (T_h - 273.15)^2 + 0.0638 \times (T_h - 273.15) - 1.9012 \quad (27)$$

Here, T_h (K) is the hot-side solid temperature. Similarly, the TEG internal resistance R_{TEG} (Ω) is approximated by the following empirical relationship:

$$R_{TEG} = -0.00001087 \times (T_h - 273.15)^2 + 0.008726 \times (T_h - 273.15) + 3.922 \quad (28)$$

The TEG output electrical power P_{TEG} (W) is then:

$$P_{TEG} = \frac{\left(-0.00004 \times (T_h - 273.15)^2 + 0.0638 \times (T_h - 273.15) - 1.9012\right)^2}{-0.00001087 \times (T_h - 273.15)^2 + 0.008726 \times (T_h - 273.15) + 3.922} \quad (29)$$

For rapid screening analyses, the efficiency of a TEG module (η_{TEG}) can be expressed as [39]:

$$\eta_{TEG} = \frac{T_h - T_c}{T_h} \cdot \frac{\sqrt{1 + ZT_m} - 1}{\sqrt{1 + ZT_m} + \frac{T_c}{T_h}} \quad (30)$$

The first right-hand side term corresponds to the Carnot efficiency, and T_c is the cold source temperature. ZT_m is a dimensionless parameter that characterizes the TEG material performance at an average temperature T_m (K):

$$T_m = \frac{T_h + T_c}{2} \quad (31)$$

$$Z = \frac{\sigma \alpha^2}{\kappa} \quad (32)$$

where α ($V.K^{-1}$) is the Seebeck coefficient, σ ($1.m^{-1}.\Omega^{-1}$) is the electrical conductivity, and κ ($W.m^{-1}.K^{-1}$) is the thermal conductivity of the TEG material.

3. Results and Discussion

Table 1 lists the input parameters to compute the output thermal power and mass flow rates of the current PEMFC model. The latter are then used as input data for TEG module simulations, to investigate the recovery performance along with the flow behavior and spatial temperature distribution in the TEG unit.

Table 1. Input parameters for PEMFC numerical modeling [40,41].

Parameter	Value
H_2 permeation in agglomerate (k_{H_2})	2×10^{-11} m^2
O_2 permeation in agglomerate (k_{O_2})	1.5×10^{-11} m^2
Channel length L	100 mm
Footprint area of the cell (A_{cell})	100 cm^2
Limiting current density (i_l)	1.4 A.cm^{-2}
Molar fraction of H_2 in the anode (M_{H_2})	0.5
Molar fraction of O_2 at the inlet ($M_{O_2}^{inlet}$)	0.21
Molar fraction of O_2 at the outlet ($M_{O_2}^{outlet}$)	0.095
Number of cells (N_{cell})	90
Operating pressure (anode and cathode side) (P)	1 atm
Operating temperature (T)	353.15 K
Reference temperature (T_{ref})	298.15 K
Anodic/cathodic symmetry charge transfer coefficient ($\alpha_{a,c}$)	0.5

3.1. Model Validation

The aim of the current developed PEM fuel cell model was to predict the output temperature and the mass flow rate of the water in the cooling channels, which enters the TEG heat exchanger. In this regard, to validate the output results of the PEMFC model, the I-V characteristic curves of the model should be compared with the verified results in the literature at both low and high current densities. Once the validation is performed, the model is suitable to predict the needed output values. The PEMFC model predictions were tested with both experimental and simulation results by Ticianelli et al. [42] and Mohammedi et al. [43], respectively. It is noteworthy to mention that the input model parameter values for the present verification were not those listed in Table 1, but correspond to the values reported in Mohammedi et al. [43] to allow for direct comparison. Figure 9a shows good agreement between both models and the experimental results up to at least 600 mA/cm^2, for conditions corresponding to a fuel utilization of 0.75, a mass flow rate of 6×10^{-7} kg s$^{-1,}$ and surface specific flow of 6.67×10^{-7} kg.s^{-1}.m^{-2} with a molar fraction of 0.5/0.5 H_2/H_2O at the anode side while having the mass flow rate of 5×10^{-6} kg.s^{-1}.w and surface specific flow of 5.56×10^{-7} kg.s^{-1}.m^{-2} with a species molar fraction of 0.1785/0.15/0.6715 $O_2/H_2O/N_2$ at the cathode side. Figure 9a presents the validation results compared to the experiments by Ticianelli et al. [42] and simulations by Mohammedi et al. [43].

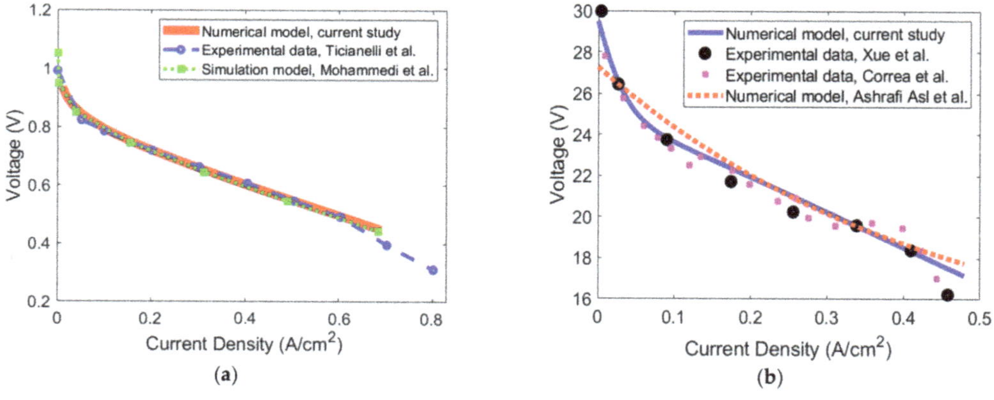

Figure 9. Validation results of the developed PEMFC model at low current densities: (**a**) model comparison with the experimental data of Ticianelli et al. [42] and simulations of Mohammedi et al. [43]; (**b**) model comparison with the 500 W BCS stack using the experimental data of Xue et al. [44], Correa et al. [45], and simulation data of Sharifi-Asl et al. [46].

Furthermore, the current model was also validated with three different sets of experimental and numerical data at low current densities. Figure 9b compares the output of the current PEM fuel cell model with the 500 W BCS stack manufactured by BCS technologies [44]. The number of cells for the 500 W BCS stack [44] was 32, with an active area of 64 cm^2 operating at 333 K. It should be noted that in this stack, the resistance to the electron flow was 0.0003 Ω, while the limiting current density was 0.469 $A.cm^{-2}$. Accounting for the high current densities, Figure 10 illustrates the I-V characteristic curve of the current model with the output performance of the single-cell Ballard Mark V PEM fuel cell [44] at the operating temperature of 343 K. In this case, the limiting current density was 1.5 $A.cm^{-2}$, the active area of the cell was 50.6 cm^2, and the resistance to the electron flow was 0.0003 Ω.

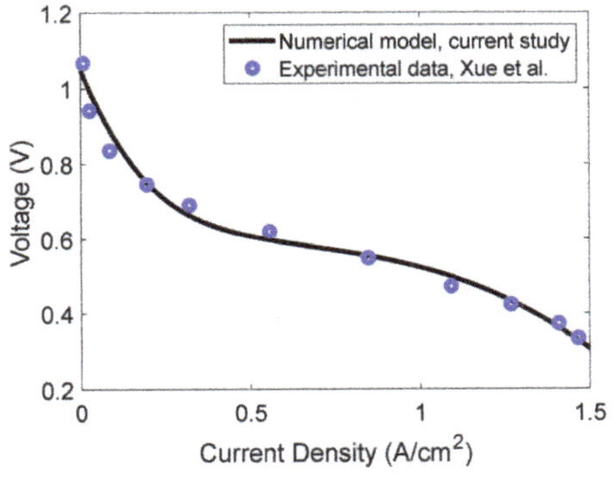

Figure 10. Validation results of the developed model at high current densities with the Ballard Mark V PEM fuel cell given by Xue et al. [44].

The heat transfer verification of the TEG model consisted of the comparison of pressure drops computed by Fernández-Yañez et al. [28] (see Table 2), which was a follow-up of the investigation in ref. [47]. The published dataset included nine stationary modes of a light-duty diesel engine at different torques. In that article, simulations were performed to obtain the corresponding temperature and pressure drop of the exhaust gases of the diesel engine. The present TEG model verification used a subset of four of the modes reported in [26] (A, I, G and D), which means four different input temperatures and mass flow rates. The agreement shown in Table 2 indicates that the present model pressure drop predictions were in line with the mixed numerical and experimental study by Fernández-Yañez et al. [28].

Table 2. Validation of the simulation data obtained by the present TEG model and that by Fernández-Yañez et al. [28].

Engine Mode	Engine Rotation Speed (min^{-1})	Torque (N.m)	Temperature (K), Present Study	Temperature (K), Fernández-Yañez et al. [28]	Pressure Drop (Pa), Present Study	Pressure Drop (Pa), Fernández-Yañez et al. [28]
A	1000	10	388	387	110	119
I	2400	120	663	691	876	890
G	1000	120	583	658	211	253
D	1000	60	475	484	156	159

In every CFD analysis, the values of the output results should be independent from the size of the cells in the mesh structure. In this regard, a grid independency analysis is needed to prove the suitability of the simulation model for the TEG unit. As the validation of this unit (see Table 2) was conducted with the given data by Fernández-Yañez et al. [28], the grid independency of the TEG thermal model was also performed using the temperature corresponding to engine mode D. Figure 11 shows that grid independency of the temperature simulated at 475 K was reached starting around 4.2×10^6 cells. In this condition, the average surface area of the cells was 1.57×10^{-3} m^2, the minimum edge length was 1×10^{-3} m, and the maximum cell edge size was 0.1132 m, with highest skewness of 0.9. Figure 12 shows the mesh structure of the developed model, only when the number of cells is lowest (1×10^6), for illustration reasons.

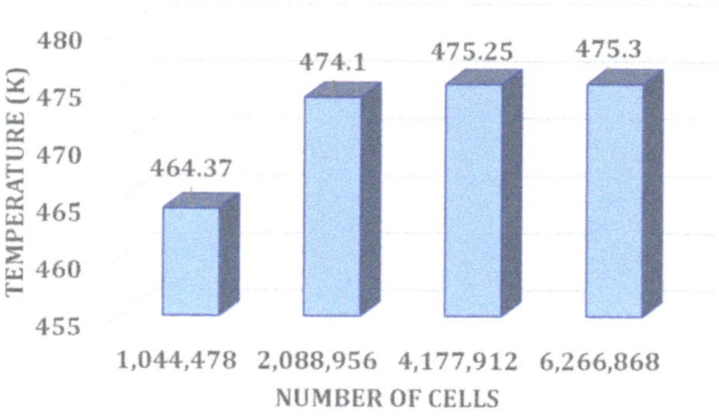

Figure 11. Grid independency study of the three-dimensional CFD model of the TEG unit considering the temperature in the engine mode D given by Table 2 as the verification value.

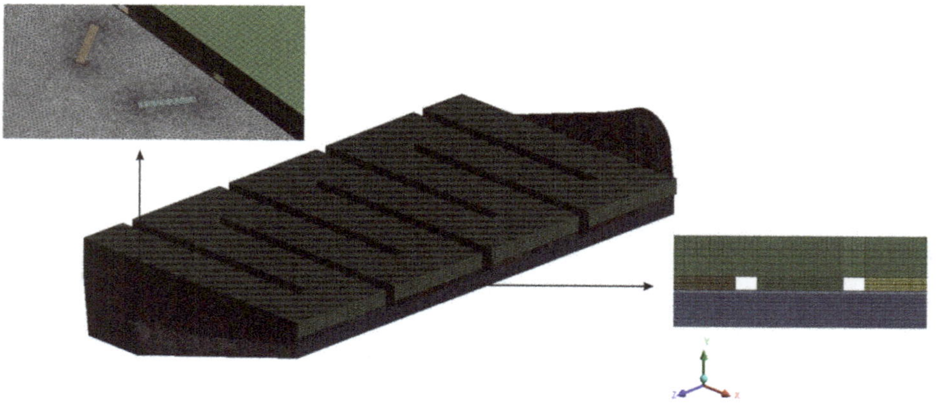

Figure 12. The mesh structure of the computational domain for the simulation of the TEG unit.

3.2. PEMFC Modeling

In this study, the mass flow rates and temperature of the exhaust water from the PEMFC's coolant channel were used as inputs for the TEG unit simulations. Figure 13a,b shows the computed characteristic curves of the PEM fuel cell with the input parameters listed in Table 1. The open circuit voltage at low current density decreased because of fuel crossover. Activation losses caused the pronounced non-linear decrease in voltage at low current density. Ohmic losses resulted in a close to linear dependence until the onset of mass transport limitations became dominant at high current densities and fuel conversion.

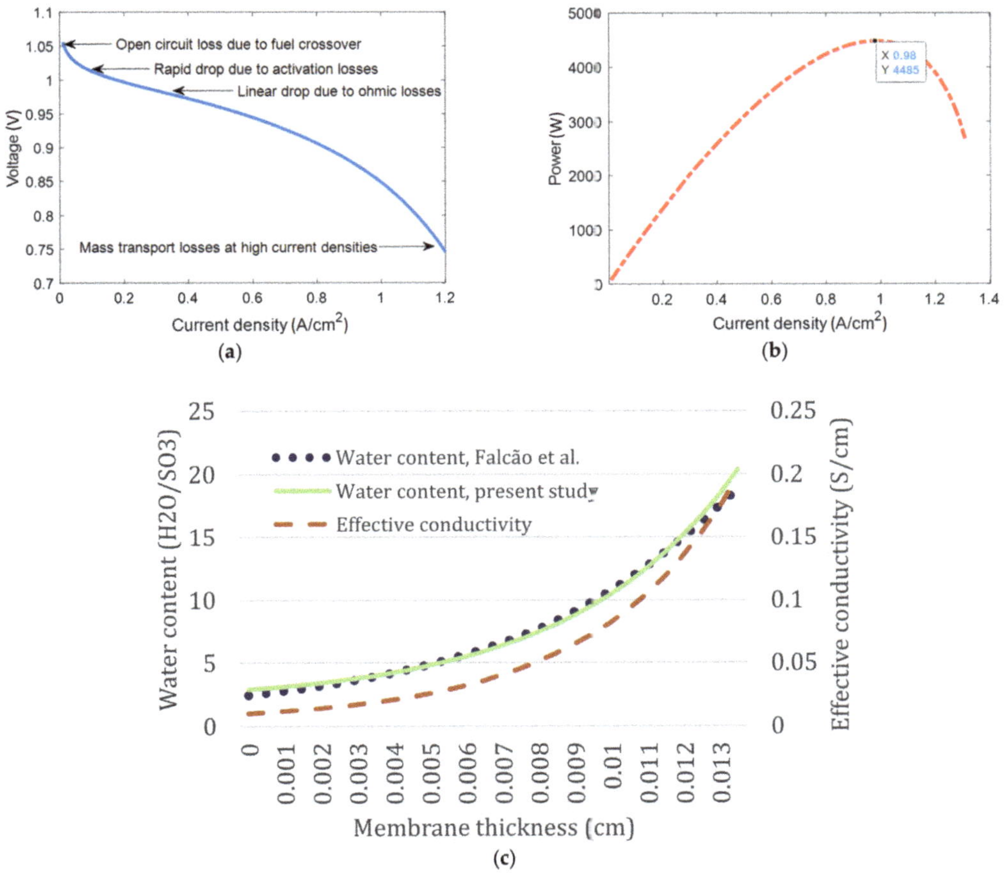

Figure 13. Performance characteristics and membrane properties simulated with the input parameters in Table 1 considering fuel utilization of 78% and 800 mA/cm^2; (**a**) I-V, (**b**) P-I, (**c**) simulated dependence of the water content and effective conductivity on the thickness of the membrane.

The effects of selected model parameters relevant to thermal and water management were further evaluated using published data, as a further verification after the I-V curve comparison in Section 3.1. The one-dimensional numerical study by Falcão et al. [48], which was compared to experimental data [49,50] and a three-dimensional model [51], evaluated differences in water content and local transport properties, depending upon the membrane thickness using a one-dimensional model. Their evaluation considering four thicknesses (0.0051 cm, 0.0127 cm, 0.0178 cm, 0.003 cm) concluded that low thicknesses result in lower ionic resistance and voltage loss, hence better cell performance. However, the fabrication of MEA with thin membranes is a challenge, and other factors such as mechanical integrity of the membrane and manufacturing cost should be considered. Figure 13c illustrates the variations in water content and effective conductivity computed for different membrane thicknesses, showing qualitative agreement.

The purpose of the present agglomerate PEMFC model was to calculate the output thermal power. It should be noted that all types of heat transport and transfer by conduction, radiation, and convection were at play in the electrodes. The sources of heat in the polarized CLs were ohmic losses induced by charge transport, entropy changes, and activation overpotentials due to electrochemical reactions. The effectiveness factor (Equation (19)) is a metric for the assessment of the cell performance. An effectiveness factor equal to one corresponds to the limiting case where all the catalyst surface area is active under polarization. It can be determined for both cathode and anode sides after calculation of the Thiele modulus (Equation (20)), which is the ratio of the reaction rate on the catalyst surface to the reaction rate without any transport losses. Figure 14 illustrates the changes in the effectiveness factor and the flux density of hydrogen upon variations in current density. Figure 14b shows that transport losses increase with increasing current, leading to an effectiveness factor approaching zero. Figure 14a indicates that hydrogen crossover increases drastically at current densities higher than 0.8 A/cm^2, which is detrimental for the cell's performance and is known to increase the degradation. A current density of 0.8 A/cm^2 was, therefore, the selected operation point for the coupling with the TEG module. The corresponding simulated electrical power and efficiency were 3.34 kW and 39%.

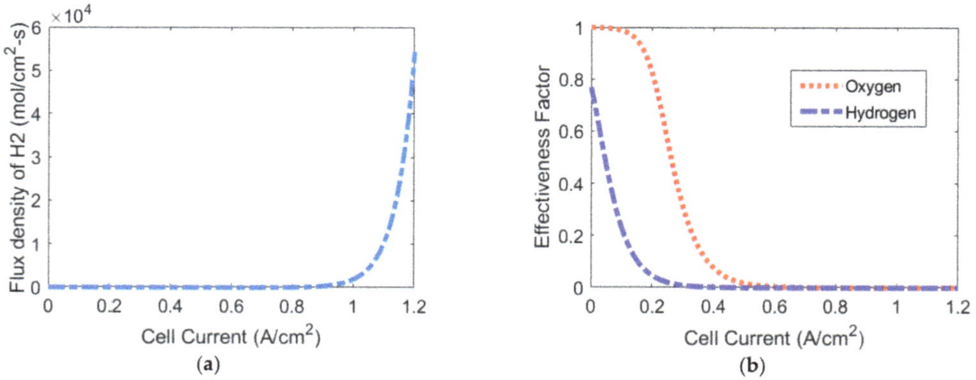

Figure 14. Simulated CL behavior as a function of current density: (**a**) surface-specific flux of H_2; (**b**) effectiveness factors on the cathode and anode sides.

3.3. TEG Simulation

As mentioned, the TEG unit will not have a direct impact on the performance of the PEMFC unit, although the integrated system of the PEMFC-TEG will have higher output power in comparison to the standalone PEMFC. The simulated output flow of water from the PEMFC's coolant channel (3 kg/s at 330.5 K, at 0.8 A/cm^2) is used as the hot-side water exchanger inlet condition for the analysis of the TEG unit for waste heat recovery. Prescribed boundary conditions are the pressure outlet of the TEG unit and a temperature of 30 °C at all exterior walls of the PEMFC and TEG coolant domains, except those in contact with the TEG modules. A uniform convection heat transfer coefficient of 15 $\frac{W}{m^2K}$ is assumed. The cold side coolant exchanger (see Figure 6) is fed with 1.37 kg/s of water at 30 °C. The reason for the mentioned flow rates of water in the TEG heat exchange module is to validate the results with the given data by Fernández-Yañez et al. [28]. Ahn and Choe [52] had already suggested the minimum mass flow rate of 0.8 kg/s for the PEM fuel cell in the coolant channel. The inlet temperature at the hot side of the water exchanger has been calculated using the agglomerate model for the PEMFC while the inlet temperature at the cold side of the water exchanger is based on the study developed by Liso et al. [53]. The high resolution Rhie-Chow [54,55] scheme with a convergence criterion of 10^{-5} is applied for the coupling between pressure and velocity. Additionally, the 40 elements forming the TEG unit are assumed to be thermally and electrically insulated from each other.

Recovering the output flow of the PEMFC by the TEG unit, Figure 15 shows the corresponding contours of temperature, voltage, and power density in the TEG unit both on the water and coolant exchanger sides. The cooling effect of cold water on the flow of water from the PEMFC entering the water exchanger of the TEG unit is as expected in Figure 15. The regions with higher temperature in Figure 15b also result from back flow pressure on the edges, leading the water to flow around a circle. In the water exchanger, the presence of a boundary layer on the sides of the heat exchanger causes lower temperature compared to the other parts of the heat exchanger. Consequently, the temperature on the side of the coolant exchanger is comparatively lower as well.

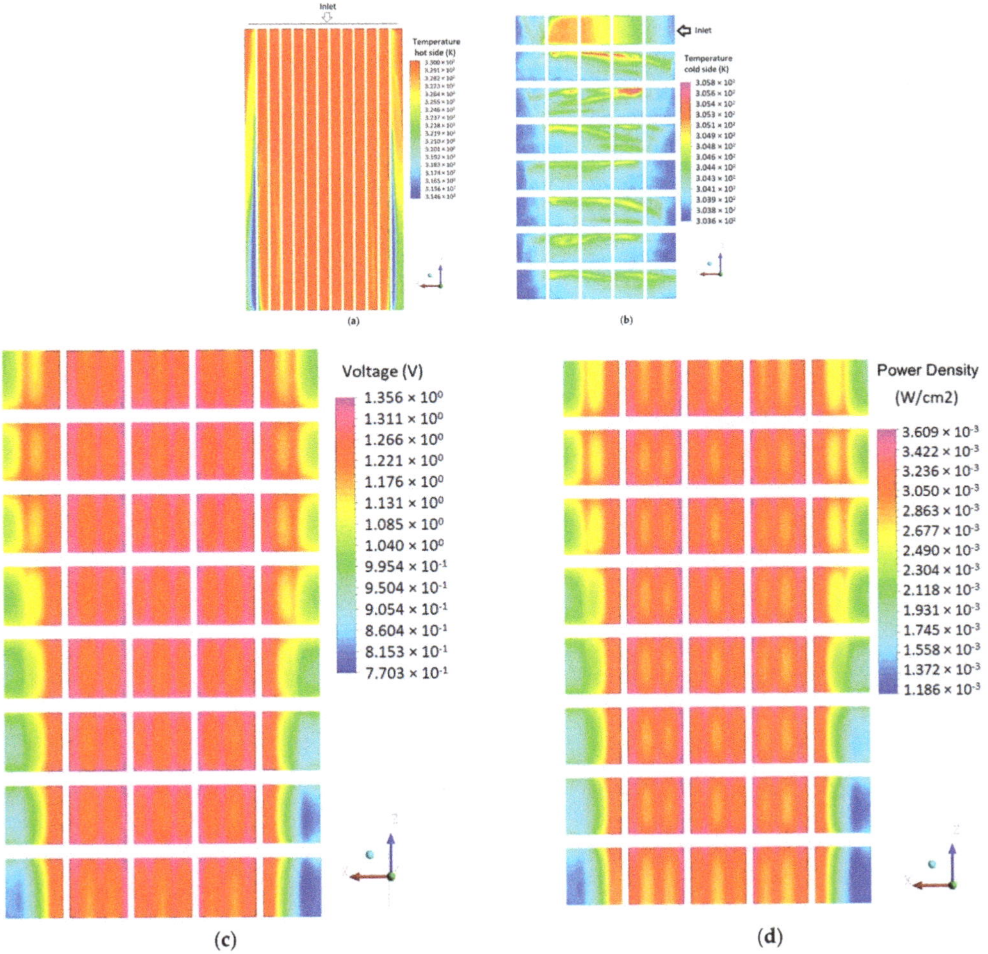

Figure 15. The contours simulated by the TEG unit model: (**a**) temperature on the hot side of the TEG that is in contact with the water flow from the PEMFC's; (**b**) temperature on the cool side of the TEG fed with cooling water at 30 °C; (**c**) voltage distribution in the TEG; (**d**) spatial distribution of electrical power in the TEG considering the surface are of 117 cm^2 for each module of the TEG unit.

The voltage of the TEG units computed by Equation (27) is dependent upon the solid hot water exchanger side temperature, leading as expected to a potential distribution such as shown in Figure 15c. The maximum voltage is 1.36 V, which is much lower than values around 8 V in ref. [28]. The reason is the higher hot source temperature in the light-duty diesel engines due to higher exhaust heat. The cold-side temperature varies by approximately 2 °C, showing that the assumption of constant cold-side temperature in Equations (27)–(29) was acceptable for the present study, even though it does not allow the analysis of the effects of flow patterns in Figure 15b on the output electrical power. Figure 15d shows the contour of produced electrical power in footprint mesh cell of the TEG unit. It indicates that the maximum local power in the module is 3.61×10^{-3} W/cm^2, compared to the total power of 37.7 W and the averaged power density of 0.00806 W/cm^2.

The corresponding value of the TEG surface-specific power density for the whole unit was also 3.41 W/cm^2.

Other parameters to be considered were the PEMFC's current density, which was set at 0.8 A/cm^2 in this study, and the PEMFC water cooling flux. Under the corresponding thermal conditions, the visual inspection of the temperature profile indicated that the TEG modules produce up to roughly twice less electrical power than the central ones. Moreover, the temperature difference between the cold and hot side is almost constant along the central lines in Figure 15a. This suggests that higher efficiency recovery may be achieved with a larger, yet more expensive heat exchanger module and/or improved heat exchanger designs. PEMFC operation at higher current density produces more excess heat. In the case where the stack temperature is kept at 353 K (80 °C) and the coolant water flow is kept constant, the temperature of the latter will increase, compared to 330 K (57 °C) in Figure 15a at the inlet. Therefore, the TEG output electrical power is expected to increase. Herein, the beneficial effect on TEG performance may well compensate for the heat exchanger limitations and lead to higher recovery efficiencies. Alternatively, the recovery efficiency in the present case of 0.8 A/cm^2 can be increased by decreasing the PEMFC cooling water flow, which was not the case in the current study, as the coolant flow of PEMFC was in a separate water loop of the PEMFC's stack, and the water mass flow rate was assumed to be constant. As expected, the TEG module and PEMFC must be designed and operated in synergy for viable recovery efficiencies over a sufficiently large window of operation conditions for automotive applications.

4. Conclusions

This study investigated the relevance of TEG technology to recover the waste heat from the PEMFC. It was inspired by the idea of producing electricity from the exhaust heat of light-duty diesel engines by TEG. Herein, an agglomerate one-dimensional PEMFC model was developed to estimate the maximum output heat from the fuel cell stack. The validation of the I-V characteristic curves of the current model with six different series of experimental and simulation data proved the suitability of the developed model. Under the considered PEMFC operating conditions and assumptions, the simulated output water temperature and flow rate from the PEMFC's coolant channel were 330.5 K and 3 kg/s, respectively. The CFD thermal simulations with cooling water at 30 °C combined with post-processing based on empirical relationships for the local TEG electrical performance resulted in a modest total output electrical power of 37.7 W. The voltage was in the range of 1.25 V and showed non-uniformity (0.77–1.36 V), suggesting potential for improvement of the TEG heat exchanger design, as well as of the PEMFC operating conditions for waste heat recovery. The figures were, therefore, modest. Although this study presented the integration of the TEG unit as a cooling method for a medium-scale size of the PEMFC, further research can be performed for future studies:

- This study only evaluated the possibility of using a TEG unit as a cooling method in a medium-scale PEMFC stack by calculating the output power and the voltage of the TEG unit. However, exergy and economic aspects of the system could be assessed through exergo-economic and thermo-economic analyses to evaluate the suitability of this integration.
- The design of the current TEG unit was inspired from the geometry given in ref. [28], which was mainly for diesel engine applications. Due to the novelty of this topic, there is a lack of optimized geometry for the application of PEMFC. In this regard, further studies on the geometry of the TEG unit, the needed number of TEG modules in the TEG unit, and the required mechanical/physical properties of the unit for maximum performance are needed.
- The current modeling was performed in the steady-state condition for both the PEMFC numerical model and the TEG simulation model. However, performing a dynamic and real-time analysis would be of value to better investigate the possibility of the current integration.

- The materials in the TEG modules and their figures of merit play a crucial role in the performance of the TEG unit and the waste heat recovery. This study assumed Bi_2Te_3 as the base material of the TEG modules, while many other alternatives can be analyzed and evaluated to reach the highest waste heat recovery.

Author Contributions: Conceptualization, H.P.; Formal analysis, H.P.; Methodology, H.S.; Project administration, J.V.h.; Resources, H.P.; Software, H.P. and H.S.; Supervision, J.V.h.; Validation, H.P. and H.S.; Visualization, H.S.; Writing—original draft, H.P. All authors have read and agreed to the published version of the manuscript.

Funding: This project received funding from the European Union's Horizon 2020 research and innovation program under the Marie Sklodowska-Curie grant agreement No. 754354.

Data Availability Statement: The data presented in this study are available on request from the corresponding author.

Acknowledgments: The authors would like to thank Arata Nakajo for the constructive comments to improve the quality of the current article. The CFD analysis of this study was carried out with the commercial software ANSYS CFX 19.2 developed by ANSYS Inc.

Conflicts of Interest: The authors declare no conflict of interest.

References

1. Farias, C.B.; Barreiros, R.C.; da Silva, M.F.; Casazza, A.A.; Converti, A.; Sarubbo, L.A. Use of Hydrogen as Fuel: A Trend of the 21st Century. *Energies* **2022**, *15*, 311. [CrossRef]
2. Shahabuddin, M.; Mofijur, M.; Shuvho, M.; Ahmed, B.; Chowdhury, M.A.K.; Kalam, M.A.; Masjuki, H.H.; Chowdhury, M.A. A Study on the Corrosion Characteristics of Internal Combustion Engine Materials in Second-Generation Jatropha Curcas Biodiesel. *Energies* **2021**, *14*, 4352. [CrossRef]
3. Luciani, S.; Tonoli, A. Control Strategy Assessment for Improving PEM Fuel Cell System Efficiency in Fuel Cell Hybrid Vehicles. *Energies* **2022**, *15*, 2004. [CrossRef]
4. Choi, J.; Sim, J.; Oh, H.; Min, K. Resistance Separation of Polymer Electrolyte Membrane Fuel Cell by Polarization Curve and Electrochemical Impedance Spectroscopy. *Energies* **2021**, *14*, 1491. [CrossRef]
5. Cigolotti, V.; Genovese, M.; Fragiacomo, P. Comprehensive Review on Fuel Cell Technology for Stationary Applications as Sustainable and Efficient Poly-Generation Energy Systems. *Energies* **2021**, *14*, 4963. [CrossRef]
6. Nguyen, H.L.; Han, J.; Nguyen, X.L.; Yu, S.; Goo, Y.-M.; Le, D.D. Review of the Durability of Polymer Electrolyte Membrane Fuel Cell in Long-Term Operation: Main Influencing Parameters and Testing Protocols. *Energies* **2021**, *14*, 4048. [CrossRef]
7. Nishimura, A.; Toyoda, K.; Kojima, Y.; Ito, S.; Hu, E. Numerical Simulation on Impacts of Thickness of Nafion Series Membranes and Relative Humidity on PEMFC Operated at 363 K and 373 K. *Energies* **2021**, *14*, 8256. [CrossRef]
8. Mariani, M.; Peressut, A.B.; Latorrata, S.; Balzarotti, R.; Sansotera, M.; Dotelli, G. The Role of Fluorinated Polymers in the Water Management of Proton Exchange Membrane Fuel Cells: A Review. *Energies* **2021**, *14*, 8387. [CrossRef]
9. Rashidi, S.; Karimi, N.; Sunden, B.; Kim, K.C.; Olabi, A.G.; Mahian, O. Progress and challenges on the thermal management of electrochemical energy conversion and storage technologies: Fuel cells, electrolysers, and supercapacitors. *Prog. Energy Combust. Sci.* **2022**, *88*, 100966. [CrossRef]
10. Hmad, A.A.; Dukhan, N. Cooling Design for PEM Fuel-Cell Stacks Employing Air and Metal Foam: Simulation and Experiment. *Energies* **2021**, *14*, 2687. [CrossRef]
11. Woolley, E.; Luo, Y.; Simeone, A. Industrial waste heat recovery: A systematic approach. *Sustain. Energy Technol. Assess.* **2018**, *29*, 50–59. [CrossRef]
12. Nishimura, A.; Kojima, Y.; Ito, S.; Hu, E. Impacts of Separator Thickness on Temperature Distribution and Power Generation Characteristics of a Single PEMFC Operated at Higher Temperature of 363 and 373 K. *Energies* **2022**, *15*, 1558. [CrossRef]
13. Blanco-Cocom, L.; Botello-Rionda, S.; Ordoñez, L.C.; Valdez, S.I. A Self-Validating Method via the Unification of Multiple Models for Consistent Parameter Identification in PEM Fuel Cells. *Energies* **2022**, *15*, 885. [CrossRef]
14. Kravos, A.; Kregar, A.; Mayer, K.; Hacker, V.; Katrašnik, T. Identifiability Analysis of Degradation Model Parameters from Transient CO2 Release in Low-Temperature PEM Fuel Cell under Various AST Protocols. *Energies* **2021**, *14*, 4380. [CrossRef]
15. Zhang, G.; Jiao, K. Multi-phase models for water and thermal management of proton exchange membrane fuel cell: A review. *J. Power Sources* **2018**, *391*, 120–133. [CrossRef]
16. Mu, Y.-T.; He, P.; Gu, Z.-L.; Qu, Z.-G.; Tao, W.-Q. Modelling the reactive transport processes in different reconstructed agglomerates of a PEFC catalyst layer. *Electrochim. Acta* **2022**, *404*, 139721. [CrossRef]
17. Pourrahmani, H.; Siavashi, M.; Moghimi, M. Design optimization and thermal management of the PEMFC using artificial neural networks. *Energy* **2019**, *182*, 443–459. [CrossRef]

18. Pourrahmani, H.; Moghimi, M.; Siavashi, M.; Shirbani, M. Sensitivity analysis and performance evaluation of the PEMFC using wave-like porous ribs. *Appl. Therm. Eng.* **2019**, *150*, 433–444. [CrossRef]
19. Pourrahmani, H.; Moghimi, M.; Siavashi, M. Thermal management in PEMFCs: The respective effects of porous media in the gas flow channel. *Int. J. Hydrogen Energy* **2019**, *44*, 3121–3137. [CrossRef]
20. Kwan, T.H.; Wu, X.; Yao, Q. Bidirectional operation of the thermoelectric device for active temperature control of fuel cells. *Appl. Energy* **2018**, *222*, 410–422. [CrossRef]
21. Kwan, T.H.; Wu, X.; Yao, Q. Multi-objective genetic optimization of the thermoelectric system for thermal management of proton exchange membrane fuel cells. *Appl. Energy* **2018**, *217*, 314–327. [CrossRef]
22. Kwan, T.H.; Shen, Y.; Yao, Q. An energy management strategy for supplying combined heat and power by the fuel cell thermoelectric hybrid system. *Appl. Energy* **2019**, *251*, 113318. [CrossRef]
23. Shen, Y.; Kwan, T.H.; Yao, Q. Performance numerical analysis of thermoelectric generator sizing for integration into a high temperature proton exchange membrane fuel cell. *Appl. Therm. Eng.* **2020**, *178*, 115486. [CrossRef]
24. Shen, Y.; Zhao, B.; Kwan, T.H.; Yao, Q. Numerical analysis of combined air-cooled fuel cell waste heat and thermoelectric heating method for enhanced water heating. *Energy Convers. Manag.* **2020**, *213*, 112840. [CrossRef]
25. Tuoi, T.T.K.; Toan, N.V.; Ono, T. Theoretical and experimental investigation of a thermoelectric generator (TEG) integrated with a phase change material (PCM) for harvesting energy from ambient temperature changes. *Energy Rep.* **2020**, *6*, 2022–2029. [CrossRef]
26. He, J.; Tritt, T.M. Advances in thermoelectric materials research: Looking back and moving forward. *Science* **2017**, *357*, eaak9997. [CrossRef] [PubMed]
27. Jouhara, H.; Żabnieńska-Góra, A.; Khordehgah, N.; Doraghi, Q.; Ahmad, L.; Norman, L.; Axcell, B.; Wrobel, L.; Dai, S. Thermoelectric generator (TEG) technologies and applications. *Int. J. Thermofluids* **2021**, *9*, 100063. [CrossRef]
28. Fernández-Yañez, P.; Armas, O.; Capetillo, A.; Martínez-Martínez, S. Thermal analysis of a thermoelectric generator for light-duty diesel engines. *Appl. Energy* **2018**, *226*, 690–702. [CrossRef]
29. Zorbas, K.T.; Hatzikraniotis, E.; Paraskevopoulos, K.M. Power and Efficiency Calculation in Commercial TEG and Application in Wasted Heat Recovery in Automobile. In Proceedings of the 5th European Conference on Thermoelectrics, Odessa, Ukraine, 10–12 September 2007; p. 4.
30. Arasteh, H.; Mashayekhi, R.; Ghaneifar, M.; Toghraie, D.; Afrand, M. Heat transfer enhancement in a counter-flow sinusoidal parallel-plate heat exchanger partially filled with porous media using metal foam in the channels' divergent sections. *J. Therm. Anal. Calorim.* **2020**, *141*, 1669–1685. [CrossRef]
31. Spiegel, C. *PEM Fuel Cell Modeling and Simulation Using Matlab*; Academic Press: Cambridge, MA, USA; Elsevier: Amsterdam, The Netherlands, 2008.
32. Li, S.; Yuan, J.; Xie, G.; Sundén, B. Effects of agglomerate model parameters on transport characterization and performance of PEM fuel cells. *Int. J. Hydrogen Energy* **2018**, *43*, 8451–8463. [CrossRef]
33. Thosar, A.U.; Agarwal, H.; Govarthan, S.; Lele, A.K. Comprehensive analytical model for polarization curve of a PEM fuel cell and experimental validation. *Chem. Eng. Sci.* **2019**, *206*, 96–117. [CrossRef]
34. Siegel, N.P.; Ellis, M.W.; Nelson, D.J.; von Spakovsky, M.R. A two-dimensional computational model of a PEMFC with liquid water transport. *J. Power Sources* **2004**, *128*, 173–184. [CrossRef]
35. Ge, M.; Zhao, Y.; Li, Y.; He, W.; Xie, L.; Zhao, Y. Structural optimization of thermoelectric modules in a concentration photovoltaic–thermoelectric hybrid system. *Energy* **2022**, *244*, 123202. [CrossRef]
36. Ge, M.; Li, Z.; Wang, Y.; Zhao, Y.; Zhu, Y.; Wang, S.; Liu, L. Experimental study on thermoelectric power generation based on cryogenic liquid cold energy. *Energy* **2021**, *220*, 119746. [CrossRef]
37. Zhao, Y.; Fan, Y.; Li, W.; Li, Y.; Ge, M.; Xie, L. Experimental investigation of heat pipe thermoelectric generator. *Energy Convers. Manag.* **2022**, *252*, 115123. [CrossRef]
38. Module Specifications of the Thermoelectric Generator. Available online: https://thermoelectric-generator.com/wp-content/uploads/2014/04/TEG1-4199-5.3_C_R_T_CBH1.pdf (accessed on 10 April 2014).
39. Shen, Z.-G.; Tian, L.-L.; Liu, X. Automotive exhaust thermoelectric generators: Current status, challenges and future prospects. *Energy Convers. Manag.* **2019**, *195*, 1138–1173. [CrossRef]
40. Zhang, G.; Kandlikar, S.G. A critical review of cooling techniques in proton exchange membrane fuel cell stacks. *Int. J. Hydrogen Energy* **2012**, *37*, 2412–2429. [CrossRef]
41. Mohamed, W.A.N.W.; Kamil, M.H.M. Hydrogen preheating through waste heat recovery of an open-cathode PEM fuel cell leading to power output improvement. *Energy Convers. Manag.* **2016**, *124*, 543–555. [CrossRef]
42. Ticianelli, E.A.; Derouin, C.R.; Srinivasan, S. Localization of platinum in low catalyst loading electrodes to attain high power densities in SPE fuel cells. *J. Electroanal. Chem.* **1988**, *251*, 275–295. [CrossRef]
43. Mohammedi, A.; Sahli, Y.; Moussa, H.B. 3D investigation of the channel cross-section configuration effect on the power delivered by PEMFCs with straight channels. *Fuel* **2020**, *263*, 116713. [CrossRef]
44. Xue, X.D.; Cheng, K.W.E.; Sutanto, D. Unified mathematical modelling of steady-state and dynamic voltage–current characteristics for PEM fuel cells. *Electrochim. Acta* **2006**, *52*, 1135–1144. [CrossRef]
45. Correa, J.M.; Farret, F.A.; Canha, L.N.; Simoes, M.G. An electrochemical-based fuel-cell model suitable for electrical engineering automation approach. *IEEE Trans. Ind. Electron.* **2004**, *51*, 1103–1112. [CrossRef]

46. Asl, S.M.S.; Rowshanzamir, S.; Eikani, M.H. Modelling and simulation of the steady-state and dynamic behaviour of a PEM fuel cell. *Energy* **2010**, *35*, 1633–1646. [CrossRef]
47. Fernández-Yáñez, P.; Armas, O.; Gómez, A.; Gil, A. Developing Computational Fluid Dynamics (CFD) Models to Evaluate Available Energy in Exhaust Systems of Diesel Light-Duty Vehicles. *Appl. Sci.* **2017**, *7*, 590. [CrossRef]
48. Falcão, D.S.; Rangel, C.M.; Pinho, C.; Pinto, A.M.F.R. Water Transport in PEM fuel cells. In Proceedings of the II Iberian Symposium on Hydrogen, Fuel Cells and Advanced Batteries (Hyceltec 2009), Porto, Portugal, 13–17 September 2009; p. 8.
49. Falcão, D.S.; Oliveira, V.B.; Rangel, C.M.; Pinho, C.; Pinto, A.M.F.R. Water transport through a PEM fuel cell: A one-dimensional model with heat transfer effects. *Chem. Eng. Sci.* **2009**, *64*, 2216–2225. [CrossRef]
50. Falcão, D.S.; Rangel, C.M.; Pinho, C.; Pinto, A.M.F.R. Water Transport through a Proton-Exchange Membrane (PEM) Fuel Cell Operating near Ambient Conditions: Experimental and Modeling Studies. *Energy Fuels* **2009**, *23*, 397–402. [CrossRef]
51. Falcão, D.S.; Gomes, P.J.; Oliveira, V.B.; Pinho, C.; Pinto, A.M.F.R. 1D and 3D numerical simulations in PEM fuel cells. *Int. J. Hydrogen Energy* **2011**, *36*, 12486–12498. [CrossRef]
52. Ahn, J.-W.; Choe, S.-Y. Coolant controls of a PEM fuel cell system. *J. Power Sources* **2008**, *179*, 252–264. [CrossRef]
53. Liso, V.; Nielsen, M.P.; Kær, S.K.; Mortensen, H.H. Thermal modeling and temperature control of a PEM fuel cell system for forklift applications. *Int. J. Hydrogen Energy* **2014**, *39*, 8410–8420. [CrossRef]
54. Zhang, S.; Zhao, X. General Formulations for Rhie-Chow Interpolation. In Proceedings of the ASME 2004 Heat Transfer/Fluids Engineering Summer Conference, Charlotte, NC, USA, 11–15 July 2004; Volume 2, pp. 567–573. [CrossRef]
55. Zhang, S.; Zhao, X.; Bayyuk, S. Generalized formulations for the Rhie–Chow interpolation. *J. Comput. Phys.* **2014**, *258*, 880–914. [CrossRef]

energies MDPI

Article

A Holistic Consideration of Megawatt Electrolysis as a Key Component of Sector Coupling

Bernd Emonts [1,*], Martin Müller [1], Michael Hehemann [1], Holger Janßen [1], Roger Keller [1], Markus Stähler [1], Andrea Stähler [1], Veit Hagenmeyer [2], Roland Dittmeyer [3], Peter Pfeifer [3], Simon Waczowicz [2], Michael Rubin [3], Nina Munzke [4] and Stefan Kasselmann [5]

1 Institute of Energy and Climate Research-Electrochemical Process Engineering (IEK-14), Forschungszentrum Jülich GmbH, 52425 Jülich, Germany; mar.mueller@fz-juelich.de (M.M.); m.hehemann@fz-juelich.de (M.H.); h.janssen@fz-juelich.de (H.J.); r.keller@fz-juelich.de (R.K.); m.staehler@fz-juelich.de (M.S.); a.staehler@fz-juelich.de (A.S.)
2 Institute for Automation and Applied Informatics (IAI), Karlsruhe Institute of Technology (KIT), 76344 Eggenstein-Leopoldshafen, Germany; veit.hagenmeyer@kit.edu (V.H.); simon.waczowicz@kit.edu (S.W.)
3 Institute for Micro Process Engineering (IMVT), Karlsruhe Institute of Technology (KIT), 76344 Eggenstein-Leopoldshafen, Germany; roland.dittmeyer@kit.edu (R.D.); peter.pfeifer@kit.edu (P.P.); michael.rubin@kit.edu (M.R.)
4 Institute of Electrical Engineering (ETI), Karlsruhe Institute of Technology (KIT), 76344 Eggenstein-Leopoldshafen, Germany; nina.munzke@kit.edu
5 Technical Infrastructure-Projects and Personnel (TB-P), Forschungszentrum Jülich GmbH, 52425 Jülich, Germany; s.kasselmann@fz-juelich.de
* Correspondence: b.emonts@fz-juelich.de; Tel.: +49-2461-61-3525

Citation: Emonts, B.; Müller, M.; Hehemann, M.; Janßen, H.; Keller, R.; Stähler, M.; Stähler, A.; Hagenmeyer, V.; Dittmeyer, R.; Pfeifer, P.; et al. A Holistic Consideration of Megawatt Electrolysis as a Key Component of Sector Coupling. *Energies* **2022**, *15*, 3656. https://doi.org/10.3390/en15103656

Academic Editors: Bahman Shabani and Mahesh Suryawanshi

Received: 17 March 2022
Accepted: 8 May 2022
Published: 16 May 2022

Abstract: In the future, hydrogen (H_2) will play a significant role in the sustainable supply of energy and raw materials to various sectors. Therefore, the electrolysis of water required for industrial-scale H_2 production represents a key component in the generation of renewable electricity. Within the scope of fundamental research work on cell components for polymer electrolyte membrane (PEM) electrolyzers and application-oriented living labs, an MW electrolysis system was used to further improve industrial-scale electrolysis technology in terms of its basic structure and systems-related integration. The planning of this work, as well as the analytical and technical approaches taken, along with the essential results of research and development are presented herein. The focus of this study is the test facility for a megawatt PEM electrolysis stack with the presentation of the design, processing, and assembly of the main components of the facility and stack.

Keywords: renewable electricity; PEM electrolyzers; megawatt electrolysis system; large-scale test facility; water splitting; living lab; simulation platform; coupling of energy-related technologies

1. Introduction

The systematic conversion of the global energy industry to a renewable basis poses a challenge to national economies and is referred to in Germany as the 'energy transition' (Energiewende). In addition to the supply of electricity from wind, solar, biomass, and hydro- and geothermal sources, an affordable and demand-oriented energy supply is essential for a sustainable and functioning economy. Therefore, there is a need for flexible, efficient, and scalable energy storage for hydrogen, a chemical storage medium, as a product of electrolysis [1–3]. To this end, eroded salt domes with large volumes are suitable for underground compressed gas storage. In the future, adapted or repurposed pipeline networks will be available to transport and distribute hydrogen from storage sites to various points of end use. There will be a focus on the direct use of hydrogen as a fuel for fuel cell-electric passenger cars, vans, buses, trucks, and trams. Additional downstream possibilities for hydrogen will be offered by the steel industry as a reducing agent in the production of pig iron, by future fuel producers via synthesis with carbon dioxide from bionic or industrial point sources [4], or even captured from air (direct air capture) [5] to produce liquid fuels [6] equivalent to petroleum, diesel, and kerosene for trucks, aircraft, and ships. The demand-oriented reconversion of hydrogen in appropriate gas turbines into electricity will round out the portfolio of hydrogen usage. The results of analyses by Jülich scientists have demonstrated that the use of power-to-X (PtX) technologies and the use of hydrogen for transport and industrial applications will lead to significant hydrogen demand in the future. The CO_2-free generation of electricity is an essential prerequisite to this. A H_2 infrastructure encompassing generation, storage, and transport will be required for its implementation [7]. Due to its diverse range of applications, hydrogen plays an especially important role in scenarios that assume complete carbon neutrality. Thus, hydrogen demand in 2050 will amount to more than 12 million tons. Over 50% of this demand will be covered by domestic production and the rest will be supplied by imports. This level of demand means that water electrolysis will be a key technology for a future sustainable energy industry.

2. An Overview of the Electrolyzer State of the Art

Work on the material and technological improvement of water electrolysis began at Forschungszentrum Jülich in the early 1980s. Initially, the focus was on increasing the efficiency of alkaline electrolysis through new, innovative approaches to cell composition and structure. As a result of the high foreseeable demand for highly efficient electrolyzers, research work at Jülich was subsequently expanded to include PEM electrolysis technology.

The decomposition of water, or water-splitting, by means of electrolysis involves two partial reactions separated by an ion-conducting electrolyte. Three technically-relevant water electrolysis processes result from the species of ion being used:

- High-temperature electrolysis with a solid oxide as an electrolyte;
- Alkaline electrolysis with a liquid alkaline electrolyte;
- "Acidic" PEM electrolysis with a proton-conducting solid-state polymer electrolyte.

At present, reversible high-temperature electrolysis is only being pursued to a limited extent due to the ceramic materials used in research and industry, and there are currently very few commercial products with a performance range relevant to the energy industry. An experiment involving high-temperature electrolysis in the kW range is being conducted using the network of facilities at Forschungszentrum Jülich, which are described in further detail below [8]. However, as this article focuses on MW application, the technology itself is not described any further.

Low-temperature electrolysis processes based on alkaline and PEM technologies can be used to establish large, powerful systems (1–100 MW) [9]. In contrast to PEM electrolysis, a number of alkaline electrolysis designs have been used for several decades at various scales. Their operating temperature is typically around 80 °C and systems of up to 750 $Nm^3{}_{H2}/h$ are available. These systems use an aqueous solution of potassium hydroxide (KOH) as an electrolyte with a typical concentration of 20–40% and achieve current densities of 0.25–0.45 A/cm^2. Rectangular and circular electrodes and cells with an active area of up to approximately 3 m^2 have also been used. The plant engineering firm Lurgi manufactures pressurized electrolyzers that supply hydrogen and oxygen below 30 bar. Lurgi's pressurized electrolyzer produces 760 Nm^3/h of hydrogen, which corresponds to an electrical output of approximately 3.6 MW [10]. Lifetimes of up to 90,000 h are achieved by the stack, with the electrodes and diaphragms needing to be replaced after this period. According to the NOW study by the Fraunhofer Institute for Solar Energy Systems (ISE, Freiburg), the voltage efficiency of the stack amounts to 62–82% in relation to the higher heating value (HHV) [11]. At less than 20%, the lower partial load operation is especially critical for its flexible application in combination with renewable energy sources. This is a result of the diaphragms that are used, which facilitate the mixture of hydrogen and oxygen due to diffusion, which in turn leads to safety-related switch offs. Another disadvantage of alkaline electrolyzers as compared with PEM electrolysis is the costly gas treatment of the product gases, for which expensive noble metals must be used.

In contrast to alkaline water electrolysis, PEM electrolysis with proton-conducting membranes uses platinum group metals for the electrodes. Due to the use of dense membranes as an electrolyte and the possibility of integration with recombination catalysts, the systems can be operated at 0–100% power; however, in technical facilities the lower threshold is limited to approximately 5% of nominal power due to the internal consumption of peripheral components [10].

In particular, the high overvoltage at the oxygen electrode is a challenge for the development of materials and is partially responsible for the energy loss of the systems. Studies have shown that RuO_2 and IrO_2 are especially suitable catalysts for oxygen electrodes [12–14]. These metal oxides exhibit a high level of activity, adequate long-term stability, and low performance losses [15–24]. For this reason, IrO_2 is often used as an anode catalyst for PEM electrolyzers [25]. Despite its low specific activity compared to RuO_2, IrO_2 is especially suitable due to its low overvoltages [22,26] and excellent electrochemical stability compared to RuO_2 [22].

In current commercial systems, approximately 2 mg/cm^2 iridium is required for the anode and approximately 1 mg/cm^2 of platinum is required for the cathode. At the given operating conditions, these systems run at voltages of about 2 V, current densities of approximately 2 A/cm^2, and operating pressures of up to 30 bar [27]. This is equivalent to the voltage efficiency of alkaline electrolysis (approximately 67–82%), but with much higher current densities (0.6–2.0 A/cm^2) [28].

Other studies have investigated the level of iridium loading that is advantageous for function and stability, and identified 1–2 mg/cm^2 as the ideal range [29]. However, other studies have investigated much lower iridium loading and found that this also results in reasonable operating times [30]. When using noble metals which are limited in terms of availability, this automatically raises the question as to whether they will be available in sufficient quantities. In [31], it was investigated whether the extraction of iridium could sufficiently cover demand for the increasing expansion of PEM electrolysis. It was found that noble metal loading would need to be significantly reduced in the next 15 years to ensure a sufficient supply of noble metals. The target loading should reach 50 mg/kW in 2035. Based on a power density of 6 W/cm^2, this corresponds to a catalyst loading of 0.3 mg/cm^2. However, it was found that recycling of the catalyst is an essential prerequisite for this. Carmo et al. describe how this can be implemented in a simple and environmentally-friendly manner, despite the use of solvents [32].

All large PEM electrolysis manufacturers are working on the development of MW systems with various stack concepts. Hydrogenics, for instance, recently developed and built a 1 MW PEM electrolyzer with a single stack and a nominal power of 1 MW [33]. For the most part, the commercially-available stacks only operate with current densities < 2.0 A/cm^2. In these systems, approximately 6 mg/cm^2 of iridium or ruthenium is required for the anode, and approximately 2 mg/cm^2 of platinum is required for the cathode [34]. In contrast to alkaline electrolysis, the lifetime of PEM electrolysis stacks is estimated at <20,000 h. However, Proton Onsite has already achieved a lifetime of more than 50,000 h for stacks used in the PEM electrolyzers of the HOGEN C series [35].

In terms of thermodynamic considerations, the hydrogen-producing electrode should ideally be operated under pressure [36]; the oxygen-producing electrode can be operated under atmospheric conditions, as the oxygen being produced is not typically stored under pressure. Tjarks et al. [37] showed that an electrolyzer operating pressure of up to 20 bar can improve the system's efficiency.

One of the technical limitations of this differential pressure operation is the mechanical stability of the membrane electrode assembly (MEA) and the sintered body. In addition, the design of modern stacks must account for the fact that in the future, thinner membranes will be used that significantly increase the hydrogen production of an electrolyzer. The influence of membrane thickness on cell performance has been simulated many times and analyzed in the literature [36,38–40]. Stähler et al. [41] were able to achieve current densities of 11 A/cm^2 at 2 V in a single cell with a membrane thickness of approximately 20 μm. If the membrane thickness is reduced, H_2 permeation through the membrane and the mechanical stress of the membrane, particularly during pressure operation, must be taken into account [36,39].

Another important aspect is the adjustment of compression and contact pressure for the porous transport layers (PTLs) being used. Stähler et al. [42] showed that by increasing compression, the performance of the electrolysis cell can be enhanced, although excess compression of the gas diffusion layers (GDLs) leads to an increase in hydrogen permeation through the membrane. When designing the stack, it is important to avoid increased compressive and shear stresses [43]. Borgardt et al. conducted analyses on membrane and stack component mechanics in terms of their relation to contact pressure, efficiency, and creep behavior [44,45].

In addition to the noble metals required for the electrodes, another challenge of system development is posed by titanium-based separator plates and current collectors, to which coatings must be applied due to hydrogen embrittlement, the formation of oxide layers, and the associated increase in contact resistance. According to publications by Ayers et al. (Proton OnSite), the separator plates (including the current collectors) represent around 48% of the stack's costs [46].

On the anode side, porous transport layers made of titanium are typically used to enable the transport of water to the electrode and electrical contact with the flow field. Liu et al. [47] showed that an iridium coating is important for ensuring long-term stability and low contact resistance of PTLs. A crucial means of limiting costs is the increasing integration of stainless steel materials instead of titanium [48], which requires a coating without any defects [49]. In order to reduce costs, the use of grade 316 stainless steel is being investigated with various coatings [50,51]. A potentially inexpensive and stable coating is comprised of Ti/TiN, and was investigated in greater detail by Rojas et al. [52].

In the long term, PEM electrolysis can play a significant role in providing operating reserve, as its better dynamics in relation to alkaline electrolysis make it interesting for larger applications with systems > 1 MW. However, the electrode or cell area must be scaled up to 600–2000 cm^2 in order to reduce the high level of investment costs (>2000 €/kW [10]). This is confirmed in the Plan-DelyKaD study, which demonstrated that PEM electrolysis has a moderate cost advantage over alkaline electrolysis in terms of investment costs [53]. As part of a study by Bertuccioli et al. [48], the manufacturers of such systems were asked about anticipated reductions in costs. The study came to the conclusion that in the long term, similar costs to alkaline electrolysis are expected. A current list of the most significant manufacturers of PEM electrolysis systems is given in Table 1. PEM electrolysis is currently in a phase of development in which the absolute power of the systems is to be further increased. The most powerful systems are currently offered by Siemens and ITM.

Table 1. Commercially-available electrolysis systems.

Manufacturer	Type	Pressure Level	Production Rates	Energy Consumption	Power Range	Reference
		bar	m^3/h (i.N.)	kWh/m^3 (i.N)	%	
Giner	OGP	207	13	5.4		[54,55]
	LPE	55	15	5.4		
h-tec	ME100/350	15–0	15–46.3	4.9	32–100	[56]
	ME450/1400		42–210	4.8	20–100	
Hydrogenics	HyLyzer®	2.75–5.5	1–2	6.7	HyLyzer®	[57]
ITM Power	HGAS1SP	20	122	5.8		[58]
	HGAS3SP	30	400	5.1		

Table 1. *Cont.*

Manufacturer	Type	Pressure Level	Production Rates	Energy Consumption	Power Range	Reference
		bar	m^3/h (i.N.)	kWh/m^3 (i.N)	%	
NEL hydrogen	C Series	30	10–30	6.2–5.8	0–100	[59]
	M Series		246–492	4.5	10–100	
Siemens	Silyzer 100	50	22–44	5.4	5–100	[60–63]
	Silyzer 200	35	225	4.3		
	Silyzer 300	low	1112–22,246	4.7		

In a study that focused on optimizing the technology and reducing the costs of electrolyzers, the factors influencing the levelized cost of hydrogen (electricity costs, operating costs, and control and investment costs) were investigated for the various electrolysis technologies [64]. The study showed that for optimized PEM electrolyzers in the MW range in the year 2030, investment costs of € 585/kW (roughly four times lower than the current costs) can be expected. To this end, PEM electrolysis will need to be further scaled up to ensure that it is economically viable in a higher MW range. A reduction in noble metal loading in the electrodes can help reduce costs, whereas additional stack components such as bipolar plates and sintered bodies also offer further potential for reducing costs.

In addition to systems development work, the integration of technology is vitally important. From a technological standpoint, it is crucial for the required system dynamics; from an economic perspective, it is important for evaluating value creation. Electrolyzers can be used to serve the various operating reserve requirements of the energy market [65]. In a study conducted at Energiepark Mainz, the operation of an electrolyzer for an energy system participating in the energy market was presented as an example. The associated pressure operation requirements of an electrolyzer were also shown [66]. There are numerous other possible approaches to producing hydrogen on a large scale under as realistic conditions as possible [67–69]. However, all of these projects share an emphasis on the use of commercially-available systems and the scaling of performance; the focus is not on making a leap in technological development or the possibility of realizing new approaches in a dimension that enables all energy and mass flows to be evaluated.

With the development described in the following sections, we therefore address this vital link in terms of energy supply and aim to show how a 400 kW next-generation electrolysis stack can be integrated into a highly dynamic, intelligent energy system of the future and what results can be expected. We have a unique infrastructure at our disposal that, for the first time, enables validation of the systems that have been almost exclusively modeled to date.

3. Research and Experimental Work for Next-Generation PEM Electrolysis

Electrolysis research generally focuses on understanding the mechanisms that are essential to improving the components required for water electrolysis. Starting with the catalyst material and emphasizing suitable support and distribution structures, the various approaches and specialized methods involve the research, development, production, and assembly of the electrolyte, MEA, bipolar plates, media, and power supplies at scales ranging from micrometers to meters. The following sections will describe this work in further detail and focus on the development of a 400 kW electrolysis stack.

The focus of this investigation is the electrolysis stack, which was planned, designed, produced, and assembled within the scope of electrolysis research and is described in detail in the following sections. The MW electrolysis system is an instrument of research infrastructure that enables the industrial-scale use of electrolysis for technological research and development. The close-to-realistic conditions, in combination with the testing facilities, were aimed at fundamental and application-oriented areas for improving electrolysis technology and its integration into a sustainable energy supply pathway:

- Electrolysis research—the study of electrolysis stacks equipped with cell components with large areas that were developed using novel approaches to materials, design, and production;
- Energy Lab 2.0—research infrastructure for a demand-oriented hydrogen supply to be used directly and indirectly in the transport sector;
- Living Lab Energy Campus—research infrastructure for the demand-oriented supply and storage of hydrogen consisting of compressed gas cylinders, dynamically-operable pipelines, and LOHC storage systems.

In the following sections, three fields of work will be presented in which the MW electrolysis system is the key component for various research and development tasks (see Figure 1). Electrolysis research employs innovative approaches to tackle fundamental issues with the aim of improving cell components for PEM water electrolysis and developing efficient, long-lasting, and economical electrolysis stacks. The Energy Lab 2.0 and Living Lab Energy Campus (LLEC) projects are focused on integrating PEM water electrolysis into application-oriented energy pathways for methanation, use in a micro gas turbine, and the synthesis of liquid electrofuels, as well as temporary storage as a compressed gas or in the form of liquid organic hydrogen carriers (LOHCs). For the various uses of stored hydrogen, the LLEC intends to utilize an alkaline fuel cell in which the oxygen from water electrolysis can also be used to generate electricity. In addition, hydrogen can be added to natural gas for combined heat and power internal combustion engines.

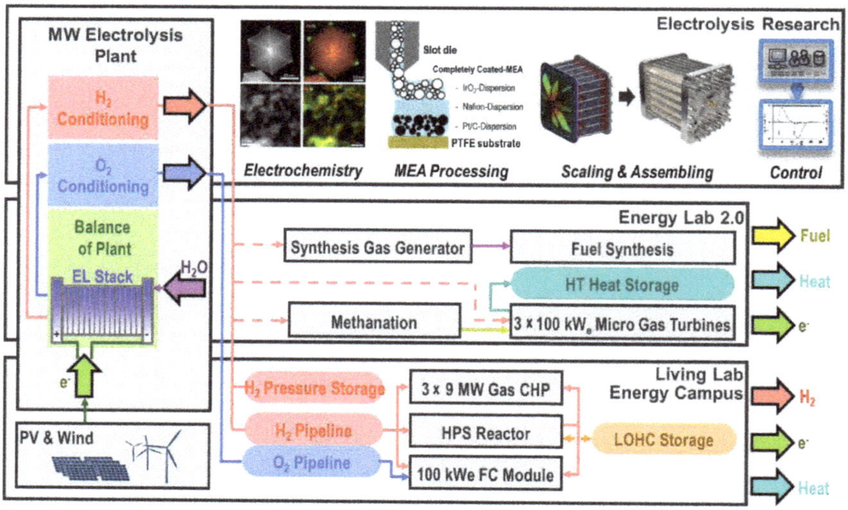

Figure 1. The MW electrolysis system as the focus of research.

3.1. Reproducible Production Technology for MEAs

In PEM electrolyzers, the MEA is situated between the gas diffusion layers and consists of two porous, gas-permeable electron- and proton-conducting catalyst layers (anode and cathode) that comprise a gas-tight proton-conducting polymer membrane. Iridium oxide and platinum are common catalyst materials. Nafion, a perfluorinated alkyl sulfonic acid, is predominantly used as a proton-conducting material. Nafion is also used as an electron-isolated proton-conducting material for the membrane itself.

For the production of MEAs in the MW scale, it is important that the manufacturing processes used meet the requirements for scalability and production speed. The transfer or decal method outlined in Figure 2 meets these requirements. The process steps used, such as coating a substrate with a catalyst dispersion using a doctor blade or slot die and the subsequent drying, can be carried out in a laboratory or a roll-to-roll process. This also applies to the subsequent assembly process where the dried catalyst layers are transferred from substrate to a membrane with a hot pressing process.

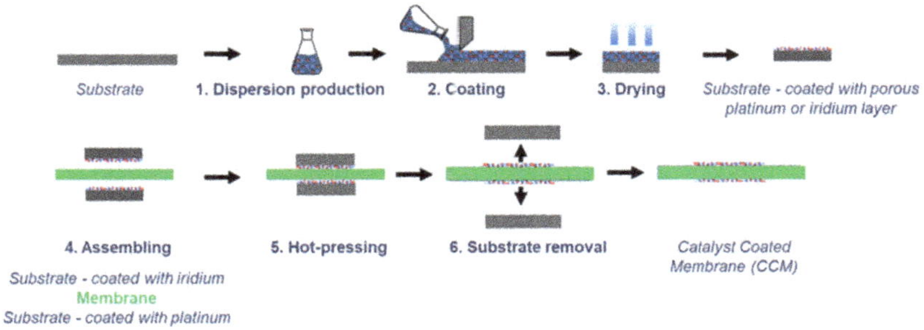

Figure 2. Process steps for producing an MEA using the decal process.

The production of MEAs for PEM water electrolysis faces the challenge of increasing the electrochemical efficiency of the MEAs and structuring the entailed processes in such a way that they are scalable and can be mass-produced. The electrochemical efficiency of MEAs partially depends on the thickness of the membrane used, as the protons flowing through the membrane during cell operation induce ohmic losses. This requirement can be fulfilled using the aforementioned decal process in which the thinnest possible membranes are used during assembly. However, a compromise must be made between membranes that are as thin as possible but still easy to handle in the production process.

Scalable processes, such as slot die or blade coating, are ideal for the industrial-scale coating of a substrate. A disadvantage of blade coating (a form of self-metered coating) is that the homogeneity of the coating cannot always be precisely controlled [70], whereas slot dies (which allow for a pre-dosed coating) can produce highly homogeneous coatings in combination with homogeneous substrates [71]. The possibility of pre-dosing means that the catalyst loading of the electrodes can be precisely adjusted by means of the dosing rate of the pump and the coating speed, provided that the dispersion composition is known. As the MEAs in the stack are connected in series for PEM electrolysis, it is highly important that the electrodes are identical and that the rolls of electrodes being produced from which the electrodes are subsequently cut are also homogenous. The specifications shown in Table 2 were applied for the catalyst loading of the MEAs being produced for the MW electrolysis system.

Table 2. Materials used in the construction of the MEAs.

Anode	Membrane	Cathode
Ir coating 1.0 mg/cm^2	Nafion NR212 (51 µm)	Pt coating 0.25 mg/cm^2

A Premion (Alfa Aesar) IrO_2 catalyst was used as the anode catalyst and a Pt/C 60% high surface area (HAS) Ketjen Black (Fuel Cell Store) catalyst was used as the cathode catalyst. The Nafion NR212 membranes were purchased from Chemours. An additional specification was the use of a recombination catalyst to be inserted into the MEA to break down the hydrogen permeating through the membrane with oxygen and thus minimize the safety risk presented by oxy-hydrogen formation during operation of the stack. As previously published methods of use of such recombination systems [72,73] could not be scaled to the required production size, a new method was developed and used for the production process described in the following sections [74].

Dispersions were produced from the catalyst powders that were applied to a fiberglass-reinforced polytetrafluoroethylene (PTFE) roll in the roll-to-roll coating system (Figure 3a) by means of a slot die. During drying, it must be ensured that the combination of solvent evaporation and the formation of layers does not lead to layer defects and the flaking of dried layers [75,76]. The specific machine settings were determined in preliminary tests. After drying, the coated material was rolled up (see Figure 3b) and the electrodes were cut from the roll with dimensions of 32.5 × 32.5 cm^2 (see Figure 3c).

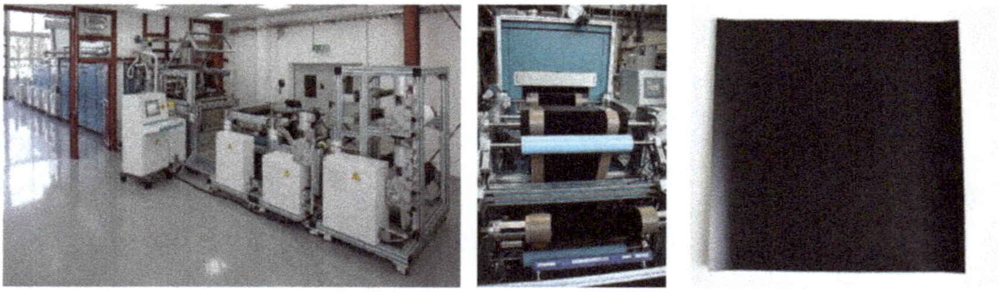

(a) (b) (c)

Figure 3. (**a**): Roll-to-roll coating system; (**b**): The coated substrate was dried and subsequently unrolled; (**c**): The electrode was cut from the roll with dimensions of 32.5 × 32.5 cm^2.

After producing the anodes and cathodes, they were assembled together with the Nafion NR212 membranes using hot presses from P/O/Weber. After pressing, the substrates were then removed from the cooled MEAs, which were then completed (see Figure 4).

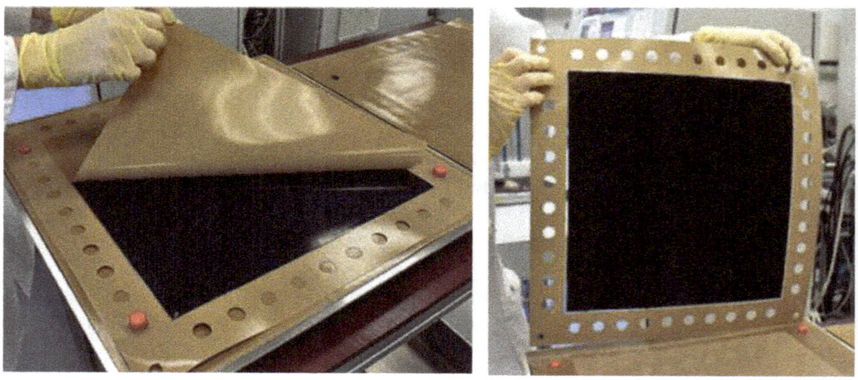

(a) (b)

Figure 4. (**a**): After assembly, the substrate could be removed from the electrode; (**b**): A complete and isolated MEA.

For the purpose of quality assurance, samples were taken from the rolls of electrodes to produce MEAs for laboratory cells and for the electrochemical characterization of the MEAs.

Figure 5 shows a polarization curve plotted from the characterization measurements, an efficiency curve, and a graph demonstrating the proportional volume of hydrogen in the anode gas. The latter clearly shows that safe operation can be achieved with the MEAs produced, even with small current densities. Within the scope of measurement uncertainty, no hydrogen could be detected as a result of the recombination catalyst being used.

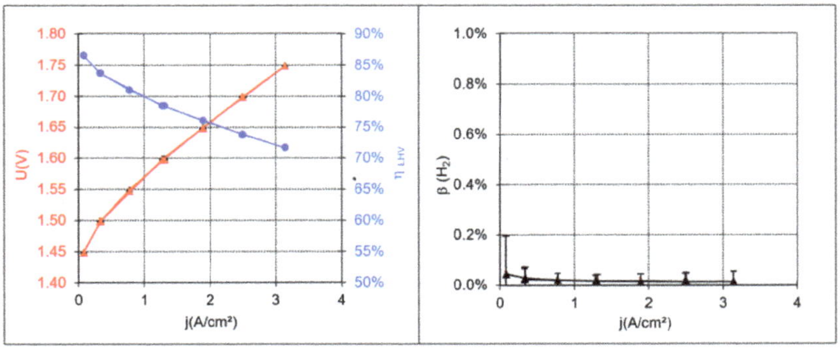

(a) (b)

Figure 5. Results from the characterization measurements of an MEA sample. (**a**): Polarization curve (red) and cell efficiency (blue) in relation to the lower heating value $\eta_L = 1.25\ V/U_{Cell}$; (**b**): H_2 volume fraction in O_2.

3.2. Compact, Efficient, and Robust Cell Components

The previously considered MEA is a component that plays the main role in producing hydrogen in electrolyzers. It is here that the electrochemical water splitting reactions and the generation of heat occur due to operational and material-related overvoltages. However, as indicated by the setup of the MEAs, additional cell and stack components are required to produce hydrogen in MW electrolyzers on an industrial scale. The entire assembly of the electrolyzer must support the following process functions:

- Introduction of electrical current and feed water;
- The homogeneous distribution of current and feed water to all active areas;
- Safe removal of produced hydrogen (H_2) and oxygen (O_2);
- Safe separation of H_2 and O_2 spaces;
- Sealing of the media from their surroundings;
- Heat dissipation in the entire area of operation.

On the left-hand side of Figure 6 can be seen an exploded view of an electrolysis stack from the Ekolyser project (funding reference no. 03ESP106A) being funded by the Federal Ministry for Economic Affairs and Energy (BMWi). The stack is closed off by two massive end plates through which the media are added and removed and through which the power connection is made. The combination of end plates and tie rods also ensures that the inner components are pressed together. A defined contact pressure helps to ensure that the components are sufficiently sealed and that they have electrical or mechanical contact. The contact pressure is adjusted via the tractive forces in the tie rods. A distinction must be made between two areas: the active cell area and the sealing area. Although the optimal contact pressure on the active cell area is 2–3 MPa [44], the required pressure on the sealants is dependent on the sealing concept. The inner components are the repeating units of the electrolysis stack. A repeating unit always consists of one bipolar unit and one MEA. In Figure 6a, three bipolar units are shown. The MEAs (not shown here) are each arranged between the bipolar units. The active cell area is 300 cm^2 (10 cm × 15 cm). In Figure 6b, an assembled electrolysis stack comprising 27 cells can be seen. The stack is designed for an electrical power supply of 50 kW with a maximum operating pressure of 50 bar.

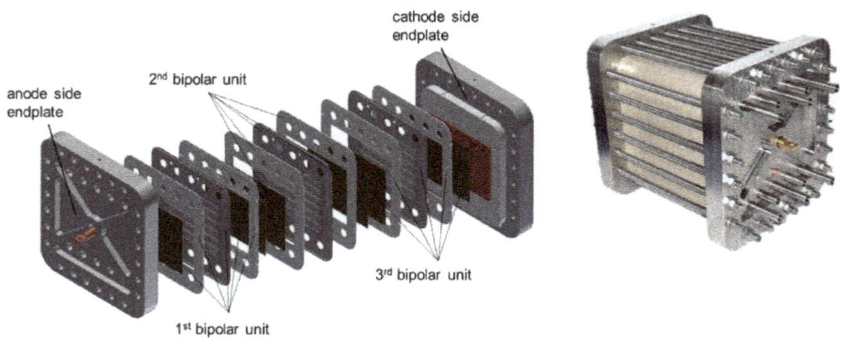

(**a**) (**b**)

Figure 6. Electrolysis stack from the Ekolyser project; (**a**): Exploded view with three bipolar units; (**b**): 27-cell stack, 50 kW_{el}, 50 bar.

Developed as part of the Ekolyser project, the bipolar unit with seven individual components is shown in Figure 7. The central aspect of the bipolar unit is the bipolar plate, which separates the anode and cathode sides of the two adjacent electrolysis cells in a gas-tight manner. The bipolar plate is made of stainless steel 1.4404 (316L) and also features flow distributor structures that are produced by hydroforming to distribute the feed water equally across the active cell area and to remove the produced gases—oxygen on the anode side and hydrogen on the cathode side—with surplus water. The bipolar plate is enclosed on both sides by a polyether ether ketone (PEEK) frame. The frames feature distribution and collector structures for the inflow/outflow of media (water and two-phase flow) from pipes installed at the end plates (see Figure 6b) to the respective cell level. Additional covers made of stainless steel are required to provide mechanical support for sealants (not shown) in the distribution and collector structures. The frame cutouts admit fine distribution structures of varying porosity on the anode and cathode sides. The anode side features a layer of sintered Ti powder. A perforated stainless steel plate is used on the cathode side. In addition, a carbon fiber layer can be found on the MEA side, which is not shown in Figure 7.

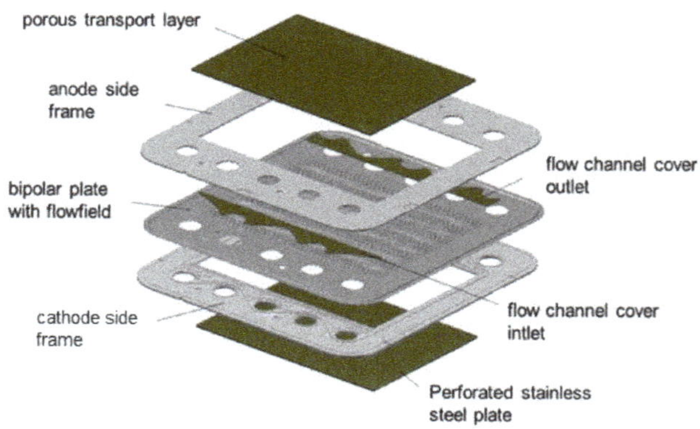

Figure 7. Bipolar unit consisting of several parts from the Ekolyser project.

A significant drawback of this bipolar unit is the fact that all components must be individually laid on top of each other when assembling the stacks. For the 27-cell stack shown above, the individual parts (including O-rings) amount to around 600 for the repeating units alone. The assembly of the stacks is extremely time-intensive and there is also the potential for sources of error, which can lead to malfunctions in stack operation. These include leaks outside and within the stack, as well as inhomogeneities in contact pressure distribution. The higher the number of individual components, the greater the likelihood of errors occurring during assembly. The dimensional tolerances of the components and the shifting of components during assembly play a key role in ensuring the quality of the entire assembly.

To counter this problem, the development of a "one-component bipolar unit" was initiated as part of the Energy Lab 2.0 and LLEC projects. Another aspect to consider was that only cost-effective, readily available raw materials should be used. The active cell area also had to be significantly increased to approximately 1000 cm^2 in order to advance development towards MW electrolysis. The starting components for the newly developed bipolar unit are shown in Figure 8. A distinguishing feature of the bipolar unit is that all of its components are comprised of Ti. According to Lædre et al. [77], Ti is a material that exhibits good corrosion stability under electrolysis conditions. To counteract degradation effects, an additional coating may need to be applied to the contact areas of the MEA. The setup shown in Figure 8 reveals a simple central plate with drill holes and elongated holes for the supply of media. The double-sided simple frames merely contain holes that enable media to pass through. Three-part expanded metal sandwiches are inserted into the frame interiors. The elongated holes in the center plate allow the media to flow in and out of these areas. The expanded metal sandwiches serve as flow distributors. A Ti fleece acting as a fine distribution structure can be found on the anode side of the bipolar unit.

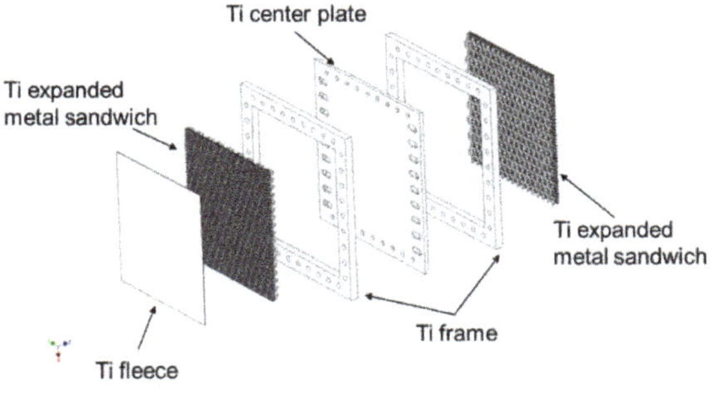

Figure 8. Exploded view of the bipolar plate unit comprising multiple parts from Energy Lab 2.0 and LLEC prior to the joining process. The active cell area of its basic quadratic design amounts to 1056 cm^2.

All previously described components are connected to each other in an interlocking manner by means of diffusion welding. This process takes place in a vacuum oven in which the stacked components are heated by thermal radiation up to a temperature close to enabling phase transition (882 °C for Ti). When the desired temperature is reached, a defined pressure is applied to the components through a mobile stamp in the oven.

As is shown in Figure 9, the stack assembly can be handled manually and easily by two people. Simple assembly aids in the form of plastic rods enhance the positioning accuracy of the components. The repeating units now comprise the following components: a bipolar unit, flat gasket, carbon fiber sheet, and an MEA. In Figure 9c, a fully assembled short stack can be seen. The performance and long-term behavior of the stack in operation has yet to be determined and will be documented and discussed in a later publication. The nominal power of a stack with 70 cells of this type will be in the range of 400 kW.

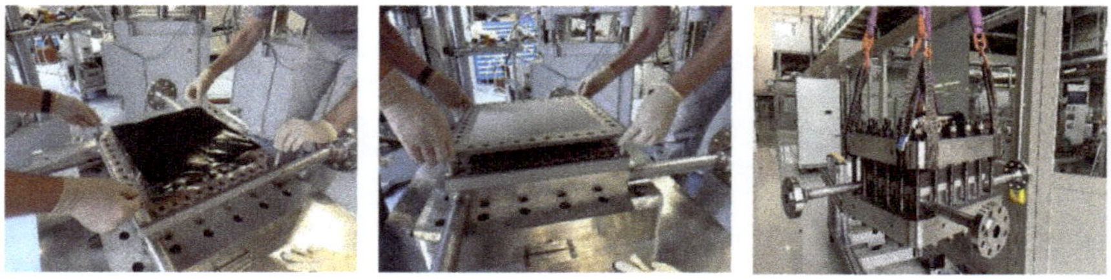

(a) (b) (c)

Figure 9. Manual stack assembly with one-part bipolar plate; (**a**): MEA; (**b**): bipolar unit; (**c**): interlocked stack.

3.3. Design and Setup of the Test Facility

On the basis of interdisciplinary approaches, Jülich researchers and engineers work on electrolysis technologies at various stages of development. A relatively new technology for industrial-scale application is polymer electrolyte membrane water electrolysis, which features dynamic operation and a high overload capacity. A new test facility, which is unique in this way, was developed in close collaboration with the Canadian company Greenlight, which is schematically illustrated in Figure 10. The test facility allows for the comprehensive characterization of PEM electrolysis stacks with an electrical output of up to 500 kVA with current intensities of up to 4000 A_{DC}/10,000 A_{DC}. The pressures can be flexibly adjusted between 4 bar and 50 bar for the anode and cathode circuits. Characterization can thus be flexibly performed at various pressure levels with balanced pressure or differential pressures.

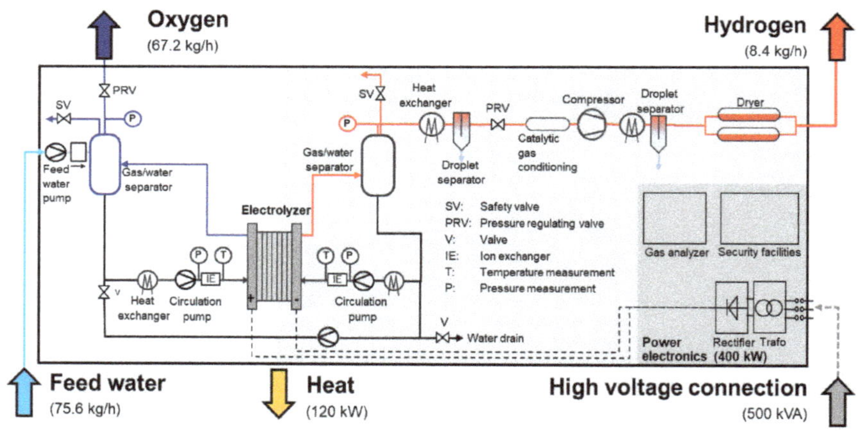

Figure 10. Process overview of the MW electrolysis system.

The main features of the test facility are as follows: power electronics enable the conversion of 400 VAC mains electricity into direct voltage up to 125 VDC. The anode and cathode are each fitted with their own gas/water circuits, which are used for heating and cooling the stack as well as supplying it with water. The gas separators also separate the converted water from the produced gas in the circuits and can regulate the produced gases at pressures of up to 50 bar by means of membrane pressure regulators. Therefore, the gas can be made available for consumers or temporary storage. For stable and disruption-free electrolysis operation, the regulation of pressure to minimize the differential pressure between the two gas circuits is very important. The deionized fresh water required for electrolysis is supplied by an in-house system and introduced to the test stand using a booster pump in accordance with the current operating pressure. In addition to regulating pressure, the temperature is also optimally regulated, which has a significant influence on achieving optimal operation and helps to compensate for highly dynamic load changes and the resulting waste heat as quickly as possible.

Comprehensive recording and processing of the measurement data allows for automated operation under predefined load profiles and ensures the transfer of data to overarching control systems of EnergyLab and LLEC living labs, which are described in the following sections. Data are connected to the control systems of the LLEC, which enables optimal electrolysis operation in terms of energy efficiency and performance in the entire network of systems by means of a specially initiated data exchange in the Jülich virtual local area network (VLAN). The protocol used for the data exchange is based on the message queuing telemetry transport (MQTT) network protocol. The MQTT is very well suited for large networks of systems with high data transfer rates, as is predominantly the case with automation technology. Therefore, it is important that all units in the LLEC are equipped with an MQTT protocol. All required data are collected and evaluated centrally with the LLEC's control systems. To enable optimized operation in the entire network of systems, the desired values must be transferred to the available systems in addition to the evaluated data. This again takes place using the MQTT protocol to ensure optimized control at the highest control level. The electrolyzer itself is then capable of regulating the desired values and can implement the provisions of the LLEC's control system. Safety-based limits for the LLEC's desired values are integrated into the test stand control system and help protect the personnel and systems.

The test facility shown in Figure 11a enables the integration of various PEM electrolysis stacks up to an electrical power input of 400 kW for an active cell area of up to 1200–3000 cm^2 if the power electronics reserve is expanded for currents of up to 10,000 ADC.

Power	kW_{el}	400	73 cells, 1056 cm^2			
H_2 production	m^3_N/min	2	50 μm Nafion membrane	Voltage	V_{DC}	Max. 125
O_2 production	m^3_N/min	1		Current	A_{DC}	4000 @ 100 V_{DC}
Pressure	bar_g	2–50	One-piece bipolar plates			
Temperature	°C	≤95				
(a)			(b)	(c)		

Figure 11. Electrolysis facility with first stack in the pilot plant; (**a**): PEM electrolysis test facility (GREENLIGHT); (**b**): 400 kW PEM stack (EK-14, FZJ); (**c**): IGBT-based AC/DC converter (AIXCOM).

From a safety perspective, the legal and operational provisions were implemented, adherence to which was confirmed by successful inspections of the Technical Supervisory Association (TÜV), particularly in the areas of explosion protection and pressure equipment. From a technical standpoint, many of the required inspections were performed with the aid of a safety-oriented programmable logic controller (PLC).

The entire electrolysis system is part of a technical center that is equipped with the necessary facilities for the supply of media and energy, as well as the forward transport, conditioning, and removal of all products from the electrolysis process. The technical center also features the technical facilities required for ventilation and the safe removal of media, as well as a control room in which all measurements are recorded and processed.

3.4. Energy Lab 2.0: Real-Life Laboratory and Simulation Platform for the Energy-Related Testing of Sustainable Conversion and Storage Technologies

Energy Lab 2.0 [78] is a large-scale research infrastructure of the Helmholtz Association. Its mission is to develop technological solutions for a smart and integrated energy system for a defossilized future. Energy Lab 2.0 allows for technology-oriented research on a demonstration scale in order to successfully integrate renewable energy (RE) into the power grid.

In particular, all relevant components of a future energy system are considered, namely RE generation, storage and grid integration, and power-to-X conversion technologies (X = gas, liquid, heat)—all complemented with a comprehensive system analysis. At Karlsruhe Institute of Technology (KIT), the majority of the components have been set up. Moreover, certain components are located at the German Aerospace Center (DRL) in Stuttgart (Germany) and Forschungszentrum Jülich (Jülich, Germany), respectively (see Figure 12).

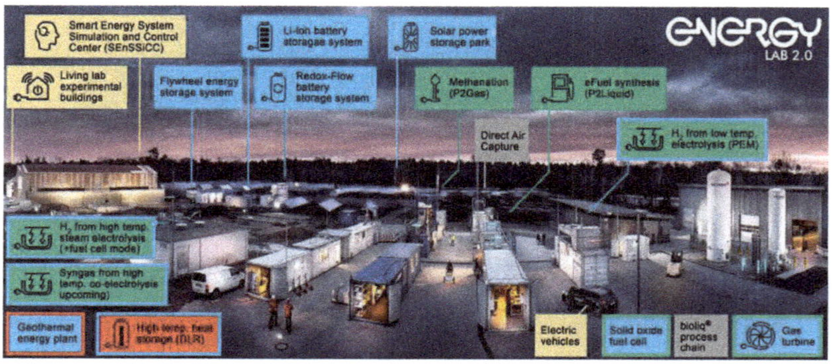

Figure 12. Energy Lab 2.0 at Karlsruhe Institute of Technology (KIT), Karlsruhe, Germany; Photo: M. Breig and A. Bramsiepe (KIT).

For the Karlsruhe site, research at the Energy Lab 2.0 can be subdivided into three major areas: (1) the solar power storage park mainly focuses on short-term energy storage via different types of batteries, whereas the plant network (2) targets chemical energy carriers (power-to-molecules), and (3) the Smart Energy System Simulation and Control Center (SEnSSiCC) focuses on power hardware, the power grid, and consumers/prosumers both experimentally and via simulations. In order to investigate and demonstrate the coupling of the different components, plants are interconnected and can be operated within a smart grid design of interest (see Figure 13). Furthermore, power-to-heat is focused on at the DLR (Stuttgart, Germany), whereas major research is conducted on electrolysis at Forschungszentrum Jülich (Jülich, Germany).

3.4.1. The Solar Power Storage Park of the Energy Lab 2.0

A solar power storage park was built at KIT back in 2014. This storage park includes a 1 MW_p photovoltaic (PV) system, as well as smaller storage systems. In addition, a 1.5 MWh Li-ion battery and an 800 kWh redox flow storage system were integrated into the park as part of the Energy Lab 2.0 project. Storage systems on a large scale will only be used when their deployment is economically-viable. Economic viability is not only influenced by the investment costs but also by the lifetime and design of the systems, the system control, and system design, including their overall efficiency. The impact and optimization potential of these aspects in terms of the economic viability of the systems are being investigated as part of various projects centered around the Energy Lab 2.0. Efficiency losses in such storage systems can occur, for instance, due to the high cooling and heating requirements. In order to reduce these losses, an innovative cooling concept was developed for the 1.5 MWh Li-ion battery, which was integrated into a concrete structure. The thermal component activation of the concrete structure and the use of groundwater to control the temperature of the batteries by indirect water cooling ensure minimized system operating costs (through increased efficiency) and a long lifetime for the battery modules as a result of the enhanced temperature control. There also exists the potential to optimize redox flow batteries (RFBs) in terms of efficiency [79]. In addition, research on intelligent operating strategies is being conducted on RFBs and Li-ion batteries. In the majority of cases, the operator of a Li-ion storage system is provided with information on the state of charge. However, the actual state of charge and the battery's usable energy depend on the system's discharge capacity. A model that is capable of self-learning during operation is currently being developed and would be able to provide information on the state of energy of the entire storage system. Furthermore, one of the smaller storage systems is being operated to replicate a Li-ion storage system in a multifamily home and is virtually connected to a part (30 kW_p) of the 1 MW_p PV system.

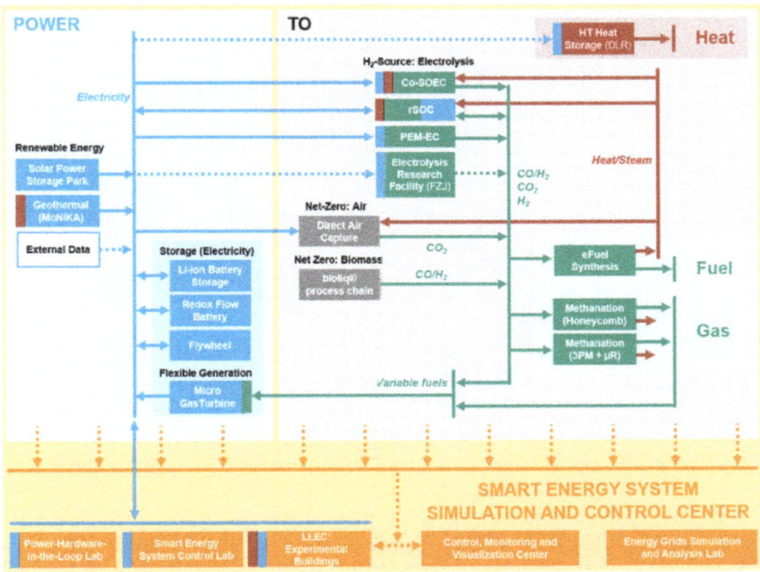

Figure 13. Interacting components: power generation, transformation, and storage in the Energy Lab 2.0 plant network (power-to-X) coupled with energy systems analysis on the theoretical and experimental levels in the Energy Lab 2.0 Smart Energy System Simulation and Control Center. Solid lines—physically-implemented coupling in the Energy Lab 2.0; dashed lines—data and model exchange; blue lines—electricity; green lines—chemicals; red lines—heat; yellow lines—data and models.

Li-ion batteries are subject to particularly strong degradation if they are in a high state of charge for a long period of time. This can be reduced through an intelligent operating strategy [80,81]. PV and load forecasts are both vital components for arriving at such intelligent operating strategies. Currently research is being conducted to determine which type of PV forecasting methods are best suited for intelligent charging strategies [82].

3.4.2. Power-to-Molecules in Decentralized and Highly Efficient Plants

In the future energy system, "power-to-molecules" technologies will be an important piece of the overall puzzle, not only as an option for chemically-storing electrical energy, but also as a source for CO_2-neutral fuels and chemical feedstocks. Within the plant network of the Energy Lab 2.0, there is a focus on the synthesis of hydrocarbons—namely methane and Fischer–Tropsch-based fuels—from CO_2 and H_2. Although green hydrogen can be obtained through electrolysis, non-fossil CO_2 must be captured from the air (direct air capture—DAC), generated from biomass, or separated from unavoidable industrial point sources that cannot be decarbonized, for example, by electrification. At the Energy Lab 2.0, a PEM electrolysis unit (100 kW_{AC} nominal power, operating pressure of up to 47 bar) and a high-temperature reversible steam electrolysis/fuel cell unit (electrolysis mode: 150 kW_{AC}; fuel cell mode: H_2: 25 kW_{AC}, CH_4: 20 kW_{AC}) are used to investigate integration into certain process chains. Within the Kopernikus Project "P2X", funded by the German Federal Ministry of Education and Research, a high-temperature co-electrolysis system will be installed. In order to capture CO_2 from ambient air, DAC is integrated into the plant network, for example, within the framework of the Power-Fuel project funded by the German Federal Ministry for Economic Affairs and Energy.

Predetermined by the local and temporal availability of both non-fossil CO_2 and green electricity in a future defossilized energy system solely based on renewable energy, power-to-molecule process chains must be tolerant to input fluctuations (not only in terms of minutes but also on a day/season scale) and intensified, also allowing for decentralized application. Dynamic operation and process intensification are the major objectives investigated in the synthesis plants at the Energy Lab 2.0, namely the three-phase methanation [83] (output equivalent to 100 kW) and eFuel synthesis plants (1 bpd Fischer–Tropsch products).

While conventional plants for eFuel synthesis, for example, via the Fischer–Tropsch (FT) route, are designed for steady-state operation due to several hurdles such as hot-spot formation; in contrast, process intensification enables the reactor's volume and plant's complexity to be reduced [84]. This is why, in the eFuel synthesis plant (INERATEC GmbH) of the Energy Lab 2.0, a modular microchannel-based reactor is used. Projects utilizing the Energy Lab 2.0's infrastructure investigate the benefits of process-intensified and modular equipment on the design of the process chain. For example, in the PowerFuel project, the required tank size for buffering the hydrogen produced from renewable sources was investigated. The results show that it is possible to drastically reduce the intermediate tank size due to the ability of the microchannel-based FT reactor and of the eFuel synthesis plant in general to operate under reduced feed within a response time of a few minutes [85].

In a study conducted at KIT, the eFuel synthesis plant of the Energy Lab 2.0 was simulated in AspenDynamics within the framework of the aforementioned PEM electrolyzer plant (160 kW_{AC} overload situation), the 50 m^3 hydrogen tank (which corresponds to up to 2300 Nm^3 H_2) and an assumed 320 kW_p PV field with experimentally-determined power profile data of the solar power storage park (two weeks in the spring of 2015). The electrolyzer in this scenario was assumed to instantaneously follow the PV profile (with a 10 min temporal resolution). Only the ramp-up was capped in the model, as at least 40% of the nominal load was required to run the system experimentally. It was shown that the control strategy applied for the synthesis plant significantly influences the number of required shutdowns enforced by the unsteady supply of electrical energy and, thus, hydrogen in such a scenario. When running the synthesis plant under steady-state conditions, its operating time can be as low as 35% and 48% when applying constant H_2 feed equivalent to 150 kW_{el} or 115 kW_{el} electrolyzer load, respectively. In contrast, when running the synthesis plant dynamically, depending on the hydrogen production rate or a desired tank pressure level (see Figure 14), the operating time can be significantly increased to 70% or even 83%, respectively.

For dynamic synthesis operation, it was also shown via experiments on a relevant lab scale (H_2 feed equivalent to 500 W_{el} electrolyzer load) that the ability of micro-structured reactors to cope with feed fluctuations does not influence the product quality, nor does it induce complete hydrogen consumption in the FT reactor, which could harm the catalyst due to the presence of the remaining CO. This is affected by the possibility of effectively controlling the reactor temperature [86].

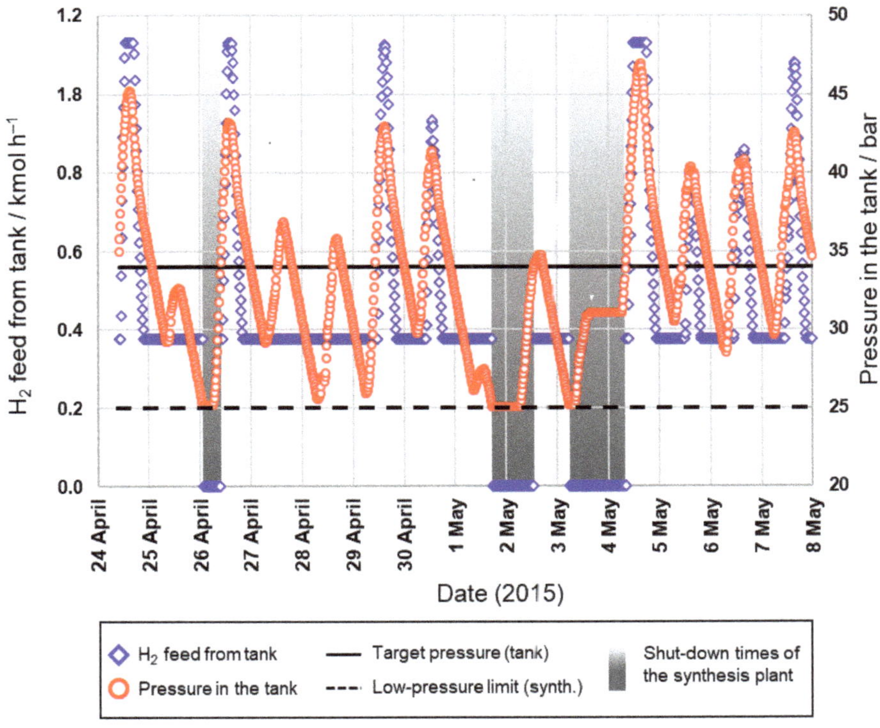

Figure 14. AspenDynamics results for hydrogen supply from the 50 m^3 hydrogen buffer tank and the pressure in the tank as a function of time with underlying PV data of two weeks in spring 2015. The scenario considers the dynamic operation of the electrolyzer and pressure-dependent control of the synthesis plant. Low pressure limit for running the synthesis plant: 25 bar.

3.4.3. The Smart Energy System Simulation and Control Center

The necessary coupling of various energy sectors in future and the fluctuation in the generation of power from renewable energy sources present an enormous challenge regarding the control and operation of future energy systems. In order to perform control and monitoring tasks under the most realistic conditions possible, the Smart Energy System Simulation and Control Center (SEnSSiCC) [87] was established as part of the Energy Lab 2.0 project. The SEnSSiCC brings together work on information technologies and the corresponding research aspects of the Energy Lab 2.0 (see Figures 13 and 15).

The following sub-labs form the SEnSSiCC: the Smart Energy System Control Laboratory (SESCL), which acts as a representation of the real power grid of the future where the most important energy systems are flexibly interconnected through a busbar matrix, ensuring that experimental configurations can be quickly changed; the Energy Grids Simulation and Analysis Laboratory (EGSAL), where the topography of future energy grids is simulated with the virtual integration of components [88] that are not available at KIT's Campus Nord; the Energy Lab 2.0's Control, Monitoring and Visualization Center (CMVC), where software tools are being developed for the control room of future energy grids [89]; and the power hardware-in-the-loop (PHiL) laboratory test environment, which enables real hardware components to be integrated into a simulated 1 MVA grid and subjected to stress tests, as in [90]. The SEnSSiCC is complemented by the Living Lab Energy Campus (LLEC, see the next chapter), also a research infrastructure of the Helmholtz Association. Three experimental buildings with identical constructions are connected to the SEnSSiCC in terms of electrical technology and data and thus expand the portfolio of energy technology systems used for experiments.

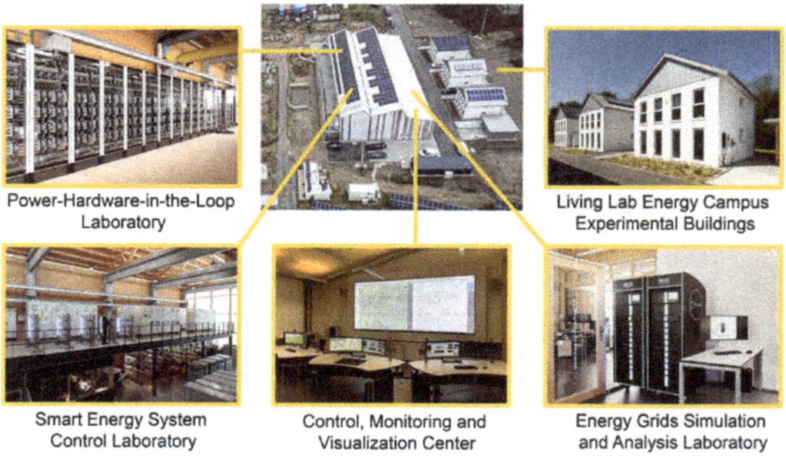

Figure 15. Smart Energy System Simulation and Control Center within the Energy Lab 2.0. Photo: M. Breig (KIT).

The Smart Energy System Control Laboratory focuses on the support of developments and innovations in the field of smart grids under realistic conditions. The SESCL is a laboratory in which state-of-the-art energy technology systems and new control algorithms are developed and tested [91]. This laboratory is galvanically isolated from the public grid, so that the control algorithms can be approved and investigated in limit ranges (frequency, voltage). In addition, it is possible to control operating points approaching the limits of stability. Such experiments would not be possible using the public grid, as the risk of a complete grid failure is too high. In order to conduct experiments on low-voltage networks under realistic conditions, the laboratory provides a number of energy systems, such as electricity generators (e.g., PV systems, wind power units), electrical machines, generators, inverters, storage systems (e.g., Li-ion batteries), and charging stations, that can be very flexibly interconnected by taking into account the physical properties of the connecting lines to form a micro-grid with a variable topology. To model the operating behavior of the cable/overhead lines, the laboratory features real means of transmission and line replicas that physically simulate this behavior using discrete resistor–inductor–capacitor (RLC) components. In terms of the interconnection, flexibility is ensured by a matrix of alternating and direct current busbars and contactors that are controlled by a centralized automation system. Load shedding and the addition of further producers and consumers, or a combination of the two (prosumers) can be achieved very easily.

3.5. *Living Lab Energy Campus: Integrated Research Platform for Energy-Related Coupling of Pioneering Conversion, Distribution, and Storage Technologies*

The Living Lab Energy Campus (LLEC) project is an integrated research platform (see Figure 16) for the coupling of electrical, thermal, and chemical energy flows through an intelligent sensor and control system. Various production, distribution, and storage systems are integrated into the energy supply of the research campus and are then monitored and optimally controlled according to predefined constraints by an adaptive, cloud-based, and model-predictive IT platform. In addition, digital models have been developed for different energy grids, energy demonstrations, and various building types. The demonstrations include various photovoltaic systems, two large Li-ion batteries, a low-temperature waste heat network, and a hydrogen sector with a high level of sector coupling. A large number of sensors collect data via the fluctuating energy flows at the research site. A special significance is attached to achieving an optimal interaction between humans and technology.

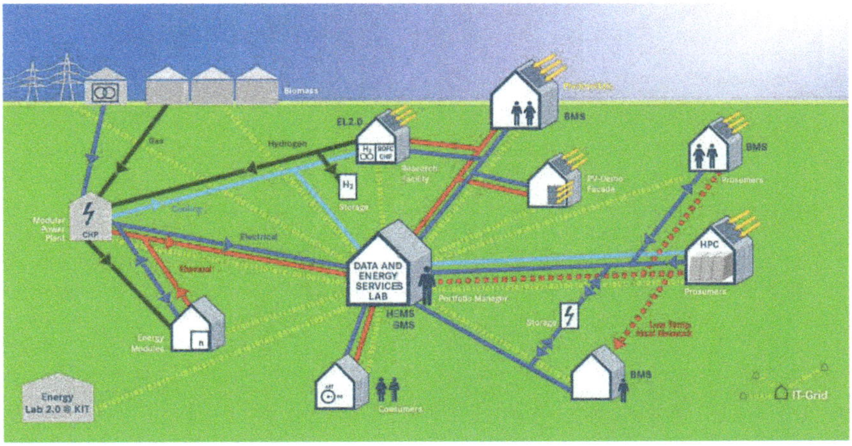

Figure 16. Living Lab Energy Campus: living lab and research platform.

A thermal energy center (WVVZ) designed as a modular system supplies the campus with electrical energy and heat. To this end, gas mixtures comprised of natural gas with biogas and hydrogen are used in the installed gas engines. An additional low-temperature network uses waste heat from the JUWELS supercomputer to supply heat to the surrounding buildings by means of heat pumps and simultaneously investigate the electrical and thermal utilization of the network, as well as potential resource savings. Various photovoltaic systems form the basis of the renewable energy supply and are integrated into the energy system in the form of conventional rooftop installations, free-standing modules, and modules fixed to the building façade. The fluctuations in the energy feed that occur are offset by two large Li-ion batteries. At the same time, these batteries serve as an uninterruptible power supply for a selection of sensitive research infrastructures. The LLEC was recently expanded to encompass research issues surrounding vehicle-to-grid (V2G) systems. Various bidirectionally operable charging infrastructures for electric research vehicles are currently being established.

One system- and process-related focus of the LLEC is the long-term and seasonal storage of renewable energy using hydrogen from the MW electrolysis system. As is shown in Figure 1, hydrogen and oxygen from the water electrolysis process each flow through a pressure-resistant pipeline, with only the hydrogen flowing into a pressure tank as additional storage. The hydrogen can also be stored in LOHCs. The loading and unloading of the LOHC takes place in a novel one-reactor system, which for the first time, demonstrates both process steps in a system with a performance of 300 kW_p. The level of heat required for hydrogenation is approximately 300 °C and is obtained from the waste gas of the gas engines. The heat that is generated during dehydrogenation is supplied to the heat supply system on campus. Significantly higher rates of storage efficiency are achieved through this type of sector coupling. All forms of storage enable the stockpiling of gases over many weeks and months before hydrogen can be used to generate electricity and heat in the WVVZ's gas engines, and hydrogen and oxygen are able to do the same in an alkaline fuel cell.

As they are incorporated in a climate-neutral office complex, all of the energy systems can be monitored in the Data and Energy Services Lab, where researchers are also able to plan and evaluate future energy system experiments together with their colleagues from the technical infrastructure department.

4. Conclusions

This article describes the detailed design and unique capabilities of the test facility for large-scale electrolysis stacks that produce substantial quantities of hydrogen with a high degree of efficiency and virtually interact with various other facilities in a future renewable energy system.

In contrast to the methods using ceramic or alkaline electrolytes, electrochemical water splitting in cells with PEMs is distinguished by the fact that large systems can be established on an MW scale and operated in a broad power range between 5%, 100% and beyond. To limit performance losses and degradation effects, iridium oxide (IrO_2) and platinum (Pt) are used as catalysts for the electrodes in PEM electrolyzers. Intensive efforts are being made to reduce the current catalyst loading to 0.3 mg/cm^2 at 6 W/cm^2 by 2035. Implementing measures for highly efficient, stable and durable H_2 production in high-MW technical systems should help reduce costs fourfold, to around €585/kW by 2030.

For the reproducible production of 1056 cm^2 MEAs, a catalyst-containing dispersion was applied in a controlled manner to a PTFE substrate using a slot die. The solvent of the dispersion layer was subsequently expelled from the roll-to-roll coating system as part of a continuous drying step, ensuring that no layer defects occurred. The coated and dried sheets were then rolled up and cut. The PTFE-supported anode and cathode layers were subsequently formed into an MEA in a discontinuous hot press using a Nafion membrane. Characterization measurements on samples from the produced electrode sheets revealed a cell voltage of 1.66 V and a cell efficiency rate of 76% at a current density of 2 A/cm^2. The proportion of hydrogen measured in the oxygen was less than 0.05% across the entire current density range.

The 400 kW stack was designed for the MW electrolysis system based on an active cell area of 1056 cm^2 and comprised a three-part bipolar unit. Two distribution structures were attached on both sides to a central plate with drill holes that enabled the supply of media. The two structures were comprised of a frame, which in one option consists of an expanded metal sandwich and in another, an expanded metal sandwich with an additional Ti fleece. In addition to a bipolar unit, the easy-to-assemble repeating unit consisted of a flat gasket, carbon fiber sheet, and an MEA.

A new test facility, which is unique in its ability to investigate the performance, dynamics and durability of PEM electrolysis stacks with a cell area of up to 1200–3000 cm^2, enables comprehensive operational testing at an electrical power input of up to 500 kVA and a current strength of up to 4000 A_{DC}, or even up to 10,000 A_{DC} in a subsequent expansion step. Highly sensitive pressure and temperature control ensures safe operation during highly dynamic load changes that correspond to the operating characteristics of renewable power sources. The recording and processing of measurement data enables automatic operation with predefined load profiles and ensures the transfer of data to the control systems of the EnergyLab 2.0 and LLEC living labs on the basis of a message queuing telemetry transport (MQTT) network protocol.

The Energy Lab 2.0 is used by pertinent researchers and technicians as a real-life laboratory and simulation platform for the energy-related testing of sustainable conversion and storage technologies on a technological demonstration level. This includes all relevant aspects, such as renewable electricity production and energy storage, grid integration, and power-to-molecule technologies, which are needed for the sustainable operation of a renewable-based energy system in future. Experimental research and development task work is complemented by extensive systems analyses. Using a 1.5 MWh Li-ion battery and an 800 kWh redox flow storage system, positive effects on the efficiency and lifetime of systems are achieved and further developed with innovative cooling concepts, the optimization of operating strategy, and the use of self-learning models. With the synthesis of methane or Fischer–Tropsch-based (FT) fuels consisting of hydrogen and CO_2, there is a focus on enhancing the process chain efficiency and dynamic operation to cope with intermittent power supplies. Operation of the microchannel-based FT plant was successfully simulated under stationary and dynamic operating conditions with real data from the PV field interconnected to electrolysis and the hydrogen buffer. On a relevant laboratory scale, the suitability of micro-structured reactors was demonstrated. Due to their efficient temperature control for FT synthesis, even with a fluctuating supply of hydrogen, neither the quality of the product nor the stability of the catalyst is affected. In the Energy Lab 2.0, the Smart Energy System Simulation and Control Center (SEnSSiCC) simulates the coupling of energy sectors with fluctuating energy producers of the future and supports developments and innovations under realistic conditions for a pioneering smart grid. The SEnSSiCC brings together numerous laboratories and institutions that will help depict the energy network of the future. Alongside the development and testing of new control algorithms, control strategies under critical operation conditions and variable network topologies can be safely and extensively investigated.

The Living Lab Energy Campus (LLEC) is an integrated research platform through which innovative production, distribution, and storage systems are investigated while integrated into the energy supply of the research campus, as well as being monitored and controlled by an adaptive, cloud-based, and model-predictive IT platform. Through connection to the LLEC electrolysis system in terms of data and gas, the aim is to investigate and further develop disruption-free processes and process efficiencies for the coupling of electricity and hydrogen generation, as well as for the production and storage of hydrogen in various forms, and H_2 storage and reconversion.

Author Contributions: Conceptualization, B.E.; methodology, B.E.; software, R.K., V.H. and S.W.; formal analysis, B.E. and M.M.; investigation, M.H., H.J., R.K., M.S., A.S., S.W., M.R. and N.M.; resources, B.E., V.H., R.D. and S.K.; writing—original draft preparation, B.E.; writing—review and editing, M.M., M.H., H.J., R.K., M.S., A.S., P.P., S.W., M.R., N.M. and S.K.; visualization, B.E., M.H., H.J., R.K., M.S., A.S., P.P., S.W., M.R., N.M. and S.K.; supervision, B.E. and M.R. All authors have read and agreed to the published version of the manuscript.

Funding: This research received no external funding.

Institutional Review Board Statement: Not applicable.

Informed Consent Statement: Not applicable.

Data Availability Statement: Not applicable.

Acknowledgments: All of the authors would like to thank the unnamed colleagues from the participating institutes of Jülich, KIT, and DLR for their fruitful discussions and excellent cooperation as part of the Energy Lab 2.0 and Living Lab Energy Campus projects. The Energy Lab 2.0, the Living Lab Energy Campus (LLEC), and the MW electrolysis system are funded by the Federal Ministry of Education and Research (BMBF), the Federal Ministry for Economic Affairs and Energy (BMWi), the Baden-Württemberg Ministry of Science, Research and the Arts (MWK-BW), and the Helmholtz Association of German Research Centers (HGF). All work related to the Solar Power Storage Park of the Energy Lab 2.0 contributes to the re-search performed at (KIT-BATEC) KIT Battery Technology Center and CELEST (Center for Elec-trochemical Energy Storage Ulm-Karlsruhe).

Conflicts of Interest: The authors declare no conflict of interest.

References

1. Falcone, P.M.; Hiete, M.; Sapio, A. Hydrogen economy and sustainable development goals: Review and policy insights. *Curr. Opin. Green Sustain. Chem.* **2021**, *31*, 100506. [CrossRef]
2. Capurso, T.; Stefanizzi, M.; Torresi, M.; Camporeale, S.M. Perspective of the role of hydrogen in the 21st century energy transition. *Energy Convers. Manag.* **2022**, *251*, 114898. [CrossRef]
3. Caglayan, D.G.; Weber, N.; Heinrichs, H.U.; Linßen, J.; Robinius, M.; Kukla, P.A.; Stolten, D. Technical potential of salt caverns for hydrogen storage in Europe. *Int. J. Hydrogen Energy* **2020**, *45*, 6793–6805. [CrossRef]
4. Gao, W.; Liang, S.; Wang, R.; Jiang, Q.; Zhang, Y.; Zheng, Q.; Xie, B.; Toe, C.Y.; Zhu, X.; Wang, J.; et al. Industrial carbon dioxide capture and utilization: State of the art and future challenges. *Chem. Soc. Rev.* **2020**, *49*, 8584–8686. [CrossRef] [PubMed]
5. Erans, M.; Sanz-Pérez, E.S.; Hanak, D.P.; Clulow, Z.; Reiner, D.M.; Mutch, G.A. Direct air capture: Process technology, techno-economic and socio-political challenges. *Energy Environ. Sci.* **2022**, *15*, 1360–1405. [CrossRef]
6. Rego de Vasconcelos, B.; Lavoie, J.-M. Recent Advances in Power-to-X Technology for the Production of Fuels and Chemicals. *Front. Chem.* **2019**, *7*, 392. [CrossRef]
7. Robinius, M.; Markewitz, P.; Lopion, P.; Kullmann, F.; Heuser, P.-M.; Syranidis, K.; Cerniauskas, S.; Reuß, M.; Ryberg, S.; Kotzur, L.; et al. *Kosteneffizient und Klimagerechte Transformationsstrategien für das Deutsche Energiesystem bis Zum JAHR 2050 (Kurzfassung)*; Forschungszentrum Jülich GmbH: Jülich, Germany, 2019.

8. Forschungszentrum-Jülich. Living Lab Energy Campus. Available online: https://www.fz-juelich.de/llec/ (accessed on 31 December 2021).
9. Smolinka, T.; Wiebke, N.; Sterchele, P.; Lehner, F.; Jansen, M. *Studie_IndWEDe-Industrialisierung der Wasserelektrolyse in Deutschland: Chancen und Herausforderungen für Nachhaltigen Wasserstoff für Verkehr, Strom und Wärme*; NOW GmbH: Berlin, Germany, 2018.
10. Smolinka, T.; Günther, M.; Garche, J. *NOW-Studie: Stand und Entwicklungspotenzial der Wasserlektrolyse zur Herstellung von Wasserstoff aus regenarativen Energien*; NOW GmbH: Berlin, Germany, 2011.
11. Winter, C.-J.; Nitsch, J. (Eds.) *Wasserstoff als Energieträger*, 2nd ed.; Springer: Berlin, Germany, 1989; p. 384. [CrossRef]
12. Marshall, A.; Borresen, B.; Hagen, G.; Tsypkin, M.; Tunold, R. Preparation and characterisation of nanocrystalline IrxSn1-xO2 electrocatalytic powders. *Mater. Chem. Phys.* **2005**, *94*, 226–232. [CrossRef]
13. Marshall, A.; Tsypkin, M.; Borresen, B.; Hagen, G.; Tunold, R. Nanocrystalline IrxSn(1-x)O2 electrocatalysts for oxygen evolution in water electrolysis with polymer electrolyte-Effect of heat treatment. *J. New Mat. Electr. Syst.* **2004**, *7*, 197–204.
14. Trasatti, S. Electrocatalysis in the Anodic Evolution of Oxygen and Chlorine. *Electrochim. Acta* **1984**, *29*, 1503–1512. [CrossRef]
15. Andolfatto, F.; Durand, R.; Michas, A.; Millet, P.; Stevens, P. Solid Polymer Electrolyte Water Electrolysis-Electrocatalysis and Long-Term Stability. *Int. J. Hydrogen Energy* **1994**, *19*, 421–427. [CrossRef]
16. Millet, P.; Andolfatto, F.; Durand, R. Design and performance of a solid polymer electrolyte water electrolyzer. *Int. J. Hydrogen Energy* **1996**, *21*, 87–93. [CrossRef]
17. Yamaguchi, M.; Okisawa, K.; Nakanori, T. Development of high performance solid polymer electrolyte water electrolyzer in WE-NET. In Proceedings of the IECEC-97 Thirty-Second Intersociety Energy Conversion Engineering Conference (Cat. No.97CH6203), Honolulu, HI, USA, 27 July–1 August 1997; pp. 1958–1965. [CrossRef]
18. Ledjeff, K.; Mahlendorf, F.; Peinecke, V.; Heinzel, A. Development of Electrode Membrane Units for the Reversible Solid Polymer Fuel-Cell (Rspfc). *Electrochim. Acta* **1995**, *40*, 315–319. [CrossRef]
19. Rasten, E.; Hagen, G.; Tunold, R. Electrocatalysis in water electrolysis with solid polymer electrolyte. *Electrochim. Acta* **2003**, *48*, 3945–3952. [CrossRef]
20. Ma, H.C.; Liu, C.P.; Liao, J.H.; Su, Y.; Xue, X.Z.; Xing, W. Study of ruthenium oxide catalyst for electrocatalytic performance in oxygen evolution. *J. Mol. Catal. A-Chem.* **2006**, *247*, 7–13. [CrossRef]
21. Hu, J.M.; Zhang, J.Q.; Cao, C.N. Oxygen evolution reaction on IrO2-based DSA (R) type electrodes: Kinetics analysis of Tafel lines and EIS. *Int. J. Hydrogen Energy* **2004**, *29*, 791–797. [CrossRef]
22. Song, S.D.; Zhang, H.M.; Ma, X.P.; Shao, Z.G.; Baker, R.T.; Yi, B.L. Electrochemical investigation of electrocatalysts for the oxygen evolution reaction in PEM water electrolyzers. *Int. J. Hydrogen Energy* **2008**, *33*, 4955–4961. [CrossRef]
23. Nanni, L.; Polizzi, S.; Benedetti, A.; De Battisti, A. Morphology, microstructure, and electrocatalylic properties of RuO2-SnO2 thin films. *J. Electrochem. Soc.* **1999**, *146*, 220–225. [CrossRef]
24. de Oliveira-Sousa, A.; da Silva, M.A.S.; Machado, S.A.S.; Avaca, L.A.; de Lima-Neto, P. Influence of the preparation method on the morphological and electrochemical properties of Ti/IrO2-coated electrodes. *Electrochim. Acta* **2000**, *45*, 4467–4473. [CrossRef]
25. Pham, C.V.; Escalera-López, D.; Mayrhofer, K.; Cherevko, S.; Thiele, S. Essentials of High Performance Water Electrolyzers–From Catalyst Layer Materials to Electrode Engineering. *Adv. Energy Mater.* **2021**, *11*, 2101998. [CrossRef]
26. Siracusano, S.; Baglio, V.; Di Blasi, A.; Briguglio, N.; Stassi, A.; Ornelas, R.; Trifoni, E.; Antonucci, V.; Aricò, A.S. Electrochemical characterization of single cell and short stack PEM electrolyzers based on a nanosized IrO2 anode electrocatalyst. *Int. J. Hydrogen Energy* **2010**, *35*, 5558–5568. [CrossRef]
27. Smolinka, T.; Rau, S.; Hebling, C. Polymer electrolyte membrane (PEM) water electrolysis. In *Hydrogen and Fuel Cells*; Wiley-VCH Verlag GmbH & Co. KGaA: Weinheim, Germany, 2010; pp. 271–289.
28. Carmo, M.; Fritz, D.L.; Mergel, J.; Stolten, D. A comprehensive review on PEM water electrolysis. *Int. J. Hydrogen Energy* **2013**, *38*, 4901–4934. [CrossRef]
29. Bernt, M.; Siebel, A.; Gasteiger, H.A. Analysis of Voltage Losses in PEM Water Electrolyzers with Low Platinum Group Metal Loadings. *J. Electrochem. Soc.* **2018**, *165*, F305–F314. [CrossRef]
30. Alia, S.M.; Stariha, S.; Borup, R.L. Electrolyzer Durability at Low Catalyst Loading and with Dynamic Operation. *J. Electrochem. Soc.* **2019**, *166*, F1164–F1172. [CrossRef]
31. Minke, C.; Suermann, M.; Bensmann, B.; Hanke-Rauschenbach, R. Is iridium demand a potential bottleneck in the realization of large-scale PEM water electrolysis? *Int. J. Hydrogen Energy* **2021**, *46*, 23581–23590. [CrossRef]
32. Carmo, M.; Keeley, G.P.; Holtz, D.; Grube, T.; Robinius, M.; Müller, M.; Stolten, D. PEM water electrolysis: Innovative approaches towards catalyst separation, recovery and recycling. *Int. J. Hydrogen Energy* **2019**, *44*, 3450–3455. [CrossRef]
33. Hydrogenics. Hydrogenics Awarded Energy Storage System for E.ON in Germany. In *World's First Megawatt PEM Electrolyzer for Power-to-Gas Facility*; Hydrogenics: Mississauga, CA, USA, 2013; Available online: http://www.hydrogenics.com/about-the-company/news-updates/2013/04/08/hydrogenics-awarded-energy-storage-system-for-e.on-in-germany (accessed on 8 August 2013).
34. Sheridan, E.; Thomassen, M.; Mokkelbost, T.; Lind, A. The development of a supported Iridium catalyst for oxygen evolution in PEM electrolysers. In Proceedings of the 61st Annual Meeting of the International Society of Electrochemistry, Nice, France, 26 September–1 October 2010.
35. Ayers, K.E.; Dalton, L.T.; Anderson, E.B. Efficient Generation of High Energy Density Fuel from Water. *ECS Trans.* **2012**, *41*, 27–38. [CrossRef]

36. Scheepers, F.; Stähler, M.; Stähler, A.; Rauls, E.; Müller, M.; Carmo, M.; Lehnert, W. Improving the Efficiency of PEM Electrolyzers through Membrane-Specific Pressure Optimization. *Energies* **2020**, *13*, 612. [CrossRef]
37. Tjarks, G.; Gibelhaus, A.; Lanzerath, F.; Müller, M.; Bardow, A.; Stolten, D. Energetically-optimal PEM electrolyzer pressure in power-to-gas plants. *Appl. Energy* **2018**, *218*, 192–198. [CrossRef]
38. Tijani, A.S.; Rahim, A.H.A. Numerical Modeling the Effect of Operating Variables on Faraday Efficiency in PEM Electrolyzer. *Procedia Technol.* **2016**, *26*, 419–427. [CrossRef]
39. Schalenbach, M.; Carmo, M.; Fritz, D.L.; Mergel, J.; Stolten, D. Pressurized PEM water electrolysis: Efficiency and gas crossover. *Int. J. Hydrogen Energy* **2013**, *38*, 14921–14933. [CrossRef]
40. Müller, M.; Carmo, M.; Glüsen, A.; Hehemann, M.; Saba, S.; Zwaygardt, W.; Stolten, D. Water management in membrane electrolysis and options for advanced plants. *Int. J. Hydrogen Energy* **2019**, *44*, 10147–10155. [CrossRef]
41. Stähler, M.; Stähler, A.; Scheepers, F.; Carmo, M.; Stolten, D. A completely slot die coated membrane electrode assembly. *Int. J. Hydrogen Energy* **2019**, *44*, 7053–7058. [CrossRef]
42. Stähler, M.; Stähler, A.; Scheepers, F.; Carmo, M.; Lehnert, W.; Stolten, D. Impact of porous transport layer compression on hydrogen permeation in PEM water electrolysis. *Int. J. Hydrogen Energy* **2020**, *45*, 4008–4014. [CrossRef]
43. Liang, P.; Qiu, D.; Peng, L.; Yi, P.; Lai, X.; Ni, J. Structure failure of the sealing in the assembly process for proton exchange membrane fuel cells. *Int. J. Hydrogen Energy* **2017**, *42*, 10217–10227. [CrossRef]
44. Borgardt, E.; Giesenberg, L.; Reska, M.; Müller, M.; Wippermann, K.; Langemann, M.; Lehnert, W.; Stolten, D. Impact of clamping pressure and stress relaxation on the performance of different polymer electrolyte membrane water electrolysis cell designs. *Int. J. Hydrogen Energy* **2019**, *44*, 23556–23567. [CrossRef]
45. Borgardt, E.; Panchenko, O.; Hackemüller, F.J.; Giffin, J.; Bram, M.; Müller, M.; Lehnert, W.; Stolten, D. Mechanical characterization and durability of sintered porous transport layers for polymer electrolyte membrane electrolysis. *J. Power Sources* **2018**, *374*, 84–91. [CrossRef]
46. Ayers, K.E.; Anderson, E.B.; Capuano, C.; Carter, B.; Dalton, L.; Hanlon, G.; Manco, J.; Niedzwiecki, M. Research Advances towards Low Cost, High Efficiency PEM Electrolysis. *ECS Trans.* **2010**, *33*, 3–15. [CrossRef]
47. Liu, C.; Carmo, M.; Bender, G.; Everwand, A.; Lickert, T.; Young, J.L.; Smolinka, T.; Stolten, D.; Lehnert, W. Performance enhancement of PEM electrolyzers through iridium-coated titanium porous transport layers. *Electrochem. Commun.* **2018**, *97*, 96–99. [CrossRef]
48. Luca Bertuccioli, A.C.; Hart, D.; Lehner, F.; Madden, B.; Standen, E. *Study on the Development of Water Electrolysis in the EU*; Fuel Cells and Hydrogen Joint Undertaking: Brussels, Belgium, 2014; p. 69.
49. Langemann, M.; Fritz, D.L.; Müller, M.; Stolten, D. Validation and characterization of suitable materials for bipolar plates in PEM water electrolysis. *Int. J. Hydrogen Energy* **2015**, *40*, 11385–11391. [CrossRef]
50. Gago, A.; Ansar, A.; Gazdzicki, P.; Wagner, N.; Arnold, J. Low Cost Bipolar Plates for Large Scale PEM Electrolyzers. *ECS Trans.* **2014**, *64*, 1039–1048. [CrossRef]
51. Lettenmeier, P.; Wang, R.; Abouatallah, R.; Saruhan, B.; Freitag, O.; Gazdzicki, P.; Morawietz, T.; Hiesgen, R.; Gago, A.S.; Friedrich, K.A. Low-Cost and Durable Bipolar Plates for Proton Exchange Membrane Electrolyzers. *Sci. Rep.* **2017**, *7*, 44035. [CrossRef]
52. Rojas, N.; Sánchez-Molina, M.; Sevilla, G.; Amores, E.; Almandoz, E.; Esparza, J.; Cruz Vivas, M.R.; Colominas, C. Coated stainless steels evaluation for bipolar plates in PEM water electrolysis conditions. *Int. J. Hydrogen Energy* **2021**, *46*, 25929–25943. [CrossRef]
53. Giner, Inc. World-Leading Specialty Electrolyzer Stacks. Available online: https://www.ginerinc.com/specialty-electrolyzer-stacks (accessed on 31 January 2021).
54. Giner, Inc. *Military and Aerospace Electrolyzers (OGP/LPE)*. Available online: http://www.ginerinc.com/products.php?a=OGP (accessed on 31 January 2022).
55. h-tec. Wasserstofferzeugung mit PEM-Elektrolyse. Available online: www.h-tec.com/fileadmin/content/sys/Flyer/HTEC_EL30_dt_web.pdf (accessed on 31 January 2022).
56. Hydrogenics. OnSite Hydrogen Generation HyLYZER® PEM Electrolysis Technology. Available online: http://www.hydrogenics.com/docs/default-source/pdf/2-1-1-1-hylyzer-1-223F620871645.pdf?sfvrsn=2 (accessed on 31 January 2022).
57. ITM Power. HGAS. Available online: http://www.itm-power.com/product/hgas (accessed on 31 January 2022).
58. Protononsite. Hydrogen C Series. Available online: http://protononsite.com/resources/technical%20brochures/c_series_spec_rev_c.pdf (accessed on 31 January 2022).
59. Siemens. SILYZER 100. Available online: http://www.industry.siemens.com/topics/global/de/pem-elektrolyseur/silyzer/silyzer-system/seiten/silyzer-100.aspx (accessed on 31 January 2016).
60. Siemens. SILYZER 200 Basis. Available online: http://www.industry.siemens.com/topics/global/de/pem-elektrolyseur/silyzer/silyzer-system/seiten/silyzer-200-basis.aspx (accessed on 31 January 2016).
61. Waidhas, M. Elektrolyseure zur H2-Erzeugung in Fluktuierenden Energiesystemen. In Proceedings of the VDE Workshop Elektrochemie 2015, Duisburg, Germany, 5–6 May 2015.
62. Siemens. Silyzer 300. Available online: https://assets.new.siemens.com/siemens/assets/public.1524040818.abae9c1e48d6d239c06d88e565a25040ed2078dc.ct-ree-18-047-db-silyzer-300-db-de-en-rz.pdf (accessed on 30 April 2022).
63. Christoph, N.; Fabian, B.; Schwan, H.S.; Philipp, L.; Svenja, K.; Stefan, B.; Josef, K.; Andreas, F.K.; Thomas, P.; Karl-Kiên, C.; et al. *Plan-DelyKaD-Studie über die Planung einer Demonstrationsanlage zur Wasserstoff-Kraftstoffgewinnung durch Elektrolyse mit Zwischenspeicherung in Salzkavernen unter Druck*; DLR: Stuttgart, Germany, 2015; p. 282.

64. Mergel, J.; Carmo, M.; Fritz, D.L. Status on Technologies for Hydrogen Production by Water Electrolysis. In *Transition to Renewable Energy Systems*; Stolten, D., Scherer, V., Eds.; Wiley-VCH: Weinheim, Germany, 2013; pp. 425–450.
65. Michaelis, J.; Junker, J.; Wietschel, M. Eine Bewertung der Regelenergievermarktung im Power-to-Gas-Konzept. *Z. Energ.* **2013**, *37*, 161–175. [CrossRef]
66. Kopp, M.; Coleman, D.; Stiller, C.; Scheffer, K.; Aichinger, J.; Scheppat, B. Energiepark Mainz: Technical and economic analysis of the worldwide largest Power-to-Gas plant with PEM electrolysis. *Int. J. Hydrogen Energy* **2017**, *42*, 13311–13320. [CrossRef]
67. Westküste 100. Available online: https://www.westkueste100.de/ (accessed on 31 January 2022).
68. H2-Wyhlen. Reallabor H2 Wyhlen. Available online: https://www.energiedienst.de/kraftwerke/wasserstoff/reallabor-h2-wyhlen/ (accessed on 31 January 2022).
69. Energiepark-Lauchstädt. Energiepark Bad Lauchstädt. Available online: https://energiepark-bad-lauchstaedt.de/ (accessed on 31 January 2022).
70. Stähler, M.; Friedrich, I. Statistical investigations of basis weight and thickness distribution of continuously produced fuel cell electrodes. *J. Power Sources* **2013**, *242*, 425–437. [CrossRef]
71. Burdzik, A.; Stähler, M.; Friedrich, I.; Carmo, M.; Stolten, D. Homogeneity analysis of square meter-sized electrodes for PEM electrolysis and PEM fuel cells. *J. Coat. Technol. Res.* **2018**, *15*, 1423–1432. [CrossRef]
72. Bessarabov, D. Membranes with Recombination Catalyst for Hydrogen Crossover Reduction: Water Electrolysis. *ECS Trans.* **2018**, *85*, 17–25. [CrossRef]
73. Klose, C.; Trinke, P.; Böhm, T.; Bensmann, B.; Vierrath, S.; Hanke-Rauschenbach, R.; Thiele, S. Membrane Interlayer with Pt Recombination Particles for Reduction of the Anodic Hydrogen Content in PEM Water Electrolysis. *J. Electrochem. Soc.* **2018**, *165*, F1271–F1277. [CrossRef]
74. Stähler, A.; Stähler, M.; Scheepers, S.; Lehnert, W.; Carmo, M. Scalable Implementation of Recombination Catalyst Layers in Membrane Phases for Gas Crossover Reduction in PEM Water Electrolysis. *J. Electrochem. Soc.* **2021**, *169*, 034522. [CrossRef]
75. Scheepers, F.; Stähler, A.; Stähler, M.; Carmo, M.; Lehnert, W.; Stolten, D. Layer Formation from Polymer Carbon-Black Dispersions. *Coatings* **2018**, *8*, 450. [CrossRef]
76. Scheepers, F.; Stähler, A.; Stähler, M.; Carmo, M.; Lehnert, W.; Stolten, D. Steering and in situ monitoring of drying phenomena during film fabrication. *J. Coat. Technol. Res.* **2019**, *16*, 1213–1221. [CrossRef]
77. Lædre, S.; Kongstein, O.E.; Oedegaard, A.; Karoliussen, H.; Seland, F. Materials for Proton Exchange Membrane water electrolyzer bipolar plates. *Int. J. Hydrogen Energy* **2017**, *42*, 2713–2723. [CrossRef]
78. Available online: https://www.elab2.kit.edu (accessed on 31 January 2022).
79. Grün, T.; Mast, M.; Schwarz, B. Großspeicherwahl. *Nachr. Aus Chem.* **2020**, *68*, 37–39. [CrossRef]
80. Munzke, N.; Büchle, F.; Smith, A.; Hiller, M. Influence of Efficiency, Aging and Charging Strategy on the Economic Viability and Dimensioning of Photovoltaic Home Storage Systems. *Energies* **2021**, *14*, 7673. [CrossRef]
81. Verma, B.; Munzke, N.; Hiller, M. Performance of In-House Li-Ion Battery Storage System Based on Various Strategies. In Proceedings of the 35th European Photovoltaic Solar Energy Conference and Exhibition, Brussels, Belgium, 6 November 2018; pp. 1739–1743.
82. Starosta, A.; Kaushik, K.; Jhaveri, P.; Munzke, N.; Hiller, M. A Comparative Analysis of Forecasting Methods for Photovoltaic Power and Energy Generation with and without Exogenous Inputs. In Proceedings of the 38th European Photovoltaic Solar Energy Conference and Exhibition, Online, 6–10 September 2021; pp. 938–945.
83. Held, M.; Schollenberger, D.; Sauerschell, S.; Bajohr, S.; Kolb, T. Power-to-Gas: CO_2 Methanation Concepts for SNG Production at the Engler-Bunte-Institut. *Chem. Ing. Tech.* **2020**, *92*, 595–602. [CrossRef]
84. Loewert, M.; Hoffmann, J.; Piermartini, P.; Selinsek, M.; Dittmeyer, R.; Pfeifer, P. Microstructured Fischer-Tropsch Reactor Scale-up and Opportunities for Decentralized Application. *Chem. Eng. Technol.* **2019**, *42*, 2202–2214. [CrossRef]
85. Pfeifer, P.; Biffar, L.; Timm, F.; Böltken, T. Influence of Power-to-Fuel Plant Flexibility Towards Power and Plant Utilization and Intermediate Hydrogen Buffer Size. *Chem. Ing. Tech.* **2020**, *92*, 1976–1982. [CrossRef]
86. Loewert, M.; Riedinger, M.; Pfeifer, P. Dynamically Operated Fischer–Tropsch Synthesis in PtL—Part 2: Coping with Real PV Profiles. *ChemEngineering* **2020**, *4*, 27. [CrossRef]
87. Hagenmeyer, V.; Kemal Çakmak, H.; Düpmeier, C.; Faulwasser, T.; Isele, J.; Keller, H.B.; Kohlhepp, P.; Kühnapfel, U.; Stucky, U.; Waczowicz, S.; et al. Information and Communication Technology in Energy Lab 2.0: Smart Energies System Simulation and Control Center with an Open-Street-Map-Based Power Flow Simulation Example. *Energy Technol.* **2016**, *4*, 145–162. [CrossRef]
88. Çakmak, H.; Erdmann, A.; Kyesswa, M.; Kühnapfel, U.; Hagenmeyer, V. A new distributed co-simulation architecture for multi-physics based energy systems integration: Analysis of multimodal energy systems. *At-Automatisierungstechnik* **2019**, *67*, 972–983. [CrossRef]
89. Düpmeier, C.; Stucky, K.-U.; Mikut, R.; Hagenmeyer, V. *A Concept for the Control, Monitoring and Visualization Center in Energy Lab 2.0.*; Springer: Cham, Switzerland, 2015; pp. 83–94.
90. Karrari, S.; De Carne, G.; Noe, M. Model validation of a high-speed flywheel energy storage system using power hardware-in-the-loop testing. *J. Energy Storage* **2021**, *43*, 103177. [CrossRef]
91. Wachter, J.; Gröll, L.; Hagenmeyer, V. Adaptive Feedforward Control for DC/DC Converters in Microgrids-A Power Hardware in the Loop Study. In Proceedings of the 2021 9th International Conference on Smart Grid (icSmartGrid), Setubal, Portugal, 29 June–1 July 2021; pp. 49–56.

energies

MDPI

Article

Physical Properties of $Ti_{45}Zr_{38}Fe_{17}$ Alloy and Its Amorphous Hydride

Antoni Żywczak [1,*], Łukasz Gondek [2], Joanna Czub [2], Piotr Janusz [3], Nivas Babu Selvaraj [4] and Akito Takasaki [5]

1 Academic Centre for Materials and Nanotechnology, 30-059 Krakow, Poland
2 Faculty of Physics and Applied Computer Science, AGH University of Science and Technology, 30-059 Krakow, Poland; lgondek@agh.edu.pl (Ł.G.); joanna.czub@agh.edu.pl (J.C.)
3 Faculty of Drilling, Oil, and Gas, AGH University of Science and Technology, 30-059 Krakow, Poland; pjanusz@agh.edu.pl
4 CICECO, Department of Materials and Ceramic Engineering, University of Aveiro, 3810-193 Aveiro, Portugal; snivbabu@gmail.com
5 Department of Engineering Science and Mechanics, Shibaura Institute of Technology, Toyosu, Kotoku, Tokyo 135-8548, Japan; takasaki@shibaura-it.ac.jp
* Correspondence: zywczak@agh.edu.pl

Abstract: The alloys based on Ti-Zr are considered an excellent candidate for hydrogen storage applications. In this communication, we report the results of Fe substitution for Ni in the well-known $Ti_{45}Zr_{38}Ni_{17}$ compound. The parent and related compounds can be obtained as amorphous powders, transforming into the quasicrystalline phase (i-phase) after annealing. The amorphous $Ti_{45}Zr_{38}Fe_{17}$ phase is transformed into the icosahedral quasicrystalline state, and it is a quasi-continuous process. The i-phase is well-developed close to 500 °C. At higher temperatures, the quasicrystal structure transforms into the other phase: the w-phase (an approximant to the crystalline phase) and another crystal phase with a small addition of the $FeZr_3$ and the $Fe_2(ZrTi)_3$. The amorphous $Ti_{45}Zr_{38}Fe_{17}$ phases can be hydrogenated while maintaining the amorphous nature, which constitutes another very fascinating research field for our group. The investigated alloy shows a good capacity for gaseous H_2 at level 2.54 wt.% at elevated temperatures. The ferromagnetic signal of the amorphous TiZrFe comes from magnetic nanocrystallites in the amorphous matrix. After heating, the magnetic signal significantly decreases due to the lack of long-range magnetic ordering in the i-phase of the $Ti_{45}Zr_{38}Fe_{17}$ alloy.

Keywords: hydrogen-storage materials; amorphous alloys; quasicrystalline alloys; magnetic properties; neutron diffraction

Citation: Żywczak, A.; Gondek, Ł.; Czub, J.; Janusz, P.; Selvaraj, N.B.; Takasaki, A. Physical Properties of $Ti_{45}Zr_{38}Fe_{17}$ Alloy and Its Amorphous Hydride. *Energies* **2022**, *15*, 4236. https://doi.org/10.3390/en15124236

Academic Editor: Giovanni Esposito

Received: 29 March 2022
Accepted: 30 May 2022
Published: 9 June 2022

1. Introduction

Hydrogen can be a promising energy carrier for future energy economy [1,2], which, of course, involves changes in the transport infrastructure [3,4] and the adaptation of the entire energy system to the properties of hydrogen. One of the most important challenges is hydrogen storage [5]. In this context, the solid-state alloys that allow for reversible hydrogen storage [6] belong to the class of materials that can be suitable for application.

The titanium-zirconium-based alloys are the second largest class of solids, for which quasicrystallinity was found, e.g., TiZrNi [7] or TiZrFe [8–10]. The Ti-Zr compounds are promising candidates for many applications such as biomedical [11–13], filler metal [14], medium entropy alloys [15], bulk metallic glasses [16], dental implant [17], shape memory alloys [18], and high-entropy alloys [19]. The TiZrNi quasicrystal seems to be especially suitable for hydrogen storage due to its large hydrogen uptake capacity [20,21]. In this case, the hydrogen atoms are situated preferentially near Ti and Zr atoms in the quasicrystal lattice.

The chemical-physical properties of the $Ti_{45}Zr_{38}Ni_{17}$ compositions can be changed by substituting other elements such as Ag [22], Pd [23–25], V [26], Co [27], Cu [28], Li [29], Ce [30], Mg [31], and Fe [32] in order to increase hydrogen sorption.

Since the discovery of quasicrystals, magnetic properties have not been studied much due to the lack of translational symmetry required for establishing long-range magnetic order. However, some quasicrystals exhibited spin-glass-like behavior [33]. Recently, the quasicrystals showing both ferromagnetic and antiferromagnetic behavior have been studied [34–38]. The magnetism in the TiZrNi quasicrystals was highly demanded [39]. The ferromagnetic signal for the $Ti_{45}Zr_{38}Ni_{17}$ originated from nickel nano-cluster precipitations in the quasicrystalline alloy [40].

The present work aims at tracking the transformation of the $Ti_{45}Zr_{38}Fe_{17}$ from the amorphous to the quasicrystalline/crystalline phases by the in-situ neutron diffraction technique and monitoring the magnetic properties. Additionally, our goal is to obtain the amorphous $Ti_{45}Zr_{38}Fe_{17}$ alloy with the highest hydrogen capacity.

2. Experimental

The $Ti_{45}Zr_{38}Fe_{17}$ nanopowders were synthesized by mechanical alloying (MA), which was performed in the Frisch Pulverisette 7 planetary mill. Commercially available titanium (99.9%), zirconium (99.9%), and iron (99.9%) powders were used as starting materials. A mixture of the starting elements corresponding to the chemical composition of the $Ti_{45}Zr_{38}Fe_{17}$ was placed in stainless steel vials (45 mL), which contained stainless steel balls (14 mm in diameter). An initial mass of the powder mixture before MA was 8.5 g, with the ball-to-powder weight ratio equal to 8:1. The vials containing the powder mixture and balls were evacuated by a rotary pump and then refilled with argon gas (99.999%) in a glove box. Then, gaseous argon was pumped out several times in order to extract all spurious gases from the operating atmosphere. The final argon pressure was maintained at 0.1 MPa. The ball acceleration was 15 g, while the maximum alloying time was 40 h. To avoid a temperature increase during MA, alloying periods of 0.5 h were alternated with rest periods of 0.5 h. After the first 20 h of milling, the vials were opened in a glove box and the powder was mixed. Then, the powder was subsequently alloyed for 20 h under an argon atmosphere.

The morphology of the amorphous nanopowder was studied using FEI Versa 3D scanning electron microscope (SEM). The differential scanning calorimetry (DSC) measurements were conducted under an argon gas flow at a heating rate of 5 K/min^{-1}.

The sample was tested using neutron diffraction to check if transformation from the amorphous to the crystal structure occurred. The amorphous phase was hydrided. Hydrogen gas was introduced at a pressure of 4 MPa, and the reaction chamber was heated to 163 °C to initiate hydrogen uptake. The X-ray powder diffraction (XRD) patterns were collected before and after the hydrogenation of the amorphous sample to check its quality. All the above steps were performed following the procedures reported in Ref [27].

Magnetization as a function of temperature from 127 °C to 927 °C was measured using a LakeShore Model 7407 vibrating sample magnetometer (VSM) equipped with an oven under an argon atmosphere (6N) heating rate of 5 $°C/min^{-1}$ in the presence of a magnetic field of 1 Tesla. The sample was mounted with Thermeez 7020 ceramic putty on a quartz rod. The characteristic and Curie temperatures were estimated as the maximum of the first derivative of the curve.

3. Results and Discussion

Figure 1 presents the submicrometer agglomerates of primary <100 nm particles, which were observed in the amorphous specimen.

Figure 1. The SEM images of the amorphous $Ti_{45}Zr_{38}Fe_{17}$.

The phase changes for the amorphous $Ti_{45}Zr_{38}Fe_{17}$ during the heating in the temperature range from 200 °C to 800 °C recorded with the DSC are shown in Figure 2.

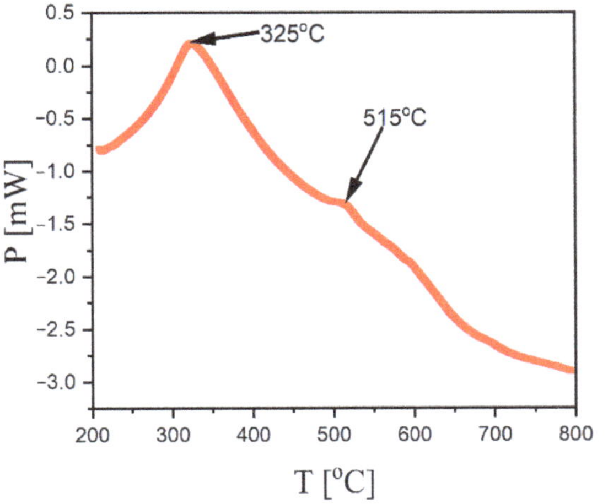

Figure 2. The DSC curve for the $Ti_{45}Zr_{38}Fe_{17}$ amorphous powders was obtained by MA for 40 h.

The broad exothermic shoulder is visible between 200 °C and 700 °C. The energy introduced by mechanical alloying is released during heating, which is visible in the DSC shoulders.

Between 325 °C and 515 °C, the structure changes from amorphous to quasisrystalline with some addition of the crystal phases such as the $FeZr_3$ and the $Fe_2(ZrTi)_3$. The minor peak appears at about 515 °C, which indicates that the quasicrystalline phase is fully formed.

Figure 3 shows the evolution of the neutron diffraction patterns for the amorphous structure during heating. As is apparent from Figure 3a, broad maxima at about 60° of 2θ is changing due to the start of thermal diffusion above roughly 300 °C. The change is connected with the gradual forming of crystalline and quasicrystalline (i) phases. The maxima broadens splitting into small reflections that at 500 °C develop rapidly into well-defined reflections of icosahedral and metastable $FeZr_3$ [41] phases. Also at this temperature, the $Fe_2(ZrTi)_3$ phase appears roughly at 40° of 2θ. The DSC results are consistent showing maxima at 325 and 515 °C, which is understandable as the DSC measurement has a much higher temperature ramp than neutron diffraction measurements. The i-phase can be evidenced by several well-defined reflections, which appears at about 53.5°, 58.8° of the 2θ angle. The results of the neutron diffraction experiment show the quasi-continuous character of the transition from the amorphous phase to the quasicrystalline phase. The icosahedral structure of the $Ti_{45}Zr_{38}Fe_{17}$ starts to evolve into the w-phase above 525 °C. At higher temperatures (750 °C), an additional transition from the w-phase into the cubic phase was evidenced. The transitions at 525 °C and 750 °C are not strongly reflected in DSC as both are not associated with the significant rearrangement of the constituting atoms.

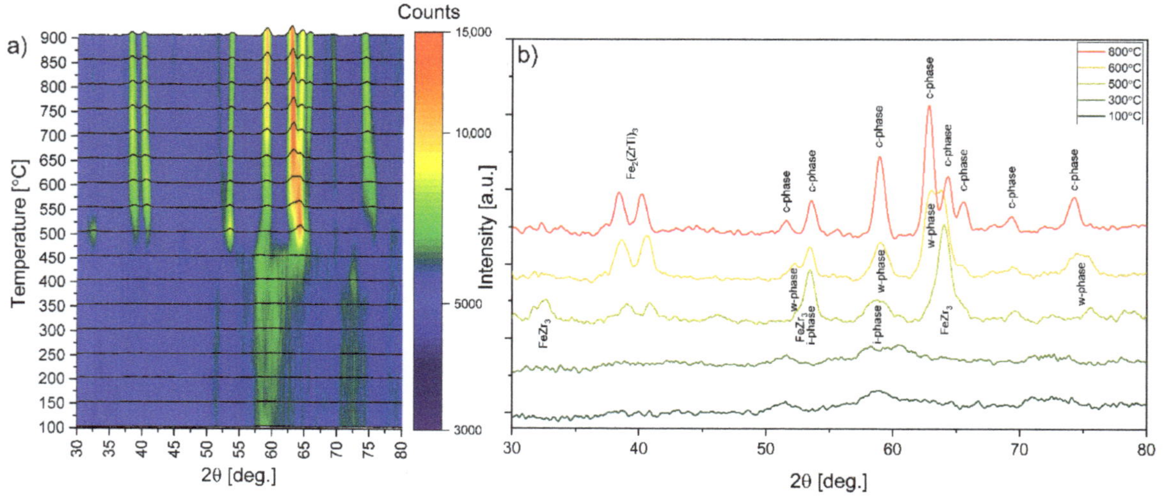

Figure 3. The transformation from the amorphous to the crystal phase for the $Ti_{45}Zr_{38}Fe_{17}$ alloy as seen by neutron diffraction (**a**); the in-situ neutron diffraction patterns at some chosen temperatures (**b**).

The pattern for the cubic phase can be indexed in the same crystal phase as the w-phase. The reflections are in similar positions; however, the primary reflections become significantly narrowed. The cubic phase is stable while cooling down to room temperature. It is worth noting that the $Fe_2(ZrTi)_3$ phase is stable to the highest investigated temperature of 900 °C and is also present after the cooling of the specimen.

The magnetic properties measured in the temperature range of 127 °C to 927 °C for the amorphous $Ti_{45}Zr_{38}Fe_{17}$ powder are shown in Figure 4. A systematic decrease in the magnetization of $Ti_{45}Zr_{38}Fe_{17}$ is observed as a function of increasing temperature, and the transition of the ferromagnetic phase into the paramagnetic phase is observed. The amorphous phase changes its structure into some crystal phases during the first heating.

The amorphous phase is not expected to be magnetic. The magnetic properties of the amorphous material originate probably from small amounts of the magnetic nanocrystals not successfully observed by neutron diffraction. The characteristic temperature with the Curie temperature (T_C) was defined as the peak position in the dM/dT-T curve, as shown in Figure 4b. It is equal to 267 °C/300 °C, 352 °C, 468 °C, and 737 °C, respectively. From 127 °C to 267 °C/300 °C, a slight increase in the magnetic signal was observed, which can be explained by the ordering of the magnetic moments of the magnetic nanocrystallites in the amorphous matrix under the influence of magnetic field. The transformation of the amorphous phase started above 300 °C. The temperature value 352 °C obtained from the

VSM measurement is correlated with the value of 325 °C derived from the DSC measurement, in the results of which one can observe the evolution of the amorphous phase to the i-phase. The formation of the quasi-phase is noticed both as peaks in the DSC (515 °C) or the VSM (468 °C) measurement, which can be additionally confirmed by the neutron diffraction at the temperature of 500 °C. The start of the formation of the i-phase is observed probably at 468 °C, and the end is observed at 515 °C. The Curie temperature (737 °C) is observed during the transition of the amorphous phase into the crystalline phases and during the subsequent heating of the crystalline phase. This temperature is close to the value of 770 °C known for pure iron [42]. This means that apart from an occurrence of the similar crystalline phase close to the w-phase, the sample also contains a small number of iron atoms that have not reacted.

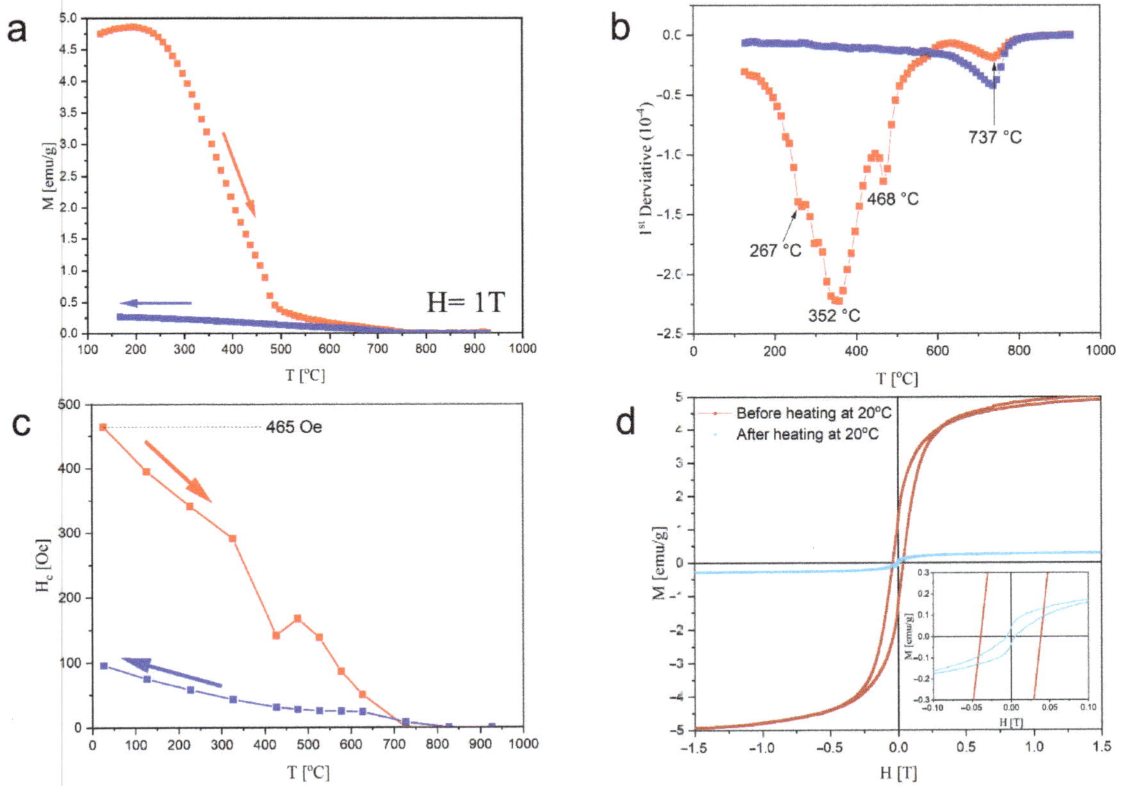

Figure 4. The temperature-dependent magnetization at 1 T for the amorphous phase (**a**); the derivative of the temperature-dependence used for the determination of the characteristic temperature (**b**); the thermal evolution of the magnetic coercivity (**c**); and the isothermal magnetization curves (**d**) before (red) and after heating (blue).

Figure 4c,d show the evolution of the magnetic coercivity vs. temperature and the magnetization vs. magnetic field curves measured at 20 °C before and after heating. The high values of the magnetic coercivity (H_C = 46.5 mT) and the magnetic saturation (Ms) (M_S = 4.9 emu/g at 1.5 T) are observed before heating; they can be associated with magnetic nanocrystals dispersed in an amorphous matrix. The abnormal behavior of the magnetic coercivity between 328 °C and 480 °C is related to the changes in the structure. It is correlated with the earlier results of the DSC and neutron diffraction measurements. After heating, the magnetic coercivity and the magnetic saturation change significantly (H_C = 9.6 mT Oe, M_S = 0.3 emu/g at 1.5 T).

The amorphous phase with some nanocrystals showed ferromagnetic properties. In the i-phase formation, most of the ferromagnetic behavior disappears due to a lack of long-range magnetic order in the quasicrystal structure. The w-phase and the new c-phase showed a weak magnetic signal, which came from the small addition of unreacted iron, which was confirmed by the Curie temperature after annealing.

After introducing hydrogen into its structure, the TiZrFe alloy remains amorphous. It creates an exciting opportunity to observe and research the behavior of hydrogen in the amorphous matrix.

Bringing hydrogen into an amorphous alloy leads to the creation of simple nanocrystalline hydrides. The hydrogen atoms are bonded to their particular constituents. The diffraction patterns for the base and hydrided alloy obtained from the XRD measurements are typical for amorphous materials. The broad central maxima observed for the hydrided sample are shifted toward the lower angles than for the base alloys (see Figure 5a), which can result from an increase in the mean metal–metal distances. Interestingly, we optimized the thermodynamic conditions so that formations of simple hydrides are not observed, which is favorable at higher temperatures or pressures.

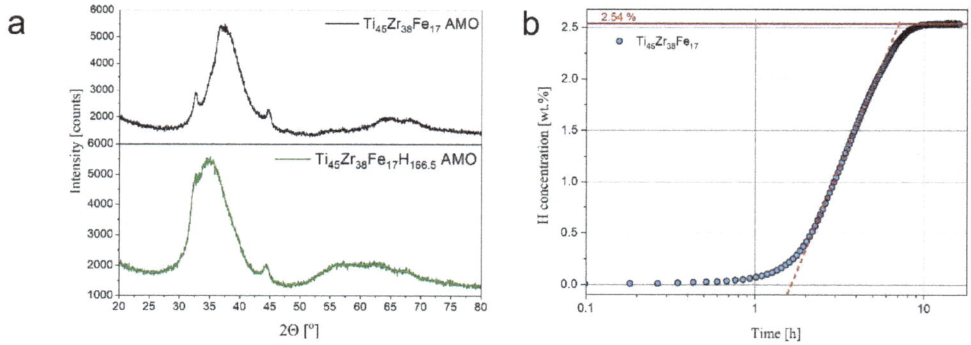

Figure 5. The diffraction patterns for the amorphous phase before (black) and after (olive) hydrogenation (**a**); hydrogen mass absorption vs. time for the amorphous TiZrFe (**b**).

The isothermal hydrogen absorption kinetic curves for the TiZrFe compounds are shown in Figure 5b. Precisely at 163 °C, the investigated alloy can readily absorb hydrogen under initial 4 MPa H_2 pressure and reach a hydrogen storage capacity of about 2.54 wt.%. The activation process energy for hydrogen atoms at some hydrogen storage alloys is a long process [43]. Hydrogen atoms must pierce the oxide layer surface of the nanoparticles before creating metal hydrides during the activation process. In the initial hydrogenation process, hydrogen atoms must penetrate the previously formed hydride layer to hold the hydrogenation reaction. Hydrogen atoms diffuse rapidly into the amorphous matrix from grain boundaries and phase interfaces between the amorphous phase and nanocrystals, which keep accelerating the hydrogenation process rate. After some time related to the processes mentioned above, the reaction is described with a single exponential function. It is probably related to hydrogen diffusion into the grains (bulk diffusion) after creating possible diffusion paths through the surface.

4. Conclusions

The amorphous phase is stable below 300 °C, and it is slowly evolving into the quasicrystal phase above this temperature., The i-phase structure is well formed close to 515 °C, with the addition of the $FeZr_3$ and the $Fe_2(TiZr)_3$. The TiZrFe compound demonstrates the quasi-continuous character of the transformation from the amorphous into the quasicrystalline (+some other crystals) phase. The structure of the $Ti_{45}Zr_{38}Fe_{17}$ is evaluated from the i-phase to the w-phase above 500 °C, close to 600 °C, at which temperature the reflections of the w-phase observed in the XRD diffraction patterns are well developed with some amount of the $Fe_2(ZrTi)_3$ with the $FeZr_3$. The ferromagnetic signal of the amorphous TiZrFe comes from magnetic nanocrystallites in the amorphous matrix. The Curie temperature is close to 740 °C, and it shows a small number of iron atoms that have not reacted. The amorphous phase with a hydrogen capacity exceeding 2.54 wt.% is still stable after the hydrogenation process, which is exciting for the materials designed for hydrogen storage.

Author Contributions: Conceptualization, A.Ż.; Data curation, Ł.G. and J.C.; Formal analysis, A.Ż.; Resources, A.T.; Visualization, N.B.S.; Writing—original draft, A.Ż.; Writing—review & editing, Ł.G. and P.J. All authors have read and agreed to the published version of the manuscript.

Funding: This research received no external funding.

Conflicts of Interest: The authors declare no conflict of interest.

References

1. Reuß, M.; Dimos, P.; Léon, A.; Grube, T.; Robinius, M.; Stolten, D. Hydrogen Road Transport Analysis in the Energy System: A Case Study for Germany through 2050. *Energies* **2021**, *14*, 3166. [CrossRef]
2. Capurso, T.; Stefanizzi, M.; Torresi, M.; Camporeale, S. Perspective of the role of hydrogen in the 21st century energy transition. *Energy Convers. Manag.* **2022**, *251*, 114898. [CrossRef]
3. Clees, T.; Baldin, A.; Klaassen, B.; Nikitina, L.; Nikitin, I.; Spelten, P. Efficient method for simulation of long-distance gas transport networks with large amounts of hydrogen injection. *Energy Convers. Manag.* **2021**, *234*, 113984. [CrossRef]
4. Kuczyński, S.; Łaciak, M.; Olijnyk, A.; Szurlej, A.; Włodek, T. Thermodynamic and Technical Issues of Hydrogen and Methane-Hydrogen Mixtures Pipeline Transmission. *Energies* **2019**, *12*, 569. [CrossRef]
5. Oliva, D.; Fuentes, M.; Borzone, E.; Meyer, G.; Aguirre, P. Hydrogen storage on LaNi5−xSnx. Experimental and phenomenological Model-based analysis. *Energy Convers. Manag.* **2018**, *173*, 113–122. [CrossRef]
6. Dematteis, E.M.; Barale, J.; Corno, M.; Sciullo, A.; Baricco, M.; Rizzi, P. Solid-State Hydrogen Storage Systems and the Relevance of a Gender Perspective. *Energies* **2021**, *14*, 6158. [CrossRef]
7. Takasaki, A.; Kelton, K. High-pressure hydrogen loading in Ti45Zr38Ni17 amorphous and quasicrystal powders synthesized by mechanical alloying. *J. Alloys Compd.* **2002**, *347*, 295–300. [CrossRef]
8. Kim, W.J.; Kelton, K.F. Icosahedral and related phase formation in rapidly quenched Ti-Zr-Fe alloys. *Philos. Mag. A* **1995**, *72*, 1397–1408. [CrossRef]
9. Kim, W.J.; Gibbons, P.C.; Kelton, K.F.; Yelon, W.B. Structural refinement of1/1bcc approximants to quasicrystals: Bergman-typeW(TiZrNi)and Mackay-typeM(TiZrFe). *Phys. Rev. B* **1998**, *58*, 2578–2585. [CrossRef]
10. Żywczak, A.; Gondek, Ł.; Figiel, H.; Żukrowski, J.; Czub, J.; Takasaki, A. Structural and hyperfine properties of Ti48Zr7Fe18 nano-compounds and its hydrides. *J. Alloys Compd.* **2011**, *509*, 3952–3957. [CrossRef]
11. Ji, P.; Chen, B.; Li, B.; Tang, Y.; Zhang, G.; Zhang, X.; Ma, M.; Liu, R. Influence of Nb addition on microstructural evolution and compression mechanical properties of Ti-Zr alloys. *J. Mater. Sci. Technol.* **2020**, *69*, 7–14. [CrossRef]
12. Ivanova, A.; Surmeneva, M.; Shugurov, V.; Koval, N.; Shulepov, I.; Surmenev, R. Physico-mechanical properties of Ti-Zr coatings fabricated via ion-assisted arc-plasma deposition. *Vacuum* **2018**, *149*, 129–133. [CrossRef]
13. Jhou, W.-T.; Wang, C.; Ii, S.; Chiang, H.-S.; Hsueh, C.-H. TiNiCuAg shape memory alloy films for biomedical applications. *J. Alloys Compd.* **2018**, *738*, 336–344. [CrossRef]
14. Liu, S.; Miao, J.; Zhang, W.; Wei, R.; Chen, C.; Wang, T.; Zhao, W.; Jiang, Z.; Li, F. Interfacial microstructure and shear strength of TC4 alloy joints vacuum brazed with Ti–Zr–Ni–Cu filler metal. *Mater. Sci. Eng. A* **2020**, *775*, 138990. [CrossRef]
15. Nguyen, V.T.; Qian, M.; Shi, Z.; Song, T.; Huang, L.; Zou, J. A novel quaternary equiatomic Ti-Zr-Nb-Ta medium entropy alloy (MEA). *Intermetallics* **2018**, *101*, 39–43. [CrossRef]
16. Liens, A.; Ter-Ovanessian, B.; Courtois, N.; Fabregue, D.; Wada, T.; Kato, H.; Chevalier, J. Effect of alloying elements on the microstructure and corrosion behavior of TiZr-based bulk metallic glasses. *Corros. Sci.* **2020**, *177*, 108854. [CrossRef]
17. Cordeiro, J.M.; Faverani, L.P.; Grandini, C.R.; Rangel, E.C.; da Cruz, N.C.; Junior, F.H.N.; Almeida, A.B.; Vicente, F.B.; Morais, B.R.; Barão, V.A.; et al. Characterization of chemically treated Ti-Zr system alloys for dental implant application. *Mater. Sci. Eng. C* **2018**, *92*, 849–861. [CrossRef]
18. Zuo, S.; Wu, R.; Pang, G.; Yang, Y.; Jin, M. High temperature internal friction in Ni50.3Ti29.7Zr20 shape memory alloy. *Intermetallics* **2019**, *109*, 174–178. [CrossRef]
19. Mishra, S.; Yadav, T.; Srivastava, O.; Mukhopadhyay, N.; Biswas, K. Formation and stability of C14 type Laves phase in multi component high-entropy alloys. *J. Alloys Compd.* **2020**, *832*, 153764. [CrossRef]
20. Lee, S.-H.; Huq, A.; Yang, W.; Kim, J. Analysis of local sites of deuterium in Ti53Zr27Ni20 alloys. *Phys. B Condens. Matter* **2018**, *551*, 33–36. [CrossRef]
21. Liu, Y.; Zhu, Z.; Liu, F.; Xu, J. Nonlinear dynamic characteristics and bifurcation analysis of Ti–Zr–Ni quasicrystal as hydrogen storage material. *Int. J. Hydrogen Energy* **2021**, *46*, 16667–16675. [CrossRef]

22. Lee, S.-H.; Park, T.; Yi, W.; Kim, J. Structure and Electrical Conductivity of Ag-Doped TiZrNi Quasicrystals. *J. Nanosci. Nanotechnol.* **2016**, *16*, 10532–10534. [CrossRef]
23. Jo, Y.; Lee, S.-H.; Shin, H.S.; Kim, J. Analysis of Structure and P–c–T Curve of Hydrogenated Ti53Zr27−xNi20Pdx Quasicrystals. *J. Nanosci. Nanotechnol.* **2013**, *13*, 7959–7962. [CrossRef] [PubMed]
24. Zhai, X.; Li, Z.; Zhou, X.; Liu, H.; Sun, J.; Su, Z.; Liu, W.; Zhao, J. Improved electrochemical hydrogen storage properties of Ti49Zr26Ni25 quasicrystal alloy by doping with Pd and MWCNTs. *Int. J. Hydrogen Energy* **2019**, *44*, 29356–29364. [CrossRef]
25. Liu, B.; Zhang, Y.; Mi, G.; Zhang, Z.; Wang, L. Crystallographic and electrochemical characteristics of Ti–Zr–Ni–Pd quasicrystalline alloys. *Int. J. Hydrogen Energy* **2009**, *34*, 6925–6929. [CrossRef]
26. Lee, S.-H.; Kim, J. Structure and hydrogen absorption properties of Ti53Zr27Ni20(Pd,V) quasicrystals. *Int. J. Hydrogen Energy* **2018**, *43*, 19130–19140. [CrossRef]
27. Żywczak, A.; Rusinek, D.; Czub, J.; Sikora, M.; Stępień, J.; Gondek, Ł.; Takasaki, A.; Hoser, A. Amorphous hydrides of the Ti45Zr38Ni17−xCox nano-powders. *Int. J. Hydrogen Energy* **2015**, *40*, 15534–15539. [CrossRef]
28. Liu, B.; Wu, Y.; Wang, L. Kinetic and electrochemical properties of icosahedral quasicrystalline Ti45Zr35Ni17Cu3 powder. *Int. J. Hydrogen Energy* **2006**, *31*, 1394–1400. [CrossRef]
29. Liu, C.; Duan, Y.; Ouyang, Z.; Shang, J.; Liu, K.; Xing, C.; Fu, Y.; Liu, W.; Wang, L. Effect of Li on structure and electrochemical hydrogen storage properties of Ti55V10Ni35 quasicrystal alloy. *Int. J. Hydrogen Energy* **2015**, *40*, 3015–3022. [CrossRef]
30. Liu, W.; Duan, Q.; Liang, F.; Lin, J.; Jiang, D.; Wang, L. Effect of Ce on electrochemical properties of the TiVNi quasicrystal material as an anode for Ni/MH batteries. *Int. J. Hydrogen Energy* **2013**, *38*, 14810–14815. [CrossRef]
31. Lin, J.; Liang, F.; Wu, Y.; Liu, W.; Wang, L. Hydrogen storage properties of Ti1.4V0.6Ni + x Mg (x = 1–3, wt.%) alloys. *Int. J. Hydrogen Energy* **2014**, *39*, 3313–3319. [CrossRef]
32. Żywczak, A.; Shinya, D.; Gondek, Ł.; Takasaki, A.; Figiel, H. Hydriding of Ti45Zr38Ni17−xFex nanocompounds. *Solid State Commun.* **2010**, *150*, 1–4. [CrossRef]
33. Ibuka, S.; Iida, K.; Sato, T.J. Magnetic properties of the Ag–In–rare-earth 1/1 approximants. *J. Phys. Condens. Matter* **2011**, *23*, 056001. [CrossRef] [PubMed]
34. Hiroto, T.; Gebresenbut, G.H.; Gómez, C.P.; Muro, Y.; Isobe, M.; Ueda, Y.; Tokiwa, K.; Tamura, R. Ferromagnetism and re-entrant spin-glass transition in quasicrystal approximants Au–SM–Gd (SM = Si, Ge). *J. Phys. Condens. Matter* **2013**, *25*, 426004. [CrossRef]
35. Tamura, R.; Muro, Y.; Hiroto, T.; Nishimoto, K.; Takabatake, T. Long-range magnetic order in the quasicrystalline approximant Cd6Tb. *Phys. Rev. B* **2010**, *82*, 220201. [CrossRef]
36. Ishikawa, A.; Fujii, T.; Takeuchi, T.; Yamada, T.; Matsushita, Y.; Tamura, R. Antiferromagnetic order is possible in ternary quasicrystal approximants. *Phys. Rev. B* **2018**, *98*, 220403. [CrossRef]
37. Inagaki, K.; Suzuki, S.; Ishikawa, A.; Tsugawa, T.; Aya, F.; Yamada, T.; Tokiwa, K.; Takeuchi, T.; Tamura, R. Ferromagnetic 2/1 quasicrystal approximants. *Phys. Rev. B* **2020**, *101*, 180405. [CrossRef]
38. Goldman, A.I.; Kong, T.; Kreyssig, A.; Jesche, A.; Ramazanoglu, M.; Dennis, K.W.; Bud'Ko, S.L.; Canfield, P.C. A family of binary magnetic icosahedral quasicrystals based on rare earths and cadmium. *Nat. Mater.* **2013**, *12*, 714–718. [CrossRef]
39. Shin, H.; Lee, S.-H.; Jo, Y.; Kim, J. Effects of hydrogen on the magnetic properties of TiZrNi quasicrystals. *J. Korean Phys. Soc.* **2012**, *61*, 1541–1544. [CrossRef]
40. Czub, J.; Przewoźnik, J.; Żywczak, A.; Takasaki, A.; Hoser, A. On magnetism in the quasicrystalline Ti 45 Zr 38 Ni 17 alloy. *J. Non-Cryst. Solids* **2017**, *470*, 108–111. [CrossRef]
41. Liu, W.; Feya, O.D.; Debela, T.T.; Hester, J.R.; Webb, C.J.; Gray, E.M. Experimental and computational modelling study of Ni substitution for Fe in Zr3Fe and its hydride. *J. Alloys Compd.* **2018**, *781*, 131–139. [CrossRef]
42. Chun, Y.; Kang, S.; Lee, D.; Cho, S.; Jeong, Y.; Żywczak, A.; Rhee, C. Development of Zr-containing advanced reduced-activation alloy (ARAA) as structural material for fusion reactors. *Fusion Eng. Des.* **2016**, *109–111*, 629–633. [CrossRef]
43. Liu, Y.; Chabane, D.; Elkedim, O. Intermetallic Compounds Synthesized by Mechanical Alloying for Solid-State Hydrogen Storage: A Review. *Energies* **2021**, *14*, 5758. [CrossRef]

MDPI
St. Alban-Anlage 66
4052 Basel
Switzerland
www.mdpi.com

MDPI Books Editorial Office
E-mail: books@mdpi.com
www.mdpi.com/books

www.ingramcontent.com/pod-product-compliance
Lightning Source LLC
Chambersburg PA
CBHW061548220926
51850CB00005B/279
9783036593005